SAHARASIA

THE 4000 BCE ORIGINS OF CHILD ABUSE,
SEX-REPRESSION, WARFARE AND SOCIAL VIOLENCE
IN THE DESERTS OF THE OLD WORLD

THE REVOLUTIONARY DISCOVERY OF A
GEOGRAPHICAL BASIS TO HUMAN BEHAVIOR

Revised Second Edition, with New Evidence

James DeMeo

For the most recently updated information on
the *Saharasia Discovery and Controversy*, visit:
http://www.saharasia.org

Books by James DeMeo:

* *The Orgone Accumulator Handbook: Wilhelm Reich's Life-Energy Discoveries and Healing Tools for the 21st Century, with Construction Plans.* Revised and Expanded 3rd Edition. Natural Energy Works, Ashland, OR 2010.

* *Preliminary Analysis of Changes in Kansas Weather Coincidental to Experimental Operations with a Reich Cloudbuster,* with the Appendix, *Evidence for a Principle of Atmospheric Continuity* Orgone Biophysical Research Lab, 2010.

* (Editor) *On Wilhelm Reich and Orgonomy.* Orgone Biophysical Research Lab, 1993.

* (Editor) *Heretic's Notebook: Emotions, Protocells, Ether-Drift and Cosmic Life Energy, with New Research Supporting Wilhelm Reich.* Orgone Biophysical Research Lab, 2002.

* (Co-Editor with Bernd Senf) *Nach Reich: Neue Forschungen zur Orgonomie – Sexualökonomie, Die Entdeckung der Orgonenergie.* Zweitausendeins Verlag, Frankfurt, Germany 1997.

* For a full listing of Dr. DeMeo's publications: www.orgonelab.org/demeopubs.htm

SAHARASIA

THE 4000 BCE ORIGINS OF CHILD ABUSE,
SEX-REPRESSION, WARFARE AND SOCIAL VIOLENCE
IN THE DESERTS OF THE OLD WORLD

THE REVOLUTIONARY DISCOVERY OF A
GEOGRAPHICAL BASIS TO HUMAN BEHAVIOR

Revised Second Edition, with New Evidence

James DeMeo

Natural Energy Works
Orgone Biophysical Research Lab
Greensprings Center
Ashland, Oregon, USA
1998, 2006

Symbol of Orgonomic Functionalism
Embracing the Earth

Published by, and Distribution to public and booksellers through:
Natural Energy Works
PO Box 1148
Ashland, Oregon 97520 USA
Tel/Fax: 541-552-0118
E-mail: demeo@mind.net info@naturalenergyworks.net
http://www.naturalenergyworks.net

and

Lightning Source / Ingram Distribution

ISBN: 978-0980231649   0980231647

Library of Congress Catalog Card Number: 98-091255

Second Revised Edition, with New Evidence

110512

Cover Art:

Front and rear cover design and artwork produced on
Macintosh computer by Sam Doust of Sydney, Australia.

The symbol at the bottom of the rear cover is a Sumerian *Ama-gi*, dating to c.2,500 BCE, the first known inscription of the word *freedom*.

Acknowledgments and Credits

I wish to thank a number of individuals who stimulated my research and thinking in one or another direction, and provided encouragement and constructive critique. Robert Nunley, my mentor and friend, deserves a special note of thanks. He opened the doors for my work at the University of Kansas, and defended my right to do this controversial work against not a small number of upset critics, often at a personal cost to himself. In addition to providing intellectual stimulation over the years, he educated me — originally a computer skeptic — in the usefulness of portable computers, a tool which was centrally necessary to unearth the spatial relationships presented in this work. Other professors at KU — Robert McColl, Pete Shortridge, John Augelli, Felix Moos, Ann Cobb — provided important constructive criticism and ideas, forcing me to dig deeper for answers to their challenging questions. A very big thanks goes to Robert Morris, my friend and mentor in Miami, who provided many ideas and a long-standing personal support. He firstly introduced me to Reich's writings many years ago, and provided a constant emphasis on the important role of social institutions in creating and maintaining irrational human behavior. His counsel at times when critics sought to put an end to my research and professional career, was crucial. Thanks also to the Illinois State University Geography Department, for assistance in production of my original Pacific-centered (*DeMeo Projection*) world base map used throughout this work, and to the Interlibrary Loans departments at the University of Kansas and Illinois State University, who chased down the many obscure publications cited in this work. A special thanks also to Theirrie Cook, who provided valuable ideas and stimulus as well as proofreading of the text, and to Bernd Senf of Berlin, whose many invitations to lecture on this topic at the *Fachhochschule für Wirtschaft* provided a wonderful encouragement and opportunity for refining the presentation of this complex material.

Regarding the various quotations and illustrations used in *Saharasia*: My thanks to Evelyne Accad for permission to use the lengthy quote from her book *Veil of Shame*. All other quotations are fully cited, being within the limits of fair-use copyright laws or from public-domain sources. My thanks to Debora Carrino for the drawing of a swaddled infant, to Angelos Rethemiotakis for various photos of artifacts in the Heraklion Museum, Crete, to Bernd Senf for the photos of open desert near Tassili N'Ajjer, Sahara, and to F. Monckeberg for the photos of malnourished infants. Thanks also to the editors of *Aramco World Magazine* for their generous reproduction policies of various Saharasian ruins and artifacts. And finally, a big thanks to Sam Doust for his cover design for this edition of *Saharasia*; he has produced a real work of art. All other illustrations and photos have been identified as to their sources to the best extent humanly possible. In most cases unattributed photos come from my own field expeditions and visits to museums. However, many years of shifting employment, requiring several long-distance moves from one end of the country to the other, has left many of my files incomplete and some illustrations may not be properly credited. In such cases, when so informed we will be happy to add such credits, or to remove the item from subsequent printings.

James DeMeo, Ph.D.
Greensprings, Oregon 1998

Contents

Page No.

Page No.

Tables, Figures & Maps

Preface to the Revised Second Edition

In the first edition of *Saharasia*, I presented comprehensive evidence for the first-origins of social violence and war within the large desert belt of the Old World, starting around c.4000 BCE or a bit earlier. The arguments were founded upon a very robust global cross-cultural and geographical evaluation of anthropological data, with many world maps of social and behavior variables, revealing a heretofore unknown geographical pattern in human behavior. The mapped patterns were traced back in time, using historical and archaeological materials, strongly suggesting the genesis of emotional armoring and violence with the appearance of harsh desert conditions across Saharasia. My study drew upon archaeology, anthropology, history, psychology and the health sciences. It was both a solid discovery of a geographical dimension to human behavior and warfare, and a larger body of theory which attempted to explain the geographical discovery, founded primarily upon the controversial findings of the late Wilhelm Reich.

Since that publication, in 1998, additional new information has come to my attention, clarifying and adding support to the findings and conclusions. In 2002, I published the article "Update on Saharasia"[1] which brought much of this new evidence together; an expanded version of the article appears in this second printing as the new Appendix B. Also, the 9-11 attacks by Islamo-fascists upon the USA prompted a renewed interest in my findings, and my own review of the first edition with "fresh eyes" and new insights into the dynamics of modern-day violence around the world.

A weakness in the first edition, by my own self-criticism, is that too-little attention was placed upon 1400 years of Islamic despotism as the major force by which the Saharasian patterns emerged on my world maps. New books by various scholars have clarified the enslaved situations of once-independent but now conquered and subordinated "dhimmi" populations of non-Muslims within Islamic Saharasia.[2] This finding prompted a re-evaluation and correction of some paragraphs and sentences in the first edition which were, in retrospect, too harshly critical of contemporary Western and American society, and too lenient and overlooking of the more extreme patrist and totalitarian cultures and empires dominating Saharasia and Saharasian-border regions in the modern times. For example: My original analysis, which ended at c.1900, clearly demonstrated the pre-Columbian New World cultures of the "American Indian" were *generally* more peaceful and social than were the Old World cultures — and the new Appendix provides even more evidence on this point. However, as "politically incorrect" as it may be, *this same general contrast between the Old and New World cultures holds true for modern times — the present-day inhabitants of Oceania and the Americas appear generally less-armored, less-patristic and less-violent as compared to Old World cultures, especially those within Islamic Saharasia.* One only has to consider the strutting blood-soaked dictators, the genocides and gigantic wars across Europe, Africa and Asia during the 20th Century, and into which nations of the New World were drawn only reluctantly, or not at all. Or more dramatically, to consider how Islamic Saharasia continues to harbor the most abundant extreme-patrist behaviors as documented in this work, which is not the case for the contemporary New World nation-states.[3]

This latter point is of considerable importance in understanding and confronting violence in the modern-day world. Islamic Saharasia today has become a fountainhead of international terrorism, targeting "infidel" captives and civilians for the most foul murder — burnings-alive, slow beheadings,

1. J. DeMeo, "Update on *Saharasia*: Ambiguities and Uncertainties about *War Before Civilization*", in *Heretic's Notebook (Pulse of the Planet #5)*, 2002, p.15-44. A revised and expanded version appears in Appendix B, page 423.

2. Ye'or, B., *The Dhimmi: Jews and Christians Under Islam,* Fairleigh Dickinson Univ. Press, 1985; Ye'or, B., *Islam and Dhimmitude: Where Civilizations Collide,* Fairleigh Dickinson Univ. Press, 2002; Bostom, A., *The Legacy of Jihad: Islamic Holy War and the Fate of Non-Muslims,* Prometheus Books, NY 2005; Spencer, R.: *The Politically Incorrect Guide to Islam (and the Crusades)*, Regnery Pub., Washington, DC, 2005.

3. North-Western Europe and a few other regions within the Old World always had a more matristic, lesser-armored cultural makeup, in part due to geographical distance from Saharasia. Southern and Eastern Europe after World War II and the defeat of Nazism and Communism has likewise softened as compared to the situation during the mid-20th Century. However, Europe is once again being challenged by an expansionist wave of new immigrants from Saharasia, who bring with them their more hardened character structures, a reduced and often nearly-enslaved status for women, and various extreme sex-negative, life-negative social institutions. In a 2004 interview with the German press, Middle-East scholar Bernard Lewis commented: *"Europe will be part of the Arab west, the Maghreb. Migration and demography indicate this. Europeans marry late and have few or no children. But there's strong immigration: Turks in Germany, Arabs in France and Pakistanis in England. They marry early and have many children. Following current trends, Europe will have Muslim majorities in the population by the end of the 21st century at the latest."* ("Europa wird am Ende des Jahrhunderts islamisch sein," *Die Welt*, July 28, 2004). Also see: Ye'or, B., *Eurabia: The Euro-Arab Axis,* Fairleigh Dickinson Univ. Press, 2005.

mass-murder, poison-gas attacks directed at civilian populations, etc. — equal to the butchery of any ancient "god-King" one might mention, and even using their own young people as "bomb delivery systems" — something which is most rationally understood in the context of *human sacrifice*, comparable to the *child-sacrifice* of ancient Carthage. Such butchery, directed against both infidels and fellow Muslims, has unfortunately become a "fingerprint" of fully Islamic societies, either through the active role of Jihad-warriors, or indirectly by ordinary Muslims who support Jihad butchery by silent agreement or actively donating money, and often by literally "dancing in the streets" at the news of the latest atrocities.[4] This book has exposed, in glaring detail, the deeply anti-sexual foundations of such social violence and chaos, formed via cruel child-rearing practices, and hateful attitudes regarding heterosexuality and females.

These same character traits, laden with sadistic impulses, have not merely been isolated to Muslims, however. They have overflooded into the Saharasian borderlands over the centuries, leaving tracks into modern times as seen in the bloody regimes of the Nazis, the Soviet Union, Red Chinese, North Koreans, Imperial Japanese, Southeast Asian Reds, and spilling into sub-Saharan Africa as well. *Nothing within modern New World history gets anywhere close to what the 20th Century records for the Old World.* One must dig back to the ancient Aztec or other bloody Meso-American empires, or to the era of black slavery, a foul practice which was brought to the New World by Europeans starting with Columbus, but which was ended more than 140 years ago — often in wars of liberation led by the lesser-armored descendants of those same Europeans. The World Wars I and II, the Korean and Vietnam wars, the Cold War, and now the war against Islamo-fascism, may be viewed in a similar historical perspective — as defensive expressions of the lesser-armored, freedom-aspiring peoples of the Western-oriented liberal democracies fighting back against the outward-directed expansionist invasions and attacks of the dominant totalitarian Saharasian empires, and their more hardened character structures and ideologies (Islamism, Communism, Nazism, Japanese Fascism).

Primarily in Appendix B, this second edition addresses both the newer archaeological findings on ancient peace and violence, and the above-mentioned sentence-changes on *"Saharasia Since 1900"*, which will demand an entirely new volume to do the subject justice. Some of this material was already included in Chapter 7 on "Expressions of Saharasia in Contemporary Demography", and in Chapter 11 on "Saharasia Today", both of which appeared in the original first edition. Where a few changes or additions have been made, I have placed the notation **New Material** into the margin, or **Note 2005** to identify those parts, to assist the reader who is already familiar with the first edition.

Also, to allow for the new "Update on Saharasia" in Appendix B, I have removed the Appendix giving criticism of the HIV theory of AIDS,[5] not because I have changed my views on that subject, but merely for space considerations, and also because it has been published elsewhere, with many excellent books on this subject available. They are cited where appropriate.

Aside from these considerations, the present edition remains unchanged from the original.

James DeMeo, Ph.D.
Greensprings, Oregon
January 2006

4. Mainstream news media do not make any significant reporting on the daily death tolls and terror incidents as carried out by what appears to be an accelerating Islamic Jihad directed against the rest of the world, as well as against less-extreme Muslim elements within Saharasia — including the awful spectacle of in-the-street "honor murders" of young women and girls. Global Islamic conquest, "cleansing Islam" of the "evil and dirty" Western influences, making the Islamic world *Judenrein* (cleansed of Jews, similar to Nazi goals), and restoration of a singular Islamic Caliphate — as with the bloody and despotic Ottoman Caliphate — these are their openly-stated goals. The reader who makes a periodic review of those websites which closely track Islamo-fascist attacks will get quite an education on materials which ordinary news-media rarely touch, possibly for reasons of "political correctness", but also due to political biases and even emotional alliances with the anti-sexual, anti-female and anti-freedom ideology of the Saharasian Islamist plague. The reader who doubts the accuracy of my statement is challenged to review the following websites every day for a month, and then to explain why only a small fraction of this material ever appears within mainstream news outlets:

www.thereligionofpeace.com
www.jihadwatch.org
www.dhimmiwatch.org
www.memri.org
www.camera.org
www.honestreporting.com

5. See the "articles" section of the author's website: www.jamesdemeo.org

New Material

Preface to the First Edition

The present work, *Saharasia*, was the product of seven years of systematic research on the genesis and geographical basis of human behavior. The basic findings were developed from a series of preliminary cultural maps completed in 1980, thereafter presented at academic conferences and in a few specialized publications, and finally accepted in 1986 as my doctoral dissertation at the Geography Department of the University of Kansas. It was not an easy work to complete, nor will it be easy to read, dealing as it does with so many terrifying historical events and pain-inflicting rituals. One is not made happier about the state of humankind while measuring the great depths of misery and depravity into which our species has collectively fallen. But for the reader who really wants to know how things came to be this way, *Saharasia* will provide answers.

It has taken 12 years for *Saharasia* to finally be published. Over this period, the basic findings — of a major geographical pattern imbedded in the archaeological, historical and anthropological literature as found in every major university library the world over — have found additional support and no serious challenges. However, in spite of the profound significance of *Saharasia* for the entire scope of behavior sciences, until now it has been impossible to publish the bulk of the findings, except as short summary articles in a few specialized journals. The female editor of one major academic publisher was positively excited to publish the work in the late 1980s, but the male professors of her advisory board expressed outrage at the implications and conclusions — while fully ignoring the supporting data and maps. The many publishers and book agents I contacted were uniformly uninterested, or only slightly veiled in their contempt, even if they had previously published many books on the subject of child abuse or "goddess" culture. *"Get rid of the citations by Wilhelm Reich"* stated one, *"How dare you misrepresent prostitution"* declared another (who later confessed to routinely visiting prostitutes), *"Interesting maps, but what's it got to do with all this sex stuff?"* said one uncomprehending professor. *"I won't approve of anything that's got Wilhelm Reich's name in it"* said another. And so it went. My attempts to openly undertake research into the even more controversial biophysical findings of Wilhelm Reich, on a revolutionary new method for *desert-greening*,[1] finally created such an uproar of opposition that I was *excommunicated* — informally but factually *blacklisted* from academic teaching in the USA. While theoretically top-heavy arguments on the genetic or innate origins of human violence, with far less supporting data, were routinely published in academic journals and books throughout the 1980s and 1990s, *Saharasia* was treated with a deafening silence.

Aside from the genuinely supportive and constructively-critical openness given to my work at the University of Kansas, and the very engaged interest expressed by a few scholars in Germany (where I have been repeatedly invited to lecture), there has been nearly no interest in *Saharasia* within the American academic world. For these reasons, this first edition is being published privately, through my own institute, following the tradition of heretics throughout the ages. *Saharasia* has nevertheless been treated to an intensive critical peer review, not only for the dissertation version (which stimulated many hot debates and critical revisions prior to its acceptance), but also in many shorter chapter articles and summaries published in different journals, and in lecture presentations to groups of scholars at the *Association for American*

1. DeMeo, J., *Preliminary Analysis of Changes in Kansas Weather Coin-cidental to Experimental Operations with a Reich Cloudbuster*, University of Kansas, Geography-Meteorology Dept., Thesis, 1979. DeMeo, J. & R. Morris, "Preliminary Report on a Cloudbusting Experiment in the Southeastern Drought Zone, August 1986", *Southeastern Drought Symposium Proceedings*, March 4-5, 1987, South Carolina State Climatology Office Pub. G-30, Columbia, SC, 1987. DeMeo, J., "OROP Arizona 1989: A Cloudbusting Experiment to Bring Rains in the Desert Southwest", *Pulse of the Planet #3*, J. DeMeo, Editor, p.82-92, 1991. DeMeo, J., "OROP Israel 1991-1992: A Cloudbusting Experiment to Restore Wintertime Rains to Israel and the Eastern Mediterranean During an Extended Period of Drought", in *On Wilhelm Reich and Orgonomy, Pulse of the Planet #4* , J. DeMeo, Editor, p.92-98, 1993. DeMeo, J., "Green Sea Eritrea: A 5-Year Desert Greening CORE Project in the East African Sahel-Sahara", in *Heretic's Notebook: Emotions, Protocells, Ether-Drift and Cosmic Life Energy, Pulse of the Planet #5*, J. DeMeo, Editor, p.183-211, 2002.

Geographers, at several *International Symposia on the Sexual Mutilation of Children*, to various groups of clinicians, and at invited lectures in German and Japanese universities. *Since the final form of Saharasia was worked through in the 1980s, there has not been a single occasion where any central tenet or finding has been seriously challenged;* only the most ambiguous and isolated bits of counter-evidence have been raised in opposition, as well as a few reactive arguments which held no substance. Notably, my articles and lectures more often elicited numerous supportive observations from other scholars, strengthening the overall findings. *The Saharasian patterns on the behavior maps are real, and constitute a serious challenge for every major theory on the origins of human behavior, save for the sex-economic theory of Reich which constituted the major basic starting point.*

The present book version contains several chapters and side-bar items which were not included in the original dissertation, many new and controversial items, and I alone bear the responsibility for the work and conclusions. Saharasia will hopefully — and should — provoke an open public discussion, in part given its testable and refutable character, but also because of the implications for various widespread beliefs and social institutions. A harsh spotlight is cast upon many "social customs" and "beliefs", and this tends to make people squirm who strongly adhere to them. For this, I make no apologies, as the future of the human species is dependent upon our making a direct and open challenge to the cruel manner in which we treat our infants and children, and suppress our sexuality. It should be clear that patristic authoritarianism and social violence is today no longer confined to just Saharasia and its immediate borderlands, but is nearly a global phenomenon. People living in or near to Saharasia today will find plenty of discussion in this book about the cultural sins of the West, which historically have been as bad or worse than any Old World desert tribe.[2]

It may be that someone will eventually spot a fatal flaw in either my logic or theoretical framework, though I am presently confident that this will not be the case. However, even if this is the case, the distributions of various beliefs, behaviors, and social institutions given on my maps are *new phenomena* not previously available to the scientific, academic, and health-care communities. Whatever explanation or criticism is leveled at *Saharasia* must, if it is to be penetrating, explain the distributions according to some *other* theory or mechanism. It is not acceptable for only the theoretical grounding, or only the conclusions of this work to be attacked while leaving the new distributional information and maps untouched. The theoretical assumptions and starting-points of *Saharasia* are considered *proven*, by scientific logic, given the highly structured patterns in world archaeology, history and anthropology — confirmed repeatedly in the orthodox published literature of those disciplines — which appear on the maps. The patterns are real, and constitute a major new, bona-fide discovery. Arguments about this or that small point of theory, pointing to isolated ambiguous evidence of social violence prior to 4000 BCE, expressions of anger because I criticize circumcision and other cruel tribal customs here and there, or expressions of contempt towards Wilhelm Reich will not constitute a "scientific critique" by any means, and will not erode the conclusions in the least.

There are several aspects of this work on Saharasia where the reader may feel more needs to be said, notably regarding the cultural sins of the West as compared to those of Saharasian cultures, and on the important subject of *what can be done* to restrain and oppose socially-dangerous patrism. I would partly agree with such criticisms, though both are addressed in various places. There are limits to what can be covered even

2. **Note 2005:** More recent research nevertheless demonstrates Saharasia and its immediate borderlands still carry the most highly armored peoples, with the most extremely violent and patristic social institutions existing on planet Earth. This is seen most dramatically within Islamic Saharasia, in the upsurge of suicide/homicide bombings, mass-murdering of civilians deemed to be *untermensch,* with a 50-year's war against the Jews in Israel — historically this is correctly viewed as an extension of Jewish-extermination politics which exploded with fury across Europe during WW-II. The current epoch of Muslim terror attacks against "infidel" civilian populations also falls into this same category of hatred and scapegoating of those considered "inferior": the attacks against infidel Americans on 9/11, and Europeans on 3/11 and 7/7, against Western tourists in Bali and Egypt, against Russians in Moscow and Besylan, etc., and even against fellow Muslims deemed "insufficiently pure", with even greater slaughters in Algeria, Sudan, and Iraq, where even mothers celebrate the wasting of their own children's lives as homicide-suicide bombers. Violence and warfare within Saharasia since c.4000 BCE has been unrelenting, never-ending and on-going down into modern times, especially with 1400 years of Islamic slaughter and jihad-conquest following Muhammed's visions. Over this same period, Europe and America have undergone dramatic social changes with the Reformation, Enlightenment, Renaissance, the democratic revolutions and ending of kings, feudalism and slavery, the civil rights movement, the women's movement, the sexual revolution, and the alternative medicine movement, with a dramatic spreading of democratic social institutions and freedom to ever larger parts of the world. While often chaotic and certainly imperfect, these are nevertheless significant changes towards less-armored and less-violent conditions which are scarcely mirrored, or not to be found at all, within the Saharasian heartland. A new work under development by the author, *Saharasia Since 1900,* will address these concerns more specifically. See my short discussion in the *Preface to the Second Edition* of this work, as well as the discussions on life in Islamic regions as covered in Chapters 3 and 11.

in 450+ pages. Some elements of modern-day patristic conditions in both the West and in Saharasia are discussed, however, and certainly there is plenty in this work which will likely offend people in both cultural regions. Nevertheless, there surely is a pressing need for *more* to be written *and done* about those social institutions within our own Western culture which continue to destroy new life and re-create armoring in the young — as well as about similar social institutions within Saharasia itself, and other world regions. As mentioned in many places, armored patrism today has nearly a worldwide distribution, though some areas are certainly more deeply infected than others. The present volume is dedicated to the issue of *ultimate origins*, on *how armored patrism got started in the first instance*, and on that matter, the reader should not be disappointed.

Additionally, I considered to include detailed geographical maps by which the reader could identify various landforms, rivers, political boundaries, cities, archaeological sites and migratory patterns. This task was too large for our small institute, however, so it may be helpful to keep handy a good world atlas.

It is my deepest hope the new knowledge about how patrism was generated and spread globally will allow its modern day expressions to be more easily restrained and eliminated. It is past time for various social scientists and health professionals to begin wrestling with those damaging social institutions which disturb behavior in whole societies. Therapy treatment of individual armored and distressed patients as done in the West is well and good, where needed, but *therapy is currently undertaken by so few people that, by itself, it will change little or nothing socially.* So long as our families, hospitals, schools and work-places are predominantly *factories* for the production of highly armored, sex-frustrated and angry people, the next generations will be just as damaged as the last, and those who give or undertake emotionally-transforming therapy will, in old age, be confronted with a world just as sick, or worse than what existed in their youth. *Saharasia* will hopefully provide some assistance and encouragement for those with the energy and strength to confront damaging, armor-producing patristic practices in the home, schools, hospitals and in more public institutions and the political arena as well. The research and discovery of Saharasia has been an incredible learning and transforming experience for me personally, not always pleasant, but certainly illuminating. Hopefully, what is written here will be similarly eye-opening and transforming to others.

James DeMeo, Ph.D.
Greensprings, Oregon
January 1998

"Man is born free; and everywhere he is in chains. One thinks himself the master of others, and still remains a greater slave than they. How did this change come about?"
Jean Jacques Rousseau, *The Social Contract*

"Those who cannot remember the past are condemned to repeat it."
George Santayana

Part I:

A Survey of New Territory:

Basic Starting Assumptions

and Theory

1. Introduction and Overview

We start with ten important questions, which must weigh heavily upon the concerned individual; they are the starting points and guiding themes for the research findings presented in this work:

1. **What are the causes and ultimate sources of human violence and war?**
2. **Why, if everyone talks about "world peace" and "loving thy neighbor", is there so much hatred and killing around the world?**
3. **In an era of exaggerated sexuality, why is there so much sexual misery and so little love, and why is natural sexual functioning so often cloaked and hidden?**
4. **Why do so many political and religious leaders behave in such a hypocritical manner, and why are the most religious nations often the most bloody and violent?**
5. **What are the roles of politics, religion, and of the ordinary person, in the cultural dynamics which produce violent societies?**
6. **Are humans innately violent, burdened with "original sin" or "violent genes"? Is violence simply "learned", or are there other reasons related to traumatic childhood experiences?**
7. **Do truly peaceful societies exist? Did they ever exist?**
8. **Is there any truth to the idea, as reflected in various mythologies and religions, that there existed an ancient time of widespread peaceful social conditions?**
9. **If so, what were those peaceful cultures like, and where were they located?**
10. **What specifically happened to change the face of the world so dramatically for the worse, to produce the big mess in which so much of humanity finds itself today?**

Over 15 years ago, as a university student of geography and environmental science, I undertook several unusual, previously untried but very productive methods of investigation, aimed at answering some or all of these questions. Nearly seven years of work ensued, requiring long periods of seclusion within several large university libraries, additional years of work at the computer, and extended field work in the deserts of Egypt, Israel, and the American Southwest. My approach involved a combination of geographical, cross-cultural, archaeological, and historical techniques, to identify specific regional differences in levels of warfare, sex-repression, and child-abuse. Most importantly, I investigated and focused upon a smaller number of lesser-known, but well-documented peaceful and cooperative cultures. Of primary interest was the expression or inhibition of love in its male-female and maternal-infant aspects.

Early in my study, a major climatic expression was observed in the regional distributions of human behavior and social institutions: *a general association was observed between desert environments and emotional inhibition, with a corresponding association between rainforest regions and a more fluid, expressive emotional life.* Further investigation gradually revealed this simplistic climatic correlation to be only partially true — correlation does not equal causation — but something quite profound had been uncovered in my approach. Additional work clarified the relationships. When standard cross-cultural data on human behavior were reviewed in a geographically explicit manner — that is, with the data actually plotted on maps, rather than being reviewed only in tabular, numeric form — several startling, but clear-cut patterns appeared on the world maps I had created: *Human violence appeared to have a specific time and place of origins on the Earth; antisocial violence was not distributed world-wide at all times in the past!* Furthermore, it was learned, *the origins of violence was precisely timed to a major historical epoch of climate change from relatively wet towards dry conditions.* [1]

1. DeMeo, J.: *On the Origins and Diffusion of Patrism: The Saharasian Connection*, U. Kansas Geography Dept. Dissertation, 1986. University Microfilms, Ann Arbor, MI 1987.

3

My research blended the cross-cultural approach of the anthropologist with the spatial analytical techniques of the geographer, emphasizing historical processes and natural environment. With this approach, a global geographical pattern was uncovered in human behavior and social institutions, specifically those related to family and sexual life; moreover, this global pattern strongly correlated to well-known patterns of desert climatology, and also to the migrations of peoples. The social variables I studied included infant and child treatments, adolescent rites, female status, type of marriage, kinship, inheritance rules, male/female relations, sexual behavior, cooperative or competitive tendencies, destructive aggressions, social hierarchies, religion, and so on. Cultures which tended to inflict pain and trauma upon infants and young children, punish young people for sexual expression, manipulate them into arranged marriages, subordinate the female, and otherwise greatly restrict the freedoms of young people and older females to the iron will of males also tended to possess high levels of adult violence, with various social institutions designed for expression of pent-up sadistic aggression. Such cultures I call the *patrist* groups. It was found that, at various times in their past histories, all patrist cultures once possessed (or were connected by migration-diffusion to) large warrior kingdoms, with extremely authoritarian and cruel cultural expressions such as divine kingship, ritual widow murder (*suttee*, or *mother murder*), human sacrifice, and sadistic ritual tortures of enemies, heretics, social rebels and criminals.

These patrist groups contrast in almost every manner with the peaceful *matrist* cultures, where child treatment and sexual relationships were of an entirely different character, being very gentle and pleasure oriented. Matrist cultures are also democratic, egalitarian, sex-positive, and possess very low levels of adult violence. The cultural histories of matrist groups are additionally devoid of the extreme expressions of patrism, as given above. Table 1, presented here, summarizes the various differences between patrist and matrist behaviors, attitudes, and social institutions.

These considerations have been, for the most part, ignored or even actively censored by mainstream sociology, psychology and psychiatry, given their critical challenge to the cherished beliefs of our own violence-prone and patristic popular culture. While a few other scholars and researchers have previously dealt with these issues in a coordinated manner, my work provides an interdisciplinary treatment using geographical, cross-cultural, and historical approaches combined. Patrism, for instance, is synonymous with the *armored character structure* and society as discussed by Wilhelm Reich in his various works,[2] a culture type which today more or less typifies the overwhelming majority of people on Earth. Matrism is likewise synonymous with Reich's descriptions of the *unarmored*, or *genital character*, a minority culture type today, but one which once was dominant worldwide. In this work, the terms *armored patrist* and *unarmored matrist* replace the less specific "patriarchal" and "matriarchal", which have been widely used without any uniform or explicit definition.[§]

The global validity of the matrist/patrist schema of character structure and social organization in Table 1 was cross-culturally tested and verified through a variety of sources and techniques, which are fully discussed and cited in subsequent chapters. From such sources, primarily the works of Reich, a *testable* climatic/geographical theory was developed, linking areas of harsh desert landscape to very specific anti-child, anti-female, and anti-sexual behaviors, attitudes, and social institutions.

I subsequently plotted the actual locations of nearly 1200 different matrist and patrist cultures on the world map, taken from the global

2. Reich, W.: *Character Analysis*, Farrar, Straus & Giroux (FS&G), NY, 1971; *Function of the Orgasm*, FS&G, 1973; *The Sexual Revolution*, Octagon Books, NY, 1971; *Reich Speaks of Freud*, FS&G, 1967; *People in Trouble*, FS&G, 1976; *Mass Psychology of Fascism*, FS&G, 1970; *Invasion of Compulsory Sex-Morality*, FS&G, 1973; *Children of the Future: On the Prevention of Sexual Pathology*, FS&G, 1983; *Genitality in the Theory and Therapy of Neuroses*, FS&G, 1980.

§ After completing and publishing my research findings between 1980-1986, I subsequently learned about Riane Eisler's work *The Chalice and the Blade* (Harper & Row, 1987), which argued for similar cultural transitions in Europe and the Mediterranean. Eisler's *dominator* culture type holds many similarities to the armored patrist culture described here, while her *partnership* culture type is generally comparable with unarmored matrist culture. While Eisler has yet to explicitly define her terms regarding childbirth practices, adolescent sexuality and genitality, her terms appear similar enough that such a comparison is generally valid. In her more recent book *Sacred Pleasure* (Harper Collins, 1996, pp.92-94 and pp.101-102) Eisler embraces much, but not all parts, of my work on *Saharasia*, regarding the origins of patristic-dominator cultures.

Table 1: DICHOTOMOUS BEHAVIORS, ATTITUDES & SOCIAL INSTITUTIONS

	Armored Patrist	*Unarmored Matrist*
Infants & Children:	Less indulgence	More indulgence
	Less physical affection	More physical affection
	Infants traumatized	Infants not traumatized
	Painful initiations	Absence of pain in initiations
	Dominated by family	Children's democracies
	Sex-segregated houses or military	Mixed sex children's houses or age villages
Sexuality:	Restrictive, anxious attitude	Permissive, pleasurable attitude
	Genital mutilations	Absence of genital mutilations
	Female virginity taboo	No female virginity taboo
	Vaginal intercourse taboos	No intercourse taboos
	Adolescent lovemaking severely censured	Adolescent lovemaking freely permitted
	Homosexual tendency plus severe taboo	Absence of homosexual tendency or strong taboo
	Incest tendency plus severe taboo	Absence of incest tendency or strong incest taboo
	Concubinage/prostitution may flourish	Absence of concubinage or prostitution
	Pedophilia exists	Pedophilia absent
Women:	Limits on freedom	More freedom
	Inferior status	Equal status
	Vaginal blood taboos (hymenal, menstrual & childbirth blood)	No vaginal blood taboos
	Cannot choose own mate	Can choose own mate
	Cannot divorce at will	Can divorce at will
	Males control fertility	Females control fertility
	Reproductive functions denigrated	Reproductive functions celebrated
Culture, Family, Social Structure	Patrilineal descent	Matrilineal descent
	Patrilocal marital household	Matrilocal marital household
	Compulsive lifelong monogamy	Noncompulsive monogamy
	Often polygamous	Rarely polygamous
	Authoritarian	Democratic
	Hierarchical	Egalitarian
	Political/economic centralism	Work-democratic
	Military specialists or caste	No full time military
	Violent, sadistic	Nonviolent, absence of sadism
Religion:	Male/father oriented	Female/mother oriented
	Asceticism, avoidance of pleasure, pain-seeking	Pleasure welcomed and institutionalized
	Inhibition, fear of nature	Spontaneity, nature worshiped
	Full-time religious specialists	No full-time religious specialists
	Male shamans/healers	Male or female shamans/healers
	Strict behavior codes	Absence of strict codes

anthropological data base developed by the late George P. Murdock at the University of Pittsburgh.[3] My geographical study, which constituted the first-ever mapping of Murdock's aboriginal cultural data, demonstrated that *the most extreme, harsh patrist peoples lived in, or were strongly influenced by peoples who originated within the most extreme, harsh desert environments*. Temperate and wetland regions close to the harshest deserts were likewise very patrist in character, owing to the migrations of patrist peoples out of drought-afflicted or desert regions, often (literally) for "greener pastures". However, other temperate and wetland regions at great distance from the harshest deserts, as well as various *semiarid* regions around the globe, possessed only very low levels of patrism, and were predominantly matrist. Indeed, with adjustments made for diffusion of culture, the actual mapped distributions of human behavior suggested that *patrism — violent, war-making, sex-repressive and child-abusive homo sapiens — had originated first and only within the harshest of hyperarid desert environments, and then, only around 6000 years ago.*

The above conclusion has been solidly established by a rigorous scientific methodology, involving the spatial and temporal (space and time) analysis of relatively recent anthropological data describing native aboriginal populations, and by an independent analysis of ancient historical and archaeological materials. The data employed are overwhelmingly drawn from those previously peer-reviewed, critiqued, and published in standard scholarly research journals and books, as found in university libraries around the world. What is new and different in my approach was, and is, the quite simple, though often difficult and time-consuming task of placing and reviewing the large mass of cultural data on world maps, and reviewing them against the backdrop of climate and history.

This volume consists of three major sections, each of which constitutes an independent body of evidence supporting the general conclusions.

Part I focuses upon the emotional and sexual life of the human animal, to include family and general social structure. The matrist/patrist schema of character structure and social organization, given in Table 1, is a basic starting assumption for the development and interpretation of the various maps, and also for my conclusions. Consequently, some effort was given in Part I to cross-cultural testing of these assumptions. Specific mechanisms were also identified which clarify exactly how the desert landscape could influence character structure over the generations — converting peaceful unarmored matrist peoples towards armoring and violent patrism — and how the newly-formed armored character structure, with its necessary patrist social institutions, could subsequently be carried out of the desert to appear and persist in places far removed. Here, the observations made by famine researchers were central.

During extreme drought and desertification, food supplies dwindle and famine sets in; as this happens, children tend to suffer most severely from protein-calorie deficiencies, such as marasmus and kwashiorkor, and have a high mortality and morbidity. Surviving children will not recover to full physical or emotional vigor once food supply is restored, and will suffer lifelong physical and emotional effects. It does not take much imagination to envision the cruel and hard effects of starvation on infants, children, families, and entire social groups — our television broadcasts bring the images into our living rooms on a regular basis, from Ethiopia, Somalia, and other drought- and war-ravaged lands.

Sadly, *the emotional responses of a child to famine and starvation are similar to those stemming from maternal rejection or isolation-rearing, factors which are known to have powerful disturbing effects upon later adult behavior.*

3. Murdock, G.P.: *Ethnographic Atlas*, U. Pittsburgh Press, 1967.

The child's own contractive emotional responses to drought and famine persist, more or less, depending upon the length and severity of the deprivation. As adults, these individuals who have suffered through severe famine during childhood will raise their own children differently from prior generations, even during times of plenty. Other drought-induced trauma, such as infant cranial deformation and swaddling are discussed as accidental products of the change from a settled to nomadic existence. These also disturb the maternal-infant bond within social groups across entire regions, to radically affect behavior thereafter. Both starvation trauma and cranial-deformation/swaddling trauma caused a radical shrinkage and permanent contraction within the emotional structures and central nervous systems of the peoples so affected.

Emotional and physiological effects of famine and starvation trauma, it will be argued, persist within a culture irrespective of subsequent climate, food supply, or settlement patterns. This persistence of culture-shock takes place by virtue of *altered behavior and altered social institutions*, which adjust to the new drought-famine-starvation conditions. In regions which experience repeated widespread droughts, with significant long-term declines in vegetation, wildlife, agriculture and domesticated animals, famine and starvation can persist for years or even periodically through generations. Under such conditions, many people die, family ties are shattered, and mass migrations take place. With so much death and displacement, family life is gradually or even radically diverted away from prior emotionally-rich and pleasure-oriented patterns; new patterns emerge, focused upon basic survival, and with little or no emphasis upon pleasurable emotional bonding or social living. Social conditions become disturbed and emotionally diminished, much in keeping with the surrounding landscape, which is drying out and withering away.

Once so anchored into social institutions, the new drought- and famine-derived behavior patterns reproduce themselves in each new generation, irrespective of subsequent turns in climate towards wetter conditions. Warfare, which often appears as a secondary effect of drought and famine starvation, compounds the damage to family life and social structure, promoting even more disturbed maternal-infant and male-female relations. And like a major war, or the invasion of a conquering, marauding horde, extreme drought and desertification (desert spreading) promotes famine, starvation, and the premature death of loved ones — in the worst cases, the majority of individuals within entire tribes, villages, or social units will starve, leaving behind only a small number of "shell-shocked" survivors. Combined with other traumas related to forced migration, these environmental forces are quite sufficient to initiate the process of armoring among previously unarmored peoples. When viewed against climatic changes towards drought and increased desert spreading over the last 6,000 years, the above facts have deep and profound social and historical significance.

Part II presents a series of world maps, prepared from contemporary environmental and anthropological data. These maps constitute part of the central new evidence supporting my conclusions, demonstrating that *the region of greatest preponderance of patrist character traits was the Old World desert belt stretching across North Africa, the Near East, and into Central Asia*. World maps from various sources likewise identify this same swath of aridity as *the harshest of global environments, with the least amount of precipitation and vegetation, with the most stressful temperature extremes*. A striking degree of spatial correlation exists between climatic, biological, and cultural characteristics within this large and unique region I call *Saharasia*.

Saharasia: *...the region of greatest preponderance of patrist character traits ... the Old World desert belt stretching across North Africa, the Near East, and into Central Asia.*

Figures 1 and 2 demonstrate the profound spatial correlation which exists in *relatively recent* anthropological and climate data. These and several dozen similar maps are presented with more detail in Chapters 3, 4, 5, 6 and 7, with a full discussion of their methods of preparation and significance.

Part III presents a coordinated survey of paleoclimatic, archaeological, and historical sources, confirming that patrism has its deepest roots in Saharasia. In fact, my research demonstrates *a coordinated change in both environment and culture across Saharasia, from wetland to extreme desert, and from matrist to patrist, around 4000-3500 BCE.* § Human armoring and patristic social institutions did not significantly exist in Saharasia, nor within any other region on planet Earth, prior to that time — though a few small exceptions dating back to c.5000 BCE do exist, corresponding to temporary episodes of regional drought which "prove the rule" of the Saharasian generality. These are also discussed in Part III. Nevertheless, patrism makes its first clear and lasting appearance on Earth only after c.4000-3500 BCE, within Saharasia at the same times when vast stretches of Africa and Asia finally and awfully converted from relatively wet to extremely dry conditions. Patrism subsequently was carried to borderland regions by mass migrations, and in later centuries, to a few regions at great distance from Saharasia. According to my independent review, *the archaeological and historical evidence shows no clear, unambiguous traces of patrism, anywhere on Earth, before approximately 5000 BCE, and no significant, lasting traces until around 4000 BCE.* Only peaceful, unarmored matrist traits can be inferred from the oldest, deepest layers of archaeological materials.

Climatologists, for instance, have demonstrated that Saharasia was once wetter and greener prior to around 6000 years ago. The Arabian and Central Asian cores of Saharasia began to dry up around 5000-4000 BCE. Other parts of Saharasia began to dry up shortly thereafter, in an oscillatory manner. Whole regions were devastated and abandoned as desiccation set in. Droughts, famines, and starvation occurred, and mass migrations ensued. Out of this social chaos, armored patrism emerged and unarmored matrism declined.

Regarding those wetter regions adjacent to Saharasia, the social histories of Europe, sub-Saharan Africa, India, and Eastern China all demonstrate profound similarities regarding the onset of patrist conditions — in each case, these regions (which are wetlands) lost their original matrist character under the influences of migrating and invading patrist peoples, who were abandoning the drier portions of Saharasia. A similar process occurred to defeat the original matrist character of peoples living on secure water sources within Saharasia itself, on the Nile, Tigris-Euphrates, and Niger Rivers.

Also of interest are the studies by arid lands climatologists, who show that the Saharasian desert lands are significantly drier and larger, with scantier food resources than deserts elsewhere. My own comparative field observations in Northern and sub-Saharan Africa, and in Israel, Namibia, and the arid Great Basin of North America have repeatedly confirmed this important difference. Long-term epochs of drought across the vast expanses of Saharasia have more certainly resulted in widespread famine and starvation than within other arid or semi-arid regions. This explains why other deserts did not independently generate a permanent armoring and patrism among their inhabitants. It also explains why the wetland regions close to Saharasia possess an intermediate level of patrism, while other wetland *and* dryland regions distant from Saharasia largely retained their original matrist character. Many of these small regional

§ Secular Chronology

BCE refers to *Before the Current Era*, with a chronology identical to "BC" or "Before Christ".

CE or *Current Era* is likewise substituted for "AD" (from the Latin "anno Domini" meaning "year of our Lord" — as in *dominator* or *conqueror*).

Figure 1: The World Behavior Map (Detailed in Chapters 3, 4 and 5)

For the period roughly between 1840 and 1960, as reconstructed from aboriginal cultural data given in Murdock's *Ethnographic Atlas* (1967), with minimal historical interpretation

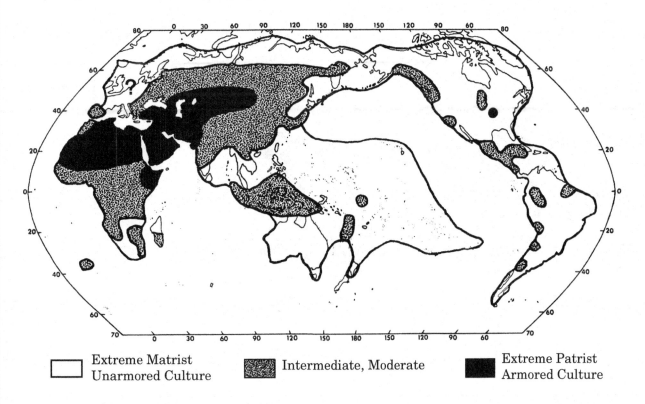

☐ Extreme Matrist Unarmored Culture   ▨ Intermediate, Moderate   ■ Extreme Patrist Armored Culture

Figure 2: The Budyko-Lettau Dryness Ratio (Detailed in Chapter 4)

Contrasting the relative dryness of different arid lands.

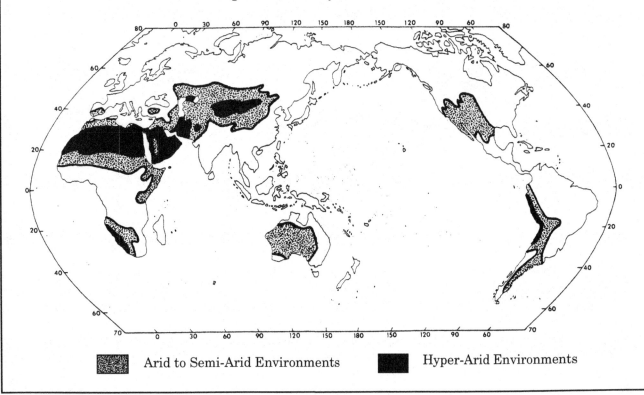

▨ Arid to Semi-Arid Environments   ■ Hyper-Arid Environments

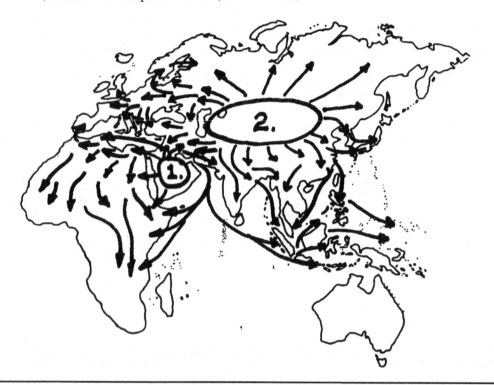

Figure 3: Generalized Paths of Diffusion of Armored Human Culture (Patrism) in the Old World, *after* c.4000 BCE.
1. Arabian Core Region   2. Central Asian Core Region
(Detailed in Chapters 4 and 5, and Part III)

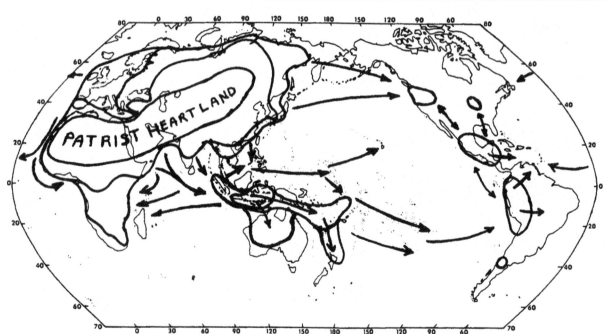

Figure 4: Suggested Patterns of Diffusion of Patrism Around the World
Starting after c.4000 BCE. (Detailed in Chapter 5 and Part III.)
The earliest migratory diffusion of culture, across the Bering Strait and into the New World from
c.18,000 - 8,000 BCE, as generally accepted, is assumed to have been entirely matrist in character,
and constitutes the "background" over which the more recent patristic migrations occurred.

Woodcut of 15th Century Chinese ocean-navigating junk, estimated at over one hundred meters (328') length. Smaller Chinese ships had been sailing into the open Pacific, to India and elsewhere, since the Han Dynasty, 2nd Century BCE, or earlier.

Seaworthy Tartessian ship of Lebanon, fully wind-powered and similar to Celtic ships of the 1st Century Roman Era, capable of long ocean voyages, approximately 50 meters (170') length.

Christopher Columbus' flagship, *Santa Maria*, no more than 31 meters (100') in length, a small boat compared to those produced by the ancient divine-king patrist states of Asia and the Mediterranean.

differences find a common resolution in the more general patterns of out-migrations from Saharasia, which began around 4000 BCE and persisted over millennia. Saharasian migrations persisted even into the epoch of early ship-building and navigation of the seas, which allowed patrism to be carried across the Pacific and Atlantic Oceans, into the New World prior to the arrival of Columbus, but much later than the interglacial or postglacial migrations across the Bering Strait. The European Colonial period was but the last of a series of near-global ocean voyages, which transplanted Saharasian and Saharasian-borderland culture around the globe. Figures 3 and 4 present a summary of the generalized migratory patterns, as derived and presented in detail in Part III of this work.

As will be demonstrated, the total combined evidence presented in this volume leads to a set of inescapable though controversial conclusions: *Human armoring, with its violent antisocial and institutionalized patristic cultural conditions, appeared and persisted on planet Earth for the first time about 4000-3500 BCE, in Saharasia, after it dried up.*

Pre-Columbian Contact Theory

According to the best-available evidence, during the last Ice Age global sea levels declined dramatically as water was taken up into the vast ice sheets covering the near-polar latitudes of North America, Europe, Asia and Antarctica. This decline in sea level exposed a number of areas of shallow sea-bottom, allowing migrations of humans and other animals across natural land bridges. One such natural land bridge existed between far north-eastern Asia and western Alaska, connecting the two coasts of the modern-day Bering Strait. It was during this glacial period, between c.25,000 to 8,000 BCE, that the first migrations took place between the Old World and the New World. Humans are thought to have arrived in North and South America by long-distance migrations across "Beringia", into Alaska, and southward from there.

When the Ice Age ended around 8,000 BCE, sea levels rose and flooded Beringia to produce the present-day Bering Strait. Migrations were then ended, it is popularly believed, and no additional contact took place between the Old and New Worlds, until the much later arrival of Christopher Columbus in 1492. However, archaeological and cultural evidence exists strongly suggesting occasional or even relatively continuous *Pre-Columbian contacts* between the Old and New Worlds throughout much of this lengthy period, to include artifacts, rock carvings, architecture, crops, linguistic similarities and other lines of evidence. The present work on Saharasia provided unanticipated support for Pre-Columbian contact theory, as discussed and fully cited in Part III.

Armored patrism gradually intensified thereafter among peoples living in Saharasia, who later migrated or invaded away from the Saharasian core to affect other borderland regions, and a few regions at very great distances. These findings, derived from history, archaeology, and relatively contemporary anthropological and climatic sources, strongly support or even prove — so far as proof of any ancient historical event is possible — that the innate, or primary, core components of human behavior and social life are unarmored and matrist in character. The maps and data which support these conclusions, and which are presented in sometimes tedious but necessary detail in the following chapters, are primarily derived from classical sources, reflecting the field observations of hundreds of anthropologists, archaeologists, geographers, and historians over the last several hundred years.

Why Has the Saharasian Connection Not Been Observed Before?

If the above points are correct, then why haven't historians, anthropologists, or geographers previously observed this quite profound and apparent *Saharasian pattern* within their own data? The fact is that a number of scholars in various disciplines have elaborated upon *parts or components* of the overall Saharasian discovery. Indeed, without such pioneering work, the connections demonstrated here would not have been possible. For various reasons, however, the full global pattern — of armored patrism's historical roots within the Saharasian desert region — has not been previously identified and is an entirely new finding.

One major barrier has been the disciplinary approach itself, which tends to inhibit synthesis, insuring that scholars will primarily focus upon their own individual specialty, region, or culture, with less emphasis given to regional or cross-disciplinary comparisons. However, Homo sapiens is simultaneously a biological, psychological, and social creature, living within an environmental and cultural dynamic, working, eating, reproducing, loving, interacting, and migrating across broad areas of landscape. Only a broader approach can encompass the full sweep of such interaction.

The work presented here is also, to my knowledge, the first attempt to geographically map the very large anthropological data bases which have been developed in recent decades. G. P. Murdock of the University of Pittsburgh, as mentioned above, spent decades compiling a massive data base on many different variables for some 1200 different native, subsistence-level cultures from around the world. His *Ethnographic Atlas* [4] became available for other scholars to use only within the last 20 years. A computer-readable version of his research data was developed by Murdock's students only after 1970, and most of my cultural maps were prepared from that version of his data. The microcomputer used to read the massive data base and print the maps was also a more recent innovation. It allowed me to make maps in minutes which otherwise would have taken months. (Although, it required nearly one year to perfect the computer program which printed the one-minute maps!) Many new and important archaeological and paleoclimatic field studies also became available only within recent years, providing crucial evidence on both climate and culture change across the Saharasian region.

A number of methodological difficulties also existed in the overspecialized academic disciplines which heretofore blocked the discovery of the Saharasian connection. For example, there is an almost systematic exclusion of study of the family and sexual life of peoples among scholars who have the skills necessary for making global cultural maps. Sexolo-

4. Murdock, 1967, ibid.

gists and sex-historians are well acquainted with materials on the family and sexual life of Homo sapiens, but they rarely review their own data in a cross-cultural manner — or, if so, they do not pay much attention to geography, and do not make maps of their data. Psychologists and sociologists, I learned, also mostly used the anthropological data in anecdotal form, and rarely viewed the natural environment, or climate, as bearing a fundamental role in behavior (even if they do acknowledge the "environment" of the child as a developmental factor). One can search the behavior science journals and textbooks and not find a single map anywhere, as if history, geography, and natural environment have nothing to do with contemporary behavior! And among human or cultural ecologists, who do emphasize natural environment, factors regarding the family and sexual life of Homo sapiens are generally ignored, as if humans only eat, wear clothes, build houses, migrate, or compete against each other. So far as I can tell, the combined geographical and cross-cultural approach used here, with an emphasis upon children, sexuality and antisocial violence, is unique.

It should also be mentioned that some of the most highly specialized, disciplinary approaches to the question of the genesis of human family life, sexual behavior, and violence (e.g., *sociobiology*) have prematurely concluded that patrism is genetically determined. Some have attempted to root aspects of sexism, human psychopathology, and destructive human aggression in the genes, a form of biological "original sin" which labels the cultural status quo of child-beating, sexually-disturbed, war-making *Homo Normalis* as unavoidable biological fact.[5] Of course, these theories uniformly assume that our earliest primitive ancestors, and the earliest settled peoples were sadistic, warlike brutes, even though the archaeological and historical evidence does not support such an assertion. Instead, the presence of peaceful matrist societies is indicated. I argue that such theories, about vicious "naked apes" and greedy "selfish genes", are in reality nothing more than *psychological projection in "scientific" cloth, where the behaviors, beliefs and attitudes within the guts of various psychologists and academics are being inappropriately transferred to cultures and peoples to whom they do not belong.* We have seen this tendency at work constantly in history, wherein peaceful societies are classified as "barbarians", "pagans", "cannibals" or "savages", just before gun-boat missionaries or jihadi-nomads arrive to enslave them and steal their land. I argue, such theories on the "innate nature of repression and violence" can only persist so long as the observed real-world histories and behaviors of people are ignored. A systematic, cross-cultural, and geographically explicit review of anthropological, historical, and archaeological data, as presented in this volume, utterly shatters all theories on the supposed genetic origins or innate nature of the armored warlike human condition, or of the various patrist social institutions.

Another important starting point from which this work proceeded, and which has been largely and unjustifiably ignored (or slandered) within academic-professional circles, was the sex-economic theory of Wilhelm Reich. Of those within Freud's "inner circle", Reich remained the most focused and insistent upon the real nature of childhood trauma and sexual repression. His findings and writings put him at odds with the prevailing social order of Europe and the USA during the period 1920-1957; he was one of the first to state clearly and openly that our treatment of infants, children, and adolescents was both cruel and damaging in the extreme, and was totally responsible not only for the chaotic, violent and self-destructive behavior of many young people, but also for the destructive violence in the adult world, alternatively acted out in uncontrolled sadistic outbursts, and in regimented rituals of nationalistic

5. Wilson, E.O.: *Sociobiology: The New Synthesis*, Harvard U. Press, Cambridge, 1975; Ardrey, R.: *The Territorial Imperative*, Antheneum, NY, 1966; Morris, D.: *The Naked Ape*, McGraw-Hill, 1967; Goldberg, S.: *The Inevitability of Patriarchy: Why the Biological Difference Between Men and Women Always Produces Male Domination*, Wm. Morrow, NY, 1973.

or "holy" wars. He boldly opposed the torturing of infants and children which occurred, and continues to occur, under the banner of "medical obstetrics", childrearing pedagogy and sex-negative obedience training. He championed the right of the adolescent to a healthy love life, well before marriage, something few other natural scientists or social reformers have so straightforwardly and uncompromisingly done.

Reich's works have always been explicit and robust enough for evaluation and testing, though rarely has this occurred. The fact that this study was one of the first to take his observations as reasonable starting points for research appears as another reason why so many new and unexpected connections between culture and natural environment were observed. Reich's work was a major foundation-stone for the Saharasian connection, and the Saharasian connection stands as an independent confirmation of his general theory on human behavior. Reich was the first natural scientist to recognize the close association between the environmental desert and the *emotional desert* of humankind. My research stands upon the shoulders of his prior good work.

Patristic "Civilization" Versus "Uncivilized" Cultures

Before proceeding, it is necessary to focus some sharp criticism upon the central concept of *civilization*, which as popularly viewed is entirely self-serving. How do we define "Civilization"? Traditionally, this word was reserved for "High Cultures", which had developed agriculture, animal domestication, writing, monumental architecture, transportation methods, and technology. The pyramid-building Egyptians and Babylonians were "civilized", the textbooks say. However, little is said about the "primitive" villagers living on the Nile and Tigris-Euphrates prior to the first pyramid or ziggurat; the implication is that these simple villagers were "not civilized". Accordingly, our own "Western Civilization", or the "Eastern Civilization" of the Orient, and even the Inca or Maya are all classified as well-developed "civilizations". Following from this logic, the native peoples of the Pre-Columbian Americas, of Pre-Christian or Pre-Islamic regions, or of tropical areas where little clothing was worn, have historically been labeled with various pejorative terms which themselves are more revealing about those who point the accusing finger: "primitive", "uncivilized", "heathen", "pagan", etc. It should be apparent from this short discussion that something is radically wrong with the above-given definition of civilization, which is equated mainly with technology and the central state. A more revealing definition can be formulated if we focus upon the word itself, which implies *civil behavior and peaceful social conduct*.

From this latter definition, how does the modern world of the 20th-Century stack up? How would we rate the cultures which produced Hitler, Hirohito, Stalin, and Mao Tse'Dung, with their military conquests, death-camps, gulags, institutionalized tortures, and widespread suppression and murdering of both foreigners and citizens? All of these cultures possessed a well-developed technology with strongly hierarchical central states, religious and medical priesthoods, a military caste, and bureaucrats authoritatively ministering to the moral, spiritual, health, and ideological "needs" of their peoples. Even our own American culture, so widely admired and emulated around the world today, has produced its own small share of serial rapists, child-killers and other murderers. Historically, European immigrants to America subordinated and occasionally committed acts of mass-murder against the aboriginal population, and their descendants engaged heavily in the African slave trade —

though to be fair, later generations of Americans fought wars to end slavery, eventually nourishing successful civil-rights and women's rights social movements — and history records no institutionalization of violent criminal elements within America, nor in *most* of Europe, in the formation of brutal "security police" or "religious cops" as seen today within totalitarian Islamic Saharasia. While modern American and post-WW-II European culture contrasts very favorably against the most extremely aggressive and cruel patrist cultures of history, we unfortunately do not look quite so healthy as compared to the truly peaceful matristic cultures recorded in anthropology, and about which Chapter 3 gives many details.

While the impulses to engage in mass murder may lay dormant within a given culture for decades, history shows that many have periodically been *compelled en-masse,* by deep and unconscious psychological urges, to undertake wars for "national glory", religious or ideological supremacy, or more blatantly for land, resources, property, or slaves. If "civilization" is defined only by high technology and the (secular or religious) central state, then we have a great deal of cruelty and bloodshed to explain.

An additional element at work is the *kind and form* of technology an armored, patristic culture finds useful and gives enthusiastic support for, as compared to those technologies which it rejects. Often as not, technological innovations have been misapplied to control us, to reduce our pleasure, vitality and health, and to restrict our choices and freedoms. Only the most blind cheerleader for industry would claim that television, automobiles, nuclear power plants, computers and biomedical pharmacology have unambiguously "set us free", without serious ecological, health and social side-effects, as commercial propaganda would have it. Indeed, the creation of faceless controlling bureaucracies, so essential to the authoritarian central state, have in some cases been amplified and speeded along by certain politically-favored forms of technology.

If a particular technology can be readily applied to building the social pyramid even higher — for controlling people, stealing their wealth, crushing dissent, and exploiting nature — armored patrism will strongly favor and put such a *hard technology* into widespread use. On the other side of the equation, armored patristic cultures tend to avoid or even actively oppose *soft* or *appropriate technologies*, which treat nature and life in a gentle way. For decades, as a culture we collectively chose hard, inefficient nuclear power and unrestricted imported oil, over soft and efficient solar energy and wind-power systems, with disastrous consequences. Harsh and generally ineffective chemotherapy, radiation therapy and surgery was culturally selected for cancer treatment over more health-promoting nutritional, life-energetic and preventative approaches; as was hard hospital obstetrics over softer homebirth and midwifery; hard chemical pesticides and herbicides over softer ecological agricultural methods; hard chemical/shock psychiatry and penal institutions over softer social and sex-economic reforms; and mechanistic genetics and bioengineering over preservation of biodiversity.

On the sexual scene, around the world and also in the West, there is much confusion and misery. Heterosexual sex-repression appears as a causal factor within the same cultures afflicted with child rape, pedophilia, bestiality, genital mutilations and other sexual sado-masochistic rituals — these are sometimes spoken of as "alternative sexualities" by a few university academics[6] who appear motivated to "normalize" their own sexual disturbances. Sexual violence and perversity is also admitted into public movie theatres and onto internet, while the healthy heterosexual norm, and especially adolescent sexuality — as depicted in Shakespeare's *Romeo and Juliett* — is sometimes treated as a crime equal to pedophilia

6. For a discussion with specific citations, see: DeMeo, J.: "Editor's Postscript", *Heretic's Notebook (Pulse of the Planet #5),* p.65, 2002.

or child-rape. Meanwhile, others try to legitimize the pedophiles! Billions are spent to propagandize blatantly unscientific antisex hysteria about "AIDS", *a non-infectious lifestyle-related immune disorder linked to specific high-risk behaviors (i.e., high usage of toxic pharmaceutical and illegal street drugs), which has never been conclusively demonstrated to be caused by a virus, nor to be spread sexually, or in blood.* [7] Those scientists and clinicians pushing for a reappraisal of infectious-HIV theory, or who advocate a healthier sexuality, or who simply point out this confused situation with some clarity are badly mistreated. Censorship, firings, blacklisting, public attacks, social shunning, ridicule, media disinformation and other measures are still supported and rationalized by the elites of nearly every political, journalistic or scientific-medical group, to suppress those who stand on the side of life, and challenge patrist orthodoxy on any level — especially by those who make the biggest fuss about "social justice", "academic freedom" or "freedom of the press." Meanwhile, the deadly orthodox are given full official permission and vast sums of wealth by which to carry forth their pleasure-hating, life-destroying, and socially destructive activities.

Wilhelm Reich
Orgonon, Maine, 1954

However, it is within the Islamic and communist core of Saharasia that one today sees the most socially bleak of situations, murderous towards children and women, hysterically and harshly sex-repressive, intolerant to all forms of independent thought and behavior, particularly if it leads to heterosexual expression or the erosion of the power of the mullahs, or of the party bosses. The average person — brainwashed by state- or mosque-controlled media, and dumbed-down by authoritarian educational systems and demands for conformity to religious and/or political ideology — either acts as hysterical cheerleader to whatever the Glorious Leader or Supreme Ayatollah has to say, or supports their particular system of patristic armoring indirectly via silent consent and emotional-numbness (apathy). While a few brave souls may struggle towards genuine love, and to maintain their freedom, only a social disaster will awaken the larger collective psyche to the tragi-comical *Grand Theatre*.

Whatever the problems within Saharasia today, most certainly all other societies world-wide have their own problematic emotional armoring to contend with. The deeper emotional and sexual needs of men, women, adolescents and children have been badly trounced and distorted, and are often under the domination of those who do not have the best interests of humankind at heart. Our species has generated too much collective sadism, including or especially the hidden sadists in positions of political, religious and even medical power, who stand in the way and block the path to human happiness in virtually all of these soft and life-affirming directions. There also is much financial profit to be made from the economics of patrism, from the war machines, hard technology, ecological destruction and centralized monopoly, all of which further complicates matters and makes change towards the softer, less-armored and matristic alternatives all the more difficult. But again, in this work we are primarily interested in the basic sexual-emotional factors which are the *foundations* upon which these complex institutional and economic patterns thrive.

Within this book we therefore dispense with the unwarranted, unproven, *disproven* popular definition of "civilization". Technology and the central state are insufficient and meaningless in our efforts to define what constitutes a civil social order. Instead, we will view *civil conduct and peaceful, sex-positive, life-enhancing social conditions* as evidence of civilization, no matter what the level of technology or social organization. For most of the contemporary world, we agree with Wilhelm Reich, that *"Civilization has not yet begun"*.

7. Duesberg, P.: *Inventing the AIDS Virus*, Regenery, NY 1996; Duesberg, P.: *Infectious AIDS: Have We Been Misled?*, North Atlantic Books, Berkeley, 1996; Duesberg, P.: *AIDS: Virus- or Drug-Induced?*, Kluwer Academic, NY 1996; Adams, J.: *AIDS: The HIV Myth*, St. Martin's, NY, 1989; Lauritsen, J.: *The AIDS War: Propaganda, Profiteering and Genocide from the Medical-Industrial Complex*, Asklepios, NY 1993; Rappoport, J.: *AIDS Inc.: Scandal of the Century*, Human Energy Press, San Francisco, 1988; Root-Bernstein, R.: *Rethinking AIDS: The Tragic Cost of Premature Consensus*, Free Press, NY 1993; Hodgkinson, N.: *AIDS: The Failure of Contemporary Science. How a Virus That Never Was Deceived the World*, Fourth Estate, London, 1996.

2. Wilhelm Reich's Discovery of Human Armoring: The Link Between Environment, Biology and Social Violence

The sex-economic findings of Wilhelm Reich constituted starting points and major basic assumptions for my own work on Saharasia. Most readers, however, will scarcely have heard his name, or if so, often in the most negative manner. The reasons for this, of why Reich's findings are never mentioned, was raised by R.D. Laing in 1968, but his words could just as well have been written today:

> "It is as though he had never existed. Few medical students, if any, will have heard his name so much as mentioned in medical school, and will never come across him in their textbooks. It is not that his views are less scientific than many of those taught today — which are no more scientific than those clinical dogmas of even 50 years ago that we are now pleased to ridicule or patronize. Reich's proposals as to the social influences on the functions of sympathetic, parasympathetic, and central nervous systems, and on our biochemistry, are testable, but are never tested, as with much else that is really important.
>
> Whether or not one agrees or disagrees with this or that of Reich's theory and practice, it is inescapable that he was a great clinician, with an unusually wide range... He understood the mess we are all in — hysteric, obsessional, psychosomatic _Homo normalis_ — as very few have done. Yet one will look through a hundred journals in the Royal Society of Medicine without coming across one mention of him. Why is he never mentioned?" [1]

The answer to Laing's question requires some discussion of Reich's differences with his mentor, Sigmund Freud, differences which also put him in opposition to the general trend of thought in the behavioral sciences. Up until around 1933, Reich was an active member of Freud's "inner circle", and a regular contributor to psychoanalytic journals. His writings on _Character Analysis_ [2] were considered a breakthrough in analytical technique, and his book is often still used as a standard text for analytical training. Freud was so impressed with the young Reich that he invited him to full membership within the Vienna Psychoanalytical Society and began referring patients to him for treatment _only one year_ after meeting him, at a time when Reich was still a medical student at the University of Vienna. Reich eventually became the Assistant Director of the Berlin Psychoanalytic Polyclinic, and was charged with the responsibilities of training all new analysts. However, he was eventually forced into an open conflict with the psychoanalytic establishment, as he increasingly focused upon the real, physical traumas and genital frustrations experienced by people, and as his sex-political work brought him into public opposition to the growing Nazi influences in Germany.

Today, with the perspective of history, we observe Reich's views about German fascism were far more telling and accurate than were the views of the other analysts of his day, including or especially the views of Freud and Carl Jung — both of whom are much better known and acknowledged within the behavioral sciences today than is Reich. Also, from the work of writers such as Jeffrey Masson, we have solid archival documentation about Jung's open and unapologetic collaboration with the Nazis, [3] and

1. Laing, R.D.: "Liberation by Orgasm", _New Society_, 28 March 1968, p.464; Reprinted as "Why is Reich Never Mentioned?", _Pulse of the Planet_ 4:76-77, 1993.

2. Reich, W.: _Character Analysis_, Orgone Institute Press, NY, 1949; Reprinted by Farrar, Straus & Giroux, NY 1972.

3. Masson, J.: "Jung Among the Nazis", _Against Therapy_, Atheneum, 1988.

additional facts regarding Freud's abandonment of his own early findings on trauma and repression.[4] What is rarely appreciated is the fact that, among the psychoanalysts, it was Wilhelm Reich who made the first and strongest public challenge to the Nazi movement in Germany, and also to Freud regarding the ignoring and de-emphasis of traumatic experience and sexual repression as factors in the genesis of neuroses. Here is a brief review of the historical development of Freud's psychoanalysis and Reich's later character analysis. These are central to understanding how the real traumas of drought-induced famine and starvation could powerfully affect a peaceful, unarmored culture, pushing them towards violent, armored, patrist social conditions.

Freud's Early Work, and Subsequent Betrayal of Truth

Whatever his later faults, as a young man Freud brought to the world a dynamic view of psychic processes and of the child, a unique view which connected the experiences of childhood to the attitudes and behaviors of the adult. His theory of the unconscious, which expressed itself in dream content and free associations, brought about a revolution in the understanding of neurotic and antisocial behavior.[5] Freud did not see the child as a *tabula rasa*, a blank slate upon which anything could be "written", in the manner of modern day behaviorism. Neither did he believe that the seed of the fully formed adult character, replete with both social and antisocial elements, was present in the infant from birth onward. His early work clearly indicated a belief that, while the human animal did come newborn into the world with specific instincts and drives, those impulses tending towards sadism, masochism, warring, brutality, and the like developed later in life, as a response to traumatic early childhood experiences. While Freud clearly retreated from this position later in his life, it was his young student, Wilhelm Reich, who became the most uncompromising and vocal champion of Freud's own early viewpoint.

Schooled in the vitalism of Charcot (one of whose teachers was the vitalist Mesmer, who claimed the discovery of a specific *animal magnetism*), Freud perceived the newborn infant and child as being charged with *libidinal energy* which was invested in certain *pleasurable* biological functions. The infant would naturally reach out for pleasure, but not for pain; there was, indeed, no primary reason to believe an organism would deliberately reach out towards painful stimuli, although this clearly could be seen in the case of the human animal, in various neurotic and self-destructive expressions. Freud searched for the answer to *why* this was so. He elaborated upon the major sensory pleasures of childhood: breast-feeding, defecation, and genital excitation. And he observed that the child passed through distinct growth phases, during which its interests were focused upon these same biological functions, and their corresponding anatomical areas. If the maturing child was repressed or inhibited in meeting its early needs, Freud observed, its psychic development would be halted at that particular stage; later adult behavior and thought would be laden with a preoccupation or "taint", derived from that same unsatisfied stage of development.

Freud viewed neurotic and psychotic, destructive behavior basically as the product of inhibition or blocking of the libidinal energy of the drives. He saw the libido as primarily a sexual energy, which could be invested by the child in various pleasurable activities. Or, libidinous impulses could be repressed, sometimes in the natural course of its development, or sometimes pathologically giving rise to neurotic elements in the character structure. The sexual etiology of the neuroses was central

"...no neurosis is possible with a normal vita sexualis"
Sigmund Freud (1898)
From *Sexuality in the Aetiology of the Neuroses* written when Freud was a young man; contrast this view with those of the elder Freud, at right.

4. Masson, J.: *The Assault on Truth: Freud's Suppression of the Seduction Theory*, Farrar, Straus & Giroux, NY 1984.

5. For instance, see "Biology and Sexology Before Freud", in Reich, W.: *The Function of the Orgasm*, Noonday Press, NY, 1961, pp.3-19.

Sigmund Freud, c.1930

"Civilization has been built up on the denial and control of instinctive urges, and every single individual should repeat the history of the ascent of the human species on his path through life with understanding and resignation. Psychoanalysis has shown that the instinctive urges which must go through the transmutation into civilization are mainly — though not solely — those of sex."

Sigmund Freud (1930)
Civilization and its Discontents

6. Freud, S.: *Sexuality in the Etiology of Neuroses* (1898); *Three Essays on the Theory of Sexuality* (1905); *"Civilized" Sexual Morality and Modern Nervous Illness* (1908), *The Standard Edition of the Complete Psychological Works,* Early Psychoanalytic Publications, London, 1961.

7. Freud, S.: *The Aetiology of Hysteria* (1896), *Standard Edition*, Vol.III, 1961.

8. Masson, J. *Assault on Truth*, ibid.

9. Malinowski, B.: *Sex and Repression in Savage Society,* Routledge & Keegan Paul, London, 1927; Malinowski, B.: *The Sexual Life of Savages*, Routledge & Keegan Paul, London, 1932.

10. Reich, W.: *People in Trouble*, Orgone Inst. Press, Rangeley, Maine, 1953; Reich, W.: *The Sexual Revolution*, Octagon Books, NY, 1971. Also see: Rackelmann, M.: "Was War die Sexpol? Wilhelm Reich und der Einheitsverband für proletarische Sexualreform und Mutterschutz", *Emotion* 11:56-93, Berlin, 1994.

to Freud's early thinking, and his early writings revealed a concern for the effects of sex-repression on the health and behavior of the adult.[6] However, Freud also saw some aspects of sex-repression as being socially beneficial, paving the way to civilization. He argued that childhood sexual *latency* (where the child loses interest in the opposite sex, or develops a negative attitude) was a biologically-demanded stepping-stone for healthy sexual development, with sexual impulses re-emerging later in adolescence, to be affirmed in marriage afterward. He saw no social benefit to adolescent or premarital sexuality, and never advocated a loosening of social restraints upon adolescent sexuality. Freud also saw sadism as being attached to the sex impulse in some unknown way. The germs of these ideas were present in the earliest of his papers, though it must be said that the young Freud more clearly articulated the negative effects of sex-repression and childhood trauma more than any supposed social benefits.[6] This emphasis would change as he grew older.

For instance, in his 1896 work *The Aetiology of Hysteria*,[7] Freud openly discussed the real childhood sexual trauma experienced by an unknown, but higher-than-expected percentage of girls (and boys also) within "normal" society. However, after presenting these early findings, he had been mercilessly rejected by establishment doctors and co-workers in Vienna. When even more solid and alarming evidence for sexual abuse during childhood emerged — that is, violent rape by a father or uncle — Freud increasingly drew back from interpreting this evidence so straightforwardly. He later changed his mind about the reality of the materials presented by his patients, viewing them instead as unconscious wishes and fantasies. As documented in the works of Masson,[8] Freud evaded his own evidence through rather contorted logic, and argued that such women had not, in fact, been raped as children — instead, he argued, unconscious childhood "Oedipal desires", childhood wishes to sleep with the parent of the opposite sex, had merely surfaced during the analysis, and were being misinterpreted by the patients. After changing his findings in this manner, and asserting that no real sexual abuse was taking place, Freud received the public applause formerly withheld. Of course, the emotional and physical health of his female patients were sacrificed on the altar of "social approval".

While there is a factual basis to the existence of childhood sexuality, and children do express a powerful desire to be held and loved by parents in a manner which includes a strong sexual component (but without any "wish" or capacity for genital intercourse), Freud refused to confront the facts regarding real parental assaults upon the sexuality of children. He rejected the observations of his student Reich, that the compulsive nuclear family contained a great deal of sexual frustration and bottled-up sexual tension, and that this tension was the source of the familial frustrations and childhood assaults. Freud also rejected the observations of the psychoanalytically-trained anthropologist Bronislaw Malinowski, that — in sharp contrast to Nineteenth Century Europe — sex-positive, matrilineal cultures did not exhibit an "Oedipal Complex" nor any traces of "childhood sexual latency".[9] Rather than directly and openly address these concerns, Freud and the overwhelming majority of the psychoanalysts simply denied any reality to the reports of trauma coming from their female patients, insisting that the Oedipal situation and fantasy/wish were the sources of the reports. Freud became entrenched in his views, and was destined to clash with workers like Reich, who had established sexual hygiene clinics in working-class neighborhoods, and witnessed first-hand the desperate social and economic situations, and often dysfunctional sexual lives of ordinary people.[10]

As a new social storm gathered over Europe, the elder Freud, having already lived through one major World War, openly expressed his disillusionment with the human condition in *Civilization and Its Discontents* (1930),[11] developing themes on a hypothetical "death instinct" which previously had been asserted in *Beyond the Pleasure Principle* (1920).[12] Whereas Freud once saw hope for humanity, by the 1930s he fully reversed his position, expressing great pessimism, and arguing that humans carried the seeds of their own destruction in a truly independent, destructively aggressive instinct. One only needs to look at Freud's picture from this time, to see the disappointment and resignation in his face. By contrast, his younger student Reich was full of energy and engaged in a battle to transform society, to steer it away from the quite obvious social catastrophe which lay ahead.

Reich's Sex-Political Work, and Break With Psychoanalysis

Reich was the de-facto leader of a small group of younger psychoanalysts who were directly engaged in social reforms — giving lectures on healthy childrearing and sexuality, distributing contraceptives in violation of local laws, fighting against the Nazi political agenda, etc. — and together they developed ties to various sex-reform and political movements. Indeed, psychoanalysis was, in those early days, viewed as a radical or even socially dangerous doctrine.[13] Freud, however, did not go as far as the younger analysts in advocating or actively pursuing sexual and social reforms. Of his younger students who did so, Wilhelm Reich was the dominant figure.[14] As Assistant Director of the *Berlin Psychoanalytic Polyclinic* for 7 years, he was charged with the task of training the younger German analysts, and therefore had a great influence. With help from many of these younger analysts, Reich gradually moved from clinical work towards sex-political reforms, wedding psychoanalysis to socialism.[15]

Reich viewed psychoanalysis as a force which, when applied socially, would liberate society faster than either psychoanalytic or socialist-economic reforms alone. Many of the younger analysts in Reich's circle worked with either the Social Democrats or the German Communist Party to oppose the growth of fascist National Socialism in pre-Hitler Germany. National Socialism, the Nazi movement, was documented by Reich and his followers as a sex-negative reactionary political movement, a dangerous expression of mass neurosis with central components of paranoia and psychosis. He exposed its anti-Semitic roots, well before the period of the concentration camps and Nazi street murders, warning of the dangers to come. The socialists and communists were, at that time, the only political groups openly opposing Hitler's National Socialism, and Reich moved into these circles in attempts to better introduce psychoanalytically-based reforms. Reich and his followers worked politically to have psychoanalytic principles written into party platforms, in a grand attempt to help move the masses towards health, in a collective manner, rather than only through a small number of individuals in therapy.

In addition to his clinical work, Reich gave lectures and wrote articles and pamphlets calling for abolishment of laws prohibiting sex-education or the distribution of contraceptive information and devices; he fought for the exemption of consenting adolescents from statutory rape laws, and for reform in family law to allow easy divorce and full legal rights for illegitimate children. He advocated equal rights and pay provisions for female workers, as well as the setting up of child care facilities in work places. Reich also believed prostitution and homosexuality would end once their roots in the sexual frustrations of the masses, and the poverty of lower class women, was eliminated. However, he also worked for an end to

"Both through his published work and in the personal contacts he [Reich] has impressed me as an original and sound thinker, a genuine personality, and a man of open character and courageous views. I regard his sociological work a distinct and valuable contribution to Science. It would, in my opinion, be the greatest loss if Dr. Reich were in any way prevented from enjoying the fullest facilities for the working out of his ideas and scientific discoveries...
— Bronislaw Malinowski"

(From an Open Letter dated 12 March 1938, defending Reich from malicious attacks in the Scandinavian newspapers. Reprinted in: *Reich Speaks of Freud*, Farrar, Straus & Giroux, NY 1967, p.219)

11. Freud, S.: *Civilization and its Discontents*, W.W. Norton, NY, 1961.

12. Freud, S.: *Beyond the Pleasure Principle*, Standard Edition, London, 1961.

13. Jacoby, R.: *The Repression of Psychoanalysis*, Basic Books, NY, 1983.

14. Otto Fenichel, Erich Fromm, Edith Jacobson, and Karen Horney were among the analysts active in social and political reform movements, and were generally allied with Reich; cf. Sharaf, M.: *Fury on Earth*, St. Martin's/Marek, NY, 1983, p.160; Jacoby, *Repression of Psychoanalysis*, ibid.

15. Reich, W.: *Sex-Pol Essays, 1929 to 1934*. L. Baxandall, Ed., Random House, NY, 1966; Reich, W.: *Early Writings*, Farrar, Straus & Giroux, NY, 1975; Rackelmann, 1994, ibid.

persecution of prostitutes and homosexuals. Working with teams of doctors and volunteers, he set up clinics for sexual hygiene counseling and for bringing psychoanalytic therapy to the masses, with many functions similar to, or broader in scope than those advocated today by Planned Parenthood. In Austria, it was the *Socialist Association for Sex Hygiene and Sexological Research* (1929), and later in Berlin the *German Association for Proletarian Sex-Politics* (1931). This latter Association was an umbrella organization encompassing other social-reform groups with a total of around 40,000 associated members, and for a brief period became a political force within the anti-fascist movement in pre-war Germany.[16]

Although it is important to note that many of the reforms sought by Reich have been achieved in the West today, such reforms were not then present, and were only remotely possible in Europe of the 1920s and 1930s. Sexual education and hygiene work was undertaken in the face of great opposition from both Church and State, and reformers were often thrown into prison, or forcibly exiled. The sex-political reforms were vigorously opposed by much of the conservative German population, including the Nazis, who strongly defended the patriarchal authoritarian family structure, and viewed ideas about free contraceptives, legalized abortion, and adolescent sexual love with total horror. While the elder Freud never supported the fascist movement, neither did he support the sex-political efforts of his younger students. Indeed, it has been said that Freud, with two young daughters, was not prepared to grant sexual freedom within his own household.[17] While psychoanalytical theory stated that blocked or repressed sexual energy would result in neurosis, Freud did not confront the social implications of his own theory head-on, as did Reich. Reich opposed Freud on the supposed benefits of adolescent sexual sublimation, on the universality of childhood sexual latency and Oedipal conflict, and later, on the genesis of sadism and masochism.[18]

As fascism grew in Germany during the 1930s, and as Reich's anti-fascist, sex-political activities increased, Freud distanced himself from Reich. Some of the analysts, primarily Carl Jung, even began to collaborate with the Nazis, openly condemning "Jewish psychology". For a time, Jung was editor of the Nazi-supported *Allgemeine Aerztliche Gesellschaft für Psychotherapie*, becoming an official representative of National Socialism. For instance:

> *"In 1933, Jung took over the presidency of the <u>Allgemeine Aerztliche Gesellschaft für Psychotherapie</u> and the editorship of its organ, the <u>Zentralblatt für Psychotherapie</u>. Speaking about the major goal of the <u>Zentralblatt</u>, Jung stated in the first number edited by him: 'That there are actual differences between Germanic and Semitic psychology has long been known to intelligent people. These differences are no longer going to be obliterated, which can be only to the advantage of science.' This was completely in line with the simultaneous statement of the <u>Reichsfuhrer of the Deutsche Allgemeine Aerztliche Gesellschaft für Psychotherapie</u>, Prof. Dr. M. H. Göring, that 'the society (for psychotherapy) assumes of all members who are active as writers or speakers that they have worked through in all scientific earnest Adolf Hitler's fundamental work <u>Mein Kampf</u> and that they acknowledge it as the basis of their work. The society wants to cooperate in the work of the Volkskanzler, to educate the German people to a heroic, self-sacrificing attitude.'"* [19]

This alliance of Jung and other psychoanalysts with the Nazi movement (or at very best, unapologetic capitulation) was solidly documented by Masson, but others have previously noted this same widely-ignored connection.[20]

16. Sharaf, *Fury on Earth*, ibid., pp.133, 162; Reich, *People in Trouble*, ibid.; Rackelmann, 1994, ibid.

17. Reich, W.: *Reich Speaks of Freud*, Farrar, Straus & Giroux, NY, 1967, pp.45, 50f,57-8,129-30. Anna Freud subsequently became an ardent spokesperson for adolescent sexual repression.

18. Reich, W.: *Genitality in the Theory and Therapy of Neurosis*, Farrar, Straus & Giroux, NY, 1980; Reich, W.: *Children of the Future, on the Prevention of Sexual Pathology*, Farrar, Straus, & Giroux, NY, 1983.

19. T. Wolfe, translator's footnote, in Reich, *Function of the Orgasm*, ibid., p.127f.

20. Masson, "Jung Among the Nazis", ibid.; Also see Cocks, G.: *Psychotherapy in the Third Reich*, Oxford U. Press, 1985, and "Psychoanalysis and the Nazis: A Collaboration?", in *Tempo* section of *Chicago Tribune*, 29 August 1984; Wolfe, Ibid.

Reich's 1931 Sexpol Proposals:

1. Free distribution of contraceptives to those who could not obtain them through normal channels; massive propaganda for birth control.
2. Abolition of laws against abortion. Provision for free abortions at public clinics; financial and medical safeguards for pregnant and nursing mothers.
3. Abolition of any legal distinction between the married and the unmarried. Freedom of divorce. Elimination of prostitution through economic and sex-economic changes to eradicate its causes.
4. Elimination of venereal diseases by full sexual education.
5. Avoidance of neuroses and sexual problems by a life-affirmative education. Study of principles of sexual pedagogy. Establishment of therapeutic clinics.
6. Training of doctors, teachers, social workers, and so on, in all relevant matters of sexual hygiene.
7. Treatment rather than punishment for sexual offenses. Protection of children and adolescents against adult seduction.

Reich's 1933 work, *The Mass Psychology of Fascism*, was a penetrating analysis and criticism of National Socialism, acting as an irritant to the Nazis, attracting fascist attacks, including episodes of book-burning. In that work, Reich exposed the roots of the German fascist philosophy and political movement in the clear-cut repressive sexual pathology of the German patriarchal authoritarian family.[21] Shortly after the book was published, Reich was formally but secretly *expelled* from the International Psychoanalytic Association (IPA) — at a time when various psychoanalysts, including Freud himself, were quietly lobbying Jewish members of the IPA to formally resign, so as to spare the IPA from attack by the Nazis.[22] (None of these facts have been included within the "official history" of psychoanalysis, of course.) During the IPA Lucerne Conference of 1934, Reich was notified of his expulsion, and nearly blocked from presenting a paper — he was finally allowed to present as "guest". This unethical political move was accepted without apparent protest by both the German and American psychoanalysts in attendance.[23] Reich was also banished from future publishing in psychoanalytical journals. Soon afterward, he was attacked in the Nazi press and fled to Scandinavia.

The points Reich made in *Mass Psychology of Fascism* are important enough for the basic assumptions to be restated here: *State structure is determined by family structure.* The male-dominated German family, by demanding strict obedience from wife and children, he argued, was a prerequisite for the development and maintenance of the fascist character and authoritarian state apparatus. Children were made docile and obedient not only through various threats and physical abuses, but also by strict requirements for absolute denial of sexual feelings and emotions. The German Church and schools likewise structured children so as to oppose almost all forms of independence in thought or behavior. The child was crushed, through compulsive rules and harsh punishments, so as to eliminate spontaneous and self-regulated inclinations, which were subor-

21. Reich, W.: *The Mass Psychology of Fascism*, Orgone Institute Press, NY, 1946.

22. Nitzschke, B.: "Wilhelm Reich, Psychoanalyse und Nationalsozialismus", *Emotion*, 10:183-190, 1992; cf. "Freuds Ungeduld wuchs – Psychoanalyse und Nationalsozialismus", *Die Zeit*, 5.10.1990.

23. Karl Menninger, a strong advocate of compulsive morality, sex-repression, and psychiatric shock therapy was at the Lucerne IPA meeting as the official representative of American psychoanalysis. He apparently approved of Reich's expulsion; in later years, he would help to spread malicious lies about Reich in his journal, *Bulletin of the Menninger Clinic.* (For example, the poisonous article by free-lance journalist M. Brady in: *Bull. Menninger Clinic,* 12(2):61-67, March 1948.)

Wilhelm Reich in Tyrifjord, Norway, 1939.

dinated entirely to the prevailing authoritarian status quo. Children produced under the German patriarchal authoritarian family system, he argued, were timid and cowed, afraid of their own inner feelings, and incapable of openly reaching out to satisfy their deeper emotional needs. When grown to adults, they fell easy prey to various mystical philosophies, which promised in a mystified "cosmic" way, to satisfy their ungratified inner longings. National Socialism, as Reich documented, always alluded to a coming new era of deep emotional and sexual fulfillment, though only in the context of racial purity, and with open legitimizing of sadistic outbursts towards Jews, Gypsies, Slavs and foreigners. The unsatisfied sexual longings of the masses, and the associated inner biological tensions, Reich observed, were being capitalized upon and harnessed into a mass movement by the political outlaws of National Socialism. With a charismatic posture and authoritarian doctrine, the Nazis vowed to resolve people's anxieties and problems at once, to "restore lost German Pride", and, like a familiar and comfortable strict father, take the heavy yolk of freedom and responsibility from their shoulders. A child whose biological needs had been met, Reich argued, which grew up in a loving, self-regulated environment, who was not sexually repressed in adolescence, and who consequently established their own sexually-gratifying love-life, would never fall prey to fascist doctrines. With views such as these, Reich became a thorn in the side of the psychoanalysts, particularly those who still felt they could work "within the system" of Hitler's Germany.

The German communists also became upset with Reich, claiming that he was "too preoccupied with the sex thing" and did not pay enough attention to Marx's economic prescriptions. As Reich described it, his public lectures on sexual hygiene would draw ten times as many people as did the dry lectures on Marxist economic principles given by stiff party bureaucrats. Many priests and puritanical elements had also dominated the Communist Party in the Soviet Union, where all the family reforms which briefly existed following the Russian revolution were being undone. These events in the Soviet Union clearly affected the German communist movement, where the Party leadership eventually broke with the sex-political platform developed by Reich and his followers. The patriarchal authoritarian family structure was to remain the norm under the banners of both the Communist Party and National Socialism. For the communists, the breaking point especially came in 1933 with the publication of Reich's *The Sexual Struggle of Youth*.[24] This was a pamphlet aimed at young people, presenting clear and concise information about reproductive physiology, sexual hygiene, contraception, abortion, and about the emotional aspects of sex. This was too much for the communists, who in short banished him and his co-workers from using their lecture halls and clinics, and ordered his books out of their storefronts. After these events, which came only a few years after his earlier disappointing visit to Stalinist Russia in 1932, Reich publicly condemned and refuted the communists, proclaiming their leadership to be power hungry, without any genuine interest in human freedom. He called the communists "red fascists", with little functional difference from the "black fascists" of Nazi Germany.[25] His complete and open break with the communist movement in 1934 came nearly 20 years earlier than that of other Western socialists.[26]

While always careful to acknowledge his debt to Freud, after his formal expulsion by the IPA, Reich entirely removed his working approach from the umbrella of psychoanalysis, calling it by a new name: *Sex-Economy*. He left Germany for Denmark just before the war, evading border guards who had his name on their lists for immediate arrest, and probable execution. From 1934 to 1939, Reich moved to Sweden, Denmark, and later to Norway at the invitation of analysts in those nations, many of whom openly embraced his work.[27] In 1939, Reich came to the United States at the invitation of Theodore Wolfe, a pioneer in psychosomatic medical research.[28] In America, he taught briefly at the *New School for Social Research* in New York City, but eventually established his own private institute for research, education and clinical training of his new American students. Reich's work increasingly focused upon bioenergetic phenomena he previously discovered during his research in Europe.

As history has it, psychoanalysis has today largely abandoned its original central elements: the sexual etiology of the neurosis, the libido energy theory, and the primacy of the pleasure drives. Psychoanalysis may be counted as but one of many mystical "original sin" doctrines popular among contemporary behavioral "scientists" who, like Catholic priests or Iranian Mullahs, refuse to ever consider the idea that sexual frustration plays a central role in the genesis of violence or neurosis. The modern psychoanalytic philosophy regarding child-rearing and sexual repression do not seriously challenge the existing patriarchal social order or status quo; the child is today seen by psychoanalysis as "unruly", needing to be "civilized" or "socialized", very much in keeping with the repressive, anti-scientific doctrines of the Church. One only needs to quote from the elder Freud himself:

> *"Civilization has been built up on the denial and control of instinctive urges, and every single individual should repeat the history of the ascent of the human species on his path through life with understanding and resignation. Psycho-analysis has shown that the instinctive urges which must go through the transmutation into civilization are mainly — though not solely — those of sex."* [29]

Reich's sex-economy, however, took the reverse position, which once was partly embraced by psychoanalysis itself: the existing social order is sick and, to the extent that it opposes biological urges and needs, creates pathos in people who otherwise would remain healthy; the child is spontaneously civil, decent and social, not harboring innate violent, destructive,

24. Reich, W.: *Der Sexuelle Kampf der jugend (The Sexual Struggle of Youth)*, Sex-Pol Verlag, Berlin, 1932.

25. Reich, *People in Trouble*, ibid.

26. Grossman, R., Ed.: *The God that Failed*, Harper & Bros., NY 1949.

27. Siersted, E.: *Wilhelm Reich in Denmark*, English translation in *Pulse of the Planet* 4:44-69, 1993.

28. Wolfe was a former husband of Flanders Dunbar, another pioneer in psychosomatic medical research. Dunbar never embraced Reich's work, however, and the two later divorced.

29. Quoted by M. Hodan, *The History of Modern Morals*, Wm. Heinemann, London, 1937, p.303.

or sadistic drives. These only appear if the individual is damaged by severe trauma, neglect and repression.

As history has demonstrated, those analysts who had long-term and intimate contact with Reich have tended, in a small way at least, to echo Reich's views on the child and human sexuality, and on the role of repressive social institutions in creating neurosis.[30]

Clinical and Experimental Aspects of the Sex-Economic Theorem

Let's review some of the specific clinical findings of Reich, for clues about how a culture-change might develop from a radical change in environment, and how, when culture opposes biology, problems develop.

1) Discovery of the Muscular and Character Armor

Of all the psychoanalysts, Reich was the most biologically oriented; in addition to having Freud as teacher, he engaged in postgraduate medical studies at the University of Vienna under the Nobel laureate neuro-psychiatrist Wagner von Jauregg. His study brought him into contact with other famous scientists of the period, such as Kammerer and Steinach, and he assimilated the vitalistic writings of Kant, Bergson, Driesch, and others.[31] Hence, it was not surprising that Reich came to view Freud's "libido" as a real energetic force, the emotions and sexual excitation as real, powerful phenomenon, capable of mobilizing behavior and muscular activity, or immobilizing a patient with severe emotional conflicts. Freud had hoped for eventual verification of psychoanalysis in the biological sciences, but not along the vitalistic lines of reasoning developed by Reich. It was Reich, though, who initiated some of the first clinical and laboratory experiments to test the energetic aspects of behavior and emotion, the "libidinal" forces of psychoanalytic theory.

Reich's studies on the genesis of sexual pathology and other forms of neurotic behavior demonstrated related psychic and somatic components. Every psychological problem possessed a corresponding somatic aspect, he observed, and both were underlain by the common denominator of blocked or repressed sexual/emotional energy. When emotions are repressed, be it sadness, terror or rage, those specific muscular groups which ordinarily participate in the *expression* of an impulse become swollen or spastic; the energy of the impulse or emotion literally becomes *trapped within the organism's nervous system and muscles*. The mental process of the individual is also affected, with a preoccupation and "taint" in the psyche corresponding to the painful events which led to the original blocked impulse or feeling. With this blocking of emotional energy, Reich observed clear somatic and motor influences within the expression, mannerisms, and posture of the individual. The way people walked or talked, their mannerisms, how they stood or gestured revealed a great deal about their personal history, the kinds of feelings they were either allowed or forbidden to express as children, and how their childhood experiences were affecting their adult existence. The psychic components of the repressed emotion, he observed, appeared within the character structure as defensive attitudes and resistances to feeling, or as resistances to any behavior or social situation which had the potential to lead to increased feeling. As Reich argued, both character structure and body, psyche and soma, were affected by past experiences of trauma and repression, and participated in

30. Horney, K.: *New Ways in Psychoanalysis*, Norton, NY, 1966; Fromm, E.: *Escape from Freedom*, Hold, Rinehart & Wilson, NY, 1963; Perls, F.: *In and Out the Garbage Pail*, Bantam, NY, 1969.

31. Sharaf, *Fury on Earth*, ibid., pp.55,66-7; Reich, *Function of the Orgasm*, ibid., pp.22-24.

Functioning of the Autonomic Nervous System (After Reich 1934)

Organ System	Sympathetic System *Anxiety Reaction*	Parasympathetic System *Pleasure Reaction*
Iris of eye	Dilation of pupils	Constriction of pupils
Tear glands	Inhibition, "dry eyes"	Stimulation, "glowing eyes"
Salivary glands	Inhibition, "dry mouth"	Stimulation, "mouth watering"
Sweat glands	Stimulation, skin wet "clammy"	Inhibition, dry skin
Peripheral blood vessels	Constriction; pallor, cool skin	Dilation; flushed skin; warm
Piloerector muscles	Hair bristles, "goose flesh"	Smooth skin
Bronchial muscles	Dilation, relaxation	Constriction, tension
Heart muscle	Excitation, accelerated pulse	Inhibition, slows pulse
Digestive system	Inhibition of peristalsis	Stimulation of peristalsis
Adrenal	Stimulation of secretion	Inhibition of secretion
Urinary bladder	Inhibition of expulsion	Stimulus of expulsion
Bladder sphincter	Sphincter closes, inhibition	Sphincter relaxes, urination
Male genital: penis	Flaccidity, withdrawal	Enlargement, erection
Male genital: scrotum	Excites, tightens	Inhibits, relaxes
Female genital	Contraction, dryness	Expansion, moist

Reich found the healthy organism, depending upon its circumstances, was capable of experiencing full excitation of the parasympathetic (pleasure) nervous system. However, people who have grown up and continue to live under long-term painful and repressive social situations may suffer from chronic over-stimulation of the sympathetic (anxiety) nervous system — and gradually become incapable of full pleasurable parasympathetic excitation. Such is the neuromuscular basis of human armoring, a phenomenon which affects both psychic and somatic aspects of life. Culture institutionalizes such expressiveness or inhibition.

The Basic Antithesis of Vegetative Life Functions

**Only one of the branches can be fully excited at a given moment.
Armor results from chronic over-excitation of one or another branch.**

Parasympathetic
Pleasure - Expansion
Towards the World

Sympathetic
Anxiety - Contraction
Away From the World

**Autonomic Nervous System
Bioenergetic Impulses**

the subsequent repression of emotional and sexual impulse energy.

Speaking more and more in energetic terms, Reich abandoned the terminology of psychoanalysis — id, ego, and the like — as the biophysical expressions of emotional energy gained in importance. For example, Reich observed that almost all neurotic patients displayed a chronic inhibition of respiration, or holding of the breath, a stiffening of the pelvis related to sexual repression, and an inability to express deep emotion such as crying, sadness, rage, or terror, even though they had suffered miserably, and had much to cry or rage about. Instead, emotions and sexual longing remained unexpressed and "held down", as it were, by chronically stiffening and tensing those muscle groups which normally participated in the expression of a given emotion. Evidence of neurosis or psychosis could be found etched in the face and body, in the posture, skin tone, and eyes, which revealed as much or more than the free associations or dreams produced in psychoanalytical sessions. Reich argued that a patient's entire being participated in the neurosis, which included not only repressed emotions, inhibited breathing, retracted pelvis, tight stomach and other stiff muscles, but also disturbed thought patterns, and pathological changes in skin pallor, muscle tone, and contactfulness of the eyes. In this manner, and more so than any other researcher before or since, his work demonstrated *the unity and functional identity of both psyche and soma.*[32]

In a milestone research paper, *The Basic Antithesis of Vegetative Life Functions*,[33] Reich identified the basic biochemistry, neurology and biological organ systems by which the individual would expand and reach out to satisfy their needs, or alternatively by which inhibition would be accomplished. The individual could reach out *towards the pleasurable world* if they were not punished or threatened into inhibition of their impulses. Alternatively, following a lifetime of early trauma and repression, an individual might become totally incapable of reaching out emotionally or sexually, moving towards, or working towards satisfaction of their deeper needs. Under such circumstances, the predominant mode of expression was *away from the painful world*. An organism could, depending upon the life circumstances, either emotionally grow and expand, or contract into emotional and sexual stasis.

Reich used the term *armoring* to describe the way in which individuals would chronically wall themselves off from a painful outer world, and also from contact with their own hurt and unsatisfied feelings. Repressed emotions and undesired sensations could be warded off and submerged, while impulses were blocked and inhibited, a feat accomplished through both censorious character attitudes and chronic neuromuscular contractions of specific body areas. The armored character thereby presents to the world a stiff protective facade which, to the perceptive eye, reveals their personal emotional history. One individual, who chronically received physical punishment in childhood and swallowed their sadness and rage, resigning to the punishment, might as an adult chronically hunch the shoulders, breathe shallowly and have very frightened eyes, as if still expecting a surprise blow; another individual experiencing a similar childrearing might possess a stiff, puffed-out "defiant" chest and jaw, holding back enormous bottled-up rage; another still pulls the facial muscles upward, into a false smile, to hold back the misery which would otherwise flood the organism if those muscles were to relax downward, into a forbidden cry. The biophysical stiffness of heavy armor might give rise to contactless "dead" eyes, rubbery or cold, clammy skin, chronic stiffness in the walk and posture, and other all-too common somatic expressions.

For the child, the character and somatic armor at one point served a rational and necessary protective survival function, against the severity of

32. Reich, Ch*aracter Analysis*, ibid.; *Function of the Orgasm*, ibid. By contrast to Reich, most contemporary popular discussions on the "mind-body" problem appear absolutely shallow and insipid. The most public and vocal champions of the "mind-body" relationship have never, to my knowledge, cited Reich's priority on this important question.

33. Reich, W.: "Der Urgegensatz des vegetative Lebens (The Basic Antithesis of Vegetative Life Functions)", *Zeitschrift für Politische Psychologie und Sexualokonomie,* 1(1-2):4-22, 1937. Reprinted in *Pulse of the Planet* 4:5-19, 1993.

external assaults, much like the armor of the medieval knight. But, as punishments and irrational demands upon children are usually repetitive, and often chronic, so too does the armor become chronic in the older child and adolescent. As the child is emotionally crushed and sexually castrated, biological impulses are more and more subdued and repressed, spontaneity is reduced, and deeper feelings extinguished. The passions of love and enthusiasm for life and work are thereby gradually or radically eliminated from adult experience, often being supplanted with a feeling of "having lost something important", or a feeling of resignation. Emotionally, Reich argued, we live as mere shadows of our deeper selves, the armor sapping our strength, pleasure and joy in life.

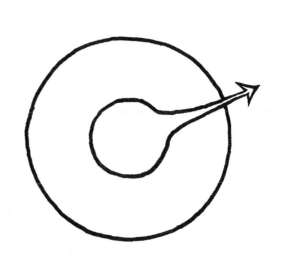

In the Absence of Chronic Armoring Primary drives are expressed directly from core through periphery toward environment, without significant or chronic restraint or inhibition.

2) Primary Versus Secondary Drives and the Social Facade

Reich's clinical work identified a set of deeper *primary drives*, or impulses, which were exclusively of a pleasure-seeking, social, and nonviolent character. The primary drives were present in the infant from the start: they were not "learned". The primary drives included those impulses such as the urge to suck in the newborn, feelings of pleasure associated with child-hood defecation or genital play, and the sexual longing of the adolescent and adult. The primary drives also include the spontaneous reaching out to the world and trusting qualities often seen in small children. However, *secondary drives* developed with the armored blocking of the primary drives, and with the consequent build-up of undischarged tension. Repression of impulses does not simply "do away with" the energy of the original impulse — the original impulse will still appear in surface behavior, but only in a distorted manner, tainted with varying degrees of anxiety, fear, or anger. In addition, when armor leads to a severe blocking of sexuality, impulses of an antisocial, violent character also may emerge, when sufficient inner tension has built up which can no longer be held back.

Whereas primary drives emerge from the deepest layers of the character structure, from the biological core, secondary drives are contained within the individual's peripheral musculature, within the armor itself, where suppressed emotional energy mis-shapes the posture and external appearance, and impedes the movement of the organism. Secondary drives include all those antisocial, sadistic, destructive, and deceitful impulses which have heretofore been wrongly attributed to "death instincts", "bad genes", "original sin", the "devil", and so forth. They obtain their energy and specific form directly from the inhibition and blocking of the primary biological drives.[34]

Psychic health, from this perspective, becomes a matter of whether or not a particular pleasure-directed drive or emotion is expressed or held back. A child reaching for the mother's breast, if given the breast, might be content and satisfied. Likewise, an easy and anxiety-free defecation, or pleasurable masturbation would feel quite satisfying to the child, and holds no harm at all. However, if punishment is meted out to a child for expressing such natural biological impulses, then the child will react with initial shock and pain. It might cry, and also express anger. The parent might then punish the child for crying or throwing a tantrum, requiring that these feelings also be blocked. If the punishments are repeated, the

Sadism and Masochism
Sex-Economic Viewpoint:
Reich viewed the organism as a "sack of energy" which, in situations of health, would periodically accumulate and then discharge inner tension via work or play activity, emotional expression and sexual orgasm. Early trauma and sex-repression blocks these normal outlets, especially the capacity for orgastic discharge, leading to armoring and a build-up of undischarged bioenergetic tension. Whereas ordinary character neurosis may lead to pleasure-starvation and sexual dysfunctions the individual lives with in quiet resignation (as per Freud, on p.19), the sadist holds a considerable amount of inner rage which is mobilized to "break out" and "pierce" their armor whenever the deeper push of sexual excitation is experienced — as readily seen in the case of the wife-beater. In extreme cases, sexual excitation and climax (not to be confused with full orgasm) may only be experienced in the ex-
(next page, sidebar)

34. Reich, *Function of the Orgasm*, ibid.

With Chronic Armoring Primary Drives are blocked, leading to pent-up inner tension and the development of Secondary Drives which are diverted from their original goals, or significantly changed in character. Some of the energy of the original core impulse is diverted inward to form the block, to create and maintain the armoring, and contain the forbidden and socially-dangerous secondary impulses. An outer layer or "facade" develops by which the painful and explosive trauma-created emotions constituting the armoring are made "unconscious" and hidden from casual view.

child eventually reacts to its own biological impulses with a fear of subsequent punishment. This brings about a chronic internal tension, as the child armors against the original biological impulse, and also against its own natural desires to protest the external restraints and punishments.

Once the child's character is so restructured and armored against its original, natural biological impulses, those impulses become disguised within a complex layering of secondary emotional and character activity. Various "civilized" layers, or a *social facade* develops within the character structure. Extreme politeness, sneakiness or deceitfulness might appear, or the child might simply become dulled, lose its spontaneity, become "quiet", "shy", or "obedient" — a "good boy" or "good girl", sitting quietly on its unexpressed fear, anger, or resentment. Behavioral problems usually remain submerged or at a low level until the energetic push of puberty occurs. As such a child matures, the armor would be maintained, but usually without being perceived as foreign to the self. The armor produces a reduction in sensory capabilities; in addition to reducing respiratory volume and sexual vitality, the spastic muscular contractions necessary to anchor the repressed impulses also reduces physical sensation. People forget the original reasons for their misery, and eventually lose the awareness that they are holding back painful emotions. They resign to the armor, losing awareness of its existence, but are affected powerfully, nonetheless. Indeed, Reich felt the armor constituted the biophysical basis of what Freud called the "unconscious".[35]

haustive brutalization of another person, as with concentration-camp guards or mass-murderers who ejaculate only when committing horrible crimes. The masochist similarly is armored with a considerable inner rage and bottled-up tension, expressed internally, towards the self in various ways. In extreme cases, sexual excitation occurs only when the skin is being physically attacked and scourged, allowing some temporary discharge of bioenergy through broken skin. They seek "to be pierced" in a physical manner. Neither the sadist nor the masochist can experience a *full genital orgasm* — hence, the brutal sexual savagery of extreme patristic nations during warfare, and the "peacetime" whips, bondage, body-cutting and piercing of sadistic-masochistic sexuality.

Reich's ideas developed from clinical experience with neurotic patients in therapy, where he found that the elimination of neurosis would not always occur if only old memories were elicited through free association. However, when *buried feelings and emotions* connected with an old trauma were liberated and expressed, then neurotic symptoms diminished, indicating that it was bound-up emotional energy, and not the memory trace per se, which constituted the core of the neurosis. As with chronic breath-holding, Reich also observed the uniform inability of neurotic patients to achieve full sexual gratification. He felt that childhood trauma left greater or lesser disturbances within the sexual life of the adult, and he emphasized this aspect. He redefined "sexual potency" into the more precise term *orgastic potency*, which included the ability to surrender fully to both the sensory-erotic and emotional-romantic aspects of genital lovemaking.[36]

Reich discovered that specific emotions could often be triggered into expression by applying a gentle manual pressure to those muscle groups which participated in the holding back of emotion. To better facilitate these discoveries, he radically modified psychoanalytic procedure, moving the analyst's chair from behind the couch to directly aside it, where he could make direct eye contact with the reclining patient. He would observe the patients breathing, their posture and skin pallor to ascertain their biophysical condition. He not only analyzed patients to uncover their emotional history and talked with them to encourage abandonment of self-destructive tendencies, but he also encouraged the uninhibited expression of buried feeling in the therapy room. Depending

35. Reich, *Function of the Orgasm*, ibid.
36. Reich, *Function of the Orgasm*, ibid.

upon the particular feeling a given patient would be holding back, they were encouraged to breathe deeply, and to shout or hit out at the couch, to bite hard with anger on a rag, or to cry openly.

If done properly, the repressed emotion would come pouring forth, often to the patient's surprise, and sometimes with a memory of the original traumatic events associated with the repression of the feeling. With time, once the patient fully surrendered to the emotion and expressed it, the muscles would relax, posture would change, and breathing would spontaneously deepen. Assuming that blocking of impulses and emotions did not continue, neurotic symptoms and destructive secondary behavior would cease, and the therapy would be ended.[37]

3) Sexual/Emotional Energy, and Function of the Orgasm

Sex-economic therapy became a matter of liberating blocked or repressed emotions, bringing people more in touch with their feelings, and restoring a fully gratifying genital love life. Reich redefined the function of the orgasm, as a basic regulatory-discharge function and determinant of emotional and physical health. The sexual life of an individual was both a determinant and diagnostic tool for understanding their general condition.[38] He observed that patients who could establish a healthy love life recovered in therapy far quicker than those who could not, and noted that patients who had long felt sexually "dead" would "come to life" again once other chronically repressed feelings were liberated. In many cases, however, this restoration of sexual feeling often placed the individual into even greater conflict with their social surroundings. Bad marriages, angry parents, poor living and working conditions, and even the legal system often stood in the way of people finding a satisfying sexual partner. Familial and social circumstances often had to be resolved and altered to accommodate an individual's restored capacity for pleasure in life. In this respect, Reich increasingly wrote about the pathology of "normal" society, of both the deadly political and legal systems, and the anti-life Church, which systematically worked in a authoritarian manner to destroy people's pleasure in life.[39]

Later in his research, Reich sought to objectify the sexual/emotional energy he observed in the body, by using sensitive millivoltmeters. He was the first researcher to demonstrate that the bioelectrical potential or charge at the skin surface would change with emotional state. To summarize, Reich observed that a strong bioelectrical charge (electronegative polarity) would register at the skin surface only during pleasurable experiences. Antithetically, when anxiety was experienced, the millivoltmeter would record a lowering of electronegativity, or lowering of bioelectric charge. Reich undertook many control procedures to assure himself that these readings were real, and not mere artifacts. He later interpreted the differing bioelectrical readings as *different directions of movement of a measurable vegetative "current" within the body.* In states of pleasure, this energy expanded outward from the core to the periphery, *towards the world,* giving rise to a stronger charge at the skin surface. During pain, fear or anxiety, a bioenergetic contraction or withdrawl of biological energy would occur, from the periphery back into the core, *away from the world.* Reich additionally observed that strong energetic currents could be found at the skin surface of emotionally fluid individuals with a full, deep respiration, and who otherwise satisfied the clinical diagnosis for a lack of significant armor. In more highly armored, chronically anxious patients with a shallow respiration, the reverse was observed; where the psyche was governed by chronic anxiety, tension,

Orgastic Potency
Sex-Economic Viewpoint

"...the capacity for the surrender to the flow of biological energy without any inhibition, the capacity for complete discharge of all dammed-up sexual tension through involuntary pleasurable contractions of the body. Not a single neurotic individual possesses orgastic potency; the corollary of this fact is the fact that the vast majority of humans suffer from a character-neurosis"
Wilhelm Reich,
Function of the Orgasm (1927)

37. Reich, W.: "Der Orgasmus als Elektrophysiologische Entladung (The Orgasm as an Electrophysiological Discharge)", *Zeitschrift für Politische Psychologie und Sexualökonomie,* I, 1934; Reich, W.: *Experimentelle Ergebnisse uber die elektrische Funktion von Sexualitat und Angst (Experimental Investigation of the Electrical Function of Sexuality and Anxiety),* Sexpol Press, Copenhagen 1937. Both articles reprinted in: *The Impulsive Character and Other Writings,* New American Library, NY, 1974; *The Bioelectrical Investigation of Sexuality and Anxiety,* Farrar, Straus & Giroux, NY, 1982; *Pulse of the Planet,* 4:5-43, 1993. Also see Reich, "The Basic Antithesis" 1934, ibid., and *Function of the Orgasm,* ibid.

38. Reich, *Function of the Orgasm,* ibid.

39. Reich, *People in Trouble,* ibid; *The Sexual Revolution,* ibid.

fear of pleasure, or sex-negative doctrines, the bioelectric measurements would demonstrate a consistent, uniformly low charge at the skin surface, without significant swings or alterations in amplitude, corresponding to a chronically contracted emotional condition. The low charge resulted from the absence of strong emotional, energetic charging of the organism's periphery, and a general inability or unwillingness to experience strong pleasurable excitation.[40]

Reich's bioelectric experiments straightforwardly quantified fundamental emotional phenomena, demonstrating a build-up or concentration of charge at the skin surface when the organism expanded outward to pleasurably embrace objects in its environment. Antithetically, charge was lost at the skin surface, being concentrated at the interior, when the organism anxiously shrank back, away from the outer world. Reich also measured the electrical potentials of erogenous zones, finding them to possess a higher bioelectrical charge than other skin surface tissues.

He also confirmed the discharge function of the orgasm through such bioelectrical measurements. Sexual excitation during lovemaking resulted in an accumulation of both physical tension and bioelectric charging of the erogenous zones, as determined with the millivoltmeter. During orgasm, both the accumulated physical tension and bioelectric charge was *discharged*, being experienced by the individual as pleasurable gratification. Sexual excitation and all pleasurable sensation would appear on the millivoltmeter as a distinct increase in bioelectrical charge. A full orgasm would show a sharp decline in bioenergetic charge to levels lower than the original base level, indicating a *net loss of energy*. Sexual dysfunctions, such as the absence of capacity for sexual excitation and/or gratification, all had correlates in the millivoltmeter curves, which in any case matched the stated feelings of the lab volunteers.[41]

Based upon his experimental findings, Reich felt he had quantified Freud's metaphorical libido, as a specific bioelectrical force which underlay both emotions and sexual excitation. His bioelectric experiments on human subjects, which were run between 1934-35, significantly predate those of Masters and Johnson and others on such questions by decades. Only in 1968 did others dare to make bioelectric measurements on the erogenous zones,[42] and then using the less sensitive and artifact-producing galvanic skin response (GSR) technique; Reich measured electrical potentials only, using sensitive vacuum-tube millivoltmeters with a very high input resistance, which neither added nor removed energy from the organism.

Reich's emphasis upon the pleasure/anxiety aspects of bioelectric skin potentials have, to my knowledge, still not been approached by any other sex researcher, given the social taboos which still surround issues of sexual pleasure and anxiety. His work indicates that emotional feelings of both tender love *and* erotic excitement are *prerequisites* for complete orgastic discharge. His clinical and laboratory research demonstrated that where emotional contact was not present between sexual partners, or where intercourse was characterized by a general lack of feeling, or by fear, anxiety, anger or pushy aggression, strong orgastic release or "letting go" was not possible. In the laboratory, Reich had quantitatively objectified why the sexual act would not result in full gratification if the partners were not comfortable with or attracted to each other. Without a unity between erotic and tender feelings, a full and complete orgasm is simply not possible. This latter point is central to understanding why the "Don Juan" "macho man" and the criminal rapist or violent impulsive character obtains no deep gratification or psychic benefit from their sexual experiences. Likewise the controlling female (or male) who runs from tender feelings of surrender, and seeks to dominate their partner.

40. Reich, "The Orgasm as an Electrophysiological Discharge" and *Experimental Investigation of the Electrical Function of Sexuality and Anxiety,* ibid.

41. ibid.

42. Zuckerman, M.: "Physiological Measures of Sexual Arousal in the Human", *Psychological Bull.*, 75:297-329, 1971.

Reich never advocated sexual license or "free love", but was a critic of compulsive abstinence, and of the sex-negative preachings of the Church. He viewed both the moralizer and pornographer as opposite sides of the same coin, that character attitudes of either a moralistic or pornographic nature would effectively kill feeling, and thwart one's love life. Each of these two aspects, he observed, was neurotic and inhibitory in character, having been formed by similar childhood traumas, punishments, and verbots. In both cases the attitude toward the opposite sex was laden with great anxiety, fear, and/or hatred, having been produced by a family and social structure which thwarted the natural heterosexual instincts of the child and adolescent.

Reich also identified homosexuality as being predominantly neurotic, founded upon feelings of *fear and hatred of the opposite sex*. His work makes understandable the often-frenzied promiscuity of many contemporary homosexuals, who — like the heterosexual Don Juan — never obtains full sexual gratification from their experiences, and therefore continues on a never-ending search for additional sexual partners. However, it is not true that Reich was a "homophobe". He was surely critical that homosexuality in the majority of cases was developed from heterosexual repression of youth, but also fought openly for the decriminalization of homosexuality.[43]

It therefore would be completely inaccurate and wrong to conclude that Reich advocated a loveless sexual free-for-all, as some have claimed.[44] Sexual functions and behavior were foremost among his concerns, but the precepts of general emotional health, to include the treatments of infants and children, and the socioeconomic conditions under which working-class families lived, were also a primary focus.

Sex-economy demonstrated a mechanism whereby early childhood experiences of a chronically painful and traumatic nature could disturb sexual attitudes and potency later on in life, generating neurotic or psychotic elements in the character structure. Orgastic disturbance in turn would increase the accumulation of undischarged sexual tension, fueling the formation of destructive secondary drives. Contrarily, where children would be allowed to grow naturally, without severe traumas or chronic fears and punishments, and when sex drives were allowed full expression, emotional blocking and repression would not become internalized, nor would secondary drives appear.

4) Reich on Infants and Children

Reich did not unrealistically argue that childhood had to be completely free of all trauma, fear, or pain for psychic health to exist. All parents set necessary limits to protect the life and limb of their children. However, the degree to which parental prohibitions trespass into biologically necessary functions for purely cultural reasons, and the degree to which the child chronically internalizes the prohibitions in spite of its own biological needs and impulses to the contrary, is the degree to which armor, neurosis and secondary drive formation will take place.

Reich was also one of the earliest medical writers to observe the perceptive, sentient, and highly sensitive nature of the newborn infant. He expanded upon psychoanalytic discussion of infantile sexuality and drives, viewing the infant and child as capable of responding to a much broader variety of stimuli than was thought at the time. He discussed a variety of modern influences which opposed the biological needs of the child, and led to neurotic behavior. These included questionable routine treatments: spanking the newborn to generate its first breath, separation

43. Reich, R.: "The Problem of Homosexuality", in *The Sexual Struggle of Youth*, 1932, ibid; Reprinted in *Journal of Orgonomy*, 20(1):5-8, 1986.

44. For instance, Chesser has given an error-filled, distorted, and downright nasty reaction to Reich's works. (Chesser, E.: *Salvation Through Sex: The Life and Works of Wilhelm Reich,* William Morrow, NY, 1973). Such irrational, unscientific and hostile commentary on Reich's works were widespread in the 1940s and 1950s, exacerbated by Reich's orgone energy experiments, which were also attacked and distorted in the popular media. Similar unscientific, malicious attacks unfortunately continue today, mostly by the organized "skeptic groups". See the discussion of these attacks in: Wolfe, T.: *Emotional Plague Versus Orgone Biophysics*, Orgone Institute Press, NY, 1948; Greenfield, J.: *Wilhelm Reich Vs. the USA*, W.W. Norton, NY, 1974; DeMeo, J.: "Response to Martin Gardner's Attack on Reich and Orgone Research in *The Skeptical Inquirer*", *Pulse of the Planet*, 1:11-17, 1989.

from the mother immediately after birth, use of stinging antiseptic eyedrops, circumcision, timetable feeding, and hostile feelings by aggressive adult caretakers. Reich argued that parents and doctors, who themselves had suppressed various emotions related to their own childhood traumas, were incapable of either recognizing or sympathizing with the emotional traumas or needs of infants and children. He considered it axiomatic, that *a person cannot recognize in someone else an emotion which they have learned to repress, and can no longer feel themself.*[45] This, he felt, partly explained the compassionless, brutal and horrific treatment of infants by adults.

These, then, were the major reasons why armored parents, doctors, and nurses, even with the best of conscious intentions, were incapable of recognizing and eliminating trauma directed toward their own offspring, or children in their care. Indeed, after around 1940, Reich became convinced that there was little hope in changing social structure by only working with neurotic patients in therapy, and that the only successful approach to the problem of human neurosis was to work with mothers and babies, to prevent neurosis from occurring in the first instance. His last work more clearly focused upon such concerns in the establishment of an *Orgonomic Infant Research Center.*[46]

Infants who were chronically traumatized, and children who were constantly reprimanded and punished, Reich argued, would view both parents and the outer world with suspicion, anxiety, fear, or hostility. Here was the source of the child's hatred of the disciplining same-sex parent, in real prohibitions, punishments, and traumas, and not in psychoanalytic "fantasy" or "wish". The child's appeasing behavior had more to do with avoiding a beating or emotionally-crushing experiences than with development of "wish impulses" or "fixations", though the child could be confused by its own conflicting feelings of love, fear, dependency and anger toward the parent.

According to the sex-economic view, emotionally healthy parents are unconsciously propelled by their own instinctual pleasure drives to meet the biological needs of their children, much in the same manner that the animal in the wild spontaneously and instinctively attends to the needs of its offspring. However, where parents were themselves raised with a great deal of trauma and punishment, as adults they remain fearful and anxious, or inhibited regarding their own pleasure in life, or regarding the expression of their own feelings. When confronted by their own children's impulses, such as the urge to suck, desire to be held, to freely defecate or express sexual feeling, they are often *driven* to unconsciously thwart the expression of those impulses. This is done usually for the stated purpose of "helping" or "training" the child, but in reality is done to kill the child's expressive feeling and energy, *so that the parent will no longer feel anxious.*

Reich called this driven reaction of the armored caretaker to repress the uninhibited life of the unarmored child *pleasure-anxiety.* In extreme cases, pleasure anxiety can become an overwhelming mode of behavior, which dominates the individual's psyche, driving them to devote their energies, on a full-time basis, to suppressing the uninhibited life and energy of infants, children, and of other lesser-armored adults in their own cultural group, or in other cultural groups.

In every armored patrist culture, there are small armies of adult caretakers who spend considerable energy making sure that all children are "properly" swaddled into immobility, circumcised, denied the breast, abandoned to "cry themselves to sleep", fed by timetable, and hit, spanked or crushed down in a thousand different ways. In so doing, the pleasure drives of youth are opposed, anxiety is implanted, and secondary drives

"...the organism can perceive only what it itself expresses. The armored physician cannot hear the screams of the infant, or else he takes them for granted because he has stifled the screams inside himself; and because his own organism can no longer perceive what another organism reveals to him." Wilhelm Reich 1949. (*Ether, God & Devil*, Farrar, Straus & Giroux edition, 1973, p.69)

45. Reich, W.: *Ether, God and Devil*, Farrar, Straus and Giroux, NY, 1973, p.69.
46. Reich, *Children of the Future*, ibid.

are repetitively created, in each new, succeeding generation.

Reich's findings have had some, albeit very limited effect upon society. While *Character Analysis* is today used in some medical schools for psychiatric students, his views on adolescent sexuality and the function of the orgasm have been received by the psychiatrists much like water on a duck's back. (More frequently, ordinary people have read his works, and been emotionally touched.) Some who disavow Reich's work on sex-economy, or the biological orgone energy,[47] view *Character Analysis* as a major contribution to our understanding of human behavior.[48] His views on human behavior are today discussed in a few college textbooks on clinical work and psychotherapy, and in a few traditional psychotherapy journals.[49] A number of theses and dissertations have also been accepted in recent years which contain sections devoted to, or which are based upon Reich's works.[50] Several scientific research journals also exist which are fully devoted to the questions raised by Reich.[51] However, these are but small cracks in an immense wall of social and professional opposition. In America, there is still an enormous wall of denial regarding the horrific treatment of infants in our hospitals, the current epidemic of sexual mutilations of women by surgeons and obstetricians, and the similar repressive treatments meted out to children in most schools — it is all widely ignored or rationalized. "Sex" is still an unplanned anxiety-filled event that *happens to* most young people, who are "educated" much more about unproven "infectious AIDS" scare propaganda and ineffective condoms than about basic sexual hygiene, contraceptives and love. Our public media presentations and sexual education classes oftentimes resemble the "antisex education" described in George Orwell's novel *1984*.

Public discussion and research into Reich's sex-economic findings have been blockaded now for over six decades. Papers are still rejected from academic journals for citing Reich's works; university students and professors with a research interest in Reich's findings often publish their findings under pseudonym, and medical students even hide Reich's books at home, lest their medical mentors find out about their interest and expel them. Reich's books were, of course, burned by the Nazis, Stalinists, and, in as recently as the 1950s by the US government.[52] This latter insult to Reich, and to the US Constitution, has been as widely ignored by journalists and civil libertarians as are his scientific findings ignored by the academics. Laing was correct. Almost no one wants to mention Reich.

Other Research Supporting Sex-Economy

Over the last hundred years, a revolution has occurred in the West regarding family law, sexuality, female status, and child treatment. Women have achieved not only the right to vote, but also to divorce, obtain contraceptives, acquire professional educations, work outside of the home, and so on. Children have become increasingly independent of parental absolutism, through laws against child physical and sexual abuse. Wife-beating has also been recognized as a criminal offense. Books on sex-education and sexual hygiene exist in most libraries and book stores, and contraceptives are openly available at any drug store, without prescription or marriage license, obtainable by anyone with money. Many of these reforms, particularly the decriminalization of abortion and homosexuality, have been and still are hotly contested to be sure, but contrasted to the era pre 1940, the change has been enormous.

Many of the legal reforms sought by the radical psychoanalysts of the 1920s and 1930s have been achieved, even though it is clear that a lot of

47. Reich, *Function of the Orgasm*, ibid., p.339; Also see sections IV-VII in Reich, W.: *Selected Writings*, Farrar, Straus & Giroux, NY, 1973.

48. Edwards, P.: "Wilhelm Reich", *Encyclopedia of Philosophy*, Macmillan/Free Press, NY, 7:104-15, 1967.

49. Corey, G.: *Theory and Practice of Counseling and Psychotherapy*, Brooks/Cole, Monterey, CA, 1982, pp.51-5; Konia, C.: "Orgone Therapy: A Case Presentation", *Psychotherapy Theory, Research, & Practice*, 12(2):192-7, 1975; Nelson, A.: "Orgone (Reichian) Therapy in Tension Headache", *Am. J. Psychotherapy*, 30(1):103-11, 1976.

50. A listing of these dissertations and theses is given in DeMeo, J.: *Bibliography on Orgone Biophysics*, Natural Energy Works, 1986, PO Box 1148, Ashland, Oregon 97520.

51. For example: *Pulse of the Planet, Journal of Orgonomy, Annals of the Institute for Orgonomic Science, Emotion, Lebensenergie, Sciences Orgonomiques*.

52. For documentation on the FDA'S banning and burning of Reich's books in the 1950s, see Greenfield, 1974, ibid., and J. DeMeo, Introduction to *The Orgone Accumulator Handbook*, Natural Energy Works, Ashland, 1989.

sexual confusion and misinformation exists. Each reform mentioned above also possesses a literature of its own, quite independent of sex-economy, to warrant its adoption into law. It will provide additional support for the basic assumptions of the Saharasian thesis to cite those researchers and findings which were stimulated by Reich and sex-economy, or which otherwise independently support his work and theories.

1) Prenatal Influences

There is growing evidence that physiological and emotional retardation of a child may result from emotional stress and trauma experienced by the mother during the period of gestation. Reich discussed such maternal influences on the fetus in the context of the "spastic uterus" of mothers, who had clamped down against sexual feeling during adolescence, and subsequently carried a great deal of tension in their pelvis and abdomen.[53] One interesting confirmation on this subject comes from Ashley Montagu, who put together an entire volume on *Prenatal Influences*, including a chapter on "Maternal Emotions" which in some ways echoes Reich's view.[54] A mother who is under great stress during pregnancy, who is pregnant as a result of rape or otherwise does not wish to have a baby, will be at greater risk for a premature baby, a baby with deformities or physical problems, or an emotionally irritable and agitated baby.

"Maternal attitudes, whether of acceptance, rejection, or indifference to their pregnancy, may well spell the difference, in some cases, between adequate and inadequate development of the fetus..."[55]

Verny, in *The Secret Life of the Unborn Child*,[56] has gathered more recent evidence supporting the points raised by Montagu. From this standpoint, emotional trauma and stress experienced by the mother could be transmitted to her offspring, to condition in subtle ways both the psychic-emotional and somatic-physical condition of the child. If culture-wide, such trauma and stress would have major significance.

2) Life-Negative Effects from Birth Trauma in Modern Hospitals

The role of various treatments of infants in shaping adult character structure has been systematically studied by modern clinicians. Practices such as separation of the infant from the mother immediately after birth, use of stinging eyedrops to allegedly "disinfect" the babies eyes, neonatal circumcision, timetable feeding and bottle feeding, have come under increasing criticism as being scientifically unsupportable and detrimental to the long term physical and emotional well-being of the child. Studies have even indicated that human mothers are more likely to emotionally reject and distance themselves from their own babies if they are separated from them during the first hours after birth, underscoring our similarities with other mammals.[57]

In every controlled study where these relationships have been addressed, the relationship between painful trauma and biological or psychic damage appears. Indeed, the demand for the abandonment of painful and traumatic practices aimed at infants and children, so far coming from only a tiny minority of the general public and health-care establishment, has been spurred by these same clinical and laboratory findings. Some circles in modern medicine, notably obstetrics and gynecology, have taken

53. Reich, *Reich Speaks of Freud*, ibid., p.91.

54. Montagu, A.: *Prenatal Influences*, Charles Thomas, Springfield, IL, 1962, pp.169-216.

55. Montagu, ibid., p.215.

56. Verny, T.: *The Secret Life of the Unborn Child*, Summit Books, NY, 1981.

57. Klaus & Kennell, *Maternal-Infant Bonding*, C.V. Mosby, St. Louis, 1976; Peterson, G.H. & L.E. Mehl: "Comparative Studies of Psychological Outcome of Various Childbirth Alternatives", in Stewart & Stewart, *21st Century Obstetrics*, ibid., Vol. 1:209-38; Rice, R.: "Maternal-Infant Bonding: The Profound Long-Term Benefits of Immediate, Continuous Skin & Eye Contact at Birth", in Stewart & Stewart, *21st Century Obstetrics*, ibid., Vol. 2:373-86; Montagu, A.: "Social Impacts of Unnecessary Intervention and Unnatural Surroundings in Childbirth", in Stewart & Stewart, *21st Century Obstetrics*, ibid., Vol. 2:589-610; Klaus, M., et al.: "Maternal Attachment: Importance of the First Postpartum Days", *New England J. of Medicine*, 286:460-3, 1972; Greenberg, M., et al.: "First Mothers' Rooming-In with Their Newborns: It's Impact upon the Mother", *Orthopsychiatry*, 43(5):783-788, 1973; Klaus, M., et al.: "Human Maternal Behavior at the First Contact with Her Young", *Pediatrics*, 46(2):187-92, 1970; Stechler, G.: "Newborn Attention as Affected by Medication During Labor", *Science*, 144:315-7, 1964; Kron, R., et al.: "Newborn Sucking Behavior Affected by Obstetrical Sedation", *Pediatrics*, 37:1012-6, 1966; Korner, A.: "Visual Alertness in Neonates as Evoked by Maternal Care", *J. Experimental Child Psychiatry*, 10:67-78, 1978.

patently murderous anti-female and anti-infant positions, dictated partly by concerns for maintenance of their economic monopoly, and partly by negative emotional attitudes towards women and female sexuality: i.e., the unscientific, religiously-rooted ideology of the inherently "dangerous" nature of the vagina, and of childbirth (eg, vaginal blood taboo, "birth as an illness", "the wages of sin", etc.).[58]

Those who doubt the accuracy of this charge should consider the shocking 25%+ Caesarean rate present in American hospitals (50%+ in many teaching hospitals), or the hundreds of thousands of unnecessary hysterectomies and breast-mutilation surgeries performed each year, including thousands of so-called "preventative mastectomies" performed upon perfectly healthy women, to "cure" them of their dangerous breasts! (See pages 129-132 for more detailed documentation). Perhaps 90% or more of all Caesareans and hysterectomies, and an unknown percentage of all mastectomies fall into the category of *unnecessary surgery*, being undertaken mainly to compensate for the physician's general ignorance — and sometimes for their bungled interferences with natural biological processes — regarding basic matters of sexuality and biology.[59]

There is a clearly documented correlation between childbirth mortality and morbidity to the various routine, compulsory forms of surgery and medications which are practiced in the modern hospitals; *the more classical medical "training" the birth practitioner has, the worse is the outcome for both mother and infant.* Ob-Gyn specialists have the worst statistical record for mortality and morbidity among women and infants, seconded by general practitioner MDs. Nurse-midwives are generally safer than MDs, but it is the *lay midwives,* who are trained by peer-groups and never set foot into a medical college, who have the safest record of all groups.[60] A problem here is the fact that medical journals have often published unsupportable sham statistics in an attempt to conceal the facts, masking the true dangers of hospital birth — spontaneous abortions, births in taxicabs en-route to the hospital, and even births in hospital elevators have been known to be included within the "homebirth" category, all in effort to deceptively distort the true safety record of homebirth.[61]

Meanwhile, policemen have been marshalled by the doctor's organizations such as the American Medical Association and the American College of Obstetrics and Gynecology to crush the grassroots midwifery/homebirth movement, and indeed, to crush or co-opt the entire alternative health movement. Midwives and doctors who do home births are generally harassed, losing hospital privileges (at least), sometimes being jailed and subject to loss of medical license.[62] The increasing pressure upon midwives to "conform" to the existing sickness-industry is producing a form of "nurse-midwife" who is increasingly indistinguishable from their sadistic male counterparts — today, it is not unusual to see stiff-faced and emotionally-hardened nurse-midwives sternly advocating unnecessary hospitalization, induced labors, episiotomies, circumcision, and drugs. These emotionally hardened midwives are fully acceptable to the existing medical-social system, while the more emotionally soft (and safe) lay midwife, who generally practices an art based upon empirical observation, inexpensive natural remedies, intuition and empathy, is not.

In spite of the above difficulties, there is a slow effort at work to rediscover what mothers in some cultures have instinctively known about the natural history of sexuality, birth, and childrearing. A grassroots social movement did grow in the United States during the 1960s and 1970s calling for a return to gentle birthing techniques, the abandoning of harsh and traumatic hospital neonatal practices, and a return to breast-feeding.[63] Nevertheless, childbirth conditions remain generally worse in the USA than in many European nations, where midwifery is often an

58. Arms, *Immaculate Deception*, ibid.; Stewart, *Five Standards of Safe Childbearing*, ibid.; Stewart & Stewart, *21st Century Obstetrics*, ibid.; Brackbill, et al., 1984, ibid.; Corea, G.: *The Hidden Malpractice: How American Medicine Mistreats Women*, Wm. Morrow, NY, 1977; Mendelsohn, R.: *Male Practice: How Doctors Manipulate Women*, Contemporary Books, Chicago, 1982.

59. Cohen, N.W. & L.J. Estner: *Silent Knife: Caesarean Prevention and Vaginal Birth After Caesarean*, J.F.Bergin, South Hadley, Mass, 1983; Corea, G.: "The Caesarean Epidemic—Who's Having This Baby Anyway — You or the Doctor?", *Mother Jones*, July 1980; Hufnagel, V.: *No More Hysterectomies*, Plume, NY, 1989; Stokes, N.M.: *The Castrated Woman: What Your Doctor Won't Tell You About Hysterectomy*, Franklin Watts, NY, 1986; West, S.: *The Hysterectomy Hoax*, Doubleday, NY 1994.

60. Stewart, *Five Standards of Safe Childbearing*, ibid., particularly Chapter 6, "Skillful Midwifery: the Highest & Safest Standard of All"; also see: Montgomery, T.: "A Case for Nurse-Midwives", *Am. J. Obstet. Gynec.*, 105(1):3, 1969; Levy, B., et al.: "Reducing Neonatal Mortality Rates with Nurse-Midwives", *Am. J. Obstet. Gynec.*, 109(1):50-58, 1971; Haire, D.: *Childbirth in the Netherlands: A Contrast in Care*, monograph, Intern'l Childbirth Education Assoc., 1973; Mehl, L., et al.: "Outcomes of Elective Home Births: A Series of 1146 Cases", *J. Reproductive Medicine*, 19:281-99, 1977; Devitt, N.: "The Transition from Home to Hospital Birth, USA", *Birth & Family J.*, 4(1):47-58, 1977; Mehl, L.: *Scientific Research on Childbirth Alternatives & What It Tells us about Hospital Practice*, NAPSAC Internat'l Publications, 1978; Adamson, G. & D. Gare: "Home or Hospital Births?", *J. Am. Medical Assoc.*, 243(17):1732-6, 1980; McQuarrie, H.: "Home Delivery Controversy", *J. Am. Medical Assoc.*, 243(17):1747, 1980; Burnett, C., et al.: "Home Delivery and Neonatal Mortality in North Carolina", *J. Am. Medical Assoc.*, 244(24):2741-5, 1980; Sullivan, D. & R. Beeman: "Four Years Experience with Home Birth by Licensed Midwives in Arizona", *J. Am. Public Health Assoc.*, 73(6):641-5, 1983; Weitz, R. & D. Sullivan: "Licensed Lay Midwives in Arizona", *J. Nurse-Midwifery*, 29(1):21-8, 1984; Hinds, M., et al.: "Neonatal Outcome in Planned vs. Unplanned Out-of-Hospital Births in Kentucky", *J. Am. Medical Assoc.*, 253(11):1578-82, 1985.

61. Stewart, D., *Five Standards of Safe Childbearing*, ibid., particularly Appendix C, "Statistics, Definitions & Ambiguities" & "A Rebuttal to Negative Homebirth Statistics Cited by ACOG".

62. Stewart, D., "The Conspiracy of Doctors Against Doctors", *NAPSAC NEWS*, 6(1), Spring, 1981; "The Conspiracy of Doctors Against Midwives", *NAPSAC NEWS*, 6(3), fall, 1981; also see various issues of *Birth Gazette, Mothering, Compleat Mother* magazines (addresses in Appendix C).

63. Becker, E., et al.: *Midwifery and the*

independent profession, not subordinated to medical doctors, and where special childbirth clinics (not in the hospital) and homebirths are more common. Whether or not this social movement will survive and continue to grow, or be subverted and crushed by policemen, is an open question. Most doctors in the USA unapologetically or ignorantly continue with their medically-disguised sadistic aggression towards women, children, and midwives.

3) Primate Studies

Harlow's studies with monkeys further confirm the above patterns in the origins of disturbed behavior.[64] Rhesus monkeys separated from their mothers immediately after birth and raised with a warmed soft cloth dummy "mother" developed a host of emotional disorders, but not nearly as many disorders as similar baby monkeys raised with a cold metal wire dummy. Importantly, infant monkeys raised without their mother, or without sensory stimulus (being held, played with, etc.) turn out to be rather cold and abusive parents themselves. Mother monkeys, who were themselves raised without a mother, are unmoved by the cries of their own offspring, refusing to pick them up and comfort them. This trait of maternal coldness is entirely absent from mother monkeys who were themselves nurtured by a healthy mother. Maternally and sensory deprived monkeys also become aggressive and violent towards other monkeys, including their own offspring, when forced to share the same cage. Monkeys raised by a healthy mother lacked this kind of "don't touch me" aggressiveness. A monkey which is deprived of nurturing love and body contact in infancy, does not tolerate much affection or touching in adulthood, and does not seek to provide touching or body contact to their own offspring. Is not the same true with humans? There are three general patterns:[65]

1) Maternally-deprived and isolation-reared infant monkeys grow into adults who react with anxiety and violence to normally behaving monkeys and to the natural maternal-affection needs of their own monkey offspring.

2) Maternally-deprived and isolation-reared infant monkeys grow up to be adults whose anxiety, aggression, psychotic and depressive states mirror those of humans with a similar history of isolation, abuse, and maternal deprivation.

3) Maternally-deprived and isolation-reared monkeys also become anxious and disturbed during normal mating behaviors, the effects of early trauma being generalized into the sexual sphere of life.

As discussed by Prescott:

"...it is the deprivation of body contact and body movement — not deprivation of the other senses — that produces the wide variety of abnormal emotional behaviors in these isolation-reared animals. It is well known that human infants and children who are hospitalized or institutionalized for extended periods with little physical touching and holding develop almost identical abnormal behaviors, such as rocking and head banging." [66]

Interestingly, at least one primatologist observed that males appear to be more traumatized than females from adverse rearing conditions.[67] Such a possibility brings to mind Montagu's well-documented work on *The Natural Superiority of Women*.[68]

Law, Mothering Magazine, Santa Fe, 1982; P. Simkin, *Directory of alternative Birth Services and Consumer Guide*, NAPSAC, Marble Hill, MO 1978.

64. Harlow, H.: *Love in Infant Monkeys,* San Francisco, 1959, reprinted in *Scientific American*, 200(6):68-, June 1959; *The Human Model: Primate Perspectives*, V.H. Winston, Wash., D.C., 1979; cf. Mitchell, G.: "What Monkeys Can Tell Us About Human Violence", *The Futurist*, April 1975, pp.75-80; Heath, R.: "Maternal-Social Deprivation and Abnormal Brain Development: Disorders of Emotional and Social Behavior", in *Brain Function and Malnutrition*, J. Prescott, et al., Eds., J. Wiley & Sons, NY, 1975, pp.295-310.

65. Harlow, *Love in Infant Monkeys,* ibid., *The Human Model*, ibid.; Prescott, *Brain Function and Malnutrition*, ibid.; Mitchell, "What Monkeys Can Tell Us...", ibid.

66. Prescott, J.: "Body Pleasure and the Origins of Violence", *The Futurist*, April 1975, p.65.

67. Mitchell, "What Monkeys Can Tell Us...", ibid., p.75.

68. Montagu, A.: *The Natural Superiority of Women*, Macmillan, NY 1953.

These studies provide stark and compelling evidence that monkey behavior, even among those groups living in the wild, cannot automatically be classified as innate or genetic in origins. Monkey species which normally are exceptionally gentle and nurturing as parents will, if deprived of nurture during their own infancy, show signs of sexual anxiety and violence. The violence is acted out toward their own offspring, who then develop the same anxious and violent tendencies, along with consequent disturbed sexual behavior. Maternally-deprived and isolation reared monkeys also express violence towards the opposite sex, particularly during expressions of mating behavior.

These observations of monkeys, by Harlow, Prescott and others, are in complete agreement with Reich's prior observations on the effects of infant trauma, which he argued was the first step in a series of blows by which the sexually anxious, neurotic, love-fearing and pleasure-hating character structure of *Homo normalis* would be formed. The monkey observations also support Reich's arguments that early trauma regarding pleasure is generalized into sexuality, to later form the core of adult genital and orgastic disturbances. Humans who were sensory deprived or abused as children would feel tremendous anxiety about matters of sex and touching, even into adulthood. As a means to control their anxiety, such previously-traumatized humans would also crush down any expressions of love and pleasure in their own children. Much of this *pleasure-anxiety* was, as Reich demonstrated, organized and institutionalized into culture, displacing previously softer and more pleasure-oriented forms of social living. As parents, such deprived humans would adopt or develop various forms of pedagogy designed, at the basic level, to deprive or deny their own children the physical body pleasures they themselves were denied, and could no longer experience due to chronic armoring.[69] While parents, priests, shamans, and various obstetricians and "mental health specialists" the world over give high-sounding excuses for inflicting pain and trauma on infants and children, and for crushing their sexuality, in reality they all are acting out disguised and rationalized feelings rooted in their own pleasure-deprived and frozen emotional structures, in a manner no different from Harlow's and Prescott's sensory-deprived, motherless monkeys.

4) Sexual and Physical Abuse of Children

As previously discussed, one central aspect of Reich's criticism of Freud was that the father of psychoanalysis was inclined in his later years to dismiss the real sexual and physical traumas of infancy and childhood as but "fantasy" or "wish". Freud would not, perhaps could not, straightforwardly deal with such sexual pathos in the family histories of his own patients. Nor could he fully address the social implications of such pathos.[70] In recent years, an abundance of documentation has surfaced about the real nature of sexual trauma.

For instance, the analyst Miller has discussed the widespread nature of child sexual and physical abuse, and she has called into question Freud's rejection of his own early observations on the subject.[71] The psychiatric social worker Rush has similarly argued against the "Freudian cover-up" of child sexual abuse.[72] The historian Sulloway also noted the problem.[73] But the most damning critique of Freud's shift in thinking (since Reich) has come from the analyst Masson, as discussed above. As Projects Director of the Freud Archives, Masson had access to unpublished documents which revealed how Freud had deliberately ignored corroborated, bona fide reports of incest in the childhoods of his female

69. Reich, *Function of the Orgasm*, ibid., p.7.

70. Reich, *Reich Speaks of Freud*, ibid.

71. Miller, A.: *Thou Shalt Not Be Aware, Society's Betrayal of the Child*, Farrar, Straus & Giroux, NY, 1984; Miller, A.: *For Your Own Good: Hidden Cruelty in Child-Rearing and The Roots of Violence*, Farrar, Straus & Giroux, NY, 1983.

72. Rush, F.: *The Best Kept Secret*, Prentice Hall, 1981.

73. Sulloway, F.: *Freud: Biologist of the Mind - Beyond the Psychoanalytic Legend*, Basic, 1979.

hysteria patients, and instead favored theories which would bring him professional applause from the prevailing patriarchal authoritarian social structure.[74] Like Reich, Masson was severely attacked by the psychoanalysts for his revelations, and was fired from his post.[75] Nonetheless, a large literature has accumulated which demonstrates a greater incidence of child sexual abuse than has previously been admitted by the traditional psychiatric community.[76]

Regarding the beating of children, in recent years evidence has been accumulating which clearly demonstrates that, contrary to popular myth, beating of children at home and in the schools does not work to "civilize" them, or to develop a more social morality. Instead, it invokes fear, docility, and conflicting emotions, reducing spontaneity and general joy of life, while actually increasing the violent and antisocial tendencies of a minority of high-energy, strong-willed, and independent-minded children. The clinical psychologist Welsh, for instance, was struck by the fact that a large number of delinquent young people reported severe parental punishments as part of their childhoods. His controlled study of over 2000 children led him to construct the "belt theory of juvenile delinquency", and clearly demonstrated that children who are frequently beaten, with belts, lampcords, paddles, fists, or other devices more stringent than an open hand to a padded bottom, are most likely to become delinquent and violent themselves. Children who were the most aggressive in the classroom, as determined by peer interviews, tended to have parents who most frequently beat them, with parental beatings being a powerful predictor of disruptive, rebellious, or aggressive behavior.[77]

Regarding the schools themselves, a number of studies have demonstrated how corporal punishment actually increases disruptive and destructive tendencies in schoolchildren.[78] One Oregon study further demonstrated that schools which had the highest rates of corporal punishment also had the highest rates of theft and vandalism, with the lowest rates of theft and vandalism occurring in schools where corporal punishment was never or rarely used.[79] These studies underscore the relationship between violent treatment of the child by the adult, and the development of violent tendencies in the child or adolescent later on.

Indeed, recent studies have identified a "child abuse syndrome", wherein children are beaten by parents who themselves were beaten when they were little, the beaten child maturing into an adult who is in turn violent towards its own infants and children. These modes of behavior tend to be transmitted from one generation to the next by virtue of duplication of specific traumatic, antipleasure, and antisexual modes of childrearing. Adult sexual confusion and pathos is in turn generated by such methods of upbringing, which are often turned against the sexuality of the next generation of children. The evidence suggests that a high proportion of children raised in such a fashion will become either child abusers or abuse seekers.[80]

Miller has addressed these problems, and discussed the life histories of several small-time and big-time mass-murderers, such as Charles Manson and Adolf Hitler. The pattern which emerges is that of a traumatic, cold, and heartless childhood for the mass murderer, where physical affection was absent, and outlets for emotion were effectively blocked. The mass murderer was an abused and neglected child with essentially no one to turn to for consolation. As adults, these hardened individuals heaped their vengeance upon the world, many times over.[81]

Most of the studies cited here have concentrated primarily upon child abuse, or the sexual abuse of children by adults. What they almost uniformly neglect to discuss, however, is the amplifying influences of later sexual repression of the adolescent. An abusive childhood, all by itself, is

74. Masson, *Assault on Truth*, ibid.

75. Malcolm, J.: *In the Freud Archives*, Knopf, 1984; cf. Masson, J.: "The Persecution and Expulsion of Jeffrey Masson, as Performed by Members of the Freudian Establishment & Reported by Janet Malcolm", *Mother Jones*, December 1984; Travis, C.: "How Freud Betrayed Women", *Ms.*, March 1984; "Assault on Freud", *Discover*, April 1984.

76. Rush, *Best Kept Secret*, ibid.; Herman, J.L.: *Father-Daughter Incest*, Cambridge, 1982; Butler, S.: *Conspiracy of Silence: The Trauma of Incest*, New Glide Press, San Francisco, 1978; Adams, P.L.: *Carnal Aggression and Abuse*, Louisville, 1978.

77. Welsh, R.: "Severe Parental Punishment and Delinquency: A Developmental Theory", in *Psychology and the Problems of Today*, M. Wertheimer & L. Rappoport, Eds., Scott, Foresman & Co., Glenview, IL, 1978; cf. earlier version of same title in *J. Clinical Child Psychology*, 5(1):17-21, 1976.

78. Special issue on "Corporal Punishment in the Schools", *Inequality in Education*, #23, Center for Law and Education, Gutman Library, Cambridge, MA, September 1978.

79. "Reaping the Whirlwind: Vandalism and Corporal Punishment", monograph, EVAN-G, 977 Keeler Ave., Berkeley, CA 94708.; cf. "Corporal Punishment in the Schools", ibid., p.25.

80. Leavitt, J.E., Ed.: *The Battered Child: Selected Readings*, General Learning, 1974; Helfer R.E. & C.H. Kempe, Eds.: *The Battered Child*, U. Chicago Press, 1974; Herbruck, C.C.: *Breaking the Cycle of Child Abuse*, Winston Press, 1979.

81. Miller, *Thou Shalt Not Be Aware*, ibid.; *For Your Own Good*, ibid.

generally insufficient to allow for accurate predictions about who will, and who will not, become a violent adult. Many abused children, as adults, do not act out violently towards their children or spouses, although they may have been significantly affected emotionally. A major determinant of who does, and who does not turn violent, is the presence or absence of a gratified sexuality during later adolescence and early adulthood. Given the social taboos surrounding adolescent genitality in the West, this variable has rarely been included in studies or discussions on social violence. However, at least one cross cultural study, undertaken by Prescott, has included variables for both child abuse and sexual repression, demonstrating a significantly increased capability for predicting social violence.[82] Later chapters will greatly expand on this same cross-cultural evidence.

While the studies discussed above have been focused upon behavior in the USA, Europe or Soviet Union, it would be a mistake to think that the problem was only isolated in those regions. In fact, the problem is global, though without a uniform geographical or historical distribution. Traumatic infant and child treatments, sexual repressions, and social violence are not evenly distributed around the globe. In subsequent chapters, this point will be fully and repeatedly documented. For example, exceptionally important and valuable information has been compiled by a number of authors on child abuse throughout history, but without discussion of regional differences, long-term historical, migratory or environmental changes, nor emphasis upon the *natural sexual interests* of the older child and adolescent.[83] One such author, deMause, observed:

> *"The evidence which I have collected on the methods of disciplining children leads me to believe that a very large percentage of the children born prior to the 18th century were what would today be termed <u>battered children</u>."* [84]

We in the West cannot take comfort in the fact that social violence is global. Social violence and pathos is epidemic within our society, and — aside from a recent welcomed emphasis to "stop child abuse" — there remains a larger measure of "officially recommended" methods of abusive "child treatment", supported by government, medicine, church, and families. Various methods of abusive treatment of children (to be documented in forthcoming Chapters) as seen in Western culture, have been traced back in time at least to the days of Imperial Rome,[85] although this study will argue their ultimate source was in Saharasia, c.4000 BCE.

5) A.S. Neill and Summerhill School

Reich often drew from the work of the educator A.S. Neill, a close friend, for evidence on the self-regulatory potentials of children.[86] Neill likewise drew from Reich's work for support of his *Summerhill School* in rural England, which was run on democratic principles.[87] At Summerhill, children were given almost complete freedom regarding choice of classes, attendance, dress, etc. School rules were developed at regular school assemblies, where each teacher and student, regardless of age, had one vote. Any student or teacher, regardless of age, could also address the entire assembly in attendance, on any issue. Neill demonstrated that under such conditions of nearly complete freedom, anarchy did not predominate, though a harmless chaotic playfulness often did. Neill also demonstrated that, even though class attendance was noncompulsory, children would, after an initial phase of rebellion (his pupils were often

82. Prescott, "Body Pleasure and the Origins of Violence" ibid.; Prescott, J.: "Deprivation of Physical Affection as a Primary Process in the Development of Physical Violence", in D. Gill, Ed., *Child Abuse and Violence*, AMS Press, NY, 1979, pp.66-137.

83. deMause, L.: "The Evolution of Childhood", in *The History of Childhood*, L. deMause, Ed., Psychohistory Press, NY, 1974; Schatzman, M.: *Soul Murder: Persecution in the Family*, Random House, NY, 1973, p.175.; Radbill, S.: "A History of Child Abuse and Infanticide", in Helfer & Kempe, *The Battered Child*, ibid.

84. deMause, *History of Childhood*, ibid., p.40.

85. Radbill, *History of Child Abuse and Infanticide*, ibid., p.3.

86. Neill, A.S.: *Summerhill: A Radical Approach to Child Rearing*, Hart, NY, 1960; Neill, A.S.: *Freedom, Not License!*, Hart, NY, 1966.

87 Reich, W. & A.S. Neill: *Record of a Friendship: The Correspondence of Wilhelm Reich and A.S. Neill*, B. Placzek, Ed., Farrar, Straus & Giroux, NY, 1981. Summerhill School, Leiston, Suffolk, 1P16 4HY, Great Britain.

"rejects" from authoritarian schools) attend class quite regularly, on their own initiative. Neill viewed the spontaneous curiosity of children and their self-regulating independent-minded tendencies as positive attributes to be nurtured, rather than crushed — as is the usual case in the modern school system. Moreover, Neill observed his students finished their curricula quicker, and with higher test scores, than the students in the nearby compulsory state schools.[88]

Summerhill demonstrated the fallacy of the assumption that children are innately "sinful", needing to be "tamed" or "civilized". Like Reich, Neill saw the coercive, punitive, and sex-repressive families and schools as responsible for "making a muck" of children's lives. He did not believe in a "problem child". His early books were titled *The Problem Teacher*, *That Dreadful School,* and *The Problem Parent,* emphasizing just where one found the main cause of children's misery and difficulties.[89] Neill had come to his own conclusions independent of Reich, being originally stimu-

88. Neill, *Summerhill*, ibid; *Freedom, Not License,* ibid.

89. Neill, A.S.: *The Problem Family,* Herbert Jenkins, London, 1948; Neill, A.S.: *That Dreadful School,* Herbert Jenkins, London, 1937; Neill, A.S.: *The Problem Parent,* Herbert Jenkins, London, 1932.

Problem Teachers, Problem Schools, Problem Families

Approximately 1,000,000 American schoolchildren, mostly boys, are being given the toxic drug *ritalin,* for "treatment" of "disorders" which themselves have never been scientifically constructed (ie., "Attention Deficit Disorder", "Attention-Deficit Hyperactivity Disorder", "Learning Disorders", etc.) America currently consumes 90% of the entire global production of ritalin, used as a chemical restraint against schoolchildren who in most cases may only be reacting to toxic environmental factors (such as environmental chemicals and low-level electromagnetic radiation from powerlines, television, computers, fluorescent lights, etc.). Or it is used against those who biologically rebel against an authoritarian family or compulsory school system.§ In one recent ritalin-advocacy article in *Parade Magazine*, a journalist blandly reported a mother putting her child on ritalin for not being able to sit still in church, and how "good" he behaved afterward. The long-term side-effects from such drugs are generally unknown, but may include later addiction to stimulants, and other serious health problems. The approximate 25% dropout rate in US schools (approaching 50% for some urban areas) is another startling confirmation that school children, particularly "problem children", cannot be forced to "learn" or be "taught" by increasingly compulsive and authoritarian methods. The problem is predominantly with the school structure itself, with contactless stressed-out parents and teachers, often of an autocratic and sadistic nature, who if dealing with adults would be quickly jailed for treatments they so readily visit upon children. Those who doubt the sadistic or contactless nature of some parents or grade school teachers are either looking at the problem with a blind eye, or have simply never cultivated the trust of children such that they would honestly tell them what often goes on at home, or in the schools. The most vile abuses of parents and teachers are hushed up for long periods, until some small or large disaster finally brings it to the attention of the community. It is typical that adults will anxiously joke about the abuses they experienced themselves at the hands of a sadistic teacher or parent, considering themselves to have somehow "benefited" from the harsh treatment. But whereas one child might simply learn to "obey", others respond to such treatment by becoming withdrawn, suicidal, throw themselves into drugs or alcohol, or become violently rebellious and hateful. The banding together of school-age dropouts into aggressive (and protective) street gangs should be viewed in this context for a deeper understanding of their origins. Children do not become violent and aggressive, contactless and hyperactive, or drug and alcohol addicted or suicidal by virtue of genetics or "original sin"; they are driven in these directions by the adult world at nearly every turn.

§ Peter Breggin: *Toxic Psychiatry: Why Therapy, Empathy, and Love Must Replace the Drugs, Electroshock, and Biochemical Theories of the "New Psychiatry",* St. Martin's Press, NY, 1991; Peter Breggin: *Talking Back to Ritalin,* Common Courage Press, Monroe, ME, 1997.

lated by Homer Lane's *Little Commonwealth*,[90] a reform school which was converted to democratic principles and demonstrated that even "delinquent" children would positively respond to true freedom.

The work of deMause, who has studied the history of methods of childrearing and child abuse, also demonstrates that a child who is raised in a non- authoritarian and sex-positive manner "...is gentle, sincere, never depressed, never imitative or group oriented, strong willed, and unintimidated by authority"; "The child" he says "knows better than the parent what it needs at each stage of life."[91] Both Reich and Neill drew additionally upon Malinowski's work on the Trobriand Islanders as confirming evidence on the social benefits of a self-regulated childhood and adolescent sexual freedom.[92] Chapters 3 and 9 provide additional information on the Trobriand Islanders, and similar peaceful cultures.

6) Confirmation of the *Mass Psychology of Fascism*

Reich's views on the rooting of right and left wing fascism in the psychosexual pathology of the patriarchal authoritarian family structure — as detailed in *The Mass Psychology of Fascism* and *People in Trouble* — have also received some recent attention and confirmation. Again, Reich felt that sexual pathology and cruelty to children were primal sources of the fascist character structure and totalitarian state apparatus, as well as of the sadistic elements present within the fascist state. Such elements could only be formed, he felt, through particular modes of harsh and antisexual child treatment. Frankl has written supportively on this question for the Soviet Union,[93] as did the emigre Russian psychiatrist Stern regarding the social-sexual conditions before and after the Russian revolution.[94] As stated by Stern:

"...there is a real connection between the bloody outbreaks of terror that have periodically stained the pages of Soviet history and the profound abnormalities of Soviet sexual life."[95]

"Stalinist ideology officially banished sex from Soviet territory. The 'Soviet Man' was meant to be a sort of superman of irreproachable morals whose amorous activities, reduced to a chaste minimum, were only intended to strengthen the 'Red soviet family' and the socialist economy. When a woman made an appearance in the official iconography, she was always brandishing either a rifle or a sickle, and if she should make so bold as to uncover a breast, you could be certain that such immodesty could only serve the greater good of suckling a future Young Pioneer of the socialist motherland."[96]

Puritanism (and alcoholism) was widespread in the USSR up into 1980s, with the moral authority of the "fatherland" state taking the place of the patriarchal High God. As documented by Frankl and Stern, young lovers might have their pictures, names and addresses posted on public bulletin boards, that proclaimed their "immorality", or be denounced as "moral offenders" in the press. The average Soviet citizen would often send in postcards to the police denouncing women who had male guests.[97] Stern noted that the Soviet Union was:

"...a country where lovers are harassed and pursued like common criminals, where the individual has lost control over his private life. ...the moral fervor of the mob, already ingrained as an instinct, still hunts down immorality in the most secluded doorway."[98]

90. Lane, H.: *Talks to Parents and Teachers*, Allen & Unwin, London, 1928; Also see Croall, J.: *Neill of Summerhill, the Permanent Rebel*, Pantheon, NY, 1983; Hemmings, R.: *Children's Freedom: A.S. Neill and the Evolution of the Summerhill Idea*, Schocken Books, NY, 1973.

91. deMause,*The History of Childhood*, ibid., pp.52-4; cf. Stallibrass, A.: *The Self-Respecting Child*, Penguin, 1977.

92. Malinowski, B.: *Sexual Life of Savages*, Routledge & Kegan Paul, 1932; Elwin, V.: *The Muria and their Ghotul*, Oxford U. Press, Bombay 1942; Elwin, V.: *The Kingdom of the Young*, Oxford U. Press, Bombay 1968; Prescott, "Body Pleasure and the Origins of Violence", ibid; Prescott, "Deprivation of Physical Affection", ibid.

93. Frankl, G.: *The Failure of the Sexual Revolution*, Kahn & Averill, London, 1974.

94. Stern, M.: *Sex in the USSR*, Times Books, NY, 1979.

95. Stern, ibid., p.xvi.

96. Stern, ibid., p.viii.

97. Stern, ibid., pp.79-81.

98. Stern, ibid., pp.80-1.

Regarding women, according to Stern and others,[99] the communist "revolution" has scarcely touched them in any positive way:

"In the USSR, the sexual repression of women is so extreme that a woman might well faint at the sight of a nude body. Sexual frustration, feminist ideas picked up piecemeal from the West, and profound sexual inhibitions — such is the curious amalgam that is the pseudo-liberation of the Soviet woman." [100]

"I am convinced that this phenomena of the battered wife exists on a scale that simply cannot be compared with anything in the experience of such countries as France or the United States..." [101]

Stern pointed out that, while sexual disorders afflicted an estimated 60% to 90% of the population in the USSR (a figure not too different from the USA or Western Europe), "neither the scientific discipline of sexology nor the clinical practice of sex therapy exists..."[102] Moreover, and of great interest for this study, Stern cites evidence for the greater severity of such repressive characteristics in Southern Russia, closer to Saharasia. There, ritual murder of females to preserve "family honor" continues, much in the manner of the Moslem Arab, Turkic, and Persian Near East. Up into the 1960s, baby girls were still considered a misfortune, and young girls were sold into arranged, polygamous marriages, facing death if they ran away. In these regions of the Soviet Union, adulterous wives were by "custom" buried alive by their husbands into the 1950s.[103]

Stafford made similar observations in the former communist satellites of the Soviet Union, and in communist China as well.[104]

"Totalitarianism provides an unparalleled opportunity for sexual deviation, especially for sadism... they [sadists] have flourished and are still flourishing in every communist country."[105]

Stafford's observations mirror those of Reich, who argued that emotional plague characters (sadistic antichild, antifemale, antisexual persons) had gained political power and become institutionalized in the extreme left and right wing nations, a characteristic not seen to the same extent in the established Western democracies, and not seen at all in cultures such as the Muria, Pygmy, or Trobriand. The factual character of these observations regrettably finds additional support in recent historical events which followed the break up of the former Soviet Union and Yugoslavia. In those regions bordering upon or lying within Saharasia, civil wars with the deliberate murdering of civilians have developed. The term "ethnic cleansing" has crept into our language, along with regular television images of artillery shelling of civilian populations, with the deaths and mutilations of women and children. Nothing like this has happened in the European-Asian region since the Nazi era — and yet, from a sex-economic standpoint, this kind of explosive reaction to the ending of external authority was predictable.[106]

Schatzman has also brought new evidence on the incredibly cruel and soul-crushing child treatments which were championed by German parents of the pre-Nazi era, and by Soviet parents in more recent years. He observed "totalitarian societies from Sparta to Soviet Russia have practiced [these] principles of child-rearing... especially the emphasis on obedience and discipline."[107]

Along similar lines, deMause has cited studies on German parents of

99. Hansson, C. & K. Liden: *Moscow Women: Thirteen Interviews,* (G. Bothmer et al., translators), Pantheon, 1983; Also see Hochschild, A.: "Is the Left Sick of Feminism?", *Mother Jones,* June 1983, pp.56-8; Pollit, K.: "Muscovites And Owenites", *Mother Jones,* August 1983, pp.55-6.

100. Stern, ibid., p.70.

101. Stern, ibid., p.208.

102. Stern, ibid., pp.ix-x,88,97.

103. Stern, ibid., pp.106,116,235-9.

104. Stafford, P.: *Sexual Behavior in the Communist World,* Julian Press, NY, 1967.

105. Stafford, 1967, ibid., p.3.

106. See Chapter 11, "Saharasia Today", for elaboration on these points.

107. Schatzman, M.: *Soul Murder: Persecution in the Family,* Random House, NY, 1973, p.175.

the 1960s, 80% of whom admitted to beating their children, 35% using canes; he also discussed the predominant type of German schoolmaster, a despot with a whip who would beat and slap small children for the slightest disobedience.[108] Fromm has also written on the question of human freedom, essentially echoing some of Reich's views on the relationship of political structure to family structure.[109]

For additional support on this particular issue, the reader is referred to historical sections of this work. There, I have drawn upon the works of a number of authors with a variety of research backgrounds, further demonstrating the roots of the despotic authoritarian state apparatus in the antisexual, antifemale, and antichild patriarchal authoritarian family.[110]

Summary of the Sex-Economic Viewpoint

The studies cited above document patterns of infant somatosensory deprivation, child abuse, adult violence, and sexual dysfunction which occur within families of humans and other primates, being "inheritable" not through genetic mechanisms, but through *similar repeating modes of harsh and antisexual childrearing*. A behavior cannot be called "genetic" just because it preferentially duplicates itself within a given branch of a family tree, or just because it occurs among primates other than Homo sapiens. As demonstrated in this chapter, *the familial and social environments of the infant, child, and adolescent are the crucial factors which determine and predict whether or not subsequent generations will be abused, or whether high or low levels of violence and destructive aggression will be present in adult behavior and society.*

In sex-economic theory, Reich gave all the necessary ingredients whereby a culture free from inhibition of innate, biologically mandated pleasure drives would maintain itself in a social, cooperative, and loving manner, without the presence of secondary behaviors of a violent or destructive nature. Male domination, the subordination of females and children, repressive sexual taboos, compulsive marriages, and tendencies towards infant neglect, child abuse, and social violence would not spontaneously occur in such a society, appearing as it were "from nowhere". They would have to come from *outside* the culture.

Sex-economic theory — which provides a dynamic and scientifically testable alternative to mechanistic genetic determinism for the passing on of behavior traits from one generation to the next — rests upon the existence of a spontaneously honest, loving, social and peaceful core to human nature, characterized by strong maternal-infant bonds, and strong bonds of both sexual and romantic love between males and females, including adolescents and the unmarried. As such, it demands some <u>outside force</u>, some powerful, prolonged and widespread anxiety-provoking trauma, to destroy a peaceful and loving social group and drive it towards pleasure-anxiety and social violence. Whatever this outside force is, or was, it would also be required to initiate generalized changes within the sphere of sexual behavior and genital function, to create chronic sexual frustration and undischarged emotional-energetic tension — the "fuel" of pleasure-anxiety, armoring, and sadistic aggression which is turned back into the social group, organized into social institutions and "traditions", for the socially-approved attack upon and repression of sexual functions and pleasurable activity in the next generation. Once the armor is formed and generally exists culture-wide, it becomes a force for its own self-perpetuation, this time wholly from <u>within</u> the culture. After the outside force had worked its traumatic damage, social structure would change, and institu-

108. deMause, *History of Childhood,* ibid., pp.41-3.

109. Fromm, E.: *Escape from Freedom,* Holt, Rinehart & Wilson, NY, 1963; Fromm, E.: *The Sane Society,* Holt, Rinehart & Wilson, NY, 1963.

110. Bullough, V.: *Sexual Variance in Society and History,* John Wiley, NY, 1976; Mantegazza, P.: *The Sexual Relations of Mankind,* Eugenics Pub. Co., NY, 1935; Lewinsohn, R.: *A History of Sexual Customs,* Harper Brothers, NY, 1958; Hodin, M.: *A History of Modern Morals,* AMS Press, NY, 1937; Tannahill, R.: *Sex in History,* Stein & Day, NY, 1980; Brandt, P.: *Sexual Life in Ancient Greece,* AMS Press, NY, 1974; Kiefer, O.: *Sexual Life in Ancient Rome,* Barnes & Nobel, NY, 1951; Van Gulik, R.: *Sexual Life in Ancient China,* E.J. Brill, Leiden, 1961; Levy, H.S.: *Sex, Love, and the Japanese,* Warm-Soft Village Press, Washington, D.C., 1971; Stone, M.: *When God Was a Woman,* Dial, NY, 1976; May, G.: *Social Control of Sex Expression,* Geo Allen & Unwin, London, 1930; Gage, M.J.: *Woman, Church & State,* Persephone Press, Watertown, Mass., 1980; Taylor, G.R.: *Sex in History,* Thames & Hudson, London, 1953.

tionalization of trauma would occur. Newborn infants, mothers, and children would then be attacked and repressed by ritual traditions, by distorted and damaged social institutions which would rationalize a thousand different methods for the denial of pleasure and infliction of pain, implanting new barriers of fear, compulsion, and anxiety into male-female relations. Outside forces would then no longer be necessary to re-create the trauma in subsequent generations. People thereafter would become their own traumatic oppressors, and would hence only "feel comfortable" with neighbors and leaders who would likewise support the continuation of the same harsh and repressive ways of living. The softer individual with a new idea or new way of living, would then be the odd person, the "heretic" or "provocateur" who — by virtue of their unbearable softness — would be a constant reminder of what had been lost, and what was missing from the lives of the majority. Likewise, the emotionally-soft and fluid cultural group would be viewed with much suspicion and anger by the more hardened social group. Of necessity, as a means to control within-group anxiety (family, school, professional group, village, or nation) the softer persons or groups would, through some convoluted and illogical rationalization, be compulsively attacked, driven out or even totally destroyed by the more hardened and rigidly armored individuals. [111]

Sex-economic theory thereby provides a major, testable mechanism for the observed and recorded behaviors and events of culture and history. It also has predictive value. Biological impulses can be allowed expression, or be crushed, by families and social institutions, leading to very different social consequences. State structure mirrors family structure in the average household, as expressed in various social institutions. Rarely do social leaders, or even professional groups with a proclaimed interest in such matters, confront the average citizen with facts about their self-destructive, life-destructive behavior, or push to implement any real social reforms (i.e., the abolishment of circumcision within a given hospital, or institution of democratic methods in schools, allowing the children to have a say in running things, or to encourage and legalize midwifery and home-birth, or to openly defend the sexual rights of adolescents). Mostly, our leaders and professionals are supporters of social orthodoxy, with whom they are compelled to form socio-economic and emotional alliances. Otherwise, they would not become "leaders", and would be rejected by professional certifying committees, as being "too unorthodox". The system thereby becomes self-replicating, admitting only the most orthodox into the inner sanctums of power, and centrifuging out dissenters without hesitation. The Western democracies are reformed sufficiently that they no longer burn heretics, as is still the case in many places around the world, but they are rarely otherwise tolerated.

While there are notable exceptions, in large measure, mainstream education, medicine, psychology and law have continued this historical pattern. Reich's penetrating social analysis in *The Function of the Orgasm* and *The Mass Psychology of Fascism* are hardly known in today's universities. Nor have the professionals been much interested in the works of various cross-culturalists, such as Malinowski and Prescott, or the various other reformers, on the institutional roots of neuroses and violence in the physical and sexual traumas of infancy and childhood. Instead, the various professions are largely preoccupied with certifying and legitimizing, or even *amplifying* the existing armored, patristic social structure and *status quo*. Significantly, they willfully rely upon police power to enforce their monopolies, to prevent the growth and development of alternative social structures which would, if allowed to exist, quickly render them irrelevant and obsolete. [112]

The often-stated goal of most "mental health" professionals is to

111. This phenomenon, of the impulse to attack that which is softer and more emotionally alive than one's self, Reich termed the *emotional plague*. It is unfortunately a widespread social phenomenon, at work within political organizations, professional groups, churches, families, and other places where basic living functions are of primary concern. See Reich, W.: *Selected Writings*, Farrar, Straus & Giroux, 1973, p.467-513; Reich, *Character Analysis*, ibid., 1961, p.504-539. Reich, W.: *The Murder of Christ*, Farrar, Straus & Giroux, 1971.

112. Peter Breggin (*Toxic Psychiatry: Why Therapy, Empathy, and Love Must Replace the Drugs, Electroshock, and Biochemical Theories of the "New Psychiatry"*, St. Martin's Press, NY, 1991) has written an eye-opening account of the leading role played by modern psychiatry in this process, reinforcing what has been said before about how the secular "medical priesthood" has taken over the former role of the Church in maintaining and profiting from the sick social order, and in using emotionally-distressed people and "social misfits" as an outlet for sadistic aggression. Historically, the alliance of the medical profession with Church power and the authoritarian state apparatus is well documented. Behavioral conformity in Nazi Germany and Stalinist Russia was also enthusiastically supported by most MDs, particularly psychiatrists, with social reformers and political dissidents often being identified as "mentally disturbed". (ie., Ehrenreich, B & D. English: *Witches, Midwives and Nurses: A History of Women Healers*, Feminist Press, NY 1973; Lapon, L.: *Mass Murderers in White Coats: Psychiatric Genocide in Nazi Germany and the United States*, Psych. Genocide Research Inst., Springfield, MA 1986.)

better fit hurting and upset people into a sick and self-destructive social order. The root causes of social problems, as listed in this chapter, are generally not addressed, not given very much attention. The overwhelming amount of attention is, instead, directed towards repressing the symptoms of distress and irritation people display from living in the sick social order; they will help one "adjust" to life "within the trap" or "social cage", as Reich would say. Symptoms of distress, from living within the social cage, are repressed chemically, by shock-torture, by surgical mutilation, and/or by simple incarceration — but only rarely are the underlying causes examined or addressed. For the record, I do not refer here to bonafide criminals, serial murderers and child-molesters who must be restrained and incarcerated, or to those volcanic character structures who suddenly take a gun and murder numerous strangers without warning. These individuals must be restrained, without question — but they are, for the most part, quite "normal" in their outer appearance and behavior before they are arrested for their crimes (which itself is revealing of the pathos of much "normality"). I refer instead to the thousands of young people who are incarcerated every year for "crimes" no more severe than disobedience to adult authority, sleeping with their girlfriend, smoking pot, staying out late at night, or other generally victimless activities. Completely healthy or harmless behaviors, such as the independent mindedness and sexual impulses of young people, may be branded as "criminal" or "antisocial" by a larger society which, in fact, is far more criminal and destructively antisocial in its deadly "normality".

Where a young person's biological needs conflict with family, school, church, or state, then such needs are generally "restructured", "socialized", "sublimated", or otherwise done away with through either mild or severe obedience training. If this does not work, the courts or a clinical specialist may step in to pronounce the individual as "deviant", to forcibly institutionalize them, drug them, or both, the underlying clash with the sick social structure being either glossed over, moralized away, or fully ignored. What is rarely done, by contrast, is for the professionals or the courts to make changes in the existing social situations — to drug, jail or repress the social misfit or rebel is accepted; demands for social change or the establishing and significant funding of new life-positive social institutions proposed by reformers are almost always ignored.

Sex-economic theory informs us that the social tensions and sadistic impulses which predictably develop within an infant-traumatizing and sex-repressive social order, will in large measure be channeled into special authoritarian social institutions — the military caste, police or security forces, certain forms of sado-medicine, the priesthood, etc. — where they receive the "official approval" and blessings of society for discharge towards some other nation, or more generally towards minority groups, non-conforming individuals, women and children within one's own national boundaries, or against the poorer classes. Civil unrest or even international wars are the inevitable consequences for the armored cultures, as is easily demonstrated historically. Social conditions may remain calm and subdued for a period, only to explode a generation later.

The social pathology described above is today applicable to nearly every nation on Earth, although in some, the forms of expression are more highly developed and well-organized than in others. Our goal here is not to elevate one cultural group over another, but rather to review the important historical changes and regional differences in such conditions, with the goal of determining their ultimate origins. With such new knowledge, we gain important tools for changing all cultures towards a more peaceful, emotional healthy, and life-positive future.

Aged Ignorance,
Perceptive Organs Closed
(William Blake)

3. Basic Assumptions, Observations and Probable Mechanisms for the Genesis and Global Diffusion of Armored Patrism

The variables selected for producing the behavior maps presented in later chapters include factors of major consequence to the pleasure and well-being of people: the manner in which infants and young children are treated; the way in which adolescent sexual drives are handled; the rights of young people to spontaneously and instinctively choose their own mates; the rights of a woman to determine her own fertility, to freely divorce if desired, to maintain legal status and property as a full member of society; and general social structure in the context of individual freedoms. Of great interest also was the degree to which the above factors correlate with the expression of violence and destructive aggression on the social scene.

The thrust of my research was to produce a geographical study of cultural adaptations and behavior in deserts around the world. But, before moving on to the various behavior maps, we must ask: *Why should variables on family life, women, children, and sex be selected as indicators of response to arid environments? Why the specific ordering and theoretical linkages between the variables, namely the linking of destructive aggression and sadistic violence to infant trauma and sex-repression? And why would such variables, in turn, possess a link to deserts?* These questions were addressed indirectly in the prior Chapter, outlining Reich's sex-economic findings, but the sections below provide additional structural evidence and considerations. They are additional foundation stones in the overall development of the Saharasian thesis.

ar-mor (är´mer) *n.*
1. A defensive covering, such as chain mail, worn to protect the body against weapons.
2. Any tough protective covering, such as the body of tanks or warships, or wire sheathing.
3. Anything serving as a safeguard or protection.
4. The armored vehicles of an army.

Reich's Speculations on the Origins of Human Armoring

Between the 1920s and 1950s, Wilhelm Reich developed the *sex-economic* theory of human behavior and social functioning, which attributed mass human destructive aggression and warfare to a complex of specific culturally-demanded trauma and repressions, primarily aimed at the infant, child and adolescent, and being reproduced through social institutions supported by the average individual. Reich's views on sex-economy provided a specific, testable mechanism whereby an individual or culture possessing antichild, antifemale, and antisexual behaviors and beliefs, could, through the existence of specific social institutions, transmit such behavior from one generation to the next.

The process of human *armoring* (as Reich called it) against impulse, feeling, and emotion, begins in the cradle as the child is purposefully (though often unconsciously) traumatized by contactless and emotionally deadened adult caretakers, who themselves had been subject to similar trauma in infancy. Additional traumas and repressions, particularly those of a sexual nature, were heaped upon the older child and adolescent, by which time the fully armored character structure crystallizes. Examples of the specific traumas I studied, and which were mapped and discussed in both anthropological and archaeological/historical settings, are infant cranial deformation, swaddling, genital mutilations, the bride price, and various other pleasure-censoring or very painful childrearing and adolescent rituals. Reich's discussions did not, however, provide much more than clues as to how such harsh, painful, and traumatic practices and institutions could have started in the first place. How and why does an originally unarmored, matrist people armor up in the first instance, to chronically harden its previously soft and fluid emotional responses to basic biological urges and feelings, and the pleasures of life?

It can be successfully argued that a culture could armor against its feelings and needs by being *physically forced* to adopt the social institutions and behaviors of some other, more powerful invading culture which already possessed such traits themselves. Indeed, history is replete with examples of such forced change in social structure, in the form of warfare and subjugation. Similarly, a culture already armored might become even more hardened, and "discover" or "invent" new methods of traumatizing infants and children. In such a case, new traumas would gradually become more firmly institutionalized by popular will, further reinforcing and amplifying previously existing tendencies. For example, the invention and institutionalization of new forms of infant trauma can be observed in the field of medical obstetrics, discussed in the later sections on "Couvade" and "Genital Mutilations". But elucidation of the process whereby armoring spreads from one culture to another, or intensifies in modern times, still does not solve the riddle of how, where, or under what circumstances the earliest cultures began to develop the first characteristics of emotional armoring. In 1951, Reich posed the question as follows:

> "If nothing exists beyond the confines of natural processes, why does the armoring of the human species exist at all, since it contradicts nature in man at every single step and destroys his natural, rich potentialities? This does not seem to make sense. Why did nature make a 'mistake'? Why only in the human species? Why not also in the deer or in the chipmunk? Why just in man? His 'higher destiny' is, clearly, not the answer. The armor has destroyed man's natural decency and his faculties, and has thus precluded 'higher' developments. The twentieth century is witness to this fact." [1]

Reich first theorized that trauma and repression originated when adults, for economic motives, began to manipulate the marriage choices and sexual lives of their children. In his work *The Invasion of Compulsory Sex-Morality* (1931),[2] he drew upon the works of Morgan, Engels, Bachofen, Freud, Malinowski, and Roheim to argue the origins of sexual repression in the compulsory, polygamous marriage and marriage gift. He demonstrated that wealth and power would accumulate in the hands of those males who manipulated the marriage choices of their children. In particular, he discussed the arranged, cross-cousin marriage which predominated among the polygamous Trobriand chiefs; they received yearly gifts from the brothers of each wife, and because of their higher status also acquired gifts of greater value than the common man.

> "...the clans have different ranks, and the chief, who belongs to the uppermost clan, has the right to polygamy. Therefore, the flow of marriage gifts, which otherwise would be leveled out by intermarriage, is deflected one-sidedly to the chief of one clan and his family."[3]

> "...the institution of the marriage tribute alters the balance of power in favor of the father and chief... father-right, and with it the patriarchal polygamous family, grows out of the primeval matriarchal organization and the kinship clans. For the chief, possibilities and rights come as a result of his power (and of his obligations) — eg., the right of polygamy and the beginnings of a feudal power to command over his wives' brothers and other relatives, who are obliged to pay tribute to him."[4]

1. Reich, W.: *Cosmic Superimposition*, Farrar, Straus & Giroux, NY, 1973, p.288.

2. Reich, W.: *Invasion of Compulsory Sex Morality*, Farrar, Straus & Giroux, NY 1971.

3. Reich, *Invasion of Compulsory Sex Morality*, ibid., p.54.

4. Reich, ibid., p.61.

Additionally, a favored cross-cousin marriage existed between the chief's son and niece (his sister's daughter). Under such an arrangement, the chief's yearly gifts to his sister's family would partly return to his own son, who was under his control and tutelage. While this form of cross-cousin marriage was considered "desirable", other forms of cousin marriage, between the chief's daughter and nephew (his sister's son) were "not looked upon favorably". In this latter case, the marriage gift accrued to the chief's sister's sons, who were not under his control.

Through these various means, power and wealth accumulated in the chiefly clans of higher status, and to chiefs or others who could control the marriage choices of their children. But how could such control be accomplished, given that Trobriand culture allowed a great sexual freedom for children who, above all else, would not be inclined to marry someone "picked" for them, whom they did not love?

In fact, only the children of the chiefs, as opposed to the average child of Trobriand society, were pressured and molded to accept the parental choice of marriage partner. Consequently, the chief's children were subject to a set of sexual taboos to which other children in the villages were not. Reich argued that such a strict upbringing, which included mandatory sexual abstinence, forced the children of the chiefs to accept the marital decisions of the father, and in general made such high-caste children docile, meek, able to censor their own inner feelings and desires, and capable of tolerating a compulsive marriage.

> *"With one exception, the genital love life of the Trobrianders before marriage not only is completely free since childhood, but in addition is socially sanctioned. The one exception is those children who are destined for a cross-cousin marriage; social custom demands of them premarital chastity, and they must refrain from the otherwise usual and eagerly indulged-in genital activities. Malinowski merely records this fact under the heading 'Ceremonies of Infant Betrothal' without drawing any connection."* [5]

> *"...the social purpose of the demand for asceticism in youth and for sexual suppression in early childhood is...the insuring of man's capacity for lifelong compulsory marriage. That is openly expressed in many publications of the church and of undisguisedly reactionary moralists."* [6]

To my own thinking, this work of Reich's still did not get to the origins of the armoring process. The *Invasion of Compulsory Sexual Morality* revealed the way in which armoring and human pathos could spread into a culture which already possessed weak tendencies in that particular direction. It also exposed the shift in socioeconomic structure which occurred during a transition from an unarmored to armored condition. However, it did not get to the ultimate origins of such tendencies, nor discuss processes whereby one armored culture could influence or "infect" a previously unarmored culture with its own social institutions and behaviors. [7] By 1950, Reich had given up the economic thesis as a primary cause of armoring, stating that the question of the origins of armoring remained an unsolved mystery.

> *"We know it is mostly socio-economic influences (family structure, cultural ideas on nature versus culture, requirements of civilization, mystical religion, etc.) that reproduce the armor in each generation of newborn infants. These infants will, as grown-ups, force their own children to armor, unless the chain is broken somewhere,*

5. Reich, *Invasion of Compulsory Sex Morality*, ibid., p.69.

6. Reich, ibid., p.71.

7. With respect to the Trobriand Islands, this question is discussed in detail in Chapter 9.

sometime. The present-day social and cultural reproduction of the armor does not imply that when armoring first began, in the faraway past of the development of man, it was also socio-economic influences that set the armoring process into motion. It seems rather the other way around. The process of armoring, most likely, was there first, and the socio-economic processes that today and throughout written history have reproduced armored man were only the first important results of the biological aberration of man...

Still, the question of how the human animal, alone among the animal species, became armored remains with us, unsolved, over-shadowing every theoretical and practical step in education, medicine, sociology, natural science, etc. No attempt is made here to solve this problem. It is too involved. The concrete facts that possibly could provide an answer are buried in a much too distant past; reconstruction of this past is no longer possible." [8]

Reich later speculated that armoring began when the human animal first became consciously aware of itself, as the perceptive apparatus turned inward, the emotions becoming withdrawn and stuck.[9] But even this speculation carries with it the assumption that the natural process which gave rise to the human species carried with it the seeds from which armoring could spontaneously arise. In this sense, his speculation is not too far removed from a kind of "death instinct" or "Original Sin", albeit one which only needed to appear temporarily, only once within the course of human history to initiate the process. Also, assumptions regarding the lack of consciousness or self-awareness in nonhuman animal life appear unwarranted, particularly given the more recent studies on dolphin and primate intelligence. Aside from a single paragraph, cited above, Reich never elaborated or dwelled upon his speculation regarding a "stuck perceptive apparatus", and it is clear that he was still searching for a concrete demonstrable answer to the question.

In one of his last writings, Reich articulated an observable connection between armoring and deserts, between the physical, geographic desert landscape, and what he called the *emotional desert*. Following a period of extended atmospheric research in the deserts of Arizona, in 1954-1955, he discussed how deserts "suck dry" the juiciness of plant and animal life, which developed a thick and prickly outer covering against the hostile environment. This was similar, he argued, to the way harsh and trau-matic methods of infant and child treatment, and antisexual attitudes, dry up the softer emotional aspects of life, giving rise to a particularly acid, dry, and/or prickly character structure:

"When a desert begins to develop, when the natural, original vegeta-tion gradually falls prey to and perishes under the strain of drought, lack of dew in the morning, progressive parching of the land under a burning sun...life still fights on. A new type of life, a secondary vegeta-tion, adapting itself to the harsh conditions of existence in the desert, arises. It is an ugly, poorly equipped life. The stems of the chollas or cactus or palo verdes are not solid as the stem of an oak or a birch. The stem consists of single, narrow strands which are and remain brittle, and have no connection, show no fusion with each other. The whole plant is covered with bristles, reminding us, in analogy, of the prickly outer behavior of human beings who are empty and desert-like inside. This is not a mere analogy. The simile goes very far, indeed. The desert plants either grow leathery prickly leaves as does the cactus plant, or as in the cholla, the chlorophyll-bearing structure is re-stricted to the outermost ends of the branches. It is characteristic of

8. Reich, *Cosmic Superimposition*, ibid., pp.288-9.

9. Reich, ibid., pp.289-93.

desert life that even animals have a bristly, prickly surface or sharply pointed organs to kill: the scorpion, the rattlesnake, the Gila monster. ... This vegetation slowly replaces the last remnants of the primal vegetation, until, with the progress of desert development to the last stage, the Sahara sands desert, the secondary vegetation, too, dies out, and nothing remains but sand dunes.

With the spreading of the global desert, civilizations go under, life perishes completely in the affected realm, man either tries to escape or he too adjusts to the life in the desert on rare spots of green, called oases.

The continuous presence of death...and the ever-present dull awareness of the inevitable end is characteristic of both life in the desert and life in armored man. The deadness of emotion, the dehydration of tissues alternating with puffy swelling, fatty flabbiness, or inclination to edema or disease which causes edema, alcoholism which serves to stimulate what is left from an original sense of life, crime and psychosis and the last convulsions of a thwarted, frustrated, badly maltreated life are only a few of the consequences of the emotional desert." [10]

To my knowledge, Reich never elaborated beyond the above paragraphs on the relationship between the geographical and emotional deserts, nor on the origins of armoring.[11] Still, his thinking clearly was moving towards a view of the desert as a primary agent for shaping human character structure, a view which today finds clear support in the studies discussed below, and in my own findings. Consequently, the original suggestion of a specific role of the desert in the genesis of human armoring belongs to Reich.[12] Formulating the correct question to ask is often the most difficult part of any research effort. I took Reich's suggestion at face value, systematically tested the idea with rigor in a number of ways, and thereby solved a long-standing riddle of history and behavior. *A very specific historical/geographical relationship was discovered and observed, on the genesis of armoring within certain hyperarid desert regions around 4000-3500 BCE, and the subsequent diffusion of armoring out of these same deserts to distant lands.*

My research involved a review of materials which Reich did not have access to, namely more recent environmental, archaeological, and anthropological studies, which I set into a geographical and historical framework. Indeed, my work indicates a parallel development of warfare, the chiefly institution, sex repression, child abuse, female subordination, and other various forms of human pathos, all of which were rooted in the physical processes of desertification, famine, and forced migrations. The following sections outline several lines of reasoning and evidence, other than those given by Reich, which also suggested a causal relationship between desertification and certain armored, patrist forms of behavior and culture.

Cross-Cultural and Geographical Factors: Armored Patrism versus Unarmored Matrism

Some of my earliest observations on cultural variation between different environments were rather simplistic, with a number of honest, but incorrect assumptions which had been made by others as well. Nevertheless, these early observations proved to be a key to the relationship between the Saharasian desert belt and patrism. For instance, I originally postulated that there was a specific "desert culture" type, with a contrary

10. Reich, W.: "The Emotional Desert", in *Selected Writings*, Farrar, Straus & Giroux, 1973, pp.461-3.

11. Reich's research came to an abrupt halt in 1956, when his books and journals were banned and burned in incinerators under a fraudulently-obtained Food and Drug Administration court injunction; Reich died in jail in 1957 after being found in technical violation of the injunction, which was clearly aimed at ending his controversial research; cf. Greenfield, J.: *Wilhelm Reich vs. the USA*, W.W. Norton, NY, 1974; M. Sharaf, *Fury on Earth*, St. Martin's/Marek, NY, 1984; DeMeo, J.: *Orgone Accumulator Handbook*, Natural Energy Works, Ashland, Oregon, 1986.

12. Immanuel Velikovsky, an unfairly maligned psychoanalyst-historian, made some of the first connections between major environmental catastrophes and social change, arguing the Earth and all human cultures were thrown into upheaval by the close passage of a large comet in ancient historical times. He did not, however, attribute the social changes to desertification *per se*. See Chapter 8, pp.214-215, for more discussion on this point.

"rainforest culture" of almost exact opposite qualities, the former being armored and patrist, and the latter being unarmored and matrist. My research on "desert" culture had primarily been focused upon the peoples of the arid Near East. And, as it turned out, my desert-patrist generalizations were largely correct for that area of the world. For the "rainforest" culture type, I had focused upon the Trobriand Islanders of Oceania, the Pygmy of the African rainforest, and the Muria of the Indian rainforest. And these cultures likewise supported my rainforest-matrist generalizations. Later, a cross-cultural and location-specific geographical testing method was developed based upon the two culture types. The tests clarified a number of important criticisms which had been leveled at my premature generalizations, namely the fact that there were some desert peoples who were relatively unarmored and matrist, and some rainforest peoples who were relatively armored and patrist. The resolution to this apparent contradiction was achieved when the World Behavior Map was developed. That map, and other lines of evidence, demonstrated the rainforest regions closest to Saharasia had become armored and patrist by virtue of culture contacts with armored peoples who had migrated from the desert regions. Rainforests and even other semiarid regions most distant from Saharasia (in Oceania and the New World) remained relatively unarmored save for a few specific cultures who likewise had a historical migratory connection to Saharasia. This latter point constitutes a controversy within itself, which will be discussed later in the Chapter on "Oceania and the New World". Before elaborating upon these factors, amplification of the two culture types, matrism and patrism, is in order.

A) Matrism: Unarmored, High-Pleasure, Low-Violence, "Rainforest" Cultures

The Trobriand Islands of Oceania are inhabited by a high-pleasure, low-violence culture living in a lush tropical island setting. As documented by Malinowski,[13] the Trobrianders did not repress the budding sexuality of their adolescents, yet neither sexual anarchy nor perversion existed (excepting among the children of the chiefs[14]). Neither did any "Oedipal conflict" exist; the mother's brother invoked needed instruction and discipline of children (which by most standards would hardly be called discipline), and the child's budding sexual interests were not bottled up within the nuclear family. Neither sexual "sublimation" nor "latency" existed. Indeed, Trobriand children joined general age groups starting around five years of age, playing all sorts of games, including sexual ones. A special institution known as the *bukumatula* existed where the children met on their own, developed their own friendships and love-matches, without adult interference or supervision.

Groups of adolescent boys and girls would travel in age groups from one village to the next, seeking to meet with other groups of children. Girls displayed the same sexual interests, assertion, vigor, and unabashed forwardness as did the boys, and virginity was seen as something rather peculiar or pitiful. Indeed, the adults looked upon the affairs of the children and adolescents with fond memories of their own childhoods.[15] Pregnancies among unmarried girls were very rare, probably due to the use of effective contraceptive herbs known to the women.[16]

From the generally fluid and changeable love-matches of the Trobriand children grew the more permanent love affairs of the older adolescents and, eventually, stable marriages. Marriages, then, were developed and maintained out of strong emotional bonding and mutual sexual attraction and genital gratification. They were not compulsive, nor

13. Malinowski, B.: *The Sexual Life of Savages*, Routledge & Kegan Paul, London, 1932; Malinowski, B.: *Sex and Repression in Savage Society*, Routledge & Kegan Paul, London, 1927.

14. Reich, *Invasion of Compulsory Sex Morality*, ibid., 1971.

15. Malinowski, *The Sexual Life of Savages*, ibid.; *Sex and Repression in Savage Society*, ibid.

16. See Chapter 6 of this work for full documentation.

dictated by the adults (except among the chiefly families), and could be dissolved by either the wife or husband, as they wished; re-marriage could then take place with equal ease. Males did not dominate the basic life decisions of females and children. Adults remembered the time of youth and the *bukumatula* as one of much happiness, and both children and adults behaved in a lively, spontaneous fashion, without inhibition of emotion. According to Malinowski, Trobriand adults exhibited a high degree of emotional health, and a lack of neurotic, antisocial behaviors or sexual pathology;[17] they were democratically inclined, monogamous (except for chiefs), and had a relatively high status for women.

As previously mentioned, the few exceptions where Trobriand children were manipulated and controlled by parents demonstrate even more clearly the truth of Reich's sex-economic findings. Children of chiefs were subject to arranged marriages which would work to the economic gain of the chief's clan. Such children were not allowed to play in the age groups with the other children, nor participate in the *Bukumatula*. As Malinowski observed, they were docile, obedient, and meek by comparison to other village children. Indeed, his work suggested that the few episodes of familial violence which did occur in Trobriand society were related to these unusual traces of compulsiveness which had invaded the family structure. Where social unrest or neurosis did occur, it was confined to the chiefly classes who repressed their children, or due to the antisexual influences of various missionaries. Hence, these events constituted the exceptions which proved Reich's point that such behavior had *invaded into* Trobriand society *from some unknown source outside of itself*. It is illustrative to quote at length from Malinowski's study of Trobriand society; these materials will later be contrasted against descriptions of patrist cultures from the heart of Saharasia.

On Infants and Children:

"A man of any clan would often, in speaking of his family relations, expatiate on the number of his sisters and of their female children as being a matter of real importance to his lineage. Thus girls are quite as welcome at birth as boys, and no difference is made between them by the parents in interest, enthusiasm, or affection. It is needless to add that the idea of female infanticide would be as absurd as abhorrent to the natives." [18]

"...the husband fully shares in the care of the children. He will fondle and carry a baby, clean and wash it, and give it the mashed vegetable food which it receives in addition to the mother's milk almost from birth. In fact, nursing the baby in the arms or holding it on the knees...is the special role and duty of the father. It is said of the children of unmarried women who...are without a husband...that they are 'unfortunate' or 'bad' because 'there is no one to nurse and hug them...The father performs his duties with genuine natural fondness: he will carry an infant about for hours, looking at it with eyes full of such love and pride as seldom seen in those of a European father. Any praise of the baby goes directly to his heart, and he will never tire of talking about and exhibiting the virtues and achievements of his wife's offspring." [19]

"Children...enjoy considerable freedom and independence. They soon become emancipated from a parental tutelage which has never been very strict. Some of them obey their parents willingly, but this is entirely a matter of the personal character of both parties; there is no

17. Malinowski, *The Sexual Life of Savages*, ibid., pp.382,395-8.
18. Malinowski, ibid., p.25.
19. Malinowski, ibid., p.17.

idea of a regular discipline, no system of domestic coercion... A simple command, implying the expectation of natural obedience, is never heard from parent to child in the Trobriands. ...the idea of definite retribution, or of coercive punishment is not only foreign, but distinctly repugnant to the native... Such freedom gives scope for the formation of the children's own little community, an independent group, into which they drop naturally from the age of four or five and continue til puberty. As the mood prompts them, they remain with their parents during the day, or else join their playmates for a time in their small republic. And this community within a community acts very much as its own members determine, standing often in a sort of collective opposition to its elders. If the children make up their minds to do a certain thing, to go for a day's expedition, for instance, the grown-ups and even the chief himself...will not be able to stop them..." [20]

The Children's Democracy, Trobriand Islands, c.1918
(Malinowski, *Sexual Life of Savages*)

On Childhood Sexuality:

"The child's freedom and independence extend also to sexual matters. To begin with the children hear of and witness much in the sexual life of their elders." [21]

"There are plenty of opportunities for both boys and girls to receive instruction in erotic matters from their companions. The children initiate each other into the mysteries of sexual life in a directly practical manner at a very early age. A premature amorous existence begins among them long before they are able really to carry out the act of sex. They indulge in plays and pastimes in which they satisfy their curiosity concerning the appearance and function of the organs of generation, and incidentally receive, it would seem, a certain amount of positive pleasure." [22]

"The attitude of the grown-ups and even of the parents toward such infantile indulgence is either that of complete indifference or of complacency — they find it natural, and do not see why they should scold or interfere." [23]

20. Malinowski, *The Sexual Life of Savages*, ibid., pp.44-6.
21. Malinowski, ibid., p.46.
22. Malinowski, ibid., p.47.
23. Malinowski, ibid., p.48.

"It is important to note that there is no interference by older persons in the sexual life of children... There is certainly no trace of any custom of ceremonial defloration by old men, or even by men belonging to an older age class." [24]

On Adolescent Sexuality:

"As the boy or girl enters upon adolescence the nature of his or her sexual activity becomes more serious. It ceases to be mere child's play and assumes a prominent place among life's interests. What was before an unstable relation culminating in an exchange of erotic manipulation or an immature sexual act becomes now an absorbing passion, and a matter for serious endeavor. An adolescent gets definitely attached to a given person..." [25]

"During this intermediate period love becomes passionate and yet remains free. As time goes on, and the boys and girls grow older, their intrigues last longer, and their mutual ties tend to become stronger and more permanent. A personal preference as a rule develops and begins definitely to overshadow all other love affairs. It may be based on true sexual passion or else on an affinity of characters. Practical considerations become involved in it, and, sooner or later, the man thinks of stabilizing one of his liaisons by marriage. In the ordinary course of events, every marriage is preceded by a more or less protracted period of sexual life in common." [26]

Trobriand girl in front of Bukumatula adolescent dormitory, c.1918. (Malinowski, *Sexual Life of Savages*)

On Marriage & Adult Sexuality:

"The natives...order their marriages as simply and sensibly as if they were modern European agnostics, without fuss, or ceremony, or waste of time and substance. The matrimonial knot, once tied, is firm and exclusive, at least in the ideal of tribal law, morality, and custom. As usual, however, ordinary human frailties play some havoc with the ideal. The Trobriand marriage customs again are sadly lacking in any such interesting relaxations as jus primae noctis, wife-lending, wife-exchange, or obligatory prostitution. The personal relations between

24. Malinowski, ibid., pp.50-1.
25. Malinowski, *The Sexual Life of Savages*, ibid., p.54.
26. Malinowski, ibid., pp.57-8.

the two partners...do not present any of those 'savage' features..." [27]

"*The natives regard such practices as bestiality, homosexual love and intercourse, fetishism, exhibitionism, and masturbation as but poor substitutes for the natural act, and therefore as bad and only worthy of fools. Such practices are a subject for derision, tolerant or scathing according to mood, for ribald jokes or funny stories. Transgressions are rather whipped by public contempt than controlled by definite legal sanctions. No penalties are attached to them, nor are they believed to have any ill results on health. Nor would a native ever use the word taboo (bomala) when speaking of them, for it would be an insult thus to assume that any sane person would like to commit them... The Trobriander's contempt for any perversion is similar to his contempt for the man who eats inferior or impure things in the place of good, clean food or for one who suffers hunger because there is nothing in his yamhouse.*" [28]

"*Many natives are, under the present rule of whites, cooped up in gaol, on mission stations, and in plantation barracks. Sexes are separated and normal intercourse made impossible; yet an impulse trained to function regularly cannot be thwarted. The white man's influence and his morality, stupidly misapplied where there is no place for it, creates a setting favorable to homosexuality. The natives are perfectly well aware that venereal disease and homosexuality are among the benefits bestowed on them by Western culture.*" [29]

"*...flagellation as an erotic practice is entirely unknown; and the idea that cruelty, actively given or passively accepted, could lead, of itself alone, to pleasant detumescence is incomprehensible, nay ludicrous, to the natives.*" [30]

Trobriand Family, c.1918
(Malinowski, *Sexual Life of Savages*)

On Male and Female Relations:

"*The frank and friendly tone of intercourse, the obvious feeling of equality, the father's domestic helpfulness, especially with the children, would at once strike any observant visitor. The wife joins freely in the jokes and conversation; she does her work independently, not with the air of a slave or a servant, but as one who manages her own department. She will order the husband about if she needs his help. Close observation, day after day, confirms this first impression. ...she is the owner of separate possessions in the house; and she is — next to her brother — the legal head of her family.*" [31]

"*The water-hole is the woman's club and centre of gossip, and as such is important for there is a distinct woman's public opinion and point of view in a Trobriand village, and they have their secrets from the male, just as the male has from the female.*" [32]

Malinowski's study of Trobriand society has long stood as a rebuttal to Freud's contention that the Oedipal conflict and childhood sexual latency were innate and biological in character. However, his writings had minimal effect upon the scholarly community with professed concerns on such matters; Freud's views are generally accepted without reform, or rejected wholesale.[33] However, Malinowski's work did influence Reich.[34]

27. Malinowski, ibid., p.65.
28. Malinowski, *The Sexual Life of Savages*, ibid., pp.395-6.
29. Malinowski, ibid., p.398.
30. Malinowski, ibid., p.400.
31. Malinowski, ibid., p.15.
32. Malinowski, ibid., p.17.
33. See the prior Chapter for more detailed discussion.
34. The reverse appears equally true. Malinowski and Reich became friends, and communicated on a first-name basis. See the sidebar quotation in this work, p.19.

As Reich viewed it, Trobrianders were free from infant and child traumas and adolescent sexual repression, because the adults had never been raised with significant trauma or repression themselves. The lack of adult neurosis was the expected consequence of a happy infancy and childhood, and a healthy love life.

A number of studies have noted the existence of cultures where babies, children, and adolescents were raised with a great amount of physical affection, and where a high level sexual affection was allowed among children and unmarried adolescents. Examples are the Muria, studied by Elwin,[35] and the Pygmies, studied by Hallet[36] and Turnbull.[37] Like the Trobrianders, these cultures also possessed very low levels of adult violence. Their very existence collectively refutes all theories on the supposed biological or genetic basis of the Oedipal conflict, patrism, and destructive aggression.

Murian Girls, Bastar, India, c.1920s.
(Elwin, *Muria and Their Ghotul*)

Indeed, Hallet argued that the behavior of the gentle Pygmy peoples, who lived deep in the rainforests of Africa, reflected the original, uncontaminated state of behavior of Homo Sapiens. Elwin likewise argued that the Muria, a Dravidian-speaking tribe dwelling in one of the most isolated parts of India, possessed attributes of nonviolent innocence which reflected a most ancient condition, and he devoted himself to protecting them from, and educating them about the outside patristic world. Both the Pygmy and the Muria were characterized by the presence of a high level of body pleasure, for infants, children and adults, with a low level of trauma, anxiety, psychopathology, and violence, precisely as forecast by Reich's sex-economic principles. As was the case with Trobriand society,

35. Elwin, V.: *The Muria and Their Ghotul*, Oxford U. Press, Bombay, 1942; *The Kingdom of the Young*, Oxford U. Press, Bombay, 1968.

36. Hallet, J.P. & R. Relle: *Pygmy Kitabu*, Random House, NY, 1973.

37. Turnbull, C.: *The Forest People*, Simon & Schuster, NY, 1961.

the Muria and Pygmy, in spite of the free conditions, did not display evidence of "unbridled promiscuity", "group marriage" or other fantasies conjured up by Westerners accustomed to conditions of sex-repression and sex-starvation.[38]

The *Ghotul* of the Muria, for instance, was a house and social institution set up and maintained by children of both sexes, and from which adults were generally barred from entering. According to Elwin, it was a central aspect of Muria society, even more so than are the sex-segregated, adult-dominated "youth huts", dormitories or military barracks of armored patrist cultures. The exclusion of adults from interference in the affairs of Murian youth stands in stark contrast to the most extreme patrist cultures, where young boys and girls are segregated from each other, dominated by adults, and where adolescent sexual expression

38. Elwin, *The Muria and Their Ghotul*, ibid., p.452.

Mother and Child, Bastar, India, c.1920s.

"*I was greatly struck by the differences in the incidence of crime between the Muria and the great Maria tribe to the south... The Muria differ from the Maria in the quite extraordinary absence of jealousy among them... I believe this difference is largely due to the existence among the Muria of the Ghotul or village dormitory, in which boys and girls of the tribe grow up...*"
— Verrier Elwin
(*Maria Murder and Suicide*, pp.xxi-xxii.)

Abandonment of the *Ghotul*, or *Children's Dormitory*, with subsequent loss of adolescent and premarital sexual freedom led to increased suicides and murder.

Sexual jealousy was minimized among the Muria, who retained the Ghotul and its sexual freedom, but increased among the Maria who were pressured to abandoned the Ghotul. Consequently, suicides and murders increased among the Maria, mostly regarding issues of sex-frustration, compulsive marriages and the paternity of children, as is the case in other patrist cultures.
(V. Elwin, *Maria Murder and Suicide*, Oxford U. Press, Bombay, 1942)

may carry the penalty of death. These latter institutions are certainly related to the virginity taboo, and also to various "military" or "headman" hierarchies, and their associated sexually painful initiation rites. Elwin noted that the Ghotul was once part of the social tradition of other cultures who lived near to the Muria, such as the *Maria* peoples. However, under the pressures of surrounding puritan Hindu culture, the Maria had given up the Ghotul, thereafter developing characteristics of severe sex-repression. Antisocial violence and psychopathology subsequently became widespread among these latter groups, similar to that found in the rest of Hindu and Moslem India.[39] This observation of Elwin supports the argument being developed in this work, that the few existing sex-negative influences which did affect both the Trobriand Islanders and the Muria had come from *specific geographical regions and cultural groups outside* their own culture.

From the field reports of Malinowski, Hallet, Turnbull, Elwin and others, it appears the unarmored peoples of high-pleasure, low-violence cultures possessed very low levels of sadistic anger or sexual jealousy, quite a different situation from the patriarchal authoritarian cultures. They also appeared incapable of even conceptualizing violent treatment of their children. They in fact never "beat" them in a cold, calculated manner. One anecdote underscoring this point appeared in the *Wall Street Journal*, regarding the peace mission of an 1800s North American Nez Perce Indian Chief. While riding through a white settler's camp, the Chief saw a soldier beating a child. Reining in his horse, he is reported to have said:

> *"There is no point in talking peace with barbarians. What could you say to a man who would strike a child?!"* [40]

Oftentimes, at lectures I am asked: What has happened to these peaceful, sex-positive cultures today? Unfortunately, they are rapidly changing towards increasingly patristic, armored conditions, being caught into the socio-economic web of 20th Century life. Nearly every report which comes in the newspapers or in travel magazines or academic publications suggests the Trobriand Islands of today are quite different as compared to Malinowski's time. The former period of sexual freedom and *Bukumatula* appears largely gone; the natives identify themselves as Christians, and a growing social violence has been documented, frequently related to newly-introduced soccer games. The Muria have also come under assault, being considered the "lowest of the low" castes in India. The former Prime Minister of India, Indira Ghandi, took a deep personal offense at the Muria, attempting to coerce and intimidate them into adopting puritanical Hindu ways. Elwin gathered many reports of Murian groups coming under such outside pressure, distorting their original traditions, and actually published a map which showed the different geographical regions from which the sex-negative influences were coming.[41] Tribes located closest to Hindu areas were generally affected the most. The Pygmy are likewise under assault, migrating as ragged paupers into the slums of African cities as their protective forest is decimated by European and Japanese lumber firms. It is not a pretty picture, suggesting the final decimation of the last true human beings of planet Earth.

39. Elwin, V.: *Maria Murder and Suicide,* Oxford University Press, Bombay, 1942; also see Elwin, *The Kingdom of the Young,* ibid., and Elwin, *The Muria and Their Ghotul,* ibid.

40. Chase, N.F.: "Corporal Punishment in the Schools", *Wall Street Journal*, 11 March 1975.

41. Elwin, *Muria and Their Ghotul,* ibid., p.13.

B) Patrism: Armored, Low-Pleasure, High-Violence, "Desert" Cultures

In addition to readings on nonviolent, sex-positive cultures, such as the Trobrianders, Muria, and Pygmy, I also studied the behavior of cultures in the desert regions of North Africa, the Near East, and Central Asia. I had numerous discussions with university students from those areas, both in the United States and Near East, with observations made "on the street" in Egypt and Israel (presented in Chapter 11). Of primary concern was the interaction between males and females in day-to-day living, a facet which I found discussed only occasionally in scholarly, "mainstream" publications and news sources.[42]

The character structure in the Old World desert belt is entirely different in almost every possible way, from that found in the tropical island and rainforest regions I have studied. These desert belt cultures are antifemale, antichild, and antisexual in almost every imaginable manner, losing no opportunity to structure the entire society in a hierarchy subordinate to the oldest, most powerful male. In terms of family structure, the father, husband, and sons rule over females, whether as daughters, wives, or widows. Enormous sexual anxiety prevails over even the most mundane and innocent of "unauthorized" social contacts between young men and women.

One of the best general descriptions of such a highly armored, patrist social structure has been given by Evelyne Accad, from a chapter on "The Social Position of Women in North Africa and the Arab World", in her book, *The Veil of Shame*.[43] Her contemporary account is in good agreement with those given by other scholars,[44] and is generally valid for most of the Saharasian region. Quotations from her chapter follow:

Infants and Children:

"...the birth of a child is a time for rejoicing—providing that the child is a male. When a boy is born there is a celebration in the house and...a special pudding...is prepared... If on the other hand, the child is a girl, her birth will be greeted differently: the mother is likely to be sad at the prospect of her daughter repeating her own fate; the father is annoyed because he must bear the burden of guaranteeing the girl's (and the family's) honor and must find her a husband. The family system is 'programmed' only to accept the birth of male children, and a baby girl is a severe impediment... As a result, the mother's attitude toward a daughter is likely to be grudging, compared to her attitude toward a son. This is borne out by the Arab lactation customs; boys are traditionally nursed twice as long as girls. And in general, the position of the male child in the family is central, while that of the female child is peripheral. The boy quickly learns that the family system exists to serve his every whim; the girl discovers that she serves but is not served."[45]

Later Childhood and Adolescence:

"Prior to adolescence, a girl is allowed a certain amount of freedom of movement; she may go out about town, provided she is accompanied by her father or a brother; she may attend school, provided it is nearby and is not co-educational. But even during this time of relative freedom, the girl does not escape the constant knowledge that she is a second class citizen in a man's world. In the home she occupies a servant-like position: schooling and other activities outside the home

42. An excellent source of contemporary information on the legal and social status and treatment of children and women in various countries can be found in *Women's International Network News* (187 Grant St., Lexington, MA 02173); *Ms. Magazine* also carries regular reports and articles on these subjects.

43. Accad, E.: *Veil of Shame*, Editions Naaman, Sherbrooke, Quebec, 1978.

44. Beck, L. & N. Keddie, Eds., *Women in the Muslim World*, Harvard University Press, Cambridge, Mass, 1978; Huston, P.: *Message from the Village*, Epoch B Foundation, NY, 1978; Whyte, M.: *The Status of Women in Preindustrial Societies*, Princeton U. Press, NJ, 1978; Stephens, W.: *The Family in Cross-Cultural Perspective*, Holt, Rinehart & Winston, NY, 1963; Paige, K. & J. Paige: *The Politics of Reproductive Ritual*, U. California Press, 1981; S. de Beauvoir, *The Second Sex*, Vintage, NY, 1952.

45. Accad, *Veil of Shame*, ibid., p.20.

Aramco World

The full-length black veil, a requirement for women in the most extremely patristic regions of Saharasia, and its affected borderlands. While men dress for climate, women must dress for "morals". Her face may be exposed so long as foreigners and men are not nearby, in which case, the loose fabric covering the head is pulled sideways across the face. Complete and full-time facial veiling is required for women to travel in public in some Saharasian regions, as even the eyes are considered too "sexually provocative" for men to withstand. Only the hands, but not the arms or even a lock of hair may appear in public view. If so, the woman can expect to be approached, sternly lectured or even arrested by the duly-authorized "religious policemen", who take their orders directly from the mullahs. In other parts of Saharasia, rules may be more relaxed, with only a headscarf required.

46. Accad, *Veil of Shame,* ibid., pp.20-23.

are of minor importance compared to housework and waiting on the males in the household. Furthermore, if the girl is from the lower social classes, it is likely that she will be circumcised before or during adolescence. Unlike male circumcision, which is an occasion of rejoicing... female circumcision is done in secret and largely as an act of degradation. Depending on local custom, it may involve anything from simple perforation of the clitoris to complete excision of the external genitalia and suturing of the vaginal opening. The intent in all cases seems to be to reduce or preclude the girl's sexual desires, in part to ensure that she will arrive at her marriage bed an intact virgin and in part to make her completely passive as a sex partner...

The social life and education of young girls often leaves much to be desired, by any standard. Girls are often locked in the house (as are their mothers) and are allowed out only if accompanied by their fathers or brothers. The strictures regarding contact between the sexes are so severe that when Sheikh Ahmed Abdal Samad (a religious leader) was asked his opinion of friendship between boys and girls, he replied simply that it was forbidden... ...the question of whether a girl will be educated is not one that is long argued: when she reaches puberty it is almost a foregone conclusion that she will be hastily withdrawn from school and placed under virtual house arrest until she can be safely married off. Once she has reached puberty there is a pathological fear that she will 'dishonor' the family if she is allowed any opportunity to get out into the world.

From early childhood, a girl is brought up in constant fear of losing her virginity. There are several psychosocial explanations for this virginity mania... In this culture a man is only convinced that he made a wise choice if his bride brings an intact hymen to the marriage bed; he considers it evidence of exclusive possession, proof that the merchandise is brand new. Further, it means that the man is assured that his wife has had no prior sexual experience and thus will not be able to compare his performance unfavorably to that of another man.

Beyond these personal considerations, however, virginity has an intrinsic value in that it represents the 'honor' of the girl and, more importantly, of her family. In the event that a girl brings dishonor to her family by losing her virginity, it is considered normal for her brother to murder her in order to avenge the family's honor. In 1966 Germaine Tillion found this custom to be common in parts of Morocco, in Kabylie, the Arab states and in Lebanon. Public opinion permits such ritual murder and in the courts it is considered only a misdemeanor. So far, no court has ever handed down a sentence of capital punishment for a man judged guilty of having murdered a woman for 'misconduct'. Once it is established that the victim had led a 'disorderly life', the sentence does not exceed four years at forced labor. Those who avenge their 'honor' therefore know in advance that the law will be lenient.

There is heavy social pressure on a brother to take such vengeance: he is made to feel that he is 'unmanly' if he does not wash the family honor with blood. In most cases it is the brother rather than the father or husband of the girl who is considered to have the responsibility for avenging honor, but in some cases the whole village turns out to stone the girl in a display of collective social responsibility." [46]

Marriage:

"The desire to ensure that the bride will be an intact virgin often leads men to take very young wives. Conversely, the fear that their daughters

may be deflowered prematurely compels parents to marry off their daughters at an early age. The result is a predominance of young marriages... ...in rural areas where cloistering of girls is more difficult (they are often needed to work in the fields) and the temptations consequently more numerous, marriage ages tend to be even younger and girls are often married off before they have achieved physical maturity. The result is that they are often nursing their first children at an age when girls in other cultures are still playing with dolls. Predictably, infant mortality is high and death in childbirth frequent among these young mothers.

..marriage is unlikely to be a joyous affair for the bride, first because she will have no part in choosing her future husband and second because marriage becomes, for most women, a state of oppression and imprisonment as severe as that which they experienced as adolescents in their own homes.

Opportunities for rebellion against arranged marriages are few, and usually open only to the privileged. The writer Fadela M'Rabet

Peasant women on the way to the market, Aswan, Egypt, c.1981

estimates that in modern-day Algiers a girl commits or attempts suicide every two days to escape an arranged marriage. The majority, however, acquiesce and suffer in silence. ...probably 99% of all marriages are arranged by the parents.

The [young man's] parents will...point out a girl of 16 or 17 who would make a 'good' match for him. In most cases she will be in some way related to the family. If the parents of the girl agree to the match, she will be allowed to see her fiance before the wedding day, but only in the presence of a chaperone. Usually this period of 'dating' is limited to six months or less in order to reduce the chance of a broken engagement, which endangers the girl's 'honor' and makes it difficult for her to find another suitor. In some countries, such as Iraq, Saudi Arabia, Yemen and Jordan, the 'dating' is virtually nonexistent; a young man who has not seen his cousin for ten years may get to see her for only a few minutes before their wedding night.

...the man must pay a certain amount of money (mahr), which constitutes the girl's dowry...[but] in most cases, the full amount need only be paid in the event that the husband divorces his wife.

...The price of a girl depends on her personal qualities and the

ASIPA Press

Shiite women in Beirut (1985) celebrating the Ayatollah Khomeini, whose hysterically anti-sexual fascistic social movement dramatically reduced the status of urban women in Iran, turning the clock back to the Middle Ages. Khomeini reversed the few social reforms enacted by his predecessor, the Shah, engaged in genocide against minority groups such as the Kurds and Bahais, and ordered the execution of prominent Western-oriented Iranian women as a "lesson" to the nation, creating a widely unreported flight of thousands of women west through Iraq and Turkey into Europe. Today, power-drunk "religious police" patrol the streets of Iranian cities (also in Saudi Arabia and elsewhere in Islamic Saharasia), monitoring that every woman is properly veiled, checking the marriage licenses of young couples, and searching out birth control devices, which can bring the charge of "prostitution" — and thereafter a death sentence. (See Chapter 11)

47. Accad, *Veil of Shame*, ibid., pp.23-27.

affluence of her family, as well as on the social and financial position of her suitor and, at times, on the general condition of the local market. In Kuwait, for instance, there is a chronic shortage of marriageable girls and consequently they generally fetch high prices. A few years ago this situation led a group of enterprising businessmen to import girls from Lebanon for sale in Kuwait. This was eventually exposed as a major scandal by the Lebanese press, presumably on the theory that some commodities are best hoarded for local consumption. Prices also rise when there is a large difference in the ages of the parties. One 65-year-old who bought the hand of a 14-year-old Syrian girl had to pay a mahr of no less than $2500.

When the day of the leilat-al-dokhla (literally, 'night of penetration') arrives, a ritual of preparation takes place. The ritual varies according to the country and social class, but the spirit and intent are everywhere the same: they seek to turn the woman into an object which conforms to certain social standards of desirability. The ballana (beautician) prepares the halawa, a thick sugar paste which she uses as a depilatory to remove every trace of hair from the bride's body. In this way she will be totally naked before her husband; further, it is believed, the removal of the pubic hair removes the bestial stigma from the female genitalia.

If the bride does not have an intact hymen, the local daya (midwife) is called in to remedy this lack. She will either introduce into the girl's vagina a small sack of animal blood mixed with water, or she will cut the surface of the labia minora with a razor blade.

On the marriage night the relatives generally gather outside the nuptial chamber and, after the consummation, the mother-in-law (and sometimes others as well) troops in to view the sheet or handkerchief stained with the hymenal blood (or its substitute).

Nor is it sufficient that the girl be physically a virgin: she must be one psychologically as well. She must not show an awareness of any sexual sensations which her husband does not want her to feel. For this reason, the husband is likely to prefer that his wife be young and uneducated: she is less likely to know more than he does in this, as in other areas. Furthermore, she should be sexually passive, not overly thin and ready to satisfy him whenever he wants. Numerous reasons can be given for this desiderata. The Arab male wants very badly to preserve his dignity before what he considers to be an inferior being; he sees women and the sexual act as basically unclean (a conception reinforced by Judaism, Christianity and Islam); and he would in any case have great difficulty satisfying his wife sexually, particularly if she has been circumcised and in any case because of her tender years and repressed sexuality." [47]

Childbirth:

"In marriage, the prime role of an Arab woman is to bear as many children (preferably male) as possible. An Arab whose wife has not delivered a child, or is not at least pregnant, within the year following his marriage sees it as a reflection on his honor, and his prestige in the community may well be diminished. A typical fellahah (peasant woman) will deliver seven to ten times in her lifetime and it is quite likely that she will die in childbirth. On the other hand, a man whose wife fails to produce male children — or any children — can divorce her very quickly...while a woman has no such rights over a sterile husband... ...the procreation of numerous children is a *religious virtue, a divine gift granted to women. Their fulfillment of this role enhances their prestige and importance. ...women are treated as factories for producing the yearly offspring...[and] women unable to assume this function are regarded with disdain and contempt. Their integrity is questioned."* [48]

Polygamy, Concubinage, Divorce, and Female Obedience to the Male:

"The woman's position in marriage is rendered even more precarious and unsatisfying by the institutions of polygamy, concubinage and divorce by repudiation. Since in most countries in North Africa and the Middle East there was no distinction between secular and religious law, it was legal for a Moslem male to have as many as four wives and an unlimited number of concubines. The more socially conscious leaders of the Arab world recognize the need to curb polygamy and, with it, the runaway birth rate, but the changes are slow.

Divorce is often so ridiculously simple that some writers have made comic material of it. In certain Sunni sects the man need only repeat to his wife three times the phrase, 'You are repudiated'. This act then acquires the force of law. The wife can then, if she has any means, appeal to the local tribunal for a small indemnity. But it is almost impossible for a woman to get a divorce: only if she has wealth of her own can she secure one. The law allows a wife the right to be separated from her husband if she pays a sum of money — but the amount is determined by the husband! For both Christians and Moslems, the man almost always receives custody of the children, no matter who is to blame for the divorce. It is legal for ...[the man] to acquire as many concubines as he wishes. In general, the concubines are slaves — often black Africans — purchased on the open market, although they may also be acquired in the same manner as a Western mistress. In either case the concubine generally leads a miserable existence, subject to the will of her master but lacking even the scanty legal protection and social status accorded a wife. If there are several wives and concubines on the scene, the concubine may only have sexual relations with the man two or three times a year. The rest of her time is spent in enforced continence at the mercy of the wives who, being mistreated and unhappy themselves, lose no opportunity to take out their frustrations on the concubines... If a man tires of a concubine he can usually resell her for nearly her original price. There is no mahr to be forfeited, no indemnity to be paid.

On the other hand, there is no institutional or social sanction for a woman who, unsatisfied with her husband, decides to take a lover. This practice is promptly labeled adultery and a woman guilty (or even suspected) of it will be sent back to her family by her husband. Once home, her brother (or father or uncle) may go to the extreme of killing her to regain the honor of the family.

Even if a woman avoids or survives all of these pitfalls and manages to remain a wife, she still must practice ta'a, the submission of the woman to her husband's will and desires even if she hates him and even if he oppresses or degrades her. She does not have the right to leave his house (or the conjugal bed) even if she feels revulsion toward her mate. He, on the other hand, has the legal right to keep her under his roof and make her submit to his will. He can appeal to the police and have them bring his wife back

48. Accad, *Veil of Shame*, p.27.

by force if she has left his house without permission. In many countries, a woman needs her father's or husband's permission to travel or leave the country. If she has not received such permission and her father or husband informs the authorities of it, she will not be allowed to board the ship or plane.

In the quest for obedience, ta'a allows the Arab male to beat his wife with a stick. This custom was quite widespread as evidenced by the general sensation when, in 1955, an Egyptian religious tribunal imposed regulations as to the size of the stick and distributions of the blows. [A similar discussion was held by the Kenyan government in 1980. J.D.]

" ...It would seem natural that having generally suffered a long list of social oppressions — circumcision, enforced ignorance, an arranged marriage, subservience to her husband and sons — she would not be a staunch supporter of the status quo when it comes to determining the course of her daughter's lives, but this is not so...

...most older women in the cultures under consideration curse the fate which caused them to be born female, but do not see the need to improve their daughters' lot even when they have acquired a certain measure of power and are capable of instituting certain reforms: 'Why shouldn't she go through it? ...didn't I? ...it will make a woman out of her.' Thus older women are often the firmest holders of traditional customs."[49] [See the account of female genital mutilation in Chapter 5 for a shocking confirmation on this point. J.D.]

Widowhood:

"Young widows constitute a numerous class since old men tend to marry young girls, and their fate is not a pleasant one. Not only are they usually unable to remarry (since they are no longer virgins) but widows also find themselves cursed as the objects of malediction, their widowhood being taken as a sign of divine punishment for their sins. In some places widows are considered so accursed that they are cloistered for forty days... ...tradition often requires a man to marry the widow of his brother, in order to provide her with economic support and to serve as a father to the children.

Nor does death of her husband liberate the widow where sexual conduct is concerned: her sons are now the guardians of the family honor in this respect. Even in the cities of the more liberal Arab countries a widow cannot seek sexual relations with other men for fear of her sons: there are records of Lebanese teenagers who have murdered their mothers for so slight an offense as going out on a date."[50]

A barely-teenage Palestinian girl, all dressed up in traditional garb and prepared for forced-arranged rape-marriage.

Aramco World

Accad's account was confirmed by my own 1980 field observations in Egypt and Arab Israel.[51] A deep, culture-wide anxiety, fear and hatred of sexual pleasure, female reproductive functions, and the maternal figure exists, and affects everyday life in a most profound manner. Children are initiated into the extreme patrist social system by special childbirth practices, followed by specific antisexual and antifemale strictures later in childhood and adolescence. Females are made obedient to males by specific legal codes and economic sanctions, and through threats of death which are carried out from time to time. They are made dependent upon males in the first instance by virtue of emotional castration in childhood, and later by a nearly impenetrable wall of socioeconomic, legal, and educational disenfranchisement. The genitals of young men and women are physically attacked and mutilated with razors, without anesthetics,

49. Accad, *Veil of Shame,* ibid., pp.28-30.

50. Accad, ibid., pp.30-31.

51. See my discussion in Chapter 11, "Saharasia Today".

65

just at the age when their sexual feelings are most powerful, in early adolescence. In most cases, a heavy wall of taboo — enforced by a very real death-penalty — prevents even the most casual and innocent of social relations to occur between young men and women.

In most of Saharasia, however, such conditions are not static, but are undergoing transformation by virtue of contacts with Western culture and technology. Such changes have taken the form of an exceedingly slow reform of those customs which act against the interests of women and children, or as a reactionary outburst against previous small reforms. As Accad has pointed out, stories of harem intrigues and emotional/sexual grief are a common topic of the literature in the region. "Honor killings" appear to be all too commonplace in North Africa and the Near East; they continue even in Palestinian (and Jordanian) areas where, due to Israeli influences, progressive social reforms are advanced beyond other Arab or Moslem regions. According to prevailing "custom", a daughter who loses her virginity before marriage has so totally dishonored and soiled the family name, that family "honor" can only be regained by spilling the girl's blood on the ground. The hapless girl may have her throat slit, or be burned or buried alive by her own brother, father or uncle. The girl's own mother and other female relatives in most cases *demand* the murder most emphatically, ridiculing the males if they fail to carry out the deadly act[§]

While ritual or impulsive murder of females is today nearly a global phenomenon, the Arab, Turkic, and Persian parts of Saharasia stand out as regions where such ritual murder of females is given social, and even legal approval. In some parts of Saharasia, to be pinpointed later, sensitive areas of a young girl's genitalia, the clitoris, labia majora and labia minora, are amputated without anesthetic, by older women. Only then is she marriageable. The stated purpose and goal of the mutilations is to "stamp out the female sex drive" which would otherwise "ruin society", or lead to "anarchy".[52] Indeed, for a woman of these regions to be called "uncircumcised one", or a man, the "son of an uncircumcised woman" is an insult of the most extreme proportions.

Another set of observations which I have made on these subjects centers on Iran and Egypt, where popular reactionary movements deposed two Western-oriented leaders. The Shah of Iran, for example, was no lover of democracy and was generally ruthless with his political enemies. Still, he enacted many reforms in family law, largely at the urging of his unveiled, Western-oriented wife. The same was generally true in Egypt, where the family law reforms of Anwar Sadat were largely promoted by his unveiled Western-oriented wife, Jihan Sadat. However, the legal reforms were passionately opposed by the Mullahs, and by much of the population. Sadat was ridiculed in public by ordinary Egyptians as a "cuckold", someone bossed by his wife, and the family law reforms were derisively called "Jihan's Law". His portraits, hung in public places in Cairo and other cities, were routinely defaced.

The reactionary movements in Iran and Egypt gained in strength and momentum with each new social reform enacted. Both societies were, from the 1950s through the 1980s, being invaded by less restrictive, relatively matrist Western ideas regarding females, sexuality, childrearing, and secularism. Western books and magazines were easily available, and Western influences were easily heard in music and radio broadcasts. In particular, Western films and movies translated into Arabic and Persian were shown in the major cities and on television stations, with large advertising billboards on the street presenting unveiled, sexual women. These influences prompted mixed feelings among the average person, some rejecting the influences as "blasphemy", and others becoming quite engaged. Still, most ordinary women continued to wear the veil,

"...in the West Bank and Gaza Strip nearly all 107 Arab women killed as suspected Israeli informers during the six-year Palestinian uprising were in fact victims of honor killings."

[§] Scheherezade Faramarzi, "Honor Killings Persist Among Druze", Associated Press article, 20 Dec. 1995, *Medford Mail Tribune*, p.8-A.

"According to Dr Shalhoub-Kevorkian, a criminologist from Hebrew University, the real figures are much higher with almost all murders [of Arab females] in the West Bank and Gaza most likely to be honor killings. In a two-year period between 1996 and 1998, Shalhoub-Kevorkian uncovered 234 suspicious deaths in the West Bank alone, which she believes were honor killings. Palestinian police do not record these deaths as murder but as deaths due to "fate and destiny." Shalhoub-Kevorkian believes the real number of honor killings may in fact be 15 times higher than the official figures."

[§] Sharon Lapkin, "Palestinian 'Honor'," *Frontpage Magazine*, 19 Jan. 2006.

52. Another aspect of such practices is to inhibit the amount of pleasure the male experiences during intercourse, by insuring that the female will not become too passionate; See my account on Egypt in Chapter 11, and the section on "Genital Mutilations" in Chapter 5.

and the repression of female sexuality continued unabated. By stark contrast, the outward appearances of the ruling elites and upper classes, who often had been educated in the West, were nearly indistinguishable from Europeans.

As history shows, the Shah was eventually deposed, and Sadat assassinated. Among the first official acts of the new leaders, Hasni Mubarak and Ayatollah Khomeini, was the repeal of their respective Family Reform Acts. After the Iranian "revolution", when the Mullahs were installed in power, reformers began to be arrested and executed. Mobs of young men and young women alternatively appeared in the streets to demand a return to fundamentalist values and a reduction in the amount of freedom which they once possessed, along with "death" to this or that Western leader. Masochistic frenzies of self-flagellation also appeared in the streets. Iranian "prostitutes", meaning young unmarried women caught in possession of a birth control device, were increasingly harassed, or even arrested and raped by special "religious police", or kidnapped, tortured and executed by the gangs of young men assembled by the Mullahs.[53]

Young couples caught walking together in the streets of Iran today are imprisoned unless they can produce a marriage license, and the veil has become mandatory for all women in public. Similar conditions today exist across much of the Muslim world, though in most areas the existence of either secular dictators or Monarchs has checked some of the power of the Mullahs. Even so, special "religious police" exist in many Muslim nations, including Saudi Arabia, where lengthy jail sentences, public whippings, and executions for "moral crimes" are not unusual. Even after the introduction of thousands of British, French and American soldiers in Saudi Arabia during the Gulf War, there was little or no change in the extremely sex-negative, anti-female nature of Saudi society. These factors are clearly indicative of severe sexual anxiety generalized to all forms of pleasure, what Reich called *pleasure anxiety*, and a particularly virulent and politically organized expression of such attitudes, which he called the *emotional plague*.[54] While we are most familiar with the emotional plague in the context of the extreme political left or right (eg. Stalinism and Nazism), in Iran and a few other parts of Saharasia we see examples of an entirely new form, perhaps the most extreme on Earth today, with exceptionally deep roots that unite the most dogmatic aspects of local religion with sociopolitical and economic elements. The long-term, widespread use of hashish and qat, potent psychoactive drugs, helps to maintain the strong emotional contraction, pleasure-anxiety, psychosis and paranoia within these same regions.

From the perspective of the female, and of youth, it would be entirely inappropriate, and highly unscientific, to simply call such behavior only a "custom" or "trait" of a particular culture, similar to predominant food-stuffs, clothing, or house type. We do not call the butchering of Jews by the Nazis a "cultural trait of the German people", even though some aspects of German character structure once leaned toward public advocacy of the "übermench" and organized outpourings of violence toward Jews. Such outpourings of violence, whether directed toward an external cultural group, an internal ethnic minority, or toward females or children, are properly identified as sadistic, emotional pathology. It is no less so for large groups of peoples than for individuals. In both cases, the charismatic sway of an authoritarian leader may persuade an emotionally thwarted individual to become completely absorbed by one or another mystical, antifemale, and antisexual philosophy or doctrine, so long as it provides that individual with some form of non-genital excitation for their dammed-up biological tension, and a socially-affirmed pathway for

53. See the depressing account of M. Coyne, "Iran Under the Ayatollah", *National Geographic*, July 1985, pp.108-35; Also my account in Chapter 11.

54. Reich, W.: *Character Analysis*, Farrar, Straus & Giroux, NY, 1971, p.248 (First published in 1933).

discharge of sadistic energy.

Once again, it was Reich who first identified these mass psychological connections, demonstrating various antisexual, antichild, and antifemale factors at work in the social structure of patriarchal authoritarian states.[55] I observed a similar mass characterological pathos at work in the various reactionary movements gripping the various Saharasian nations in recent years. Such pathos is acted out on the social scene in epidemic proportions in a manner similar to smallpox or the Black Plague. However, with the *emotional plague*, there is no microbe at work, only special emotionally contagious doctrines and philosophies which are rooted in the childhood traumas and pent-up genital frustrations of people.

Irrespective of the present conditions, the archaeological and historical record, to be given later, clearly reveals the regions now characterized by such armored, plagued conditions once were entirely different, being more like the gentle Trobriander or Pygmy. But if this is so, then how, and under what circumstances could such a drastic change have taken place? What was the first, original source of such virulent hatred of the child, female, and natural sexuality?

C) A Preliminary Cross-Cultural Comparison

James Prescott's cross-cultural study of childhood trauma, physical affection, and adult violence is most telling on these issues.[56] Using a regionally balanced sample of 400 cultures from around the world, Prescott demonstrated that cultures which had high levels of physical affection or *body pleasure* for their infants and children, including high levels of adolescent sexual expression, also had low levels of adult violence; cultures with low levels of physical affection and body pleasure were observed to have high levels of adult violence. To my knowledge, Prescott's work was the first independent and systematically-derived indication that Reich's sex-economic thesis was in harmony with human behavior as expressed in many different cultures around the world.

While the arguments of Reich and Prescott were convincing enough on the validity of the variables previously given in Table 1 (see page 5), the matrist/patrist schema, and while my own observations and readings on the matter appeared in good agreement with their arguments, it was important to undertake an independent cross-cultural evaluation of the cultural relationships. The variables in Table 1 needed to be rigorously evaluated, to prove its internal logic and legitimacy before it could be used to evaluate human cultures around the world, or to make global maps of human behavior, or to assess historical trends in culture.

I firstly developed a large *Correlation Table of Sex-Economic Factors* (fully reproduced in Appendix A) from 63 different dichotomously coded cultural variables, as taken from Textor's large compendium on the subject, *A Cross Cultural Summary*.[57] Textor's *Summary* is, essentially, a massive computer printout of all existing statistically-significant correlations, whether positive or negative, within a group of 500 different cultural variables as recorded for a geographically balanced set of 400 different cultures. It is a widely used and easily consulted source of cross-cultural data for social theory testing. The 63 variables selected for the *Correlation Table* closely corresponded to the extreme patrist variables previously given in Table 1 (page 5). Each variable in the list of 63 was independently checked against all other variables to see if a correlation existed, and if so, whether the correlation was positive or negative in character. The *Correlation Table* was constructed such that armored,

"...cultures which had high levels of physical affection or body pleasure for their infants and children, including high levels of adolescent sexual expression, also had low levels of adult violence; cultures with low levels of physical affection and body pleasure were observed to have high levels of adult violence."

55. See the section on "Confirmation of the *Mass Psychology of Fascism*" in Chapter 2, p.42.

56. Prescott, J.: "Body Pleasure and the Origins of Violence", *Futurist*, April 1975; also in *Bull. Atomic Scientists*, Nov. 1975, and *Pulse of the Planet*, 3:17-25, 1991.

57. Textor, R.: *A Cross-Cultural Summary*, HRAF Press, New Haven, CT 1967.

patrist modes of behavior were cross-checked against each other. Sex-economic theory predicted that positive correlations would exist between the entire set of patrist variables. For example, cultures which had strong virginity taboos, who manipulated their children into arranged marriages with a high bride-price were predicted to possess, significantly more often than not, a high level of infant trauma, the presence of genital mutilations, patrilineal favored inheritance, caste systems or slavery, emphasis upon military glory, and so forth, for all 63 variables.

Sex-economic theory, and the general matrist-patrist schema in Table 1, was strongly confirmed in the *Correlation Table*, in that 500 separate positive correlations were found in the data, with only 20 negative correlations. The positive correlations constituted fully 95% of all observed correlations between the various armored patrist behaviors, attitudes, and social institutions. Statistically insignificant correlations (where p>0.10) were not included within Textor's work, and hence were automatically excluded from the *Correlation Table* — this is the reason for the blank places in the *Table* (again, reproduced in Appendix A). Indeed, all correlation probability coefficients ranged from the 0.10 level, to below the 0.001 level, indicating that the recorded correlations were either "significant" or "highly significant" in nature.

Reich's sex-economic theory predicted that only positive correlations would exist between the given patrist variables, and 95% of all observed correlations were, indeed, positive in nature. If the relationships between the variables were determined entirely by chance, such that the specific traumatic and sex-repressive treatments of infants and children *did not* drive adult behavior towards despotism and violence, then the correlations would have been closer to a chance distribution of 50% positive and 50% negative. The small number of negative correlations can also be explained as due to either a high level of missing data for certain variables, or to improper or unavailable codings of a given variable.

For example, the variable "Segregation of Adolescent Boys" is clearly a patrist social institution designed to prevent the interaction of boys with girls. However, boys are not segregated in areas characterized by the even harsher practice of *female seclusion*, or *purdah*, which exists across most of Saharasia. Given that some of the most extremely armored, patrist peoples practice female seclusion, but not the "segregation of boys", some negative correlations appeared for that variable. Unfortunately, there was no variable in the *Cross Cultural Summary* addressing the seclusion of females. These and a few other quirks in the data are addressed in Chapter 5.

Cross-cultural testing, of course, provides data on tendencies only, requiring logical theoretical grounding. Such has been provided, however. Unlike so much social theory testing where multivariate or correlation analysis suggests one or another theory *a-posteriori*, the strong unidirectional correlations observed here were anticipated, *a-priori*, as predicted from a robust theory of human behavior, *sex-economy*, which has garnered independent support from other directions over the past 50 years.

The *Correlation Table* suggests that all the identified patrist traits bear a positive relationship to one another which is not due to chance alone. By inference, all the opposite unarmored, matrist behaviors and social institutions must also bear a positive relationship to one another. The cross-cultural data, systematically derived from classical anthropological and ethnographical sources, clearly do not support the notion of any cultural benefit from repressive child "socialization", strict childrearing pedagogy, obedience training, "tough love", or other authoritarian modes for "civilizing" a child's instincts. Such usually means sex-repression, pain-inflicting punishments, and lists of specific "do-not's". It is highly significant that such strong positive correlations exist between infant and childhood trauma, adolescent sex-repression, male dominance, destructive aggression and warfare.

The above tests therefore confirmed the findings of Reich on the legitimacy of the sex-economic formulation, summarized in the patrist-matrist schema in Table 1, and also the prior cross-cultural evaluations of Prescott.[58] These cross-cultural findings, combined with specific suggestions from archaeology, history and climatology, suggested that the data, when analyzed geographically, would yield additional geographical correlations to regions of desert.

D) Geographical Analysis of Cross-Cultural Data

From the above starting assumptions, and with the confirmation found in the larger *Correlation Table*, I undertook the first careful steps towards making a geographical study of the cross-cultural data. These steps are most clearly spelled out in the original dissertation,[59] but here we can summarize the most important findings which underlay the more significant World Behavior Map (previously presented in the Introduction, and discussed throughout this work).

Using the same global 400-culture sample from Textor's *Cross-Cultural Summary*, which was also used in the *Correlation Table*, each of the 400 cultures was evaluated individually and given a "percent patrist" numerical value, depending upon how it fared in accordance with the 63 variables contained in the *Correlation Table,* which mirror the general matrist-patrist schema given in Table 1. If all of a given culture's 63 social institutions and behaviors were patrist in character, that culture was given a score of 100%. If none of the culture's 63 social institutions and behaviors were patrist, then it got a score of 0%. In reality, however, few or none of the cultures possessed such extreme conditions. Most clustered in the middle of this scale, ranging from between 10% to 85% patrism. A Histogram of Regional Behaviors was constructed from the Textor data, as given in Figure 5, which is broken down into continental regions. The mean values are also identified. The Textor data suggests the most extreme patrist cultures are found in Africa and Asia, followed by intermediate values in both Europe and Oceania, with the most extreme matrist (the least patrist) conditions found among the original native groups in North and South America.

These regional differences were significant, and using a tedious hand-calculation and plotting method, I used the same 400-culture data sample to evaluate the differences in behavior in even smaller regional units, composed of 5° x 5° blocks of latitude and longitude, which were then plotted on the world map. While those hand-plotted maps left much to be desired, they confirmed the powerful and strong difference in behavior between the different continents, and furthermore showed even more significant climate-based regional differences. Unfortunately, because the Textor data was composed from only 400 different cultures, when distributed across the world map, many regions were left unrepresented, or under-represented.

I then turned to the larger cultural data base of G.P. Murdock, his *Ethnographic Atlas*,[60] which was composed of data on 1170 different cultures, nearly three times as many as used in the Textor data base. The Murdock data posed its own special problems, however. While Murdock studied more cultures, he only included information on about 50 different cultural variables. Textor, by contrast, provided data on approximately 500 different cultural variables. In spite of this difficulty,

58. Prescott, "Body Pleasure..", ibid.

59. DeMeo, J.: "On the Origins and Diffusion of Patrism", U. Kansas Dissertation, Geography Dept., 1986, pp.125-140.

60. Murdock, G.P.: *Ethnographic Atlas*, U. Pittsburgh Press, PA, 1967.

the Murdock data base had far less missing data, and because the data was in a computer-readable format, overall it was more amenable to analysis and mapping of the data by computer.

The Murdock data was provided on a set of nearly 1200 IBM-cards, and my work on the subject took place before the era of convenient and fast desk-top microcomputers. The IBM-cards were given to the University of Kansas computer center, which read the data onto a large reel of magnetic tape, and from there the tape was recorded into the KU mainframe computer. I had recently purchased an Osborne portable computer – a heavy, bulky machine weighing as much as a large sewing machine – and by patch-cords was able to connect the Osborne to the KU mainframe computer, from which the Murdock data was transferred and stored onto floppy disks. I then had in my hands, the first such set of cultural data on easy-to-use and portable floppy disks.

Specific latitude and longitude coordinates were obtained for each of the 1170 cultures, and added to the data base. I then set about to write a computer program which would compute and plot the data for the different tribes onto a world base-map. It took about a year to write the necessary computer program, and draw the special base-map which was centered upon the Pacific Ocean. Most base maps center on the Atlantic, but the historical and archaeological research I was undertaking strongly suggested a much more active diffusion of culture across the Pacific, which possessed many island groups. Oceania truly was a rather "contiguous region", even though the islands were often separated by hundreds of miles of open ocean. The above steps being completed, the first World Behavior Maps were printed out in 1981 and 1982.

As mentioned above, the Murdock database contained information on a larger number of cultures, 1170, but with fewer cultural variables being addressed. In fact, only 15 of the 50 Murdock variables were relevant to issues of matrism or patrism. Nevertheless, the 15 variables were revealing, and were used to construct the individual cultural evaluations. The 15 variables are summarized in Table 2.

As in earlier preliminary efforts, each of the 1170 cultures in the *Ethnographic Atlas* was evaluated by the computer program for the 15 variables in Table 2, and given a simple numerical rating according to the percentage of variables with a patrist orientation. A second histogram was constructed, this time using the Murdock data, and is presented in Figure 6. As before, the distributions range from between 10% to 90%, with higher patrist values in Africa and Asia than in the Pacific or North and South America. While the cultural data contained in the 400-culture Textor data base is also contained within the larger 1170-culture Murdock data base, and while the Murdock data base involved the evaluation of only 15 different variables as compared to the 63 variables in the Textor-derived *Correlation Table*, the two histograms are nearly identical. This close identity further confirms the validity of the methodology.

Using the new Murdock data, each of the 1170 evaluated cultures was then located within its given 5° x 5° regional block of latitude and longitude, from which a regional-block "behavior average" was extracted. The block-averages were then plotted on the new Pacific-centered base map, which formed the basis of the World Behavior Map. The original computer-plotted data is reproduced here, in Figures 7A - 7H. A visual scan of this map sequence reveals a clustering of extreme patrist cultures in Saharasia, with a dispersion from there outward towards surrounding regions where patrism undergoes a gradual dilution with the matrist background. Regions most distant from Saharasia possess the most extreme matrist conditions, which are spread rather evenly in North and South America, and also in Oceania. The cross-cultural pattern which was firstly described clinically and socially by Wilhelm Reich, identified in cross-cultural data by James Prescott, confirmed and extended in the present *Correlation Table*, and then plotted on two histograms constructed from two separate but similar sets of global cultural data, is finally seen as the expression of a global-historical pattern in human behavior with roots in Saharasia. The final World Behavior Map, presented in the Introductory chapter of this work (and elsewhere), was developed directly from Figures 7A - 7H, with some historical interpretations for a few small regions of North and Central Asia, and Northern North America, where data was scarce or absent.

These histograms and figures were the first historical demonstration of the Saharasian geographical pattern, as determined through mathematical aggregation of the behavior data. Chapter 5, which follows in Part II, shows the same behavior data with each variable being independently mapped, a step with further reinforces and supports the overall Saharasian thesis.

Figure 5: Histograms of Regional Behaviors
Textor Data (400 cultures, 63 variables)

Table 2: *Ethnographic Atlas* Variables Used to Construct the World Behavior Map

	Patrist	Matrist
Female Premarital Sex Taboo	present	absent
Segregation of Adolescent Boys	present	absent
Male Genital Mutilations	present	absent
Bride Price	extreme	low or absent
Family Organization	polygamy	monogamy
Marital Residence	patrilocal	matrilocal
Post-Partum Sex Taboo	present	absent
Cognatic Kin Groups	absent	present
Descent	patrilineal	matrilineal
Land Inheritance	patrilineal	matrilineal
Movable Property Inheritance	patrilineal	matrilineal
High God	present	absent
Class Stratification	present	absent
Caste Stratification	present	absent
Slavery	present	absent

Figure 6: Histograms of Regional Behaviors
Murdock Data (1170 cultures, 15 variables)

Number of cultures with a given percent patrist value

▬▬ = mean value

Percent	Africa	Circum Mediter.	East Eurasia	Insular Pacific	North America	South America
Extreme Patrist 94						
92						
90	2	4				
88	5	9	1			
86						
84	5	19	1			
82						
80	10	12	3			
78	17	17	6	2		
76						
74	51	14	5	2	1	1
72						
70	57	19	5	1		
68	72	15	10	6		1
66						
64	50	8	10	4	2	2
62						
60	34	5	11	8	1	1
Percent 58	41	4	8	8	5	1
Patrist 56						
54	24	3	8	11	2	1
Behavior 52						
1170 50	16	5	9	12	13	7
Cultures 48						
46						
44	9	2	3	13	17	4
42						
40	4	5	6	19	21	5
38	1	1	1	10	23	4
36						
34	2	2	1	9	19	7
32						
30	1	5	3	9	27	12
28		1	4	8	26	9
26						
24	3	2	4	2	25	10
22						
20		2	2	2	19	12
18		1	2	8	26	7
16						
14	1	1		2	15	4
12						
10			2		20	5
8					5	2
6						
Extreme 4					3	1
Matrist 2						
0						

73

Figures 7A - 7H: Average Regional Percent-Patrist Values
Each black dot represents a 5° x 5° block of latitude
and longitude with the given percent-patrist averages.

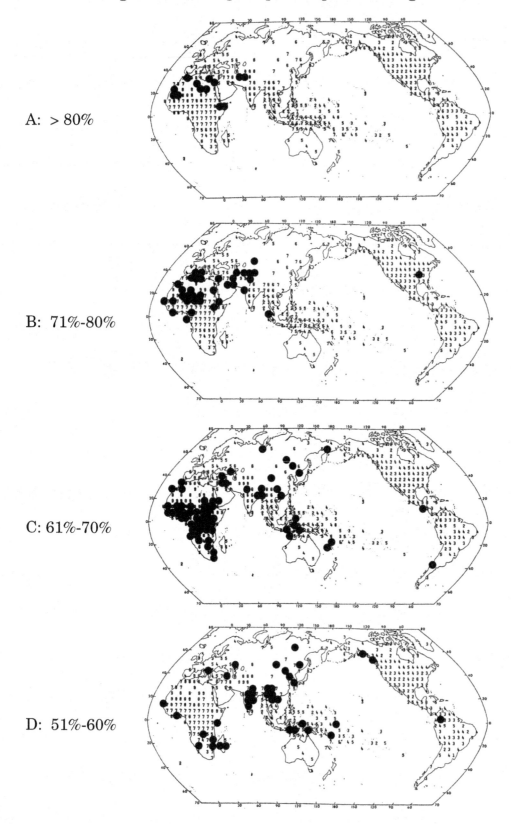

A: > 80%

B: 71%-80%

C: 61%-70%

D: 51%-60%

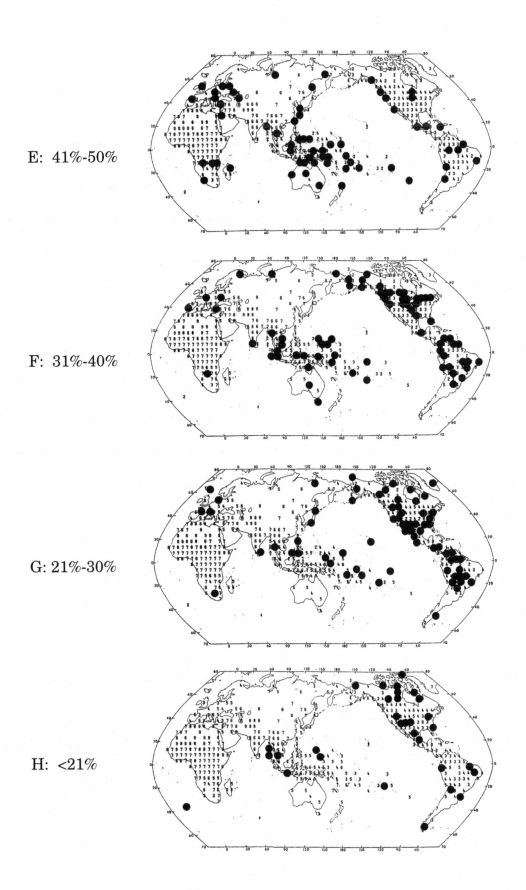

E: 41%-50%

F: 31%-40%

G: 21%-30%

H: <21%

Changes in Ancient Climate, Landscape and Archaeology

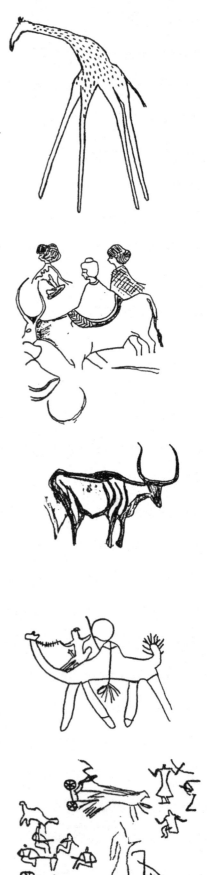

Another major consideration which suggested a desert-culture connection was the evidence found in archaeological sites throughout Saharasia, indicating social changes towards armored patrism simultaneous with ecological changes toward hyperaridity. The evidence for these coordinated changes required nearly seven years of research to assemble and review and was derived from more than 100 separate definitive regional compendiums. From that evidence, presented in detail in Part III, a new time- and location-specific historical/archaeological data base was created. Evidence from various archaeological and paleoclimatic studies demonstrates the great desert belt of modern-day Saharasia has experienced major climatic shifts over the centuries, and was for many thousands of years prior to c.4000-3500 BCE, a semi-forested grassland, abounding with streams, lakes, and rivers, teeming with life. Further, existing evidence strongly suggests the peoples of this wet, lush period were almost entirely unarmored, and matrist in character. Indeed, from the evidence gathered and presented in Parts II and III, I concluded there is only the most fleeting, short-term and ambiguous evidence for the existence of significant armoring or patrism on Earth prior to c.4000 BCE. And where such fleeting evidence exists, it is always connected to temporary episodes of aridity and famine.

These inferences and conclusions are made in part from the *presence* of certain artifacts from the earliest times of rainfall abundance: the sensitive and careful burial of the dead, irrespective of sex, with a relatively uniform grave wealth; anatomically realistic female statues; naturalistic, sensitive artwork on rock walls and pottery which emphasized women, children, music, the dance, animals, and the hunt. In later centuries, some of these same unarmored peaceful peoples would progress technologically, and develop large, unfortified agrarian and/or trading states, notably in Crete, the Indus Valley, and Central Asia. The inference of matrism in these early times is also made from the *absence* of archaeological evidence for chaos, warfare, sadism and brutality, which becomes quite evident in more recent strata, after long-term drought set in and Saharasia dried up. This latter archaeological evidence includes: weapons of war; destruction layers in settlements; massive fortifications, temples and tombs devoted to big-man rulers; infant cranial deformations; ritual murder of females in the tombs or graves of generally older men; ritual foundation sacrifices of children and slaves; mass or unkept graves with mutilated bodies thrown in helter-skelter; and cast stratification, slavery, extreme social hierarchy, polygamy, and concubinage, as determined from grave goods, mortuary arrangements, and architecture. Artwork style and subject matter of the later, dry period also changed to emphasize mounted warriors, horses, chariots, battles and camels. Scenes of women, children and daily life vanish. Female figurines simultaneously become abstract, unrealistic, or even fierce, losing their former gentle, nurturing, or erotic qualities; or they disappear entirely, to be replaced by male gods. Artwork quality as well as pottery and architectural styles decline for Old World sites at such times, to be followed in later years by monumental, warrior, and phallic motifs. These and other aspects of my findings will be discussed in detail and fully referenced in Parts II and III.

Physiological, Behavioral, and Social Effects of Prolonged Drought and Famine

A) Somatic and Emotional Effects

Another important connection between desertification and culture change was presented most clearly by Turnbull in his heartbreaking account of the Ik, in *The Mountain People*.[61] Here was an East African hunter-gatherer people whose ancestors, unlike the Pygmy, had not retreated into the rainforest or fled south in the face of invasions by more aggressive Bantu peoples. The Ik adopted some of the ways of the Bantu, but being smaller and not warlike, always gave ground to the militant, taller people. Still, they remained a happy, lively culture, with much spontaneity, at least before a resettlement program and drought forced them into a semiarid highland region with few resources and a low carrying capacity. Under such conditions, as dramatically described by Turnbull, the Ik began to slowly starve. And as they starved, their social structure broke down entirely. A passive indifference to the needs or pain of others manifested itself, and hunger, feeding of the self, became their all consuming passion. They sat in the midst of great heat and drought, thin and exhausted, losing interest in the pleasures of life, to include both pleasurable social interaction with members of their family, and sexual pleasure. The very old and young were abandoned to die. Brothers stole food from sisters, and husbands left wives and babies to fend for themselves. While the maternal-infant bond endured the longest, eventually mothers abandoned their weakened infants and children. Older children gathered into gangs dedicated to stealing food, even from their own younger siblings, or older, weakened kinfolk.

Turnbull's account of the starvation and hunger of the Ik is among the most emotionally shocking and heartbreaking ever recorded. Their experience demonstrated that famine is a major mechanism whereby a severe trauma of external origins could disturb the family bonds and social structure of a previously cooperative and cohesive, relatively unarmored people. The end result of the process is similar to armed invasion by a merciless, brutal enemy, who kills the majority of the population, scatters the survivors and steals all resources and property of value. Given the fact that Saharasia had been much wetter in the ancient past, the role of desertification-induced famine loomed as a primary mechanism for initiating profound and potentially lasting cultural change. A review of the literature on famine, and the physiological and psychological effects of starvation confirmed this possible mechanism. Turnbull's accounts of massive social effects were, as given below, mirrored in accounts of famine in other regions.

Studies of famines have indicated a wide variety of associated cultural shocks and traumas: migrations, wars, revolutions, insurrections, riots, crimes, and economic problems.[62] Further, it is drought, and its effects upon food supply, which is the number one natural cause of famine. Floods and insect damage are the second and third most important factors, and these often occur in conjunction with droughts or climatic instability.[63] Among larger populations dependent upon agriculture, economic and social factors could also result in famine, such as from warfare, blockades, epidemics, or gross economic mismanagement. However, such dependency upon a national or international food economy infrastructure would have been absent among early subsistence level

61. Turnbull, C.: *The Mountain People*, Simon & Schuster, NY, 1972.

62. Cahill, K., Ed.: *Famine*, Orbis Books, Maryknoll, NY, 1982; Sorokin, P.: *Hunger as a Factor in Human Affairs*, Univ. Florida Book, Gainesville, 1975.

63. Carlson, D.: "Famine in History: With a Comparison of Two Modern Ethiopian Disasters", in Cahill, 1982, ibid., p.6; Aykroyd, W.: *The Conquest of Famine*, Chatto & Windus, London, 1974, p.7.

hunter-gatherer, pastoral, or agricultural peoples. During the fourth millennium BCE, the effects of drought would have been a major, and probably the only major mechanism which would lead to famine and starvation, given the normally self-reliant condition of early peoples.

The physiological effects of starvation upon the human organism include wasting, emaciation, and edema, plus nutritional disorders such as marasmus and kwashiorkor. Parasitic infestations and infectious diseases, such as malaria, measles, and tuberculosis, quickly spread among the weakened population.[64] The mortality and morbidity is unequally spread among the population, with infants, children, the elderly, and pregnant and lactating women being most severely affected.

> *"During a famine infant mortality rises further as diseases become more widespread, as birth weights decline, and as protein calorie malnutrition increases... The rise in mortality among breastfeeding infants is...much less than that among older children...[who are] hardest hit by famine... Low calorie and protein intake stunts growth and allows disease to take its toll. The age group from one to three years, following weaning from breast milk, is especially at risk... The heavy additional nutritional demands made on the woman's body during pregnancy and breastfeeding are extremely difficult to meet during periods of food shortages. As a consequence, there is a sharp drop in the average birth weight of babies born during famines and breastfeeding and pregnant women become rapidly emaciated. However, breastfeeding can and does continue during the early phases of famines as long as adequate nipple stimulation to the breast is maintained by the suckling infant. This greatly benefits the infant, but puts a heavy strain on the mother. ...starvation per se is often not the main cause of death. Instead, there is a large increase in mortality from endemic infectious diseases to which the nutritionally deprived population has greatly reduced resistance... A severe famine produces a substantially altered population age composition, with more individuals in the central age groups and fewer in the oldest and youngest age groups."[65]*

The physiological aspects of famine have been studied more than social or behavioral changes per se, but it is clear that the two generally go hand-in-hand, with powerful behavioral and social changes occurring both during and after the famine period.

> *"Psychological effects [of starvation] are less obvious but important: after an initial period of increased activity and agitation, a general apathy and lethargy prevails, with loss of interest in normal human concerns except for finding food. Interpersonal relationships may be broken down so completely that parents devour their own children; cannibalism has been reported during famine times in almost every part of the world, although apparently less frequently in China and India. Infants, children, and the elderly suffer first; young and middle-aged adults are usually able to travel to food sources, leaving the young and the old to fend for themselves and possibly die. The animal world can also become ravenously hungry and leave their usual habitats to encroach upon human territory in towns and cities; animals may attack humans in broad daylight without apparent fear. Social chaos is evident everywhere, as refugees flock to cities or roads seeking food. Almost all normal functions are reduced or halted. Economic confusion is pronounced,*

64. Cahill, K.: "The Clinical Face of Famine in Somalia", in Cahill, 1982, ibid., pp.41-3; Carlson, 1982, ibid., p.6.

65. Bongaarts, J. & M. Cain: "Demographic Responses to Famine", in Cahill, 1982, ibid., pp.46-7.

as hoarding occurs, and costs of staple foods rocket to three, five, or one hundred times the normal price." [66]

66. Carlson, 1982, ibid., pp.7-8.

67. Zeitlin, M. et al.: *Nutrition and Population Growth: the Delicate Balance,* Oelgeschlager, Gunn & Hain, Cambridge, Mass., 1982, pp.7-13.

68. Monckeberg, F.: "The Effect of Malnutrition on Physical Growth and Brain Development", in *Brain Function and Malnutrition*, J.W. Prescott, et al., Eds., John Wiley, NY, 1975, pp.23-9.

69. Winick, M. & P. Rosso, "Malnutrition and Central Nervous System Development", in Prescott, et al., 1975, ibid., p.47.

70. From Monckeberg, 1975, ibid.

Maternal and infant malnutrition have long been recognized as factors in infant mortality and morbidity. Low birth weight babies and retarded physical growth will occur when the pregnant mother or infant is malnourished.[67] However, the neurological impact of malnutrition on the growing infant can also be long-term in character. In humans and other animals, the brain is the fastest growing organ during the first months of life. Not only can brain growth be inhibited, but atrophy can set in which eventually leaves a space between the brain and skull which fills with cerebrospinal fluid. Malnutrition before the seventh year of life has a permanent retarding effect upon growth and development, while later malnutrition effects are repairable.[68] The first year of life is most important, when malnutrition will not only inhibit the size of brain cells, but also their absolute number.[69] Figure 8 shows some of the tremendous physical differences between healthy, well-fed babies, and babies suffering from marasmus due to chronic malnutrition.[70]

From the above, a question is naturally raised about the long-term psychological or behavioral effects of malnutrition among peoples inhabit-

Figure 8: Normal Versus Marasmatic Infants[68]

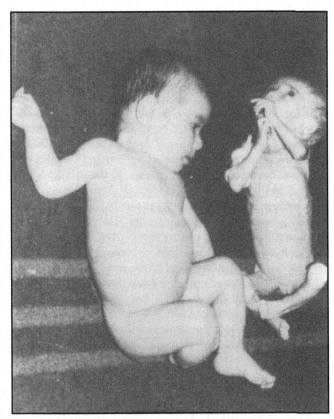 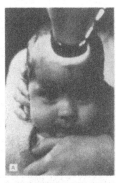  

Left Infant: 5 months old, healthy; Right Infant: 7 months old, marasmatic.

Photos courtesy of F. Monckeberg

Transillumination of the skull of A. normal (upper left), B. malnourished (upper right) and C. marasmatic (lower left) infants. The skull is illuminated in proportion to the amount of brain atrophy, and fluid-filled space between the brain and skull.

ing regions of repetitive drought and famine. Unfortunately, studies have indicated that psychological impairment accompanies the retarded physical growth common during times of famine.

> "Pediatricians and nutritionists had repeatedly observed that apathy was one of the most constant signs in severely malnourished children. From the very first description of protein-calorie malnutrition it was clear that psychological disturbances were prominent features of this syndrome. The malnourished, severely ill patient seems to have lost the curiosity and desire for exploration that characterizes the normal preschool child. This unresponsiveness is so marked that renewal of interest in the environment is considered a premonitory sign of recovery and is clinically judged as one of the most reliable signs of improvement." [71]

The malnourished child emotionally contracts and withdraws from its painful, unresponsive, and meager environment. The fullness of its recovery depends upon its age at the time of deprivation, and the length and intensity of deprivation. Improvement of nutrition will bring the child back to activity, but not to the same physical or emotional developmental level as other well-nourished children of the same age group. In other words, long term drought and famine can, through the physiological effects of malnutrition, impart negative emotional, behavioral effects upon the child which will later be expressed in adult behavior.

Monckeberg, whose study summarized years of work on the role of malnutrition in child development, expressed it so:

> "...malnutrition per se not only affects the expression of genetic potential in physical development, but acts upon intellectual development. If it is true that in developing countries undernutrition is the principal cause of premature deaths, the damage in the survivors is still worse." [72] [emphasis added]

This statement becomes all the more shocking when one considers that *fully half of the infant deaths on Earth are due to malnutrition, with [in the early 1980s] over 200 million children suffering from its effects.* One third of all peoples in the "developing" nations are identified as malnourished, with about 30 million being severely malnourished in the "developed" countries. [73] Indeed, malnutrition has been called "the most widespread disease in the world". [74] From the above discussion, it is seen how this malnutrition has long term social implications, and political consequences.

The number of field studies in drought regions on behavioral effects of famine are relatively small, given the lack of facilities in the field. Understandably, the major concern during famine is to provide food to starving people, with scientific measures being a secondary concern. For instance, Inhelder and others studied the specific effect of malnutrition on early cognitive development in infants and children of the Baoule group in Sahelan Africa. [75] Contrasted to better nourished children in their own region, the Baoule children with moderate malnutrition showed signs of marginal retardation in psychological development measures by several months, particularly those behavioral aspects concerned with the "active exploration of the environment". Within the Baoule culture, which suffered from serious anemia, repeated infections and parasites, children were observed to move rather slowly, hesitating to undertake acts which they were quite capable of doing, calling upon their mothers for help with even the slightest tasks. These observations confirm the overall pattern of

71. Cravioto, J. & E. DeLicardie: "Neurointegrative Development and Intelligence in Children Rehabilitated from Severe Malnutrition", in Prescott, et al., 1975, ibid., p.53.

72. Monckeberg, 1975, ibid., p.37.

73. Garcia, R. & J. Escudero: *The Constant Catastrophe: Malnutrition, Famines and Drought*, Vol. 2 of the *Drought and Man* series, IFIAS Project, Pergamon Press, 1982, p.30.

74. Garcia, R.: *Nature Pleads Not Guilty*, Vol. 1 of the *Drought and Man* series, IFIAS Project, Pergamon Press, 1981, p.103.

75. Inhelder, B.: "Early Cognitive Development and Malnutrition", in Garcia & Escudero, 1982, ibid., pp.24-9.

slightly slowed rates of physical and psychological development during times of mild malnutrition, with a persisting element of deep emotional withdrawl and contraction. Aykroyd had this to say about the deadly affect of malnutrition on children:

> *"If you walk down a children's ward in a hospital in the tropics, you can immediately pick out the children with severe kwashiorkor by their appearance of extreme misery. These children are withdrawn into a dim wretched world of their own and take no interest in toys and people. After proper treatment for a few weeks, often by 'drip feeding' with skim milk, they will suddenly and unexpectedly smile; the nurse and doctor then know that all is well and that the child will recover. All this shows that lack of protein has a striking effect on the brain and mind of a small child, though the physiological and bio-chemical changes underlying these remarkable phenomena are not yet understood. But it is important to note that the child with kwashiorkor who recovers has suffered not only from deficiency of protein; it has also passed through a period, perhaps several months in duration, when it was almost unaware of external stimuli. Such a period of withdrawl, at a receptive time of life when the growing brain should be receiving daily a multitude of new impressions, may perhaps in itself result in subsequent retardation in mental development.*
>
> *Much attention is now being given to the question whether dietary deficiency during the early stages of life may check the development of the brain and perhaps cause life-long mental impairments."* [76] [emphasis added]

Here, we see a pattern of famine-induced emotional contraction and contactlessness at work. Such deprivations fall heavily upon the emotional needs of the child, and will leave scars which will affect adult character structure and behavior, just as other forms of severe infant and childhood trauma do. During such a period of environmental and nutritional stress, the maternal-infant bond becomes all the more important, particularly for infants whose only source of food may be breast milk. Yet it is precisely during the period of intense drought, with its oppressive, stifling heat and lack of food and water, that adults, including mothers, become inert, lethargic, and inattentive to the needs of infants and small children. Under such conditions, adults would necessarily become less attentive caretakers, and dependent infants and children would thereby suffer not only malnourishment, but maternal and sensory deprivation. As discussed by Prescott:

> *"...it is important to indicate that other sources of impaired brain-behavioral development may exist during the formative periods of ontogeny which interact with and confound the effects of malnutrition upon development. When discussing the effects of impoverished environments it is clear that malnutrition is only one characteristic of such an environment and that usually, if not invariably, malnutrition is accompanied by other conditions of deprivation. Maternal-social deprivation may be considered as one of these 'other conditions' and when interpreted as a form of sensory deprivation, it is possible and meaningful to search for the biological consequences of such early insults upon the developing brain and behavior. This search becomes particularly relevant when certain emotional-behavioral characteristics reported as a consequence of experimental malnutrition are similar to those reported consequent to maternal-social deprivation."* [77] [emphasis added]

76. Aykroyd, 1974, ibid., p.123.

77. Prescott, J.W., Editor's note in R. Heath, "Maternal-Social Deprivation and Abnormal Brain Development; Disorders of Emotional and Social Behavior", in Prescott, et al., 1975, ibid., p.306.

The effects of somatosensory deprivation of infants are carried into adulthood, and can be transmitted to the next generation: Mothers and fathers, deprived of maternal affection and physical touching during their own infancy and childhood, will as adults raise their own children in a similarly affectionless, cold, and uncaressing manner. Moreover, as demonstrated in the primate studies of Harlow and Prescott (previously discussed), these individuals will additionally exhibit a general intolerance and anxious aggressivity towards the basic biological expressions of others, especially to include expressions of touching and body contact. Such contractive emotional responses to touch are invariably generalized into the individual's sexuality. It is therefore significant that, in addition to malnourishment-related physical and emotional retardation, the effects of famine on infants also include maternal-somatosensory deprivation. Both infants and adult caretakers suffer the effects of physiological malnutrition and starvation lethargy, along with emotional shocks from the death of loved ones, and from deteriorating familial and social conditions. In this manner, prolonged famine and starvation produce profound disturbances in the capacity for uninhibited and mutually-pleasurable sexual expression, in both maternal-infant and male-female aspects. Additionally, a major portion of these emotional/sexual responses would persist to affect behavior later on, through the development of biophysical armoring (as defined by Reich). This would be so, even after food supplies would be restored to a famished group, as following the end of drought, or when the survivors of a devastated social group would migrate from a drought/desert region to another region of greener pastures.

To summarize: The child who suffers through the miseries of drought-induced famine and starvation undergoes a mild to severe emotional withdrawl and contraction, or shrinking away from its harsh, painful, and unsupportive environment. The *emotional shrinking* is accompanied by a *physiological withering*, in an organic, neurological sense, as protein-calorie intake is severely restricted. Damage then becomes chronic and likely permanent to both the psyche and soma. Indeed, this shrinking and withering of the child, during a most important phase of its emotional and physical development, parallels Reich's bioenergetic findings on the effects of chronic pain and anxiety upon the autonomic nervous system, conditions which clearly lead to the subsequent appearance of respiratory blocking, pleasure-anxiety, and character armor.[78]

From the observations of Sorokin, a Russian emigre who experienced the famine of 1919-1921, we also have first-hand evidence suggesting that, as starvation intensifies and hunger dominates the interests and activities of the individual, the energetic intensity of the sexual urge is progressively extinguished.[79] Of major importance is that recovery of sexual functioning may not be complete following a period of prolonged starvation.[80] The evidence Sorokin cites on this last point is restricted to measures of sperm and egg viability, and to observations of the degeneration of sexual organs in animals; still, he feels that the "experimental data can easily be applied to human beings".[81] Aykroyd also discussed the loss of sexual potency in starving men. Some experiments he cited had shown a return of sexual feeling after food was again provided, but others indicated that starvation would cause young females to suffer afterward from a delay in puberty, underscoring at least temporary biological effects on sexuality following recovery from starvation.[82]

Sorokin also discussed the role of extended fasting in diminishing the sex drive, a practice which has been adopted for this expressed purpose by various religious sects over the centuries. On the other hand, he gives a number of arguments on the relationship between satisfaction of hunger and the subsequent onset of the sexual urge, a relationship which is

"The child who suffers through the miseries of drought-induced famine and starvation undergoes a mild to severe emotional withdrawl and contraction, or shrinking away from its harsh, painful, and unsupportive environment. The emotional shrinking is accompanied by a physiological withering, in an organic, neurological sense, as protein-calorie intake is severely restricted."

78. Reich, W.: *The Biological Investigation of Sexuality and Anxiety*, Farrar, Straus & Giroux, NY, 1982.

79. Sorokin, 1975, ibid., pp.117-29; Bongaarts & Cain, 1982, ibid., p.48.

80. Sorokin, 1975, ibid., pp.122-7.

81. Sorokin, 1975, ibid., p.121.

82. Aykroyd, 1974, ibid., p.18.

acknowledged in popular myths and warnings given to young girls, such as: "all boys think about is food and sex".[83]

These latter observations are most important from a sex-economic viewpoint, which predicts a gradual increase in pent-up sexual bioenergetic tension and destructive, antisocial behavior within cultures where genital functioning and orgastic discharge are disturbed, for whatever reason.

Other evidence for the physiological effects of aridity and famine come from Oliver and Lee, primatologists who recorded a decline of interactions between male and female baboons, with increased levels of male dominance, in dry regions as opposed to wetter regions.[84] Juvenile baboons in the dry savanna habitat were observed to spend less time playing with each other, or with younger individuals, as opposed to juvenile baboons in a wetter, forested region. Grooming patterns were also different: juvenile males and females groomed each other in the wet, but not the dry regions, where male juveniles spent more time grooming their male peers. In the drier areas, dominance/supplicant interactions increased as the availability of food declined.

Viewed in conjunction with Harlow's studies on isolation-reared and maternally deprived primates,[85] these baboon observations suggest the existence of sexual anxiety, mild armoring, and a "patrism" of sorts, among animals in the wild, related to aridity and scant food supply. If so, Reich's discussion on the process of armoring would, in a limited way at least, have to be extended to some nonhuman primates which have suffered through repeated phases of drought and famine.

Taken together, the observations suggest that both human and monkey respond to prolonged aridity and famine in a similar manner, developing similar behavioral disorders and social responses. It is further suggested that such climatological patterns may be at the root of conflicting observations regarding primate behavior in the wild: Do primates who live for generations in more arid regions, having been subject to starvation trauma, behave in a more disturbed manner, a bit closer than we may realize to that of "civilized" *Homo normalis*?

B) Competition for Food and Water During Prolonged Drought and Famine

More should be said about the role of competition for scarce resources. Observations of animals during times of drought provide reasonable models for response patterns among subsistence peoples. The biological principle of *carrying capacity* is a major determinant of the number and size of animals which can successfully live within a given natural ecosystem. The application of this principle to human populations forecasts stress, greatly heightened competition, starvation, or migration during times of severe drought, when carrying capacity is lower than normal.

Grinnell's principle of competitive exclusion is seen to apply to different human cultures under situations where, due to drought, the carrying capacity of the land is adjusted downward.[86] This principle states *A:* No two species are on equal footing in terms of extracting energy from a given environment, and *B:* With the limitation of a vital resource, only one of two competing species will ultimately survive. If this principle is extended to cover competing human cultures, it suggests that those cultures most successful in competing for food or water, possibly by using violent means, would be at an advantage over those who were less competitive in character.

This suggests that even people who were nonviolent in character

83. Sorokin, 1975, ibid., pp.119-21.

84. Oliver, J. & P. Lee, "Comparative Aspects of the Behavior of Juveniles in Two Species of Baboon in Tanzania", Recent Advances in Primatology, Vol. 1: Behavior, D. Chivers & J. Herbert, eds., Academic Press, 1978, pp.151-3.

85. See section on "Primate Studies" in Chapter 2 for discussion and full references.

86. Himes, N.: *Medical History of Contraception*, Williams & Wilkins, Baltimore, 1936, p.451.

might be compelled, through raw thirst and hunger alone, to steal or fight over water and food. However, we cannot infer that this would extend to the premeditated killing of other people to obtain their food and water. The historical record is filled with examples of one culture fighting against another to secure food and water supplies during times of drought, famine and displacement. But these are uniformly reported from already-armored, moderately to severely patriarchal authoritarian cultures, with an already-existing military caste or other hierarchical elements. One cannot casually infer that a similar warlike response would automatically appear — as if from nowhere — among unarmored and peaceful social groups suffering from similar drought and famine. Unarmored peoples might simply lay down and die in the desert, rather than do the unthinkable and use their hunting tools to kill other human beings. Clearly, some additional factor must work its damage (such as discussed above, regarding the physiological response to famine) to turn a peaceful group towards organized social violence. But it is clear that, once such a turn towards armored social conditions got started, those who were capable of the greatest violence would soon dominate remaining water and food resources, driving away or subjugating those who maintained nonviolent or less-violent behaviors.

Those individuals best adapted for stealth or combat, or who were better at hunting or otherwise procuring food would likely become leaders of a cultural group. They would likely be the larger, stronger males, who could run fastest, farthest, and possessed the greatest hunting and weapons-making skills, which would put them at advantage in any fight. Under persisting drought and famine conditions, strongman leadership would be favored in a manner which, over the generations, might result in more permanent change in social structure. Indeed, once armoring sets in and people were prepared to use violent means to defend or occupy a region with secure water or food resources, then those with the greatest ferocity and mercilessness, or those with better weapons or larger numbers would likely prevail.

C) Migration and Nomadic Adjustments

History shows that droughts and famine have driven people out of the desert in large groups organized around strongman leaders, towards regions possessing secure food and water supplies. The archaeological/historical sections of this work cite a number of examples of such a process. In particular, Huntington discussed the evidence on these points,[87] identifying cyclical climate change toward aridity as a major factor in promoting land abandonment, migrations, and conflicts.

> *"The histories of wars of primitive peoples to secure salt, the wars of ancient Egypt, Assyria, Babylonia, Persia, Greece, Rome, and the Hebrews, the wars of the Middle Ages, and those of our times often record the capitulation of people suffering from starvation. They also tell of desperate forays, attempts to break sieges, and of many who perished in the end. Such behavior indicates that hunger had overpowered the determinant of self-preservation and forced people to perform acts which were detrimental to their health and life. Had it not been for hunger, such acts never would have been performed."* [88] [emphasis added]

As discussed in a recent United Nations volume on desertification, the "effects of desertification on man appear most dramatically in the mass

87. Huntington, E.F.: *The Pulse of Asia*, Houghton Mifflin, NY, 1907; E.F. Huntington, *Palestine and its Transformation*, Houghton Mifflin, NY, 1911.
88. Sorokin, 1975, ibid., p.102.

exodus that accompanies drought crisis".[89] Under persisting drought conditions, migration to new regions and a shift towards a nomadic existence would be favored over the sedentary lifestyle, given the relative persistence of small grasses and shrubs which could provide fodder for cattle, sheep, goats, and camels. As people began to migrate, herds, tents, and entire villages would be forced to move, usually traveling through territories already inhabited by other people. A graphic and recent example of such land abandonment and turn toward nomadism in Turkey has been given by Helms:

> "During the early 1970s water sources had become dry and the villages depending upon them had to be abandoned. The people — easily 1000 — with all their livestock had to move and by the time we saw them they had been on the road for several months. To keep alive they would approach water sources and grazing land owned by more fortunate villagers along their way and negotiate. It was a scene of much historical precedent: of a people dispossessed — by nature this time, not by man — of a people made refugees and nomads who now pitched their tents on the sown lands not their own and asked for terms.
>
> Often they were told to move on after the three days of traditionally free grazing. The flocks would survive by passing from place to place and eating enough at each to stay alive until the next confrontation, which could end in violence.
>
> The cyclical pattern of dispossession, movement and confrontation along established trade routes in the Near East can be traced back through the archaeological record..." [90]

Such widespread migrations would have several major effects. The migrations themselves would, initially at least, bring the migrating peoples into conflict with the prior inhabitants of the moister regions into which they moved. Conflicts and wars might break out, probably followed by displacement of the conquered group toward some new territory. A long-distance, mass migration might originate in a drought-stricken region, but migrations toward lands adjacent to a drought zone may stimulate additional cultural displacements, creating a chain of migratory invasions and displacements stretching over thousands of miles, with effects lasting for hundreds of years.[91]

History also shows that the most potent weapons, cavalry, and chariots developed among nomadic peoples of the drought-afflicted Asian steppe. Their military advantage was often translated into dominance over peoples along the Saharasian borderlands, or dominance of its remnant water sources. As will be demonstrated, the powerful nomadic warrior cultures of Central Asia and Arabia have played a prominent role in the genesis of kingly states, military alliances, and political history in both Saharasia and its moister borderlands.[92]

A number of authors have noted the more recent development of pastoral nomadism in the history of human affairs, a development which has been called "a blind alley in human progress".[93] Phillips, who studied the nomad hordes of Central Asia, observed that "...the pattern of life of the nomad contrasts at every point with that of the farmer or townsman".[94] In the case of Central Asia, such contrast included a significant component of destructive, sadistic aggression which periodically irrupted from the dry steppe to wreak havoc upon surrounding moister regions. These steppe nomads, as well as those from Arabia, revealed more violent tendencies than either the settled pastoralists, farmers, or hunter-gatherers living in surrounding territories.

89. "Desertification – An Overview", in *Desertification: Its Causes and Consequences*, UN Conf. on Desertification, Nairobi, 1977, Pergamon Press, 1977, p.41.

90. Helms, S.: *Jawa: Lost City of the Black Desert*, Cornell U. Press, Ithaca, NY, 1981, pp.57-8.

91. Sorokin, 1975, ibid., pp.200-2; Huntington, 1907, ibid.; Also see the various historical sections of Chapter 8.

92. See Chapter 8.

93. Phillips, E.: *The Royal Hordes*, McGraw Hill, London, 1965, p.10.

94. Phillips, 1965, ibid., p.56.

"Among the [nomadic] herders we get the beginnings of a much more dominating, aggressive form of ritualized action by which supernatural forces are controlled and ordered to the will of man." [95]

Another major aspect of nomadism which would work to transform character structure towards violent, antifemale tendencies is that of infant cranial deformation, a severe trauma to the infant which appears to have originated among nomadic peoples of Central Asia. It is known that infant cranial deformation occurs when a baby is unmovingly secured against a cradle board, similar to those employed by the Asian nomads. The newborn baby's head might be tied against the cradle board, to prevent it from flopping about as the mother walked over irregular terrain. The child's arms and legs would be wrapped and also securely tied against the cradle board. Such children would spend great amounts of time in the cradle, particularly as migrations were under way; and the drier the environment, the longer the migration time. Not only would a child's movement be inhibited, frustrating its need to explore its world, but the pressure against the head would deform the skull, in a painful, often life-threatening manner.[96]

When on the move, the child strapped to the cradleboard would not be breast-fed as regularly, nor allowed out for regular waste functions. If a mother stopped to attend the infant too often, she might fall behind the rest of the group, which would be dangerous to survival. Once raised in such a deforming manner, adults would not only carry the psychic trauma in their character structure, with a hostile and anxious view of the maternal figure, but also a head with a biologically unnatural shape. Physiological damage to the squeezed brain organ on a massive scale also appears possible, as such deformations were often severe and widespread.

As discussed in more detail later on, cranial-deforming cultures viewed the head deformity as a "special mark" of the tribe which, like circumcision, distinguished them from other undeformed cultures. Given the extreme painfulness of this procedure, and the manner in which it would interfere with the child's emotional needs and interests in its' mother, the child's view of the maternal figure would likely become contaminated with anxiety and rage. This would be so, even though the culture in question would view the deformity in a positive light. If a cranial-deforming culture conquered another culture which did not engage in the practice, the deformity might be seen as a mark of ruling class distinction, to be emulated by the lower classes of conquered people. Ill treatment of the infant, and negative attitudes towards the female, would thereby spread; inadvertently at first, purposefully later on, as migrations and invasions of cranial-deforming peoples spread out of the desert.

As I argue later on, there appears to be a parallel development of painful and restrictive *infant swaddling* along with the infant cranial deformations, occurring first among the same desert-dwelling Central Asian nomads (starting around c.3000 BCE), appearing later in other regions adjacent to the Central Asian deserts — the swaddling being used to "still" the child and immobilize it in the painful cradle, and to keep it from pushing off the painful cranial tourniquet.[97]

D) Direct Effects of the Desert Atmosphere

Prolonged drought and aridity might also work to change culture through mechanisms other than famine, infant malnutrition and trauma, or competition for food and water alone. The heat, dust, and low humidity

95. Turnbull, C.: *Man In Africa*, Doubleday, 1976, p.18.

96. See the section on "Infant Cranial Deformation and Swaddling" in Chapter 5.

97. ibid.

of the desert atmosphere could increase the lethargy, irritability, and likelihood of physical conflict within a given group of people, via a *biometeorological* effect. That conflicts appear during drought and famine is beyond question. A separate question is raised, however, as to the long-term biological influences of living in a climate characterized by extreme air temperatures, scanty rains, high dust and haze levels, and low relative humidities.

To my knowledge, there have not been any studies addressing the question of behavioral effects from prolonged exposure to atmospheric conditions characteristic of deserts. Homo sapiens evolved in wetlands with both trees and grasses, but did not, like the Gila monster or horned toad, evolve in deserts among sand dunes and cactus. A general observation is that hot, dry conditions lead peoples to engage in water-conserving, low body-heat activities, to shelter within protected environments or under protective clothing during the heat of the day, and to follow the diurnal cycles of other animals. This is a general biological reaction, present whether an individual is native to desert climates or not. However, such mild, shelter-seeking activities would not in and of themselves disturb the maternal-infant or male-female bond, and turn a gentle culture into killers. But other aspects of the desert atmosphere *might* work toward such an effect, especially where emotional tendencies leaning in the same direction already existed.

Early in this Chapter, Reich's speculations on the origins of armoring were covered, and he identified a similar biometeorological effect – though his arguments on the mechanism of this effect were made within the context of his *orgone energy* discoveries, of a plasmatic *life-energetic force* at work in the atmosphere and in living organisms. Under certain conditions, Reich argued, this aether-like phenomenon developed a stagnant, immobilized, and "dead" quality across entire regions, dehydrating organisms and disturbing weather systems as well. Even if one is skeptical about Reich's ideas on orgone energy, there is considerable support for this basic observation, of a functional identity between life and land.[98]

Some studies have demonstrated physiological responses to what are called, by classical atmospheric scientists, "electrically charged small air ions". Distinct behavioral influences have been identified for such "ions", and tend to be distributed according to the local microclimate, seasonal air circulation patterns, extensiveness of plant cover, and availability of water. For instance, *negatively* charged air ions are generated in quantity by waterfalls and lush green plants, and have a stimulating, invigorating effect upon individuals, generally making them feel more alive, alert, and contactful, with a heightened sensory awareness. By contrast, *positively* charged air ions predominate in the "bad winds" which regularly blow within and out of desert regions (sharav, hamsin, sirocco, Santa Ana), where life and flowing water are generally absent or at low levels. Such positive ions can generate feelings of irritability, restlessness, malaise, or even migraine, nausea, vomiting, edema, conjunctivitis, and respiratory congestion in an individual.[99] These observations, like those of Reich, suggest a direct bioenergetic effect from the atmosphere of the dry, dusty desert which would not be present in the atmosphere of the rainforest or grassland. However, the larger social and cultural influences of such an atmospheric effect, over long time-periods, can only be speculated about. It does not appear sufficient to totally destroy the gentle and loving qualities of any single individual, much less of whole cultural groups — severe drought with famine-starvation and deaths of large numbers of people in a given village appears as a far more traumatic, dramatic and culture-changing influence than biometeorology alone. It clearly remains a question, ripe for further investigation.

98. Reich argued for the existence of two antithetical qualities of a unique and quantifiable *atmospheric and biological orgone energy*: one of an oppressive, stagnant quality predominating in desert regions, and another of a invigorating and lively quality, predominating in wetter, forested regions. See the various citations on "deadly orgone (dor)" indexed in the online *Bibliography on Orgonomy*:
http://www.orgonelab.org/bibliog.htm In a related manner, Elsworth Huntington also developed an "Ozone Hypothesis", suggesting a direct influence of the atmosphere upon human behavior and cognition. (See: McGregor, K.: *Evaluation of Huntington's Ozone Hypothesis as a Basis for his Cyclonic Man Theory*, unpublished Thesis, Univ. of Kansas, Geography Dept., 1976.)

99. Sulman, F.: *The Effect of Air Ionization, Electric Fields, Atmospherics and Other Electric Phenomena on Man and Animal*, Charles Thomas, Springfield, IL, 1980; Persinger, M.: *The Weather Matrix and Human Behavior*, Praeger, 1980, pp.213-26; Tromp, S.: *Biometeorology: The Impact of the Weather and Climate on Humans and Their Environment*, Heyden, Philadelphia, 1980. It should be noted that Reich used some of the same techniques to measure orgone energy in the atmosphere that air ion researchers subsequently used for identification of "air ions" (ie., electroscopical discharge rates, a principle underlying many modern air-ion meters). This suggests a logical parallel between his findings on *dor-ish* desert atmospheres and *positive air ions* identified in more recent times within desert regions. Reich's writings on atmospheric orgone and dor appear to pre-date those of the air ion researchers by at least a decade.

Chapter Summary: *Basic Assumptions, Observations and Possible Mechanisms for the Genesis and Global Diffusion of Armored Patrism*

We have demonstrated the stark and dramatic cultural differences between matrist and patrist cultures, differences which impinge upon basic social structure, sexual life, and family relationships. In turn, it has been demonstrated that these differences have important continental and regional aspects to their global distribution, with the Old World (Europe, Africa, Asia) having a more armored, patristic cultural make-up than either Oceania or the pre-European New World. Furthermore, when the cross-cultural data is mapped, it yields an even more explicit Saharasian-Patristic geographical pattern, as given on the World Behavior Map.

From here, a number of lines of argument converged to independently support the hypothesis that hyperarid, desertified environments — specifically, hyperarid *Saharasia* — would be the source-region for development of contracted, armored behaviors, which were previously unknown to human beings. Some of these arguments, such as the historical and archaeological evidence for major climate change in Saharasia c.4000 BCE, remain to be developed in subsequent chapters. Additional evidence will be presented in those chapters to fill in detail regarding the geographical dimensions of Saharasia, its characteristics, and the timing of its genesis and growth. But from what has been already presented, a general outline sketch can be made with some confidence.

The cumulative cultural effects of the drought-famine factors discussed in this chapter would be severe, disturbing the emotional relationships between infant and mother, and between young men and women, and husbands and wives. Under a constant, generations-long pattern of recurrent drought and famine, where carrying capacity of the land declined, cultures which were initially peaceful and unarmored might be decimated, the majority dying off, family units and social living arrangements shattered, with survivors scattered to the winds in an independent search for food and water. All survivors would have lived through hell. All surviving children would have experienced severe trauma and deprivation. Under such pressures, the child's view of the female would turn from being that of the "Great Nurturing Mother", who protected and provided for all needs, to that of a distant, unresponsive, nonprotecting figure, blurred in early memory against a background of parching heat, thirst, dust, and flies.

During times of severe famine and starvation, a child would be subject to deprivation and trauma, in spite of the parent's best intentions. As starvation lethargy would set in, parents would understandably respond less and less to the needs of the infant or child. Similarly, a mother under the pressure of an incessantly nomadic condition, where food grubbing on the move would occupy almost all energies, would probably not do much about the crying of the newborn strapped into the back-pack cradle. If she was lethargic and hungry herself, she would become increasingly oblivious to the crying. And, a mother who was herself raised without attention to infant needs would not be prone to look after the needs of her own offspring. As seen in Harlow's studies of motherless mother monkeys, and in the infant wards of modern hospitals, the crying infant would be ignored, or forcibly hushed, by adult caretakers who were armored against their own sad feelings, and made anxious by emotional expressions they could no longer experience themselves. As Reich observed: *An individual cannot recognize the expression of an emotion in someone else which they have learned to inhibit and repress in themselves.* The parent, where such environmental patterns persisted over generations, would thereby develop a hardened attitude, rooted in the traumas of their own childhood experience.

A frustrated, anger-laden view of the maternal figure would develop, culture-wide, and maternal functions would be viewed increasingly with a sense of horror and severe anxiousness. Such natural functions as menstruation, childbirth, breastfeeding, heterosexual intercourse, micturition and defecation would be burdened with all forms of taboo, anxious jokes, and rituals of absolution. The crying of infants would thereafter stir up unwelcome feelings in the adult, and numerous methods would be developed to keep it "hushed", such as swaddling, or various gags and "pacifiers". The "good" child would be increasingly defined in terms of the *absence* of biological need, and the *absence* of emotion or feeling.

The adult would not know why, or probably even question, any feelings of anger or disgust directed toward the natural functions in question. However, such feelings, when culture wide, would be translated into new rules, religious verbots, and institutions. Originally developed under severe environmental pressures, the new traumatic and repressive treatments of infants and attitudes towards females would be incorporated into social structure, and thereby become self-duplicating and perpetuating.

The thwarting of the biological nature of the maternal-infant bond would also negatively affect the relations between male and female, and lead to increased violence and sadistic anger, quite apart from any competitive or "territorial" tendencies associated with food-gathering in a region of dearth. Such destructive impulses would primarily be directed toward the sources of anxiety: children and women, procreative functions and the sexual organs, and other races or cultures who behaved in a more emotionally fluid, sexually healthy manner.

These secondary drives, as Reich called them, would eventually develop compulsory institutionalized expressions, through various taboos, rituals and rites. They would displace any previously existing social institutions which once affirmed or supported the natural instincts and pleasures, and impose new modes of male dominance, female subordination, child obedience, sex restriction and sex inhibition. Once so institutionalized, such practices would remain as a force to shape and modify the child's instinctual nature, implanting armoring as "culture", and continuing to do so irrespective of natural environment. The new behaviors could then be exported to regions outside of the desert.

Migration toward wetter regions would most certainly occur as the desert became drier and drier, and the peoples who migrated would increasingly be characterized by anti-infant, antifemale, antisexual, and violent, warlike tendencies. Nonviolent peoples of the regions adjacent to the desert would thereby be displaced or conquered, perhaps to adopt the practices and institutions of their conquerors, thereby eventually developing similar behavior characteristics themselves.

The above scenario is more than mere hypothesis: It is confirmed in the archaeology, history, and anthropology of Saharasia, as previously summarized in this work, with much more detail to be provided in Parts II and III. From these perspectives, it can be seen how prolonged drought, with its accompanying malnutrition, famine and starvation, provides an important *triggering influence* whereby massive cultural changes can be initiated, particularly when drought is widespread and incessant, lasting from one generation into the next. As will be shown in the following sections, such a widespread, incessant drought condition characterized much of Saharasia during its transition from wet to dry conditions, between c.4000 to c.3000 BCE. And it was precisely during this period of climatic transition that the first widespread evidences of similar social trauma, armoring and patrism appeared within human culture.

"...prolonged drought, with its accompanying malnutrition, famine and starvation, provides an important triggering influence whereby massive cultural changes can be initiated, particularly when drought is widespread and incessant, lasting from one generation into the next."

89

PART II:

The Recent Historical Dimensions of Armored Patrism:

Mapping the Environmental and Cross-Cultural Evidence

**Saharan Sand Dunes
near Tassili N'Ajjer, Algeria**

Photo courtesy of
Prof. Bernd Senf

4. The Saharasian Desert Belt

The great belt of arid lands encompassing
North Africa, the Near ("Middle") East, and
Central Asia is the largest single contiguous land region of generally similar climate
on the face of the Earth. Like a great
planetary wound where rainfall is scanty,
temperatures generally high, and vegetation relatively scarce or nonexistent, this
belt of aridity stretches from Port Etienne
on the Atlantic Ocean in West Africa
eastward almost to Beijing, China, near the
Pacific. Averaging 1600 kilometers in width
and nearly 13,000 kilometers long (1000 x
8000 miles), this arid zone has played a
central role in the natural history of life on
Earth. It encompasses some of the largest
individual true desert regions: the Sahara
Desert, the Arabian Desert and the Rub'al
Khali or "empty quarter", the Negev and
Syrian Deserts, the Iranian Desert, the
Rajasthan and Thar deserts, and the
Turkestan, Takla Makan, and Gobi deserts
of Central Asia. These various desert
regions are each surrounded by belts of
semiarid lands which merge them together
into a generally contiguous zone or belt of
aridity, stretching nearly halfway around
the Earth. All other arid and semiarid
regions of the world, if combined together,
would not equal the area of this large
contiguous arid zone.[1] I call this large
global swath of aridity *Saharasia*, to assist
in identifying its distinct environment and,
as will be demonstrated, unique cultural
characteristics.[2]

1. McGinnies, W.G., et al., Eds., *Deserts of the
World*, U. Arizona Press, Tucson, AZ, 1968, p.6.

2. While the term 'Saharasia' is my own, the region
itself has been discussed before by others: cf. Huzayyin,
S.: "Changes in Climate, Vegetation, and Human Ad-
justment in the Saharo-Arabian Belt, with Special Ref-
erence to Africa", in *Man's Role in Changing the Face of
the Earth*, Vol. 1, W.L. Thomas, Ed., U. Chicago Press,
Chicago, 1973; Von Wissmann, H.: "On the Role of
Nature and Man in Changing the Face of the Dry Belt of
Asia", in Thomas, 1973, ibid.; Wadia, D.N.: *The Post
Glacial Desiccation of Central Asia: Evolution of the Arid
Zone of Asia*, National Institute of Sciences of India,
Delhi, 1960; Wendorf, F. & A.E. Marks, Eds.: *Problems
in Prehistory: North Africa and the Levant*, SMU Press,
Dallas, 1975.

Climatic Aspects of the Saharasian Environment:

Saharasia stands out from other arid regions of the world in terms of size, intensity of aridity, and other criteria used to identify or contrast arid zones. Deserts and drylands are generally defined according to either climate data or soils and vegetation data. The maps in Figures 9 and 10 are fairly standard and widely-used for identification of desert lands. Figure 9 was taken from a climate classification system developed by Trewartha,[3] which was based upon a system developed by Koppen.[4] Koppen's early work mapping vegetation types proceeded at the turn of the century, a time when measured meteorological data were generally unavailable for many parts of the world; his system served as a foundation for Trewartha's climate classification, which incorporated precipitation amount and seasonality. The map in Figure 10 identifies all world regions with rainfall less than 250mm (10") per year.[5] Figures 9 and 10 appear very much alike. The arctic polar "deserts" also show up on the maps, given their small amounts of precipitation. Taken by themselves, these two maps suggest more similarities than differences between the various arid zones. From them, we might conclude that the intensity of aridity in Saharasia was nearly identical to that found in other arid zones, such as in Australia, Southern Africa, or the Southwestern USA. However, a more detailed analysis of environmental data, to include other climate classification systems, reveals that such an identity is false and misleading.

Meigs, for instance, developed a climate classification system emphasizing factors of specific importance to arid zones,[6] such as season of rainfall, maximum temperature of the hottest month, and minimum temperature of the coldest month. His arid zone map, reproduced in Figure 11, possesses more restrictive criteria for the most stressful "extremely arid" environments. Meigs' system also embraced the concept

Both Figures 9 and 10 provide widely-used but misleading contrasts and comparisons between the Earth's different arid lands.

3. Trewartha, G.T.: *Introduction to Climate*, 4th Ed., McGraw Hill, 1978; Espenshade, E.B. & J.L. Morrison, Eds.: "Climatic Regions Map", in *Goode's World Atlas*, 15th Edition, Rand McNally & Co., Chicago, 1978, pp.8-9.

4. Koppen, W.: "Das Geographische System der Klimate", Vol. 3 of W. Koppen & R. Geiger, *Handbuch der Klimatology*, Gebruder Borntrager, Berlin, 1936.

5. Griffiths, J.F. & D.M. Driscoll: *Survey of Climatology*, Charles Merrill Pub. Co., Columbus, Ohio, 1982, p.164.

6. Meigs, P.: "World Distribution of Arid and Semi-Arid Homoclimates", in *Reviews of Research on Arid Zone Hydrology*, UNESCO, Paris, Arid Zone Programme, 1:203-9, 1953.

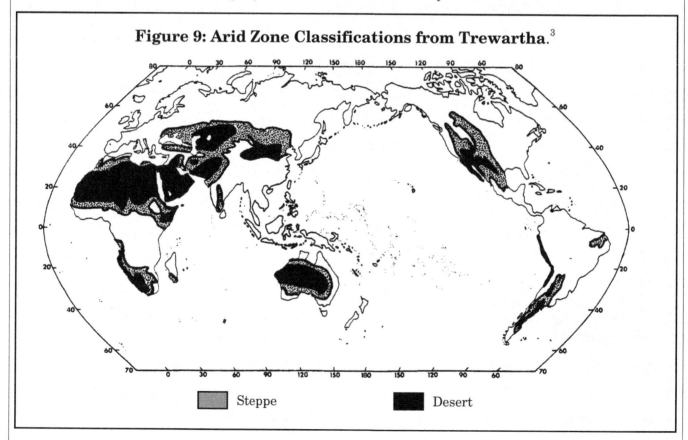

Figure 9: Arid Zone Classifications from Trewartha.[3]

Steppe   Desert

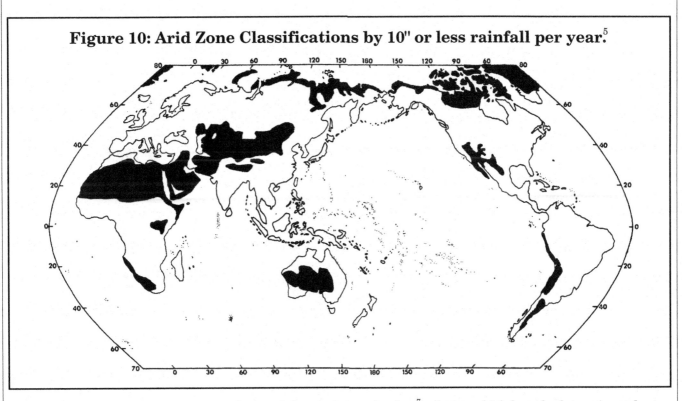

Figure 10: Arid Zone Classifications by 10" or less rainfall per year.[5]

7. Thornthwaite, C.W.: "An Approach Toward a Rational Classification of Climate", *Geographical Review*, 38(1):55-94, 1948.

of *potential evapotranspiration*,[7] a factor which largely determines plant stress and growth. This factor identifies *precipitation effectiveness*, in the sense that plants can make more effective use of available precipitation when air temperature and evaporative stress are lower. His classification system is therefore a more sensitive indicator of biological stress, and more accurately reflects the distributions of plants and other living

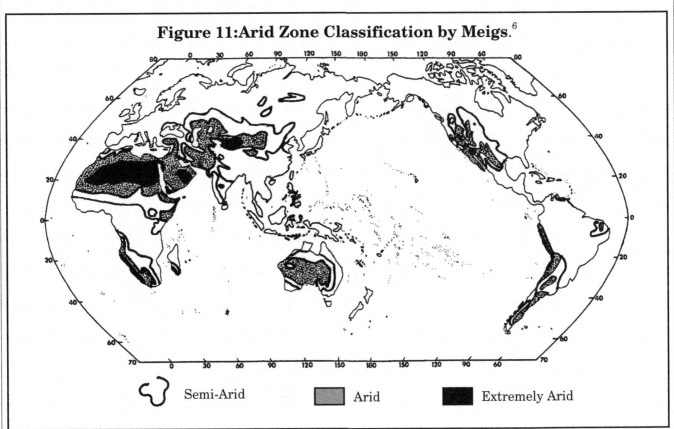

Figure 11: Arid Zone Classification by Meigs.[6]

Semi-Arid   Arid   Extremely Arid

organisms in arid regions. In Meigs' map, Saharasia contains almost all of the world's extremely arid environments. In fact, the only extremely arid environments identified outside of Saharasia are comparatively small regions: the narrow coastal Namib Desert of South Africa, the Atacama Desert of South America, and a few narrow basins & confined valleys in North America, such as the environment surrounding the Great Salt Lake, Death Valley, and the arid zone near Yuma, Arizona, at the northern end of the Gulf of California.

The uniqueness of the Saharasian environment is also confirmed by the Budyko-Lettau Dryness Ratio, presented below in Figure 12.[8] Both of these latter maps present a much clearer and accurate comparison between the different arid zones of the Earth.

The Dryness Ratio roughly indicates the amount of evaporative energy available in a given environment relative to the amount of water available as precipitation. Accordingly, Dryness Ratios of various arid zones are given at right in Table 3. From this Table, and from the map given in Figure 12, it is noted that Egypt and Libya are the driest regions on Earth. Only the Namib and Atacama, both narrow coastal deserts, achieve Dryness Ratios approaching those within Saharasia.

A number of other parameters can be identified which reveal the difference between Saharasia and other arid zones. Figure 13 presents a map of Precipitation Variability. Precipitation in many arid zones may be low, but even so, if it comes regularly, once per year during a given rainy season, it can be effectively used by plants to support their growth cycles, and to replenish water supplies in small pools. Animal life in such deserts with a regular rainy season can be quite abundant. Saharasia, however, possesses the largest single region of high precipitation variabil-

In Figures 11 and 12, a more accurate contrast and comparison is made between the Earth's different arid zones.

8. Budyko, M.I.: *The Heat Balance of the Earth's Surface*, N.A. Stepanova, trs., US Dept. of Commerce, Washington, D.C., 1958.

Figure 12: Arid Zone Classification by the Budyko-Lettau Dryness Ratio.[8]

Contrasting the relative dryness of different arid lands. Values reflect the ratio between precipitation and evaporative energy; values of 2 receive twice as much evaporative solar heat as moisture from precipitation, while values of 10 receive ten times as much evaporative energy.

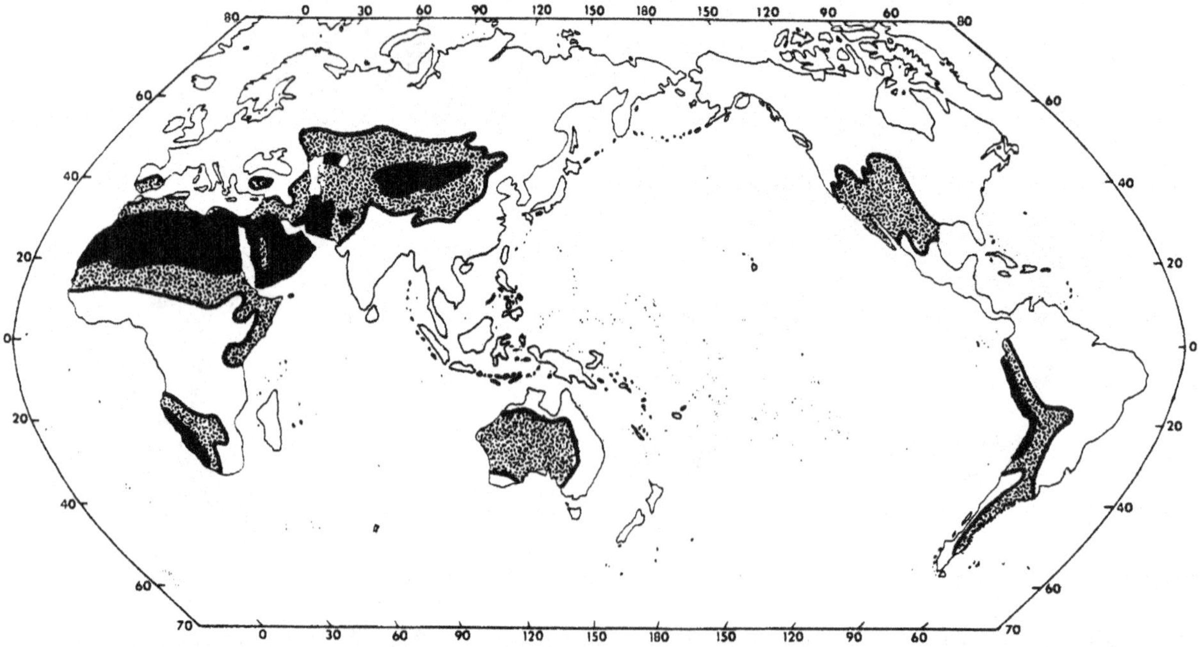

Arid to Semi-Arid Environments (values of 2 - 10)	Hyper-Arid Environments (values of > 10)

Table 3: Contrast of Dryness Ratio in Various Regions[8]
Higher numbers indicate more arid and drier, harsher landscapes.

WITHIN SAHARASIA

Arid Region	Dryness Ratio
Egypt & Libya	200+
Rub'al Khali	20 - 50+
Other North Africa	20 - 50+
Takla Makan & Gobi	7 - 20+
Other Arabia	10+
Jordan, Syria, Iraq	7 - 10+
Caspian Sea/L. Balkhash	7 - 10+
Pakistan, Afghanistan	3 - 10+
Iran	2 - 10+

OUTSIDE OF SAHARASIA

Arid Region	Dryness Ratio
Atacama	10 - 50+
Namib	10 - 50+
Kalahari	3 - 7+
Central Australia	3 - 7+
North American Basin	3 - 7+
Argentina	3 - 5+

HUMID ZONES

W. Europe & N. Asia	<1 - 2+	Amazon & C. America	<1 - 1+
E. USA & Canada	<1	Congo Basin	<1
Oceania	<1		

Computation of the Dryness Ratio

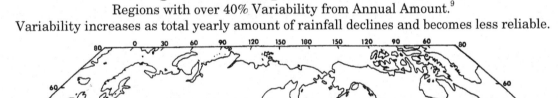

$$\text{Dryness Ratio} = \frac{\text{Mean annual net solar radiation}}{\text{Mean annual precipitation} \ \times \ \text{Latent heat of evaporation}}$$

ity found on Earth.[9] This great precipitation variability, from a biological point of view, makes annual precipitation data nearly meaningless. Near the desert margins some regular seasonal rainfall might occur, but in the heart of the desert signs of regularity, or atmospheric pulsation, cease. Rains may come only once every several years without any seasonal pattern, and then in the form of a destructive torrent which might last

9. "Precipitation Variability Map", in Espenshade & Morrison, 1978, ibid. p.14.

Figure 13: Precipitation Variability
Regions with over 40% Variability from Annual Amount.[9]
Variability increases as total yearly amount of rainfall declines and becomes less reliable.

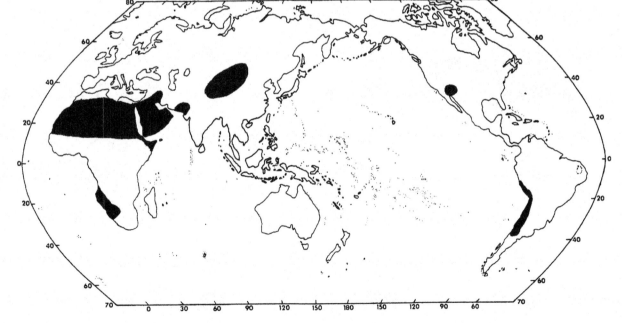

hours or even days.[10] Such infrequent rains may wash away earthen homes and structures built by desert peoples, scattering their herds, and wear away the exposed landscape. Remnant soils and plant cover may be washed away, even as rains pool and temporarily replenish groundwaters. During my 1980 visit to the Aswan region of Egypt, I interviewed several native informants in their early 20s, who gave testimony to the great precipitation variability: they could not remember ever having seen rain falling at that location. One of these young men said he had never seen rain at all! While many plants and animals may be adapted to survive a single year's drought, a greater hardship to life is experienced with each additional year that passes without rains.

Saharasia also clearly stands out in terms of its thermal characteristics, as seen with respect to highest mean monthly maximum temperature,[11] given in Figure 14. From this Figure it is observed that most of the Earth's hottest mean monthly maximum temperatures, of 37.8°C (100°F) or more, are found in Saharasia. Exceptions here are parts of Central India, which luckily experience more abundant seasonal rains, and a narrow strip across Northern Australia, also a desert. Additionally significant is the fact that the North African portion of Saharasia experiences mean monthly maximum temperatures exceeding 43.3°C (110°F), a condition not recorded anywhere else on the face of the Earth. Indeed, the highest temperature ever recorded under standard meteorological conditions (shaded conditions one meter above ground surface), a scorching 58°C (136.4°F), was in Aziza, less than 50km (31mi) south of Tripoli, Libya. And this is certainly not in the heart of the Sahara! Relative humidities in this area are frequently below 10% during the hottest months, a most inhospitable set of conditions for any living organism. Summer mid-afternoon temperatures at the exposed ground surface in such regions will often reach a roasting 82°C (180°F)![12] The exceedingly

10. Walton, K.: *The Arid Zones*, Aldine Publishing Co., Chicago, 1969, pp.25-7.

11. Griffiths, J.F. & D.M. Driscoll: *Survey of Climatology*, Charles Merrill Pub. Co., Columbus, Ohio, 1982, p.153.

12. Walton, 1969, ibid., pp.27-30.

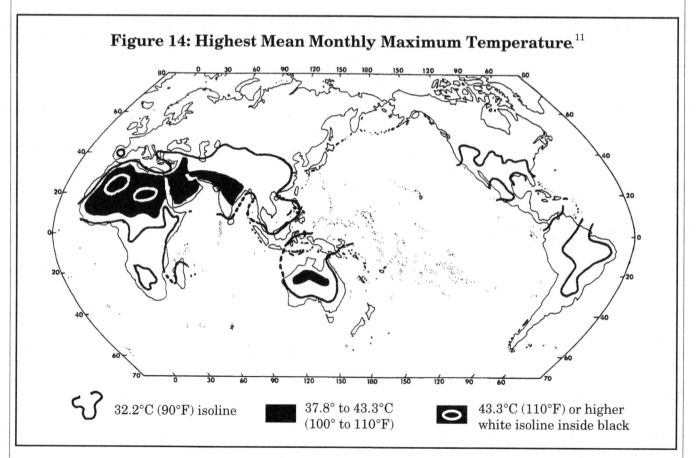

Figure 14: Highest Mean Monthly Maximum Temperature.[11]

⌇ 32.2°C (90°F) isoline   ▬ 37.8° to 43.3°C (100° to 110°F)   ◉ 43.3°C (110°F) or higher white isoline inside black

harsh conditions of the Libyan Desert extend eastward into the Western Desert of Egypt, which receives an average of less than 1mm (0.01") per year,[13] with an extreme variability: 10 years or more may lapse between significant rain showers.[14]

Deserts of Arabia suffer from not only scorching high temperatures, but also very cold winter temperatures. Frost and snow routinely occur in winter, except in the southeast. The interior southern portions of the Iranian Desert similarly experience cold winters and hot summers, with temperatures of 54.5°C (130°F). Combined with strong winds and scanty rains, the area is rendered generally unfit for human habitation. The Turkestan Desert, north of the Iranian Plateau and Hindu Kush, does not get as hot in summer (only 46°C or 115°F!), but may drop to -40°C (also -40°F) in winter. Indeed, most of the Central Asian deserts are mountain-rimmed basins open toward the north, and they experience subzero freezing conditions during winter. The Aral Sea, lower Syr Darya, and Lake Balkhash are frozen for several winter months, with the latter occasionally freezing over completely. Rains in Turkestan average 70mm to 200mm (3" to 8"), but as many as ten consecutive summers have passed without a drop of rain being recorded.[15]

The miserable conditions of the Iranian, Afghan, and Central Asian Deserts are compounded by the presence of seasonally persisting high winds of very low humidities, and a high content of choking dust and haze. Violent sandstorms may reach up to 160kmph (100mph); these occur in a belt across Iran, Afghanistan, and into Pakistan.[16] A tremendous haze hangs in the air for much of the year stretching from the Eastern Sahara, across Arabia, Jordan, Syria, Iran, Afghanistan, and Pakistan. An estimated 60 million tons of dust is borne eastward out of the Saharan and Arabian Deserts into Pakistan and India each year, a quantity equal to 1/3 of the yearly sediment transport of the formerly undammed Nile River.[17] Eventually, the haze cloud reaches north into Soviet Central Asia. Here one finds, once again, seasonal and quite miserable hot, dry, high-velocity winds carrying a fine dust, called *Sukhovei* by local people.[18]

Data for the Takla Makan and Gobi Deserts are hardly more appealing. The driest portions of the Takla Makan and Turfan Depression range between 50mm to 100mm (2" to 4") of rainfall yearly, with extremely low winter temperatures (6 to -19°C or 43 to -2°F for January). However, summer temperatures regularly exceed 50°C (122°F) in the shade, and a sand surface record of 78.8°C (174°F) has been recorded. The Gobi Desert ranges from 50mm (2") rainfall in the west to 170mm (7") in the east. The growing season averages a scant 165 days, and temperatures range from -25 to 44°C (-13 to 111°F). The central portions of these large basins are generally uninhabited, and only small populations live on their fringes.[19]

From Figures 13 and 14 it is noted that only Saharasia, specifically its North African, Arabian, and Middle Eastern portions, can be identified as a region of both highest mean monthly maximum temperature and greatest precipitation variability combined. Compared to Saharasia, other arid regions possess significantly lower mean monthly maximum temperatures (South Africa, North America, and South America, in Figure 14), or they receive a more regular, albeit minimal, precipitation (Australia, in Figure 13).

13. Wendorf, F., et al.: "The Prehistory of the Egyptian Sahara", *Science*, 193:103, 1976.

14. Wendorf, F. & R. Schild: *Prehistory of the Eastern Sahara*, Academic Press, NY, 1980, pp.12-4.

15. McGinnies, et al., 1968, ibid., pp.419,424,427-30; Walton, 1969, ibid. pp.29-30,35-7.

16. Fairservis, W.A.: "Archaeological Studies in the Seistan Basin of SW Afghanistan and E. Iran", *Anthro. Papers Am. Museum of Natural History*, 48(1):14-16, 1961; Dupree, L.: *Afghanistan*, Princeton U. Press, 1973, pp.4,31.

17. Bryson, R. & D.A. Baerreis: "Possibilities of Major Climatic Modification and their Implications: NW India, A Case for Study", *Bull. Am. Meteorological Society*, 48(3):136, March 1967, pp.137,140; Hare, F.K.: "Connections Between Climate and Desertification", *Environmental Conservation*, 4:96, 1977; Bryson, R. & T. Murray: Climates of Hunger, *U. Wisconsin Press*, 1977, p.112; cf. Bryson, R.A.: "Is Man Changing the Climate of Earth?", *Saturday Review*, 1 April 1967, p.112.

18. Kovda, V.A.: "Land Use Development in the Arid Regions of the Russian Plain" in *A History of Land Use in Arid Regions*, L.D. Stamp, Ed., UNESCO, 1961, p.184.

19. McGinnies, et al., 1968, ibid.; Walton, 1969, ibid.

Biological Aspects of the Saharasian Environment

Kuchler's maps on natural vegetation[20] demonstrate the harsh conditions in many subregions of Saharasia through identification of places where vegetation is largely or entirely absent, as seen in Figure 15. The extremes of thermal and moisture stress present in Saharasia are further demonstrated in Figure 16, which identifies world regions of lowest biological productivity,[21] a factor which roughly translates into carrying capacity — that is, the quantity of plants and animals produced by a given piece of land in an average year. Saharasia appears similar to the icecap and extreme permafrost regions of the arctic in terms of its very low biological productivity, as do a few parts of central Australia. Figure 17 reveals another facet of the harsh conditions in deserts, the presence of desert soils.[22] Such soils are not restricted to Saharasia, but it is apparent from Figure 17 that desert soils are most widespread across Saharasia.

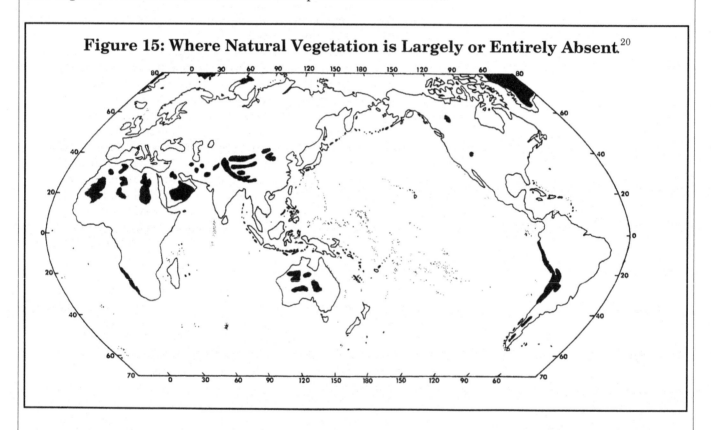

Figure 15: Where Natural Vegetation is Largely or Entirely Absent.[20]

Cultural Aspects of the Saharasian Environment

In terms of its inhospitableness and outright hostility to life, plant or animal, Saharasia, as seen in these preceding figures, stands out from among all the world's regions. But what do the various harsh extremes described above mean in terms of human occupation, subsistence, and settlement? Most of the social and behavioral characteristics which define the Saharasian desert belt are discussed at great length in subsequent chapters. However, a number of the more direct responses to climate may be identified here.

It is apparent from what has already been given that plants and animals in Saharasia must not only be capable of withstanding great temperature extremes; they must also be capable of surviving off of less

20. After Kuchler, W.: "Natural Vegetation Map", in Espenshade & Morrison, 1978, ibid., pp.16-17.

21. After Lieth, in Golley, F.B.: "Productivity and Mineral Cycling in Tropical Forests", in *Productivity of World Ecosystems, Proceedings of a Symposium: August-September, 1972*, National Research Council/National Academy of Sciences, Washington, D.C., 1977, pp.106-15.

22. After USDA, in Sears, P.B.: "Climate and Civilization", in *Climatic Change, Evidence, Causes, and Effects*, H. Shapley, ed., Harvard U. Press, Cambridge, 1953, p.39.

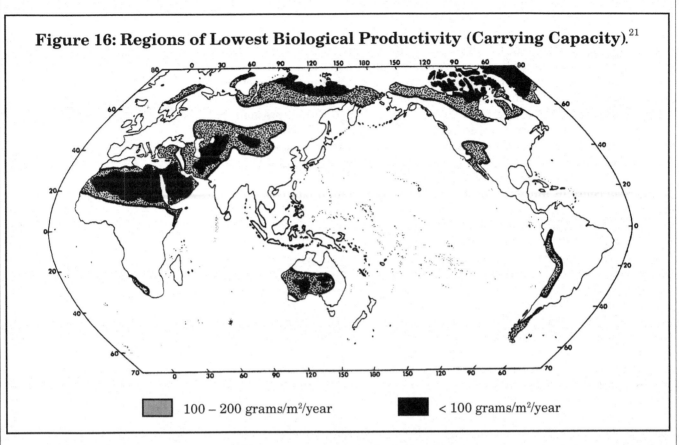

Figure 16: Regions of Lowest Biological Productivity (Carrying Capacity).[21]

100 – 200 grams/m²/year   < 100 grams/m²/year

regular or dependable water supplies. Desiccating, low relative humidity winds and dust storms, powerful erosive forces, and blistering heat from high insolation (sunlight), high albedo (glaring reflection from the ground surface), low moisture landscapes, as well as severe freezing conditions in

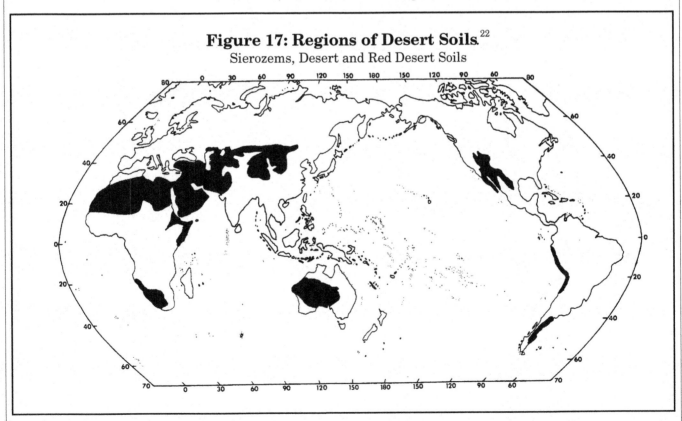

Figure 17: Regions of Desert Soils.[22]
Sierozems, Desert and Red Desert Soils

some areas must be tolerated. A greater adaptation to these various physical, thermal, and moisture stresses must occur, or, as is typical in many subregions of Saharasia, life simply will cease to exist.

Can it be different for the human animal? The answer is a qualified no: human life can exist in this stressful environment, but only in specific places where the micro-environment is not quite so harsh, where water supplies are secure on a year-round basis, and given specific adaptations and technology. Some areas are so hostile to human life that, without a constant supply of food and water from the outside, or from deep wells, they have remained essentially uninhabited. The world's uninhabited regions (around the turn of the century) are identified in Figure 18; Saharasia contains almost all of the world's nonarctic uninhabited lands.[23]

Another prominent cultural adaptation in Saharasia is that of nomadic herding, the distribution of which is given in Figure 19.[24] Human populations engaged in nomadic herding essentially live in a symbiotic relationship with their herd animals, moving from one pasture region to another, often following seasonal shifts in skimpy rains along the border of the desert. In Saharasia, these animals are primarily the camel, goat, sheep, and on the moister fringes, cattle. As such, nomadic herding may be seen as a survival adaptation to a low carrying capacity. It should be noted that in most of the central regions of Saharasia nomadic herding is a tenuous and risky business at best; regions with barely adequate forage for livestock encompass and surround the various vegetation-barren or uninhabited regions, identified in Figures 15 and 18.

Human populations did not first discover and enter Saharasia in its present desert condition, however. When Neolithic hunting, herding, and agricultural peoples were spreading out across the globe, around c.8000 BCE, Saharasia possessed a relatively moist and lush environment, thick with grasses, trees, lakes, rivers, swamps, and a plethora of large and small animals. In its moist state, Saharasia was settled by various

23. After "Population Density Map", Espenshade & Morrison, 1978, ibid., pp.20-21.

24. After "Major Agricultural Regions Map", in Espenshade & Morrison, 1978, ibid. pp.28-29.

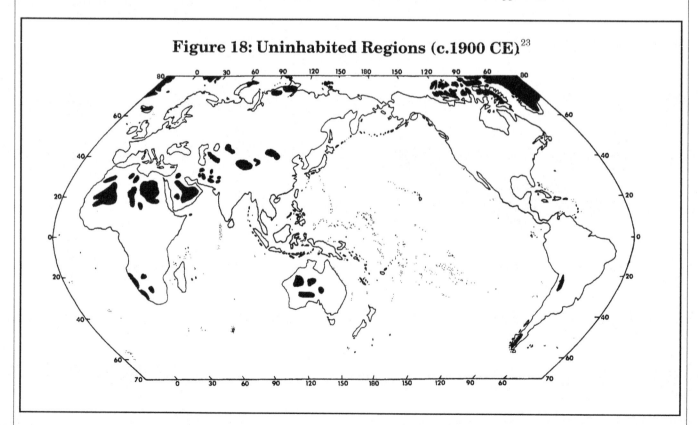

Figure 18: Uninhabited Regions (c.1900 CE)[23]

Figure 19: Regions of Nomadic Herding (c.1900 CE)[24]

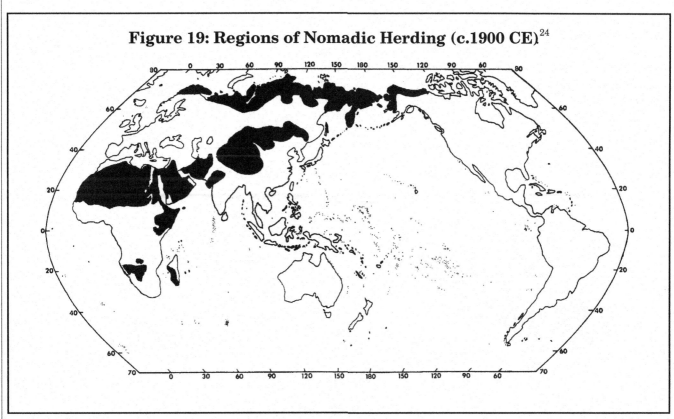

Neolithic groups who established small and large settlements based upon hunting, animal herding and limited agriculture; they traded with each other and lived peacefully. By c.4000-3500 BCE, however, the Saharasian environment began to dry up. Permanent settlements were abandoned, except where water supplies remained secure. A shift towards nomadic adaptations occurred. As drought and desertification intensified, famine, starvation, mass migrations and conflicts inevitably developed, intensified, and have continued ever since.[25]

As will be detailed in Part III, two major Saharasian subregions from which repeated out-migrations occurred were Arabia and Central Asia. As Arabia began drying out after c.4000 BCE, various warrior-nomad Semitic groups repeatedly irrupted and invaded territories with secure water supplies, or wetland regions bordering the desert. The last irruption from Arabia was that of the Islamic Arab armies around 640 CE and 1000 CE. The regions influenced by Arab armies, or those which adopted the Islamic social institutions and faith under their influence, are shown in Figure 20.[26] Central Asia also began drying out after c.4000 BCE, and various warrior-nomad Indo-Aryan groups (eg., Kurgan, Battle-axe peoples) repeatedly irrupted from the region, also invading and conquering regions with secure water supplies, or wetland regions bordering the desert. Later irruptions of peoples from the deserts of Central Asia included those of the Scythians, Sarmatians, Shang, Chou, Huns, Mongols, and Turks. Figure 21 identifies regions which, at one time or another since 540 CE, were under military domination of Mongol/Turkish peoples and their social institutions.[27] *Significantly, Figures 20 and 21, taken together, encompass 100% of Saharasia, spilling out beyond the borders of the arid zone into moister regions.*

Semitic and Indo-Aryan peoples, to include their more recent descendants, the Arabs and Mongol/Turks, historically dominated the Saharasian landscape and its borderlands. Importantly, the beliefs, social institutions, and behaviors of these warrior-nomad peoples were

25. See Part III for details.

26. After Jordan, T. & L. Rowntree, *The Human Mosaic*, Harper & Row, NY, 1979, p.187.

27. After Pitcher, D.E.: *An Historical Geography of the Ottoman Empire*, E.J. Brill, Leiden, 1972, Map V.

significantly different from those of the peoples inhabiting moister regions outside of Saharasia, particularly regarding infant and child care, female status, and sexual matters. Practices such as swaddling, infant cranial deformations, virginity taboos, genital mutilations, the bride price, polygamy, concubinage, female seclusion, castes, slavery, divine kingship, and human sacrifice occurred first among the warrior-nomad peoples of Saharasia, or within the despotic states they would later erect on the

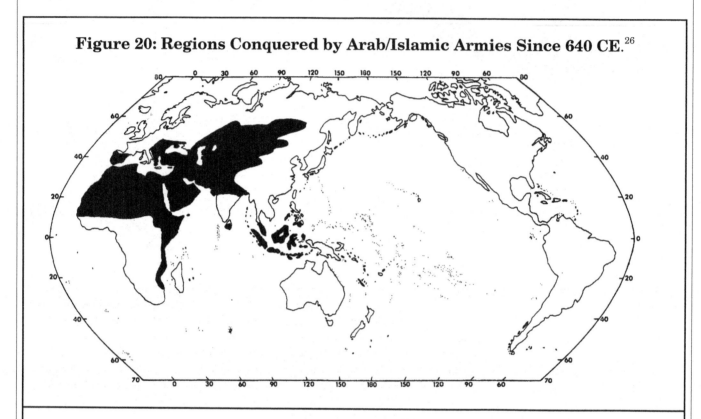

Figure 20: Regions Conquered by Arab/Islamic Armies Since 640 CE.[26]

Figure 21: Regions Conquered by Turkish/Mongol Armies Since 540 CE[27]

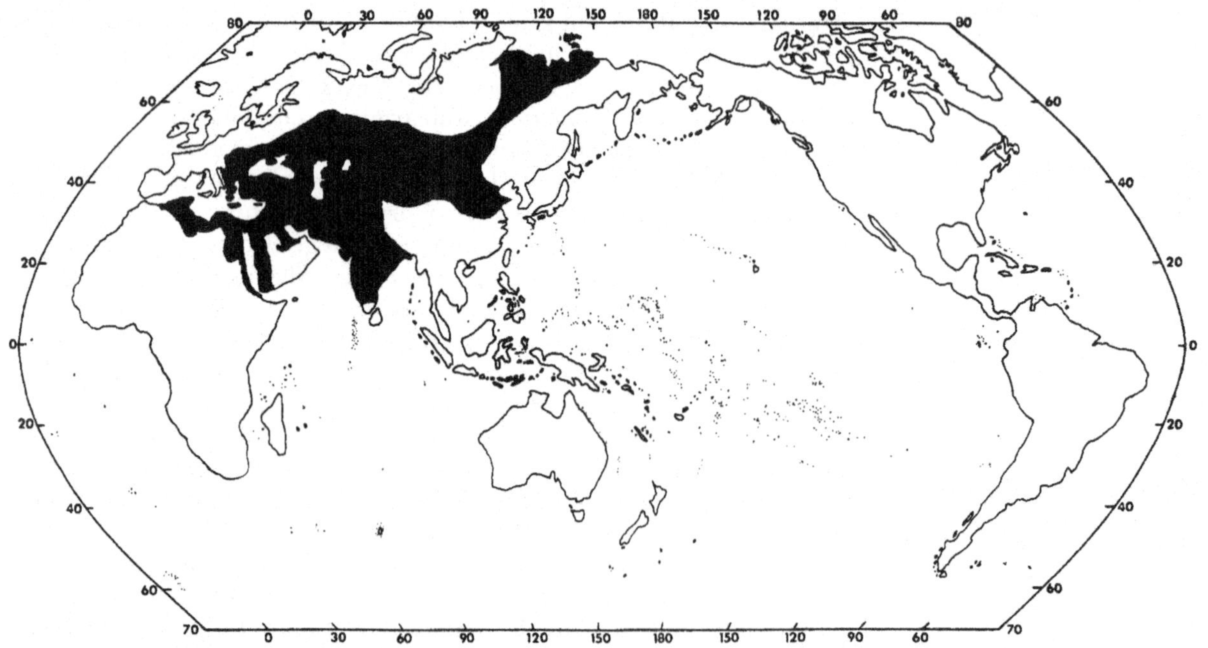

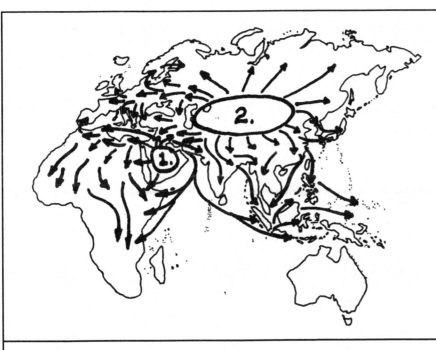

**Figure 22:
Core Spreading Centers
(1. Arabia, 2. Central Asia)
for Origins of Patrism
Within Saharasia**
(see Part III for details)

Figure 23: Diffusion of Patrism From Saharasia to Other World Regions
(see Part III for details)

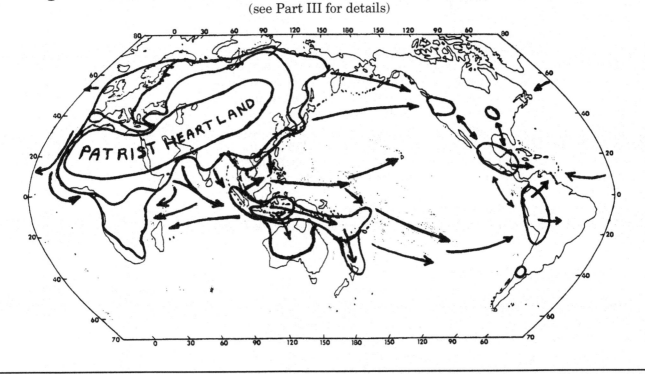

moister fringes of the desert. This process, which began with massive famine and migration, was set into motion at the same point in time that major desiccation of Saharasia began. From Saharasia, these behaviors and social institutions spread to other non-arid regions along its borders through the mechanisms of migration, borrowing, conquest, and/or forced adoption. Figures 22 and 23 present, once again, two maps which are fully derived and developed in Part III, showing the migratory pathways from which these earlier episodes of armored patrism spread outwards, to affect the Saharasian borderlands, from regions of origin *within* Saharasia.

The armored patrist behaviors and social institutions existed in their most extreme form in the Central Asian Khanate and Persian (and Arab) harem system, which was a pathological form of "family" organization that originated among peoples who had led centuries of nomadic life in the traumatizing, famine-prone deserts of Saharasia. The fully-blown harem system developed after these same peoples acquired the wealth and captive females of various conquered territories. In these social systems, women and children were chattel, their major life decisions (love, marriage, divorce, residence, education, fertility, etc.) were decided by one or another dominant male (father, brother, husband), who also could use socially approved physical punishment, including murder, to enforce his will. All men were subordinated in similar fashion to other males in a rigid military caste system where a single dominant male leader was given unquestioning obedience, in god-like manner. Male and female children were segregated to prevent sexual play and exploration, which would in any case have been severely punished. Young men were initiated into all-male, military or quasi-military organizations, where they were punished and threatened into a system of unquestioning obedience to older males; young girls were similarly segregated, threatened and coerced. In such regions, pleasure was generally eschewed or avoided, and pain generally endured or actively sought out. Infants were regularly subjected to confining wraps and other pain-inflicting appliances. The genitals of young children were also routinely cut and mutilated by elders of the group, in rituals associated with childbirth or marriage. Sadistic impulses generated from severe traumas and dammed-up, distorted sexual urges, were consequently widespread, institutionalized, and ritually directed towards infants, children, women, minorities, conquered enemies, slaves, or neighboring peoples. From a sex-economic point of view, it can be seen that the sadistic impulses had to go somewhere, and simply could not remain submerged forever. While it is more usual to speak about these matters in a smaller-scale context, of pent-up anger spilling outward within a single individual, the "outbreak" of pent-up violence and sadistic rage within an entire cultural group always has a geographical context over which it is acted out — the physical landscape. More typically the issues of landscape and territory are discussed in the context of "geopolitics", but such "purely political" matters are, in fact, underlain by deeply emotional motivations which require careful analysis if the phenomenon of war is ever to be understood and finally ended.

The World Behavior Map

The Saharasian distribution of armored, patrist societies is most clearly seen in Figure 24, the World Behavior Map (duplicated here from Chapter 1). This map was derived through a systematic analysis of cross-cultural data from within a global sample of over a thousand different cultural groups, worldwide, as developed in G. P. Murdock's *Ethnographic Atlas* data base — all the variables conformed to Wilhelm Reich's sex-economic theory, as previously detailed.[28] Murdock's data base focused upon indigenous, aboriginal native peoples in each geographical region, and was compiled from the field reports of hundreds of ethnographers and anthropologists around the world, whose studies were undertaken roughly between 1840 to 1960 CE. Murdock's data allowed a significant number of the sex-economic variables (see Table 1) to be systematically and simultaneously addressed for a large number of cultures around the world. As previously discussed in detail, each of the 1170 cultures in the *Ethnographic Atlas* was evaluated according to its percent-patrist

28. Murdock, G.P.: *Ethnographic Atlas*, U. Pittsburgh Press, 1967; cf. Chapter 2 and the section on "Geographical Analysis of Cross-Cultural Data" in Chapter 3.

Figure 24: The World Behavior Map

For the period roughly between 1840 and 1960, as reconstructed from aboriginal cultural data given in Murdock's *Ethnographic Atlas* (1967), with minimal historical interpretation.

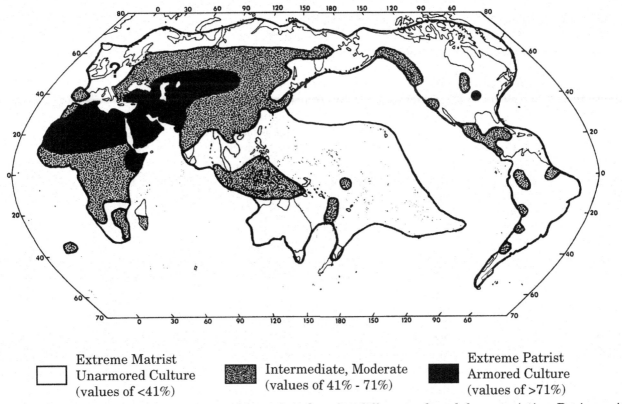

☐ **Extreme Matrist** **Unarmored Culture** (values of <41%)	▒ **Intermediate, Moderate** (values of 41% - 71%)	■ **Extreme Patrist** **Armored Culture** (values of >71%)

The Percent-Patrist values were composed from an index of 15 different cultural characteristics. Regions with a high degree of patrism are defined by the presence of: Female Premarital Sex Taboo, Segregation of Adolescent Boys, Male Genital Mutilations, High Bride Price, Polygamous Family Organization, Patrilocal Marital Residence, Patrilineal Descent (without Cognatic Kin Groups), Patrilineal Land Inheritance, Patrilineal Movable Property Inheritance, High God, Class Stratification, Caste Stratification, and Slavery. Regions possessing few of these traits were defined as predominantly matrist in character, while intermediate regions possessed a mix of both matrist and patrist traits. See Chapter 3 for details.

characteristics. Using latitude and longitude coordinates for each culture, regional behavioral averages were extracted. World regions characterized by a high percentage of cultures with a high degree of armored-patristic, warlike social institutions and behaviors scored relatively high on the scale, while those of a predominantly peaceful unarmored-matrist character were given an appropriately low percentage score. The average percentage of patrist behavior was determined for all cultures residing within a 5° x 5° regional block of latitude and longitude. The various regional block scores were plotted on the world map by computer, after which they were divided into low, intermediate and high patrist categories, and coded by grey-scales. The computer-generated maps formed the basis of the World Behavior Map, with some additional information derived from specific historical factors expressed in the individual cultural maps presented in the next Chapter. The World Behavior Map reflects behavior among native, subsistence-level peoples, thereby minimizing the effect of European-derived culture in Oceania and the Western Hemisphere. For example, Murdock's cultural data for Asia, Africa, Oceania, North America, and South America are almost entirely derived from ethnographical studies of native aboriginal peoples in those regions. His European cultural data was derived from culturally isolated groups, or reconstructed from historical data, thereby minimizing somewhat the effects of 20th Century modernization and conflicts in Europe. Today, much of Europe and the Americas clearly would deserve classifications of "intermediate" to "high" on the patrist scale — Murdock's data simply informs us that native aboriginal peoples and culturally-isolated groups, even within the relatively recent past, expressed a gentler and softer, more loving, peaceful and cooperative form of social organization, but *only with great geographical distance from Saharasia.*

The World Behavior Map thereby provides a first-time global view of the more recent spatial-historical dimensions of human behavior, as recorded in the observations and writings of hundreds of scholars over the last several hundred years.[29] The identified Saharasian spatial pattern is very real, a heretofore previously hidden pattern deeply imbedded within the collected anthropological and ethnographical scholarly literature, as found within every university library around the world. In spite of whatever problems may be posed by the generalities and final classifications, or the data base from which it is composed, the World Behavior Map is a robust approximation of a very real pattern of recent history, a still portrait of a specific time-period, revealing an ongoing historical process. A data base composed of twice as many cultures and related variables, analyzed in a similar systematic manner, would predictably look very similar.

The spatial correlations between the armored patrist cultural characteristics of Figure 24 and the environmental factors previously given are an additional striking factor which cannot be the product of some grand coincidence. The spatial-temporal relationships are much more than coincidental — there is a powerful causal element at work. It will be a major thrust of this work to examine these relationships.

A number of specific questions on environmental and cultural change force themselves. For instance: To what degree has the Saharasian environment changed over the years? Clearly, it was not always so dry and harsh, being much wetter and filled with plants and animals in the ancient past. Where did the aridity begin? Where can the earliest traces of armored patrism be found? What do archaeology and history have to say about human responses to this dramatic environmental change towards aridity?

As shall be demonstrated, there is a reasonably large amount of historical and archaeological evidence demonstrating that patrism originated, persisted and spread within Saharasia at the same time when the arid conditions started to intensify, and that patrism later diffused outward to affect borderland regions. The smaller regions of patrism in Oceania and the Americas are likely also connected to events in Saharasia, through the phenomenon of pre-Columbian contacts between the Americas and the ocean-navigating patrist states of the ancient historical period. In summation, *human armoring and the earliest patrist social institutions and cultures appeared and persisted on Earth for the first time precisely within Saharasia, and at the same moment in prehistory when Saharasia was transformed from lush grassland into harsh desert. As the great belt of desert landscape called Saharasia appeared, so too appeared the emotional desert of the human social landscape.*[30]

"...human armoring and the earliest patrist social institutions and cultures appeared and persisted on Earth for the first time, precisely within Saharasia, and at the same moment in prehistory when Saharasia was transformed from lush grassland into harsh desert. As the great belt of desert landscape called Saharasia appeared, so too appeared the emotional desert of the human social landscape."

29. Murdock, 1967. ibid. See Chapter 3 for more details on the Murdock data base, and how it was used to construct the World Behavior Map.

30. The phrase "emotional desert" was coined by Wilhelm Reich, to describe the emotionally dead and contactless manner in which modern "civilized" humans generally behave; included here would be the abrasive, hostile, and hateful treatments often meted out to vibrant, freely moving, pulsating life, as present in the newborn infant, the lively child, in youthful sexual behavior, the unconfined spirit of wild nature, etc.; cf. Reich, W.: "The Emotional Desert", in *Selected Writings*, Farrar, Straus & Giroux, NY, 1973, pp.461-6.

About the Maps and *Correlation Tables* Presented in This Chapter

The maps and tables presented in this Chapter were predominantly drawn up from two major sources of data: the *Ethnographic Atlas* of George Peter Murdock (U. Pittsburgh Press, 1967), and the *Cross-Cultural Summary* of Robert Textor (HRAF Press, New Haven, 1967).

Murdock was one of the founding fathers of American anthropology, and a pioneer in cross-cultural testing methods. During his tenure at the University of Pittsburgh, Murdock tirelessly reviewed hundreds of books and reports from field researchers in his efforts to compile a data base on different human cultures from around the globe, to allow researchers a simple and easy way to review a sizable number of behavior traits and social variables for a given culture, or to contrast and compare different cultures, without having to go through the tedious process of consulting a dozen or so different texts for each culture of interest. His decades of labor produced the *Ethnographic Atlas*, which did not contain any maps, but which was a concise tabular coded listing of over 50 different variables on 1170 different human cultures. His data base has been constructively criticized and reviewed by the anthropological community, and is considered to be one of the most accurate sources available for making cross-cultural determinations. I used Murdock's data to construct the World Behavior Map, as well as most of the maps in this chapter. For the *Ethnographic Atlas* variables mapped here, each has been arranged into dichotomous categories, according to the "matrist/patrist" schema previously given in Table 1 (on page 5). The percentage of cultures in each 5° by 5° block of latitude and longitude with a *patrist* orientation was then determined and mapped. The small numbers given on some of the maps, multiplied by ten, constitute the upper end of a ten-point numerical range representing the percentage of cultures within that block region possessing the cultural trait under investigation (the maps provide more details on this point).

The smaller *Correlation Tables* presented in this Chapter are extracted from a much larger *Correlation Table of Sex-Economic Factors* composed of 63 different variables; it is presented in Appendix A, and more fully discussed in Chapter 3 in the section on "A Preliminary Cross-Cultural Comparison". All variables presented in the larger and smaller *Correlation Tables* were derived from Textor's 400-culture subset of the Murdock data base. Textor added significantly more variables for those 400 cultures than did Murdock, and used a computer to look for all possible statistically significant correlations (positive or negative) between the variables. His *Summary* is a computer print-out of those correlations, and is routinely consulted by behavioral scientists for social theory testing. My *Correlation Tables* were drawn directly from Textor's computer analyses. The significance of these *Tables* lies in the fact that some 95% of all the observed correlations between different patristic variables were positive in nature, with only a few negative correlations, discussed shortly. Additional details can be found in Chapter 3 and Appendix A.

5. The Global Geography of Social Institutions As Derived from Recent Cultural Information

The Saharasian thesis and the patterns previously given on the World Behavior Map (Figure 24) may be independently verified by a review of the geographic distribution of a number of individual, specific behavioral maps and variables. The following sections present maps of individual social institutions and behavior variables, reflecting relatively recent social and cultural conditions, which further support the Saharasian distribution of armoring and patrism. The data used for preparation of most of these maps was taken from the same *Ethnographic Atlas* data base used to construct the World Behavior Map, but — as detailed below — many other sources were consulted. In a few cases, certain social institutions are discussed without presentation of any map, given the difficulty of obtaining suitable data. In those cases, specific predictions are made as to what such a map would eventually look like, and these predictions may eventually act as additional tests of the overall accuracy of the Saharasian thesis.

Infant and Child Treatment

1) Infant Cranial Deformation and Swaddling

The variables of infant cranial deformation and swaddling were not included in Murdock's *Ethnographic Atlas*, nor in the composition of the World Behavior Map; but they repeatedly appeared in the historical record as occurring among Central Asian nomads, including the Hittites, Scythians, Sarmatians, and Huns.[1] History and geography suggest that infant swaddling and cranial deformations are related, complementary practices, having been invented by these same Central Asian nomadic peoples, firstly by historical accident through use of the nomadic back-pack cradleboard. Once established, however, the deformations became an identifying mark of the militant nomads, and were purposefully produced wherever they roamed.

Infant swaddling and cranial deformation are severely traumatic practices, having all the characteristics necessary to alter radically the psyche of the child and to contaminate the view of the maternal figure with tremendous anxiety and anger. They are severe, often life-threatening practices which could, by themselves, produce a fearful submission to external authority, thwarted adult male-female relations, and an impulsive, enraged, violent streak in adult behavior. If this sounds like an exaggeration, consider these eyewitness reports gathered by Dingwall from travelers in the Eastern Mediterranean and West Coast North America, where babies were observed undergoing both swaddling and artificial cranial deformations.[2]

Pacific Northwest, North America
(Dingwall: *Artificial Cranial Deformation*)

"The child is kept in the cradle for long periods of time and is not even removed for the purposes of suckling or for the natural needs. Ingenious contrivances are used in order to avoid removing the child. The urine is caught in a little iron or earthenware receptacle placed beneath the cradle. The apparatus is arranged that a tube leads from the genitals to the pot and a similar arrangement serves for defecation. Washing the child is usually done once a week, and thus for seven days at a stretch the child remains fastened in the

1. See the appropriate sections in Chapter 8 for additional discussion, particularly for Central Asia.

2. Dingwall, E.J.: *Artificial Cranial Deformation*, J. Bale, Sons & Danielson, Ltd., London, 1931.

cradle, the monotony only being broken by the weekly wash which probably does not take any considerable time. The head rests upon a small but very hard sort of bolster upon the middle of which the occipital portion of the skull is placed. Arms and legs are secured by a band. Another compresses the forehead... Thus restrained, the child remains for about the first two years of its life." [3]

"The child cries and turns black, and when the mother presses on its forehead a white slimy fluid comes out its nose and ears. Thus it sleeps every night until its head has taken on the desired shape..." [4]

"As the child's head is flattened its eyes stand as prodigious way asunder and the hair hangs over the forehead in a manner like the eaves of a house... On the forehead a lump of clay was placed and tied down very tightly. Under the pressure the children suffered extremely during the early stages of the operation. Their faces turned black and whitish viscous fluid exudes from the nose, eyes, and ears." [5]

"The heads of the children and especially of the girls are pressed so tightly by means of a peculiar kind of ligature, that little by little the heads assume the shape of sugar-loaves. The pressure is so great that the noses of the children who submit to it are constantly bleeding." [6]

"One case...a woman had 18 inches of solid flesh and bone towering above her eyes before her head began to slope backwards." [7]

"Its little black eyes, forced out by the tightness of the bandages, resembled those of a mouse choking in a trap... The child's eyes protruded to a distance of half an inch and appeared both inflamed and discolored, as did also the surrounding region." [8]

"The process is attended...by a good deal of pain, and certainly the appearance of the child is shocking. Its eyes seem to start from their sockets, its mouth contorted...the noses of the children were accustomed to emit a whitish pus." [9]

There are no modern-day accounts of infant cranial deformation I know about, given that the practice has died out over the last several hundred years. Dingwall and others have gathered photographs of various adults, or their skulls, showing the end result of some of the practices. The people looked absolutely horrifying and frightening, the apparatus to mold the head appearing as devices of torture.[10] Indeed, this aspect of developing a frightening appearance reinforced the use of the practice among peoples prone to warfare, as a means of appearing more fierce in battle. Central Asian Hunnish warriors, for example, were said to appear so frightening, with towering ghoulish skulls piled up so high above the eyes as to appear inhuman, like demons of some sort — Dingwall gathers various historical accounts where the Huns routed enemy armies simply by their frightening appearance alone.

The cranial deformations took the form of a simple flattening of the back of the head, as from an inadvertent pushing the crania backwards against a cradle board. Or the crania were more purposefully flattened at both back and front, or elongated into a loaf shape. These latter deformations were generally accomplished by squashing the infant's skull be-

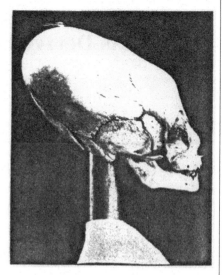

Deformed Chinook Skull
(Dingwall: *Artificial Cranial Deformation*)

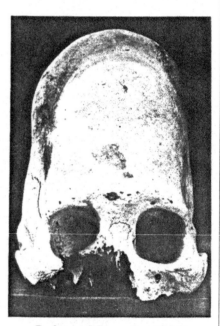

Deformed Peruvian Skull
(Dingwall: *Artificial Cranial Deformation*)

3. Dingwall, *Artificial Cranial Deformation*, ibid., p.84.
4. Dingwall, ibid., p.188.
5. Dingwall, ibid., pp.186-7.
6. Dingwall, ibid., p.178.
7. Dingwall, ibid., pp.177-8.
8. Dingwall, ibid., p.166.
9. Dingwall, ibid., pp.167-9.
10. In particular see Plates XXXIV in Dingwall, 1931, ibid. XXXVII, XLVI, LI; Also Comas, J. & P. Marquer, *Craneos Deformado de la Isla de Sacrificios*, Veracruz, Mexico, Instituto de Investigaciones Historicas, Mexico, 1969.

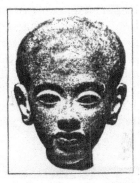

Akhenaten's Daughter
(Dingwall: *Artificial Cranial Deformation*)

Kwakiutl Woman
(Dingwall: *Artificial Cranial Deformation*)

11. Dingwall, *Artificial Cranial Deformation*, ibid., pp.146,167-9,238.

12. Dingwall, ibid., p.238.

13. Dingwall, ibid., pp.181,238-9.

14. Dingwall, ibid., pp.12-4,130,150, 238; Also see sections in Chapters 8 and 9.

15. Dingwall, ibid., pp.238-41.

tween two boards, which were then tied together, or by applying a metal or rope tourniquet to the infant's head. These methods were applied almost from the moment of birth onwards for up to several years. In order to keep the tourniquet in place, and to prevent the infant or child from interfering with the painful apparatus, some form of body restriction or swaddling was performed as an accompaniment. Both deforming tourniquet and swaddling were maintained, day and night, for years; swaddling bands often remained in place constantly, around the clock, except for an occasional bath.

Dingwall noted the association of cranial deformation with ideas of nobility, aristocracy, and chieftainship, and rightly pointed out the impossibility of any innate or instinctual aspect to the practice.[11] As he put it:

"The custom of cranial deformation is not uniform in all parts of the world. It cannot be the response to some innate human impulse, since if this were so it would be more widely distributed than it is. Moreover we can see the custom fading away, as it were, from those centers where it flourished and where succeeding generations clung to it with tenacity." [12]

He discussed its distribution in Asia, Europe, Africa, Oceania, and the Americas, pointing out that the practice breaks down as one moves away from certain central locations. These central "heartland" regions for cranial deformation were the Eastern Mediterranean, Egypt, Central Asia, Northwest China, North America's Pacific Northwest, the Great Basin, Mesoamerica, and Inca Peru.[13]

Cranial deformation among Egyptian royalty can be traced back at least to Akhenaten's time, and mudpack hair dressings persist in the Sub-Saharan Nile Valley today which resemble the trait. The Hittites, Scythians, Sarmatians, and Huns also practiced it. The Near Eastern turban itself may have had a history associated with the practice.

Asian nomads transplanted the deformations to the ruling castes in both China and India, from where they spread to other parts of Southeast Asia. The Huns introduced them into Europe, and they persisted among the French and German military castes until the late 1800s. Their distribution in the Old World appears intimately connected with the nomads of Central Asia, though they were also found in Oceania, the Pacific Northwest of the USA, in Mesoamerica, and in Peru. Dingwall believed the practices could have diffused to the New World from the Old, by ocean-navigating peoples. He cited the findings of Chinese and Korean pottery with deformed crania in archaeological digs in the Philippines, with a gradual diffusion and diminution of the trait across Polynesia.[14]

Dingwall did not directly consider swaddling in his study, but identified three steps to the dilution and diminution of the cranial deformations:[15]

1. The methods become mixed and various manual manipulations are substituted;

2. The apparatus vanishes, with only manual molding and pressing surviving;

3. These also disappear, leaving behind only body massage.

The functional association between swaddling and cranial deformations is, to my knowledge, my own, largely derived from the quotes and discussion given by Dingwall, and developed from two lines of evidence. Firstly, some form of swaddling was always used to restrain the child during the cranial deformation process, to prevent it from pushing off the painful cranial tourniquets, and also to allegedly "mold the limbs" of the child, which would otherwise "grow crooked". Swaddling also has persisted with a similar geographical distribution to infant cranial deformation, among the same peoples. While infant cranial deformation appears to have died out today in Saharasia and most areas around the world, swaddling persists as a remnant practice.

These two practices, infant cranial deformation and swaddling, appear as the most painful and traumatizing of all "childrearing" practices. They are aimed at literally molding or crushing the child's physical form to cultural expectations, and also towards the goal of developing a permanent emotional immobilization of the child. It is clear that infants and children often died from the organ-crushing aspects, or from secondary infections which developed. Gorer made a case for the negative psychological effects of swaddling among the "Great Russians",[16] but beyond this, the question appears not to have been studied to any significant extent. Also deMause, who studied the psychohistorical aspects of a wide variety of similar related practices aimed at children, says:

> "...restraints were thought necessary because the child was so full of dangerous adult projections that if it were left free it would scratch its eyes out, tear its ears off, break its legs, distort its bones, be terrified by the sight of its own limbs, and even crawl about on all fours like an animal... Swaddling was often so complicated it took up to two hours to dress an infant. Its convenience to adults was enormous — they rarely had to pay any attention to infants once they were tied up. As a recent medical study of swaddling has shown, swaddled infants are extremely passive, their hearts slow down, they cry less, they sleep far more, and in general they are so withdrawn and inert that the doctors who did the study wondered if swaddling shouldn't be tried again. The historical sources confirm this picture; doctors since antiquity agreed...and children were described as being laid for hours behind the hot oven, hung on pegs on the wall, placed in tubs, and in general 'left, like a parcel, in every convenient corner'." [17]

Some practices aimed at older children appear related to infant swaddling, namely various forms of bindings, corsets, backboards, and bonds designed to keep the child from moving about too much, "making a fuss", "slouching", or touching its genitals.[18] The following reports from earlier centuries are illustrative:

> (1230 CE) "And for tenderness the limbs of the child may easily and soon bow and bend and take diverse shapes. And therefore children's members and limbs are bound with lystes [bandages], and other covenable bonds, that they be not crooked nor evil shapen..." [19]

> (1826 CE) Swaddling "...consists in entirely depriving the child of the use of its limbs, by enveloping them in an endless length bandage, so as to not unaptly resemble billets of wood; and by which the skin is sometimes excoriated; the flesh compressed, almost to gangrene; the circulation nearly arrested; and the child without the slightest power of motion." [20]

New Material 2005: See *Update on Saharasia* (Appendix B) for presentation of new archaeological findings of mild or uncertain examples of cranial deformation, in Australia and elsewhere.

16. Gorer, G.: "Some Aspects of the Psychology of the People of Great Russia", *American Slavic and East European Review*, VIII(3):155-67, October 1949; Gorer, G. & J. Rickman, *The People of Great Russia; A Psychological Study*, W.W. Norton, NY, 1962; cf. Mead, M.: "The Swaddling Hypothesis, Its Reception", *Am. Anthropologist*, 56, 1954.

17. deMause, L.: "The Evolution of Childhood", in *The History of Childhood*, L. deMause, ed., Psychohistory Press, NY, 1974, p.37; Also see Lipton, E., et al.: "Swaddling, A Child Care Practice: Historical, Cultural, and Experimental Observations", *Pediatrics*, Supplement, 35, part 2, March 1965, pp.521-67; Dennis, W.: "Infant Reactions to Restraint: an Evaluation of Watson's Theory", *Trans. NY Acad. Sci.*, Ser. 2, Vol. 2, 1940.

18. deMause, 1974, ibid., p.11.

19. deMause, ibid., p.11.

20. deMause, ibid., p.37.

21. Drawn courtesy of Deborah Carrino, from an original photo by Allen, T. & D. Conger: "Time Catches Up With Mongolia", *National Geographic*, February 1985, pp.252-3.

Swaddled infant of Mongolia [21]

Figure 25: Infant Cranial Deformation and Swaddling.
Arrows suggest generalized path of diffusion from a possible common point of origins in Central Asia.

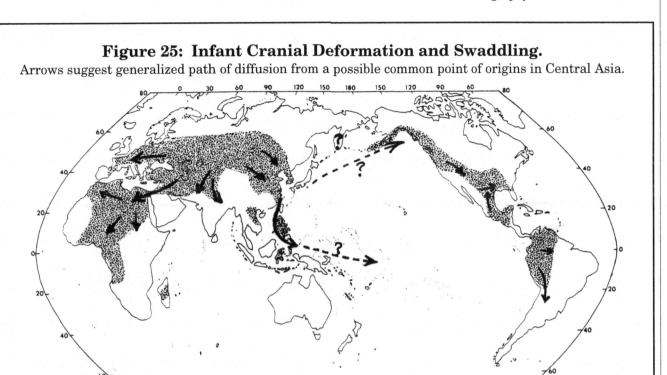

European Cranial Deformation, c.1700s.
Tall hats and cute bonnets hid an awful secret.
(Dingwall: *Artificial Cranial Deformation*)

22. Dingwall, 1931, ibid.; Gorer & Rickman, 1962, ibid.; Comas & Marquer, 1969, ibid.

A photograph of a tightly swaddled Mongolian infant, with eyes shut and oblivious to the fact that it was being handled, appeared in a recent issue of *National Geographic*.[21] The child, about six months old, appears much like a larva inside a tight cocoon, from which it emerges only infrequently. The practice still persists today across much of Saharasia, to include the Soviet Union, Eastern Europe, and China. News videos and magazine photos from hospitals in Russia and other east-European and Central Asian nations often will show dozens of similarly swaddled and inert infants lying in groups on carts or across tables.

Gorer and deMause have stated that swaddling was distributed worldwide, but maps of the actual distributions of swaddling do not support this idea. Furthermore, a casual review of the sources cited in studies on swaddling and cranial deformation indicate that most of the Old World evidence is gathered from Saharasian warrior nomad cultures, or from cultures which obtained the practices after being invaded by those same nomadic peoples. The question of swaddling and cranial deformation in Oceania and the New World is addressed in a later section, but clearly appears connected with the phenomenon of trans-oceanic, Pre-Columbian cultural diffusion.

Dingwall developed maps of the distribution of infant cranial deformation around the world, but his maps, like Gorer's discussion of the distribution of swaddling, remained very general. Still, they suggest a Saharasian connection in the Old World at least. Figure 25 is my own very rough and incomplete reconstruction of the distribution of infant cranial deformation and related practices, taken from the sources cited above.[22]

115

2) Breastfeeding / Denial of the Breast

There was no variable in the various data bases I consulted which directly addressed this important concern. Breastfeeding is obviously a most important factor in both the emotional bonding between mothers and infants, as well as for full nutrition. Breastfeeding by nature may last up to two years after birth of the child, but for various cultural reasons, may be terminated much earlier, or be denied altogether. In both economically developed and developing nations, there are trends towards denying the breast to babies, while simultaneously making the female breast a target for all kinds of sexually-erotic and sadistic impulses. Some men will pay money to watch women expose their breasts, but at the same time, policemen may arrest and incarcerate any woman who dares to expose them in public, beyond certain restricted places, such as nude beaches. A baby at the breast may be tolerated, but photos of the breast are largely defined as "pornography". Medical interest in the breast is also great — to "correct" breasts which do not fit the current fashions, and otherwise to engage in surgical mutilations of the breast, as discussed in a following section on "female mutilations".

From a sex-economic perspective, the emotional bonding between mother and baby is similar to the later emotional bonding between young men and women — the infant mouth is attracted to the breast nipple in a deeply bioenergetic, *sexual* manner. Wilhelm Reich described a specific *oral orgasm* experienced by the nursing infant, along with the pleasurable genital feelings and streaming sensations experienced by uninhibited healthy nursing mothers.[23] It is all quite natural, but reveals the very reason why, in sex-negative pleasure-denying cultures, infants are routinely denied the breast. Breastfeeding stirs up sexual anxiety in those who are sex-repressed: witness the various instances where mothers who breast-feed in public have been harassed by policemen, even arrested, and in at least one instance, the baby forcibly taken away from the mother for over a year by "social workers", solely for the reason of breast-feeding longer than the social "norm".[24]

Even before the invention of the "baby bottle", sex-negative cultures were inventing methods for denial of the breast to infants.§ In my readings on the literature of cranial deformation, swaddling, repressive couvade and other social institutions related to the crushing down of the infant (literally, in some cases), it was often mentioned that the "baby was given a moist cloth to suck on during the absence of the mother", or "baby was fed pap" or some other passing statement which demonstrated the infant was not being allowed to nurse at its mother's breast, which often was considered "filthy" (sexually). In all the cases I came across, the cultures engaged in such practices were the patristic, armored cultures. Additionally, in all those cultures which employed the "wetnurse" to breast-feed the children of the rich, little or nothing is said about the *children of the nurse*, who would often be neglected or abandoned in favor of the children of their employer.

Consequently, when a data base becomes available whereby one can identify those cultures which provide the infant immediate and uncompromised access to mother's breast, from the moment of birth onward, as compared to those cultures which deny the breast to infants —using various substitutes, for whatever reason — I predict that a map of that data will demonstrate a Saharasian distribution, with the greatest *denial of the breast* existing within Saharasia and its borderlands, and with the greatest openness and frequency of breastfeeding in Oceania, the New World, and other places far removed from Saharasia.

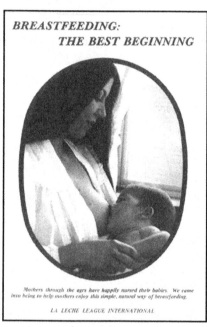

BREASTFEEDING:
THE BEST BEGINNING

Mothers through the ages have happily nursed their babies. We came into being to help mothers enjoy this simple, natural way of breastfeeding.

LA LECHE LEAGUE INTERNATIONAL

La Leche League International

23. Reich, W.: "Falling Anxiety in a Three-Week-Old Infant", in *Children of the Future: On the Prevention of Sexual Pathology*, Farrar, Straus & Giroux, NY, 1983, p.115; cf. other papers by Reich in this same citation.

24. See the case of Denise Perigo, reported in *Birth Gazette*, Spring 1992, pp.11-12; also in *Pulse of the Planet*, 4:158, 1993.

§ Michel Odent, the French physician and former student of Frederick LeBoyer (of gentle-birth fame) has gathered a lot of evidence on the cultural practice of denying babies access to the mother's first milk, called colostrum. Colostrum is an especially rich and vital substance, filled with maternal antibodies, and very healthy for the baby to ingest. However, a great many cultures are filled with superstition about colostrum, that it is a dangerous or foul substance, etc. Even in Western hospitals, such superstitions often prevail. (Info from a lecture by Dr. Odent at the *3rd Int. Symposium on Circumcision*, in Washington, DC, 1994.)

3) Male Genital Mutilations

Various mutilations have been performed upon the male genitalia, notably the foreskin of the penis. *Circumcision*, where the foreskin of the penis is surgically removed during infancy or puberty, was and is practiced across much of Saharasia, and in a number of Sub-Saharan, African, Near Eastern, Asian, and Pacific Ocean groups. *Subincision* was practiced primarily among Australian aborigines, and consisted of a cutting open of the urethra on the underside of the penis down to as far as near the scrotum; the subincision ritual was generally preceded by a circumcision ritual. *Incision*, a milder form of the mutilations, consisted of either a simple cut on the foreskin to draw blood, or a complete cutting through of the foreskin in a single place to partly expose the glans. Incision is found primarily on the East African coast, in Island Asia and Oceania, and parts of the New World. Other, more severe forms of genital mutilation existed elsewhere, as discussed below. Figure 26 is a composite map of the distribution of the various male genital mutilations, constructed primarily from the data of Murdock,[25] plus a variety of other sources.[26] Notably, the map does *not* address the more recent historical diffusion and adoption of male genital mutilations into Europe (primarily England) or North America.

Regarding the origins of the practices, the historical sources cited below and the distribution maps suggest a starting point in Arabian or North African Saharasia, with a possible spread to other world regions from there. The origin of subincision in Oceania and Australia, as well as certain analogous mutilations of the female genitalia are unclear. Two possibilities exist for their genesis, that of independent invention within the heart of the Australian desert with a subsequent diffusion outward to nearby regions of Oceania, or introduction into Oceania and Australia by

25. Murdock, G.P.: *Ethnographic Atlas*, U. Pittsburgh Press, 1967.

26. Montagu, A.: "The Origins of Subincision in Australia", *Oceania*, 8(2):193-207, 1937; Montagu, A.: "Ritual Mutilation Among Primitive Peoples", *Ciba Symposium*, October 1946; Montagu, A.: *Coming into Being Among the Australian Aborigines*, Routledge & Kegan Paul, London, 1974, pp.312-25; Byrk, F.: *Circumcision in Man and Woman*, American Ethnological Press, NY, 1934; For a discussion of circumcision and subincision in Australia, see Davidson, D.: *The Chronological Aspects of Certain Australian Social Institutions, As Inferred from Geographical Distribution* (Thesis, U. Pennsylvania), Philadelphia, 1928.

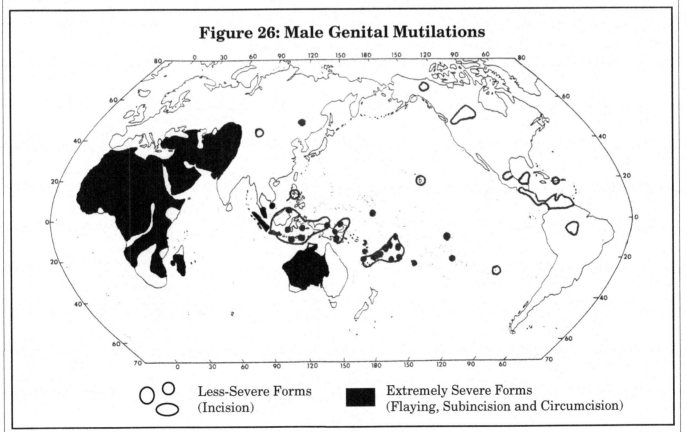

Figure 26: Male Genital Mutilations

Less-Severe Forms (Incision)

Extremely Severe Forms (Flaying, Subincision and Circumcision)

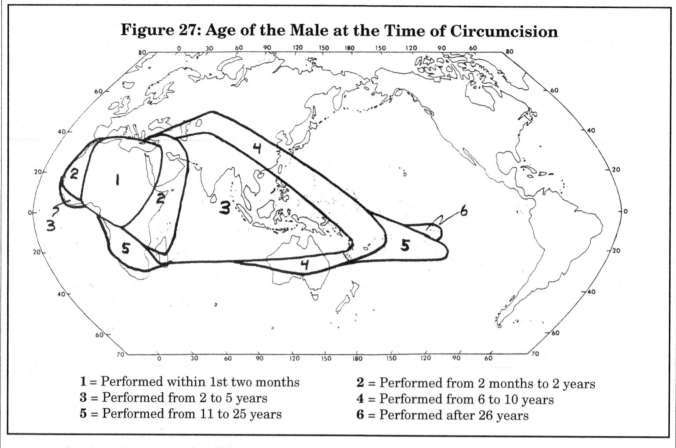

Figure 27: Age of the Male at the Time of Circumcision

1 = Performed within 1st two months
2 = Performed from 2 months to 2 years
3 = Performed from 2 to 5 years
4 = Performed from 6 to 10 years
5 = Performed from 11 to 25 years
6 = Performed after 26 years

ocean-navigating migrant peoples. Whatever, it is probable that the antisexual and antipleasure cultural attitudes by which people justify and desire the genital mutilation rituals also appeared at the same time.

Another severe form of mutilation, *genital skin stripping*, was practiced along the Red Sea coast in Arabia and Yemen at least into the 1800s. Here, in an endurance ritual performed on a potential marriage candidate, skin was completely flayed from the entire penile shaft as well as from a region of pubis. The community blessing would only be bestowed upon the young man who could refrain from screaming or fainting during the event.[27] Montagu cites the work of Elkin, who identified Northwest Australia, specifically the Kimberly region, as another location where genital skin stripping was performed.[28] Elkin believed that circumcision and subincision spread into Australia from Kimberly, diffusing east and south. Davidson, however, argued for independent development of the traits within Australia, based upon the observation that the most intense forms of subincision occurred in the center of the continent, being absent in a few border regions where only circumcision was practiced.[29]

The *Ethnographic Atlas* data also give the age at which the mutilations are customarily done. The data reveal an interesting distribution and structure, namely, an increasingly widespread distribution the older is the male at the time of the mutilation. These age distributions are seen in Figure 27. As one moves farther and farther east away from Africa, not only are the males progressively older at the time of the mutilation, but the practice itself usually becomes less harsh. In general, circumcision gives way to incision, which requires little cutting or exposing of raw flesh, as one moves eastward across the Pacific.[30] Genital mutilations were not practiced among the majority of cultures in Oceania and the Americas, and it is precisely in these regions where decorative "penis tops" were worn by native men. This suggests a similar cultural interest

27. Burton, R.: *Personal Pilgrimage to Al-Medinah and Mecca, Vol. 2*, Dover, 1962, p.110; Doughty, C.: *Travels in Arabia Deserta*, Random House, NY, 1920, p.170; Remondino, P.: *History of Circumcision*, F.A. Davis, Philadelphia, 1891, p.55; Byrk, 1934, ibid., p.137; Montagu, 1946, ibid., p.424.

28. Montagu, 1937, ibid., p.203; Also see Chapter 9.

29. The antithesis between these two contrary positions is addressed in Chapter 9; Also see Davidson, 1928, ibid.

30. Derived from examples given in Montagu, 1937, 1946, 1974, ibid.; and from Carter, N.: *Routine Circumcision: the Tragic Myth*, Londinium Press, London, 1979; Wallerstein, E.: *Circumcision, an American Health Fallacy*, Springer, 1983; Bettelheim, B.: *Symbolic Wounds*, Collier Books, NY, 1962; Loeb, L.: "The Blood Sacrifice Complex", *Memoirs, Am. Anthropological Assn.*, #30, 1923; Davidson, 1928, ibid.; Byrk, 1934, ibid.

Bas-Relief at the Tomb of Ankhmahor (6th dynasty, c.2300 BCE) at Saqqarah, Egypt. Earliest unambiguous evidence for circumcision.
(From Byrk, 1934, p.238)

in the genitalia, but only in a decorative, and presumably pleasurable sense. The presence of isolated subincision and incision in Oceania and the Amazon Basin will be discussed in detail later on. But like cranial deformation, it is extremely unlikely that these are the product of independent invention — the evidence, instead, points to their origins in those latter regions by diffusion of ocean-navigating patristic peoples from Saharasia.

The earliest unambiguous evidence for circumcision comes from Egypt's 6th Dynasty (c.2300 BCE), during the age of the pyramids. An inscription on the tomb of Ankhmahor at Saqqarah shows two young boys being subjected to the mutilation. One of them understandably has his arms being held out of the way by an older man.[31] Circumcision was fully adopted by the Hebrews only after their exodus from Egypt, and Joshua thereafter mandated it as a special mark of the Hebrew faith "to take away the reproach of the Egyptians".[32] Moses Maimonides (c.1175 CE), physician and rabbi of Cairo, articulated the reasons for circumcision which are still prevalent in the Near East:

"I believe one of the reasons for circumcision was the diminution of sexual intercourse and the weakening of the sexual organs; its purpose was to restrict the activities of this organ and to leave it at rest as much as possible. The true purpose of circumcision was to give the sexual organ that kind of physical pain as not to impair its natural function or the potency of the individual, but to lessen the power of passion and of too great desire." [33]

Such views were also adopted by the Moslems, and Mohammed proclaimed both male and female circumcision as "desirable", though not mandatory. Male circumcision spread wherever Moslem armies carried it, being nearly ubiquitous in Moslem lands, to include within Moslem parts of Oceania as a transplant (e.g., Indonesia, Bahrain). However, genital mutilations existed in some parts of Island Asia, Australia, and Oceania long before Moslem times.

Male genital mutilations were not adopted widely in Europe, Canada, European Australia, Latin America, in the Orient or among Hindus. Only a few native cultures within North and South America mutilated the genitals. The spread of the rite of infant circumcision to the United States is a recent, major diffusion which is not reflected on the maps. It gained in importance in the United States only during the 1800s, at the urging of allopathic medical doctors. In various medical journals and popular magazines of that era, circumcision was advocated as a "cure" for a whole host of childhood diseases and disorders, to include polio, tuberculosis, bedwetting, and a new "syndrome" which appeared widely in the medical literature of that time: "masturbatory insanity". Circumcision was then advocated along with a host of exceedingly harsh, pain-inducing devices and practices designed to thwart any vestige of genital pleasure in children, and to bend the child's biology and instinct completely to parental will.[34] Official medicine in the universities — and not the so-called "quacks" and natural healers — played the central role in encouraging the adoption of these mutilating "remedies"; and they continue to do so.

The distributions of male genital mutilations strongly suggest a Saharasian origin during the early Egyptian dynastic period, with several

31. Montagu, 1946, ibid.; "Passage Rites", *Encyclopedia Britanica*, 13:1050, 1982; Also see section on North Africa in Chapter 8.

32. Joshua 5:3-9.

33. As quoted in deMause, 1974, ibid., p.24.

34. Paige, K.: "The Ritual of Circumcision", *Human Nature*, May 1978.

major phases of diffusion. The adoption of circumcision by Moslems, with subsequent spread by conquest with the Moslem faith, and the more recent adoption by allopathic medical doctors in the United States, are but the most recent phases. The distribution of female genital mutilations also supports the suggestion of a Saharasian origin, as will be discussed shortly.

From the standpoint of the pain involved in circumcision as a puberty or premarital rite, the easterly dilution of the mutilations towards less painful methods, and to older ages, makes perfect sense if we also assume that the emotional attitudes and beliefs which originally mandated them were likewise diluted as they were carried eastward from a Saharasian point of origin. With the social and emotional root reasons for the mutilation becoming diluted with time and distance, less painful methods such as incision were substituted, or it was put off as long as possible, certainly well past the period just before marriage, preferably into the period of old age; or they were relinquished altogether. In some Saharasian regions, more highly institutionalized and sadistically traumatic forms of the mutilations developed, such as infant circumcision and genital skin stripping.

Male genital mutilations exist within a cultural complex where infant care is negligent, childhood anxieties are high, females are subordinated, and strong social hierarchies of one form or another are present. An overview of more recent biomedical literature, with some specific examples of the genital mutilations, will reinforce these cross-cultural findings.

In the most common male genital mutilation, circumcision, the glans-protecting foreskin is entirely removed or peeled away. In some regions (including the USA) this occurs during infancy, when the foreskin and sensitive glans penis are still bonded together with living tissue. Infant circumcision is not a "relatively mild procedure", contrary to widespread popular assertions. The foreskin must be ripped away from the entire

---

Correlation Table: Male Genital Mutilations
are positively correlated with the following other patrist variables:

By Column:
Narcissism index is high
Slavery is present
Castes are present
Class stratification is high
Land inheritance favors male line
Cognatic kin groups are absent
Patrilineal descent is present
Female barrenness penalty is high
Bride price is present
Father has family authority
Polygamy is present
Marital residence near male kin
Painful female initiation rites
Segregation of adolescent boys is high
Oral anxiety potential is high
Average satisfaction potential is low
Speed of attention to infant needs is low
Others by row: High God present, active, supportive of human morality

No negative correlations, only positive correlations, were found between "Male Genital Mutilations" and other patrist variables.

Male Genital Mutilation, Iraq c.1950

"Circumcision, although nowhere mentioned in the Koran, is generally regarded as obligatory for Moslems, following the example of the Prophet himself who was circumcised in accordance with Arab custom. No uncircumcised person may lawfully make the pilgrimage to Mecca. Among the tribes in southern Iraq, whether Madan or shepherds, the operation was often deferred till manhood, as in the present case, and was seldom performed before puberty. It was done by specialists who travelled round from village to village in the summer. Their traditional fee was a cock, but more often they charged five shillings. The examples of their work which I saw later were terrifying. They used a dirty razor, a piece of string and no antiseptics. Having finished, they sprinkled the wound with a special powder, made from the dried foreskins of their previous victims, and then bound it up tight with a rag. People living under these conditions acquire a remarkable resistance to infection, but they could not resist this, and boys sometimes took two months to recover, suffering great pain in the meanwhile. One young man came to me for treatment ten days after his circumcision, and although I am fairly inured to unpleasant sights and smells, the stench made me retch. His entire penis, his scrotum, and the inside of his thighs were a suppurating mess from which the skin was sloughing away, the pus trickling down his legs. I cured him eventually with antibiotics. In spite of the social stigma of being uncircumcised, some boys not unnaturally refused. In other cases the fathers would not allow their sons to be operated on, because there was no one else to look after the buffaloes. A few maintained that they had been circumcised by an angel at birth, a superstition that is also current in Egypt. Later I visited villages, among the Suaid and Kaulaba in particular, where I heard that hardly anyone was circumcised — almost incredible among Moslems." (W. Theisenger, *The Marsh Arabs*, E.P.Dutton, NY, 1964, pp.101-2.)

glans in a very painful procedure which leaves the glans raw and bleeding, and the infant in a state of gasping shock. The doctors generally insert a probe under the foreskin and push it around until the foreskin tissue has been peeled away from the glans. In other regions, the circumcision mutilation is performed on young boys, who must refrain from expressing emotion in a demonstration of pain-endurance. While the foreskin is usually separated from the glans by age 15, the account given above by Theisenger from the 1950s demonstrates that even pubertal circumcision (a milder practice than infant circumcision), is still accompanied by much trauma and danger.

Under conditions of sanitation, the mortality and morbidity of the mutilations declines significantly. Even so, the mutilations have potential severe and life-threatening complications, and infant boys still occasionally die from shock-trauma in modern hospitals following circumcision. According to evidence presented at a recent International Symposium on Circumcision, "routine" circumcision performed in hospitals by trained obstetricians can result in horrifying penile deformities, to include the formation of inelastic scar tissue (leading to painful or bent erections), and occasional significant or complete loss of the glans penis. Other shocking cases exist where incompetent or drunk doctors have by accident completely amputated the penis' of baby boys, leaving them without a sexual organ.[35] Such cases always prompt lawsuits, but these are nearly always settled out of court, with provisions that the full facts do not

35. NOCIRC organization coordinates public education and political action against sexual mutilations, world-wide. They publish a newsletter and maintain files on cases of extreme sexual damage or loss of the penis from botched circumcisions. NOCIRC, http://www.nocirc.org

receive exposure in the public press. Consequently, the American public rarely hears anything bad about infant circumcision.

The emotional trauma to infants and boys is also significant. Documented emotional complications can include disturbed sleep patterns, increased irritability and aggressiveness, castration complexes, and phobias.[36] One researcher clearly observed increased post-mutilation aggressiveness toward the mother.[37] Another team of researchers observed that early gender differences in response to stimuli were absent among male and female babies studied in Europe, where circumcision of male infants was uncommon. However, among American babies, where around 80% of all male infants were circumcised in the 1970s, such gender differences were present.[38]

Like the sexual abuse of children, wife-beating, or other hidden social pathologies which spring from sex-repression, the true extent of pathology and damage from "routine" circumcision has not been studied significantly.

It must be stressed that there is no scientifically valid medical indication for routine circumcision of infants or boys. This is true, even though about 60% of newborn male babies in America are currently subjected to the cruelty. Both the American Academy of Pediatrics and the American Academy of Obstetrics and Gynecology have acknowledged this fact, which has long been known in the European medical community.[39] Still, many if not most American doctors unconscionable continue to misrepresent the biomedical evidence on circumcision benefits and risks, and encourage parents to have the mutilations performed.

All theories which attempt to explain the rite based upon its assumed conferred biological or health advantages have been shown to fail completely.[40] In fact, circumcised boys are at a distinct physical and psychological disadvantage in terms of survival, particularly under the unsanitary conditions which have characterized the practices throughout most of history. Compared to other medical concerns or religious rites, very little study has been given to the motivations or psychological effects of either male or female genital mutilations, including circumcision.

Genital mutilations tend to be supported by a cultural ethos driven by powerful, but hidden psychological motives. Indeed, a dispassionate, scientific review of the question leads to a conclusion supporting nature's decision in giving baby boys a foreskin. Although the function of the foreskin may be debated, the surgical removal of it has demonstrable stressful and damaging effects upon the child. Performing the mutilations in a hospital with a surgeon's scalpel instead of in a Synagogue by the mohel's sharpened fingernails or in the bush by a "specialist" wielding a rusty razor, does not change these facts in the least. Use of anesthetic and sanitation will somewhat diminish the negative health consequences, and the age at which it is performed will have some difference in the way the trauma is perceived by the child, but none is as beneficial to the child as simply leaving the foreskin well enough alone!

Freud and other psychoanalysts have discussed male genital mutilations as inducing a form of "castration anxiety" in the child.[41] Montagu, Lewis, and Bettelheim have revealed the cognitive/social rationalizations for the male mutilations as being partly rooted in male anxiety over vaginal blood: males may imitate menstruation and the vulva through a subincision ritual, or male infants may be absolved of contact with "poisonous" childbirth blood through infant circumcision, or of contact with equally "poisonous" hymenal blood through pubertal circumcision.[42] Reich, going more directly to the core of the issue, identified genital mutilations as but one of a series of brutal and cruel acts directed toward infants and children which possessed hidden motives designed to cause

36. Brackbill, Y.: "Continuous Stimulation and Arousal Level in Infancy,: Effects of Stimulus Intensity and Stress", *Child Development*, 46:364-9, 1975; Ozturk, O.: "Ritual Circumcision and Castration Anxiety", *Psychiatry*, 36:49-60, 1973; Marshall, R., et al.: "Circumcision I: Effects Upon Newborn Behavior", *Infant Behavior and Development*, 3:1-14, 1980; Emde, R., et al.: "Stress and Neonatal Sleep", *Psychosomatic Medicine*, 33(6):491-7, 1971; Anders, T.: "The Effects of Circumcision on Sleep-Wake States in Human Neonates", *Psychosomatic Medicine*, 36(2):174-9, 1974; Carter, 1979, ibid., pp.85-93.

37. Cansever, G.: "Psychological Effects of Circumcision", *British Journal of Medical Psychology*, 38:328-9, 1965.

38. Richards, M., et al.: "Early Behavioral Differences: Gender or Circumcision?", *Developmental Psychobiology*, 9(1):89-95, 1976.

39. Thompson, H., et al.: "Report on the Ad Hoc Task Force on Circumcision", *Pediatrics*, 56(4):610-11, 1975.

40. Carter, 1979, ibid.; Paige, 1978, ibid. pp.40-8; Wallerstein, 1983, ibid.; Zimmer, P.: "Modern Ritualistic Surgery", *Clinical Pediatrics*, 16(6):503-6, June 1977; Klauber, G.: "Circumcision and Phallic Fallacies, or The Case Against Routine Circumcision", *Connecticut Medicine*, 37(9):445-8, 1973.

41. Schlossman, H.: "Circumcision as Defense: Study in Psychoanalysis and Religion", *Psychoanalytic Quarterly*, 35:340-56, 1966; Zimmerman, F.: "Origin and Significance of the Jewish Rite of Circumcision", *Psychoanalytic Review*, XXXVIII(2):103-12, 1951; Nunberg, H.: "Problems of Bisexuality as Reflected in Circumcision", *Imago*, London, 1949.

42. Bettelheim, 1962, ibid.; Montagu, 1946, ibid., pp.421-36; Lewis, J.: *In the Name of Humanity*, Freethought Press, NY, 1967.

severe genital pain, and bioenergetic-emotional contraction (armoring) of the pelvic area. Such severe genital pain and pelvic armor would, in turn, reduce the child's energy level and thwart its ability to experience full sexual pleasure later in life — thereby easing the unconscious parental/social anxiousness over sexual matters in general. Reich viewed the mutilations as a major step in the armoring process, and claimed that parents and doctors blindly advocated or performed them in proportion to their own emotional armoring and genital anxiety.[43] Parents who had been subjected to the mutilations in their own childhoods were simply incapable of tolerating intact genitals for their own children, driven by their own sadistic rage to "strike out" at their children's genitals, though generally with a thick facade of religious or medico-magical justifications. Along this line, the analyst Bettelheim has also acknowledged: "Such a strange mutilation, found among the most primitive and the most highly civilized people, and in all continents, must reflect profound needs."[44]

The psychohistorian deMause has reviewed materials on circumcision, and feels that "Such mutilations of children by adults always involve projections and punishment to control projected passions".[45] Foley has made comments of a similar nature, bolstered by observations in infant wards of modern hospitals:

> "Circumcision provides a convenient and socially acceptable outlet for the perverted component of the circumciser's libido. I have had personal experiences with the psychopathology that underlies the wish to circumcise. The pitiful wails of the suffering infant are all too often in the background for lewd and obscene commentary by the obstetrician to his audience of nurses. I have seen two medical students fight over the privilege of doing circumcision on the newborn, although these same students showed neither interest in nor aptitude for opening boils or doing other surgical tasks." [46]

Foley also recounted the actions of a brutal obstetrician who, while attending a breech (buttocks first) delivery of a male infant, actually circumcised the infant before it had cleared the vaginal canal. Of this incident, Foley remarked:

> "That obstetrician, I would say, may be capable. He may be an all-round fine fellow. But sexually I say he is a monster. And I say that one of the reasons why circumcision is so common in this country stems from the sadism of the crypto-pervert." [47]

Simply put, all forms of male genital mutilation, to include circumcision, are acts of sadistic violence against the sexuality of children. They are ancient blood rituals associated with primitive religion, and the supposed need to absolve the young male for contact with "poisonous" vaginal blood (childbirth or hymenal). The mutilations have absolutely nothing whatsoever to do with medicine, health, or science in practically all cases. The fact that so many circumcised American men and mothers, nurses, and obstetricians are ready to defend the practices in the face of contrary epidemiological evidence is a certain give-away to hidden, unconscious motives, and disturbed emotional feelings about the penis and sexual matters in general. This is not a small point, for before such a painful and traumatic blood ritual can be perceived as "good" and be championed by both high-caste priests and average people alike, certain other antisexual and antichild social factors must already be present and thriving.[48] And a number of such factors have already been identified above.

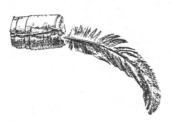

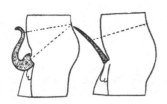

Decorative "penis-tops" worn by men in some non-mutilating cultures.

43. Reich, W.: *Reich Speaks of Freud*, Farrar, Straus & Giroux, 1967, pp.28-9.

44. Bettelheim, 1962, ibid., p.16; Recent revelations by Jeffrey Masson about Bettelheim's therapeutic sadism suggests he wrote this sentence with less than a sympathetic attitude towards the child. See Masson, J.: "Afterword to the Second Edition", *Against Therapy*, Common Courage Press, Monroe, ME, 1994.

45. deMause, 1974, ibid., p.24.

46. Foley, J.: "The Unkindest Cut of All", *Fact*, July-August, 1966, pp.3-9.

47. Foley, ibid.

48. Montagu, 1946, ibid.; Bettelheim, 1962, ibid.; Lewis, ibid.; Paige, 1978, ibid.; Carter, 1979, ibid.

4) Phallotomy and Eunuchism

Two additional aspects appear to have a relationship to male genital mutilations: phallotomy and eunuchism. Both of these practices constitute extreme examples, and involve surgical removal of the entire male genitalia. Historically, eunuchism was most often forced upon young boys who, as captured slaves, would supply the "needs" of the Arab and Turk harem system, which also predominated within Saharasia. My incomplete review of the literature indicates that phallotomy and eunuchism, while present in many regions worldwide, were most strongly institutionalized in the Near East. The effect of such practices on cultures where the harem/eunuch system prevailed must have been tremendous.[49] Young boys even casually threatened with such castration would easily have been made very obedient.

Remondino estimated, for instance, that 35,000 young African slave boys were forcibly and completely castrated each year by Coptic Monks in the Sudan to supply the harem systems of the Near East. The mortality at these eunuch factories was a shocking 90%, but the price such a eunuch boy would command in the slave markets was great enough that the practice continued.[50] The use of phallotomy or castration as a punishment, by both individuals and by the state, was widespread and institutionalized across the Near East, and appears today in these same regions as a common atrocity during times of war, or as a punishment or vengeance against males who commit sexual crimes, usually the seduction of a taboo female.

A form of modern day eunuchism also exists on a small scale today in the West, in the form of "sex-change" operations performed on homosexual men in hospitals. However, the surgical cutting and/or removal of the sexual organs of women — vagina, breasts, uterus — for highly questionable "medical" reasons, is also rampant in Western hospitals, and this may also be considered in a similar category. More details and documentation will be given below.§

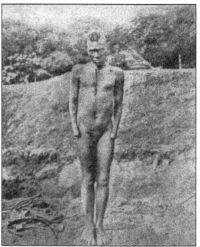

Azandeh man, mutilated as punishment for adultery.
(from Byrk, 1934)

§ Leaders and members of the recent "Heaven's Gate" mass-suicide cult were also found to have committed self-castration, a practice more widespread among both contemporary "mainstream" religions and religious cults than has heretofore been recognized. (see for example: Favazza, A. & B. Favazza: *Bodies Under Siege: Self-Mutilation in Culture and Psychiatry*, John Hopkins Univ. Press, Baltimore, 1987.)

5) Female Genital Mutilations

Female genital mutilations are generally harsher, more painful, and more life-threatening than those performed upon the male. While the ritual of male mutilation is often a doorway to a higher social status for the male, allowing him access to marriage and sexual activity, the female mutilations hold no such social benefits. Additionally, more so than male genital mutilations, the female mutilations often destroy entirely any capacity to obtain sexual pleasure whatsoever. Usually, they are specifically and consciously designed to do just this: To destroy the woman's ability to experience full sexual pleasure — the ritual's often-stated, explicit goal.[51]

Female genital mutilations are found to have a distribution very similar to, but not as widespread as male genital mutilations, but tend to occur among the same peoples, notably in North Africa, as demonstrated in Figure 28. The map of this variable was constructed primarily from data presented in Fran Hosken's work *The Hosken Report on Sexual/Genital Mutilation of Females*,[52] but with an additional number of examples taken from the anthropological literature.[53]

Unlike male genital mutilations, the history books have little to say about the female mutilations, so the discussion here will be confined to more recent observations.

49. See Chapters 8 & 9 for additional documentation. Also: Bonaparte, M.: "Uber die Symbolik der Kopftrophaen", *Imago*, 14:100-40, 1928; Coser, L.: "The Political Functions of Eunuchism", *Am. Sociological Rev.*, 29:880-85, 1964; Hopkins, K.: "Eunuchs in Politics in the Later Roman Empire", *Proceedings, Cambridge Philological Soc.*, 189, 1963; Spencer, R.: "Cultural Aspects of Eunuchism", *Ciba Symposium* 8, 1946; Byrk, 1934, ibid., pp.107-9.

50. Remondino, 1891, pp.98-9.

51. Hosken, F.: *The Hosken Report on Genital and Sexual Mutilation of Females*, 2nd Edition, Women's International Network News, Lexington, Mass., 1979; Hosken, F.: "Genital Mutilation of Women in Africa", *Munger Africana Library Notes*, 36, October 1976: Morgan, R. & G. Steinem, "The International Crime of Genital Mutilation", *Ms*, March 1980, pp.65-7,98,110.

52. Hosken, ibid.,1979.

53. Montagu, 1937, ibid.; Montagu, 1974, ibid., pp.312-25; Byrk, 1934, ibid.

Figure 28: Female Genital Mutilations[52, 53]

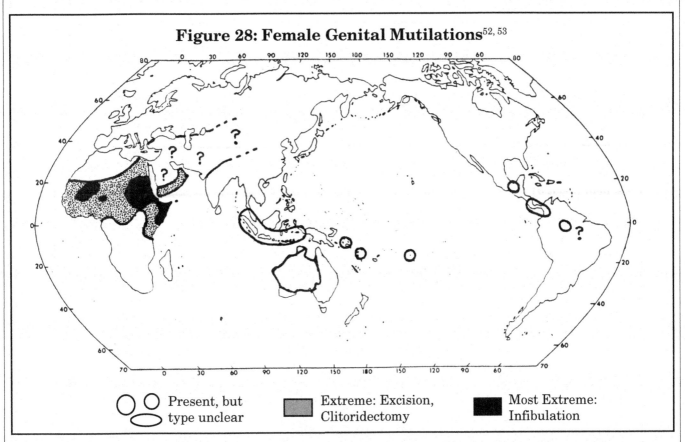

| ⬭ ◯ ◯ Present, but type unclear | ▨ Extreme: Excision, Clitoridectomy | ■ Most Extreme: Infibulation |

From many eyewitness accounts, it is known that the female mutilations are generally performed by a grandmother or *daya* (unskilled midwife) on the unanesthetized young girl. Unlike the male circumcision ritual, there is no ceremony or "coming out" ritual involved. The women of the cultural unit gather about the poor girl to chant and beat drums and tambourines, largely to drown out her hysterical screaming. She is held down by many of the village women while the daya, razor in hand, sets to work upon her genitals. I have quoted one eyewitness account at length to remove any doubt in the reader's mind of the traumatic and antisexual nature of the rite, which is not performed by males upon females, but rather by the older women upon the young girls. Indeed, the older grandmothers are among the greatest enthusiasts for the mutilations, insisting they be done "properly", meaning that the maximum amount of cutting be done, to do maximum damage to the young girl's sexuality.

With *female circumcision*, only the hood of the clitoris may be cut away, but usually the clitoris is removed as well (*clitoridectomy*). Accounts from the literature indicate that the distinctions between the two are blurred, and a good deal of additional cutting may be done if the daya is especially concerned about the "morals" of the child. With *infibulation*, the severest female genital mutilation, the clitoris, labia majora and labia minora are cut away with the razor, following which a "handful" of soft tissue may be sliced away from the exterior vaginal opening. After this, the young girl's legs are bound together with rope. She is left immobilized for months, with a hollow reed inserted for micturition, until the vaginal walls grow together to form a sheaf of skin which covers the vaginal opening. In this way the girl's virginity is ensured.

After marriage, the infibulation is cut open (*defibulation*) by the daya, just enough for intercourse; this is usually done on the wedding night when a husband's intercourse with the new bride is mandated. Subsequent sexual relations are impeded by scar tissue, a generally too-small

Village women splash water on a 7-year old girl who has just suffered through the ritual cutting out of her clitoris.
(ASIPA Press)

Female Genital Mutilation, Somalia c.1970

"The child, completely naked, is made to sit on a low stool. Several women take hold of her and open her legs wide. After separating her outer and inner lips, the operator, usually a woman experienced in this procedure, who sits facing her, with her kitchen knife pierces and slices open the hood of the clitoris. Then she begins to cut it out.. While another woman wipes off the blood with a rag, the operator digs with her fingernail a hole the length of the clitoris to detach and pull out the organ entirely. The little girl, held down by the helpers, screams in extreme pain; but no one pays the slightest attention.

The operator finishes this job by entirely pulling out the clitoris, and then cuts it to the bone with her knife. Her helpers again wipe off the spurting blood with a rag. The operator then removes the remaining flesh and digs a hole with her finger to remove any remnant of the clitoris amidst the flowing blood. The neighbor women are then invited to plunge their fingers into the bloody hole to verify that every piece of the clitoris is removed.

This operation is not always well managed, as the little girl struggles. It often happens that by clumsy use of the knife, or a poorly-executed cut, the urethra is pierced or the rectum is cut open. If the little girl faints, the women blow pili-pili (spice powder) into her nostrils.

But this is not the end. The most important phase of the operation begins only now. After a short moment, the woman takes the knife again and cuts off the inner lips (labia minora). The helpers again wipe the blood with their rags. Then the woman, with a swift motion, begins to scrape the skin from the inside of the large lips.

The operator conscientiously scrapes the flesh of the screaming child, without the slightest concern for the extreme pain she inflicts. When the wound is large enough, she adds some lengthwise cuts and several more incisions. The neighbor women carefully watch her 'work' and encourage her.

The child now howls even more. Sometimes, in a spasm, the children bite off their own tongues. The women carefully watch the child to prevent such an accident. When her tongue flops out, they throw spice powder on it, which provokes an instant pulling back.

With the abrasion of the skin completed, according to the rules, the operator closes the bleeding large lips and fixes them one against the other with long acacia thorns.

At this stage of the operation, the child is so spent and exhausted that she stops crying, but often has convulsions. The women then force down her throat a concoction of plants.

The operator's chief concern is to achieve as small an opening as possible, just big enough to allow the urine, and later the menstrual flow, to pass. The family honor depends on making the opening as small as possible because for the Somalis, the smaller this artificial passage is, the greater the value of the girl and the higher the brideprice.

When the operation is finished, the woman pours water over the genital area of the girl, and wipes her with a rag. Then the child, who was held down all this time, is ordered to get up. The women then immobilize her thighs by tying them together with ropes of goat skin. A bandage is applied from the knees to the waist of the girl, which is left in place for about two weeks. The girl must remain stretched out on a mat for the entire time, while all the excrement evidently remains with her in the bandage.

After that time, the girl is released and the bandage is cleaned. Her vagina is now closed, and remains so until her marriage. Contrary to what one would assume, death is not a very frequent result of this operation. There are, of course, various complications which frequently leave the girl crippled and disabled for the rest of her life."

(J. Lantier, *La Cite Magigue et Magie en Afrique Noire*, Librarie Fayard, Paris, 1972, pp.277-9.)

Correlation Table: **Painful Female Initiation Rites**
are positively correlated with the following other patrist variables:

By Column: Protection of infants from the environment is low
Segregation of adolescent boys is high
Male genital mutilations are present
Patrilineal descent is present
Killing, torturing, mutilation of the enemy is high

Neither Murdock's *Ethnographic Atlas* nor Textor's *Cross Cultural Summary* contained specific
data on female genital mutilations, but the *Summary* did have this one variable on "painful rites"
which is pertinent to the discussion here. No negative correlations were found between "Painful
Female Initiation Rites" and other patrist variables.

vaginal opening, the defibulation wound, and by psychological traumata on the part of both husband and wife. Childbirth for the once infibulated women is very difficult, given the amount of inelastic scar tissue which forms in the vagina. The woman in labor is generally cut open even more by the daya's knife to provide a passageway for the child; after childbirth, her vagina may be re-lacerated and her legs again bound to reform the infibulation. Women in such cultures are not only usually denied contraceptive information and devices, but even a basic education on matters of hygiene. Deprived of moral or legal rights, they live a life of total subordination on basic matters of living, punctuated by horrifying rituals of infibulation and defibulation.[54] Mortality and morbidity of such women, during childbirth and afterward, is thereby significantly increased. Infections, hemorrhage, shock, fistulas, and even death are not uncommon following such mutilations.[55]

A girl who attempts to avoid the mutilation is socially ostracized, the stigma of possessing an intact genital being enormous. Few girls escape it, as to do so would require running away and prevailing laws usually empower her relatives to bring her back by force. In the regions where the rite prevails, uncircumcised women are considered "unclean"; to call a woman "uncircumcised one", or a man the "son of an uncircumcised woman" is an insult of the most extreme proportions.

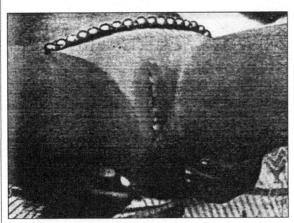

Infibulated Danakil girl.
Note the absence of normal labia
and complete closure of vagina.
(Widstrand, 1965)

The defense of female genital mutilations by native Africans, and the criticism of them by Europeans and native African reformers, have been central issues of various African political independence and social reform movements in the 1900s. Jomo Kenyata rose to power in British East Africa (renamed to *Kenya* in later years) only after the British colonials passed laws forbidding the mutilations.[56] Native Africans who could tolerate British plantations and colonial rule were outraged to open revolt over the law, which meddled intolerably in private family matters. But even after the British departed, the mutilations remained controversial. Indeed, in 1983, President Moi of Kenya finally overturned the long-standing policy of his predecessors and enacted a ban against female circumcision, ordering police to arrest and charge with murder anyone who performed a circumcision which results in death of the girl. Just prior to his action, 14 young girls had died from hemorrhages.[57] Similar laws against female genital mutilation are being proposed, or have already passed, in many sub-Saharan nations, though such reforms progress more slowly in Islamic regions.

Table 4 gives the percentage of women in various nations who were, according to the best available evidence, subjected to the mutilations in

54. Widstrand, C.: "Female Infibulation", *Studia Ethnographica Upsaliensa*, XX:95-122, 1965; Montagu, A.: "Infibulation and Defibulation in the Old and New Worlds", *Am. Anthropologist*, 47:464-7, 1945; Hosken, 1976, 1979, ibid.; Montagu, 1937, 1946, 1974, ibid.; Byrk, 1934, ibid.

55. Mustafa, A.: *J. Obstet. Gynaec. Brit. Cwlth.*, 73:302-6, 1966; Hosken, 1979, ibid.; Widstrand, 1965, ibid.

56. Murray, J.: *The Kikuyu Female Circumcision Controversy*, Diss., UCLA, 1974; Cf. Kenyata, J.: *Facing Mt. Kenya*, Vintage, NY, 1965.

57. "Mid Revolutionary Mores", *Ms*, May 1983, p.28.

the late 1970s.[58]

In the author's view, the descriptions of the infibulation of young women in North Africa are similar in character to the descriptions of forcible castration of young boys for the eunuch trade. In a similar manner, girls were once infibulated to insure their virginity when presented for the Near Eastern harem slave trade. *The regions which today practice infibulation were once primary "capture zones" for the Arab slave trade in both infibulated girls and castrated boys.* The use of eunuchs has died out over the last 100 years with the decline in large harems, but female infibulation and other forms of female genital mutilation persist in accordance with a powerful and hysterical virginity taboo.

Female genital mutilations are also found in Australia and stretch across Oceania into the Amazon, as well. Cultures that subincise men in Australia tend to also *introcise* women, that is, to make a cut into the vaginal wall, similar to an episiotomy, to enlarge the vaginal passage. This latter mutilation was also observed in parts of the Amazon.[59] Discussion of such practices in Oceania and the New World will occur later.

Postscript, 1997

Since the time when the above section on female genital mutilations was firstly researched and written, over 10 years ago, there has developed a widespread social movement across Africa and the Middle East, supported by both grassroots women's organizations and some governments, to legally abolish and end the mutilations.

Table 4: Where Females are Subjected to Genital Mutilations (Hosken c.1979)[58]

AFRICA:	% of women	number of women
Djibouti	100%	75,000
Somalia	100%	1,620,000
Ethiopia	90%	10,890,000
Mali	80%	2,520,000
Sierra Leone	80%	1,200,000
Sudan	80%	6,560,000
Ivory Coast	70%	2,550,000
Upper Volta	70%	2,270,000
Guinea Conakry	60% est.	1,350,000
Kenya	60%	4,260,000
Egypt	50%	10,000,000
Gambia	50% est.	130,000
Nigeria	50%	25,000,000
Senegal	50%	1,270,000
Ghana	30%	1,590,000
Tanzania	25%	2,500,000
Togo	25%	310,000

PRESENT BUT PERCENTAGE UNKNOWN

Algeria	Indonesia	Oman
Bahrain	Liberia	Saudi Arabia
Benin (north)	Libya	South Yemen
Cameroun (north)	Malaysia	Tunisia
Central Africa	Mauritania (south)	Uganda
Chad	Morocco	United Arab Emirates
Guinea Bissau	Niger	Western Sahara
		Zaire (north)

"I lost two little girls, one after the other. The first time, I observed the traditions, with the old women. The child bled and died. The second girl had the excision done in the hospital, but they could not stop the bleeding."

— African mother
(quoted in *People*, 6:1, 1979)

"A woman was brought into my care. In the bedroom of the newly-wed there was also an old woman; she said that the bride was not 'clean', she still had some clitoris left. The old woman cut it off, after which the husband immediately deflowered his bride. She arrived here bleeding profusely."

— Midwife, Mali, Africa
(quoted in *People*, 6:1, 1979)

"Because of the scars, which are often very hard and completely unyielding, the woman is completely torn [during childbirth] when pushing. The vagina is often obstructed by a membrane, the expulsion is very slow, the baby suffers and we have to perform a wide upward episiotomy"

— Midwife, Mali, Africa
(quoted in *People*, 6:1, 1979)

58. Hosken, 1979, ibid., pp.2-6 of section on "Geographic Overview".
59. Montagu, 1937, ibid.; Montagu, 1974, ibid., pp.312-25; Byrk, 1934, ibid.

6) Unnecessary Hysterectomy and Mastectomy: Ritual Medical Castration and Sexual Mutilation of Girls and Women in Western Hospitals

As a final note, there are practices in the West which must be mentioned here, as they possess a profound resemblance to female genital mutilations in that the sexual organs of the female are ritually cut (mutilated) or actually removed without a scientifically demonstrated health-related indication. For example, there are around 700,000 hysterectomies performed each year in the USA, the overwhelming majority of which — upwards of 90% — are not necessary, and offer no scientifically demonstrated health benefits. As such, the practice of hysterectomy in the USA can accurately be identified as a form of ritualized *female castration*.[60] Regarding childbirth, Caesarean sections are now performed on approximately 25% of all American mothers.[61] The ritual episiotomy cut, also an unnecessary and scientifically undemonstrated "necessity" of medicated and induced hospital birth, is probably performed on as many as 80% of all laboring women who deliver vaginally. Unlike in Africa, these Western forms of female genital mutilation are predominantly performed by males, within the high caste of medical obstetrics and gynecology.

As determined from controlled studies on birth outcomes by the training of the practitioner, it is known that *these and other surgical interventionist-procedures are not necessary in the overwhelming majority of cases, and have no scientific basis for their routine application* .[62] These facts, about the *lack* of a scientific or health-related basis for the obstetrical/surgical mutilations of female sexual organs in hospitals, demand an analysis and understanding going beyond the self-serving platitudes and excuses commonly offered up by the doctors themselves. Certainly the economic motivation can explain a lot— the more cutting which is done, the more surgical fees are earned by both doctor and hospital; and the incidence of these mutilations declines where medical insurance refuses to pay for them, or in poor areas where women often do not have medical insurance — but there must be other, more profound factors at work which determine why it is mainly a certain sub-group of medical doctors who are so enthusiastic to undertake this line of "work", and why so many females willingly line up to have the mutilations performed.

As with the case of the genital mutilations of infants and children of both sexes, the sexual mutilations of adult women are primarily undertaken for *hidden psychological and emotional motivations,* held by both the consenting women and the surgical specialists. The reasons are certainly rooted in culture-wide anxieties and hatreds regarding sexuality and female reproductive functions. Given that the medical group which performs these unnecessary mutilating surgeries upon women is generally the same group that performs equally unnecessary genital mutilations upon baby boys, a similar psychological motivation — deeply-buried sadistic hatred of sexual functions — appears likely. Indeed, circumcisions and hysterectomies are today the most commonly-performed surgeries in the USA.

To this consideration can also be added the "routine" amputation of the breasts of women in hospitals for the presence or *mere suspicion* of cancerous tissue — or worse (as discussed below), the newspeak Orwellian surgery of "preventative mastectomy", performed *before* a woman or girl has any signs of cancer, in order to "prevent cancer"!

While the scope of these concerns goes well beyond the present discussion — namely to the topic of alternative treatments for cancer, and the very active and socially-enthusiastic use of policemen to suppress the

60. Stokes, N.M.: *The Castrated Woman: What Your Doctor Won't Tell You About Hysterectomy*, Franklin Watts, NY 1986; Hufnagel, V.: *No More Hysterectomies*, San Diego, 1990; West, S.: *The Hysterectomy Hoax: Why 90% of all Hysterectomies are Unnecessary*, Doubleday, NY 1994.

61. Corea, G.: "The Cesearean Epidemic", *Mother Jones*, July 1980.

62. Stewart, D., Ed.: *Five Standards of Safe Childbearing*, NAPSAC, Marble Hill, MO 1981.

alternatives to medical mutilation[63] — such officially-approved but scientifically fraudulent "cancer treatments" used in modern hospitals ("preventative mastectomy") are clearly indicative of an *urge to mutilate* the sexual organs of women.

Rather than advise so-called "high-risk women" to avoid ingesting cancer-promoting tobacco products, alcohol, foods and food additives, or take jobs where workplace exposure to known carcinogens are reduced, or move to locations away from nuclear power plants or other major sources of air and water pollution, or to address their family patterns of childrearing and sexual attitudes (which probably underlie the inherited components of cancer), these smooth-talking sadistic medical butchers instead

63. DeMeo, J.: "Anti-Constitutional Activities and Abuse of Police Power by the US Food and Drug Administration, and Other Federal Agencies", *Pulse of the Planet,* 4:106-113, 1993. Also see the Editor's Notes: "Genetics Does Not Equal Heredity", "Modern Horrific Medicine", "Preventative Mastectomy", and "The Mammogram" in this same issue of *Pulse*, pp.161-164.

"Preventative" Mastectomy or Official Medical Quackery?
The Psychological Dynamics of American Sexual-Mutilation Surgeries are Indistinguishable from African Sexual-Genital Mutilations

The sadistic butchery of cancer surgeons is most clearly revealed in those specialized groups who mutilate and disfigure healthy young women, often teenage girls, who do not have a trace of cancer in their bodies — only the "theoretical" traces of unobservable "cancer genes", or textural changes in breast tissue. I speak here of the growing practice of "preventative mastectomy", in which the surgeon firstly identifies an increased tendency for breast cancer (or breast surgery) in a given family line, and secondly, based upon an unwavering faith in his genetic calculations, diagnoses those women, often young girls, as being "at high risk for breast cancer". ...

This editor first heard of "preventative mastectomy" more than 20 years ago, when it was heralded by medical journalists as an "advanced breakthrough" in the treatment of cancer. Then, it was isolated to a few doctors in training hospitals in New York City. Today, it is nationwide Big Business, with thousands of women visiting surgeons across the nation, to be evaluated by quack genetic calculations. In one major hospital, Memorial Sloan Kettering, over 150 "preventative mastectomies" are performed each year, some 20% of the total number of breast surgeries performed. Girls so evaluated as "high risk" are apparently persuaded — by the entire medico-magical "health-care" system, and by their typical sex-anxious families — into having their sexual organs amputated. No confirmed traces of cancer need be present.

Lest you think your editor is making a sick joke here, or exaggerating, I point to a recent National Public Radio "Talk of the Nation, Science Friday" broadcast (21 August 1992) wherein two surgeons, one male and one female, spoke for nearly two hours about the new "breakthrough procedure of preventative mastectomy". I was totally shocked. The doctors boasted of having "saved" several thousand young girls from the potential ravages of cancer. The doctors also claimed to be part of a larger group of "preventative" surgeons across the USA. Therefore, it appears that virtually thousands of these operations must be performed each year. And worse, about a half-dozen young women called in to the radio program, most of whom already had the mutilation performed, profusely thanking the doctors for "saving" them!!

Upon listening to the program, I was struck by how similar is our American medicine to the most sadistic, bloody and superstitious tribe you could name. Ritual female genital mutilations are occurring here, in our most modern hospitals, and all with the emotional collusion and agreement by the women upon whom it is perpetrated. Indeed, the women who had just undergone the mutilation appeared to be its biggest advocates, just as many women now demand Caesarean births and hysterectomies so they won't have to be "bothered" with natural functioning and sexual feeling. (J. DeMeo, *Pulse of the Planet,* 4:163-164, 1993)

brainwash susceptible (read: *emotionally castrated*) women into life-ruining and unnecessary surgical mutilations. On one television program, I witnessed how this butchery is socially approved and supported, in a manner which appears similar to what happens in Africa when a young girl is taken for her genital mutilations.

A 35 year-old mother, a most angry, obese and unsexual woman who previously herself had consented to a double "preventative mastectomy", was observed taking her completely healthy, but emotionally cowed and frightened 16-year old daughter into the hospital for a "heart to heart" talk with a dominating, self-assured male sadist-surgeon. Sadly, he talked the frightened young girl into having both breasts amputated. A "preventative", of course, with no other options or possibilities ever discussed! We must ask, *how different is this from the situation in Sudan or Somalia, with their female genital mutilations? Psychologically, emotionally, the situations are identical! Except that here, in the USA, our genital-sexual mutilations of women, performed by white-coated smooth sadists, are covered over with a facade of "medicine", masquerading as defendable scientific progress. That, it most certainly is not!*

In another case I personally know of, a young woman living in a small town, a student of mine where I formerly was a professor, was "diagnosed" by the town's "most highly respected" obstetrician as having a high risk of breast cancer, and advised to have a "preventative mastectomy" — mainly because her mother and sister had cancer and mastectomies, *as diagnosed and treated by this same doctor*. This woman refused the recommended "preventative" surgery, but consented to the doctor's insistence upon *multiple mammograms, given every few months!* Clearly, if she did not consent to pre-cancer surgery, he would do his best to induce a cancer in her by excessive exposure to dangerous x-radiation![66]

No doubt, the incomes of the women, or their medical insurance, was a major determinant in the diagnosis and recommendation for the medical mutilations, which are quite costly. San Francisco, for example, has one of the highest rates of "breast cancer" surgery in the USA, reflecting without question the high rates of insurance and public assistance money available to fuel the expensive surgeries. A similar argument may be leveled against the widespread encouragements to women regarding silicone breast "implants" — that the normal breast is in need of "surgical correction", in this latter case often for the most obtuse and narcissistic of reasons.

Again, it is the same sub-groups of high-caste medical priests who are advocating and performing so much of the unnecessary, life-debilitating mutilations of sexual organs. Other medical physicians may not be so directly culpable, but as a profession, and with only a few notable exceptions, they collectively participate in the *conspiracy of silence* by which the public is led to believe there are no reasonable alternatives to mutilation.

Financial greed alone cannot explain this phenomenon. In general, I have observed that the doctors who most strongly advocate cutting and surgery on women fall into two groups. Firstly, there are hardened, emotionally dried-up and sexually-dead individuals, and secondly, phallic-aggressive types, who might superficially be considered "sexually vital"

Deep-Organ Surgery on Babies... *Without Anesthesia!*

The midwifery/homebirth magazine *Birth* recently published an exposé documenting deep organ surgery routinely carried out on infants in modern hospitals, without the use of anesthetics. Since the physicians had grown accustomed to cutting up animals without anesthetics, and performing circumcisions upon struggling baby boys without anesthesia, they "reasoned" that infants had a "poorly-developed nervous system", and could therefore "withstand major surgery without distress". Deaf and dumb to the shrieks and gasps of tortured, struggling infants, strapped to special surgery-boards (eg., *torture-racks*, as with the "circumstraint" used for genital mutilations), these doctors would simply start cutting away. When the revelations were made public, prompting the rational outrage of non-physicians, both the American Academy of Anesthesiologists and the American Academy of Pediatrics firstly tried to defend these practices, but eventually they passed an admit-no-wrong resolution stating "...there now exists a theoretical consensus... that infants feel considerable pain". *Theoretical?!* The above episode did not occur in the Middle Ages, but rather, in 1988!

(*from* "Infant Medical Experiments", *Pulse of the Planet*, 4:163, 1993; cf. *Birth*, 15:36-41, March 1988.)

---

66. *"Our estimate is that about three-quarters of the current annual incidence of breast cancer in the United States is being caused by earlier ionizing radiation, primarily from medical sources"*, John Gofman, M.D., Ph.D., formerly employed by the Atomic Energy Commission, but fired in the 1970s after co-authoring the book *Poisoned Power*, which criticized the nuclear industry; Quotation from: Gofman, J.: *Preventing Breast Cancer: The Story of a Major, Proven, Preventable Cause of This Disease*, Committee for Nuclear Responsibility, 1996, PO Box 421993, San Francisco 94142.

due to a higher energy level — "sexual predator" might be a better descriptive term. Both types possess a smothered hateful anger towards natural sexuality and women which is often only barely disguised; the major difference is that the former character generally has a lower energy level, while the latter has much more energy, requiring a more developed facade, which can be very smooth or even "handsome" and seductive to the average person. Only when one takes a sharp and critical look at what these sexual monsters actually *do* on a given day — severe and life-destroying mutilations of healthy babies, women, etc. — does the gaping chasm between high-sounding rhetoric and reality, golden words and dirty deeds, become clear.

On the other hand, the doctors I have observed who advocate the least amount of cutting, or who have pioneered non-toxic alternatives to surgery, were always softer, emotionally animated, gentle individuals who had retained a loving attitude towards the opposite sex. They were not sexually inert, emotionally dead-pan, nor phallic-narcissistic nor sadistically aggressive in character.

The point here is, the same hateful and sadistic antisexual emotions underlie the genital-mutilating impulses of both the primitive, uneducated daya and the university educated "high-tech" obstetrician-surgeons. In both high-technology and "primitive" cultures, children and women can be emotionally manipulated and coerced into the subordinated position of *mutilation victim*. Later, they may become *mutilation enthusiasts*, having pushed the horrible memories into unconsciousness. Dissenters (either dissenting health practitioners or dissenting potential mutilation victims) are subjected to family and priestly-caste pressure and persuasion, to submit to them, or to support or even perform them. In the most extreme cases, dissenting victims are hunted down by policemen when they try to escape, for daring to challenge the widespread and lucrative system of culturally-mandated sexual mutilations. Both are supported or even encouraged in their sadistic butchery by the broader antisexual cultural framework in which they function. Without consent and approval from the masses for what they do, the mutilators in all nations would have long ago been put into prison for their deadly life-destructive activities.

In short, there is little fundamental difference between the sadistic butchery of sexual mutilations carried out in modern Western hospitals by obstetricians and surgeons, and the sadistic butchery of sexual mutilations carried out in the bush by primitive dayas or witch-doctors. Neither can provide credible scientific support for what they do, and both are the creators of untold human misery.

7) Scarification of the Body

In addition to the above factors already discussed, there are numerous reports in the anthropological literature of cultures which inflict painful, life-threatening wounds and cuts on their own bodies, or on the bodies of their children, in ways which interfere with important body functions: finger-amputation, tooth extraction (of healthy upper or lower incisors), tooth filing (to sharp points), deep facial cuts and other types of extreme skin scarification, extreme lip stretching, neck-stretching and so forth. In each case, the major affect of the activity was to inhibit the individual from full biological capabilities. I *do not* place these extreme scarification traits into the benign category of *body-decorations*, which includes tattooing and ear-piercing. Unfortunately, there was no time to make a systematic review of scarification traits — but for future reference, my research suggests they existed mostly in the more highly armored patrist cultures.

**Sexual Mutilations
of the American M.D.
(Short List):**

Preventative mastectomies.
Breast implants for neurotic-erotic "enhancements".
Male infant circumcision.
Female genital mutilation as performed in secret by Muslim physicians.
Female genital "shaping".
Genital piercing, nipple-rings, tongue-spikes, etc.
Routine hysterectomy.
Routine caesarean childbirth.
Episiotomy.
"Sex-change" surgery on confused young people.
Prostate surgery of older men where erectile nerves are deliberately cut.
Sexual desire-destroying or impotence-inducing "medications" given to people suffering from depression.

Some of these procedures and medications *might* be justified in a few cases, or temporarily, but overwhelmingly they are scientifically unwarranted.

Adolescents

8) Female Premarital Sex Taboo

*Children of the future Age
Reading this
    indignant page,
Know that in a former time
Love! sweet Love!
    was thought a crime.*
— William Blake

67. Murdock, 1967, ibid.

Where the arranged marriage system prevails, the sexual interests of children, particularly females, are subordinated to the cultural and economic interests of the parents. For children to accept an arranged marriage without undue protest, a high level of obedience training and sex-repression is required from a very early age. Young children and adolescents must, to accomplish this goal, be prevented from having sexual contact with each other, and adolescent romances must be crushed. Such repression nearly always falls more heavily upon the young female than the young male, who is generally granted more leniency for breaking the sexual taboo.

In sex-repressive cultures, young females are taboo objects, their virginity of considerable pre-occupation to parents; loss of virginity before marriage may be so "dishonorable" to the family in some regions the offending girl will be severely punished, or even put to death by her own brother or father. Where the father's selection or approval of the girl's marriage partner is absolute, the virginity taboo is also quite severe. The premarital sex taboo is central to the sex-economic interpretation of the origins of adult violence, as well as to the general subordination of the child to adult will, and subordination of the female to the male. Repression of the adolescent sexual drive is anticipated to occur in cultures dominated by punitive, militaristic, hierarchal and mystical tendencies, with high levels of adult violence and sexual anxiety. These predictions were clearly borne out by the cross-cultural data.

Figure 29 is a map which reflects prevailing taboos regarding sexual activity by young, unmarried females, as constructed from *Ethnographic Atlas* data on the behavior of aboriginal, subsistence-level peoples.[67] The map identifies the relative frequency of premarital sexual encounters by children, adolescents and the unmarried. Data on this variable were not

---

Correlation Table: Female Premarital Sex Taboos
are positively correlated to the following other patrist variables:

By Column: Narcissism index is high
Bellicosity is high
Class stratification is high
High God present, active, supportive of human morality
Organized priesthood is present
Patrilineal descent is present
Castration anxiety is high
Sex disabilities are present
Child sexual anxiety potential is high
Child anal satisfaction potential is low

Others by Row: Child sexual satisfaction potential is low
Child anal anxiety potential is high
Marital residence near male kin
Sex anxiety is high
Castration anxiety is high

No negative correlations were found between "Female Premarital Sex Taboos" and other patrist variables.

available for some areas of the world (notably, across much of Asia), but the following patterns are clear nevertheless: Saharasia and its adjacent borderlands contain the largest single contiguous grouping of cultures with strong taboos against premarital sexual behavior (regions marked with the number "10"). In opposition to this observation, South Asia and Africa, Oceania and the Americas have the largest groupings of cultures uniformly without premarital sex taboos (regions marked with a zero).

Block regions with a high number are characterized by a high percentage of cultures with strict virginity taboos, where premarital sexual activity by unmarried girls is prohibited, strongly censured, and in fact rare. Regions with a low or zero value are characterized by a low percentage of cultures with such taboos or the complete absence of taboos, where the premarital sexual activity of young girls is freely permitted without censure, censured only if pregnancy results, confined within monogamous trial marriages, or prohibited but weakly censured and not infrequent. The variable appropriately focuses upon rules to control the sex life of the female, given the rather ubiquitous presence of the sexual "double standard", wherein males are often not penalized for premarital sexual activity even when females are.

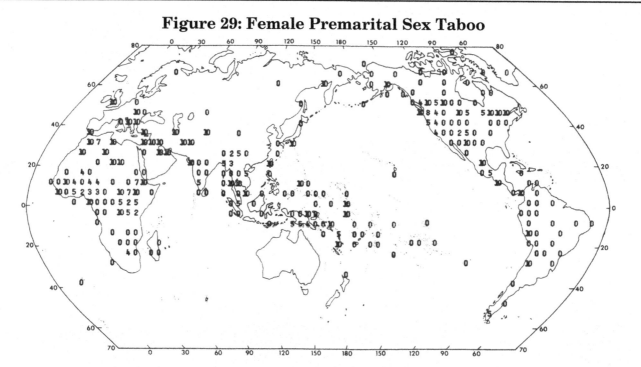

Figure 29: Female Premarital Sex Taboo

"10" identifies regions where 91-100% of cultures have a strong Premarital Sex Taboo (insistence upon virginity). Numbers indicate the percentage of cultures within a given 5° x 5° region possessing a strong taboo (number times 10 = top end of 10-point percentage category). Regions marked "0" have no cultures with such a taboo, while 1 = 1%-10%, 2=11-20%, 3=21-30%, etc. Blank areas = data not available.

Key to numbered data on maps in this Chapter:
10 = 91% - 100% of cultures in the given 5° x 5° region possess the given variable.
9 = 81% - 90% of cultures possess the given variable.   **8** = 71% - 80% of cultures possess the variable.
7 = 61% - 70% of cultures possess the variable.   **6** = 51% - 60% of cultures possess the variable.
5 = 41% - 50% of cultures possess the variable.   **4** = 31% - 40% of cultures possess the variable.
3 = 21% - 30% of cultures possess the variable.   **2** = 11% - 20% of cultures possess the variable.
1 = 1% - 10% of cultures possess the variable.   **0** = 0% of cultures possess the variable.
Blank areas = data not available.

9) Segregation of Adolescent Boys

In a society where older males dominate the sexual favors and marriage choices of women, female virginity taboos are strengthened, and the separation of young children of opposite sex (to prevent contact) becomes necessary and mandatory. Sex-segregation of children, particularly of unmarried young men approaching puberty, becomes necessary to maintain the virginity of young females. It is also mandated by sexual anxieties of parents regarding the budding sexuality of the young male. This is particularly so when the mother might be quite young herself as compared to the age of her husband, or where polygamous marriages would result in teenage sons of a man's older wife coming into contact with a young, new wife. In such cases, the overwhelming desire of one or both parents would be to segregate the young males in some social arrangement that would get them out of the home, away from sisters and younger wives, put them to useful work, and bend them to the prevailing sex-repressive social order.

Figure 30 is a map of the *Ethnographic Atlas* variable "segregation of

Figure 30: Segregation of Adolescent Boys

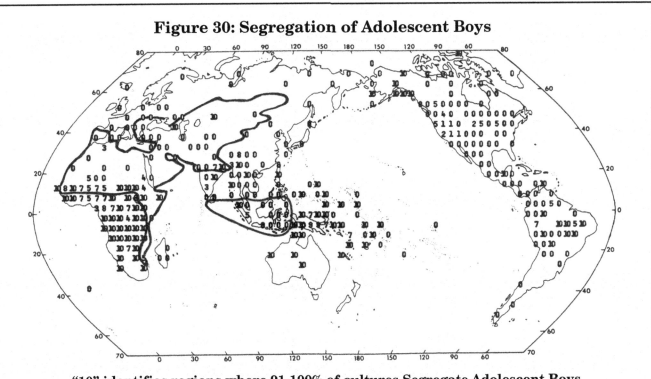

"10" identifies regions where 91-100% of cultures Segregate Adolescent Boys
from the rest of society, either in single-sex groups, or together with adolescent girls.
Drawn line around Saharasia and Island Asia demarcates areas influenced or conquered by
Arab armies since 632 CE (Figure 20), a suggested maximum distribution for female seclusion.
Numbers indicate the percentage of cultures within a given 5° x 5° region which segregate only the boys
(number times 10 = top end of 10-point percentage category). Regions marked "0" have no cultures
segregating boys, while 1 = 1%-10%, 2=11-20%, 3=21-30%, etc. Blank areas = data not available.

adolescent boys".[68] Unfortunately, from the viewpoint of sex-economic theory, there are problems with the codings on this variable which introduce certain highly patterned errors in the map. For instance, the codings on the "segregation of boys" are not restricted to purely criteria for separating young men from young women. The variable includes bachelor huts (where young women may or may not be allowed to visit), quasi-military

68. Murdock, 1967, ibid.

units (for defending cattle or organization of raids), and mixed-sex dormitories and age villages (where young people of both sexes are segregated from the adults, but allowed to live and sleep together).

Regions with a high percentage are characterized by the partial or complete segregation of boys from eating or sleeping with their natal family; boys are sent to stay in special huts, cattle sheds, dormitories, military regiments, age-villages, or with grandparents or non relatives, as retainers or apprentices. Particularly in Oceania, girls may or may not be segregated with the boys away from the community. Regions with a low percentage are characterized by the absence of segregation of boys, but this says nothing about the segregation of boys from girls through female seclusion in Islamic regions.

Sex-economic theory would predict that cultures where children were segregated only from the adults in mixed sex dormitories, to get them "out of the hair" of the adults, so to speak, without reference to male-female sexual interactions, would be of a relatively unarmored, matrist character. Contrarily, cultures where children were segregated in order to isolate them from peers of the opposite sex, in single sex "bachelor huts" or otherwise, would be of a more violent, armored, patrist character. From this perspective, two errors were anticipated:

Correlation Table: Segregation of Adolescent Boys
is positively correlated with the following other patrist variables:

By Column: Bellicosity is high
Cognatic kin groups are present
Patrilineal descent is present
Bride price is present
Grandparents have authority over grandchildren
Polygamy is present
Marital residence near male kin
Painful female initiation rites are present
Male genital mutilations are present
Anal satisfaction potential is low

This variable was found to correlate negatively with two patrist variables in the *Correlation Table*, namely "movable property inheritance favors male line" and "female status inferior or subjected". These few contradictory correlations appear to result from the problems in the codings discussed above, namely that the variable "segregation of adolescent boys" only indirectly addresses the degree to which adolescent boys and girls are segregated from each other's company, particularly in Moslem regions characterized by female seclusion.

---

1) According to the summary on adolescent dormitories by Elwin,[69] which is the most comprehensive account on the subject I have seen, most of the mixed sex dormitories of the world were concentrated in Oceania. Hence, some error will accumulate in my map for this variable in Oceania: There, cultures which have *weak or no virginity taboos and hence do not separate girls from boys, but send both girls and boys to their own mixed-sex dormitories* (to give both them and the parents some needed privacy) were classified by the *Ethnographic Atlas* the same as cultures which have *severe virginity taboos and therefore segregate boys in separate quarters, to keep them away from the girls.*

2) Another important consideration is the segregation of girls. In some regions, such as the Moslem portions of Saharasia (North Africa, Arabia, Middle East, Southeast and Island Asia), females are segregated or secluded from public view in a custom known as *purdah*, and the variable "segregation of adolescent boys" says nothing about this. Indeed, most of the Moslem territories are coded with "absence of segregation" of boys

69. Elwin, V.: *The Muria and Their Ghotul*, Oxford U. Press, Bombay 1942, pp.269-325; cf. Webster, H.: *Primitive Secret Societies*, Octagon, NY, 1968.

(zeros on the map), even though segregation of females from males is extreme and absolute in those areas, with severe penalties for transgressors. Hence, another major error is expressed in the data, and on the map in those regions where Moslem cultures which practice strict female seclusion are classified the same as cultures which do not segregate their children at all. I have consequently drawn on this map an outline area taken from Figure 20, showing the major world regions historically influenced by Arab armies, where the Moslem faith is typically practiced.

By combining Figure 20 (Arab/Moslem regions which seclude girls) and the Ethnographic Atlas data on "Segregation of Adolescent Boys", a more accurate and meaningful representation of the two opposing social institutions is revealed. The true geographical distribution of cultures which segregated boys *from* girls, and girls *from* boys, and where sex-repression is quite extreme, encompasses the combined identified regions on Figures 30.

The variable "segregation of boys" by itself appears to violate the Saharasian generality. Moslem regions of Saharasia and Island Asia stand out as regions of "zeros", which means no segregation of adolescent boys. Oppositely, regions with a high degree of segregation of boys generally lie outside the region of Arab and Moslem influences. The argument being raised here is, simply, that both factors, of "segregation of boys" and "female seclusion" are but different ways different cultures use to deal with the underlying "problem" of adolescent sexuality. A functional diagram[70] of this relationship can be given:

$$\text{Extreme Adolescent Sex-Repression} \underbrace{\Big\langle}_{} \begin{array}{l} \text{Segregation of Boys} \\ \text{(from girls)} \\ \\ \text{Seclusion of Girls} \\ \text{(from boys)} \end{array}$$

The diagram indicates that, as an extreme or efficient method to repress the sexuality of adolescents, one or the other of the two methods will prevail. Even in the West, to a limited extent, these methods are employed in single-sex schools supported by religious groups who advocate them for this very explicit reason.

While these questions are not addressed in the *Ethnographic Atlas* codings, the variable was still used in the *Correlation Tables* and also for constructing the *World Behavior Map*, under the assumption that these errors in the data classifications would tend to *bias the study against the controversial hypothesis of this work*, namely, the existence of a Saharasian-patrist correlation. It was anticipated that the mapped data would code Oceania excessively high for "segregation" and the Islamic regions excessively low; and indeed, this can readily be seen on the map in Figure 30.

The general failure of the psychological and anthropological academic scholars to seriously acknowledge Reich's sex-economic findings as a possible valid explanation for cultural variation has resulted in the creation of an explanatory vacuum. For example, one finds in the literature of ethnography some hints towards a connection between climate, "segregation of boys", and other patrist cultural variables such as circumcision, polygamy, lengthy post-partum sex taboo, and child nutrition. But these derive mostly from purely psychoanalytical or economic-determinist theories, which ignore or deny the role of virginity taboos or adolescent sex-repression as a motivating factor in the "segregation" variable. Nothing particularly insightful or revealing has developed from that work.[71] In part, this is because the behavioral data, when used in the

70. The analytical tool of *orgonomic functionalism* was developed by Reich, partly based upon his earlier interests in dialectic methods. See Reich, W.: *Ether, God and Devil*, Farrar, Straus & Giroux, NY 1973; "Orgonomic Functionalism", *Orgone Energy Bulletin*, 2:1-15, 49-62, 99-123, 1950; and 4:1-12, 186-196, 1952.

71. The relationship between mother-child and husband-wife sleeping arrangements, and other patrist variables have been previously explored in a non-geographic cross-cultural context: Whiting, J.: "Effects of Climate on Certain Cultural Practices", in *Explorations in Cultural Anthropology*, W. Goodenough, Ed., McGraw-Hill, NY, 1964; Whiting, J., et al.: "The Function of Male Initiation Ceremonies at Puberty", in *Readings in Social Psychology*, E. Maccoby, et al., Eds., Holt, Rinehart & Winston, NY, 1974; Young, F.: "The Function of Male Initiation Ceremonies: A Cross-Cultural Test of an Alternative Hypothesis", *Am. J. Sociology*, 67:379-91, 1962; Parker, S., et al.: "Father Absence and Cross-Sex Identity: The Puberty Rites Controversy Revisited", *Am. Ethnologist*, 1974, pp.687-706.

usual manner of *data tables*, and not plotted on maps for a geographical review, expresses only inexact correlations; this is particularly so when the variables under study only partially represent deeper and more primary functions. Such is certainly the case for the "segregation of boys" variable, which only partly reflects the deeper and more important function of adolescent sex-repression (which itself is underlain by the even deeper common functioning principle of *orgasm anxiety* among the adults).

Interestingly, there is one additional variable in the *Correlation Table*, taken from the *Cross Cultural Summary*,[72] which does address adolescent sex repression (aside from the previously given variable on female premarital sex taboo). It is the "adolescent sex dissociation" variable (coded as Finished Characteristic #366). This variable was not mapped as it did not appear in Murdock's *Ethnographic Atlas* data base, and also because there was a low absolute number of cultures for which data were available on the variable. Nevertheless, there was one cross-cultural correlation observed to exist between this variable and others in the *Cross-Cultural Summary*. It is telling: Where "adolescent sex dissociation" is high, "adult sex disabilities" are present.

10) Incest and Incest Taboo

Any discussion of the phenomenon of incest must firstly deal with the larger issue of childhood and adolescent sexuality, in order to free the discussion from a widely-held but scientifically unsupportable tendency which labels any and every sexual experience of the child or adolescent as "sexual abuse". On the other hand, real child sexual abuse does occur, primarily in the form of pedophilia, or child-rape, which is never, and can never be "natural" or "healthy" for any child. The term "incest", without clarifications, often mixes up the normal and healthy sexuality of children and adolescents with some highly pathological adult behaviors.

Childhood sexuality is an undeniable real and natural phenomenon. Decades ago, as discussed in Chapter 2, Reich, Malinowski and others effectively challenged the biological basis of childhood sexual "latency", and the assumed social benefits for "sublimation" of adolescent sexuality. When left alone from adult interferences, children — even toddlers — will often express an almost magical, bioenergetic attraction to the opposite sex. These attractions are perfectly natural and normal, but different cultures deal with such attractions in quite different ways.

Sex-positive cultures do not interfere with childhood attractions and sexual play, nor with adolescent lovemaking, but in fact protect them within social institutions where adults play a minimal role, or no role at all. Such is the case with the Trobriand *Bukumatula* and Murian *Ghotul*. Within such a "children's democracy", there is much innocent sexual play and sweet romance. Even so, it is quite rare for biological brothers and sisters to pair off. A brother-sister incest taboo may exist, as may a clan-incest taboo. But neither taboo is given much significance, nor is it violated with any significant frequency, for the following reason: brothers and sisters brought up together in the same household generally have a relatively weak sexual attraction to each other, even when there are no significant practical measures or sexual taboos to keep them apart. This phenomenon has been observed on some Israeli kibbutz, where children raised communally without significant sexual taboos nevertheless generally found romance with children *outside* their immediate group. Sexual excitation appears to be dulled by a too-frequent contact, a factor even married couples must confront. The absence of sexual barriers or taboos

72. Textor, 1967, ibid.

between *unrelated* children also appears crucial, as this gives each child a natural pathway by which to live out their own sexual attractions and romances, with unrelated children of the same approximate age-group.

Within sex-negative cultures, however, the situation is reversed and more complicated. Children may not be allowed open sexual expression of any kind, much less with unrelated children in their own age group — they might only have contact with their brothers and sisters. In such cases, even though the brother-sister incest *taboo* would be considerably stronger than in the sex-positive cultures, with more severe punishments, the day-to-day contact between brothers and sisters would generally increase, with a significantly heightened and increased sexual tension, given the absence of contact with other children of the opposite sex.

A related set of factors occurs regarding adult-child incest. In the sex-positive cultures, where sexual taboos do not exist, one does not find adults trying to seduce or rape little children. Love-match marriages are the overwhelming rule, given the absence of parental domination of adolescent sexual interests. Divorces also occur without difficulty, as within sex-positive matrilineal kinship systems (to be discussed shortly) each person has a greater opportunity to seek out a satisfactory love-mate. Neither is there an economic incentive in such a social framework to maintain a dead marriage — a woman cannot be divorced from her maternal kin, who contribute more to the material support of herself and her children than does her husband. Because of the natural and love-based system of non-compulsive marriages, sexual frustration is at a low level. Additionally, there are only very weak or no taboos against extra-marital sexual relationships, which may occur periodically, but without undue social uproar. Within such a sexually-satisfied culture, one does not find adults being sexually attracted to small children. Such attractions do occur, however, among the sex-negative groups, characterized by compulsive marriages and sexual taboos against adolescent and extra-marital sexuality.

In the sex-negative patristic cultures, children are subordinated to the patriarchal authoritarian family structure, and not allowed to seek out their own companions. Adults also are confined to the compulsive marriage, whether they remain in love with their husband or wife, or not. This insures that the family home is a seething inferno of suppressed and unsatisfied sexuality. Here is where the dynamics of "Oedipal" attractions are played out, and young children may in fact express one or another innocent sexual display towards the parent of the opposite sex. The usual parental reaction is one of hysterical shock, and the child gets a severe punishment, or at best is "distracted" ("here Johnny, wouldn't you rather play with this ball"). However, the budding sexuality of the child sometimes proves to be too great a temptation for sexually frustrated adults. Usually, it is the father, step-father or uncle who would use coercive words or threats to seduce or rape a young and inexperienced girl, but boys also can be the targets of such abuse. Ancient Greece institutionalized such sexual abuse of boys by older men, under social conditions of severe patriarchal authority and repression of female sexuality (see Part III for details). Children want to please their parents, to love them and be loved, but are totally unprepared for the sexual advances of a grown adult, with fully mature genitalia. Society rightfully has a "taboo" against this kind of incest, which is called *pedophilia*, *child seduction*, or *child rape* when it occurs between unrelated individuals. It is properly the domain of social workers and policemen, to protect children against pedophiles and rapists — but *social workers and police have no rational business jailing adolescents for consensual sexual play and / or intercourse with their peer group*, even when it occurs between related children. Such children and adolescents need guidance about contraceptives and sexual hygiene, and some help regarding privacy and protection from snooping puritans. They do not need the sex-negative hysterical reactions of parents and power-drunk "officials".

Sex-Positive Social Institutions:
Sexual gratification predominates
Incest is rare, weak incest taboo

Children's Democracy:
Childhood/adolescent romance within age-groups
Brother-sister incest is rare.

Non-Compulsive Marriages:
Adult romance with spouse,
Tolerance of extramarital sex.
Adult-child incest and pedophilia is nonexistent

Segregation of Boys, Seclusion of Girls:
Heterosexual premarital romance forbidden
Brother-sister sexual excitation is increased
   in the absence of other children of opposite sex

Sex-Negative Social Institutions:
Sex-frustration, misery predominates
Incest is frequent, strong incest taboo

Compulsive and/or Arranged Marriages:
Adult romance thwarted, minimized
Extramarital sex taboo, dangerous
Adult interests/preoccupations with child's sexuality
Child rape and pedophilia develop

We can recapitulate the above phenomena according to Reich's sex-economic and functional methodology:

Again, it is important to separate the issues of young boys and girls having a natural and sweet sexual attraction to each other, from the phenomenon of frustrated and sadistic adults having an unnatural sexual attraction to children. The usual definitions of "incest" do not distinguish between these clearly different phenomenon, but the distinctions are crucial and necessary to gain a better understanding of the sexual dynamics within families and small cultural groups.

In contrast to the above discussion, deMause has argued for "The Universality of Incest" in all cultures, making a plea for the defense of the child against adult seduction — but he confuses this rational and important concern with consensual childhood sexual play and adolescent lovemaking, placing them in the same category of pedophilia and rape.[73] His work does, however, suggest that adult-child incest and pedophilia might be more prevalent in the Middle East, and draws a correlation between those abuses to genital mutilations. Assuming this correlation is valid, it would by itself suggest a Saharasian distribution to violent adult-child incest, a distribution already demonstrated in this Chapter for genital mutilations.

From the above, we can make a prediction: any map which is ultimately produced of the variable "incest" should have a distribution

73. deMause, L.: "The Universality of Incest", *J. of Psychohistory*, 19(2):123-164, Fall, 1991. deMause's citations on the sex-positive matrist Muria were generally faulty and misrepresented. He claimed (p.147-148) that Murian children were a part of "family sex" activities; that toddling children were routinely sexually molested in the family bed; that adults dominated the life inside the children's dormitory or Ghotul; that toddling children were being molested in the Ghotul by adolescents and adult males such that they would "wet their beds and wake up crying". None of these bald assertions are supported by the citations and quotes attributed to Elwin (*Muria and their Ghotul*, Oxford U. Press, Bombay 1942). Elwin was clear to point out how the Ghotul was *not* under the control of adults, except among those Murian or Marian villages where Hindu puritanism had crept in. Elwin even published a map identifying the geographical source regions from which such influences were affecting the Muria (Elwin, *Muria*, p.13). deMause's portrayal of ram-

related to the patristic and matristic areas of Saharasia, but only assuming the variable is divided into two main groupings: adult-child rape and pedophilia should be identifiable mainly in the more highly armored and patristic cultures and regions, while brother-sister incest might be more globally distributed, but still less frequent in the sex-positive matristic cultures. A map of "incest taboo", however, should demonstrate a clearly patrist-Saharasian distribution for the harshest *taboo punishments*, with the weakest punishments in those sex-positive areas where adults have no desire to seduce or rape related or unrelated children.

Marriage and Family

11) High Bride-Price Marriage

The map in Figure 31 gives the percentage of cultures in a given world region where men obtain wives only through difficult or extreme means, namely bride-price, bride-wealth, or exchange of a female relative. Such practices indicate a loss of control by young people over the choice of mate. Regions marked with the number "10" are characterized by 90-100% of all cultures possessing such a high bride price. Regions characterized by a token bride price, gift exchange, bride-service, or the absence of any significant consideration are given low percentage scores on the map. As before, data was taken from Murdock's *Ethnographic Atlas*.[74]

From the given cross-cultural correlations, a high bride price appears as a cultural trait which both supports and is supported by an entire matrix of other anti-child, sex-negative, authoritarian patrist cultural institutions, attitudes, and behaviors. Sex-economic theory would predict that children who are allowed to live out their youthful romances will be far less inclined to accept the dictates of a parent over emotionally-charged basic life decisions, such as choice of marriage partner. A high bride price, where women are bought, sold, or exchanged like cattle or

pant and routine aggressive adult seduction and rape of toddling children in Murian society is false and unsupportable, though Elwin was clear to point out how many neighboring cultures, which abandoned the Ghotul or subordinated their children to adult authority were in fact very chaotic sexually. Social conditions of the Muria were thereby confused by deMause with other patristic cultural groups in geographically-adjacent areas. For example, on p.148, deMause misquotes Elwin about what was clearly an exception to Ghotul life: an old woman remembered being raped when she was a young girl — but the actual quotation came from one of Elwin's footnotes, by a third author, who discussed a rape among a completely different culture (the *Oraon*). Such incidents were falsely portrayed as "ordinary" or "typical" of life among the Muria. (p.148 of deMause, p.340 of Elwin). Likewise, the weak and often-broken "clan-incest" taboo of some Murian groups, and the generally happy and consensual sexual relationships between the unmarried adolescents, were mischaracterized as "incest" equal to old men raping small children. Clearly, cultures like the Muria and Trobriand Islanders were in the process of important cultural changes at the times when they were studied, and conditions were different from one village to the next. Elwin was clear to point this out and, as discussed in Part I, he emphasized the dramatic differences between the most isolated tribes whose social institutions reflected a more ancient situation, and the lesser-isolated tribes which had been deeply influenced by the sex-negative Hindus.

74. Murdock, 1967, ibid.

Correlation Table: High Bride Price Marriage
is positively correlated with the following other patrist variables:

By Row: Protection of infants from the environment is low
Segregation of adolescent boys is high
Male genital mutilations are present
Marital residence near male kin
Polygamy is present
Father has family authority
Patrilineal descent is present
Cognatic kin groups are absent
Land inheritance favors male line
Class stratification is high
Castes are present
Slavery is present
Warfare prevalence is high
Military glory emphasis is high
Bellicosity is high
Narcissism index is high
Insult sensitivity is extreme

No negative correlations were found between "High Bride Price" and other patrist variables.

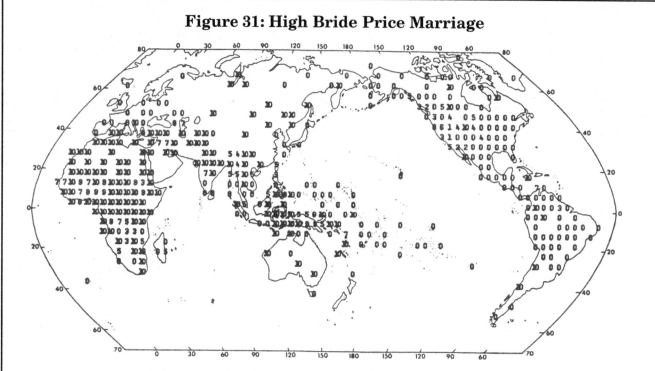

Figure 31: High Bride Price Marriage

"10" identifies regions where 91-100% of cultures have a High Bride Price

Numbers indicate the percentage of cultures within a given 5° x 5° region possessing a high bride price (number times 10 = top end of 10-point percentage category). Regions marked "0" have no cultures with a high bride price, while 1 = 1%-10%, 2=11-20%, 3=21-30%, etc. Blank areas = data not available.

camels, can only prevail where the sexuality and basic emotions and desires of young children and adolescents are totally manipulated and repressed by parents. If children were left to their own desires, the bride price would rapidly be erased, given the fact that youthful romance tends to ignore economic, class, or caste barriers. As with the song from the popular musical *South Pacific*, "you have to be taught to hate and fear, you have to be carefully taught". Cultures with strict adult control over children possess far fewer love-match marriages, as well as harsher punishments for violation of virginity taboos, more extreme measures for segregation of children of the opposite sex, plus strict and uncompromising forms of "obedience training". The high bride price thereby predominates and reproduces itself within a cultural matrix of extreme armored patrism.

The map of this variable, given in Figure 31, reveals a predominant Saharasian distribution, with diffusions of the trait to bordering regions, and to distant regions with historical connections to Saharasia (primarily, Southeast Asia and Western Oceania).

12) Marital Residence

Figure 32 is a map of prevailing modes of marital residence. Block regions of a high percentage are characterized by rules which favor the newlywed couple locating near the male kin of either spouse. This would include patrilocal (male kin of husband), virilocal (patrilocal but no patrilineal kin groups), or avunculocal (male kin of bride). The number "10" identifies regions where 90-100% of all cultures possess such marriage rules favoring location near male kin. Regions of a low percentage are characterized by rules which favor locating near the female kin of either spouse, or which do not favor male kin over female kin. This would include matrilocal (female matrilineal kin of wife), uxorilocal (matrilocal but no matrilineal kin groups), ambilocal, bilocal, or utrolocal (either parent of either spouse), neolocal (apart from all relatives), or the non-establishment of a common household. Data is from Murdock's *Ethnographic Atlas*.[75]

This variable of marital residence does not, within and of itself, appear as a causal factor in the perpetuation of armoring and patrism. It is, rather, strongly related to the issue of *family descent*, whether an individual traces their kinship down the line of their mother, grandmother, great-grandmother, etc., or through the line of their father, grandfather, great-grandfather, etc. More will be said about descent shortly, but the factor of marital residence informs us about a host of factors which either prevent or perpetuate armor, or which shift family structures towards or away from the bio-emotional aspects of the maternal-child and male-female bond.

For example, when a young couple marry and reside in a location close to the mother and matri-kin of the bride, in a matrilineal kinship organization, the woman can continue to rely upon her maternal kin for sustenance (mother, sisters, brothers, cousins), and she therefore has no economic incentives to either enter into or maintain a marriage. She cannot be "divorced" from her matrilineal kin group, as is the case with respect to her husband. The maternal-child bond is thereby secured by the maternal marital residence system, as is the emotional-love bonding between males and females (as opposed to compulsive marriages undertaken or maintained for class, caste, or economic reasons). If the couple fall out of love and decide to part, the issue of "division of property" does

75. Murdock, 1967, ibid.

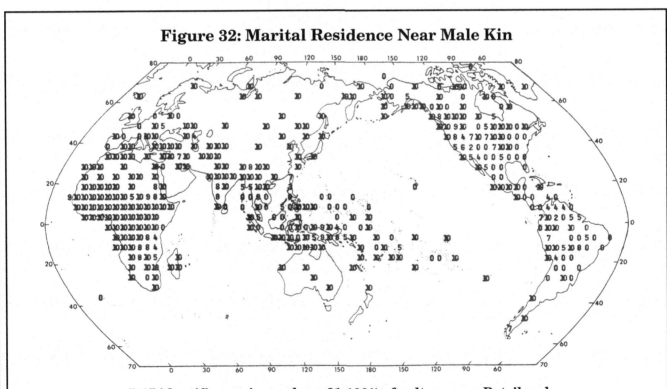

Figure 32: Marital Residence Near Male Kin

"10" identifies regions where 91-100% of cultures are Patrilocal
Numbers indicate the percentage of cultures within a given 5° x 5° region with patrilocal marital residence (number times 10 = top end of 10-point percentage category). Regions marked "0" have no cultures with patrilocality, while 1 = 1%-10%, 2=11-20%, 3=21-30%, etc. Blank areas = data not available.

Correlation Table: **Marital Residence Near Male Kin**
is positively correlated with the following other patrist variables:

By Column: Infant pain infliction is high
                       Segregation of adolescent boys is high
                       Male genital mutilations are present
                       Female premarital sex taboo is high
                       Exchange of female relatives is present
                       Bride price is present
                       Polygamy is present
                       Cognatic kin groups are absent
                       Patrilineal descent is present
                       Movable property inheritance favors male line
                       Land inheritance favors male line
                       Class stratification is high
                       Castes are present
                       Slavery is present
                       Insult sensitivity is extreme

Only one patrist variable had a negative correlation with "marital residence near male kin", that of "adolescent sex expression restrictions are present". However, this single negative correlation was a weak one, the significance of which fell between the .10 to .05 level. The positive correlation of "marital residence near male kin" to "female premarital sex taboo", given in the previous section above, is much stronger, with a probability falling between the .05 to .01 level.

---

not arise, as neither the marriage nor the divorce affects their matrilineal kinship arrangements, which secures the sustenance of both husband and wife. The wife primarily draws support from her brothers and uncles, while the husband is obligated more to his sister's children than to the children of his wife. Nobody can be divorced from their matri-kin, and divorce therefore is irrelevant to economic security. The entire matrilineal and matrilocal family unit is therefore very much in tune with deeper biological factors, such as childbirth, breastfeeding, and the absence of economic motivations or coercion regarding marriage and divorce.

All of the above factors reverse themselves in cases where the father, his brothers, sons and nephews reside collectively under the patrilocal and patrilineal kinship system, which is often organized around a military caste or other form of strong social hierarchy. In such a case, the woman is more often "obtained" by a high bride price, and has less to say about her own fate, or the fate of her children, than the males of her new family. The child is descended from the "father's line", and all bonds and social relationships revolve around him, and his family. The locating of the newlywed couple near male kin places the young bride in a position more dependent upon and controlled by her husband, to whom she must look for social support and assistance at the new patrilocal residence location. Divorce could mean a complete disenfranchisement of the woman from any means of sustenance, as well as losing her children — and this factor would reflect an increasingly compulsive marriage system, rooted more in economics than love. The requirement of patrilocality might also mean that the young husband would be expected to participate in one or another quasi-military activity, wherein men aggregated together in groups, to be bossed about by a dominant male, and certainly by older males in each individual clan or family unit. Indeed, where a high bride price existed, he might have been required to reside near to a male family member to whom he was labor-indebted for the cost of the bride price, in the manner of "securing the loan".

By the textbook, ***polygamy*** means, to have *more than one husband or wife*, while ***polygyny*** means one husband with several wives, and ***polyandry***, one wife with several husbands. In common usage outside of anthropology, however, the term "polygyny" is nearly unknown, and polygamy is generally understood to mean one man with several wives. For the discussion here, therefore, the more general usage of polygamy is applied.

76. Murdock, 1967, ibid.

Hence, where newlyweds reside with female kin, or have the freedom to choose their residence, the authority of males over females, and the elderly over the young is eroded somewhat; military caste structures appear to be eroded significantly. Matrilocality therefore suggests a completely different social structure, oriented around maternal-child bonds, and male-female bonds, with a decreased emphasis upon male dominance and military-type matters, or parental dominance of children.

These suggested relationships are strongly supported by the cross-cultural correlations, given above. As before, the area of the world where patrilocal (male-relative oriented) marriage residence predominates is Saharasia, and its African-European-Asian borderlands.

13) Polygamy

Figure 33 is a map of institutionalized unrestricted polygamy, as opposed to monogamy, or occasional or limited polygamy, developed from Murdock's *Ethnographic Atlas*.[76] Regions with a high percentage are characterized by unrestricted polygamy, irrespective of extended or nuclear form, or of kin or residence type. The number "10" identifies regions where 90-100% off all cultures practice such unrestricted polygamy. Regions with a low percentage are characterized by monogamy, or occasional or limited polygamy only. Moslem regions fall under the heading of occasional or limited polygamy, in accordance with Islamic law which limited the average man to no more than four wives, but which may in many cases allow an unrestricted number of concubines. Consequently, the map of "polygamy" presents a fairly low reading for Saharasian Islamic regions, in spite of the fact that, today, Islamic nations are

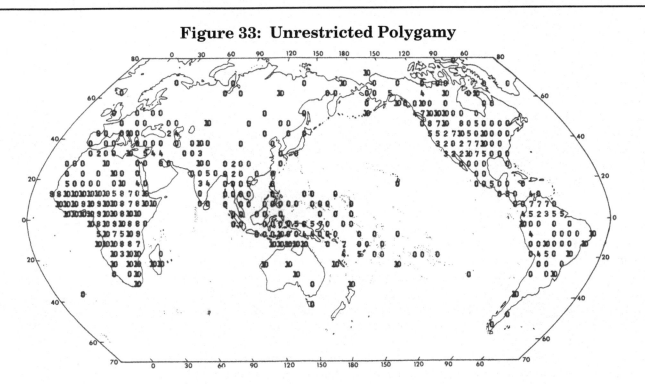

Figure 33: Unrestricted Polygamy

"10" identifies regions where 91-100% of cultures have Unrestricted Polygamy.
Numbers indicate the percentage of cultures within a given 5° x 5° region with unrestricted polygamy (number times 10 = top end of 10-point percentage category). Regions marked "0" have no cultures with the trait, while 1 = 1%-10%, 2=11-20%, 3=21-30%, etc. Blank areas = data not available.

> ## *Correlation Table:* **Polygamy**
> is positively correlated with the following other patrist variables:
>
> By Column: Infant pain infliction is high
>   Segregation of adolescent boys is high
>   Male genital mutilations are present
>   Marital residence near male kin
>   Father has family authority
>   Bride price is present
>   Exchange of female relatives is present
>   Female status is inferior or subjected
>   Barrenness penalty is high
>   Abortion penalty is severe
>   Sex anxiety is high
>   Cognatic kin groups are absent
>   Patrilineal descent is present
>   Slavery is present
>   Military glory emphasis is high
>   Bellicosity is high
>   Narcissism index is high
>   Insult sensitivity is extreme
> Other by Row: High god present, active, and supportive of human morality
>
> Two negative correlations to polygamy were present in the *Correlation Table*, to "early weaning is present" and "high god present, active, and supportive of human morality". The former of these negative correlations derives from the fact that in polygamous households children tend to be weaned later given a more extreme and lengthy post-partum sex taboo, or absence of the father for other reasons. The latter negative correlation contradicts a similar positive "By Row" correlation between polygamy and the same "high god" variable. The contradiction may be explained by the fact that while nearly all polygamous cultures possessed a high god, not all high god cultures allowed unrestricted polygamy, or viewed concubines in the same class as wives. This is clearly the case with the Islamic groups which, as mentioned, were classified as having only restricted polygamy.

among the most polygamous to be found.

The family unit based upon spontaneous, primary core feeling would tend to be, according to the sex-economic view, monogamous, but without compulsive, life-long tendencies. Therefore, where polygamy predominated, it would be caused by other intruding factors, such as changes in the male/female ratio (due to warfare or disease), or through cultural processes where the spontaneous love-relations between children and adolescents are abolished by the adults, and children lose the right to choose their own mates.

Historically, it can be seen that where the bride price and harem system prevailed, a single man could, by force of arms or wealth, accumulate a large number of "wives" or concubines. Moreover, these men would themselves have been raised under social systems emphasizing military regimentation, warfare, and slavery. These cultural factors are clear signs of armoring, sexual anxiety and genital impotence, which itself can block the individual from tender feelings towards a single partner, and drive them towards a constant seeking-out of unattainable gratification with different sexual partners. A single woman would not suffice for a man who suffered from such violence-laden sexual disturbances, and women in such cultures would similarly be conditioned to accept such a polygamous situation, and perhaps even to feel relief about the attraction of the husband towards another woman in what was most probably an arranged and passionless marriage.

The leaders of most kingdoms within or surrounding Saharasia were, in fact, gargantuan fornicators, massing harems with wives and concubines numbering in the hundreds or even thousands. These massive harems developed as major islands within a larger sea of female subordination, arranged marriages, and polygamy. The average man would not have the wealth to support so many women, though they certainly might strive to emulate the big-man leader, to whatever extent possible. Attaining high status in such a society might require a polygamous household.

Strict adherence to the sexual "double standard" generally prevails in polygamous areas, and this includes a mix of factors such as the sexual segregation of young people, female virginity taboos, vaginal blood taboos, widowhood taboos, female inferiority complexes, and generally mistrustful attitudes between the sexes. This is necessary, given the fact that loving affection and courtship ritual are not prerequisites for the development of a marriage contract — indeed, "permission" (at least, of the female) is not generally required for such arranged marriages, as the children are obligated to obey the dictates of the parents. Usually, the father has the last say in such matters.

One also finds "temporary marriages" akin to concubinage or legalized prostitution prevailing in many polygamous regions, though these factors are not addressed in the maps given here. Such temporary marriages prevailed in the Islamic regions, for example, to provide travelers and pilgrims to Mecca with a temporary female partner — such practices still persist and thrive within the Moslem world. The temporary wife is paid to cook for the man and sleep with him, but when he returns home, the contract is over and he has no further obligations to the woman, even if she bears his child.

One justification often given for polygamy is that it provides a social support mechanism for widows. However, this is true only in cultures which have blockaded women (and average men) from education, professional work, legal rights to control over money they have earned, and ownership of property. When the man dies, the widow may be left penniless, and consequently, dependent entirely upon relatives. No longer virgins, such widowed women were additionally considered sexually threatening and dangerous to society, and so were distributed to the dead man's brothers. Or, in some Saharasian border regions (India and China) widows were ritually murdered, another "cultural practice" which will be discussed shortly.

Historically, polygamy has also been linked to warfare, hierarchy, and caste, where property and wealth preferentially accumulated to those in the upper castes. Men went to battle, killed each other, and left behind a larger population of females, both at home and in conquered regions. Captured slave-women were seized by conquering soldiers as wives, concubines or harem slaves, as were the lands, wealth, and other surviving people of the conquered lands. Younger wives of older males and warriors generally outlived their husbands, leading to large numbers of widows in some cases.

All the above factors historically have reinforced the institution of polygamy, which, like concubinage and prostitution, was and still is largely dependent upon female subordination, sex-repressive childrearing, the caste system, and skewed wealth accumulation for its very existence. The various taboos placed against widows, as sexually dangerous individuals within society, as well as the practices of ritual widow murder were also functionally related. And as these social factors have declined, in most places polygamy has also declined or been subject to restrictions.

Sub-Saharan Africa is an exception where polygamy continues to thrive or even increase. This tendency is partly reflected in the map on polygamy, Figure 33. In this region, women provide most of the wealth-producing labor, for agriculture and village-level manufacturing efforts. However, women are also greatly subordinated, and both their property and money earned through their labor is usually controlled by male relatives, fathers or husbands. A man with many working wives and children can accumulate more wealth than a man with a single wife and only a few children. It is therefore not unusual to see extended families in sub-Saharan Africa composed of a single man with multiple wives, each of which having six or more children. This is quite a large labor resource, and the husband/father is given full legal rights to control both the very lives and property of the entire wealth-producing enterprise! With the historical change from a barter to cash economy in those areas, and with similar changes to cash export crops over crops for local consumption (coffee over beans, for example) the accumulation of capital became possible in a manner previously impossible, and the lure towards increasingly polygamous households was amplified. A related side-effect was that, with fewer eligible brides for each young man, the bride-price invariably increased. The status of young males in polygamous regions was thereby made increasingly more dependent upon older men, from whom they borrowed or indentured themselves for the increasingly costly bride price.

These factors have provided a difficult, and quite conscious barrier to social and political reforms in the status of women, and changes in family law which would reduce the power of the father or husband. One interesting social shift accompanying this increase in sub-Saharan "economic" polygamy has been that Christianity, which forbids polygamy, has declined in favor of Islam, which does not. Here, one can see how sex-economic factors clearly interrelate with other purely economic factors, providing a very clear, if not depressing understanding of social forces perpetuating a low status for women, dramatic population growth, environmental deterioration, and political, regional, and tribal conflicts.

We might also point to similar tendencies at work in extended polygamous Mormon families in small regions of the USA and Canada, though in this case, the prevailing laws do not give the man the same legal rights and privileges as is the case in Africa or the Islamic world. From the Mormon cases, however, it is apparent that deep emotional factors related to sex-negative, repressive childrearing are at work, and that political laws favoring the adult male are not always necessary to perpetuate the polygamous family system. Should polygamy ever gain a foothold in the West — as is now seen de-facto, in a limited way in Muslim enclaves within European and American cities — and then be legalized, it surely would dramatically increase in scope and magnitude, bringing with it all the sex-negative and patristic cultural factors upon which it is dependent, and which it creates. The extreme patrist social disaster for women which persists across Saharasia today, would then become a fact of life in the West as well, eroding whatever gains have already been made in women's rights and status over several hundred years of difficult struggle and reform. We could then quickly expect to see the return of arranged marriages, child betrothal, and "honor" murders of young women trying to escape what is basically life-long female sexual slavery.

Regarding polyandry, where one wife may have more than one husband, seven cultures out of 1170 in Murdock's data base (0.5%) were identified with such characteristics. These were located in Southern Africa, Southern India, Oceania, and other areas with a generally matrist classifications. This aspect of family organization has not been explored to any extent, however, and the small number of cultures in the sample prohibits meaningful analysis.

Pregnancy and Childbirth

14) Contraception and Abortion

The cross-cultural data bases of Murdock and Textor provided only limited information on these important and hotly-debated issues, but it is clear the anthropological literature is filled with references to both contraception and abortion. Previously, I have undertaken major cross-cultural study of the use of contraceptive plant materials by native peoples,[77] finding hundreds of references to the use of plant materials which, in some cases, had been verified by laboratory study as effective in the prevention of pregnancy. Often, the same cultures which would report the existence of a special herb or plant for contraception would also report other herbs for abortion. Chapter 6, which follows, gives a more complete discussion of the issue of contraception in native cultures around the world, but here we can address the issue of abortion.

Devereux has undertaken a comprehensive cross-cultural study of abortion in primitive societies, but his data have yet to be mapped, or analyzed in any other significant way relative to the materials in this study.[78] However, one observation has emerged from my own research:

Correlation Table: **Severe Abortion Penalties**
are positively correlated with the following other patrist variables:

By Column: Killing, torturing, mutilating of the enemy is high
    Slavery is present
    Extramarital coitus is punished
    Barrenness penalty is high
    Polygamy is present
    Adolescent sex expression restrictions are present

No negative correlations were found between "Severe Abortion Penalties" and other patrist variables.

---

whenever I came across a culture which severely punished women for abortion, that culture invariably was patrist, and possessed a number of other harsh, antifemale and antichild taboos and blood rituals. Contrarily, in the matrist cultures I reviewed, women entirely controlled whatever means were available to them for determining their own fertility; men did not have control over female procreative functions, nor did restrictive taboos or harsh penalties exist for abortion. In fact, many of the contraceptive plants discussed in the above section also acted, in larger doses, as abortifacients. This suggests, like the plant contraceptives discussed in Chapter 6, abortion knowledge once was unrestricted, though not necessarily a common practice. Cultures which allow contraceptive knowledge and abortions are characterized by planned and desired pregnancies, with few unwanted pregnancies. This conclusion is supported by more recent studies which indicate an increase in rates of abortion among societies which prohibit the practices or, and a decrease in abortion where contraception and abortion are freely permitted and available.[79] The reason for this apparent contradiction is the fact that abortion-prohibiting cultures generally suppress the rights of women, deny them basic equality of education, and additionally banish effective contraceptive methods — thereby increasing unwanted pregnancy and, hence, increasing abortion in spite of its illegality. Abortion is rarely

77. DeMeo, J.: "Herbal Oral Contraceptives: Their Use by Primitive Peoples", *Mothering*, 5:24-28, 1977; "The Use of Herbs for Contraception by Primitive People", *AAG Program Abstracts*, San Antonio 1982, Annual Meeting, Assn. of American Geographers, pp.33-34, 1982; "The Use of Contraceptive Plant Materials by Native Peoples", *Journal of Orgonomy*, 26(1):152-176, 1992; German translation: "Empfängnisverhütungsmittel bei Naturvölkern",*Emotion* 11:6-29, 1994. See Chapter 6.

78. Devereux, G.: *A Study of Abortion in Primitive Societies*, International Universities Press, NY, 1976.

79. See the discussions on "More Population Woes" and "Abortion" in *Pulse of the Planet* 3:125, 1991. Excellent summary article in *San Francisco Chronicle*, 29 May 1993. Data from Nat. Center for Health Statistics, Worldwatch Institute, and International Planned Parenthood.

prohibited all by itself. Where abortion is free, contraceptives are also freely available, as is general birth-control knowledge and information. Unwanted pregnancies (and abortions!) tend to be much lower in abortion-permitting countries than in abortion-restricting countries.

From the cross-cultural correlations, a picture emerges where the female is expected to produce children in a male-dominated, warlike, slave-owning, polygamous cultural matrix. Young people are punished for attempts to determine their own sexual lives, and women are punished if they become pregnant prior to marriage; or if they do not produce children after marriage for their husband.

Several maps addressing the issues of contraception and abortion are given in Chapters 6 and 7, constructed from data bases which were not directly comparable with the maps presented in this chapter. A true map of *abortion prevalence* should identify two cultural regions with low abortion rates, namely where abortion is freely permitted along with contraceptives, but also where abortion and contraceptive penalties are extremely severe, demanding death for the offending woman (as in some extreme patrist regions). Abortion prevalence should not, therefore, show either a clearly matrist or patrist distribution; possibly, it would be found most prevalent in transition cultures. Severe *abortion penalties*, however, predominate in patrist cultures only, as suggested by the cross-cultural data; a map of cultures with severe abortion penalties should therefore display a distinctly Saharasian distribution. Likewise, a map of contraceptive usage should also show a Saharasian distribution, though opposite in character to any map on abortion penalties.

Anti-Abortionists Promote Abortion

"Abortions are more common in countries that ban or restrict the procedure than in those where it is widely permitted."

The above conclusion comes from a report by International Planned Parenthood, and is no surprise to anyone who has studied the issue of contraception, family planning, and the problem of unplanned, unwanted pregnancy. Aggressive anti-abortion laws are successfully promoted in nations where generally antisexual attitudes predominate, and this includes attitudes against contraception. Anti-abortionists are rarely supporters of contraception. They are more likely, instead, to be promoters of antisexual propaganda, such as "abstinence", the "rhythm method", or other unreliable forms of birth control.

"In western Europe, where abortion is legal except in Ireland, there are about 14 abortions per 1000 women, the study said. In Latin America, where abortion is restricted, the rate is between 30 and 60 per 1000 women... Abortion rates are lowest in countries that not only permit the procedure, but offer family planning and sex education services. Among 22 countries that allow abortion, the rate increases as availability of family services declines... the Netherlands, which has widespread sex-education, had the fewest abortions — 5.6 for every 1000 women in 1984. The United States ranked 13th with 27.4 abortions for every 1000 women... the Soviet Union ranked last with 181 abortions for every 1000 women..." (San Francisco Chronicle, 29 May 1993)

15) The Couvade, and Similar Practices

The couvade is a cultural phenomenon wherein childbirth and labor pains, and/or post-partum fatigue and recovery are experienced by the husband during his wife's pregnancy and childbirth. For the discussion here, couvade is defined more broadly, to include any cultural practices where the focus of childbirth shifts from the infant and mother and their biological needs, to the husband and his psychological needs, or where "culture" otherwise intrudes into, disturbs, or interferes with the emotional bond between mother and infant. In effect, where the couvade is practiced (as defined here), the husband, or other male, becomes an intrusive, need-thwarting "mother" substitute, and may disturb primary care of the infant to the extent of miming breastfeeding after the birth. In extreme cases of couvade, the husband agonizes the pains of childbirth, and writhes on the floor, and is given even more comfort and attention than the birthing mother. After the child is born, the husband may demand prolonged bedrest, pampered care, and special foods. Meanwhile, the hardly-recovered mother may be sent outdoors to an isolation hut, for ritual purifications, or sent to resume household or agricultural chores. She is left to recover pretty much on her own devices, while the husband takes possession of the child, and celebrates "his" birth.[80]

A variety of theories have been advanced on the origin of couvade, none of which, to my knowledge, has used as a starting foundation the matrist/patrist dichotomy of Taylor, or the sex-economic corrections to psychoanalysis, as discussed by Reich.[81] I wish to make a few concrete suggestions on the practice from such a point of view, to include an analysis of the couvade as practiced in modern American hospitals.

The couvade may originally have had a rational, biologically necessary and pleasure-enhancing origin in matrist culture, in a mandate to a working husband to stay home and help his laboring wife through transition and recovery. However, under conditions of increasing patrism and increasing anxiety regarding childbirth, where vaginal blood taboos and post-partum taboos of various sorts began to intrude, the role of the man might have automatically increased due to the "need" to ritually "absolve" the child of the "poisonous" influences of contact with vaginal childbirth blood and the female, and due to the involvement of the mother in various purification rituals during which time the child would have to be handed over to someone else.

Under such circumstances, the mother becomes unclean and taboo, and the child must be taken from her for its own protection. In this sense, a similar practice appears in most modern hospitals, where childbirth is seen as an illness, and is also made unnecessarily dangerous by various shamanistic-obstetrical practices. In hospital deliveries, male obstetricians often boast about "their" difficult delivery, and fathers, who have traditionally been excluded from birth, engage in various nervous rituals, often becoming a major focus of attention by friends and relatives. Birth attendants are capped, gowned, and gloved ostensibly to protect the mother and baby, but perhaps unconsciously to protect themselves against contact with poisonous childbirth blood. It is still routine in many hospitals for newborn babies to be separated from their mothers "for their own protection", in a manner very little different from the most primitive of tribes.[82]

The obstetrical wards of our hospitals practice a form of "high-tech" couvade, veiled in a priestly aura of "safety" for infant and mother, which, as discussed in Chapter 2, is actually far less safe than the childbirth care given by peer-trained empirical lay midwives attending homebirths. The

80. Dawson, W.: *The Custom of Couvade*, Manchester U. Press, 1929.

81. Discussed in Chap. 2 & 3; cf. Paige, K. & J. Paige, *The Politics of Reproductive Ritual*, U. of Cal. Press, 1981, pp.34-42.

82. If infection is a primary concern, then the best place for birth is the home environment, where no hospital pathogens are present, and where the mother's immune system is already familiar with any existing germs. See the section on "Birth Trauma in Modern Hospitals" in Chapter 2 for a more complete, referenced discussion of these facts.

present-day epidemic of Caesarean deliveries is testimony to our modern couvade, and the horrific fear of allowing the child to have contact with the vagina or vaginal blood; gross economic motives — high profits and fear of lawsuits — also promote a high Caesarean rate, which cannot be justified from the standpoint of science or biological need in any case.

At a deep level such practices are an attempt by priestly-caste doctors to simultaneously dominate the childbirth function, destroy its instinctual nature, and "protect" the infant from contact with the dangerous vagina, and vaginal blood. Our society has, unconsciously perhaps, developed a system of childbirth where great birth trauma is not really considered a bad thing, with all sorts of irrational excuses being offered up as to why the trauma is allowed to persist. The spontaneous movements of transition and delivery are viewed with great fear and superstition by both mothers and birth-attendants alike. Gross interference with natural childbirth over the centuries has indeed often increased the risks of childbirth, particularly in hospitals at the hands of male obstetricians, and their often mindless, mean-spirited and subservient nurse-attendants. As is the case with the adolescent sexual drive, childbirth and the maternal-infant bond provoke much anxiety, and great effort and care is taken to prevent them from following biological, and pleasure-oriented pathways. This has included the legal harassment and imprisonment of lay midwives and doctors who deliver babies in the relaxed home environment; those who try to make childbirth more spontaneous, pleasure-oriented, and biologically natural (and safer!) do so today often at the risk of jail. As cited and discussed in more detail in Chapter 2 ("Birth Trauma in Modern Hospitals"), the medical statistics prove that the relaxed atmosphere of midwife-assisted homebirth is safer for both mother and baby than birth in "high tech" hospital obstetrical wards. This is so, even when the comparisons are made to homebirths attended by lay trained, empirical or "granny" midwives.

My own review of the couvade has not been systematic enough to construct a map, and neither of the data bases consulted included the couvade variable. However, my review of the anthropological literature on couvade suggests that the most extreme forms of the practice occur where many other forms of patrism are present. (The full *Correlation Table* given in Appendix A and discussed in Chapter 3, provides a full listing of the variables related to childbirth and infant care.) I predict that a systematically-derived geographic study on the couvade will demonstrate a Saharasian-related distribution, assuming its impact upon the maternal-infant bond, and its extremeness, are also taken into consideration.

For instance, where the husband is simply relieved of work duties and stays home to help the mother, but does not physically remove the baby from her (insuring the preservation of the maternal-infant bond), "couvade", as defined here, would not exist. On the other hand, if the husband's biological reactions are similar to the laboring mother, with labor pains and a period of "post-partum" recovery, or if the male ritually separates the child from its mother for any significant period, miming her bonding and breastfeeding functions, then the couvade would exist. As defined here, a map of couvade should demonstrate a Saharasian distribution for the practices.

The following must be stressed: without taking into account these qualitative differences in the various phenomena labeled with the heading "couvade", and its affects upon the infant, utter confusion will prevail in attempts to understand its relationship to other social institutions, or forms of adult behavior.

16) Post-Partum Sexual Taboos

During the period immediately following childbirth, an existing post-partum sexual taboo demands that husbands and wives forego sexual contact with each other for periods ranging up to several years. The reasons given for the taboo generally focus upon the dangers to the infant, or to the man, who (it is often said) would be harmed by coming into contact with dangerous vaginal blood.

It is reasonable to assume that in most cases where an excessively lengthy post-partum sex taboo exists, a strong vaginal blood taboo might also exist. In many cases such a taboo lasts up to two years or more, suggesting the marital sexual relations are greatly thwarted. "Lengthy" is defined here as that which would clearly defy any biological or medical benefits, namely periods longer than six months.

Figure 34 is a map addressing the excessively lengthy post partum sex taboos.[83] Block regions with a high percentage are characterized by post-partum sex taboos ranging from six months to more than two years. Regions with a low percentage are characterized by the absence of taboos, or taboos of less than 6 months duration. The reader will note from Figure 34 that this variable has a higher degree of missing data than most, and the missing data are disproportionately located in the Saharasian region, which other measures and approaches have identified as being largely patrist in character. Hence, it is expected that traditional cross-cultural correlation methods using this variable will not properly assess the relation of the variable to patrism until more data is provided from the Saharasian areas. Currently, few of the most extreme patrist cultures are represented. The absence of data probably reflects the past inability of

83. Murdock, 1967, ibid.

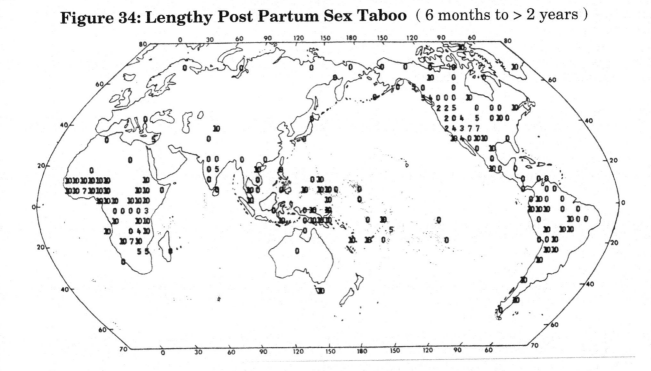

Figure 34: Lengthy Post Partum Sex Taboo (6 months to > 2 years)

"10" identifies regions where 91-100% of cultures have lengthy Post Partum Sex Taboos.
Numbers indicate the percentage of cultures within a given 5° x 5° region with lengthy post partum sex taboos (number times 10 = top end of 10-point percentage category). Regions marked "0" have no cultures with such taboos, while 1 = 1-10%, 2=11-20%, 3=21-30%, etc. Blank areas = data not available.

Correlation Table: Lengthy Post-Partum Sex Taboo (>6 months)

is positively correlated with the following other patrist variables:

By Column: Boastfulness is extreme
Narcissism index is high
Warfare prevalence is high
Slavery is present

Only one negative correlation was found to other patrist variables, and this was regarding the relationship to infant indulgence. A lengthy post partum sex taboo is correlated to a higher infant indulgence, which would be expected where men tended to stay away from wives with newborn children for lengthy periods. However, the relationship was neither strong nor uniform, as only 62 percent of the cultures in the data base shared both long post-partum taboos and high infant indulgence, with a relatively low statistical significance (from .10 to .05). It is anticipated that a data base which includes a greater quantity of cultural data from Saharasia on the post-partum sex taboo will more clearly demonstrate the positive correlation between a lengthy taboo and other patrist variables.

Western anthropologists, who have predominantly been male, to penetrate into the secluded female quarters of Arab/Asian Moslem cultures.

Adult Sexuality

17) Homosexuality

Homosexuality is often stated as being a "natural phenomenon", global and cross-cultural in its distribution — but clear and compelling scientific evidence to support such an assertion, free of the heated political pressures of the day, does not exist. Sir Richard Burton's study on homosexuality is the only one known to this author which discusses the general distribution of homosexuality in the period before the global spread of industrial society, which he felt had a geographical and climatic cause. The "sotadic zone" as he called it, included the entirety of Saharasia, plus parts of the Saharasian borderlands, specifically:[84]

1) the Mediterranean region from 43°N latitude south to 30°N latitude, which takes in all of North Africa, from Morocco to Egypt, plus Southern Spain, Italy and Greece.

2) Turkey, Mesopotamia, Syria, Iraq, Afghanistan, Pakistan and Northwest India.

3) Southern Central Asia, China, Japan, Indochina

4) unspecified parts of the "South Sea Islands", and "New World"

Unfortunately, Burton's review was not specific enough to construct a map, nor did his work, or the more recent and excellent study by Cory,[85] address the prevalence of the trait in a given culture, or its various institutionalized expressions such as eunuchism, socially-approved pedophilia, or boy-slave markets.

While it would be scientifically unwarranted to be too declarative on the question of "natural homosexuality" in such an early period of the genuine study of human sexuality, most of the evidence offered up in support of the argument for a purely innate or genetic explanation derive from examples which can be argued in the exact opposite direction. Aside from a minority of cases approximating physiological hermaphroditism, homosexuality does not appear to be innate or instinctual, but mainly derives from *social repression of heterosexual feelings of children, as well as a certain unknown percentage resultant from the adult seduction of children.*   In other words, homosexuality is predominantly the

84. Burton, R.: *The Sotadic Zone*, Panurge Press, NY, nd.

85. Cory, D.: *Homosexuality: A Cross-Cultural Approach*, Julian Press, NY, 1956; Cory's work does provide a starting point for construction of a map.

product of *heterosexual* sex-repression. Studies of unarmored, matrist cultures such as the Muria or Trobrianders suggest that homosexuals did not exist, except in those cases where missionaries or traders had erected barriers between the sexes. In such cultures, homosexuality was looked upon with puzzlement by the average person. Though it was not "taboo", with a penalty for transgressors, it had no observable prevalence given the easy and free expression of heterosexual romance within childhood and adolescence.[86] Certainly, nothing approximating or even suggestive of the homosexual bath-houses of contemporary Western cities can be found in sex-positive cultures. However, one can find example after example of such "institutions" in the histories of the most sex-repressive, anti-child, anti-female, war-making cultures, such as the imperial states of Greece, Rome and Persia. Greece, in particular, was openly homo-erotic, with socially-approved pedophilia, pairing of young boys with older men — but it was simultaneously severely repressive of adolescent heterosexuality, and was staunchly anti-female.

Homosexuality is also most prevalent in those institutions where young men and women are separated from each other, namely in single-sex schools, camps, seminaries, prisons, or on military ships — precisely within those cultures and institutions which place insuperable barriers between members of the opposite sex. Homosexuality often manifests just before or during early puberty, during the cultural artifact known as "latency"; in other words, homosexuality only appears after the repression of a child's natural heterosexual feelings.[87]

Homosexuality is not found in the animal world, except among domesticated, caged, or castrated specimens, without access to partners of the opposite sex. Some male animals will engage in a passive "display" or miming of the female when threatened with aggression, but sexual penetration does not occur in such instances. Also, some lower animals without clear sexual dimorphism have been misinterpreted as being "homosexual".[88] Such examples only serve to prove the rule of exclusively heterosexually directed instincts in wild mammals, however.

Homosexuality, in the vast majority of cases, may be explained as a result of an obedience-demanding and sex-negating childrearing and social structure, which works to physically separate boys from girls, and establish penalties for youthful heterosexual experimentation or breaking of the virginity taboo. Such penalties may include, in extreme patrist cultures, the use of severe beatings, jailing, or even death for offending children and adolescents. The powerful sexual drive cannot be repressed into oblivion, only diverted from its original purpose and goal such that it re-emerges in a form which is more easily expressed, and to which lesser social penalties are attached. In the West, moralists may object to homosexuality, but in fact they are often even more enraged by expressions of heterosexuality by children and adolescents. The following sentiments of Voltaire have likewise been expressed by Buffon, Westermarck, Havelock Ellis, Malinowski, Elwin, Reich, and others:

"The young males of our species, reared together, feeling this force that nature begins to develop in them and not finding the natural object of their instinct, throw themselves upon that which re-sembles it." [89]

Institutionalized forms of homosexuality were found in the political eunuchism of Byzantium, China, and other kingly states of the Saharasian borderlands. It appears historically connected to the bloated harem systems of patrist empires which developed either from, or as a response to, militant nomads from Saharasia. Ancient Assyria and Babylonia

86. Elwin, V.: *The Muria and Their Ghotul*, Oxford U. Press, Bombay 1942, pp.446,638,656; Malinowski, B.: *The Sexual Life of Savages*, Routledge & Keegan Paul, London, 1948, pp.377,382,394-5,397-8; cf. Cory, 1956, ibid., p.103.

87. Elwin, 1942, ibid.; Malinowski, 1932, ibid.; Cory, 1956, ibid., pp.108-9,349,398; Bullough, V.: *Sexual Variance in Society and History*, J. Wiley, NY, 1976, pp.25-7; cf. Bergler, E.: *Homosexuality, Disease or Way of Life?*, Collier, NY, 1956; Bayer, R.: *Homosexuality and American Psychiatry: the Politics of Diagnosis*, Basic, NY, 1981.

88. Cory, 1956, ibid., p.354.

89. As quoted in Cory, 1956, ibid., p.351; Also see pp.108,137.

Disturbances in Sexuality from Chemical Endocrine Disrupters?

The breakdown products of harsh pesticides and herbicides, as well as chemical residues from food-processing and packaging, often enter the food chain and act as *endocrine disrupters,* or *sex-hormone emulators*, disturbing the hormone balances of living creatures. The problem is especially acute for creatures at the top of the food chain, such as predatory birds, fish, reptiles and mammals (including humans), whose dietary habits concentrate and increase environmental levels of the chemicals. Eagles and other predatory birds long ago suffered from the effects of pesticide residues upon egg shells, leading to near extinction. More recent effects upon wildlife from the endocrine-disrupting chemicals include: Female molluscs (snails, mussels) turning into males; Production of vitellogenin (a protein necessary for egg-yolks) in some male fish, with hermaphroditism (both male and female sexual organs) in others; Reduced fertility among exposed turtles and alligators, due to underdeveloped male sex organs; Abnormal nesting behavior in birds, namely female-female pairing; Similar disturbed fertility among common and grey seals, and Florida panthers. Laboratory tests on rodents, hamsters and monkeys show a similar array of disturbances in sexual physiology and behavior, including reduced sperm counts, endometriosis, disturbed thyroid hormones, shrunken sexual organs, early-onset of puberty, persistent estrus (being "in heat" for an abnormal, prolonged time), birth defects of the penis and undescended testicles. Many of these same factors, such as endometriosis, early onset of menstruation and puberty, low sperm counts, and undescended testicles, have also been on the increase among human populations. Increases in cancers of sexual organs (breast, uterus, prostate, testicles) have also been observed among human groups with high levels of endocrine-disrupting chemical exposures. No clear alternative explanations have been offered for these phenomena by the health-medical community, which, with few exceptions, has remained generally silent on the problem of environmental toxins. To my knowledge there is no research underway addressing the clear possibility that such sex-hormone emulating pollutants might also be at work in altering the sexual physiology and behavior of humans, increasing the numbers of feminized males and masculinized females in exposed populations. This possibility may also underlie some of the *claimed* (but as yet unproven) differences in brain morphology and physio-chemistry between homosexuals and heterosexuals.
(*European Workshop on the Impact of Endocrine Disrupters on Human Health and Wildlife, 2-4 Dec. 1996 ,Weybridge UK ,Report of Proceedings* [Report EUR 17549] Copenhagen, Denmark: European Commission DG XII, April 16, 1997. Available from: European Environment Agency, Kongens Nytorv 6, DK-1050 Copenhagen K, Denmark; Also see: Theo Colborn, et al, *Our Stolen Future: Are We Threatening Our Fertility, Intelligence & Survival?* Penguin, 1996.)

demanded tribute in the form of young boys and girls for sexual slavery. Greece adopted its most widespread acceptance of homosexuality and pedophilia after the initiation of patrist tendencies, which came out of the Near East and Central Asia, to include severe taboos against female sexuality. Institutionalized pedophilia, where young boys are assigned to older boys or men to be used as homosexual partners, in Greece and elsewhere, is correlated to patterns of general child abuse and subordination to adult authority, to female subordination, severe female virginity taboos, and the military caste.[90]

Homosexuality appears to have been least prevalent among the aboriginal native populations of Southeast Asia, Oceania, and North and South America, the regions of a more matrist character. In these latter areas, it is sometimes found associated with shamanism, or the *berdache*, where a single individual of a cultural unit might act in a passive homosexual manner. Where homosexuality does occur in transitional matrist/patrist cultures, it is almost fully connected to the invasive traces of patrism, such as where a young boy is used as a sexual object by a group of warriors out on a hunt or raid, or where girls or boys are subject to adult male sexual abuse at an early age. A relationship to hermaphroditism may also exist in some of such cases, which would also explain its occasional, very isolated presence in some matrist areas.

Interestingly, it is among matrist-leaning cultures with greater sexual freedom, where the punishments for adolescent heterosexual expression are mild or absent, and where punishments for homosexuality are also absent, that homosexuality can hardly be found. It is *most prevalent* in

90. See Chapter 8 for documentation on this point. **Note 2005:** A growing body of anecdotal evidence suggests that homosexuality and pedophilia are at very high levels within the Saharasian Islamic world, where heterosexual repression is most intense. Clarification on this essential consideration must wait for a later time.

cultures with strong heterosexual taboos and punishments, and adult seduction of children (producing fear and hatred of the *opposite* sex), underscoring the genesis of homosexuality in cultural diversion of natural heterosexuality, and not in primary instinct. Indeed, the presence of severe taboos and punishments for adolescent heterosexual behavior are a giveaway for the certain presence and high prevalence of homosexuality.

No maps of homosexuality were attempted for this study, due to lack of a suitable data base. It is predicted that a geographical study on homosexuality will corroborate these general observations and demonstrate a patrist, Saharasian-related distribution, assuming the prevalence, institutionalized expressions, taboos, and punishment severities associated with the practices are also taken into account.

18) Prostitution

Prisons are built from bricks of Law; Brothels, from bricks of Religion.
– William Blake

Although prostitution has been called the "oldest profession", it would appear this is a chauvinistic insult against women, with honor of oldest profession probably belonging to the midwife, who was also the first healer and shaman. Several excellent histories and cultural summaries of prostitution exist,[91] but none have sufficient geographical orientation from which world maps can be constructed. It is commonly and erroneously assumed that prostitution was ubiquitous and present since the earliest times. However, I challenge this view for a number of reasons.

First, there does not exist, to my knowledge, any study of prostitution which addresses regional or national differences in the institution, identifying without speculative ambiguity where it exists now, in the past, or more importantly, identifying those cultures, nations, or regions in which it does not now or never did exist. Neither have data on prostitution specifically addressed its prevalence, institutionalized expressions, taboos, or associated penalties, all of which are most important for any functional analysis.

Burley and Symanski, cited above, have provided an excellent list of cultures in which prostitution appears, but no map. Neither have they identified the degree of its prevalence, or those cultures where it is known to have been absent. Still, their classification headings are themselves illustrative of the reasons for and origins of the practice: "...because of lack of, or restricted, opportunities for marriage", "Divorcees and others whose marriage has failed", "Non-virgins, unmarried girls, adulterers and other women of questionable moral character", "orphans", "diseased and outcast women", "lack of suitable husbands", and so on. All of these categories are suggestive of turbulent social conditions, economic hardship, warfare, and a generally low status for women, with classes of "tainted", "outcast" women, "divorcees", "widows" and the like who are not welcomed in society whether prostitute or not.

I take it as a basic starting point that no woman will sleep with a man who is repugnant to her, or for whom she does not have desire, unless:

1) She has lost her natural sexual feelings and has armored herself off from her desire to be emotionally loved, due to antisexual childhood traumas, or more recent sexual trauma, such as rape or sexual slavery; perhaps from an early age she was forced to unquestioningly obey the will of men, and to ignore her own feelings, viewing emotionless sexual contact as a "duty" or "obligation" to which she must submit, often at fear of her life if she were to disobey.

2) She is in dire need of money, from the standpoint of starvation or deprivation of self or children, as within a loveless marriage where the wife endures the husband out of fear of economic loss, or where a woman's

91. Burley, N. & R. Symanski: "Women Without: An Evolutionary and Cross-Cultural Perspective on Prostitution", in R. Symanski, Ed., *The Immoral Landscape*, Butterworths, Toronto, 1981; Scott, G.: *A History of Prostitution from Antiquity to the Present Day*, Medical Press, NY, 1954; Evans, H.: *Harlots, Whores & Hookers: A History of Prostitution*, Picture Library, NY, 1983.

only perceived escape from grinding poverty or starvation is prostitution.

3) She is a sexual slave, being threatened with death if she tries to escape, and has no place to turn to for help. In such cases, the pimps and local police authorities work together to trap and kidnap the usually younger female, who may be purchased for a small sum from impoverished parents, in a cultural system which in any case views the young female as a financial burden or worthless object.

The presence of very strict female virginity taboos which block sexual access to females by young men, a primary group of customers of prostitutes, must also be present before prostitution can gain in importance. Women must also be made economically dependent upon men, without equal means for taking care of themselves, before they will turn to prostitution. Older men with surplus wealth, another primary group frequenting prostitutes, must also be present in a sex-repressive social situation, with compulsive marriages and strong adult controls over children. In the above cases, prostitute-frequenting men do not have easy sexual access to women of their own class or caste, but do have sexual access via money to generally younger women of a lower, economically deprived class or caste. The relationship between slavery and prostitution has also been fundamental, and female sexual slavery still occurs today,[92] primarily in the Near East and Asia, either as a remnant from the days of the great harems, or in connection with severe poverty, female virginity taboos, and low female social status. Prostitution is generally a lower-class, poverty-driven phenomenon, dependent upon a supply of frustrated young and elderly men with excess cash to spend, and few or no willing females otherwise available to them.

The romantic side of prostitution is a money-making illusion, served up to prostitute customers, and to a sex-starved public by Hollywood producers and New York publishing houses. The woman who wrote the book *Happy Hooker* is today no doubt the "happy writer" or "happy lecturer", or maybe even the "happy housewife/mother", and no longer earns her money by the sexual practices of former times, the worst experiences of which undoubtedly were deleted from her best-selling fantasy-book. I submit, in the brothels of Asia, where virtual slavery exists for thousands of kidnapped teenage and pre-teen girls, there is little happiness or "erotic bliss" to be found.

Hence, a connection between prostitution and class, caste, and slavery is anticipated, as well as a connection to general male dominance, warfare, and economic hardship. Indeed, Sorokin provides evidence for the increase in prostitution which invariably occurs during times of war, economic depression and starvation.[93]

Also, "temple prostitution" is often pointed to as an aspect of the earliest "Goddess religions", as a means of proclaiming its innate and ancient character, and such evidence would appear to counter the argument for a Sharasian, post desiccation (c.3500 BCE) origin of prostitution as a social institution. However, there is sufficient reason to doubt that temple prostitution existed before c.3500 BCE. As discussed in Chapter 8, the great Mesopotamian temples in which such prostitution was often claimed to occur simply did not exist until the coming of the patrist states, after which a male priestly caste, divine kingship, widow murder, and other antifemale practices came into being. A female goddess may have been maintained from earlier times in such kingly states, and even had new temples constructed to her. The practice of temple prostitution, however, appears to be a vestige of male domination, arranged marriages, and the vaginal-hymen blood taboo, introduced only *after* cultural perceptions of the female and reproductive functions were radically altered and

92. Barry, K.: *Female Sexual Slavery*, Avon, NY, 1979.

93. Sorokin, P.: *Hunger As A Factor In Human Affairs*, U. Florida, Gainesville, 1975, pp.118,127-8.

laden with much anxiety. Indeed, the practice of defloration of young girls or brides by priests or strangers in a temple appears firmly rooted in the fear of the vagina and hymenal blood by the common man. This analysis is supported also by the existence of special "Christian brothels" founded by the Church of Rome in the Middle Ages. A more specific historical chronology, with cited examples of these dynamic relationships, is given in Chapter 8.

By contrast, it appears certain that fertility rites or festivities celebrating sexual pleasure did take place before the onset of patrism, but wholly as celebrations, with feasting and dancing, marking the end of a hard winter, and the planting or harvesting of crops. The character of institutionalized temple prostitution is greatly different from any such seasonal fertility rite or easy granting of sexual favors within matrist cultures. Also, the occasional giving of a gift from a young man to a young woman in sexually active cultures certainly should not be classified the same way as either institutionalized temple prostitution or other forms of sexual slavery, for the purposes of a cross-cultural or geographical study.

The subordinated organization of young girls in temples, for the sexual dominance of strangers or priests appears functionally identical to the later institution of the brothel, where young girls are similarly compelled into sexual relationships with both customers and pimps. In both cases, an economic advantage accumulates to the generally male-dominated institutions (temple, brothel) while the "recruitment" of girls was (and is) dependent upon the existence of other sex-negative, patristic social institutions, as previously discussed.

From this, it might be anticipated that a map of prostitution, which traced the migration of peoples and their interactions with neighboring groups, and which also took into account the prevalence, form, institutionalized expressions, taboos, and severity of penalties, would show a dominant Saharasian distribution, similar to class, caste, slavery, female subordination, and general adolescent sexual repression.

Kinship and Inheritance

19) Descent (Kinship)

Figure 35 is a map of patrilineal kin groupings, as opposed to matrilineal or cognatic kind groups, the data for which come from Murdock's *Ethnographic Atlas*.[94] Block regions of a high percentage are characterized by patrilineal kinship. Regions of a low percentage are characterized by matrilineal, duolateral, or quasilineage kinship, or by bilateral or ambilineal cognative kin groups. This map confirms the presence of almost uniformly patrilineal descent in Saharasia, and its borderlands.

Sex-economic theory would anticipate that an unarmored, instinctually-based family unit or kinship system would trace its ancestry either through well-remembered or mythical ancestors of the group, and/or through the female line, given the obvious functions of childbirth and in some cases due to ignorance of physiological paternity. Even where the role of the male in procreation was inferred, however, no compelling instinctual reason would have existed to change the previously existing system of female or cognatic descent. The cross-cultural correlations to patrilineality, however, imply that there are other psychological and emotional factors which turn a culture away from the maternal-child bond, towards the paternal-child relationship.

94. Murdock, 1967, ibid.

> ### Correlation Table: Patrilineal Descent
> is positively correlated with the following other patrist variables:
>
> By Column:
> - Slavery is present
> - Castes are present
> - Class stratification is high
> - Movable property inheritance favors male line
> - Land inheritance favors male line
> - Bride price is present
> - Polygamy is present
> - Marital residence near male kin
> - Female premarital sex taboo is high
> - Painful female initiation rites present
> - Male genital mutilations are present
> - Segregation of adolescent boys is high
> - Childhood indulgence is low
>
> No negative correlations were found between "Patrilineal Descent" and other patrist variables.

From the cross-cultural correlations alone, it can be inferred that, where generally negative views about the female and childbirth began to increase, tracing of ancestry through the female line would increasingly be at risk. Also, where strongman rulers dominated, as where slavery or the military caste were present, and where wealth and power was concentrated in male hands, inheritance of wealth or leadership roles would generally occur along the male line, reinforcing the pattern of male-dominated hierarchy and the warrior caste. Patrilineal kinship would

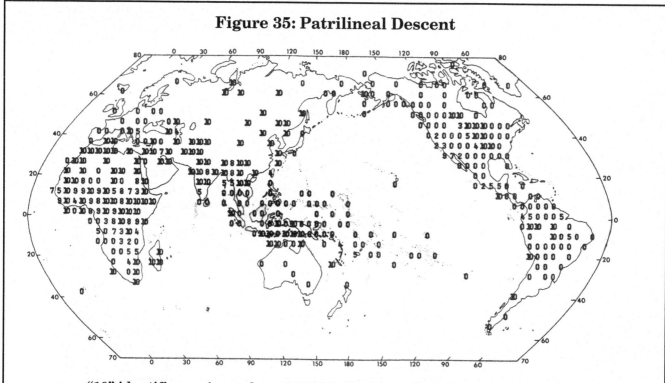

Figure 35: Patrilineal Descent

"10" identifies regions where 91-100% of cultures have Patrilineal Descent.
Numbers indicate the percentage of cultures within a given 5° x 5° region with patrilineal descent (number times 10 = top end of 10-point percentage category). Regions marked "0" have no cultures with patrilineage, while 1 = 1-10%, 2=11-20%, 3=21-30%, etc. Blank areas = data not available.

thereby be correlated with negative views about female functions, infant and childhood traumas, patterns of male dominance, and warfare.

These suggested patterns are confirmed in a review of the social factors correlated with patrilineal descent. Other related factors have already been discussed, in the context of the location of residence of the newlywed couple. Patrilineal descent, coupled with other correlated factors such a patrilocality and male inheritance patterns, works against the biologically necessary maternal-infant bond, and places the female and her children at the mercy of the husband in the event of any divorce or loss of love in marriage. In the matrilineal system, women and their children are generally cared for by a network of their matri-kin, including their brothers and maternal uncles. Their husbands, in turn, provide care for their own sisters and sororal nieces and nephews. Nobody can be divorced from their supportive maternal kin, and marriages thereby generally avoid gathering significant compulsive qualities characteristic of patrilineage, where divorce might force a woman to stay with a man she did not love, in order to keep her children, and her economic security.

20) Cognatic Kin Groups

Figure 36 is a map addressing cognatic kin groups, where ancestry is traced through kin groupings outside of blood lines, such as well-remembered or mythical heroes and ancestors. Data is, once again, from the Murdock *Ethnographic Atlas*.[95] Block regions of high percentage are characterized by a strict tracing of kin through blood line, and the absence of cognatic kin groups. Regions of low percentage are characterized by the presence of bilateral or ambilineal cognatic kin groups, without strict tracing of kin through blood line.

95. Murdock, 1967, ibid.

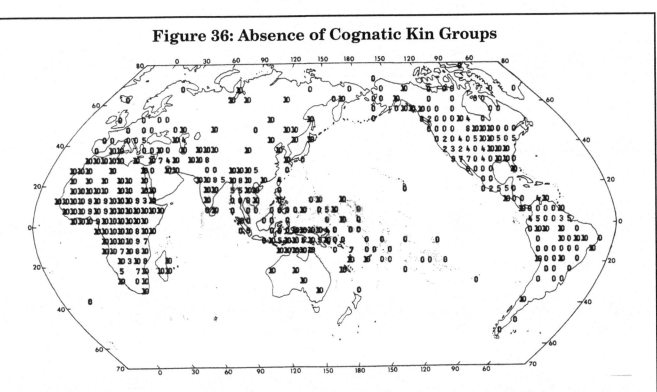

Figure 36: Absence of Cognatic Kin Groups

"10" identifies regions where 91-100% of cultures Lack Cognatic Kin Groups.
Numbers indicate the percentage of cultures within a given 5° x 5° region without cognatic kin groups (number times 10 = top end of 10-point percentage category). Regions marked "0" have no cultures lacking the trait, while 1 = 1-10%, 2=11-20%, 3=21-30%, etc. Blank areas = data not available.

Correlation Table: **Absence of Cognatic Kin Groups**
is positively correlated with the following other patrist variables:

By Column:
    Slavery is present
    Castes are present
    Isolation of pregnant women is high
    Exchange of female relatives is present
    Bride price is present
    Polygamy is present
    Marital residence near male kin
    Male genital mutilations are present
    Segregation of adolescent boys is high
    Childhood indulgence is low

One negative correlation was observed, between the "absence of cognatic kin groups" and "movable property inheritance favors male line", which I have not been able to explain. Still, both the correlations and the spatial pattern on the maps show a strong agreement between patrilineal descent and the absence of cognatic kin groups, with the latter having a slightly greater distribution than the former.

As mentioned above in the section on Descent, an unarmored, instinctually based family unit or kinship system would trace its ancestry either through well-remembered or mythical ancestors of the group, and/or through the female line, given the obvious functions of childbirth and ignorance of physiological paternity. Cognatic kin groups are not likely to prevail where children are viewed as the "property" of the parents. Where children must be bent to the parental will, through obedience training invoked by a particular guardian, or where they are seen as property of a particular parent, then blood-line kinship, either patrilineal or matrilineal, would gain in importance over more generalized social-group ancestry. Also, where matters of inheritance are important, as with land or significant movable property, and particularly where significant wealth-determined class structure existed, blood line kinship would be increasingly stressed, and cognatic kin groupings de-emphasized.

Although patrilineal kinship ties would appear to be more firmly related to a decline in female status, "matrilineal" kinship cannot completely stand as its opposite, as matrilineage often may not grant any significant advantage to women, but only to a particular woman's male relatives. Cognatic kin groupings may, however, signify a more relaxed and fluid "collective" view of children, who thereby would more likely form their own social structures, such as the Ghotul of the Muria, the Bukumatula of the Trobriander, or other forms of children's democracy.

21) Inheritance rules: Land and Movable Property

Economic control over resources translates into real social power. In cultures dominated by authoritarian males, inheritance rules tend to favor distributions of wealth to males over females. In this regard, it does not matter if the benefiting males are related patrikin or matrikin of the deceased.

Figure 37 is a map giving the percentage of cultures in a given block region where land inheritance rules favor males over females, specifically sons and/or brothers, irrespective of whether they are patrilineal or matrilineal male kin of the deceased.[96] Regions with a low percentage value are characterized by equal distributions of land between children of both sexes, or by the absence of property rights or inheritance rules.

96. Murdock, 1967, ibid.

Figure 37: Land Inheritance Favors Male Kin

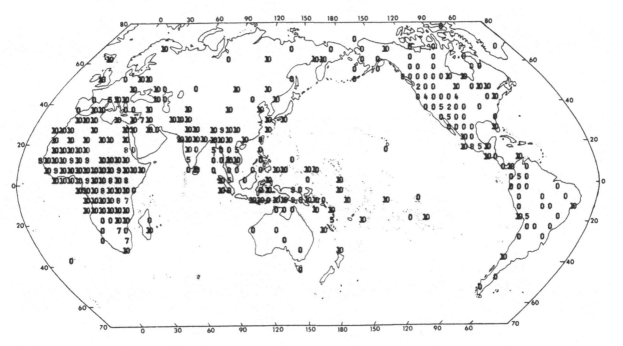

"10" identifies regions where 91-100% of cultures have Male-Favoring Land Inheritance.
Numbers indicate the percentage of cultures within a given 5° x 5° region with male-favoring land inheritance (number times 10 = top end of 10-point percentage category). Regions marked "0" have no cultures with such a trait, while 1 = 1-10%, 2=11-20%, 3=21-30%, etc. Blank areas = data not available.

Correlation Table: Male-Favoring Land Inheritance Rules

are positively correlated with the following other patrist variables:

By Column: Slavery is present
Castes are present
High God present, active, supportive of human morality
Movable property inheritance favors male line
Patrilineal descent is present
Bride price is present
Marital residence near male kin
Male genital mutilations are present
Infant pain infliction is high
Infant protection from environment is low

No negative correlations were found between "Land Inheritance Rules" and other patrist variables.

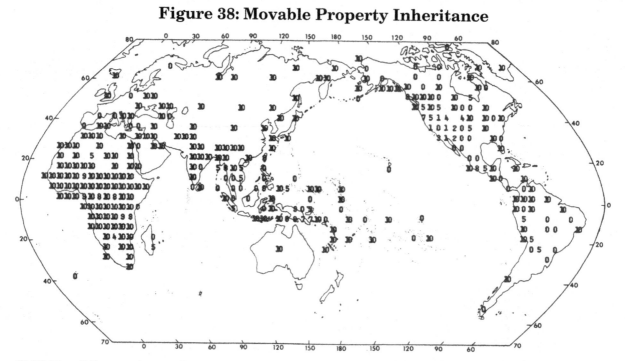

Figure 38: Movable Property Inheritance

"10" identifies regions where 91-100% of cultures have Male-Favoring Property Inheritance.
Numbers indicate the percentage of cultures within a given 5° x 5° region with male-favoring property inheritance (number times 10 = top end of 10-point percentage category). Regions marked "0" have no cultures with the trait, while 1 = 1-10%, 2=11-20%, 3=21-30%, etc. Blank areas = data not available.

Correlation Table: Male-Favoring Movable Property Inheritance Rules
are positively correlated with the following other patrist variables:

By Row:
      Infant protection from environment is low
      Marital residence near male kin
      Female status inferior or subjected
      Patrilineal descent is present
      Land inheritance favors male line
      Castes are present

Two negative correlations were found between "movable property inheritance rules which favor the male" and other patrist variables, namely to "segregation of adolescent boys" and "cognatic kin groups are absent". The error in the structure of the "segregation of boys" variable was previously discussed, but the negative correlation to "cognatic kin groups are absent" remains a mystery.

Similarly, Figure 38 is a map giving the percentage of cultures in a given block region where movable property inheritance rules favor males over females. Regions of low percentage are characterized by equal distributions of movable property, the absence of rights or rules, or the destruction, burial, or giving away of property.

Both inheritance variables demonstrate similar Saharasia (plus borderland) distributions, and statistical correlations to a variety of male-favoring, female-suppressing cultural factors, including those which assault and traumatize the infant and child.

22) Ritual Widow Murder = *Mother Murder*

A map was not produced of the widow-murder variable, but my historical reviews, summarized in Chapters 8 and 9, identified its presence primarily in those areas previously classified as extreme patrist on the World Behavior Map, particularly in Saharasia and its borderlands. It was also present among a few extreme patrist, divine-king cultures of Oceania and the New World, but never practiced there to the extent as was found, for example, in India or China. Ritual murder of females in Europe during the "witch-burning" period, rivaled or surpassed that found in China or India, though it existed for a shorter time period and was not confined to widows. A map of the various forms of female murder should therefore yield a distribution similar to that seen for extreme patrism.

Psychologically, the various forms of ritual female murder were compelled and sustained by sadistic, antisexual, and antifemale feelings of the average citizen. Widows, often young — but no longer virgins — constituted a significant sexual threat to sex-anxious males in warlike cultures whose character structures were devoted to the crushing down of tender or gentle sexual feelings. Various severe widow taboos worked to elicit "voluntary" cooperation of the women's closest relatives, and the women themselves, in the murders.

In mother-murdering cultures, women are generally economically impoverished, and widows are forbidden any significant inheritance which goes to male relatives of the deceased. Thereby, they are made destitute and additionally became an object of open scorn, even to their own blood relatives. Various eyewitness reports on the subject by travelers are clear: women who refused to cooperate in their own ritual murders would often be hog-tied and thrown on the sacrificial fire anyway. Others would be drugged or clubbed into submission, or given narcotics which often yielded a public aura of women "volunteering" to be burned or hanged.

While greed for inheritance money by male relatives must have played a role in the murders, their ultimate origins lay in the sadistic, antifemale energies of the masses. The same is true of the European witch-burnings, organized by sadistic woman-hating celibate priests, who also confiscated the wealth and property of their victims. But, one must not forget that someone had to carry out the actual murder itself, the cutting of the throat, the strangling, burning, or burying alive. The person to whom this task was assigned is most revealing of the underlying psychology behind the murders. In both China and India, the brutal acts were most often mandated to be carried out *by the woman's firstborn son*, which makes the process in reality, not "widow murder" (which implies a stranger or friend of the dead husband killing an old woman) but the even more ghastly act of *a young man murdering his own mother*.

One has to imagine what kinds of cultural practices must be in place to render an entire culture, and generations of young men, to become so cut-off from tender feelings towards their mothers as to be capable of carrying out such brutal actions — not against a stranger or a cultural enemy during warfare — but against their own mothers. Also, the ritual executions of women were usually carried out in public, with mass gatherings and a circus-like atmosphere. Without the existence of a strong element of sadistic energy within such a female-murdering culture, it would have been physically impossible for the act of public mother murder to be carried out.[97]

97. See the sections on India, China and Europe in Chapter 8 for specific discussions and citations.

Religion

23) High God Religion and the *Mother Goddess*

Figure 39 is a map of the *Ethnographic Atlas* "high god" variable.[98] Block regions with a high percentage are characterized by belief in a high god who is active in human affairs and specifically supportive of human morality. The religions of the Jews, Christians, and Moslems are most clearly characterized by such features, but authoritarian similarities are found among Hindu, Confucian, Buddhist, and other religions. Block regions with a low percentage figure are characterized by either the absence of a high god, or a high god who is inactive and/or not supportive of human morality. The map clearly indicates uniform distribution of an active, morally-supportive high god across Saharasia and its borderlands.

It is generally accepted by scholars of the subject that the earliest humans made carvings of females as a predominant art form. From this some have argued for the widespread existence of a female-centered religion, a "Mother Goddess" of some sort.[99] On the surface, this opinion is in accordance with the existence of matrist cultures at the earliest times. However, it is also clear that, whatever significance the earliest peoples gave to their female figurines, whatever kind of "religion" they developed in attempts to understand the nature of life and the universe, in no way did that early philosophy or religion mirror what we have today in the form of organized patrist religion, steeped in ritual and hierarchy.

As discussed in the historical chapters to follow, the earliest female

98. Murdock, 1967, ibid.
99. Stone, M.: *When God Was A Woman*, Harcourt, Brace, Jovanovich, 1976; cf. von Cles-Reden, S.: *The Realm of the Great Goddess*, Prentice- Hall, NJ, 1962; Goldenberg, N.: *Changing of the Gods: Feminism and the End of Traditional Religions*, Beacon Press, Boston, 1979; Murray, M.A.: *The Genesis of Religion*, Routledge & Kegan Paul, London, 1963.

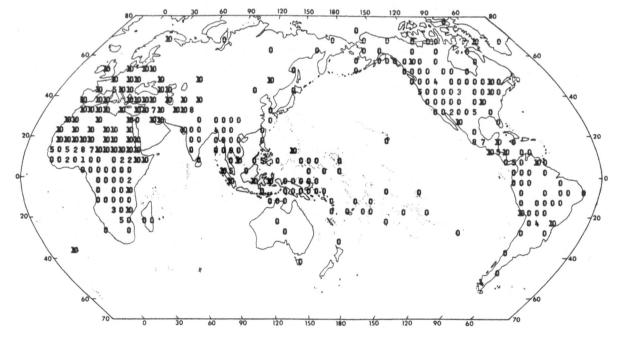

Figure 39: Presence of a High God,
Active in Human Affairs, Supportive of Human Morality

"10" identifies regions where 91-100% of cultures have a Moral/Active High God.
Numbers indicate the percentage of cultures within a given 5° x 5° region with a moral/active high god (number times 10 = top end of 10-point percentage category). Regions marked "0" have no cultures with the trait, while 1 = 1-10%, 2=11-20%, 3=21-30%, etc. Blank areas = data not available.

figurines were found in association with peaceful social conditions. Only for a brief period of history do we find female "goddesses" with temples, and these periods are transitional phases, when early matrism was being overrun by dominating patriarchal warrior cultures. Often, male priests became the dominant "servants" to such female deities, who were in some cases transformed into murderous goddesses, extracting a toll of human sacrifices from worshipers. The earliest female figurines are graceful, sexual and some are even erotic. Later female figurines, following episodes of drought, famine, starvation, migration, invasion, warfare, chaos, etc., are typically asexual, with tiny breasts, cold faces and even fierce looks. Or the female figurines vanish altogether, being replaced by male gods. Specific examples are given in the historical Chapters of Part III.

Sex economic theory predicts that extreme matrist culture, as during the earliest times, would naturally view female reproductive functions with awe — the mystery of new life, springing forth from the womb. The other great mysteries of the universe which have so preoccupied humans for so long would properly be subsumed into the same scheme or grand view of life, in which the female played a central role. It is therefore not difficult to imagine the female figurines (along with other associated material culture) as evidence of a broader matristic world view which held women in at least equal esteem to men, and perhaps even in higher esteem.

Patrist religion also does this, though in an indirect manner, by its great emphasis upon distrust of sexuality and the female — original sin, the devil, etc. — the chaste male god is constantly at war with the sexual-female devil, and virtue is defined by the avoidance of natural sexuality.

Later in early history, when patrism begins to develop across Saharasia, there do appear a few examples of domineering "Mother Goddesses", who are scarcely different from their later male counterparts. As patrism increases in power and strength, the goddesses are either married off to more powerful male gods, they become exceedingly fierce themselves, or they completely disappear from history,[100] leaving behind only the newer male gods, who fiercely dominated the social landscape.

Of interest from a sex-economic view is the degree to which a culture defines the natural environment in pleasant, comfortable terms, or in unpleasant, antagonistic terms. Such attitudes, or perceptions of the environment, appear to be largely shaped during infancy and early childhood. The qualitative aspects of early environment are of importance to religion in the context of whether or not the spirits or gods are defined as supportive, nurturing and friendly, or hostile, vicious, punitive, and angry. Another factor is the degree to which the animating forces or spirit-forces are defined as *being in* the real or natural world, the *here and*

100. Stone, 1976, ibid.

***Correlation Table:* Active, Morally-Supportive High God**
is positively correlated with the following other patrist variables:

By Column: Castes are present
Land inheritance favors male line
Polygamy is present (unrestricted form)
Female premarital sex taboo is high
Male genital mutilations are present

A contradictory, negative cross-cultural correlation existed between "high god" and "unrestricted polygamy" because, while many polygamous cultures have an active, morally-supportive high god, not all cultures with such a god are polygamous in an unrestricted manner. This question of the restrictions to polygamy in Islamic-Moslem regions, where polygamy is widespread, has been previously discussed.

now versus being divorced from the everyday world into another "plane of existence", a "heaven" or "hell", or another "dimension". Also important is the gender of the deity or spirits. If the "gods" exist in the real world, they may not need religious specialists for the average person to make contact with them, to see or feel the forces which animate the natural world, to interpret reality or intervene. If they have *departed* from the real world, then religious specialists may be required to act as intermediaries for the common person, who can not see or commune with such forces. Here, we touch upon the issue of "perception" which itself is a reflection of human armoring.[101]

A "natural religion", if such exists, would reflect a given people's own organic sense impressions. The degree to which organic sense impressions would become impaired or clouded, as from early childhood trauma or armoring, would in a general way explain the development of religions wherein the forces of nature were no longer observable or felt, having "departed from the sensible, natural world" into a "heaven" and/or "hell" which lay beyond the perception of the ordinary armored individual. It is therefore not accidental that patrist High God religion, with its spirit-versus-flesh, other-worldly qualities, demands the rigid suppression of sexual pleasure, particularly adolescent, premarital sexuality. Diverted and dammed-up sexual energy is the source from where comes armor that blocks one from having a deeper emotional contact with the real world. It is also the source of the quasi-orgiastic, often streaming and ecstatic qualities of deep religious faith. If one is having regular and full orgasms, one feels the streaming life-energetic currents in the body, and can see and feel various life-energetic expressions in nature. One feels more a part of nature and the cosmos, and ideas about god and the universe are intrinsically real-worldly, and not other-worldly.[102]

The above description is closer to animistic religion, which argues for spirit-energy forces, or the spirits of ancestors, to be residing within specific places on the landscape, in trees, streams, special places, etc. While animism in some cultures can be far more "other-worldly" and ritualistic than in others, in general, the animating forces of the cosmos are, in such a world view, a verb or an energy; a force, not a "being". The Great Spirit (verb) of the North American Plains Indians, *Wakataka,* literally means *the force which moves all things*. An energy, not a personality. Even some elements of patristic religions capture the essence of this phenomenon. The concept of *Chi energy* from Chinese medicine probably predates to the earliest periods of history, before the advent of extreme patrism, as does the similar concept of *Prana*, widespread in India. Even Christians speak about being "moved with the spirit" of the "holy ghost". But these are, for patrist high god religions, considered only secondary elements of a hierarchal pantheon, requiring spiritual experts and ritual to "make contact", for "salvation". For matristic cultures, hierarchal and ritual qualities are reduced or non-existent, and the creative forces of the universe are accessible to everyone.

In this context, we may turn once again to Wilhelm Reich for clarification of these weighty matters, for it was his discovery of the *physical and measurable, objective orgone energy, or "life energy"* — a fundamental animating natural "force which moves all things" — that unarmored matristic cultures appear to be describing. They could feel, see and touch it directly, through the senses, without the need for any "spiritual experts" to guide them into contact with this basic life force of the cosmos. Reich's orgone satisfies many of the descriptions of the older "vital force" and "aether" theories simultaneously. The energy exists as a ubiquitous ocean, filling all space, and moving in a pulsatory spiral-form manner which can be seen in the atmosphere by the naked eye,[103] under the right

Paleolithic "Venus" Figurine "Goddess" or something more ordinary (clay-doll for children)?

101. Reich, W.: *Ether, God and Devil,* Farrar, Straus & Giroux, NY, 1973.

102. Reich, 1973, ibid.

103. Reich, W.: *Discovery of the Orgone, Vol. 1: Function of the Orgasm,* Farrar, Straus & Giroux, NY 1973; *Vol.2: The Cancer Biopathy,* Farrar, Straus & Giroux, NY 1980; Also see: DeMeo, J.: *Orgone Accumulator Handbook*, 1989, Natural Energy Works, PO Box 1148, Ashland, Oregon 97520.

conditions, and assuming ones brains and ocular organs are not too contracted and spastic from armoring. Historically, as humans began to armor up in association with increasingly patristic social elements, this direct organic sense-perception of bioenergy was lost and the parental figures (ultimately, the father-figures) were elevated to god-like stature.

This above view is also supported by the findings of Hanspeter Seiler, who made a review of archaeological materials around the world, finding that the spiral form was dominant among the decorative art motifs of more peaceful, matristic groups. More angular, geometric forms replaced the spirals as armoring and patrism increased in influence (also see discussion on p.296).[104] In short, matristic peoples had a more precise and intact perceptive apparatus, and could see and feel, with their eyes and body, the physical creative forces of the universe, something which the more armored patristic groups lost due to the damaging culture-wide experiences of infant trauma and sex-repression.

Whatever the source of matrism's inspiration, patrist culture clearly has replaced it, in nearly every corner of the globe. The existence of a natural energy principle, or the female principle, as a focus of religious worship is notable only for the marginalization which has occurred over the centuries, even among animists. One may pray to a sweet, gentle Mother Mary or to the young and soft, forgiving Jesus, but the dictates of the local male priesthood (codified into law and general social philosophy) on matters of infant and adolescent sexuality, marriage and divorce, are going to have a whole lot more to say about what an individual can do with their life than anything else. Any remnants of matrism are today mostly confined only to icons, wholly defined by extremely patristic, sex-negative authoritarian ideologies.

Patrism is characterized by dominating male gods, heading up a divine family or subordinate pantheon of kindred spirits or semi-gods. Patrist gods, like the father-figures of the family structures which created them, demanded certain forms of moral behavior from their followers. The godly dictations are largely aimed at infants, children, sexual and family matters; their moral codes demand painful and pleasure-avoiding sacrifices from adherents. Cranial deformations, genital mutilations, and widow murder come to mind here. Natural functions, such as defecation, micturition, sexual intercourse, childbirth, menstruation, and so forth, become heavily burdened with rules and ritual purifications. Puberty, marriage, affirmation of offspring, and other social functions are of great concern to the god, and are also ritualized. The patrist god does not exist within nature, but is separated a distance from it, requiring full-time religious specialists to interpret and act as intermediaries. These intermediaries act on the god's behalf to invoke the moral code as the law of the land. Police and the military are the enforcers of the code. Patrist religion thereby maintains links to wealth, upper classes, caste, and other forms of social hierarchy, and historically has supported the powers of kings (or political parties) over the needs and interests of the common person.

The fact that religion plays a central role in cultural attitudes toward sexuality, and in the roles of men and women in society is straightforward enough. In cultures dominated by patrist attitudes, male gods exist alongside a male-dominated priesthood, with power links to the head of state. The moulding of children to the given social order thereby occurs in a highly structured fashion, wherein children have little to say about what happens to themselves. In many cases, certain harsh and traumatizing infant treatments are mandated by religion, such as cranial deformation or genital mutilations. It is likely that in cultures where the natural instincts of children and adults are strongly diverted into significantly narrow choices, as with the abolishment of adolescent romance and love-

104. Seiler, H.: "Spiralform, Lebensenergie und Matriarchat", *Emotion*, 10:137-167, Berlin, 1992.

match marriages, or with the abolishment of a woman's right to freely divorce or to control her own fertility, the role of a high god becomes much more supportive of human morality through the following mechanism:

Where natural feeling and biology are more harshly suppressed in the infant, child, and adolescent, the great displacement of instinctual, emotional energy causes not only disturbed, violent characterological changes when the individual grows to adulthood, but also changes the projected character of the anthropomorphic deity in a similar direction. Like the "believers", the "god" develops an asexual, pleasure-hating and violent character!

Just as the child must plea to the father for forgiveness from physical punishments and abuse, just as the adult must subordinate themselves to the warrior-king power-figures of the State, so too must the average individual, in their worship, plea for forgiveness for their "sins of pleasure" from the angry father-like deity.

Consider when individuals are called upon to commit acts which they might not normally engage in, such as the ritual murder of relatives, perhaps loved ones, as in widow murder or the "honor killing" of young girls for loss of virginity before marriage. A High God, active in human affairs and supportive of human morality, would predominate in such cultures, with human representatives on Earth holding powerful social positions. Nobody dares challenge them, because their authority flows "directly from god".

None of this, of course, says anything especially revealing about the original reform-minded intentions of the founding figures of some religions. Almost all religions superficially seek to champion "brotherhood", "peace on Earth", the "elimination of violence and the betterment of humankind", etc. Jesus brought a message of peace and universal brotherhood; likewise with Buddha and many others who championed social reforms. Even Mohammed was a reformer within the context of his own extremely repressive culture. However, to the extent that the followers and devotees of these (sometimes) reformers continue to champion various forms of infant trauma, or child and adolescent sex-repressive measures (such as masturbation phobias, genital mutilations, and the female virginity taboo), *they guarantee the creation of the very social chaos and pathology they profess to oppose.*

A survey of the Holy Bible by deMause, regarding its discussion of children, merits quotation on this important issue:

"...certainly here [Holy Bible] one should find empathy toward children's needs, for isn't Jesus always pictured holding little children? Yet when one actually reads each of the over 2000 references to children listed in the Complete Concordance to the Bible, these gentle images are missing. You find lots on child sacrifice, on stoning children, on beating them, on their strict obedience, on their love for their parents, and on their role as carriers of the family name, but not a single one that reveals empathy for their needs." [105]

The central "holy books" of other patristic, high god religions (Torah, Koran, Vedas, etc.) are no better on this point. Davies, who has extensively studied the phenomena of human sacrifice, a pathological side of extreme patrist high god religions where the "god" demands the life of human victims, has echoed the conclusions given above regarding the relationship between destructive aggression and belief in a patriarchal authoritarian high god:

105. deMause, 1974, ibid., pp.16-7.

"Paradoxically, then, the violence of Homo sapiens may be said to have arisen through the very thing that differentiates him from the beasts — his faith." [106]

Social Structure

24) Class Stratification

Figure 40 gives a map of class stratification, excluding those which are based only upon political or religious affiliation, as taken from Murdock's *Ethnographic Atlas*.[107] Block regions of high percentage are characterized by class stratification based upon hereditary aristocracies, landed or other resource-monopolizing elites, wealth, or complex social classes. Regions of low percentages are characterized by an absence of class stratification among freemen. Where social differences exist because of individual initiatives alone, such as skill, valor, piety, or wisdom, class stratification is considered to be absent. Stratifications based upon sex, castes, or slavery are not addressed.

Class stratification which is rigidly encoded into laws, customs, or inheritance, appears as a social factor working to oppose social mixing, mobility, and intermarriage, with an impact upon the lives of children and women similar to but of lesser intensity than either castes or slavery. Intermarriage across class lines can only be prevented by enactment of rules and customs which prevent children from playing with each other, or from developing inevitable romances with each other. This is only done

106. Davies, N.: *Human Sacrifice in History and Today*, Wm. Morrow, NY, 1981, p.282.

107. Murdock, 1967, ibid.

Figure 40: Class Stratification

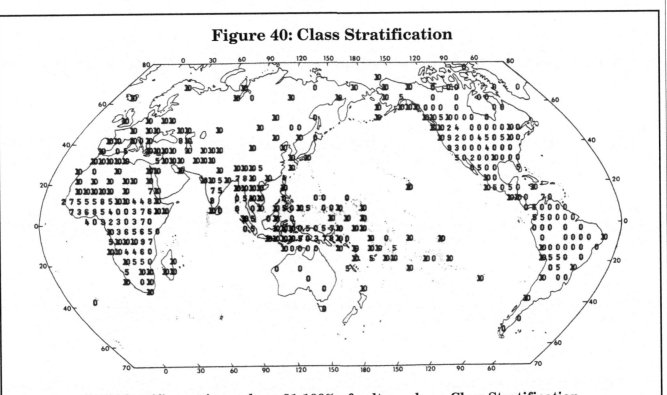

"10" identifies regions where 91-100% of cultures have Class Stratification.
Numbers indicate the percentage of cultures within a given 5° x 5° region with class stratification (number times 10 = top end of 10-point percentage category). Regions marked "0" have no cultures with the trait, while 1 = 1-10%, 2=11-20%, 3=21-30%, etc. Blank areas = data not available.

by virtue of sexual repression and other forms of "obedience training", starting with infancy. This trait appears to be the diluted remnant of possibly a stronger social barrier from an earlier historical period, when invasions, mass migrations, or slavery occurred, placing one group in a dominating position over others.

Correlation Table: **Class Stratification**
is positively correlated to the following other patrist variables:

By Column:

Insult sensitivity is extreme
Warfare prevalence is high
Slavery is present
Castes are present
Full-time religious specialists are present
Patrilineal descent is present
Bride price is present
Grandparents have authority over parents
Marital residence near male kin
Female premarital sex taboo is high
Male genital mutilations are present
Attention to infant needs is low
Protection of infants from environment is low
Infant indulgence is low

One significant negative correlation was observed, between "class stratification" and the "exchange of female relatives" for marriage; this negative correlation remains unexplained.

---

25) Caste Stratification

Figure 41 is a map which gives the distribution of caste structured societies.[108] Block regions of high percentage are characterized by the presence of one or more despised occupational groups, by hereditary occupational differentiations, or ethnic stratifications with significant barriers to intermarriage between superordinate and subordinate castes (eg., descendants of former slaves, conquered peoples, different races, etc.). Regions of low percentage are characterized by an absence of significant caste stratification. While data on the map in Figure 41 are missing from some key areas of Saharasia, the distribution of castes very definitely has a Saharasia plus borderlands distribution.

Strong caste distinctions historically appear as the result of invasions and conquering of one people by another, with a subordination of the conquered group. Strict taboos are then put in place to prevent intermarriage, or intimate contacts between upper and lower castes (except contacts without parental obligations, such as prostitution or concubinage of lower-caste women). Formalized rules, laws and religious codes generally work to maintain the deeply held cultural views.

As is the case with class stratification, children, left to their own devices, would through spontaneous friendliness, youthful romance, and love-match marriages, eventually break down the caste system. Hence, persistence of caste distinctions for many generations depends upon development of an arranged marriage system, particularly for the upper caste women. This means the moulding of the natural friendliness of children to the will of prevailing moral and cultural dogma, specifically the caste-specific censoring of a child's playmates, strict obedience training, and the abolishment or curtailing of love-match marriages. Increased

108. Murdock, 1967, ibid.

Correlation Table: Caste Stratification

is positively correlated with the following other patrist variables:

By Column: Insult sensitivity is extreme
Slavery is present
Class stratification is present
High God present, active, supportive of human morality
Movable property inheritance favors male line
Land inheritance favors male line
Cognatic kin groups are absent
Patrilineal descent is present
Bride price is present
Marital residence near male kin
Male genital mutilations are present

No negative correlations were found between "castes" and other patrist variables.

control of the female by the male, a tightening of women's privileges within the social structure, and harsher punishments for violations of the female virginity taboo inevitably occur.

Figure 41: Regions with Castes

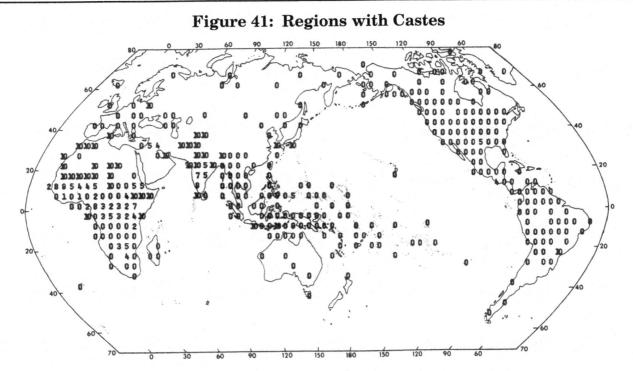

"10" identifies regions where 91-100% of cultures have Castes.
Numbers indicate the percentage of cultures within a given 5° x 5° region with castes
(number times 10 = top end of 10-point percentage category). Regions marked "0" have no cultures with castes, while 1 = 1-10%, 2=11-20%, 3=21-30%, etc. Blank areas = data not available.

26) Slavery

Figure 42 is a map of the *Ethnographic Atlas* variable on slavery,[109] specifically the form and prevalence of slave status, apart from any class or caste stratifications. Block regions with a high percentage are characterized by the presence of slavery of an hereditary, socially significant character, or the presence of incipient (temporary) or nonhereditary slavery. Block regions of a low percentage are characterized by the absence or near absence of slavery. The *Ethnographic Atlas* data on the variable slavery would appear to document the network of Arab mideastern slave-hunting, as well as some African effects from the less-intensive and shorter period of European slave-hunting, minus its New World aspects.

The picture of slave-owning cultures as given in the variables of the *Correlation Table* is striking: violent, stratified, punishing, and brutal, as might be anticipated from even a cursory review of the literature on slavery. The map does not address the presence of female sexual slavery and related forms of enforced prostitution, which are addressed under the latter heading. It is interesting to note that slavery was legally abolished only most recently in some parts of Saharasia where it was most strongly institutionalized for very long periods; for instance, in Saudi Arabia in 1962, and Oman in 1970.[110] Slavery still persists in Saharasia: on a tribal level in parts of North Africa, as female sexual slavery in some Islamic areas, and as slave labor in the gulags of communist China.

The most extensive slave-owning states historically have demonstrated high levels of violence involved in both the obtaining and treatment of slaves. Networks were established involving raiding, piracy,

109. Murdock, 1967, ibid.
110. Davis, D.: *Slavery and Human Progress*, Oxford U. Press, 1984.

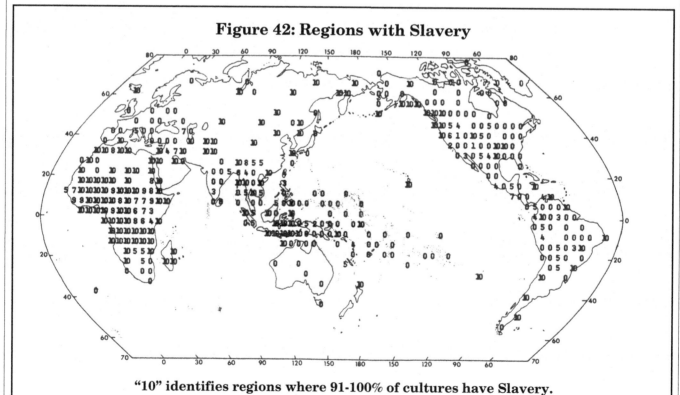

Figure 42: Regions with Slavery

"10" identifies regions where 91-100% of cultures have Slavery.
Numbers indicate the percentage of cultures within a given 5° x 5° region with slavery (number times 10 = top end of 10-point percentage category). Regions marked "0" have no cultures with slavery, while 1 = 1-10%, 2=11-20%, 3=21-30%, etc. Blank areas = data not available.

murder, and warfare. Female and young male slaves were generally sought for clearly sexual purposes, with mature male slaves valued for strength as laborers. Slavery is also associated with attitudes of cultural or racial superiority, generally reinforced by patrist religion, and given legal sanction.

Slave hunting and owning states or cultures are thereby characterized by a militant or quasi-military characteristic, either to capture or guard slaves, or to secure dominance over a subordinate group from which slaves are regularly acquired. Slavery not only insures that the children of slaves will be subject to significantly higher levels of punitive obedience training, but children of nonslaves, who otherwise would be friendly to the slaves, also would have to be molded to the will of prevailing slave-owning moral and cultural dogma. As such, slavery is anticipated to correlate positively with increased emphasis upon high levels of violence, adult dominance over children, male dominance over females, and general patrism.

The cultural tendencies where slavery is present are in a direction similar to that found in caste and class stratified cultures, but of a decidedly harsher and more severe quality. The global distributions of class, caste, and slavery also bear a clear geographical relationship to each other, as seen in Figures 40, 41, and 42.

Correlation Table: **Slavery**
is positively correlated with the following other patrist variables:

By Column:	Boastfulness is extreme
	Insult sensitivity is extreme
	Narcissism index is high
	Military glory emphasis is high
	Warfare prevalence is high
	Castes are present
	Class stratification is high
	Land inheritance favors male line
	Cognatic kin groups are absent
	Patrilineal descent is present
	Extramarital coitus is punished
	Abortion penalty is severe
	Bride price is present
	Polygamy is present
	Marital residence near male kin
	Male genital mutilations are present
	Infant pain infliction is high
	Infant indulgence is low
Others By Row:	Post partum sex taboo is lengthy

No negative correlations were found between "slavery" and other patrist variables.

27) Hydraulic Society and *Oriental Despotism*

In 1957, Wittfogel's *Oriental Despotism* was published, in which it was argued that cultures of the Near East, Central Asia, and Mesoamerica developed a despotic central state apparatus with authority and power far exceeding anything ever known in Europe, excepting perhaps Imperial Rome and Byzantium.[111] Wittfogel concluded that the tendency of Central Asian culture to lean towards such harsh authoritarianism was a result of the increased structuring necessary to maintain the large scale irrigation systems which were developed in that region. Elaborate irrigation systems, Wittfogel argued, required constant error-free attention, maintenance, and coordination to function properly, and it was towards this "need" that such "hydraulic societies" developed their increasingly authoritarian characteristics.

The cultural data presented here in spatial form confirm Wittfogel's observations regarding the more extreme forms of despotism known in those regions, and also the presence in many cases of highly developed irrigation systems. However, the correlation between despotism and hydraulic society would appear to be related to a deeper underlying variable, namely that of desertification and its direct effects upon character and social structure, as developed in this work.

As pointed out by Butzer, despotic societies existed in Egypt long before the onset of large scale irrigation schemes (as during the early Pharaonic period), and irrigation was in use prior to the development of the despotic central state in Mesopotamia.[112] In the Old World, despotism appears following invasions of militant nomads from Saharasia into moister regions, and irrigation follows later in those areas as a straightforward means of developing a dwindling water resource. Despotism would assist irrigation development, however, as a despotic state would have the means to gather forcibly a large labor force comprised of slaves. And, as Wittfogel aptly points out, in most cases this appears to be how large-scale irrigation projects were carried out. Beatings, floggings, torture, murder, and legal robbery were common tactics for gathering the needed capital and labor. Caste structured societies, comprised of wealthy elites, priesthoods, military men and peasants, also appeared in such despotic hydraulic societies. However, the hydraulic hypothesis cannot explain the genesis of sadistic violence in such cultures which previously did not have it. Despotism appears founded upon the presence of violence and sadism in a particular society, not upon irrigation.

Once established, irrigation systems might contribute towards an already established tendency of central state control. The materials gathered in Chapter 8 indicate that in Egypt, Shang China, and Tsarist Russia, the despotic state apparatus appeared first, in a sudden, fully-blown form, following invasions of militant nomads out of the desert.

Or, as was the case with the Shang-Chou transition in China, the Hyksos-New Kingdom transition of Egypt, and the Tatar-Tsar transition of Russia, a statehood of even greater despotism developed following efforts to oust the militant Asian invaders. Also following such events, large new irrigation works were constructed. Here, the development of a large army appears as a prerequisite for gathering the needed construction capital and labor force.

On the other side, a "pre-despotism" irrigation appears to have characterized a few "pre-desert" Central Asian sites, such as at Altyn Depe (Caspian region) and Indus Valley sites. These and related issues are discussed in Chapter 8, which is broken down regionally. The historical sections of this work will detail the manner in which nomadic groups

111. Wittfogel, K.: *Oriental Despotism*, Yale U. Press, New Haven, 1957.

112. Butzer, K.: *Early Hydraulic Civilization in Egypt*, U. Chicago, 1976, p.100; cf. Hoffman, M.: *Egypt Before the Pharaohs*, Alfred Knopf, NY, 1979, pp.314-5.

abandoning the deserts of Central Asia and Arabia established despotic states on the territory of the Saharasian fringe, or within Saharasia where water supplies were stable. In the Old World at least, *both the development of the despotic state apparatus, as well as the development of irrigation technology appear rooted in a deeper, common functioning principle: climate change towards desert conditions in Saharasia.*

Wittfogel acknowledged the role of the warlike pastoral nomads of Central Asia in the spread of despotism.[113] But his theory implies that human sadism and violence exists everywhere, in everyone, just waiting for the right form of social or economic organization to express itself. In this sense, Wittfogel takes a lead from Marx's economic determinism which, as Reich pointed out in the 1930s, was completely devoid of any psychological understanding of the role of childrearing trauma and sexual repression in the genesis of the patriarchal authoritarian and violent fascist (despotic) character structure.[114]

Central Asia was, at least until the most recent times, straddled by two giants of collectivism, both of whose peoples have exhibited tendencies toward historically significant sadistic mass violence (the Soviets during the Stalinist period, and the Peoples Republic of China during the Maoist period). Remnants of this sadistic mass violence continue, as seen in the outbreak of brutal "civil" warfare following the break-up of the old communist states of the Soviet Union and Yugoslavia, and in China with the continuation of slave-labor camps, routine executions of civilians for petty offenses, and the massacre of pro-democracy students in Tienamen Square. The Russian-Soviet and Chinese regions remain predominantly extreme patrist in family and social structure, and these tendencies are rooted not so much in one or another "economic system", as in prior historical periods when their respective territories were repeatedly overrun by militant and despotic Central Asian nomads (Chapter 8), changing the character structure of survivors who subsequently lived under the boot-heel of cut-throat despots for generations. The Russian Soviet and the Chinese communist states were born as part of a larger attempt to subordinate and control their old and brutal enemies, the Saharasian nomads — but after being successful in this task, their territorial ambitions spilled out into borderland regions not connected with the Asian nomads (such as Eastern Europe, and both Southeast Asia and Island Asia). As such, they followed a pattern of history which has been repeated many times.

"Oriental" despotism in Oceania and the New World is discussed in Chapter 9. As will be demonstrated, the "Oriental" characteristics in those distant regions were not coincidental, but were connected by the causal mechanism of culture contact and migrations by ocean-navigating despotic peoples, with cultural connections back in time to the patristic central-state empires of Saharasia.

113. Wittfogel, 1957, ibid., pp. 18-9, 204-6.

114. Reich, W.: *The Mass Psychology of Fascism,* Farrar, Straus & Giroux, NY, 1970; Reich, W.: *People in Trouble,* Farrar, Straus & Giroux, NY, 1976.

6. Contraceptive Plant Materials Used in Sex-Positive Cultures [§]

This Chapter will flesh out the prior discussions of what unarmored, sex-positive cultures were like before the coming of armored, patriarchal authoritarian "civilization". I hope to finally defeat any ideas the reader may have which would label sex-positive, unarmored peoples — who at the subsistence-level might go about with little clothing and who are far more gentle and affectionate towards children and the opposite sex than the typical citizen of a patriarchal central state — as somehow being "ignorant", "beast-like" or "primitive". It may also serve as a useful reference tool for the researcher who wants to identify and possibly resurrect the use of contraceptive plant materials by women in the West.

As discussed in prior chapters, the unarmored, matristic cultures around the world did not have sexual taboos against premarital and adolescent sexual union, a fact observed and recorded by various anthropologists. Social institutions in such cultures facilitated and protected the sexual feelings and desires of young people. Children in such societies spontaneously began to extend their interests outward from their maternal-infant bond to make contact with other children, at around age 2 or 3. After age 5 or 6, they joined peer groups, or *children's democracies*, and spent much of their days and nights in mixed-sex groups of children and adolescents.

The children's groups were notable for their lack of adult controls or supervision, and for the democratic, self-regulated manner in which the children ordered and determined their everyday activities. The children in such free cultures also were observed to have very strong sexual interests, and sex-games were played at a very early age. They learned about sexual matters, including intercourse, by experimentation and play with other children, in a sex-affirmative, life-positive social framework. During later childhood and adolescence, they slept together in special quarters specifically designated for the young people. Their sexual unions, which were characterized by tenderness and sweetness, eventually became quite regular, nightly for the most part, and the younger adolescents tended to engage in the greatest amount of experimentation with different partners. The period of sexual experimentation did not persist, however, and the older adolescents eventually formed more serious and stable relationships, ignoring other lovers. Eventually, love-match marriages occurred which were formulated in large measure upon mutual gratification. As documented throughout this present work on Saharasia, these cultures generally lacked many of the characteristics of "civilized" *Homo normalis*, such as homosexuality, impotence, rape, stealing, antisocial violence, and an organized military. They also possessed a high women's status, and showed great affection for their infants and children.[36, 37, 94] But the phenomena of interest for this Chapter is the general absence of pregnancy among the sexually active young girls.

The majority of these cultures do not practice *coitus interruptus*, or other mechanical contraceptive means, and some may not have made a connection between male semen and pregnancy. Here are a few quotations from the original ethnographical reports:

The Trobriand Islanders in Oceania (1929):
"Since there is so much sexual freedom, must there not be a great number of children born out of wedlock? If this is not so, what

§ First presented at the Annual Meeting of the Association of American Geographers, San Antonio, Texas, 1982, *AAG Program Abstracts*, pp.33-34. Also see: J. DeMeo, "Herbal Oral Contraceptives: Their Use by Primitive Peoples", *Mothering*, 5:24-28, 1977; "The Use of Contraceptive Plant Materials by Native Peoples", *Journal of Orgonomy*, 26(1):152-176, 1992; German translation: "Empfängnisverhütungsmittel bei Naturvölkern", *Emotion* 11:6-29, 1994.

NOTE ON CITATIONS: For this Chapter only, footnotes have been gathered as chapter endnotes, to facilitate use of the included Table as a research tool.

means of prevention do the natives employ? ...it is very remarkable to note that illegitimate children are rare. The girls seem to remain sterile throughout their period of license, which begins when they are small children and continues until they marry; when they are married they conceive and breed, sometimes quite prolifically...I was able to find roughly a dozen illegitimate children recorded genealogically in the Trobriands, or about 1%. Thus we are faced with the question: why are there so few illegitimate children? ...they never practice coitus interruptus." [73:166-167]

The Muria of India (1947):

"...here, boys and girls — in spite of various checks and disciplines — cohabit from the period of their first tentative experiments, through full sexual maturity, for 5 to 8 years until their freedom is ended by marriage. Yet here too pregnancies are comparatively rare..." [112:461]

The feelings of the Muria about coitus interruptus are summarized by the following statement of one young native man: "The sowing of the seed is the happiest moment in one's life — How should one resist it?" [112:464]

The Massim area of British New Guinea (1910):

"Wherever the confidence of the natives was gained it was admitted that abortion was induced, but the most careful inquiries failed to produce evidence that the practice was as frequent as might be expected considering the prevailing liberty. In fact, with every a-priori reason for expecting abortion to be commonly produced, I came to the conclusion that in fact it was a somewhat infrequent event." [107:500]

ANTHURIUM
Tessmannii
K. Krause

The Eddystone Islanders of Oceania (1926):

"The very free relations existing before marriage might have been expected to lead to the birth of many children and to the existence of definite regulations for assigning such children to their proper place in society. Such births seemed, however, to be extremely rare, and in the whole of the pedigrees collected by us only one such case was given, and that was many generations ago." [100:76]

The Wogeo near New Guinea (1935):

"Single girls do sometimes have children, but illegitimacy is not nearly so common as one might have expected. Just why this is so it is not possible to say. I observed one fact that bears directly upon the problem, namely that it is extremely rare for women to have children until they are, I judge, more than 21 years of age, by which time most of them are safely married. I have noticed that even when a girl is married directly after her first menstruation, which does not regularly take place until almost certainly after the 17th year, it is most unusual for her to have a child for several years." [56:320f]

The Ifuago of Oceania (1938):

"...conception results less frequently than one would expect. The surprising intensity of the sex life of the agamang period seems to create physiological conditions unfavorable for conception: Possibly the excessive intercourse keeps the female organs in a state of hyperemia." [14:11]

And so on...

This question, of how the adolescent girls can go with boys so regularly and for such long periods without getting pregnant, has been called, by the medical historian Himes,[55:29] the "puzzle of infertile premarital promiscuity".

Assuming fertile adolescent girls are sleeping regularly with their lovers, or with different lovers, pregnancy rates should be higher than those noted by the above ethnographers, which were:[§]

<1% in the Eddystone Islands, by Rivers [100]
1% in the Trobriand Islands, by Malinowski [73]
4% among the Muria of India, by Elwin [112]

A number of interesting hypotheses have circulated to explain the riddle. Infanticide and late teen abortion do not seem to qualify given the nature of the reports. Even if these practices existed, the fact that so few of the young girls did not even appear pregnant implies contraception, or use of emmenagogues or abortifacients within the first two or three months of pregnancy.

Rentoul[99] challenged Malinowski's assertions that the Trobriand natives did not know a method of contraception, stating that the native girls knew of a muscle-control technique for "expelling the male seed" following intercourse. However, Malinowski [73] dismissed the argument as "one of the typical myths which circulated among the semi- educated white residents". Hartman[52] theorized an infertile period of from 3 to 4 years following the first menstruation in young girls. However, this has not been demonstrated for any human population although irregular ovulation seems to be a characteristic of younger girls. Also, some have postulated an immunizing affect from exposure to the sperm of a variety of different men, or boys. But nothing substantive has ever been demonstrated on this last point, to my knowledge.

What, then, do the native peoples themselves have to say about their own fertility and contraception? Are there any general patterns which cut through the diversity of cultures? Himes, in his *Medical History of Contraception*, has demonstrated two important facts:[55]

1) The desire to control conception is a universal social phenomena, constant in all group cultural life;

2) Almost every society possesses some knowledge of control which use is made of, even if the methods are not always effective.

Among the different methods for conception mentioned by subsistence- level peoples is the use of orally-ingested plant substances, such as herbs, roots, and barks. These are from very specific plants and are often shrouded in great secrecy, magic, and taboo. As the ethnographers of the era 1900-1940 were a bit perplexed about the unrestrained sex lives of the youth in these societies, so too were they often hyper-critical in their assessments of native medicine. Himes flatly rejected the possibility of effective contraceptive plants.[55] His reasons for rejection of the plant contraceptive hypothesis was that medical science of his own day, the 1930s, had not produced an oral contraceptive; therefore, how could "uncivilized" peoples have had them?

Malinowski also rejected the plant contraceptive hypothesis, stating:

"Some of the herbs employed in this (abortive) magic were mentioned to me, but I am certain that none of them possess any physiological properties...equally incorrect and fantastic is the belief in mysterious contraceptives..." [73:167-168]

Malinowski's dismissal does not seem to have been based upon any

§ Ashley Montagu's study on the infertility of sexually-active adolescent girls (*Adolescent Sterility: A Study in the Comparative Physiology of the Infecundity of the Adolescent Organism in Mammals and Man*, Charles Thomas, Springfield, 1946) came to my attention after research for this Chapter was completed. Montagu surveys some of the same sexually-free cultures discussed here, providing even more statistical evidence to support the existence of a very real period of adolescent infertility. Unfortunately, he drew upon the questionable authority of the medical historian Himes (see citation 55) who was totally dismissive of the various reports coming from Oceania and the Americas suggesting the existence of effective contraceptive plants. Echoing Himes, Montagu stated "... *effective contraception is not practised among any preliterate people of whom we have any knowledge...*" (Montagu, ibid., p.71) There was no other mention of contraception in his otherwise well-researched and valuable book, published over 50 years ago. His conclusions of a period of natural infertility lasting on average about one year after a girl's first menstruation surely contributes to the phenomenon of adolescent infertility, but the cited reports contained in this Chapter demonstrate periods of infertility well beyond only one year, as well as confirmed laboratory evaluations of the contraceptive effects of many of the same plants explicitly used by native peoples for this very purpose.

empirical testing of these native medicinal plants, but he confessed to the possibility of having "...missed some important ethnological clue..." and felt "...that my information is perhaps not quite as full as it might have been had I concentrated more attention to it." Malinowski was interested greatly in the psychology of the Trobrianders, whom he felt provided a clear example to defeat Freud's theory of a universal Oedipal complex and childhood sexual latency. His interest in penetrating to the secret of the infertile period of Trobriand sex life was not great enough, or the Trobrianders themselves had not been frank with him. Malinowski may also have necessarily relied too much on male informants, as revealed in his works. As a man, he understandably would have had a most difficult time penetrating easily to the world of Trobriand female rituals and medicine. Trobriand women had among them midwives who also employed various native medicines for pregnancy, childbirth, and abortion; Trobriand male shamans worked openly, but midwives ("witches", as Malinowski called them), had to work secretly.[73:3, 39] He also admits, though derisively, that white citizens of Eastern New Guinea believed in Trobriand contraceptive and abortive herbs [73:168-169; 74]

Withholding judgement on the matter for the time being, let us review a few ethnographies regarding medical and magical secrecy. Hortense Powdermaker, Malinowski's female student, made the following observations of the peoples of New Ireland, very near to the Trobriand Islands:

> "...they [the Lesu people] think they have a method of control over birth. There are certain leaves which are supposed to have the power of making a woman sterile, and others which are said to have an abortive value. To produce sterility or abortion, several leaves are taken in succession. The leaves are chewed, the juice swallowed, and the pulp spit out. Natives swear by the efficacy of these leaves... The women are very sure of the power of the abortive leaves, and, as far as I know, no physical means of abortion is used. The knowledge of these sterility and abortive leaves is zealously guarded by those who know about them." [92:242-243]

In a similar context, the anthropologist Nicole Maxwell, in more recent times reported on her work in the Peruvian jungle, visiting the Seuene Witotos, Jivaros, Shipibos, Conibos, Campa, and Yagua tribes. She states:

> "Primitive peoples are generally reluctant to reveal their medical lore to a stranger, but with care and considerable time it is usually possible to develop a feeling of friendship and confidence. If this is done they can, as a rule, be persuaded to gather and explain the use of many plants they depend on... But this is by no means the case with plants they employ as contraceptives. Not all the tribes seem to know these plants, but those that do are extremely sensitive about them. In fact, any botanicals which they believe to affect the giving or withholding of life are surrounded by powerful taboos which prevent their disclosure to an outsider. This is not only true of the plants they take to prevent conception, but also of those used to increase fertility in women, to facilitate parturition, and to provoke abortion. Unless a feeling of intimacy, even of identification can be established, all the tribes I have known guard such secrets jealously. Any premature attempt to pry, however delicate, is apt to bring all communication to an abrupt, sullen end." [77:2-3]

Using a slow, friendly approach, and spending time with female friends in native villages, Maxwell was able to collect specimens of contraceptive plants from most of the mentioned tribes. The contraceptive plants were all similar, a sedge called the Piripiri. These were grown in small secret gardens by the women themselves, who invariably obtained them from their own mothers, from female relatives, or from friends. A tea is made from the leaves and drunk upon a young girl's first menses; one dose of this tea is considered effective for contraception for 6 or 7 years. The native women state, regarding contraception for their young, sexually active daughters: "By the time it wears off, the girl should be mature enough for motherhood." [77, 78]

Briefly, the following are reviews of a few original field ethnographies regarding contraceptive plant materials. Again, one must keep in mind that the researchers involved may not have had access to complete knowledge about the plants used, and that the natives may have hidden or purposefully distorted the process or plants actually used for contraception. In a few of these reports one will also sense the bias of the field researcher against the unencumbered, lively sexuality of some native peoples, and additional bias against the possibility of effective plant contraceptives.

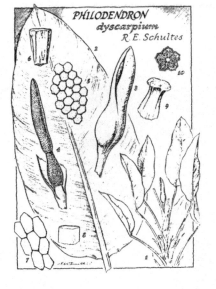

PHILODENDRON
dyscarpium
R. E. Schultes

The S.W. Pacific (1910):
"A large number...of the women have no children... The women, I believe, eat some leaves to prevent conception... It is very significant that the natives in, so far as I know, all these Pacific islands assert that abortion can be effected by administering an infusion of a certain plant, and in the cases which have come under my notice they all point to the same plant, a long straggling creeper growing on almost every sandy beach in the Pacific... The same plant is also used in cases of suppressed menstruation..." [24:33-34, 38]

The islands around New Guinea (1930s):
"A very large portion of the tribes of New Guinea and the adjacent islands (Bismarck Archipelago, etc.) believe in the possibility of producing sterility in women by means of vegetable substances taken orally. The actual species of tree or plant which supplies the berries , bark or roots so used differ in different districts. I tried to get some of them subjected to pharmacological analysis. There is evidence that at least one of these medicines does produce in a few days a shrinking of the female breast." [55:25]

The inhabitants of Dutch New Guinea (1902):
"Four obviously harmless plants are already known to us, the drinking of which produces sterility. One of these called Kakau grows by the thousands hereabout. The women eat the leaves of that plant with sago bread. Both the leaves and fruit of another plant, natunmum (making one capable of conceiving), are eaten with sago bread. Another plant, the name of which is unknown to us, is dried on the fire like tobacco, ground, made into a cigarette and smoked. The worst plant of this kind is called Lapalet. This plant has a root which is peeled, cut into little discs, and then eaten alternately with cocaonut kernels. One does not dare to chew the stuff, for its taste is too bitter; it must be swallowed whole. The poison of this plant is so destructive that it not only produces sterility, but can also kill a 3 or 4 month old foetus. Why the people wish to produce sterility and even abortion is still quite unknown to us. Even the men do not know what to say about it.

They know neither the secret means of the women, nor the art by which this childlessness is made possible. Only the fact that the women do this is known to them, and also the fact that they take plant poisons for that purpose. On the more intimate circumstances the married women, who alone are initiated into the secrets, observe an unbreakable silence." [41:383]

Natives of the Torres Straits (1908):
"Old women may give to young women the young leaves of the Argerarger, a large tree, of which the fruit is inedible, Sobe, a large tree with edible fruit, and Bok, a large shrub. The young leaves of these trees are well-chewed and the juice swallowed, until they feel that their bodies are wholly saturated with the juice. This process takes some time, but when their system is thoroughly impregnated, they are supposed to be proof against fecundity and can go with men indefinitely. Both men and women believe in the efficacy of these leaves..." [50:107]

Natives of the Fiji Islands (1887):
"Just as the Fijian midwife undertakes to rectify sterility, so, on the other hand, amusing expedients are resorted to with the object of preventing conception, and these methods are believed to be sometimes successful and sometimes not. The medicine employed for this purpose is obtained from the leaves and root of the Roga tree, and from the leaves and root of the Samalo in conjunction. The roots are first denuded of bark and then scraped. The scrapings and the bruised leaves are made into an infusion with cold water, and this, when strained, is ready for use. This herbal medicine is taken sometimes once, and sometimes twice, in order to produce the desired effect. If coitus takes place, say in the evening, the decoction is given on the following day, and this without any reference to the relation in point of time which the coitus may bear to the menstrual period. This remedy, besides being given to prevent a first conception, is also administered in the case of a woman who has had one or more children, in order to prevent future pregnancies." [18:180]

The Kurtatchi, Tabut, Baniu and Petats of Oceania (1935):
"This (contraceptive) knowledge is common to both sexes, and the practice is freely discussed... A number of different plants are used for this purpose." [17:134]

Natives of Cape York, Australia (1933):
"In the Koko Ya'o, Kanju, Yankonyu, Ompela, and Yintjingga tribes, there is a firm belief in the contraceptive (not abortifacient) properties of certain plants. I was informed of this fact by both men and women in widely separated localities, and in each case the names of two plants that are used, Ka'ata and Pi'ala, were given to me. I had striking evidence of this belief when collecting genealogies from a group of Kanju people on the Batavia River. After writing down the names of a man and his wife that occurred in one of the branches of the pedigree, I asked for the names of their children, and received the spontaneous reply: 'No got, keni yankoi'n, she shut mesel', i.e. she has not got any, she has eaten medicine, she has shut herself. Those were the exact words volunteered by one of my informants — a woman, with whom I had never discussed the question. The men, as usual, in all matters

pertaining to women, such for example as childbirth, generally disclaim any first-hand knowledge of this medicine, but they freely admit that a 'Keni belong woman' is used, and declare that they would be angry if they found their women using it. Most of my information on this subject was obtained from the old women. They stated that this Keni was 'old fashioned' and all agreed that when they used it 'piccaniny no more come out'." [111:506-507]

The Muria of India (1947):
"Certain supposedly contraceptive herbs are taken orally... another orally administered drug is supposed to confer permanent sterilization." [112:464-465]

Hill tribes of the Rif in Morocco (1931):
"...there is, in addition to the regular market, another devoted to women. All men are excluded from it and are severely punished if they enter... In it the women sell, besides the fruits, vegetables, hens, eggs, and pottery which they dispense at the regular market, magico-medical materials which are supposed to act as contraceptives and to produce abortions. The sale and possession of these is considered ample reason for divorce. Sometimes a woman who fails to bear children within a reasonable time after marriage is accused, on general suspicion, of using them and is divorced." [32:110]

Tribes of the N.W. Amazon Basin (1963):
"In the course of 12 years of field work in the N.W. Amazon Basin, I heard repeated references to the use by Indians of plants as oral contraceptives. Most of these reports were indirect or were in the nature of hearsay, and their reliability was, consequently, suspect. Some, however, were direct reports based upon personal knowledge and seemed to me to be worthy of serious investigation." [106:67]

The Canelos of the Amazon (1920):
"In order to be able to cohabit with a man without getting pregnant the Canelos women are in the habit of taking a medicine prepared from the small Piripiri plant. The root knobs of the plant are crushed and soaked in water, and the woman takes a quantity of the drink." [65:71]

Other Amazon tribes, in Brazil (1970):
"...There exists a very real control of birth. This affirmative answer is from studies and observations which we have made for several years including those by ethnologists of the Emilio Goeldi Museum. — How do they control birth? In general they use plants, whose names are secret, for baths and beverages in the form of tea." [80:10]

Tribes in North America (1891):
"We find the same tricks and crimes accompanying conception and gestation among Indians that are common everywhere. Nor is it probable that their ideas upon these matters are borrowed from civilization. Everywhere, in all grades of society, there seems to be an inherent desire with a certain number of women to avoid the cares and responsibilities of maternity. Among the Quapaws the child bearing period ends at 35 to 40. They occasionally use means to prevent conception... The Neah Bay (Washington) women drink a decoction of an herb (the name of which my correspondent did not know) to prevent conception, but the very young women are eager to

become impregnated, that they may not be compelled to go to the government school." [34:277-278]

The Navaho (1950):
"The majority of informants questioned said they were aware that certain plant medicines were effective contraceptives. But only a few admitted first-hand knowledge." [13:25]

Other North American tribes (1900):
"Thirty or forty years ago there was not an American Indian tribe among whom medicine men did not possess several kinds of magical formulae for preventing childbirth." [11:111]

Cherokee women who ate an excess of the spotted cowbane root (a type of water hemlock) would reportedly become sterile "forever". This plant resembles parsley, which is used as an abortifacient and an emmenagogue in European folk medicine. But possibly the most tested and proven effective contraceptive plant used by American tribes was *Lithospermum Ruderale*. This was used by the Shoshone as a contraceptive tea. Of Lithosperma, the following was reported in 1970:

"Preliminary experimental work showed that feeding mice with alcoholic extracts of this plant abolished the normal estrous cycle, and decreased the weights of the sex organs, thymus, and pituitary. Noble, Plunket and Taylor carried on research by administering lyophilized aqueous extracts of both the tops and roots of the plants to rats and found that the number of estrous smears decreased. Subsequent work with rabbits led them to conclude that lithosperma apparently inhibits the actions of gonadothropins in the ovary." [114:230]

Additional testing of Lithosperma has been undertaken by Breneman and Carmack[21] at Indiana University. Mice and chickens tested with extracts showed inhibitions of hormones, such as oxytocin, which suppresses blood pressure and controls womb contractions. Their more recent work has led to isolation of Lithospermic acid as the active component found in the root of Lithosperma, and closely related plants.

The above story on Lithosperma is indicative of biochemical research on only one of the contraceptive plants. Such research has greatly increased during the last decade, and several compendiums have appeared on the subject. For example, Price[95] discussed the subject, and pointed out the contraceptive, anti-steroidal substances present in a wide variety of plants. Malhi and Trivedi[72] reported on over 60 different plants used in India for influencing fertility. Brondegaard[22] compiled a list of over 60 plants used for contraceptives and related purposes by peoples all over the world. A similar list was compiled by deLaszlo and Henshaw.[35] Most recently Farnsworth, et al[43] discussed the topic from a biochemical point of view, collecting over 1600 references to laboratory tests and field reports on contraceptive plants.

The Farnsworth paper summarized the literature on fertility-affecting plant substances, reporting the existence of at least 225 different contraceptive plants and 551 abortive plants. About 20% of these plants had been subjected to laboratory or clinical analysis, and of that number tested, about 49 were known to have been used by subsistence- level peoples for contraceptive purposes. About 66% of these plants, previously used by native peoples, gave reasonably positive indications for influencing fertility.[43]

Some very specific recipes for the use of contraceptive plants come from China, in the Barefoot Doctor's Manual,[9:173-176] a source of native healing folklore and wisdom. Juices and boiled teas were made from the stalks and roots of several different plants, to make an orally-ingested decotion taken after the conclusion of the menstrual period, for five months in succession. The various prescriptions were described as being effective in preventing pregnancy for from 8 months to three years, and in different dosages were also effective for abortive or sterilization purposes. Some of the plants tested in India also gave contraceptive effects in humans for up to a year. Taken with other reports given above, it appears that long-term contraception effects from plants was not an isolated case. The South American plants mentioned by Karsten[65] and Maxwell[77, 78] which gave protection for up to 7 years, have not yet been tested, to my knowledge. Maxwell reported sending samples of these plants to pharmaceutical houses in the USA, but the male laboratory researchers simply laughed at her suggestions that they be tested for contraceptive effectiveness; the plants sat unexamined in the pharmacy offices for months, and were eventually "taken home by the secretaries".

Table 5, which follows, lists the specific cultures and locations I have gathered from the literature which mentioned used of plant substances for contraception. I should point out that, while I cannot claim to have included all references in the literature regarding contraceptive plants, nor even to having read the entire ethnographic literature for each of the different world regions, a reasonable attempt has been made to search out references to the same for different world regions and to be systematic. This listing, however, should be considered as but a subset of the general question on contraception among native peoples, world-wide. As should be apparent from the sources cited above, this question is deeply woven into the general considerations regarding female status and female sexuality.

Table 5, given on the following two pages, itemizes reports from 54 specific cultural groups and 31 specific areas where plant materials were used for contraception. Additionally, there are 14 general regions noted. One may note the preponderance of reports from Oceania, North America, and South America. A Table Summary follows which numerically organizes these reports on a regional basis, according to tribal or area designation.

Summary of Table 5 (next page)

CONTINENT	TOTAL # OF REPORTS	# AREAS REPORTED	# TRIBES REPORTED
North America	21	4	17
South & Central America	24	8	16
Oceania, India & Southeast Asia	28	15	13
Africa	12	5	7
Europe	10	10	0
Central Asia	4	2	1

Table 5: Cultures and Locations Where Contraceptive Plants Were Used
(Alphabetical listing by culture name, or place name where specific culture is unknown)

Cultural Group	Area/Location	Citations
Abyssinians	Ethiopia, Africa	16
	Africa, General	68
	Algiers, Algeria	54, 89
	America, Central	76
	America, Central + W. Indies	46
	America, North	35, 86, 96, 97
	America, South	4, 5, 27, 28, 29, 39, 67, 70, 71, 95
Apache	America, North	59
	Arabia	69
Baholoholo	Africa, Congo Basin	10
Bambundas	Africa, Central, Kasai Basin	11
Baniu	Oceania	17
Bapindas	Africa, Central, Kasai Basin	11
Barasana	Colombia	106
	Brazil	80, 93
	California, Mendocino	1, 31
Campa	Peru	77
Canelos	Peru	65, 78
Caughwawaga	Quebec, Canada	102
Chaco	America, South	8
Chehalis	Washington, USA	49
Cherokee	America, North	3, 82
Chippewa	America, North	35, 118
Conibo	Brazil	77
Creoles	S. America	20, 35
	Cuba	39, 70
	Dominica	39, 70
	Eddystone Islands	100
	Europe	2, 88. 91
	Europe, Central	23, 35, 71, 104, 121
	Europe, North	6, 15, 26, 51, 121
	Europe, South	15, 62
	Europe, West	2
	Fiji Islands	18
	Germany	2, 58
	Guadelupe	39, 70
Guarani	Paraguay	87, 105
Gunantuna	Melanesia, Oceania	109
Han Chinese	China	9
Hopi	America, North	35, 115, 117
Huichol	Mexico, North	59, 71
	Hungary	81, 90, 118, 121
Ifuago	Oceania	14
	India	38, 53, 57, 61, 72
	India, Assam	84
	India, Bastar	60
	India, Himalayas	35, 66
	India, Hindu	64
	Iran	7
Isleta	New Mexico, USA	85

Cultural Group	Area/Location	Citations
	Italy	6, 35
Jivaros	Peru	77
Karo Bataks	Sumatra	63
Kawadjii	Australia, Cape York	19, 110
Kurtatchi	New Guinea, N.E.	17
Lesu	New Ireland, Oceania	92
	Madagascar	20
	Malawi, Lake Nyassa	119
	Malaya	44
Masai	Africa	79
	Mediterranean	15
Muria	India	112
Navaho	Southwest, USA	13, 111, 113, 120
	Nevada, USA	59, 111
	New Guinea, New Britain	24
	New Guinea + Adjacent Isl.	55
Oaxaca	America, Central	76
	Oceania, Buka	17
	Oceania, "Pacific Isles"	71, 101
Opata	Southwest, USA	59, 71
Owyhee	Nevada, USA	22, 111
Paiute	Nevada, USA	111
Petats	Oceania	17
	Puerto Rico	39, 70
Quapaws	America, North	34
Quinault	America, Pacific Northwest	83
Riffians	Morroco	32
	Russia, South	89, 103
	Santa Lucia	39, 70
Seuene Witotos	Peru	77
Shawnee	America, North	12
Shipibos	Peru	77
Shoshone	America, North	114
Sinaugolo	New Guinea, British	11
	Czechoslovakia	57
	Solomon Isl., Oceania	17, 35
Songish	Northwest, USA	11
Squaxin	Washington, USA	49
Tabut	Oceania	17
Tartareans	Central Asia (?)	69
Toba	America, South	8
	Torres Straights	50, 75
Trobrianders	Trobriand Isl., Oceania	73
Tumelo	New Guinea, Dutch	41
Vaupes	Colombia	106
	Vitu Isl., Oceania	109
	Nea Bay, Washington, USA	34
	West Indies	47
Yagua	Peru	77
Yao	Africa, East	116
Zapotec	America, Central	98

Spatially, the distribution of these cultures known to have employed contraceptive plants is seen on the accompanying world map, Figure 43. This map clearly reveals a clustering of reports on contraceptive plants in the Melanesian area of the Southwest Pacific, in North America, and in South America. The areas marked with large circles possessed a large number of very vague reports on contraceptive use which were greatly mixed with magical and superstitious qualities, rendering their effectiveness highly questionable.

It is certain that some of the patterns on the world map reflect differences in data availability for different regions. Melanesia, North America, and South America, stand out as clusters of plant contraceptive use. Indeed, reports on contraception from Oceania, India, North and South America have generally been more concrete, citing specific plants involved which were used straightforwardly for contraceptive purposes. By contrast, reports from Europe, Asia, and Africa, tended to be very general, vague, and confused with much magic and superstition, diluting any potential effectiveness; or, reports were of very old remedies culled out of old botanical texts, and failed to have a first-hand report character. Hence, as seen in Table Summary, contraceptive use in the areas of the Insular Pacific, North America and South America occurred in specific tribes and peoples tied to a specific places which could be pinpointed on the map. Contraceptive use in Europe, Asia, and Africa was generally too diffuse in terms of region, and non specific, rarely coming from field reports.

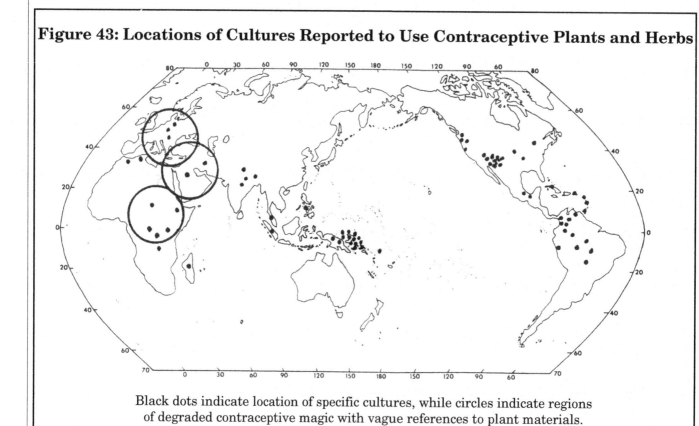

Figure 43: Locations of Cultures Reported to Use Contraceptive Plants and Herbs

Black dots indicate location of specific cultures, while circles indicate regions of degraded contraceptive magic with vague references to plant materials.

Historical Discussion

One of the earliest known documents mentioning contraceptives is the Petri or Kahun papyrus of Egypt, dating from around 1850 BCE.[45] This papyrus described pessary type contraceptives made of paste and gum-like substances, including honey, natron (a sodium carbonate compound), and crocodile dung. These methods worked through blocking the cervical opening and also via spermicidal influences. Himes[55:61-62] points out how crocodile dung, used as a contraceptive, acted similar to a sponge tampon soaked in a weak spermicidal acid; this latter method was a popular and reasonably effective technique in use before the 1900s, and is even copied today in over-the-counter products. The Ebers papyrus from c.1550 BCE describes a refinement in birth control technique using a medicated lint tampon; lactic acid and gum arabic compounds were also used.[25, 40] Interestingly, lactic acid was the active ingredient in many spermicidal jellies commercially sold in the USA and England during the 1930s, while gum arabic was generally used as a base for such compounds at that time.[55:62]

The crocodile dung-honey recipe appears in various ancient literature through to the 9th centuries CE when, in India, it is found using elephant dung as a substitute. The greater acidity of elephant dung made it an even better spermicidal agent. Such elephant dung recipes are found in Christian writings up through the 10th century and in the Arab world to the 13th century.

In Egypt, however, the recipes for contraception seem to take a distinctly magical and non-effective turn after the Hyksos period, and following Akhenaten's failed attempts to change the religion and political tone of Egypt by decree alone — Akhenaten's opponents may be viewed, in modern terms, as the "moral majoritarians" of that era, a reactionary movement. The period coincides with a loss in status for women, with a greater emphasis on male gods, and with a general increase in pharaonic and priestly authoritarianism.[36, 37] The Berlin papyrus of 1300 BCE gives purely magical and presumed worthless prescriptions, certainly a dilution in effectiveness from those of several hundred years earlier.[55:66]

Elephant dung last appears as an orally ingested, and consequently ineffective contraceptive in the texts of the Christian Dominican, Albert the Great, around 1240 CE, and similarly so in Persian texts of the 1400s. The shift from vaginal insertion to oral ingestion is a clear degeneration of the original prescription. Albert the Great also degenerates the original lint tampon, once soaked in spermicidal fluids; instead he recommends application of an oil soaked liniment which is used to anoint the temples of the head in a contraceptive ritual.[55:158-163]

This historical degeneration of contraceptive efficacy is seen in other Mediterranean area literature on the subject. In the 5th century BCE Hippocratic writings alluded to contraceptives, mostly unknown substances giving effects of up to one year of protection. Aristotle also mentions those prescriptions of his predecessors; these were reasonably effective folk methods which included herbs, tampons, and pessaries of sorts.[55:79] Later authors, such as Soranus of Ephesus (c.100 CE) and Dioscorides (c.150 CE) compiled medical works which contained many of these folk methods. By the 300s, however, following the Christianization of Rome, we find such notables as St. Jerome (Hieronymous) vociferously condemning "potion-drinking" for contraception, complaining to the Pope about the "free and immoral life" of many girls and widows. This was the start of an era which showed a rapid decline in the legal rights and social status of women, and the labeling of any sort of sexual pleasure, outside the most compulsive brand of marriage, as both sin and crime against state and church (which were identical at that time).

By the 1500s CE we find books on contraception and abortion mentioning the older remedies of Soranus and the Hippocratic authors, but now magic and amulet-wearing are also introduced. In 798 CE, Carlovingian law further repressed the use of effective contraceptives by outlawing the use of "potions".[55:93-96, 798] A short time later we enter the period of European witch hunts, where female healers and midwives of the peasant classes, the very people in whose hands plant contraceptives would have been safeguarded, were singled out by the Church for torture and execution.[42]

Hence, it is not surprising to find the European literature on contraceptives, indeed on the whole of folk medicine, to be vague, non-specific, and mixed with magic and superstition. Certainly, though, it was no less magical or superstitious than the traditional medical ideas of that time. We would expect any woman healer who had information on contraceptive or abortive plants to have hidden the fact; discovery would have meant a ghastly and gruesome death at the hands of Christian Inquisitors.

In the Islamic world, the decline of contraceptive efficacy was delayed. Al Razi, a great Persian clinician of the early 900s CE, discussed contraceptive techniques in his works. Other Islamic physicians, such as Ali ibn Abbas (c.980 CE), Ibn Sina (1000 CE), and Ismail al-Jurjani (1125 CE) discussed contraception in their texts in a straightforward manner, relaying information from earlier sources. After 1150, however, the Arabic texts begin to introduce more and more magic, and the remedies become less scientific or effective. By the 1400s Persian manuscripts discuss oral ingestion of rennet, mule's hoof, iron dust, dung, and pepper as contraceptive means. Islamic contraception reached its lowest point in the 1500s with the treatise of Dawud Al-Antaki: magical words, rites, and amulets had completely replaced the discussions on medicated or treated tampons, pessaries, and orally-ingested plant materials. His work was published in many editions all over the Islamic world, as recently as 1906.[55:137-158]

In the Christian world, the views of St. Thomas Aquinas (d. 1274) had been championed even before the Council of Trent (1545-1563) wrote them into law. According to Thomist doctrine, birth control is a vice against nature, and the wasting of male semen a crime second only to homicide. This view of birth control as a capital sin, the wasting of male seed, has dominated Catholic thinking to this very day. More recently Pope John Paul II traveled to several impoverished, overpopulated South American cities where, due to a high birth rate and grinding poverty, rampant infanticide exists; even so, he lectured the peasantry against the use of contraceptives, publicly stating that *any woman who uses a birth control device commits a sin worse than murder.*§

In South American regions politically dominated by Roman Catholic descendants of the Portuguese and Spanish colonists, the possession of a birth control device would have, in prior times, provoked a military attack against one's village, or at least the execution of the offender who knew about the contraceptives. Today, such knowledge still might result in a withholding of economic aid, or other bureaucratic pressures or political reprisals, by the largely Church-dominated political and social structure. It is no wonder that natives who retained a knowledge of contraceptive herbs have historically been extremely cautious about discussing them. In particular, any anthropologist who had connections with the Catholic Church, or who even displayed a sympathy towards its doctrines, would have simply been cut off from all discussion on the issue.

The history of contraceptive practices in Asia and Africa is less clear. Plant contraceptives are mentioned in Chinese texts from as far back as c.2700 BCE. However, a reduction in status of females, including the

§ Overheard by the author on short-wave radio, as broadcast from Brazil by the BBC, October 1997. In a similar vein, around this same time I observed a television special program on Mother Theresa at her home for unwed mothers in India, showing her giving a lecture to a group of young mothers with their babies. Abusing the totally dependent condition of these cast-away and desperate girls, who had no other place to go for shelter or sustenance, she lit into them with pious self-righteousness, pointing her finger at them threateningly, with poisonous emotions and words, and a scowl on her face, about the "sinfulness" of using contraceptives. And yet, it was the absence of contraceptives in both of these cases which had led to the desperate and miserable condition of so many women and children, and which the Catholic Church then abused in the most sanctimonious and predatory manner, to "justify" its social role in "saving women and babies." In this manner, the Church helps to create the disturbed social conditions which it later tries to resolve.

decline of female healers, shamans and midwives, immediately followed the central Asian invasions and Shang dynasty of c.1760 BCE. Female status and legal rights were further eroded during the subsequent eras of Confucian puritanism. While these changes in female status were greatest among the upper classes, such influences also penetrated to the enslaved peasantry, who adopted the patriarchal authoritarian mode of social and family structure. Much of the oriental literature remains to be translated, but a few citations can be found evidencing the use of contraceptive herbs, persisting into the modern era.[9]

In Africa, the Islamic invasions between CE 632 and 1258 certainly affected the status of women in many tribal groups. From Africa, the Orient, and Europe also come the greatest frequency of reports on *coitus interruptus*, *coitus obstructus* and *coitus reservatus*, methods for suppression of orgasm and ejaculation. One would not expect these to be practiced as methods of birth control unless other techniques, such as reasonably effective plant materials, were unavailable or suppressed. Such manipulations of male sexuality coincide generally with the presence of severe sex-repression and armoring, and coincide with practices such as the harem, polygamy, concubinage, and other forms of sexual slavery.[36, 37]

Regarding India, contraceptive reports date back to c.3000 BCE in the Susruta Samheta (Sushruta Samhita, a detailed work on surgery) and Kama castras (Kama Shastra, a book on sexual technique). This early period coincides with the early Dravidian cultures of the Indus River valley. Modern day reports on contraceptives in India occur mostly from Dravidian-descended cultures, such as the Muria, who were minimally influenced by Hindu, Moslem, and Christian brands of puritanism.

Regarding the presence of secretiveness about plant contraceptives in the face of religious sex-negativity in more recent times, consider the following reports:

[North America, 1908] *"In an 18th century Pima confessionary are the following suggestive questions: 'Have you drank, from a desire to kill the child within you, sanari [an herb] or anything else?' 'Or have you placed (with the same object in view) a very hot stone upon your abdomen?' 'Or have you lain a long time in the sun?' 'Have you abstained for a long time from eating, wishing that the baby within you dies of hunger?' 'Have you aided another woman to kill her unborn infant?' Among the Mexican Indians the writer heard more about 'medicines' and less about violence as a means of inducing abortion; but among the Mexican tribes observed, on account of their contact with whites and their adoption of Christianity, investigation of subjects of this nature is usually unsatisfactory."* [59:164-165]

[South America, 1963] *"Wherever the Indian has come into close contact with civilization, he acquires a self-consciousness and reticence about free discussion of such a topic as contraceptive agents. He may very well employ them but refuses to talk about them. This is especially true within the sphere of influencing missionaries who usually crusade against the use of contraceptives."* [106:67]

[South America, 1963] *"The present writer has come across no specific mention of contraceptive practices; apparently in recent times they were disapproved..."* [33:733-734]

Of course, Christian missionary work was not the only force responsible for global decline in plant contraceptive use, for its becoming a hidden aspect of culture, and for the secrecy surrounding it. The influences of the c.4000-3500 BCE Semitic and Indo-Aryan invasions, into China, India, Europe, the Near East, Middle East and North Africa, as well as later Judaic, Christian, Islamic, Confucian, Taoist, Hindu and Buddhist transformations generally worked to destroy the remaining elements of unarmored matristic culture — and specifically, in almost every case, reduced women to the status of property-objects whose sexuality and capacity for childbearing became an ongoing preoccupation for dominant ruling males (fathers, husbands, brothers). Under such circumstances contraceptive knowledge increasingly came under control of males, was crushed out of existence as an affront to male-domination over procreative matters, or was driven underground, only to appear vaguely as female "magic" or "witchcraft". North America, South America, and the Insular Pacific underwent such changes only mildly or only within the last 200 years, during which time some historical material was recorded. Hence, the reports from these areas are more numerous regarding contraceptives derived from native plants, and more clearly reflect a preserved empirical native knowledge about health and physiology.

In closing, the above reports left this researcher with both optimistic and pessimistic feelings. It seems clear that various native cultures around the world have long known about reasonably safe and effective means for controlling fertility and population growth, long before the arrival of western medicine, or the invention of the birth control pill. There is great hope and promise for our overpopulated planet in these facts, which should be carefully researched by those with the pharmacological or herb-growing skills. It seems pointless to hope that some drug company will come to investigate these drugs, at a time when they are making billions in profits on birth control pills, chemical spermicides and condoms. A simple herb cannot (yet) be controlled by patents or market monopolies, which undermines gargantuan profits of drug companies, but not the fair-work profits of a community-based herbarium or clinic. It is likely that only cooperative, focused work by herbalists, midwives and feminists, in concert with interested private medical specialists and clinicians, anthropologists, and native healers will rescue this ancient knowledge.

REFERENCES on CONTRACEPTIVE PLANTS

1. Aginsky, B.W.: *Am. Soc. Review*, 4:211, 1939.
2. Aigremont: Volkscrotik u. *Pflanzenwelt*, Halle: 1908, 2:19-20, 93, 134, 149.
3. Anon.: *U.S. Dispensary*, 19th Ed., 1907, p.1393.
4. Anon.: *Am. Bureau Ethnol., 44th Annual Report*, 1928, p.360.
5. Anon.: *Presse Medicale*, 1946, 54:760.
6. Anon.: *Manuale Fitoteraoia*, Inverni & Della Befa, Milan, 1951, pp.90, 491.
7. Anon.: *Qualities Plantarium Mat. Veg.*: 1959, 6:152.
8. Anon.: *Bonplandia*, 1964, 1:322, 326.
9. Anon: *A Barefoot Doctor's Manual*, The American Translation of the Official Chinese Paramedical Manual, Running Press, Philadelphia, 1977, pp.173-176.
10. Aptekar, H.: "In Anthropological Perspective", *Birth Control Review*, July, 1930, p.203.
11. Aptekar, H.: *Anjea — Infanticide, Abortion and Contraception in Savage Society*, Wm. Godwin, NY, 1931, pp.111, 121, 140.
12. Ashe, T.: *Travels in America*, London, 1808, p.272.
13. Bailey, F.L.: "Sex Beliefs & Practices in a Navaho Community", *Papers Harvard Peabody Mus. Arch. Ethn.*, 1950, 40:23-27.
14. Barton, R.F.: *Philippine Pagans*, London, 1938, p.11.
15. Berendes, J.: *Die Arzneimittellehre des Dioskurides*, Stuttgart, 1902, 1:101-109, 2:151-220 & 255, 4:376.
16. Bieber, F.J.: "Geschlechtsleben in Athiopien", *Anthropophyteia*, 1908, 5:45-99.
17. Blackwood, B.: *Both Sides of Baku Passage*, Oxford, 1935, pp.134-136, 594.
18. Blythe, D.: "Notes on the Traditions and Customs of the Natives on Fiji in Relation to Conception, Pregnancy, and Partuition", *Glasgow Medical J.*, 1887, xxviii:180.
19. Boorsman, W.G.: *Bull. Inst. Bot. Buitenzorg*, 1916, 14:20.
20. Bouton, L.: *Plantes Medicinales de Maurice*, Depuy et Dubois, Mauritius, 1864, pp.13, 131.
21. Breneman, W.R. & Carmack, M.: "Isolation & Structure Determination of Lithospermic Acid", *J. Org. Chem.*, 1975, 40:1804.
22. Brondegaard, V.J.: "Contraceptive Plant Drugs", *Planta Medica*, 1973, 23(2):167-172.
23. Brown, R.: *Bot. Soc. Edinberg, Trans.*, 1868, 9:391.
24. Brown, G.: *Melanesians & Polynesians*, Macmillan, London, 1910, p.38.
25. Bryan, C.P.: *The Papyrus Ebers*, London, 1930.
26. Caius, J.F.: "Medical & Poisonous Composites of India", *J. Bombay Nat. Hist. Soc.*, 1940, 41:607.
27. Campa, A.L.: *Western Folklore*, California, 1950, p.345.
28. Casey, R.C.D.: "Alleged Anti-Fertility Plants of India", *Indian J. Med. Sci.*, 1960, 14:590-600.
29. Ceruti, O.: *Le Plante Medicinali*, Set Turin, 1945, p.122.
30. Chaudbury, R.R.: *Indian Council Med. Res. Spec. Rept. Ser.*, 1966, 55:3.
31. Chestnut, V.K.: "Plants Used by the Indians of Mendicino County, California", *Contr. U.S. Natl. Herbarium*, Wash. D.C., 1900-1902, 7:344, 367.
32. Coon, C.S.: "Tribes of the Rif", *Harvard African Studies IX*, Cambridge, Peabody Museum, 1931, p.110.
33. Cooper, J.M.: "The Araucanians", *Handbook of S. American Indians*, NY, 1963, II:733-734.
34. Currier, A.F.: *Transactions Am. Gyn. Soc.*, 1891, XVI:277-278.
35. deLaszlo, H. & Henshaw, P.S.: "Plant Materials Used by Primitive Peoples to Affect Fertility", *Science*, 1954, 119:626-631.
36. DeMeo, J.: "On the Origin and Diffusion of Patrism: The Saharasian Connection", Doctoral Dissertation, Geography Dept., University of Kansas, Lawrence, 1986: cf. *Dissertation Abstracts Int.*, 48:457-458A, August 1987.
37. DeMeo, J.: "The Origins and Diffusion of Patrism in Saharasia, 4000 BCE: Evidence for a Worldwide, Climate-Linked Geographical Pattern in Human Behavior", *Pulse of the Planet*, 3:3-16, 1991, *World Futures*, 30(4):247-271, March-May, 1991.
38. Dutta, A. & Ghosh, S.: "Chemical Examination of Daemia extensa", *J. Am. Pharm. Assn.*, 1947, 36:250.
39. Dvorjetski, M.: "La plante sterilisante caladium seguinum st ses proprietes pharmacodynamiques", *Revue Francaise de Gyn. Obstet.*, 1958, 53: 139.
40. Ebers, G.M.: *Papyros Ebers*, Leipzig, 1875, pl.93.
41. Erdweg, P.M.: "Die Bewohnei der Insel Tumleo, Berlinhafen, Deutsch New Guinea", *Mitt. d. Anthrop. Ges. in Wein*, 1902, xxxii:383.
42. Ehrenreich, B. & English, D.: *Witches, Midwives, and Nurses: A History of Women Healers*, Feminist Press, NY, 1973.
43. Farnsworth, N.R., et al: "Potential Value of Plants as Sources of New Antifertility Agents", *J. Pharm. Sci.*, 1975, 64(4): 535-598, 64(5): 717-754.
44. Gimlette, J.D.: *Malay Poisons and Charm Cures*, London, 1915, p.71.
45. Griffith, F.L.: *The Petrie Papyri-Hieratic Papyri from Kahun and Gurob*, London, Bernard Quaritch, 1898.
46. Grosourdy: *New and Rare Drugs*, London, T. Christy & Co., 1889, p.9.
47. Guerrero, L.M.: *Phil. Bull.*, 1921, 22:149.
48. Gulil, R.H.V.: *Sexual Life in Ancient China*, Leiden, E.J. Brill, 1961.
49. Gunther, E.: *Ethnobotany of Western Washington*, 1945, pp.46-48.
50. Haddon, A.C.: "Birth Control and Childhood Customs, and Limitation of Children", *Cambridge Anthropological Expedition to the Torres Straights, Vol.VI on Sociology, Magic and Religion of Eastern Islanders*, Cambridge U. Press, 1908, pp.105-111.
51. Harms, H.: *Dutsch. Apoth. Zeit.*, 1937, p.52.
52. Hartman: *Science*, 1931, lxxiv:226-227; cf. Himes, below, p.35.
53. Hartwich, C.: *Die Neuen Arzneidrogen aus dem Pflanzenreiche*, Springer, Berlin, 1897.
54. Hilton-Simpson, M.W.: *Arab Medicine and Surgery*, 1922, p.90.
55. Himes, N.E.: *Medical History of Contraception*, Williams and Wilkins, Baltimore, 1936, pp.25, 209.
56. Hogbin, H.I.: "The Native Culture of Wogeo", *Oceania*, 1935, 5:320.
57. Hovorka, O. & Kronfeld, A.: *Vergl. Volksmed*, 1909, 2:523-525.
58. Hovorka, O.: *Vergl. Volksmed*, 1908, 1:33.
59. Hrdlicka, A.: "Physiological and Medical Observations among the Indians of Southwestern U.S. and Mexico", *Bur. Am. Ethn., Smith. Inst.*, Wash. D.C., 1908, pp.163-166.
60. Jain, S.K.: "Medicinal Plant Lore of the Tribes of Bastar", *Econ. Bot.*, 1965, 19:239.
61. Jain, S.K.: *Medicinal Plants*, National Book Trust, New Delhi, 1968, pp.99, 125, 133.
62. Janata, O.: *Res. Program on USSR*, N.Y., 1952, 21:36.
63. Joustra, M.: *Hygienische Misstanden in het Karoland*, Batak Institute, 1909, 1:287.
64. Kalyanamalia: *The Anagaranga*, India, 16:1450, 1526.
65. Karsten, R.: "Contributions to the Sociology of Indian Tribes of Equador: Three Essays", *Acta Academiae Aboensis*, Humanioria Abo, Abo Akademi, #1, 1920.
66. Kirkitar, K.R. & Basu, B.D.: *Indian Medicinal Plants*, Indian Press, Allahabad, 1918, p.2743.
67. Levi-Strausse, C.: "The Use of Wild Plants in Tropical South America", *Handbook of South American Indians*, N.Y., 1963, 6:486.
68. Lewalle, J. & Rodegem, F.M.: *Quart. J. Crude Drug Res.*, 1968, 8:1257.

69. Lewin, L.: *Die Fruchtabtreibung*, 1922, pp.216, 218, 236.

70. Madaus, G. & Koch, F.E.: *Zeit. ges. exptl. Med.*, 1941, 109:68.

71. Madaus, G.: *Lehrbuch der biologischen heilmittel*. Abteilung I: Heilpflanzen, Leipzig, 1938, 1:444, 2:1030, 3:2351, 2378.

72. Mahli, B.S. & Trivedi, V.P.: "Vegetable Antifertility Drugs of India", *Quart. J. Crude Drug Res.*, 1972, 12:1922-1928.

73. Malinowski, B.: *The Sexual Life of Savages in N.W. Melanesia*, Routledge & Keegan Paul, London, 1929, 166-168.

74. Malinowski, B.: "Pigs, Papuans, and Police Court Perspective", *Man*, art. #44, 1932.

75. Marshall, H.A.: *The Physiology of Reproduction*, Longmans, Green & Co., London, 1922, p.652.

76. Martinez, M.: *Las Plantas Med. Mex.* Publ. Botas, Mexico City, 1944, p.353.

77. Maxwell, N.: "Attitudes of Four Peruvian Jungle Tribes Toward Plants Employed as Oral Contraceptives", paper, *XXXIX International Congress of the Americanists*, Lima, 1970.

78. Maxwell, N.: "Plowboy Interview", *Mother Earth News-Lifestyle Magazine*, August, #6, 1973.

79. Merker, M.: *Die Masai*, Berlin, 1916, p.375.

80. Mierels, F.S.: in Maxwell, ibid., 1970, p.10.

81. Oefele, F.V.: *Anticonceptionelle Arzneistoffe*, Heilkunde, Vienna, 1898, 2:39.

82. Olbrechts, F.M.: *Anthropos*, 1931, 26:19.

83. Olson, R.L.: *Univ. Wash. Publ. Anthrop.*, 1936, 6:180.

84. Parry, N.E.: *The Lakhers*, Macmillan, London, 1932, p.170.

85. Parsons, E.C.: "Isleta, New Mexico", *47th Ann. Rept. Bur Amer. Ethnol.*, 1930, p.213.

86. Pflugers: *Arch. Physiol.*, 1913, 153:239.

87. Planas, G.M. & Kue, G.M.: "Contraceptive Properties of Stevia regaudiana", *Science*, 1968, 162:1007.

88. Ploss, H. & Bartels, M.: *Das Weib in der Natur. u. Volkerkunde*, Leipzig, 1899, 6:543.

89. Ploss, H.: *Das Weib in der Natur. u. Volkerkunde*, Leipzig, 1902, 1:670-671.

90. Pohl, J.: *Arch. Exptl. Pharm. Path.*, 1891.

91. Potter, J.: *Cyclopaedia of Botanical Drugs*, London, 1941, p.167.

92. Powdermaker, H.: *Life in Lesu, the Study of Melanesian Society in New Ireland*, W.W. Norton, NY, 1933, pp.242-244, 293-297.

93. Prance, G.T.: *Econ. Bot.*, 1972, 26:221.

94. Prescott, J.W.: "Body Pleasure and the Origins of Violence", *The Futurist*, April, 1975, pp.64-74; *Pulse of the Planet*, 1991, #3:17-25.

95. Price, J.R.: "Antifertility Agents of Plant Origin", in *Symposium on Agents Affecting Fertility*, Austin & Perry, Eds., Little, Brown & Co, 1965, pp.3-17.

96. Pursh, F.T.: *Flora Americae Septentrionalis*, J. Black & Son, London, 1814, 2:596.

97. Ray, V.F.: *Univ. Wash. Publ. Anthrop.*, 1913, 5:219.

98. Reko, B.P.: *Mitobotanica Zapoteca*, Tacubaya, Mexico, 1945, p.11.

99. Rentoul, A.C.: "Physiological Paternity and the Trobrianders", *Man*, 31, art. #162, 1931.

100. Rivers, W.H.: *Psychology and Ethnology*, Kegan Paul, London, 1926, p.76.

101. Roig Y Mesa, J.T.: *Plantas Medicinales: Aromaticas Venenosas de Cuba*, Havana, 1945, p.232.

102. Rousseau, J. & Raymond, M.: *Etudes Ethnobotaniques Quebecoises*, Univ. of Montreal, 1945, p.59.

103. Rud. Krebel: *Volksmed. und Volksmittel verschied. Volkerstamme Russlands*, 1858, p.134.

104. Schlickenrieder, F.G.: Dissertation, Munich, 1939.

105. Schmidt, R.: *Liebe u. Ehe im alten u Mod. Ind.*, Berlin, p.510.

106. Schultes, R.E.: "Plantes as Oral Contraceptives in the Northwest Amazon", *Lloydia*, 1963, 26(2):67-74.

107. Seligman, C.G.: *The Melanesians of British New Guinea*, Cambridge, 1910, p.500.

108. Smith, H.H.: *Ethnobotany of the Ojibwe Indians*, Milwaukee, Public Museum, 1932, p.370.

109. Sterly, J.: *Heilpflanzen der Einwohner Melanesiens*, 1970, pp.100, 107, 250-253.

110. Thompson, D.F.: "The Hero Cult, Initiation and Totemism on Cape York", *J. Roy. Anthrop. Inst. Gr. Br. & Ire.*, 1933, 63:453, 505-508.

111. Train, P., et al: "Contributions Toward a Flora of Nevada", #33, *Medicinal Uses of Plants by Indian Tribes of Nevada*, Part II, USDA, Washington, DC, 1941.

112. Elwin, V: *The Muria and their Ghotul*, Oxford U. Press, Bombay, 1942, pp.464-465.

113. Vestal, P.: "Ethnobotany of the Ramah Navaho", *Papers, Peabody Museum of Arch. & Ethn.*, Harvard U., 1952, 40:23.

114. Vogel, V.J.: *American Indian Medicine*, Ballantine, NY, 1970, p.230.

115. Voth, H.R.: *Field Columbian Museum of Anthropology*, Chicago, 1905, 6:52.

116. Weule, K.: "Wissenschaftliche Ergebdisse meiner ethnogr. Forschungsreise in den Sudosten Deutsch-Ostafrikas", *Mitt. a. d. D. Schutzgeb. Erg.*, Heft 1, Berlin, 1908, p.61.

117. Whiting, A.F.: *Mus. of N. Arizona Bull.*, 1950, 15:35, 91.

118. Williams, S.W.: *Trans. Am. Med. Assoc.*, 1849, 2:878, 890.

119. Williamson, J.: *Useful Plants of Malawi*, 1955, p.123.

120. Wyman, L.C. & Harris, S.K.: *U. of New Mexico Bull.*, 1941, 366:61.

121. Zadina, R. & Geisler, M.: *Biol. Listy*, 1950, 31:41-45.

Recent Additional Publications (not cited):

Riddle, J. & J.W. Estes: "Oral Contraceptives in Ancient and Medieval Times", *American Scientist*, 80:226-233, May-June, 1992.

Riddle, J., J.W. Estes & J.C. Russell: "Ever Since Eve: Birth Control in the Ancient World", *Archaeology*, March/April 1994, pp.29-35.

7. Expressions of Saharasia in Contemporary Demography

In this Chapter, several cultural maps will be presented representing conditions as they currently exist within the modern nation-states. The data is therefore organized by nations, rather than by individual cultural groups or blocks of latitude and longitude. This nation-state form of data presentation makes for some unique problems, given the fact that several of the larger ones (notably Russia, the former Soviet Union, and modern China) dominate various sub-regions of Saharasia. Data is averaged over these entire large nation-states, making it difficult to distinguish between the behavioral characteristics of, for example, western and eastern China, or between the northern and southern Soviet Union (before the break-up of the Soviet Union, or USSR). These problems notwithstanding, the following materials provide their own contribution to the overall Saharasian thesis, and demonstrate that the Saharasian pattern on the maps is still observable in the modern world of the nation-states, exerting a powerful influence into modern times.

The Status of Women Index and
Population Reference Bureau Demographic Maps

Andrews has prepared a world map of a computed *Status of Women Index* (SWI), composed from data on female life expectancy, literacy, and total fertility rate.[1] The Population Reference Bureau (PRB) has also prepared several individual world maps of similar data on fertility, birth rate, infant mortality, maternal childbirth mortality, contraceptive use, and other demographic variables related to family and childbirth.[2] These maps express patterns which are similar to each other, and all of them bear a functional and patterned relationship to the World Behavior Map, given in Figures 1 and 24 in Chapters 1 and 3.

Figure 44 presents the Status of Women Index (SWI) map of Andrews, based upon data from the late 1970s (before the break-up of the Soviet Union), identifying female status in low, medium and high-status categories. The areas of lowest category are characterized by short female life expectancy, low female literacy rate, and high fertility (numbers of babies per woman). Africa, the Middle East, India, and parts of SE Asia and Island Asia, as well as a few Central American and Western South American areas all stand out as regions of the lowest female status. Intermediate regions include East Europe, the former Soviet Union, and the remainder of Central and South America. High status areas are, basically, the Western democracies of USA, Canada, Western Europe and Australia. Unfortunately, data was not available for China, Mongolia, and a few smaller regions of the world. The large geographical size of China and the former Soviet Union prevent us from knowing anything about the regional differences within those very large nations.

Figure 45, a map of Contraceptive Use developed from the PRB data, presents similar data, with similar difficulties. The map identifies regions with low, medium and high levels of contraceptive use. In this case, however, data is unavailable for the Soviet Union and Mongolia, and the large size of both the Soviet Union and China prevent analysis of regional differences within those large nations.

1. Andrews, A.: "Toward A Status of Women Index", *Professional Geographer*, 34(1):24-31, 1982; Andrew's data is in good agreement with the female status data of Stewart and Winter, which are also in good general agreement with the World Behavior Map; The lowest female status occurs in Saharasia; Stewart, A. & D. Winter: "The Nature and Causes of Female Suppression", *Signs: Journal of Women in Culture and Society*, 2(3):531-53, 1977.

2. *Population Dynamics of the World*, (Maps) Population Reference Bureau and Demography Division, Office of Population, USAID, Washington, D.C., 1981; also, *World Population Data Sheet Slide Set* (Maps), Population Reference Bureau, Washington, DC, 1991.

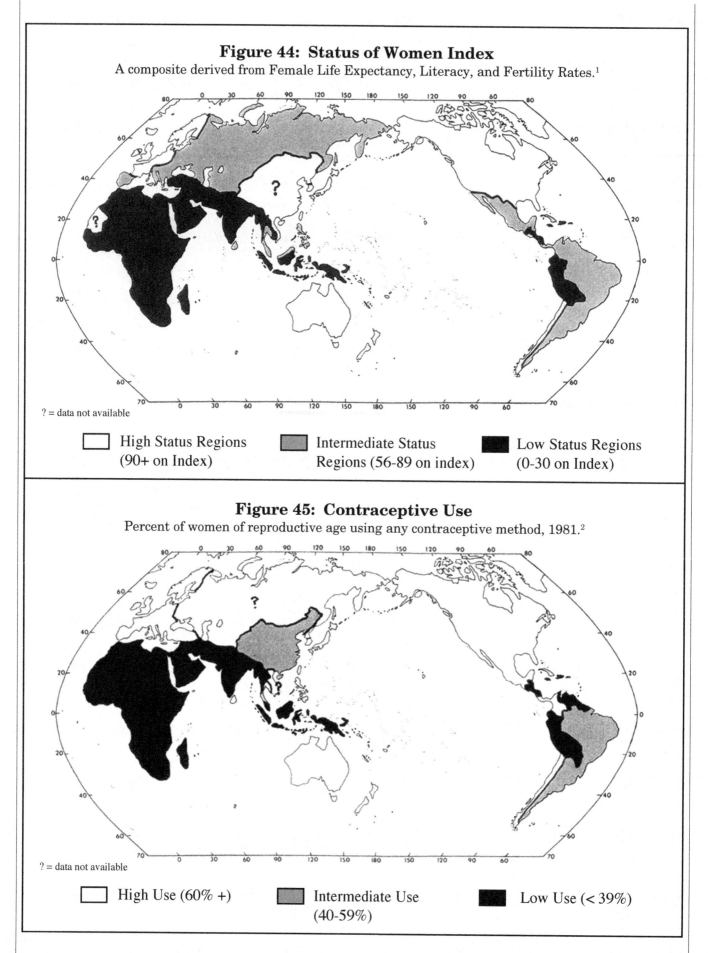

Figure 44: Status of Women Index

A composite derived from Female Life Expectancy, Literacy, and Fertility Rates.[1]

? = data not available

☐ High Status Regions
(90+ on Index)

▨ Intermediate Status
Regions (56-89 on index)

■ Low Status Regions
(0-30 on Index)

Figure 45: Contraceptive Use

Percent of women of reproductive age using any contraceptive method, 1981.[2]

? = data not available

☐ High Use (60% +)

▨ Intermediate Use
(40-59%)

■ Low Use (< 39%)

On both of these maps, Western Europe, North America, and Australia show up as having reasonably good conditions for women and children — at least in a global comparative sense. The data clearly do not address some of the life-negative and sex-negative factors presently existing in "modern" American and European culture, as exposed in previous chapters. But few educated women would argue that life in contemporary Iran or rural China or Zaire (Congo) is identical to the situation in the USA or Europe. USA and Europe have severe problems to deal with, for sure, but the last hundred years have seen incredible reforms in the status of women, and factors affecting childbirth and children. Latin America also displays an increased patrism as compared to Europe and Anglo-American regions, a factor which will be discussed momentarily.

From the maps in Figures 44 and 45, we can raise the argument that "Saharasian" patristic culture has finally and awfully encompassed the whole of sub-Saharan Africa, Middle (Near) East, South Asia, Islamic Island Asia, and likewise also afflicts most of modern China and the Soviet Union — but in the latter two cases, the averaging of cultural data for highly-patristic regions close to Saharasia with those of lesser-patrism far away from Saharasia has resulted in a gross "intermediate" classifications. But from these data alone, we cannot *prove* this argument. What is needed is data which would give information on these cultural variables allowing one to distinguish between western and eastern China, and between northern and southern Soviet Union. Also, other cultural data dealing with related issues of patrism, armoring, and tendencies towards authoritarianism would be helpful, to see if there were differences between northern and southern Africa, for example. There are several sources which we can draw upon to clarify the situation.

Stern, for instance, has provided a number of anecdotes and some data which suggest the Saharasian portions of the old Soviet Union were more armored and patristic than elsewhere within that massive nation-state. He published data on birth rates for different Soviet territories for the period of the 1970s,[3] data which was corroborated only more recently in the 1990s with Population Reference Bureau data from the newly Independent States.[4] These data are gathered and presented in Table 6.

Table 6: Relative Birth Rates, Soviet Central Asia
Organized Generally North (top) to South (bottom)

Former Soviet Union (1970s)[3]		Modern Nations (1997)[4]	
Russian SFSR	14.6	Russia	9
Latvia	14.5	Latvia	9
Estonia	15.8	Estonia	9
Ukraine	15.2	Ukraine	10
Byelorussia	16.2	Belarus	9
Lithuania	17.6	Lithuania	11
Georgia	19.2	Georgia	11
Moldavia	19.4	Moldova	13
Kazakhstan	??	Kazakhstan	15
Kirgizia	30.5	Kyrgyzstan	24
Uzbekistan	33.5	Uzbekistan	28
Tadhikstan	34.7	Tajikistan	29
Turkmenia	35.2	Turkmenistan	28

The closer one moves towards Saharasia, the higher the birth rate (and lower female status, higher patrism, etc.)

3. Stern, M.: *Sex in the USSR*, Times Books, NY, 1979, p.235; Also see pp.106,116,237-9.

4. *World Population Data Sheet 1997*, Population Reference Bureau, Washington, DC.

The Table 6 data, reflecting conditions in the 1970s and more recently in 1997, clearly indicate that the closer one is to Saharasia, the higher the birth rate. Unfortunately, we do not have similar detailed data for modern-day China, to see if the north-western, Saharasian parts are more patristic than the south-eastern parts. General news reports suggest this is so, however. The recent massacre of the pro-democracy students in Tienamen Square, was undertaken by emotionally-hardened divisions from Western China, ordered into the area specifically for this purpose by Communist Party functionaries, who were uncertain they could rely upon local troops from Eastern China to murder unarmed young people. More discussion of these regional differences is given in Part III, and in Chapter 11, on "Saharasia Today".

Political-Social Freedom and Press Freedom Maps

We may gain additional insight by examining two additional maps which relate to political and press freedoms, produced as part of a study on *The Comparative Survey of Freedom*,[5] evaluating trends towards social and political democracy as reflected in the rights and freedoms of individuals, and of the news media. Figure 46 is a map of the variable "Political-Social Freedom", as constructed from survey data on political rights and civil liberties, while Figure 47 maps the variable "Press Freedom", based upon laws, regulations, political pressures and economic influences, as well as repressive actions, directed against journalists.

These two maps are of interest for our study firstly because they directly address the issue of political authoritarianism of either left-wing, right-wing, or religious-political varieties. Secondly, the data for both freedom maps is composed from 1997 data, differentiated for all the newly-independent nation-states formerly composing the old Soviet Union. In these maps, we can see a relationship both to the prior SWI and PRB maps, and also to the World Behavior Map — Saharasia does indeed express itself in the two "freedom" maps, though the problem with the absolute size of modern Russia and China continue to conceal existing regional differences. Clearly, southern areas of the former Soviet Union, now independent nations, show themselves to be more repressive than the more northerly Russia does, a factor which is rooted in the history of the region. Similarly, many parts of Africa most far removed from the Saharan desert portions of Saharasia show up with a freer social condition, as does India. Island Asia, mostly under Islamic influences, shows up as very highly restrictive and authoritarian, while parts of Latin America fall into the "Intermediate" category. All of these factors generally agree with the World Behavior Map, especially when viewed against more recent historical migratory-diffusion patterns.

Discussion

Clearly, the four maps presented here provide additional support, once more, for the original proposition Wilhelm Reich made over 60 years ago: *state structure is rooted in family structure*.[6] We get a glimpse of Saharasia expressing itself through these maps, but there also is a clear spread of increased patrism into sub-Saharan Africa, and also (to a lesser extent) into Oceania and the Americas. The World Behavior Map was constructed from anthropological data gathered from individual, subsistence-level cultures roughly between 1840 and 1960; many of these individual tribal units reflected conditions which in most cases existed for

5. Map of "Political-Social Freedom" published in *Freedom Review*, 28(1):28-29, Jan-Feb.1997. Map of "Press Freedom" published in *Press Freedom 1997: The Press Law Epidemic, A Year of Restrictions*, pp.8-9, 1997. Both publications available from Freedom House, 120 Wall St., New York, NY 10010.

6. Reich, W.: *The Mass Psychology of Fascism*, Farrar, Straus & Giroux, NY, 1970.

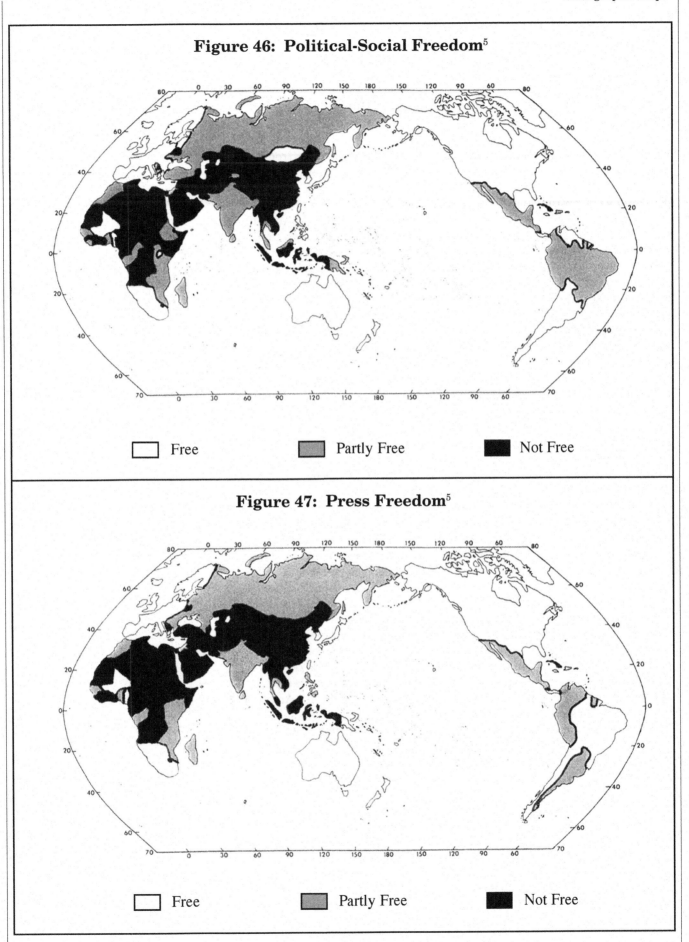

Figure 46: Political-Social Freedom[5]

☐ Free ▨ Partly Free ■ Not Free

Figure 47: Press Freedom[5]

☐ Free ▨ Partly Free ■ Not Free

hundreds of years before the coming of the ethnographer. However, the maps presented in this Chapter were constructed from data collected in the 1970s and 1990s, after many of the tribal units had been absorbed into the central state, or sharply eroded by intrusion of new migrant cultures.

The cases of Latin America, the former Soviet Union, China, and India can be considered individually:

1) Latin America: Europe underwent a slow change from extreme patrism back towards matrism over the last several hundred years, the reasons for which are discussed in Part III. Europe was never completely overrun by the extreme patrist nomad hordes of Saharasia, and remnants of matrism persisted, particularly in those parts of Europe most distant from Saharasia. Regions closer to Saharasia, such as Spain, Italy, and Eastern Europe, were pushed towards a more extreme patrism. Presently, female status in most of Europe is quite high with low fertilities and high rates of contraceptive use, even though this was not the case even a hundred years ago. Consequently, colonial era migrants to North America who came from Northern Europe carried more matrist tendencies with them, and absorbed a great many more from matrist North American natives. North American settlers also underwent an additional transition towards matrism parallel to the changes back towards matrism which occurred later in Europe. Colonial era migrants to Mesoamerica and Peru, however, came largely from Spain and Portugal, which had been greatly influenced by hundreds of years of contact with the Saharasian Moors, and also by the control of the Holy Inquisition and Church over secular affairs. Spanish and Portuguese influences in the New World were consequently relatively extreme patrist in nature, and these influences were added to the already extreme patrist character of the Mesoamerican and Peruvian empires. For these regions, the road back to matrism for Spain, Portugal, Italy, East Europe, and Latin America is longer and more difficult than for Northern Europe and North America. The Spanish Revolution against the autocracy of Church and State was late by European standards, and it was a failed effort. Small reforms which had been made regarding family matters were undone, and reformers were likewise rounded up and shot, or fled into exile. Only after Franco's death did democracy, and some limits to the authority of the Catholic Church over family matters, appear in Spain. Latin American states which were former colonies of Spain and Portugal have undergone similar very slow and hesitant transitions towards matrism; the revolution against European colonialism, for example, was only partly carried over to restrict the power and control of the Catholic Church, which continues to exert a powerful sex-negative, patristic influence in Central and South America. Additionally, in some areas of Latin America, a matrist/patrist split parallels or underlies the tensions which exist between different class or ethnic groups, such as that between some native South American peoples, and the Spanish-descended classes of wealth and power. Taken together, Mesoamerica and Peru therefore remain relatively patrist to this day.

2) The former Soviet Union, India, and China: The southern, Saharasian portions of the former Soviet Union are Moslem regions with a higher fertility and birth rate, reflecting persisting Saharasian traditions (Turk-Mongol & Arab-Moslem) regarding infants, children, and female behavior. The western parts of China and Northern India also maintain Saharasian traditions dating back at least to the Arab-Moslem and Turk-Mongol periods. Status of women may also be lower in North India and West China, with higher fertility and birth rate; but one cannot tell from

tell from the SWI, PRB, or Freedom maps because the data are also averaged for these large and diverse national territories. It is suggested that if the PRB, SWI and Freedom data from Russia, China, and India were broken down regionally, they would yield distributions which would more closely confirm the Saharasian distributions seen on the World Behavior Map.

As previously mentioned, when viewed historically, these maps suggest the continued spread of patrism around the world to areas of a greater distribution than as revealed in the World Behavior Map, which firstly demonstrated the existence of Saharasia as the largest region of acutely-intensive armored patrism. Recall that the World Behavior Map was constructed of anthropological data dating between the mid 1800s to the mid 1900s, while the Freedom maps and the PRB and SWI maps were plotted from very recent data, no more than 25 years ago. In short, *armored patrism appears to have significantly spread around the world since the turn of the century.*

Note 2005:

Since the original publication of *Saharasia*, and the subsequent collapse of the Soviet Union and its breakdown into smaller independent states, new maps have become available which validate the observations and predictions given above, for a more clearly Saharasian geographical distribution of these same demographic variables. This further supports the argument, that the World Behavior Map retains validity for understanding the social condition of the world's modern nation-states.

A new work is in preparation, *Saharasia Since 1900*, which will detail this new evidence with updated maps. Discussion will also be given in that new work on the dramatic social transformations which took place over the turbulent 20th Century, which included: The rise and fall of German and Japanese fascism, and of the Communist International Empire of the Soviet Union, Red China, and their satellite states; the end of European colonialism; the breakup of the Islamic Caliphate and its subsequent reorganization and push towards world domination via the international terrorist networks; the subsequent resurgence, as a defensive reaction, of a democratically oriented semi-patrist American-European social-economic network of largely-free nations, led primarily by the USA and United Kingdom — this network militarily confronted, then surrounded and contained the various Saharasian aggressor-warrior nations, blocking their aspirations to global empire. This defense, which played itself out during WW-I, WW-II and the Cold War (WW-III), factually preserved remnant matristic and free social institutions for most of the world outside of Saharasia, down to the present day.

Even while despotic totalitarian nations today control less territory, and the more socially- and politically-liberated nations control a greater territory than at any time in the last several millennia, a 6000-year old tragedy continues to play itself out — of expansive Saharasian Empires versus the rest of the world — now in more technologically-developed and weapons-dangerous forms.

Part III:

The Ancient Origins and Geographical Diffusion of Patrism

Mapping the Archaeological and Historical Evidence

New Material: See *Update on Saharasia* (Appendix B) for additional new archaeological findings and maps relevant to this section.

How Part III Was Developed and Organized

Approximately in 1982, when I began to observe the various patterns on the behavior maps, as presented in preceding chapters, it became clear that the maps would have to be examined in terms of specific historical events and archaeological evidence. The implications were, firstly, that specific dates could be set for each region identifying when armoring and patrism first arose, or arrived as an import from some other region. Secondly, patrism would have *earliest places of origins from which it first arose*, with *spreading centers* out from which affected peoples migrated over the centuries to influence more distant regions. From these two processes, the patterns on the behavior maps developed. And, the behavior maps implied that Saharasia would be the place from which patrism first arose, and from which those early outward migrations would have taken place. But I could not simply assume this was so. I had to *prove it*, by reviewing the archeological and historical evidence for each world region. Major climate changes had to be documented, migrations and settlement patterns traced, and any major changes in culture which were germane to the Saharasian thesis identified. This had to be done for each major world region, from the earliest times to the present.

My methodology was simple and systematic, though tedious. I read and made detailed outline note cards for several hundred major foundational articles, books and compendiums addressing climate, migrations, culture, warfare, religion, sexuality, etc. Each note card recorded a specific archaeological finding or artifact, or historical event, and was identified as to the location and date of occurrence, with citations recorded also. I paid close attention to the various facts as recorded in the "stones and bones" of archaeology and history — but I mostly ignored the myriad, often-contradictory theories being offered by the different authors to explain those facts. The Saharasian maps, derived from more contemporary anthropology, provided a completely new vantage-point from which to observe and interpret the archaeological-historical record. My primary interest was *facts*, and not *theories about facts*. If my behavior maps were truly presenting a bona-fide, real pattern, then major periods of cultural transition, from unarmored matrist to armored patrist, would be identified for each region if the appropriate data was systematically organized in a geographical and chronological manner. I read and outlined various sources, the most scholarly and up-to-date I could find (not always easy), and continued digging until the sources began to repeat themselves, suggesting I had adequately reviewed available evidence for a particular region.

Approximately 10,000 detailed note cards were thereby created, each with its own citation(s). When this exhausting two-year exercise was finished — a project which temporarily transformed me into the "hermit of the library stacks" at the University of Kansas — the note cards were organized chronologically. From this newly created *archaeological/historical data base*, which currently resides in special card files in my office, essential facts emerged which clearly demonstrate the cultural changes implied on the various behavior maps. When properly organized and reviewed, the archaeological and historical evidence thereby provides crucial independently-derived support for the overall Saharasian thesis.

The reader will note some repetition in the material as one moves from one geographical region to the next. This was unavoidable, as the entire globe could not be discussed in explicit terms simultaneously.

8. Environmental and Cultural Changes in Saharasia and its Borderlands

An Archaeological-Historical Reconstruction: The Ancient Origins of Armored Patrism in Saharasia, and the Historical-Geographical Diffusion of Patrism to the Saharasian Borderlands.

New Material: See *Update on Saharasia* (Appendix B) for additional new findings and maps relevant to the materials in this

This Chapter will present details of the environmental and cultural transitions which occurred across Saharasia, as previously discussed, on a region-by-region basis. The discussion here will focus first upon North Africa, and then move eastward across the Middle and Near East, and finally into Central Asia. Cultural changes in Sub-Saharan Africa, Europe, India, China, and Japan which were either influenced or initiated by the migrations of Saharasian peoples will also be discussed.

Note on Availability and Quality of Data:

A number of problems exist regarding the development of a comprehensive and detailed picture of Saharasian ecological, climatic, and social history, not the least of which is the sparse distribution of adequately studied archaeological sites. Blocks of territory, sometimes culturally hostile to scientists with a Western background, remain only minimally studied, or unstudied. Problems exist regarding attempts to precisely trace specific ecological or climatic phases across subregions of Saharasia; pluvials (rainy periods) in one region are not always precisely linked to wet periods in other regions. Still, there is much evidence for former wet phases across all of Saharasia, extending down even into historic times; absolute chronology of these phases, and the exact degrees of wetness can only be generally determined.

Of all the parts of Saharasia, North Africa is best known from a geographical, historical, and archaeological standpoint. The last 30 years has seen expanded research into the prehistory of the Sahara, an effort not undertaken to the same degree in either Arabia, the Near East, or Central Asia. In discussing the Arabian Peninsula, A. H. Masry, Director of Antiquities and Museums for Saudi Arabia, says "It is a land whose antiquities and early heritage have yet even to be outlined, let alone critically analyzed".[1] Butzer, writing in 1975, summarized a situation which is still generally true for almost all forms of prehistoric data in the eastern portions of Saharasia:

"The evidence...is of variable quality from region to region. In some areas, such as Turkey, Iraq, and Iran, data are limited to scattered reports of an exploratory nature. In the Levant somewhat more systematic observations are available, but there are few or no isotopic dates. In Egypt and Nubia the chronology of alluvial deposits is fairly reliable. In Transcaucasia and Arabia, there is almost no published information." [2]

Information from Central Asia is very sketchy. Knowledge of past environments there have only been researched in a systematic manner

1. Masry, A.H., Ed., *An Introduction to Saudi Arabian Antiquities*, Department of Antiquities & Museums, Ministry of Education, Kingdom of Saudi Arabia, 1975, p.13.

2. Butzer, K.W.: "Patterns of Environmental Change in the Near East During Late Pleistocene and Early Holocene Times", in *Problems in Prehistory: North Africa and the Levant*, F. Wendorf & A.E. Marks, Eds., S. Methodist U. Press, Dallas, 1975, p.393.

since around 1950, and knowledge of the region still remains very general, particularly for the Turkestan, Takla Makan, and Gobi deserts. Effective political barriers prevented most information about the arid steppes and Turkestan desert from diffusing outside the former USSR, and social turmoil following the breakup of the USSR has continued to stand as a barrier to scholarly research in the region. The Russians and Chinese have both initiated ambitious arid zone research programs, but these efforts are largely focused upon irrigation, and methods of controlling desertification or exploiting barren lands. The little that has been published and translated focuses primarily upon contemporary problems of land use, and contains only hints of conditions during prehistoric, much less historic times.

Russian and Chinese sources also appear to avoid discussion of evidence regarding climate change, favoring economic interpretations for desertification, land abandonment, and depopulation of arid zones. Ecological and cultural upsets in their arid zones may be attributed solely to the "inferiority of private land ownership" over collective farming, the "ruthless exploitation of feudalistic lords, herd owners and Kuomintang reactionaries", or the invasions of nomadic Arab and Mongol armies.[3] The environmentalist Huntington, who argued for climate change across Central Asia, was traditionally labeled by the extreme left as a "decadent bourgeois determinist". However, other Soviet scientists have more recently taken positions in agreement with Huntington's views on climate change.[4]

Current research in the Asian arid zones is restricted by a variety of other factors. For instance, it is known that the Chinese arid zone of Lop Nor is rich in archaeological sites; however, it is now a top-secret troop staging area and testing ground for nuclear weapons. Secrecy and suspicion towards foreigners has historically prevailed across both the Soviet Union and China, to the extent that one author labels it a "regional disease".[5] And from an emotional viewpoint, regarding the unconscious motivation of behavior by submerged fears and hostility, the diagnosis of a Saharasian psychological "secrecy disease" or "suspicion-of-outsiders disease" is quite precise. The following assessment of Central Asian information was made in 1968 regarding atmospheric data, which may be taken as an indicator for the availability of other forms of paleoclimatic and archaeological data:

> "The problems of language and politics have proven effective barriers to the assessment of current sources of climatic data for these...deserts... A letter from D. Tuvdendorj, director of the hydrometeorological service of the Mongolian Peoples Republic, states that although climatological observations are made, collected, and summarized, they are not published... The data available from the Takla Makan desert are at best fragmentary and the existing records are short. ...data [are] from less than 20 stations in this area, with the longest temperature record being 5 years. Isolines on maps were not extended into this region." [6]

As of 1968, the faunal knowledge for Turkestan was given a subadequate rating, while the Takla Makan and Gobi remain the least known of all deserts on Earth.[7]

A parallel can be drawn between the unavailability of research data in Saharasia and intolerance toward foreigners or foreign customs. Western researchers have more or less abandoned working in the war zones stretching from Lebanon eastward through Syria, Iran, Iraq, and Afghanistan. It is a region which has long displayed much impulsiveness,

3. For example, see Kovda, V.A.: "Land Use Development in the Arid Regions of the Russian Plain, the Caucasus and Central Asia", in *A History of Land Use in Arid Regions*, L.D. Stamp, Ed., UNESCO, 1961, and the case studies on China, in *Desertification*, M.R. Biswas & A.K. Biswas, Eds., Pergamon Press, 1980.

4. Chappell, J.E.: "Climatic Change Reconsidered: Another Look at 'The Pulse of Asia'", *Geographical Review*, 60(3):368; cf. Lamberg-Karlovsky, C.: "Afterword", in P. Kohl, Ed., *The Bronze Age Civilization of Central Asia: Recent Soviet Discoveries*, M.E. Sharpe, NY, 1981, p.388; Kohl, P.: "The Namazga Civilization: An Overview", in Kohl, 1981, ibid., p.xxxiii.

5. Tickell C.: *Climatic Change and World Affairs*, Harvard Studies in International Affairs, #37, 1977, p.52.

6. McGinnies, W., et al, Eds: *Deserts of the World*, U. Arizona Press, 1968, p.30.

7. McGinnies, et al, 1968, ibid., p.5.

oscillating between extremes of hospitality and hostility. Dangers abound for foreigners working in these regions.

The Soviet Union and China only in recent historical times have tamed marauding bands of nomads, using brutal tactics learned from the nomads themselves; these large regions have long maintained a suspicion towards foreigners, and currently tolerate only limited incursions by Western researchers. The Soviet purges and Chinese "Cultural Revolution" had a general freezing effect upon free inquiry, even or especially by native scholars. North Africa and Arabia only opened their borders to Westerners, partially, following cultural contact brought about by two World Wars and the discovery of oil. Unbelievers are nevertheless still barred from entering the holy cities of Mecca and Medina. One might politely call Saharasia, particularly its portions east of the Red Sea, a *Zone of Suspicion*.

GENERAL ENVIRONMENTAL TRANSITIONS IN SAHARASIA

Quaternary Epoch:

Period encompassing both the Pleistocene Ice Age and the more recent Holocene Era.

Pleistocene Ice Age:

Period ranging from around 1 million years ago until approximately 50,000 years ago, during which much of Canadian North America and NW Europe was covered with massive sheets of ice. These sheets of ice persisted, more or less, until around 10,000 BCE.

Holocene Era:

Period of most recent geological history, starting around 50,000 years ago, a period of shifting ice sheets and climate change, with the final melt-back of Ice Age glaciers around 10,000 BCE.

In spite of the above problems, there is an enormous body of evidence demonstrating that Saharasia changed from a relatively well-watered grassland into a hyperarid desert around 4000-3500 BCE, following a period of wetness which began during the Pleistocene. Where disagreement among scholars may exist on this question, it is only regarding the exact chronology for desiccation in the various subregions, the degree of previous wetness, the presence or absence of subphases of wetness or dryness during the general period of progressive desiccation, and the degree to which either human influences or climate change, or both, were responsible for observed changes toward desiccation. The desiccation of Saharasia as a force in shaping human prehistory has been appreciated by various scholars, but generally not in terms of its effects upon the infant, child, or female.

"The Saharo-Arabian belt of desert and semidesert has played a crucial part in the story of man on the Earth. Perhaps in no other major belt was the interaction of man and milieu more oscillating in nature, and yet uniform in pattern, than in this area. The story of human activity in these deserts and semisteppes is characterized by its immense length. It goes back into the remote past of prehistory and paleogeography. Indeed, the intricacies of human adjustment and changes in nature in this area cannot be properly understood or deeply appreciated in the light of a study limited to the present-day geography." [8]

"Three subcontinents, green with forest, are divided from one another by the vast desert belt of the Old World: Africa south of the Sahara, Monsoon Asia, and Europe, including Siberia and the Mediterranean lands. The desert barrier as well as the frost line, ...have probably been responsible for dividing humanity during parts of the Quaternary age, so that racial units could develop to some extent in isolation." [9]

"...the Saharo-Arabian belt played a particularly important role in the development of early human civilization. The remains left by man over practically all the desert are a clear indication to this effect." [10]

It is known that large ice sheets existed across Northern Europe and

8. Huzayyin, S.: "Changes in Climate, Vegetation, and Human Adjustment in the Saharo-Arabian Belt, with Special Reference to Africa, in *Man's Role in Changing the Face of the Earth*, Vol. 1, W.L. Thomas, Ed., U. Chicago Press, 1973, p.304.

9. Von Wissmann, H.: "On the Role of Nature and Man in Changing the Face of the Dry Belt of Asia", in Thomas, 1973, ibid., p.278.

10. Huzayyin, 1973, ibid., p.310.

Northwest Asia during most of the Quaternary; these ice sheets reached their maximum extent around 18,000 BCE. Following this maximum, the ice sheets shrank and by c.8000 BCE had significantly melted away. The growth, shrinking, and final melting of these ice sheets is evidence of climate changes of a global nature, particularly a transition from cooler to warmer conditions in the mid and high latitudes. Also, the end of the last glacial period coincided with the extinction of some Quaternary species of plants and animals, following which a different flora and fauna prevailed.[11]

Precipitation and evaporation also eventually changed following the close of the Pleistocene, giving rise to additional changes in physiography, flora and fauna. Childe has given the following vivid picture:

> "While Northern Europe was covered in ice as far as the Harz, and the Alps and the Pyrenees were capped with glaciers, the Arctic high pressure deflected southwards the Atlantic rainstorms. The cyclones that today traverse Central Europe then passed over the Mediterranean Basin and the Northern Sahara and continued, undrained by Lebanon, across Mesopotamia and Arabia to Persia and India. The parched Sahara enjoyed a regular rainfall, and farther east the showers were not only more bountiful than today but were distributed over the whole year, instead of being restricted to winter...We should expect in North Africa, Arabia, Persia and the Indus Valley parklands and savannas, such as flourish today north of the Mediterranean." [12]

Childe's view of Saharasia will have to be qualified somewhat, given the identification of limited aridity in some sub-regions of North Africa during the last glacial epoch.[13] Also, the period of wetness in Saharasia did not immediately end with the melting of the glaciers; wet conditions persisted and possibly even intensified in Saharasia for several thousand years following the Pleistocene. Aridity only began to grip the region in a significant way after c.4000-3000 BCE. Wadia puts it so:

> "There is a body of competent geological, biogeographical and archaeological evidence, supported by popular local legends and traditions, to prove that this desert belt is the creation of late Neolithic and sub-Recent times extending up to historic times." [14]

Figure 48 is a map which reconstructs world rainfall following the last glacial epoch, for the Altithermal period, c.6000-2000 BCE. This map, which reflects conditions at the "Dawn of Civilization", was compiled by Kellogg from over 100 references.[15] Kellogg's map details paleoclimate conditions for that early period six millennia ago, as best known at the time of its publication, in 1978. His findings were also in agreement with the map of Sarnthein,[16] depicting a generally wet Sahara at c.4000 BCE. Central Asia was given question marks by both Kellogg and Sarnthein, reflecting a relative lack of paleoclimate field study in the area, as discussed above. However, other sources I consulted indicated a wetter early Central Asia as well. As will be presented and documented in sections to follow, there is plentiful evidence for an early wet period which stretched across the whole of Saharasia.

11. Axelrod, D.I.: *Quaternary Extinctions of Large Mammals*, U. California Publications in Geological Sciences, Berkeley, 1967.

12. V. Gordon Childe as quoted in Carpenter, R.: *Discontinuity in Greek Civilization*, Cambridge U. Press, 1968, p.6.

13. Nicholson, S.E. & H. Flohn: "African Environmental and Climatic Changes and the General Atmospheric Circulation in Late Pleistocene and Holocene", *Climatic Change*, 2:313-348, 1980; Sarnthein, M.: "Sand Deserts During Glacial Maximum and Climatic Optimum", *Nature*, 272:43-46, 2 March 1978; Street, F.A. & A.T. Grove: "Environmental and Climatic Implications of Late Quaternary Lake Level Fluctuations in Africa", *Nature*, 261:385- 390, 3 June 1976; Fairbridge, R.W.: "African Ice-Age Aridity", in *Problems in Paleoclimatology*, A.E.M. Nairn, Ed., Interscience Pub., 1964, pp.356-63.

14. Wadia, D.H.: *The Post-Glacial Desiccation of Central Asia*, Nat. Inst. Sciences of India, Delhi, 1960, p.1.

15. Kellogg, W.W.: "Global Influences of Mankind on the Climate", in *Climatic Change*, J. Gribbin, Ed., Cambridge U. Press, NY, 1978, p.220.

16. Sarnthein, 1978, ibid.

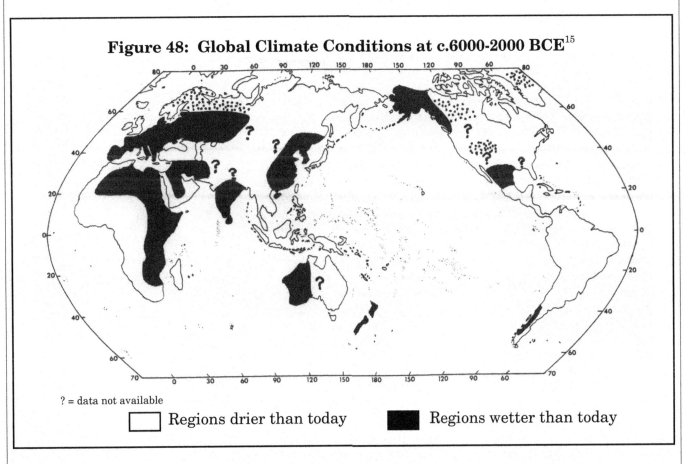

Figure 48: Global Climate Conditions at c.6000-2000 BCE[15]

? = data not available

☐ Regions drier than today   ■ Regions wetter than today

GENERAL BIO-CULTURAL TRANSITIONS ACROSS SAHARASIA

Around 1,500,000 years ago, long before any quantifiable environmental transitions had occurred in Saharasia, Homo erectus evolved from the more primitive hominoid Australopithecus. Homo erectus, as the name implies, walked upright and possessed a brain capacity half again as large (900 cc.) as its Australopithecine ancestor. Bones of Homo erectus have been found with Lower Paleolithic hand axes and other stone tools of the early Acheulean industries, which spread throughout Western Europe, Africa, Arabia, and tropical and subtropical Asia by around 500,000 BP, aided by the lowered sea levels of the Pleistocene glaciation.

Around 500,000 BP another hominoid appeared, the Neanderthal, who apparently developed under the frigid conditions of the glacial epochs of the period. Neanderthals possessed a brain capacity greater than Homo erectus, around 1350 cc. Their remains exhibit the first indications of culture as generally defined in that they buried their dead in prepared graves, leaving behind tools and offerings of food or flowers. Neanderthals spread throughout much of Europe, Asia, and Africa during the last interglacial period (Riss-Wurm) and first half of the final glacial period. However, their pebble stone tool industry, the Mousterian, was more primitive than the developing Acheulean, which by 100,000 BP had been considerably advanced by various new hominoids of a true Homo sapiens type. Although much of evolutionary history at this point remains unknown, it appears that:

"...from about 35,000 years ago Neanderthal man as a recognizable breed disappears while men of our own species, fully evolved Homo sapiens, appeared in many parts of Southwest Asia, North Africa and Europe, using skillfully made specialist tools and weapons." [17]

BP = "Before the Present"

17. Hawkes, J.: *Atlas of Ancient Archaeology*, McGraw Hill, NY 1974, p.9.

211

Between 35,000 BCE and 8000 BCE, Late Paleolithic Homo sapiens, the modern human, spread around the globe. An advanced hunting culture was developed, the tools of which included the spearthrower and the bow, and finely worked flint spearpoints and blades. The first substantial dwellings also appeared during this Late Paleolithic, as did various arts; widespread mother goddess figurines were constructed, and magnificent engravings and paintings were left behind in caves. Music, art, the dance, and oral literature also emerged during the period.

As glacial ice melted and retreated, the periglacial steppe of Europe became forested, prompting new modes of survival based upon small game stalking and fishing. Human populations expanded as the great ice sheets of the Pleistocene made their final retreat. Cultural diversity increased during this Late Paleolithic stage, during which time considerable climatic fluctuation occurred. One factor stimulating migrations during the period was a rise in sea levels of some 300 feet,[18] caused by the meltwaters from the massive continental glaciers which had shrunk considerably by c.7000 BCE. The archaeological record for these early periods is relatively incomplete, but there is no compelling evidence, or reason to believe that these changes forced any drastic, sudden, or widespread influences upon human culture which would have driven them towards an increased social violence or patrism. Where archaeological information is available from which inferences can be made, evidence for violence or patrism is almost entirely absent, with a few notable exceptions which I shall detail in the sections to follow. I therefore argue that the original innate behaviors and early cultural institutions of Homo sapiens were carried into the post-Pleistocene era intact, unchanged by any stimulus which would have driven culture away from matrism towards patrism. This argument is given substance by Fisher,[19] who has reconstructed early hominoid sexual behavior and family life from fossil evidences, demonstrating a distinctly matrist orientation. As discussed, below, the archaeological and historical evidence also supports the contention of matrism as the earliest, and hence innate, human condition.

Following the close of the Pleistocene, climatic conditions remained relatively stable across the Old World until about 4000 BCE, when desertification began to grip the Arabian and Central Asian subsections of Saharasia. Starting at c.4000-3000 BCE, and by no later than 2000 BCE, desert conditions dominated the whole of Saharasia. At that time, one of the most substantial environmental and climatic changes occurred since the close of the last glacial epoch.[20] Adaptive changes were forced upon emerging human cultures which previously thrived upon hunting and gathering of native fauna and flora, or in some areas upon mixed agriculture and animal herding. It was in Saharasia, during the desiccation phase, that irrigation agriculture and nomadic pastoralist technology first developed. Building construction and metallurgical knowledge were also first refined and advanced in Saharasia following the onset of drier conditions, as were various forms of central-state and military apparatus.

Childe discussed climate changes toward desiccation in the Near East as a major mechanism in promoting social change,[21] but his study did not focus upon the world of the infant, child, or sexuality, and defined "civilization" as being synonymous with technology and the central state, which says nothing at all regarding human civility. Other studies on the development of culture and state in the Near East have likewise defined civilization as something apart from matters related to social violence, which I see as a common trap.[22]

For example, Wittfogel noted the development of despotic tendencies in many Saharasian empires, but argued for a mechanism of cause involving development of irrigation agriculture, rather than any devastat-

BCE = "Before the Current Era" (secularized from equivalent dates "BC")

CE = "Current Era" (secularized from equivalent dates "AD")

18. Carpenter, ibid., 1968, p.20.

19. Fisher, H.: *The Sex Contract: The Evolution of Human Behavior*, Wm. Morrow, NY, 1982.

20. Lamb, H.H.: "Reconstruction of the Course of Postglacial Climate Over the World", in *Climatic Change in Later Prehistory*, A.F. Harding, Ed., Edinburgh U. Press, 1982, p.29.

21. Childe, G.: *New Light on the Most Ancient Near East*, Praeger, NY, 1952.

22. Bowden, M., et al.: "The Effect of Climate Fluctuations on Human Populations: Two Hypotheses", in Wigley, T.M.L., et al., Eds.: *Climate and History*, Cambridge U. Press, 1981, pp.479-513; Spooner, B.: *Population Growth, Anthropological Implications*, MIT Press, Cambridge, 1972; Wright, H.: "The Environmental Setting for Plant Domestication in the Near East", *Science*, 196:385-9, 1976.

ing affect of aridity upon babies, children, and sexuality.[23] His irrigation mechanism, however, has since been contradicted by archaeology.[24] As discussed in Chapter 3, more recent observations of severe drought and famine have clearly demonstrated severe and disastrous effects upon family structure and social fabric. In the modern era, famine, migration, mass mortality, conflict, warfare, and revolution invariably have followed widespread desiccation, and such certainly would have occurred during past epochs of desertification.[25]

Indeed, phases of migration and conflict are apparent in the archaeological record beginning with the early periods of extreme ecological change in Saharasia. These phases were also accompanied by famine and pestilence in many areas, similar to or more extreme than what is witnessed today during droughts in the Sub-Saharan region. Populations were exposed to severe ecological pressures, the scale of which had never before been experienced by Homo sapiens. As the region was transformed from lush grassland to desert, resident populations migrated or perished. Concurrently, populations in surrounding moist regions, or living on secure water supplies, were devastated by invasions, migrations, and conflicts.

Other scholars, notably the environmentalist Elsworth Huntington,[26] have discussed the role of climate change in prompting cultural adjustments or transitions across broad regions of Saharasia. Huntington's observations and ideas on the dynamics between climate and culture change have been discussed at length elsewhere,[27] with important supporting corroboration.

Archaeology has clearly confirmed the presence of many sites in the Near East where relatively sharp, or abrupt cultural change occurred. The archaeological strata demonstrate many cultures wherein the ages pass with only gradual, minor changes in material culture. Then a destruction layer may appear along with other evidences of mass migrations, warfare and anarchy. Major changes in material culture, technology, burial rites, social status of various classes, and other cultural transformations are evident, even after "order" is reestablished and settlements rebuilt. When these destruction layers are widespread and long-lasting, and are coincidental to the abandonment of writing, records-keeping, or the arts, they are termed "Dark Ages".

Bell has reviewed the archaeological evidence for environmental and cultural change in the Near East, and extended the ideas of Huntington and others to develop a thesis regarding the "Dark Ages of Ancient History". She argues that such major social transformations in the Near East were caused by widespread and chronic drought.[28]

"In the history of the ancient Near East two striking Dark Ages have occurred. They occurred more or less simultaneously (within the limits of current dating accuracy) over a wide area extending at least from Greece to Mesopotamia and Elam, from Anatolia to Egypt, and probably beyond." [29]

Bell identified a number of cultures and their corresponding periods of collapse, and these are identified in Table 7, below. The "Dark Ages" noted by Bell were accompanied by particularly strong climatic pendulations towards dryness inscribed upon a longer term trend towards desiccation. However, my own review of archaeological and paleoclimatic research for Saharasia suggests that the first major environmental transition to affect human culture and society in a significant way occurred across most of Saharasia around c.4000-3500 BCE. I argue that the actual creation of the male-dominated, warrior caste empires which

23. Wittfogel, K.W.: *Oriental Despotism*, Yale U. Press, New Haven, 1957; See the section on "Hydraulic Society and Oriental Despotism" in Chapter 5 for more discussion.

24. Butzer, K.: *Early Hydraulic Civilization in Egypt*, U. Chicago Press, 1976, p.100; Hoffman, M.: *Egypt Before the Pharaohs*, Alfred Knopf, NY, 1979, pp.314-14.

25. See Chapter 3 and later sections of this Chapter.

26. Huntington, E.F.: "The Burial of Olympia", *Geographical Journal*, 36:657-86, 1910; Huntington, E.F.: "Climatic Change and Agricultural Exhaustion as Elements in the Fall of Rome", *Quart. J. Economics*, 31:173-208, 1917; Huntington, E.F.: *The Pulse of Asia*, Houghton-Mifflin, NY, 1907; Huntington, E.F.: *Palestine and its Transformation*, Houghton-Mifflin, NY, 1911; Huntington, E.F.: *Asia: A Geography Reader*, Rand McNally, Chicago, 1912; Huntington, E.F. & S.S. Visher: *Climatic Changes: Their Nature and Causes*, Yale U. Press, New Haven, 1922.

27. Murphy, R.: "The Decline of North Africa Since the Roman Occupation: Climatic or Human?", *Annals, Assoc. Am. Geographers*, 41(2), 1951; Mikesell, M.: "The Deforestation of Northern Morocco", *Science*, 132:441-8, 1960; Mikesell, M.: "The Deforestation of Mt. Lebanon", *Geographical Review*, 59(1):1-28, 1969; Chu, C.: "Climatic Pulsations During Historic Times in China", *Geographical Review*, 16(2):274-83, 1926; Issar, A.: "Climatic Change and the History of the Middle East", *American Scientist*, 83:350-55, July-Aug. 1995; Chappell, 1970, ibid., pp.347-73; Wadia, 1960, ibid.; Carpenter, 1966, ibid;

28. Bell, B.: "The Dark Ages in Ancient History, 1: the First Dark Age in Egypt", *American J. Archaeology*, 75:1-26, 1971.

29. Bell, 1971, ibid., p.1.

Table 7: Old World Periods
of Social Chaos and Collapse [28]

For the period c.2200 BCE:
- Egypt: (end of VI dynasty) sudden transition from a stable society into anarchy
- Mesopotamia: Akkadian Empire disintegrates
- Levant, Palestine, Syria: Byblos, other sites destroyed by fire and abandoned
- Anatolia: Troy II destroyed by fire; end of Early Bronze II "marked by a catastrophe of such magnitude as to remain unparalleled until the very end of the Bronze Age", widespread destruction and general decline in material culture, decline of known settlements by 75%
- Greece: Lerna, other prosperous Argolid centers are burned, followed by periods of poverty
- Indus Valley: Harappan civilization collapses

For the period c.1200 BCE:
- Hittite empire collapses
- Mycenaean Greece collapses
- Egypt declines shortly thereafter
- Babylonia weakened
- Assyria weakened

---

dominated those same regions was prompted by desiccation in the first instance. Later, as desiccation became more intense and widespread, these same empires collapsed.

Other authors have claimed that major cultural transitions occurred during these times, but usually through mechanisms other than desiccation, and with a focus on cultural parameters other than infants, children, women and sexuality. For instance, Velikovsky's catastrophic thesis was probably the first to identify and address the simultaneity of environmental and social changes during the same periods identified later on by Bell. Unlike Bell, however, Velikovsky attributed the changes to the perturbation of the Earth's surface by a large comet.[30] His thesis attracted much attention, and antipathy, but credit must go to him for reopening the catastrophic hypothesis to renewed analysis.[31] Other works in recent years have also taken up the thesis of regional or global catastrophe in the shaping of human events,[32] and a number of distinctly catastrophic observations have been made by various field archaeologists.[33] In any regard, the desiccation of lands in the Near East was briefly touched upon in Velikovsky's works, as one expression of a major global cataclysm which stimulated massive culture change.[34]

The culture changes toward human pathology identified by Velikovsky did not, however, address any catastrophically-stimulated changes in the treatments of infants, children, or women, or regarding human sexual behavior. Velikovsky, a psychoanalyst, argued instead that the "collective unconscious" transmitted a species-wide "memory trace" of the catastrophe to ensuing generations, who compulsively and unconsciously repeated the disturbance in various cataclysms of warfare and mass murder.[35] Although Velikovsky has attracted the greatest amount of hostility over his reinterpretation of astronomical and geological events, the "collective unconscious" is, in my opinion, the least documented and weakest aspect of his work. For instance, if the catastrophe he identified

30. Velikovsky, I.: *Worlds in Collision*, Macmillan, NY, 1950; Velikovsky, I.: *Earth in Upheaval*, Doubleday, 1955; Velikovsky, I.: *Ages in Chaos*, Doubleday, 1952; Velikovsky placed major global cataclysms between 1500-800 BCE, citing evidence for errors in traditional radiometric dating techniques by which he claimed to resolve many puzzles in ancient chronology. However, he also stated, without elaboration (to my knowledge), that other cataclysms had occurred several millennia earlier. I do not know if he referred specifically to the same time period identified in my research, namely to c.4000-3500 BCE.

31. deGrazia, A.: *The Velikovsky Affair*, University Books, 1966; Editors of *Pensee: Velikovsky Reconsidered*, Doubleday, 1976; Ransom, C.: *The Age of Velikovsky*, Kronos Press, 1976.

32. Alvarez, L., et al.: *Science*, 208:1095-108, 1980; Kastner, M., et al.: *Science*, 226:137-43, 1984; Berggren, W. & J. Van Couvering, Eds.: *Catastrophes and Earth History: The New Uniformitarianism*, Princeton U. Press, 1984; Goldsmith, D.: *Nemesis, The Death Star and Other Theories of Mass Extinction*, Walker, NY, 1985; Sheets, P. & D. Grayson, Eds.: *Volcanic Activity and Human Ecology*, Academic Press, 1979; Sheets, P., Ed.: *Archaeology and Volcanism in Central America*, U. Texas Press, Austin, 1983; Hooykaas, R.: *Catastrophism in Geology, its Scientific Character in Relation to Actualism and Uniformitarianism*, North-Holland, 1970; Vita-Finzi, C.: *The Mediterranean Valleys, Geological Changes in Historical Times*, Cambridge U. Press, 1969.

33. A number of examples are specifically cited in this Chapter.

34. Velikovsky, 1950, ibid., pp.97,144, 169,297,384.

35. See Velikovsky, I.: *Mankind In Amnesia*, Doubleday, NY, 1984; In fairness to Velikovsky, who has been unjustifiably maligned, it must be said that my thesis on the role of desiccation in Saharasia as the major event stimulating culture change may not be wholly incompatible with his catastrophic hypothesis. Indeed, my research suggests the most abundant physical evidences for catastrophism will be found in the region stretching across North Africa, Arabia and across the Near East into Central Asia, and not, as Velikovsky believed, around the entire globe. This question could be resolved by a systematic geographical and temporal review of the archaeological and geological *evidence for catastrophism* cited in Velikovsky's writings, and in other publications on catastrophism.

was truly global, and is really at the root of human psychopathology and destructive aggression, then the distribution of human psychopathology should also be global and somewhat uniform in nature. As demonstrated in this work, such is not the case.[36]

Others have made theoretical links between global catastrophism and the origins of the repression of emotion and instinct in the human animal, but without any substantial presentation of new evidence.[37]

Jaynes similarly identified cultural changes during the Neolithic and Early Bronze Age, but also did not review such changes against environmental conditions, or the prevailing treatments of infants, children, or women.[38] Wescott's review of Jaynes' work identifies a series of authors with similar ideas on major cultural change, none of whom touch upon the specific issues of environmental desiccation, or the treatment of infants, children, and women, and the larger issues of human sexuality.[39] Even H. G. Wells observed the change in mentality during the Neolithic:

"...the...Paleolithic...hunter...killed for reasons we can still understand; Neolithic man...killed for monstrous and now incredible ideas...and... sacrificed... men, women, and children whenever...he... thought the gods were athirst." [40]

My objection to all the above works, stimulating and important as they are, is none of them take a systematic chronological and spatial review of the archaeological and historical evidence which is used to construct their theories; neither do any of them take a firm and unwavering focus upon the environment of the infant, child, or female, or include human sexuality from the standpoint first outlined by Reich. The works of Huntington, Velikovsky and Bell do emphasize the role of major environmental change as a force to initiate changes in human behavior and social structure, but are rather vague about exactly how social change proceeds, or how distant regions not influenced by climate change are also subject to profound social change. As pointed out in the introduction, discussions on environmental influences upon culture are largely found in the literature of geography and anthropology, in the context of human ecology. Yet even here, an interdisciplinary approach, emphasizing both environmental and cultural change — to include the treatments of infants, children, and women, with a straightforward discussion of the social repression of sexual impulses — has not proceeded, except in the most theoretically obtuse manner (ie., "selfish genes lead to male dominance", etc.).

The archaeological and historical literature, summarized below, demonstrates the desiccation of Saharasia coincided with a number of very specific and profound human social disasters and adjustments. These include extremely important changes in the treatment and status of infants, children, and women, with a reduction of emphasis upon sexual pleasure and shifting emphasis towards sexual pain and repression, and a growth of emphasis upon strongman leadership and the military caste. In short, matrist cultures in previously wet regions were, over the generations, driven towards patrism as conditions became drier and food supplies dwindled. Patrism eventually became a relatively permanent feature of most Saharasian cultures and was, in turn, carried out of the desert via migration and invasions, to be forced upon or copied by subordinated and conquered survivors of surrounding non-desert regions.

36. H. Fox (*Gods of the Cataclysm*, Dorset, NY, 1976) has argued for a Velikovskian-type catastrophe somewhere in the Old World, with diffusion of affected cultures thereafter across the Atlantic and Pacific Oceans. The evidence provided by Fox was not mapped, however, and neither did it include data on infants, children, women and sexuality, except in a most general sense. Still, his thesis would appear to be in agreement with the data presented here, assuming further research can document a more specific correlation between onset of aridity and catastrophe in the Old World. Unfortunately, scholars of the Velikovskian catastrophist viewpoint have not been welcomed or even tolerated within academic institutions, any more than Reichians, making it extremely difficult for these kinds of questions to be more thoroughly researched in a systematic manner. Until openness prevails, progress on these questions will be very slow in coming. See Chapter 9 for more discussion on the issue of Pre-Columbian migrations and diffusion of patrism into the New World.

37. Robert Morris, a scholar familiar with the works of both Velikovsky and Reich, first pointed out to me in 1975 the potential causal relationship of catastrophism to the origins of emotional armoring. In 1983, after my own research was essentially completed, I learned that another scholar, Theodore Lasar, had previously made a theoretical connection between Velikovsky and Reich, and had prepared a manuscript which addressed "...how, in ancient times, man was compelled by physical catastrophe to armor himself against his own natural emotions." (Lasar, T.: "The Origins of the Emotional Disaster", *Orgonomic Functionalism*, VII(1):37-62, Jan. 1961; "...Part 2", VII(2):116-154, March 1961; "...Part 3", VII(5):283-329, Sept.1961.)

38. Jaynes, J.: *The Origins of Consciousness in the Breakdown of the Bicameral Mind*, Houghton Mifflin, Boston, 1976.

39. Wescott, R.W.: "Review of: The Origins of Consciousness in the Breakdown of the Bicameral Mind", *Kronos*, III(4):78-85, 1978.

40. H.G. Wells, as quoted in Wescott, ibid, p.79.

A. NORTH AFRICA

The Sahara desert stretches across the whole of North Africa, from the coasts of the Atlantic on the west to the Red Sea on the east. Highlands of the central region are broken by elevated mountain massifs, and the northwest is dominated by the Atlas Mountains. Ancient wadis cut deep into the slopes of the highlands and plateaus, forming outwash regions across the surrounding lowlands which terminate only at the coasts or in various interior basins. Both basins and plateaus possess alternating zones of relatively flat compacted sand, drifting sand dunes, or wind-scoured zones deflated down to bedrock. Or they are strewn on the surface with sun-baked rock and pebble. Most of the Sahara receives less than 125 mm (5") of precipitation per year, and some regions get less than 25 mm (1").

The volcanic massifs of Ahaggar, Tibesti, and Darfur arc across the central Saharan plateau roughly NW to SE. Like the Atlas range, these elevated massifs obtain slightly higher rainfalls, which in turn generate slight differences in vegetation and settlement pattern. Between the Ahaggar and Atlas mountains lie the great *sand ergs* of Algeria which cover basins, sloping tablelands, and plateaus. These vast devegetated regions, such as the Grand Erg Occidental, Grand Erg Oriental, Ergs Chech and Iguidi, and El Djouf, stretch across the dry southern slopes of the Atlas range southwest into Mali and Mauritania, eventually reaching the Atlantic coast and Senegal River. The barren sand-swept tablelands of eastern Mauritania and northern Mali are nearly the same size as France, but contain little vegetation and are almost entirely uninhabited, save for a few mining towns and a very few oases at the perimeter. As desolate as they are, these regions of the Western Sahara receive more rain than areas farther to the east. Unfortunately, when rare storms do occur they are torrential or catastrophic in character, and may be mixed with heavy duststorms.[1]

Vast basins of interior drainage exist along the borders between Mali and Algeria, and between Algeria and Tunisia. Salt lakes or flats are found in the lowest spots of these basins, as well as in the valleys of the Atlas range. Only at the southern edge of these West Saharan tablelands does fresh water exist in supply, where the Niger River and its tributaries braid through alluvial swamps and lakes in an otherwise barren, rainless region. The Niger soon abandons the desert, however, deviating south towards Nigeria.

The highlands of Air and Tibesti in the Central Sahara slope south toward the Chad Basin. The northern part of the Chad basin is covered with either compacted rock and pebble *hammadas*, or drifting sand. The southerly portion of the Chad Basin is occupied by the remnants of Lake Chad,[2] as well as by salt flats and swamps. Like the Niger River, the waters of Lake Chad do not come from the desert itself, but from rivers draining from the wet subtropical regions to the south, and not from its arid north. Except for a few recently established, isolated settlements based upon the mining of deep underground water sources, used for irrigation, vast unvegetated and uninhabited regions lie north of the Chad Basin, all the way across the Sahara until one reaches Tripoli, at the Mediterranean coast.

North of Ahaggar and Tibesti, the Saharan highlands gradually slope towards the Mediterranean Sea. This region is the most arid portion of the Sahara, receiving less than 5 mm (0.2") of rain per year, and is largely unvegetated and uninhabited. Water is available only from a few springs

1. Flohn, H. & Nicholson, S.: "Climatic Fluctuations in the Arid Belt of the 'Old World' Since the Last Glacial Maximum: Possible Causes and Future Implications", *Paleoecology of Africa and the Surrounding Islands*, 12:6-7, 1980.

2. Lake Chad has nearly dried up in recent years, due to persisting droughts and overuse of its waters for irrigation.

and wells in bedrock depressions in the gently sloping terrain. The Libyan Desert lies in this section of the Sahara, on the border between Libya and Egypt. In places it is covered with vast, flat sand sheets; elsewhere drifting sand dunes or scoured bedrock exist. Several large basins of interior drainage are found here, at Qattara and other places west of the Nile River.

The Nile cuts through the far eastern edge of the Sahara, draining from the Ethiopian highlands and rainy tropical plateau regions south of the Sahara. East of the Nile, between it and the Red Sea, lie the sharply elevated highlands of the Arabian and Nubian Deserts. Like the Niger River and Lake Chad, the Nile River Valley is one of a very few places in the Sahara where water is found in copious supply throughout the year.

All in all, at present there are few places in the open Sahara where any large population of humans could exist, much less thrive. Large populations are now found only where water supplies are copious, along the Nile or Niger Rivers, or near Lake Chad. Even so, an astonishingly different landscape, fauna, and settlement pattern existed in North Africa until c.3000 BCE.

In one of the earliest comprehensive surveys of North Africa, the Egyptian geographer Huzayyin discussed changes in the landscape across the "Saharo-Arabian Belt".[3] He concluded that a wet phase or pluvial period existed as far back as the Upper Pliocene, some 2 million years ago, and persisted more or less through to the end of the Pleistocene glacial epoch, some 12,000 years ago; thereafter (he argued) desiccation gradually set in. While Huzayyin believed that periods of prolonged aridity punctuated the Quaternary, none were as severe as the one which started around 10,000 BCE, and which progressively intensified, continuing to this day.[4] Huzayyin felt that the general postglacial trend towards increased aridity was characterized by oscillations back to wetness from time to time in various subregions. However, given the lack of datable research at the time of his writing, he did not identify the exact chronology for the older pluvial (wet period) oscillations. The most recent wet period, the Neolithic wet phase, was identified at between 5500-2500 BCE. While subsequent research has refined the chronology of desiccation in North Africa, the general view of oscillatory climate change has found support.

Monod reviewed various theories on the chronology and rainfall distributions of prehistoric North Africa,[5] of which there appear to be no shortage. For instance, different theorists have alternately described Northern Sahara pluvials as occurring at both the same time and at different times than the Southern Sahara.[6] Evidence gathered within the last several decades, relying largely upon radiometric dating techniques, has resulted in a general pushing back of the start of the Neolithic wet phase to 7000-8000 BCE;[7] prior to this date, a series of wet and dry phases occurred throughout the Pleistocene.[8]

The most recent comprehensive study of climate change in North Africa is that of Nicholson,[9] who also identified moist conditions in the Sahara starting around 8000 BCE and lasting until c.3000-2000 BCE. Extreme aridity followed this humid phase, peaking out around 1500 BCE after which a number of minor humid phases occurred in subregions.

A more recent addition to the view of North African paleoenvironments is the discovery of a prolonged period of aridity during the height of the last glacial epoch. The duration, intensity, and distribution of this arid phase are still uncertain, but it was centered on 16,000 BCE. Arguments for its existence rely upon the extent of radiometrically dated ancient sand deserts,[10] as well as upon fluctuations of ancient lake levels.[11] This view of "ice age aridity"[12] has challenged the older concept of

3. Huzayyin, S. "Changes in Climate, Vegetation, and Human Adjustment in the Saharo-Arabian Belt" in *Man's Role in Changing the Face of the Earth*, Vol 1, W.L. Thomas, Ed., U. Chicago Press, 1972; cf. Huzayyin, S.: *The Place of Egypt in Prehistory, A Correlated Study of Climates and Cultures in the Old World*, Memoires, L'Institute D'Egypte, Cairo, 1941.

4. Axelrod, D.I.: *Quaternary Extinctions of Large Mammals*, U. Calif. Pub. in Geological Science, Berkeley, 1967, p.27.

5. Monod, T.: "The Late Tertiary and Pleistocene in the Sahara", in *African Ecology and Human Evolution*, F.C. Howell & F. Bourliere, Eds., Viking Fund Publications in Anthropology, #36, 1963, p.125-9.

6. Monod, 1963, ibid.; cf. Wendorf, F. & Marks, A.E. Eds.: *Problems in Prehistory: North Africa and the Levant*, S. Methodist U. Press, Dallas, 1975.

7. Haynes, C.V.: "Great Sand Sea and Selima Sand Sheet, Eastern Sahara: Geochronology of Desertification", *Science*, 217:629-33, 1982.

8. Wendorf, F.: "The Paleolithic of the Lower Nile Valley", in Wendorf & Marks, 1975, ibid., p.103.

9. Nicholson, S.: *A Climatic Chronology for Africa: Synthesis of Geological, Historical and Meteorological Data*, Diss. U. Wisconsin, 1975, pp.96-7.

10. Sarnthein, M.: "Sand Deserts During Glacial Maximum and Climatic Optimum", *Nature*, 272:43-6, 1978.

11. Nicholson S. & Flohn, H.: "African Environmental and Climatic changes and the General Atmospheric Circulation in Late Pleistocene and Holocene", *Climate Change*, 2:317, 1980; Street, F.A. & Grove, A.T.: "Environmental and Climatic Implications of Late Quaternary Lake-Level Fluctuations in Africa", *Nature*, 261:385-90, 1976.

12. Sarnthein, 1978, ibid.; Fairbridge, R.W., *Quaternary Research*, 6:529, 1976; Street & Grove, 1976, ibid.; Williams, M.A.: *Nature*, 253:617, 1975; Fairbridge, R.W.: "African Ice-Age Aridity", in *Problems in Paleoclimatology*, A.E.M. Nairn, Ed., Interscience Pub., 1964, pp.356-63.

an "ice age pluvial".[13] Even so, general agreement does exist on the existence of prolonged and extensive wet conditions in North Africa both before and after the arid phase, which itself may have been relatively short and not uniformly distributed.

Huzayyin's extensive 1941 survey of data from North Africa led him to the conclusion that "...climate was more or less favorable right through Quaternary times".[14] This view was almost echoed by Nicholson and Flohn in 1980, who, based upon an extensive review of radiometrically dated evidences, stated that:

> "...the late Pleistocene up to c.12,000 BP [10,000 BCE] was predominantly wet in North Africa, except for an arid interval c.18,000 BP [16,000 BCE]. While in some regions arid conditions prevailed also before that date, in most regions one or two arid millennia centered around that date merely interrupted a generally wet late Pleistocene." [15]

New Material 2005: See *Update on Saharasia* (Appendix B) for additional new findings and maps on "ice age aridity" in North Africa, as well as a discussion on the early massacre site of Jebel Sabaha which has now been dated to that same arid period.

Ancient Streams, Rivers, and Sands:

One type of evidence for wet periods in the Sahara is that which has been etched into the surface geology by running water. A number of very large wadis exist across North Africa, Arabia, and Central Asia, the length and extension of which indicate wetter times in the past.

The great wadis in West Africa, oriented north-south, once served as transport links between the Sahel and Mediterranean, connecting the Atlas Mountains with the Niger River System through what is now an extremely arid region. The Ahaggar Massif of the Central Sahara is also carved by numerous fossil streams which radiate in all directions, particularly to the NNE and SW. The Adrar portion of this Massif contains fossil streams which "may have formed the original upper course of the Niger."[16] Other examples are the "huge" Wadis of Tefezzezat and Igharghar in the Western Sahara, and Wadi Tilemsi north of the Niger River.[17]

Some of the fossil streams and wadis of Central and West Africa give

13. Huzayyin, 1941, ibid.; Flint, R.F.: *Glacial and Quaternary Geology*, Wiley, NY, 1971; Shaw, B.D.: "Climate, Environment and Prehistory in the Sahara", *World Archaeology*, 8(2):133-49, 1976-77.

14. Huzayyin, 1941, ibid., p.68.

15. Nicholson & Flohn, 1980, ibid., p.321.

16. Huzayyin, 1941, ibid., p.71.

17. Flohn & Nicholson, 1980, ibid., p.10; also see Pachur, H.J. & Braun, G.: "The Paleoclimate of the Central Sahara, Libya and the Libyan Desert", *Paleoecology of Africa & Surrounding Islands*, 12:351, 1980.

Large wadi adjacent to the Upper Nile River. These now-dry and salt-encrusted wadis once maintained seasonal or year-round flow, dotted with permanent lakes and swamps.

Ancient pastoral and agricultural peoples once flourished in these same areas, which today are hyper-arid.

evidence of past rainfalls in some cases approaching a torrential character. Examples are the gullies in Senegal and Mauritania, where:

> *"...ponds and lakes occupied interdunal depressions of the Majabat al-Koubra and the trend toward the present aridity did not begin until at least 4800 BP [2800 BCE]; even as late as 4450-3700 BP [2450-1700 BCE], the climate there was still much wetter than today, with many lakes existing in the Tichitti area"* [18]

Egypt and Arabia also possess a number of giant *wadis bi'lama* (waterless wadis) indicating a wetter climate in ancient times.[19] It has been estimated that permanent rivers once flowed in the now hyperarid desert center between the Tibesti Massif of Northern Chad and the Kufra Oasis of Egypt. This region now receives less than 5 mm (0.2") of rain annually, but it once must have received around 250-400 mm (10-16"), being covered by grasslands with trees, lakes, wildlife, and Neolithic hunters.[20]

The Great Sand Sea of Egypt, which lies to the west of the Nile River and Kharga Oasis, was recently penetrated by the Space Shuttle's imaging radar system (see Figure 49) to reveal intricately carved stream beds and river valleys.[21] The surface character of this 150,000 km^2 (58,000 mi^2) expanse of sand gave no hint of such an underlying character; the sand is 200 to 300 meters (650' to 1000') thick in the northern portion, near Siwa and the Qattara Depression, and gradually thins out to the south where

18. Nicholson & Flohn, 1980, ibid., p.326; cf. Munson, P.: "A Late Holocene (c.4500-2300 BP) Climatic Chronology for the Southwestern Sahara", *Paleoecology of Africa & Surrounding Islands*, 13:53-60, 1981.
19. Huzayyin, 1972, ibid., p.307.
20. Flohn & Nicholson, 1980, ibid., p.10.
21. Shuttle imagery of sand sheet; USGS Astrogeology Branch, Flagstaff, AZ, Gerald Schaber; cf. *Discover*, Feb 1983, p.6.

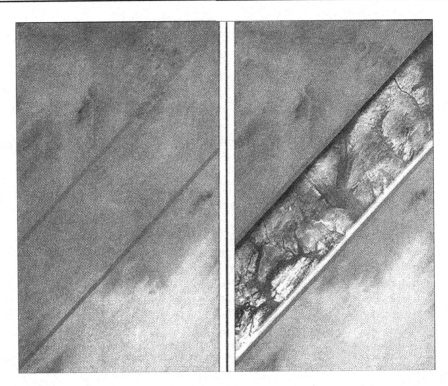

Figure 49: Ancient Riverbeds Beneath the Great Sand Sea, Western Egypt
Left: Landsat image of the surface characteristics of the sand sheet.
Right: Superimposed Space Shuttle radar image of
the ground surface underlying the sand sheet.[21]

In some areas of North Africa, ancient wadis are completely filled in and covered over with sand sheets, the surface character of which gives no hint of the underlying ancient water-cut topography. The underlying watercourses were only discovered after the space shuttle's imaging radar penetrated the sand sheets.

bedrock and gravel is eventually exposed between longitudinal dunes. Evidence of ancient lakes, fossils, carbonized rootcasts, and human artifacts, such as freshwater shells and fish spear points, appear upon the bedrock desert floor at this most inhospitable location, indicating a wetter phase which ended around 5000 years ago (c.3000 BCE).[22]

Other expanses of sand, such as the Selima Sand Sheet (south of the Great Sand Sea) is composed of flat, hard and compacted sand. The subsurface character of these sand sheets and the fossils and artifacts they may conceal are essentially unknown, save for a small amount of exploration undertaken, and the space shuttle imagery mentioned above. However, the existence of large, desiccated wadis beneath them can be inferred from the presence of the upper portions of wadis which emerge from the sand. Examples are found in the Libyan Desert south of 30°N.[23]

Physiographic evidence tells a similar story to the east of the Nile along the Red Sea coast.[24] However, the chronology of the cutting and terracing of these wadis, their human occupations, and subsequent desiccation and burial by sand sheets to present depths has been studied only superficially. The Egyptian evidence has been summarized by Wendorf and Schild,[25] who identified various wet and dry phases in the region. Evidence is similar for various sites stretching from Mauritania, across North Africa into Arabia, the Near East,[26] and Central Asia.[27] Nicholson and Flohn have summarized evidences for increased moisture in wadis across North Africa, between 8000-2000 BCE.[28]

Physiography in the Nile River valley has been discussed in detail by Butzer, who found evidence for oscillating Nile floods between 15,000 - 3000 BCE. Throughout this same period, wadi discharge into the Egyptian and Nubian Nile was greater than today,[29] and the Nile at times maintained:

> "...a host of overflow channels on a floodplain several times wider than that of today...[with] frequent levee breaching, channel shifts, and crevasses, all of which point to exceptionally violent floods."[30]

During the Nile high flood periods, the depression at Fayyoum received flow from the Nile and filled to its highest levels; these events also coincided with a wetter phase of settlement in the Western Desert of Egypt, in Kharga and elsewhere. However, by 2350 BCE Nile floods shrank to their present low levels and contemporary ecological conditions set in.[31]

Butzer similarly identified the period following 2300 BCE as one of lowered Nile levels, intensified aeolian activity, and drought. During the period, sand dunes from the Western (Libyan) Desert migrated eastward into the Nile Valley, covering the alluvium over a 175 km (110 mile) region. Only later, after c.800 BCE, did Nile floods return to their present level to cover and stabilize the invading sand dunes with Nile muds.[32] Butzer identified four major periods of dune deposition, which were characterized by aridity and drought.[33]

1) 2350-500 BCE
2) CE 300-800
3) CE 1200-1450
4) after CE 1700

These dates, marking phases of major environmental transition towards aridity, were generally reflected across the entire Saharasian desert belt, as will become apparent later.

22. Haynes, 1982, ibid.; Wendorf, F. et al: "Late Pleistocene and Recent Climatic Changes in the Egyptian Sahara", Geographical J., Roy. Geog. Soc., London, 143:211-34, 1977.

23. Huzayyin, 1941, ibid., p.118.

24. Huzayyin, 1941, ibid., p.117.

25. Wendorf F. & R. Schild, Prehistory of the Eastern Sahara, Academic Press, 1980, pp.225-34.

26. Flohn & Nicholson, 1980, ibid., p.10.

27. Huzayyin, 1972, ibid., p.307.

28. Nicholson & Flohn, 1980, ibid., pp.321-28.

29. Butzer, K.W.: "Patterns of Environmental Change in the Near East During Late Pleistocene and Early Holocene Times", in Wendorf & Marks, 1975, ibid., p.403.

30. Butzer, K.W.: Desert and River in Nubia, U. Wisconsin Press, 1968, pp.330-1.

31. Butzer, 1975, ibid., p.403.

32. Butzer, K.W.: "Studien zum vor- und fruhgeschichtlichen Landschaftswandels der Sahara und Levante seit dem klassischen Altertum. II. Das okologische Problem der Neolitischen Felsbilder der ostlichen Sahara", Abhandl. Akad. Wiss. u. Liter. Mainz, Math.-naturwiss. Kl., 1:115-6, 1958; cited in Monod, 1963, ibid., pp.162-3.

33. Butzer, K.W.: "Climatic Change in Arid Regions Since the Pliocene", in A History of Land Use in Arid Regions, L.D. Stamp, Ed., UNESCO, 1961, p.42.

Ancient Lakes:

As was the case with North African wadis, North African lakes and dried up remnants of ancient lakes attest to moister conditions in the past. Figure 50 presents evidence from a cartographic study undertaken by Street and Grove, a compilation of Carbon-14 data from 58 African basins, revealing a systematic increase in lake levels between 8000-3000 BCE as compared to the present day.[34] Figure 51 presents evidence from a similarly systematic study, by Nicholson and Flohn, who also used radiometrically dated evidence from North African lakes to identify major moist phases between 8000-3000 BCE.[35] The nearly uniform wet conditions across North Africa for the period ending around 3000 BCE has found general support; Schulz has stated, that while uncertainties regarding interpretations of various data and measuring techniques do exist, a "comparison of the various curves describing climatic developments indicates a rather high degree of uniformity for the whole Sahara".[36]

Lakes across the Southern Sahara and Sahel filled to higher levels at that time. In Niger, at Agadem and other sites in the heart of the Grand erg de Bilma, lakes of 40 m (130') depth once existed. These sites are now awash only in waves of sand, in the heart of a vast sand sea.[37] Other lakes persisted to about 700 BCE, such as those in Mauritania; in general the Western Sahara appears to have benefited from moister conditions for a longer period as compared to its eastern parts.[38] Munson estimated that arid conditions began in the Southwestern Sahara only during 2500-1500 BCE, but still being moister than today; a moisture increase was identified for c.1500 BCE in the area, with gradual desiccation to present aridity only after c.900 BCE.[39]

At Lake Chad evidence exists for higher lake levels in the form of elevated beaches and exposed sediments. The Lake, which up until most

---

Carbon-14 Dating:

A method for determining the age of organic materials containing carbon, by measuring the ratio of naturally-occurring C-12 to C-14 isotopes.

34. Street & Grove, 1976, ibid.

35. Nicholson & Flohn, 1980, ibid., p.317.

36. Schulz, E.: "Trends of Pleistocene and Holocene Research on the Sahara", *Paleoecology of Africa & Surrounding Islands*, 16:193, Boston, 1984.

37. Nicholson & Flohn, 1980, ibid., p.322.

38. Munson, P.: *The Tichitt Tradition: A Late Prehistoric Occupation of the Southwestern Sahara*, Diss., Anthropology, U. Illinois, Urbana-Champaign, 1971, pp.59-82.

39. Munson, 1981, ibid.

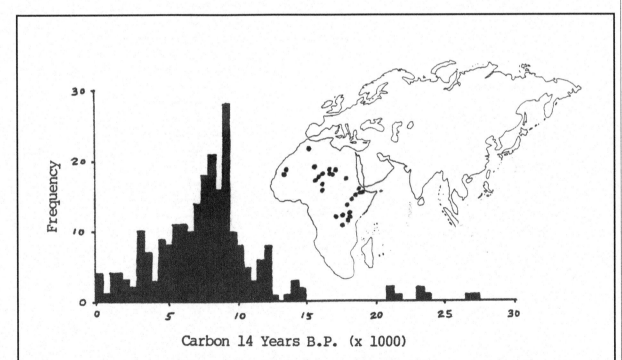

Figure 50: Histogram of North African Basin Wet Phase C-14 Dates
The period between 10,000-5,000 BP (or 8,000-3,000 BCE) had the greatest frequency of high or intermediate lake levels. Locations of the lakes are indicated on the map.[34]

recently stood at 3.8 m (12') average depth and was fresh in spite of a lack of outlet to the ocean, was fed primarily by two rivers flowing northward from the subtropics, the Logone and Shari. Without these two rivers Lake Chad would probably dry up within a period of several years; evaporation losses from the Lake reach 2000 mm (80") per year, while direct rainfall on its surface runs between 250-500 mm (10-20") inches.[40] It has undergone several expansions within the last 10,000 years, probably in concert with other climatic fluctuations in the region.[41] The peak moist phase of Lake Chad occurred between 7000-6000 BCE. The Lake was then the size of Great Britain, filling the adjacent Bodele Depression to a height of 340 m (1100'), and standing 38 m (120') above the present level. Various lacrustine and marsh deposits dating between 10,000-5000 BCE exist throughout the central highlands of the Sahara, at Ahaggar, Tibesti, Tassili, and Air. These are highland regions of slightly greater rainfall; during moist phases, considerable runoff would flow from here into lakes in surrounding lowland regions.[42]

In the Northern Sahara south of Tripoli, Libya lie a series of dried up lake terraces holding freshwater shells and Neolithic deposits, the most recent high water terraces of which have been dated to 3500-2500 BCE.[43] Other fossil lakes exist in the Libyan Desert, which were fed from increased rains and runoff from Tibesti between 6500-4000 BCE.[44]

40. Grove, A.T. & Pullan, R.A.: "Some Aspects of the Pleistocene Paleogeography of the Chad Basin", in Howell & Bourliere, 1963, ibid., pp.230-7; again, news reports from Africa state that Lake Chad is now completely dried up.

41. Huzayyin, 1941, ibid., pp.113-6.

42. Nicholson & Flohn, 1980, ibid., pp.322-3; cf. Schove, D.J.: "African Droughts and Weather History", in *Drought in Africa, Report of the 1973 Symposium*, D. Dalby & R. Harrison-Church, Eds., Centre for African Studies, U. of London, pp.29-30, 1973.

43. Petit-Marie, N. et al: "Pleistocene Lakes in the Shati Area, Fezzan", *Paleoecology of Africa & Surrounding Islands*, 12:293, 1980.

44. Nicholson & Flohn, 1980, ibid., p.324.

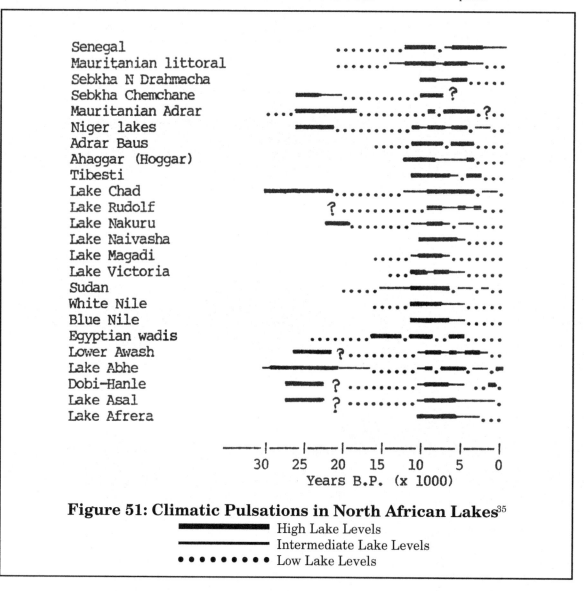

Figure 51: Climatic Pulsations in North African Lakes[35]

━━━━━━ High Lake Levels

───────── Intermediate Lake Levels

•••••••• Low Lake Levels

Nicholson and Flohn discussed the environment of the Northern and Central Sahara, which had been settled by Neolithic hunters and pastoralists:

> "...lakes were numerous in the desert of southern Libya until about 6000 BP [4000 BCE]... and from 6500-4500 BP [4500-2500 BCE] at 28° to 30° N in Libya...At that time the Sahelian summer rains may have reached to c.30° N in Libya and the Atlas Mountains... Lowland marshes developed in the northwest Sahara, as did lakes in the highland areas,.. To the southeast of Tibesti Lake Ounianga Kebir was 40 m [130'] higher than today toward 6160 BP [4160 BCE], while lakes in the northern part of the massif toward 5300 to 5100 BP [3300-3100 BCE] were surrounded by a mixture of Mediterranean and Sahelan vegetation." [45]

Farther east from here in Egypt, playas near the Qattara depression yield evidence for the presence of three separate wet subphases between 7000-3800 BCE.[46] Other depressions adjacent to the upper Nile Valley show similar evidences of ancient lakes, such as those of Jebel Aulia and Kosti, near Khartoum.[47]

And south from here in Ethiopia, the region where the Blue Nile bends from east to west was once covered by ancient Lake Yaya, formed during a wetter and cooler phase.[48] The Danakil depression of coastal Eritrea, now a desert, was filled with water between 6940-3600 BCE.[49] Butzer, et al studied paleoclimatic evidence in a series of East African lakes, to include Lake Chad, concluding that:

> "Between 10,000 and 8,000 years ago [8000-6000 BCE], it seems that lakes in many parts of tropical Africa were greatly enlarged. ... The Holocene record subsequent to the maxima...is more complex. Three basins (Rudolf, Nakuru, and Chad) show an apparently concordant, positive oscillation at some point between 6000 and 4000 years ago [4000-2000 BCE]..." [50]

It should be noted that these moist phases were not restricted only to the now arid regions of North Africa. They appear to be reflected in the subtropical and tropical regions of Africa as well, as evidenced from the simultaneous rise of lakes in the heart of the Sahara with those fed by tropical waters (Lake Chad) or which exist in tropical regions.[51] Other evidence also exists for a significantly wetter period between 10,000-5000 BCE in the semiarid and subhumid areas of tropical Africa, with similar trends in Arabia stretching northeast into the Iranian highlands.[52]

Ancient Plants and Animals:

Huzayyin's survey of faunal remains in North Africa indicated that, during the wet phase, "...half the total number of N. African fauna was shared in common with that of Europe..."[53] More recently, Mauny produced maps showing the distribution of fossil elephant and giraffe in North Africa,[54] and these are reproduced in Figure 52. McBurney also made reference to surviving colonies of freshwater fish and crocodiles of Congo affinity in waterholes in the mid-Sahara, at the Ahaggar and Tibesti massifs.[55] Monod similarly discusses surviving colonies of baboon and red monkey, isolated at the Air and Tibesti massifs.[56] Other desert fossils, which include hippopotamus, buffalo, elephant, crocodile, and giraffe, are dated to only 2,000 BCE.[57] Fossils of domestic animals have

45. Nicholson & Flohn, 1980, ibid., p.327.

46. Wendorf & Schild, 1980, ibid., pp.236-41.

47. Nicholson & Flohn, 1980, ibid., p.322.

48. Huzayyin, 1941, ibid., p.79.

49. Nicholson & Flohn, 1980, ibid., p.322.

50. Butzer, K.W. et al: "Radiocarbon Dating of East African Lake Levels", Science, 175:1074, 1972.

51. Nicholson & Flohn, 1980, ibid., pp.322-3.

52. Grove, A.T.: "Desertification in the African Environment", in Dalby & Harrison-Church, 1973, ibid., pp.33-45.

53. Huzayyin, 1941, ibid., p.66.

54. in Axelrod, 1967, ibid., pp.18-9.

55. McBurney, C.B.M.: The Stone Age of Northern Africa, Penguin Books, 1960, pp.70,76.

56. Monod, 1963, ibid., pp.163-5.

57. Esperandieu, G.: "Domestication et elevange dans le Nord de L'Afrique au Neolithique ns la protohistoire d'apres les fiions repestres", Actes 2e Cong. Pan-Afr. Prehist., Algiers, 1952, pp. 551-73.

also been found in the Sahara.[58] The migration of fishes and crocodiles, plus the presence of hippopotamus, elephant, and waterbuck clearly indicate the presence of running streams and freshwater pools, while an abundance of ostrich eggs further indicates the extension of grassland into the region.[59] Regarding aquatic fauna, Dumont's study identified the period between 3000-2000 BCE as the time when increasingly arid conditions wiped out large numbers of aquatic species in the Southern Sahara and Sahel.[60]

Climatic reconstruction of North Africa based upon forage requirements of such large mammals indicates that most of the Sahara was covered by grassland, shortgrass to tallgrass savanna, with annual precipitation of *at least* 250-380 mm (10-15"). In the upland areas of the Tibesti, Ahaggar, and Air Massifs, precipitation probably reached 500 mm (20") per year or more, based upon the past presence of Mediterranean forest and woodland of a subhumid character (Atlas cedar, Aleppo pine, juniper, oak, linden, pistachio, maple, alder). These upland forests persisted until aridity and deforestation began to take their toll.[61] Jakel also identified generally humid conditions at the northern margins of the Tibesti Massif between 14,000 BCE to around 1000 BCE.[62] Pollen analysis from Chad and Tibesti indicate 5 or perhaps 6 separate climatic oscillations between 7000 and 3000 BCE.[63]

Butzer has discussed the distribution of various fauna appearing in Neolithic "wet phase" rock art in the Eastern Sahara (c.5000 BCE to 3000 BCE).[64] Rock art in the Sahara will be discussed in detail shortly. The higher rainfall in North Africa around 6500-5000 BCE has been estimated by Hobler and Hester,[65] who drew upon Mauny's previously given distributions of large mammal fossils and Butzer's rock art distributions.

58. Monod, 1963, ibid., p.183; cf. Axelrod, 1967, ibid., p.18.

59. Huzayyin, 1972, ibid., pp.308-9.

60. Dumont, H.: "Relicit Distribution Patterns of Aquatic Animals: Another Tool in Evaluating Late Pleistocene Climate Changes in the Sahara and Sahel", *Paleoecology of Africa and Surrounding Islands*, 14:1-24, 1982.

61. Axelrod, 1967, ibid., p.19.

62. Jakel, D.: "Eine Klimakurve für die Zentralsahara", in W. Sheel, Ed., *Sahara: 10,000 Jahre Zwischen Weide und Wuste*, Museen Der Stadt Koln, c.1980, pp.382-96; cf. Paleoecology of Africa, 11, 1979.

63. Wendorf & Schild, 1980, ibid., pp.236-41.

64. Butzer, 1958, ibid., pp.1-49.

65. Hobler, P.M. and Hester, J.J.: "Prehistory and Environment in the Libyan Desert", *S. Afr. Archaeol. Bull.*, 33:127-9, 1969.

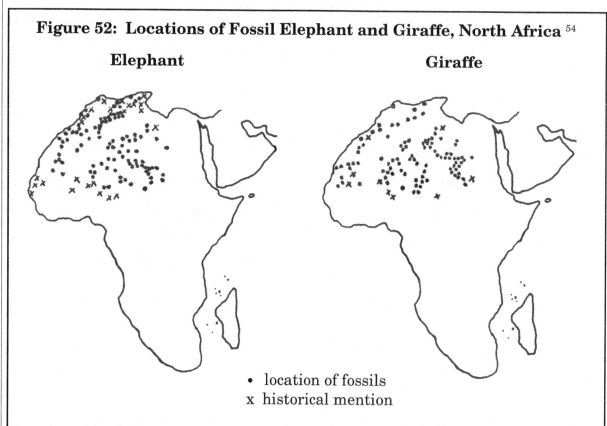

Figure 52: Locations of Fossil Elephant and Giraffe, North Africa [54]

Elephant   **Giraffe**

• location of fossils
x historical mention

Locations of fossils found in the arid zone, or historical discussion of such animals, where animals have not been found in recent times. The Eastern Sahara and Arabia were not included in the study, but other evidence cited here suggests similar conditions existed in those regions as well.

Regarding fossil evidences for the Eastern Sahara, fossil Pleistocene pollens from the Kharga Oasis in Egypt indicate the presence of species of evergreen oak, as well as varieties of Ficus, and tropical dates.[66] The deserts of Egypt possessed savanna with shrubs, bushes, scattered grass tuff and isolated trees; large numbers of tree stumps of acacia, tamarisk, and sycamore remain in the deserts to the west and east of the Nile, as well as in southern margin of the Sahara.[67]

To the south of Egypt, near Khartoum, bones of hippopotamus, elephant, antelope, giraffe, and ox have been found at depths of 18-20 m below Nile muds, indicating wetter times on the Sudan Plain.[68] According to Clark, North Africa of 10,000-2000 BCE was one where:

> *"...climatic conditions in the desert supported many favorable open habitats of grassland and cooler Mediterranean flora, and shallow lakes and swampy depressions where the large Ethiopian fauna were to be found in abundance"* [69]

Ancient Artwork:

Additional compelling evidence of the progressive desiccation of the Sahara comes from the hands of ancient artists. Rock walls covered with carvings and paintings dated as early as 8000 BCE are found scattered throughout the Sahara, the richest variety and abundance of which are found in its driest and most inhospitable portions, in the southern parts of Algeria, Libya, and Egypt.[70] Some of the most spectacular and beautiful works of rock art come from a region in southern Algeria called Tassili N'Ajjer, meaning "plateau of the waters" in Tuareg; it is a region with majestic canyons and pinnacles, carved by water. At present, of course, it is all dried up.[71]

Throughout this region of the Sahara, stone age cultures, including the Berber and Tuareg, existed into Roman times, attested to by the presence of Roman artifacts mixed with flaked and polished stone axes and other implements.

> *"These documents [rock art, artifacts] can be attributed to the successive civilizations which inhabited this huge area from Lower Pleistocene until near our Christian era. ... The presence of so many and so abundant traces of the activity of Man through hundreds of thousands of years in places now completely alien to human life and often to any form of life, testifies to the existence, in the same places, of climatic conditions completely different from those of today."* [72]

The subject of the earliest Saharan artwork was animal life; elephant, ostrich, hyena, giraffe and other animals were carved or painted by the skilled hand of artists who devoted more than a few passing moments to the work. The ancient artwork reveals the smooth and steady hand of a skilled and artistically sensitive individual. Scenes of animal life, as given in Figure 53, are among the oldest of Saharan art. They are believed to reflect the hunter-gatherer phase of c.7000 BCE, when the Sahara was a lush grassland savanna.

Later phases of North African art reveal changes toward nomadic herding, when artwork motifs emphasize cattle. Representative examples of this period are given in Figure 54. During the nomadic pastoral period, the Sahara was a meeting ground for migratory groups moving out of Arabia and Central Asia. The dress and hair styles of people in some of

66. Huzayyin, 1972, ibid., pp.307-8.

67. Butzer, 1961, ibid., pp.39-40.

68. Huzayyin, 1941, ibid., pp.119-22.

69. Clark, J.D.: "Prehistoric Populations and Pressures Favoring Plant Domestication in Africa", in *Origins of African Plant Domestication*, J.R. Harlan, et al, Eds., Mouton, the Hague, 1976, p.74.

70. Lhote, H.: *A La de Couverte des Fresques du Tassil*, B. Arthaud, Paris 1958; Many picture books exist showing these spectacular rock art, the best examples of which come from the work of Henri Lhote (cited above) whose magnificent reproductions are on display at the Musee d'Afrique in Paris. Cf. Davidson, B.: *African Kingdoms*, Time-Life, NY, 1966, pp.43-57, and Sheel, 1980, ibid.

71. Hawkes, J.: *Atlas of Ancient Archaeology*, McGraw Hill, 1974, p.22.

72. Graziosi, P.: "Prehistory of Southwestern Libya", in *Geology, Archaeology, and Prehistory of the Southwestern Fezzan, Libya*, W.H. Kanes, Ed., Petroleum Exploration Society of Libya, 11th Annual Field Conf., 1969, pp.3-4.

Figure 53: Early North African Rock Art
Neolithic Hunter/Gatherer Period c.7000 BCE

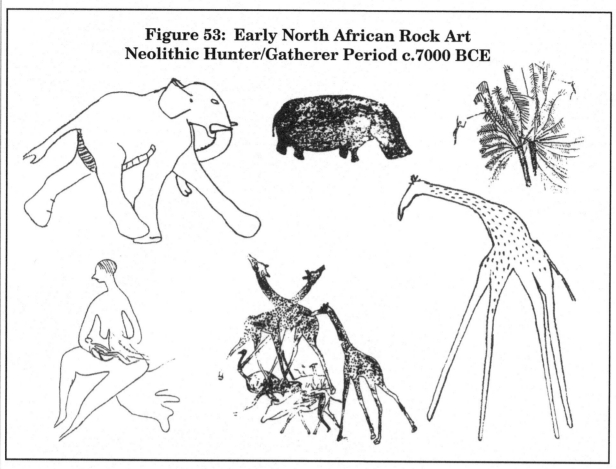

these scenes can be interpreted as influences from central and southern coastal Asia.[73] This newer phase in artwork first appeared after c.5000 BCE.[74] Butzer[75] and Rhotert[76] estimated several probable routes of entry into Africa of the pastoralists who made the rock art.

Winkler, who was first to comprehensively survey the rock art, maintained that much of it was from essentially the same group of people, with women primarily doing the painting and men primarily doing the

73. Scheel, 1980, ibid., pp.421,435.

74. McHugh, W.P.: "Late Prehistoric Cultural Adaptation in Southwest Egypt and the Problem of the Nilotic Origins of Saharan Cattle Pastoralism", *J. Am. Res. Center in Egypt*, XI:12, 1974; McHugh, W.P.: *Late Prehistoric Cultural Adaptation in the Southeastern Libyan Desert,* Diss. Anthropology, U. Wisconsin, 1971.

75. Butzer, 1958, ibid., map reproduced in Monod, 1963, ibid., p.205.

76. Rhotert, H.: *Libysche Felsbilder*, Darmstadt, L.C. Wittich, 1952; map reproduced in Monod, 1963, ibid., p.206.

Hyper-Arid Rock and Sand-Dune Landscape of the Central Sahara, near Tassili N'Ajjer, Algeria.

Photo courtesy of Prof. Bernd Senf

**Figure 54: Early North African Rock Art
Neolithic Pastoralist Period, c.5000 BCE**

engraving.[77] His views have been criticized by others, however.[78]

A general view of the sequence of rock art in the Central Sahara is given below. Though the absolute chronology is different for different sites, the general pattern indicates a westward migration of peoples.[79]

1. Hunter-Gatherer Phase c.7000 BCE: engravings of native species, with humans (both sexes) and dogs, in naturalistic styles (Figure 53);

2. Nomadic Herding or Pastoralist Phase c.5000 BCE: engravings and paintings dominated by cattle, with prominent udders and decorated, mottled coats; also present are archers and other male hunting figures, females and children, domestic scenes, and occasional wild animals (Figure 54);

3. Horse-camel group (iron age) c.2000 BCE to modern era: engravings and paintings, mounted warriors, war chariot, horses and camels, but absence of cattle, females, children and domestic scenes (Figure 55).

The rock art of the final phase, group 3, occurred after the Sahara had begun drying up. Rock art of the dry phase reflects a definite change of style and subject matter. Rock art from earlier hunter/gatherer and pastoralist stages holds a certain gentle and bold quality not present in the later arid phase. Mori, an Italian observer of the rock art, puts it so: "...the older they are...the more evident are the refinement and artistic sense with which they are executed."[80] Rock art from the camel-horse-chariot phase lacks its former artistic flavor; it appears abstract and stickman-like, without the grace and sensitivity of earlier renderings. One could argue that the arid phase rock art is really not art at all, but merely graffito, or haphazard scrawlings, indicative of a certain loss of emphasis upon art. This view is supported by a review of the principal subject matter contained in the arid phase rock art: organized warfare, armed

77. Winkler, H.A.: *Rock-Drawings of Southern Upper Egypt,* II, Sir Robert Mond Desert Expedition, Oxford U. Press, London, 1939, p.25.

78. cf. Monod, ibid., pp.117-229; McHugh, 1971, ibid.

79. McHugh, 1971, ibid.; McHugh, 1974, ibid.

80. Mori, F.: "Prehistoric Cultures in Tadrart Acacus, Libyan Sahara", in Kanes, 1969, ibid., p.21.

Figure 55: North African Bronze Age Rock Art
Warrior, Horse, Chariot, Camel Period,
After Desiccation of the Landscape, c.2000-500 BCE

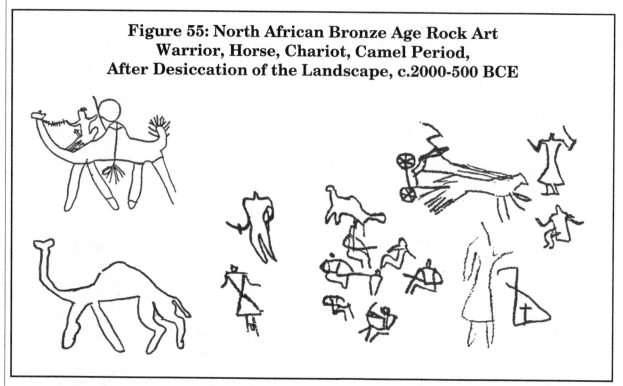

males, battles, death, chariots, horses, and camels. Davidson also discussed the rock art:

> "The peace and plenty of the Sahara's pastoral era are reflected not only in the skill of the paintings of that period, but also in other arts such as music and dancing shown in the paintings. Both men and women took part in the dances... Women played a surprisingly active role in early Saharan life compared to the veiled ladies of later North African eras... Perhaps the most striking aspect of these Saharan wives and mothers, however, is the elegance and care with which many of them are represented...[they] endow their subjects with dignity, indicating the regard in which women were held." [81]

> "...war appeared more and more frequently in paintings of their everyday lives. During pastoral times they were sometimes depicted in battle array, perhaps to settle the rights to a disputed herd. But if war was an occasional necessity for the herdsmen, it was a passion with the horsemen who appeared in the last millennium before Christ." [82]

Of the mounted camel warrior, Davidson agreed that the art of that period reflected "...a decline in Saharan art that coincided with the region's decline into desert."[83]

The humid conditions which persisted in the Northeast Sahara have also been captured on art motifs of various Egyptian dynasties. These include paintings or bas-reliefs of hunting scenes, vegetation, and wild animals. The Saharan ecology, as reconstructed from the earliest dynastic art motifs, confirms the previously given physical materials on ecological change. Elephant, giraffe, wild camel, oryx, addax, gazelle, ibex, wild ass, hippopotamus, lion, hyena, and wolf all appear in the older Egyptian art, along with bushes, grass, and trees. Butzer, who summarized the Egyptian art,[84] also identified several periods where ecological discontinuities occurred:

81. Davidson, 1966, ibid., pp.51-2.
82. Davidson, 1966, ibid., p.55.
83. Davidson, 1966, ibid., p.45.
84. Butzer, K.W.: "Die Naturlandschaft Agyptens wahrend der Vorgeschichte und der Dynastischen Zeit", idem. III, *Abhandl. Akad. Wiss. u. Liter. Mainz, Math.-Naturwiss.* Kl. 2:96-109, 1959; cited in Monod, ibid., pp.160-2.

1. c.3600 BCE: decrease of Neolithic tropical savanna fauna, increase of antelope;

2. c.2800-2600 BCE: disappearance of rhinoceros, elephant, giraffe, wild camel; diminution of lion and Barbary sheep; increase of antelope and gazelle;

3. c.2480 BCE: loss of all large animals; gazelles increase.

After the start of the New Kingdom (c.1567 BCE) paintings and bas-reliefs are dominated by Nile swamp and water fauna.[85] Viewed more broadly, the rock art implies that between 2800-2480 BCE all large game animals disappeared from the Eastern Sahara, with a loss of other species shortly afterward, as the Eastern Sahara dried out. Indeed, Butzer chose to identify c.2350 BCE, the beginning of the VI Dynasty, as the terminal date for the Neolithic wet phase.[86]

General Settlement Patterns and Culture Change, to Include Influences Transmitted into Sub-Saharan Africa:

Distributions of Neolithic human populations in North Africa parallel findings of flora and fauna, and further support the view of progressive desiccation. Neolithic settlements and remains are scattered across the Sahara, particularly along the Mediterranean coast, in the Nile Valley region, and across the Sudan or Sahel zone into the central massifs.[87]

"Not only are there numerous Neolithic settlements in places now entirely dried up, but there is also a fairly marked contrast between the distribution of the finds relating to the Early Neolithic phase and those relating to the historic cultures. Whereas the Early Neolithic settlements are generally found in the bottom of the valleys and in deltaic regions (formerly well watered), the remains of the historic cultures occur only on the tops of the highlands to which the people appear to have been forced to 'go up'"[88]

Regarding the inhospitable Libyan Desert, Ball made the following appraisal after an early expedition in the 1920s:

"I have been much struck by the wide distribution of stone implements in the desert. I have found them for instance, not only in Siwa Oasis and near the wells of Abu Mungar and Shebb, but also on the plateau between Baharia and Farafra, on the open desert between Terfawi and Owenat, and to the southwest of Owenat near the Anglo-French [Libyan-Egyptian] border. This wide range of occurrence...inclines me to think that there is scarcely any part of the Libyan Desert in which stone implements might not be found".[89]

Good evidence also exists demonstrating Neolithic communities benefited from greater groundwater resources, particularly in the Eastern Sahara:

"...at Kharga there are immense deposits of calcareous spring tufas around or in which Neolithic tools have been found in great numbers. They imply the existence over 5000 years ago of an agricultural settlement established around formerly functioning springs. Similar tufas or clay hummocks have been noted from the wells of Kurkur, Sheb, and Tarfawi. From Uweinat, Merga and Bir el Atrun there is also evidence that the static groundwater table

85. Butzer, 1959, ibid.

86. Butzer, 1959, ibid.; Monod, 1963, ibid., p.162.

87. Barker, G.: "Early Agriculture and Economic Change in North Africa", in *The Sahara: Ecological Change and Early Economic History*, J.A. Allan, Ed., Middle East & North African Studies Press, England, 1981, pp.131-45.

88. Huzayyin, 1941, ibid., p.75.

89. Ball, J.: "Problems of the Libyan Desert", *Geographical J.*, 1927, p.220.

was much higher in 'Neolithic' times — enabling nomadic cattle breeders to live there until perhaps the time of the Sixth Dynasty (2350 BCE)." [90]

The study by Camps, using 268 radiocarbon dates from Paleolithic and Neolithic sites in the Western Sahara, indicates habitation by humans between 8000-2000 BCE, with finds along Mediterranean coastal Africa extending back to 14,000 BCE.[91] These dates of occupation are in good agreement with those given previously, being a period of grasslands, lakes, rivers, and highland forests, supporting a "rather large" human population.[92]

The first agricultural settlements appear in North Africa in the Nile Valley as early as 10,500 BCE; however, the swampy character of the Valley and its high flood stages of the period are thought to have prevented any growth of these early efforts.[93] Between 6000-5500 BCE, settlements based upon plant and animal domestication spread through Egypt and Libya, altering the traditional hunting and gathering lifestyle in some regions towards cattle pastoralism.[94] By 5500 BCE, farming communities begin to clearly emerge in the Nile Valley.[95] Agricultural and pottery-making settlements appeared in the Fayum by 4500 BCE, and similar settlements were established in Merimde by 4200 BCE.[96] And by c.4000 BCE, the Badari culture of the Upper Nile region came under the influence of peoples to the east, in the Sinai, Levant, Arabia, and Red Sea coast; trade for copper and timber also proceeded.[97]

These Neolithic farming and trading communities were cooperative, productive, and peaceful in character, without significant social stratification or strongman rule. Women and children predominate in graves of the period, which contained little or no grave wealth. Their use of metals was mostly restricted to artistic pursuits, and no weapons were found.[98]

Naturalistic Female Figurine, Predynastic Egypt. Dancing Woman or "Bird Deity"? "Goddess" statues from later patristic periods are unmistakable in their rigid formality and ritual qualities. Female figurines from pre-patrist times lack such abstract or rigid qualities, suggesting they are depictions of ordinary daily life. There is a tendency for archaeologists to psychologically project our cultural preoccupation with religion and ritual onto other cultures. How would archaeologists in the year 3000 interpret hundreds or thousands of Barbie Dolls found in our archaeological strata?

Changes after 4000 BCE:

After around 4000 BCE, the region east of the Nile began to reflect the intrusion of different groups of peoples. Rock pictures of boats and strangely dressed invaders are found all over the Eastern Desert, tens of miles from the nearest shore. These pictures are of the stickman quality previously discussed, possessing a military motif not present in either the hunter-gatherer or pastoralist rock art.[99] Some Upper Egyptian settlements were abandoned under violent pressure from these invaders, and grave wealth for the first time evidenced a growing nobility, with caste and hierarchy. However, the Nile Delta region did not display these changes, though it also was engaged in trade with both Palestine and the Red Sea.[100]

The period following 4000 BCE saw a rapid spread of Bronze Age technology out from the Near East into Egypt, from which it slowly diffused westward.[101] This was a period when Indo-Aryan mounted warrior nomads irrupted into the Near East from beyond the Iranian Plateau. Later, with increasing desiccation following the end of the Neolithic humid phase, Semitic and Hamitic peoples would gradually abandon parts of Arabia and North Africa, respectively. Migrating groups swept across the Near East and into North Africa, bringing about radical changes in both technology and social structure.

By 3600 BCE, peoples of the Nile Delta were influenced by new wealth and innovations similar to those found in the Mediterranean, Mesopotamia, coastal Arabia, and possibly even the Indus Valley.[102] However, a relative lack of grave wealth in the Delta region indicates that

90. Butzer, 1961, ibid., pp.40-1.

91. Camps, G.: "The Prehistoric Cultures of North Africa: Radiocarbon Chronology", in Wendorf & Marks, 1975, ibid., pp.181-92.

92. Hays, T.R.: "Neolithic Settlement of the Sahara as it Relates to the Nile Valley", in Wendorf & Marks, 1975, ibid., pp.193,201.

93. Hoffman, M.: *Egypt Before the Pharaohs*, A. Knopf, NY, 1979, pp.89-90.

94. Hoffman, 1979, ibid., pp.177,205,238.

95. Hoffman, 1979, ibid., p.156.

96. Mellaart, J.: *The Chalcolithic & Early Bronze Ages in the Near East & Anatolia*, Khayats, Lebanon, 1966, p.54.

97. Mellaart, 1966, ibid., p.54.

98. Hoffman, 1979, ibid., pp.143, 173, 189, 329; Murray, M.A.: *The Splendor that was Egypt*, Philosophical Library, 1949, p.5.

99. Hoffman, 1979, ibid., pp.243-5.

100. Hoffman, 1979, ibid., pp.99,110,194-7.

101. Hawkes, 1974, ibid., p.268.

102. Emery, W.: *Archaic Egypt*, London, 1972, p.40; Mellaart, 1966, ibid., pp.55-57.

Table 8: Chronological Outline of Ancient Egyptian History:[105]

c.4000+ BCE: Upper Nile slowly penetrated by peoples from Red Sea/Arabia region.

c.3300-3100 BCE: Nile Valley "unified" from Delta to Nubia through conquest by militant peoples from the Upper Nile or Red Sea/Arabia region (King Menes, the first Pharaoh).

c.3100-2300 BCE: Start of Pyramid Age, minor irrigation works, military expansion and control extended over surrounding territory (Senefru, Cheops).

c.2300-2100 BCE: Nile Valley invaded by Asiatic/Semitic nomads, era of chaos, famine, locust, migrations.

c.2100-1800 BCE: Era of massive irrigation works, military expansion and control over surrounding territory, time of Karnak/Luxor, Valley of the Kings.

c.1800-1600 BCE: Nile Valley invaded by Hyksos, Asiatic nomads using horse and chariot, chaos, migration, brutal rule.

c.1600-1000 BCE: Era of Imperial Egypt (Seti, Ramses) massive military campaigns, control over surrounding regions.

c.1000+ BCE: Gradual decline of Imperial Egypt as a Mediterranean power; invasions by Kushites, Hittites, Assyrians, Libyans, Sea Peoples, Persians, Greeks, Romans, Arabs.

---

its peoples still had not copied the stratified caste system of their neighbors in the Upper Nile region.[103] South of Upper Egypt, in the Sudan, social stratification was also absent among Nubians engaged in mixed farming, hunting, and herding.[104] Table 8, above, gives a general outline of Egyptian History.[105]

The period beginning with 3300 BCE saw immense changes toward less peaceful conditions along the Nile which were gradually transmitted to the rest of North Africa. As detailed previously, this date coincides roughly with the end of the Neolithic humid phase. It was a period of increased warfare, social chaos, and class stratification, with a rise of centralized authority. A new ruling class of taller people with larger skulls appeared among the peoples of the Upper Nile. Calling themselves "followers of Horus", they herded cattle, and possessed many Semitic characteristics. The proto-kingdoms of the Upper Nile invaded and conquered the peoples of the Delta and Nubia. As determined through studies on human remains, a centralization of wealth gradually thereafter reduced the protein intake of the average farmer, as a caste system based upon divine kingship and slavery was imposed. The status of women was instantly reduced to servitude and concubinage. Grave wealth dramatically increased all over the Eastern Sahara evidencing for the first time human sacrifice, usually of widows or concubines in the graves of their deceased husbands.[106]

Influences from both Mesopotamia, the Iranian Plateau, and Central Asia were abundant in the culture of the invaders who subdued the region from the Delta to Nubia, displacing the older, relatively peaceful agricultural and pastoral communities.[107]

The date 3100 BCE marks the forcible "unification" of Upper and Lower Egypt under King Menes "the fighter", and the onset of the I Dynasty. With use of superior Bronze Age weaponry, the militant invaders subdued the Neolithic agricultural and pastoral communities of the Nile and surrounding environs.[108] Rock carvings from Menes' time are dominated by pictures of executions and war,[109] and archaeological ruins of agricultural or merchant villages of the period have occasionally yielded layers of "widespread ash and human bones".[110]

It seems clear that the invading peoples were of a pastoral nature, with similarities to better-known nomadic groups. The early Dynastic Kings of Egypt were depicted:

103. Hoffman, 1979, ibid., pp.201-9.
104. Hoffman, 1979, ibid., p.260.
105. Adapted from Murray, 1949, ibid., and Edwards, I.E.S.: *The Pyramids of Egypt*, Viking Press, NY, 1972; Hoffman, 1979, ibid., pp.15-6.
106. Hoffman, 1979, ibid., pp.116-7,152,332,340; Mellaart, 1966, ibid., pp.55-7.
107. Hoffman, 1979, ibid., pp.129,133, 211.
108. Grun, B.: *The Timetables of History*, Touchstone, NY, 1982;. Hoffman, 1979, ibid., pp.13,129,144, 200.
109. Murray, 1949, ibid., p.13.
110. Hoffman, 1979, ibid., pp.213-4.

"...in the habit of a pastoral citizen, carrying the crook and the flail-like ladanesterion, wearing an animal tail at their backs, and the beard of their goat-flocks on their chins..." [111]

Some of the earliest Egyptian tombs, precursors to the pyramids, contained heads of cattle modelled in clay, with supplied cattle horns.[112]

New social institutions were introduced into the region by the pastoral invaders, most of which were adopted by the local populations even after the invaders were absorbed or overthrown: divine kingship, slavery, concubinage, polygamy, the cult of Osiris, mummification, and the God of the Dead.[113] The older animistic, nature-based religions, with their emphasis upon sexuality and the female were either replaced or subordinated to the male dominated religion of the warrior caste and its divine king.[114] New forms of torture and mutilation, along with burning of war prisoners, were introduced.[115] And importantly, the new cultural ethos either brought with it or stimulated new painful and traumatic methods of child treatment, such as infant cranial deformation, prolonged swaddling, and circumcision.[116] These social changes initially remained a characteristic of only the upper ruling classes; over the centuries, however, they were adopted by the peasant.

Nubia also was shocked by these changes, but due to its distance from Egypt and the presence of cataracts which blocked river transportation, it remained independent from direct control by the peoples of the Egyptian Nile. Still, divine kingship, the caste system, human sacrifice, slavery, polygamy, and other social institutions of the invaders were gradually adopted by the Nubian Kings, who would in turn amplify and transmit them west and south, across the African grasslands, and into Ethiopia.[117]

Ritual widow murder became widespread among the new dynastic rulers and nobility of Egypt after the rule of Menes. The Queen of King Djer (c.3000 BCE), for instance, was murdered and placed inside the royal sepulcher after her husband's death; over 300 other people, including a host of concubines, were also murdered and placed in the tomb.[118] Other I and II Dynasty Kings were interred along with sacrificed victims, who were usually female. The tombs of these kings were unusually large, possessing massive grave wealth.[119]

By the time of the III Dynasty (c.2700 BCE) arid conditions in the Sahara appear to have perturbed the regularity of the Nile floods, and famine was widespread in Egypt, as evidenced by the following tomb inscription from King Djeser:

"I am in mourning on my high throne for the vast misfortune, because the Nile flood in my time has not come for seven years! Light is the grain; there is lack of crops and all kinds of food. Each man has become a thief to his neighbor. They desire to hasten and cannot walk. The child cries, the youth creeps along, and the old man falters; their souls are bowed down, their legs are bent together and drag along the ground, and their hands rest in their bosoms. The counsel of the great ones in the court is but emptiness. Torn open are the chests of provisions, but instead of contents there is air. Everything is exhausted." [120]

It is no wonder that writings of this period contain lamentations and skepticism about the meaning of life.[121] Only after c.2300 BCE did massive human sacrifice in graves end, after which stone statues usually, but not always, replaced live women to accompany the king to the land of the dead.[122]

As discussed previously, the period c.3000 BCE was one of increasing

King Narmer (c.3100 BCE) of Upper Egypt, depicted with war club and presumably deformed crania, preparing to bash in the skull of a captive from Lower Egypt. Cattle on the upper parts of the palet are typical decor of early Dynastic Egyptians, who came to the previously peaceful Nile Valley as invading pastoral nomads from the desiccating Red Sea region.

111. Aldred, C.: *The Egyptians*, London, 1961, p.157.

112. Edwards, 1972, ibid., p.28.

113. Murray, 1949, ibid., pp.163-7; Tannahill, R.: *Sex in History*, Stein & Day, 1980, p.64; Hawkes, 1974, ibid., p.156; Grun, 1982, ibid.

114. Stone, M.: *When God was a Woman*, Dial Press, NY, 1976, pp.18,87-8.

115. Davies, N.: *Human Sacrifice in History and Today*, Wm. Morrow, NY, 1981, pp.35-6.

116. See Chapter 5 for more discussion on this.

117. Hoffman, 1979, ibid., pp.256-8.

118. Hoffman, 1979, ibid., p.276.

119. Hoffman, 1979, ibid., pp.267,279.

120. Aykroyd, W.: *The Conquest of Famine*, Chatto & Windus, London, 1974, p.25.

121. Grun, 1982, ibid.

122. Murray, 1949, ibid., p.116; Davies, 1981, ibid., pp.36-7.

desiccation and lowering Nile flood levels. Starting with Menes, a tax-collection system was initiated, and building of levees begun.[123] Indeed, much of Egyptian civilization, as we characteristically know it, emerged after c.3000 BCE,[124] a time when the effects of desiccation would have been taking its greatest toll upon peoples living on the high deserts away from the Nile. With the start of the Pyramid Age (2686–1770 BCE) the Egyptians poured an increasing energy into digging and maintaining wells and water impoundments.[125] One massive dam was constructed in the Wadi el-Garawi, near Helwan, which is outside of Cairo. Composed of 40,000 tons of stone and 60,000 tons of gravel, the structure (known today as the *Sadd el-Kafara*, or "dam of the pagans") was washed out soon after its completion in c.2700 BCE.[126]

North Africa, particularly Egypt, continued to come under the influence of empires to the north and east. The period saw massive migrations and invasions from beyond the Iranian Plateau into the Near East, and into Europe.[127] Semitic and Hamitic groups also continued to migrate from the desiccating lands of Arabia and North Africa. Trade stretched from Egypt north to the Black Sea and beyond.[128] Similarities of pottery morphology are found for the period across the region from Egypt through the Near East to both India and China.[129] This trade, however, was generally harassed by nomadic groups, which brought periodic reprisal raids and invasions from the Egyptians.[130]

Nomadic trade-raid and military reprisal characterized Northeast Africa for the first centuries of the Dynastic period. However, as the desiccation of the region intensified, the militant character of the steppe nomads increased. Snefru (c.2600 BCE), builder of a number of early pyramids, expended much effort despoiling the prosperous pastoral peoples of Nubia and the Western Desert.[131] In Egypt itself, meanwhile, a "flogging society" was instituted, where commoners genuflected to god-kings. Enormous and magnificent structures were built at tremendous cost to house the corpse of dead kings, while the bodies of commoners and slaves who built them were interred in communal pits.[132] Military campaigns against nomads increased. Palestine, the Western Desert, Nubia, and coastal Ethiopia were invaded. Gold was imported from tropical Africa, and Pygmies appeared in the Pharaoh's court. A period of expanding military power and trade prevailed until around 2300 BCE, after

123. Turnbull, C.: *Man in Africa*, Doubleday, 1976, pp.91-6.

124. Hawkes, ibid., p.146.

125. Murray, G.W.: "Water from the Desert: Some Ancient Egyptian Achievements", *Geographical Journal*, London, 121:171-81, 1955.

126. Helms, S.: *Jawa: Lost City of the Black Desert*, Cornell U. Press, NY, 1981, p.171.

127. Hawkes, 1974, ibid., p.12.

128. Murray, 1949, ibid., p.14.

129. Gupta, S.P.: *Archaeology of Soviet Central Asia and the Surrounding Indian Borderlands*, Vol. II, B.R. Publishing, Delhi, 1979, p.238.

130. Grun, 1982, ibid.

131. Hoffman, 1979, ibid., 217.

132. Author's personal observation of worker's pit tombs near Great Pyramid of Cheops, Egypt.

Great Pyramid of Cheops (right) and Sphinx, at pyramid complex near Giza, Cairo. These and similar pyramidal structures were built across the Near East on command of divine kings who lorded over totalitarian "flogging societies".

The wives and concubines of the dead king were often sacrificed at his gravesite, to accompany his corpse to the "Land of the Dead".

which social turmoil and desiccation gripped the region. For instance, bas-reliefs on the causeway of the Sakkara pyramid of Unas, last of the V Dynasty Kings (c.2350 BCE), depict groups of emaciated famine victims, indicating that either disruptive warfare or failed Nile floods, or both, had severely reduced food supplies.[133]

The VI Dynasty saw massive invasions of the Nile Valley by Asiatic peoples from the East. Numerous punitive campaigns were launched against the "land of the sand dwellers", who were engaged in trade and raid.[134] Bas-reliefs from the VI Dynasty give the earliest known records of desert locust.[135] Around this same time, the Temehu peoples emigrated out from their homeland, which is today called the Libyan Desert.[136] A gradual breakdown of central government authority occurred, leading to a collapse of the Old Kingdom during the rule of Pepi II (c.2300 BCE).[137] O'Connor identified the Egyptian period 2160-2100 BCE as one of famine, civil strife, and migrations, with greatly increased burials, more than during any previous period.[138] It was also during this period that male circumcision became more widely practiced.[139]

Asiatics were finally driven out of the Nile delta region during the rule of Kheti III (c.2130 BCE), last of the X Dynasty Kings, and trade resumed with Palestine and Byblos. King Amenemhat III (c.2100 BCE) thereafter instituted massive irrigation works, including dams and canals at Fayum.[140] The power and authority of the ruling class was significantly increased, and "order" was restored. With time, trade resumed into the Mediterranean and Black Sea. During the period institutions such as polygamy and the marriage contract began to be practiced by the lower classes, and society in general gave a picture of greatly increased structure, mandated not from unarmored instinct, but by force of codified law. This was the era of the Old Testament Patriarchs,[141] and shortly before the time in which it has been suggested that the Hebrew Joseph counseled Pharaoh to store up grain for an impending drought and famine.[142]

A new era of Egyptian military expansion followed c.2000 BCE. Sestrosis III renewed raids above the first cataract into Nubia, which was conquered, as was Palestine shortly thereafter.[143] Another period of social unrest occurred around c.1800 BCE, described as "...a vast ethnic movement which affected the whole of Western Asia and ultimately reached Egypt."[144] During this era, the Hyksos — an Asiatic/Semitic group from the northeast — invaded the Nile Valley, instituting 200 years of brutal oppression. The Hyksos brought new iron weapons and the chariot, against which the Egyptians had no adequate defense.[145]

The Hyksos invasion was characteristic of the general type of assault which Egypt experienced over the centuries. Said to have been "thrust out by drought from the desert", the Hyksos have been described:

> *"The fearful apparition of this host, a people coming from unknown regions, strange of speech, uncouth in appearance, and bloody in act...*[146]

The Hyksos are believed to have been characteristic of other nomadic Asian/Semite invaders which had arrived previously, in 3300 BCE and 2300 BCE. However, some have argued that they were not invaders at all, but rose to power from within Egypt, possibly from homelands in the desiccating highland environment above the Nile.[147]

During this period of Hyksos rule, Nubia successfully revolted from Egyptian control. A new city was built by the Nubians at Kerma, along with large tombs where massive human sacrifice was practiced; widows and concubines were predominant among the victims, many of whom were buried alive.[148]

133. Bell, B.: "The Dark Ages in Ancient History, 1: The First Dark Age in Egypt", *Am. J. Archaeology*, 75:1-26, 1971; Aykroyd, 1974, ibid., p.25; Mellaart, 1966, ibid., pp.68-71.

134. Hawkes, 1974, ibid., p.11; Grun, 1982, ibid.; Murray, 1949, ibid., p.19; Hoffman, 1979, ibid., pp.217-8.

135. Winstanley, D.: "Desertification: A Climatological Perspective", in *Origin and Evolution of Deserts*, S.G. Wells & D.R. Haragan, Eds, U. New Mexico Press, Albuquerque, 1983, p.195.

136. Butzer, 1961, ibid., p.42.

137. Mellaart, 1966, ibid., p.71.

138. O'Connor, D.: "A Regional Population in Egypt to c.600 BCE", in *Population Growth: Anthropological Implications*, Spooner, B., Ed., MIT Press, Cambridge, 1972, pp.78-100; cf. Bell, 1971, ibid.

139. "Passage Rites", *Encyclopedia Britanica*, Macropaedia, Vol 13, 1982, p.1050.

140. Hoffman, 1979, ibid., p.182; Mellaart, 1966, ibid., p.71.

141. Hawkes, 1974, ibid., p.151; Murray, 1949, ibid., p.22; Lewinsohn, R.: *A History of Sexual Customs*, Harper Brothers, 1958, p.23.

142. Aykroyd, 1974, ibid., pp.25,28-9.

143. Grun, 1982, ibid.

144. Edwards, 1972, ibid., p.181.

145. Huntington, E.: *Palestine and its Transformation*, Houghton-Mifflin, NY, 1911, p.385; Edwards, 1972, ibid., p.181.

146. Huntington, 1911, ibid., p.386.

147. Moscati, S.: *The Semites in Ancient History*, U. Wales, Cardiff, 1959, p.87.

148. Hoffman, 1979, ibid., p.261; Hawkes, 1974, ibid., p.165.

It was against this backdrop that, around 2000 BCE, Canaanite colonists came to the relatively isolated region of Tunisia. Their agricultural settlements used existing water sources plus runoff.[149] The Mediterranean coast of Africa west of Egypt was then still populated by hunting, pastoral, and agricultural peoples.

After several hundred years of foreign occupation by the Hyksos, the Egyptians successfully revolted, around c.1567 BCE, and established the New Kingdom. In driving out the Hyksos, the Egyptians adopted the chariot and iron technology of their enemies, and Egyptian society was once again transformed. Professional soldiering gained an importance and status it had not previously held, and the Egyptians embarked upon various wars of conquest, flooding their country with captured slaves.[150] Religion was altered. Hymns to Amon made their first mention of "fatherliness" for a deity.[151] The Sun God, Book of the Dead, and animal mummification were adopted. The Gods of the upper ruling classes became, for the first time, the gods of the lower classes.[152]

The New Kingdom Pharaohs achieved power never seen before in Egypt; they ruled during an Imperial Age, and erected grand palaces in Karnak and Luxor. For 500 years, incessant war was waged with their neighbors, particularly in Palestine and Syria toward which the remnants of the Hyksos had fled.[153]

Queen Hatshepsut briefly ruled (c.1480 BCE), following the death of her husband, Thutmose II. As matron to her young son and future Pharaoh Thutmose III, she brought peace and productive trade relations with foreign ports. However, she was opposed by antifeminist elements in the military. At the end of her rule, her name was defaced from all public buildings. Thutmose III thereafter extended the Egyptian Empire into Nubia and Mesopotamia.[154] A subsequent pharaoh, Amenhotep II (c.1440 BCE), revived the custom of ritual widow murder, slaughtering four of his wives to be embalmed with him.[155] The peak of Egyptian central power, authority, and trade occurred under Amenhotep III (c.1400 BCE), when the first navigable canal was built connecting an arm of the Red Sea with the Nile delta.[156]

The Great Pharaoh's son, Akhenaten (Amenhotep IV, c.1360 BCE), attempted to restrain the military and priesthood, declaring a new monotheistic creed with a new God and capital city at El Amarna. Akhenaten called for peace and emphasized the arts, Nefertiti being his wife. During his reign, the Hittites invaded Egypt's borderlands in Palestine and Mesopotamia; the traumatic practices of infant cranial deformation and swaddling, which the Hittites also practiced, made their first clear appearance in Egypt at this time,[157] and spread from there to other parts of Africa. However, like circumcision, the practices may have been introduced at an earlier date. The contemporary mudpack hair dressings of some Sub-Saharan African cultures, such as the Hamar in Southern Ethiopia, appear to be a diluted remnant of the ancient practice.[158] After Akhenaten's death, like the despised Queen Hatshepsut, his name was defaced from buildings and stela, and El Amarna was dismantled; his burial either never occurred or has never been found.[159]

Akhenaten's successor was the boy-king Tutankhamon (c.1350 BCE), who restored the military and defeated the invading Hittites. Dying in the battle at age 18, Tutankhamon was given fabulous grave wealth which today is known the world over. His death brought in the XIX Dynasty of Seti and Ramses, the Pharaohs of Mosaic times. This was a period of almost relentless warring with powers to the northeast.

Seti considered himself a God, and constantly warred against the Hittites over Syria; Ramses II (c.1300 BCE), his successor, warred for 20 years against the Hittites. Ramses strengthened the priesthood, and

149. Evenari, M. et al: *The Negev*, Harvard U. Press, Cambridge, 1982, pp.353-5.

150. Wittfogel, K.A.: *Oriental Despotism*, Yale U. Press, New Haven, 1957, p.323; Turnbull, 1976, ibid., pp.93-6.

151. Murray, M.A.: *The Genesis of Religion*, Routledge & Keegan Paul, London, 1963, p.78.

152. Murray, 1949, ibid., pp.167,182; Grun, 1982, ibid.

153. Hawkes, 1974, ibid., pp.11,146; Hoffman, 1979, ibid., p.24; Grun, 1982, ibid.

154. Murray, 1949, ibid., pp.49-52; Lewinsohn, 1958, ibid., p.22.

155. Thompson, E.: *Suttee*, George Allen, London, 1928, p.25; Davies, 1981, ibid., pp.36-7.

156. Murray, 1949, ibid., p.53; Grun, 1982, ibid.

157. Dingwall, E.J.: *Artificial Cranial Deformation*, London, 1931, pp.102-111, 237; Murray, 1949, ibid., p.54.

158. The award-winning film *Rivers of Sand* shows these hair dressings, as well as other antifemale practices (ritual whippings of marriageable girls by their potential husbands, female scarification and tooth extractions) which probably came to the region at the same time.

159. Murray, 1949, ibid., pp.53-60; Velikovsky, I.: *Oedipus and Akhnaten*, Doubleday, NY, 1960, pp.182-3.

instituted massive taxation programs to finance his wars and grand temples. Ramses II erected the great statues of himself at Abu Simbel and Karnak, and fathered some 160 children from a small army of wives and concubines.[160] Ramses II was not alone in possessing large retinues; he merely carried it to an extreme not seen before in Egypt. Bas-reliefs of Ramses II on stelae and columns in the Royal Palaces at Karnak and Luxor graphically boast of the size of his erection, as long as his forearm, and depict the ritual catching of his semen in a cup.[161] Ramses III also warred against the Hittites, and against the Sea Peoples, Libyans, and Philistines. He died in a harem conspiracy.[162] It was a period when the male principle was seen as all-powerful, replacing entirely the female principle. This was also a period (c.1300-1000 BCE) which emphasized the role of the military, when Egypt experienced increasing military conflict at the delta, on its northeast border, and on the Libyan frontier.

The Berber-speaking Garamantians arrived in Libya around 1000 BCE, or possibly before, from unknown regions. The ruins of ancient Germa and the capital city of Garamantia lie about 800 km (500 mi) south of the modern city of Tripoli, Libya, in a most desolate region now noted for its extreme aridity and high temperatures. The peoples of Libya, who suffered under increasingly desiccated conditions, constantly pressured the Egyptians on their western frontier.[163]

Using the chariot and superior weapons, the Garamantians subdued pastoral cultures of central Sahara, occasionally clashing with the Egyptians.[164] The Garamantian culture eventually diluted itself with indigenous North African peoples, and females in Germa maintained a somewhat higher status than in Egypt proper. Daniels discussed the customs of the ancient North Africans:

"Almost every ancient author speaks of general promiscuity amongst the African tribes, e.g.: the Nasamones have many wives and their intercourse with women is promiscuous; likewise the Massagetae; the women of the Gindines wear many leather anklets because they put on one for every man with whom they have had intercourse; the Garamantes do not practice marriage but live with their women promiscuously, so that children do not know their parents, or parents their children. This sort of remark was commonly thrown by monogamous Greek and Latin authors at native peoples - Caesar says the same of the Britons - and as Bates has said it probably indicates no more than a misunderstanding of the polygamous practices of the Libyan tribes... Probably the custom of the Gindines was akin to that of the Ouled Nail [of c.1920] where girls earn a dowry by prostitution before marriage. If this is so it would suggest a considerable freedom, and no doubt status, accorded to the female sex, which may even be reflected in the wearing of male dress by a Libyan woman depicted on a tile from Medinet Habu in Egypt... Concerning dress, we have no reason to suppose the Garamantes were basically different from other tribes. Amongst Greek and Latin authors these are frequently called 'naked Garamantes' or 'nude Nasamones', but other evidence makes it clear that this was not the simple case."[165]

Feminine influences also persisted in patriarchal Egypt and surfaced after c.900 BCE, during a period of fragmentation and declining central state power. A High Priestess was established at the Temple of Ammon in Thebes, who acted as an informal consultant on all matters of government, much in the same manner as the Oracle at Delphi in Greece.[166] The practice of sacrificing children or adults under foundations of buildings

Wall carvings from Germa, a city-state which developed on a secure water-course during the period of desiccation (c.1000 BCE). The Garamantians appear to have retained *relative* sexual freedom and female influences even while in conflict with growing Egyptian power. Germa eventually declined and was abandoned; it lies in what is today a most inhospitable bone-dry environment, deep in the Sahara.

160. Murray, 1949, ibid., pp.57-9; Hawkes, 1974, ibid., p.159; Lewinsohn, 1958, ibid., p.22.

161. Author's personal observations in Luxor, Egypt. Such phallic representations are a give-away to a deep culture-wide insecurity about sexual potency.

162. Murray, 1949, ibid., pp.61-2.

163. Mellaart, J.: *The Neolithic Near East*, Thames & Hudson, London, 1975, p.269.

164. Davidson, 1966, ibid., p.55.

165. Daniels, C.M.: "The Garamantes", in Kanes, 1969, ibid., p.38.

166. Lewinsohn, 1958, ibid., p.22.

While the "God-Kings" of Egypt viewed the more naturalistic peoples of North Africa with contempt for their relative sexual freedom (i.e., the "promiscuous" and "naked Garamantes"), at home they enslaved women in enormous harems, butchered them in the tombs of their dead leaders, and engaged in absurd phallic-masturbation rituals, narcissistically boasting on bas-reliefs (c.1300-1000 BCE).

167. Davies, 1981, ibid., p.37.

168. Stone, 1976, ibid., pp.34-5; see Fox, H.: *Gods of the Cataclysm*, Dorsett, NY, 1976, pp.195-7, 201-3.

169. Evenari, 1982, ibid.. pp.353-5.

170. Hawkes, 1974, ibid., p.91.

171. Hughes, J.D.: *Ecology in Ancient Civilization*, U. New Mexico Press, Albuquerque, 1975, pp.1-2.

172. Walls, J.: *Man, Land, Sand*, Macmillan, NY, 1980, p.91.

173. Moscati, S.: *The World of the Phoenicians*, Praeger, NY, 1968, p.150; Whitaker, J.I.: *Motya: A Phoenician Colony in Sicily*, Bell & Sons, London, 1921, p.8; Scott, G.R.: *History of Torture*, Sphere Books, London, 1971, p.216; Fox, 1976, ibid. pp.131-6, 151-4.

174. Moscati, 1968, ibid., p.183.

175. as quoted in deMause, L.: "The Evolution of Childhood", in *The History of Childhood*, L. deMause, Ed., Psychohistory Press, NY, 1974.

also ended around this time.[167]

Such remnants of feminine influence in North African culture were probably at the root of the reports made by Diodorus Sicilus years later, that the female had more power than the male regarding property, inheritance, and kinship, and that Libya and Ethiopia were inhabited by "Amazons".[168] Ethiopia was, at this time, under the influence of the coastal Arabian trading state of Sabaea, or Sheba, which was ruled at least once by a Royal Queen.

Phoenician colonists also settled in North Africa, in Tunisia around 800 BCE, to build the state centered on Carthage. They introduced the grapevine and olive to the area, and established a thriving agricultural system based upon runoff accumulation.[169] Carthage grew to be a North African power based upon both agricultural exports and dominance of sea trade in the Western Mediterranean. They expanded their empire to Sicily, Sardinia, and coastal Spain, and controlled the Straits of Gibralter.[170] The forests of coastal North Africa supported their shipbuilding activities and fired huge kilns in which exquisite pottery was fashioned. Much wood was also cut for charcoal making, and for the fires from which metal implements were fashioned. Wheat, olive oil and other agricultural products were exported around the Mediterranean, primarily to the growing Roman state.[171] At this time, Berber pastoralists were still able to obtain pasture for their sheep and goats in Southern Tunisia.[172]

The Phoenicians were a Semitic people who emigrated to North Africa from the region of Syria and Palestine, and possessed a severely antifemale and antichild nature. They were ruled by despotic kings who were buried in royal tombs; ritual grave murder and child sacrifice were practiced, and they owned massive numbers of slaves. One of the bloodiest and most hated people of the Mediterranean, the Phoenicians killed their enemies by flaying them alive in a manner similar to the Aztecs; after the victim was flayed, the skin would be worn by priests like a diver's wet-suit and carried about "until it began to stink". Phoenician influences were felt throughout the Mediterranean islands, and beyond, along the West African and European coasts.[173] The impact of the Phoenicians upon peaceful Neolithic peoples of the African interior may be surmised from the following inscribed words of Carthaginian King Hanno, of c.400 BCE:

> "...*Chasing them we were unable to catch any of the men, all of whom, being used to climbing precipices, got away, defending themselves by throwing stones. But we caught three of the women, who bit and mangled those who carried them off, being unwilling to follow them. We killed them, however, and flayed them and brought their skins back to Carthage...*" [174]

Plutarch discussed the bloody nature of the Phoenician Carthaginians, whose massive sacrifices of children to the gods have been found in cemeteries of the region:

> "...*with full knowledge and understanding they themselves offered up their own children, and those who had no children would buy little ones from poor people and cut their throats as if they were so many lambs or young birds; meanwhile the mother stood by without a tear or moan; but should she utter a single moan or let fall a single tear, she had to forfeit the money, and her child sacrificed nevertheless; and the whole area before the statue was filled with a loud noise of flutes and drums so that the cries of wailing should not reach the ears of the people.*" [175]

Children not only had their throats slit, but also were burned alive, as were war captives.[176] The Phoenicians spread terror, social turmoil, and militant reactions among peoples of regions to the south, along the African coast and interior. The Phoenicians also possibly established trading relations southward along the Atlantic African coast, and into the region of the Niger River, Dahomey, and Benin. There, divine kingship, human sacrifice, and sometimes flaying of prisoners existed and more or less persisted into the late 1700s and 1800s CE.[177] The Oba of Benin, for instance, maintained absolutism and divine kingship, occasionally lapsing into massive human sacrifice.[178] Many of these characteristics were in later years also transmitted by Bantu groups into equatorial and South-eastern Africa.[179]

During the first millennium BCE, the Egyptians declined from power and were conquered and ruled by Kushite Kings (c.800-700 BCE). The Kings of Kush later established a new capital in Meroe, building large pyramids, temples, palaces, and royal tombs in the Egyptian style. The Kushites were expelled from Egypt by the Assyrians (c.700 BCE), who briefly ruled. Egyptian independence was finally curtailed by the Persians (c.500-300 BCE) under Darius III, and remained a vassal state until Persia was conquered by Alexander the Great.

Hoffman has discussed the copying of Egyptian customs in sub-Saharan Africa, and the spreading of such influences:

"If the institution of divine kingship had indeed evolved from a more general African context, then we should be able to trace its more or less parallel emergence among many fourth millennium BCE farming and herding societies along the Nile. Instead, we find a slow diffusion of Egyptian influence up the Nile beginning in Late Gerzean times (c.3300-3100 BCE) and a concomitant copying of some of the attributes and paraphernalia of Egyptian kingship by frontier and client groups over the next 2000 years." [180]

"Throughout the Napatan and Meroitic periods, till the triumph of Christianity in the fourth century AD, Nubian kings continued to imitate Egyptian royal customs. The kings of Napata and Meroe still styled themselves [in the manner of] pharaohs of Upper and Lower Egypt centuries after their expulsion from Egypt. They continued to write official decrees in the Egyptian language and to use hieroglyphs until the third century BCE and, most anachronistically, they had themselves buried in pyramids until the fourth century AD... The practice of royal pyramid burial had been abandoned for almost a thousand years in Egypt before being revived in Nubia." [181]

By 400 BCE, the regions south of the Sahara were considerably influenced by northern peoples and their iron technology, to include the empires of Egypt, Nubia, Garamantia, and Carthage. Iron-using peoples also spread across the African savanna from centers in both Meroe and Nigeria. Iron-using agriculturalists and pastoralists of a Bantu heritage gradually pushed south, displacing smaller stone age, hunter/gatherer peoples southward, where they found a relatively safe haven only within the rainforest (Pygmy), or within the less-desired semiarid southern end of the continent (Bushmen, Hottentot). While the Bantu adopted many of the cultural characteristics of the Semitic/Indo-Aryan groups who originally introduced iron technology into North Africa, these traits were diluted and mixed with local traditions.[182]

Still, Bantu females often were, and often still are, greatly subordi-

Pyramids of Kush - constructed c.800 BCE on what is now an arid plateau in modern-day Sudan (Photo credit: Francis Geius, Mission SFDAS 2001)

176. Davies, 1981, ibid., p.51.
177. Fox, 1976, ibid., pp.150-1; Davies, 1981, ibid., pp.21,104,135-8, 141-56.
178. Davidson, 1966, ibid., pp.109-12; Davies, 1981, ibid., p.153.
179. Best, A. & deBlij, H.: *African Survey*, J. Wiley, NY, 1977, p.75; Davies, 1981, ibid., pp.133-4,142-3,157,161; Scott, 1939, ibid., pp.37-8.
180. Hoffman, 1979, ibid., p.261.
181. Hoffman, 1979, ibid., p.262.
182. Hawkes, 1974, ibid., p.16; Davidson, 1966, ibid., pp.21,108-18; Hoffman, 1979, ibid., pp.257-64.

nated to their husbands, fathers, and brothers, their children strongly segregated sexually in a largely polygamous, patrilineal, and bride-price and virginity-emphasizing family structure. While the Bantu groups did not use the horse, they often maintained nomadic and militant characteristics, with strongman rulers. Slavery, divine kingship, and human sacrifice occasionally surfaced and persisted among some of the Bantu groups, and among other African cultures which came into contact with or clashed with them.[183]

Following the conquests of Alexander the Great, power in the Eastern Mediterranean passed from Persia to Greece. The port city of Alexandria on the Nile Delta became a center of learning (c.300 BCE), though the rest of Egypt was ruled with an intolerant, iron hand.[184] During the periods of Persian and Greek rule, additional Eastern concepts on family structure and sexual behavior flooded Egypt. Strict sexual segregation prevailed as women were introduced to the veil, the harem, and the system of female seclusion which the Arabs would later adopt.[185] The office of High Priestess was also abolished.[186] This was the Egypt which Herodotus described as possessing circumcision for its young men, "excision" (of the clitoris) for its young women,[187] and in which Hippocrates observed the swaddling of infants.[188]

Greek rule in Egypt was particularly harsh, and hardly democratic, with great corruption, impoverishment, insurrections, and massacres. Still, the period was more stable than under the Persians, who were much hated. The reign of the Ptolomies was characterized by brother-sister marriages, whereby Kings preserved the throne for their sons under a system of matrilineal kinship which had persisted more or less since predynastic times.[189]

Prior to the Roman era, the Garamantian kingdom spread out over the remaining Central North African grasslands, with urban settlements along specific water sources closer to the Mediterranean. Herodotus described the thriving kingdom of Germa of c.440 BCE, and the Romans recorded at least five military campaigns against the Garamantians between 19 BCE and CE 100. After this date Garamantia disappeared from the rolls of history; its ruins have so far revealed tens of thousands of graves, and hundreds of miles of foggaras (qanats), or underground water channels. A large lake appears to have existed in the center of the Wadi el Agial, in which the major Garamantian cities were built. Dates, grain, and trees existed there, as did sheep, goats, pigs, and horses or asses. These evidences speak of relatively moist conditions in a region now bone-dry.[190]

Carthage was invaded by Sicilian soldiers in the 4th century BCE, and was similarly described by Diodorus Siculus:

"The intervening country through which it was necessary for them to march was divided into gardens and plantations of every kind, since many streams of water were led in small channels and irrigated every part".[191]

During the Punic Wars, when Carthage warred with Rome (c.264–146 BCE), human sacrifice became especially important. Large pits have been excavated in which, according to historical sources, hundreds were burned alive.[192] Notably, it was only after the onset of the Punic Wars that the first gladiator fights appeared in Rome.[193] In later years, the Roman emperor Tiberius would forbid Carthaginians from sacrificing children, purchased from poor families, on the altar of Saturn.[194] At home in Rome, however, his rule was characterized by severe tortures and sadistic brutalities of the worst sort.[195] More will be said on events in Rome in the

Sack of Carthage by the Roman General Scipio the Younger, c.146 BCE.

183. Hoffman discusses various aspects of the debate on origins of such institutions in sub-Saharan Africa (1979, ibid., pp.257-64). He identifies the following groups as having possessed divine kingship: Yoruba, Dagomba, Shamba, Igara, Songhay, Katsena, Daura, Gobir Hausa, Shilluk, Baganda, & Jukun.

184. Murray, 1949, ibid., p.71.

185. Marsot, A.L.: "The Revolutionary Gentlewomen in Egypt", in *Women in the Muslim World*, L. Beck & N. Keddie, Eds., Harvard U. Press, Cambridge, 1978, p.262.

186. Lewinsohn, 1958, ibid., p.22.

187. Tannahill, 1980, ibid., pp.67-8.

188. deMause, 1974, ibid., pp.37-8.

189. Murray, 1949, ibid., pp.71,77; Lewinsohn, 1958, ibid., p.24.

190. Daniels, 1969, ibid., pp.31-51; cf. Pesce, E.: "Exploration of the Fezzan", in Kanes, 1969, ibid., pp.73-80.

191. as quoted in Evenari, et al, 1982, ibid., p.354.

192. Davies, 1981, ibid., pp.21,63,216.

193. Davies, 1981, ibid., p.47.

194. Davies, 1981, ibid., p.51.

195. Scott, G.: *A History of Torture Throughout the Ages*, Luxor Press, London, 1939, pp.141-2.

sections ahead.

After the Roman conquest of North Africa (200-50 BCE), extensive settlement based upon irrigation and runoff agriculture took place in Central Algeria, Tunisia, and points eastward. A series of Roman defensive frontier trenches and fortifications exist in Algeria which are hundreds of kilometers long, with ruins of forts, towers, terraced wadis, diversion systems, runoff fields, cisterns, and villages.[196] Pliny, the Roman historian, remarked that *"the interior of Africa as far as the Garamantes and the desert is covered with palms remarkable for their size and their luscious fruit".*[197]

Rome completely supplanted Greece in the Eastern Mediterranean during the reigns of Octavian (later titled Augustus Caesar) and Cleopatra. Octavian conquered Egypt, defeating the navy and armies of Mark Antony and Cleopatra (c.30 BCE). Antony fell on his sword, while Cleopatra died from a bite by the poisonous asp, thereby bringing to an end the long matrilineal royal blood line of Pharaonic-Ptolemaic Egypt. Egypt passed into the Roman Empire as the personal estate of the "Emperor-God" Octavian.[198]

Presently, the North African coast is littered with ruins dating from pre-Roman and Roman times, many of which were buried in sand prior to their recent rediscovery by explorers and archaeologists. The presence of these various ruins, which include cities, estates, and now dry or nearly dry water supply systems, indicates that present levels of aridity were not reached until as recent as 200 to 600 CE.[199]

Cleopatra (above) and Mark Antony (below).

Most of the oases in North Africa also give evidence of decline, having had greater amounts of water available as recently as the Roman period. Many artesian springs have ceased flowing, and groundwater levels have dropped as well. In the Libyan Desert west of the Nile, for example, are found a number of oases with drastically fallen groundwater levels. Fossil spring mounds around the Kharga Oasis indicate a prior water table 55-60 m (180-195') higher than today. Springs and wells in use during Egyptian dynastic times, at Uweinat, Gilf Kebir, Sheb, and Tarfawi, have since dried up. Wells at Bir Misaha, Bir el Atrun, and Merga have fallen 22 m, 10 m, and 5 m respectively since Roman times. Other wells in both the Western and Eastern Deserts of Egypt have fallen or gone dry since Dynastic or Roman times.[200]

Murphy surveyed the evidences for desiccation and depopulation in North Africa within the last 2000 years, and pointed out a number of factors other than climate change which have been at work transforming the landscape. He cited the Arab conquests of the seventh and eighth centuries, their hostility toward "unproductive trees", the depletion of groundwater reservoirs, and lack of technical skills as major causes for abandonment of irrigation systems, wells, and farms. While human factors certainly have played their role, much of the evidence cited by Murphy can still be interpreted as being due to climate change. In any regard, environmental change since even Roman times, has been profound:

"Large areas of the 'granary of Rome' in northwestern Africa are now a desiccated wilderness. The great amphitheatre at El Djem [Tunisia], with seats for 60,000 people, stands in the desert surrounded by a few small Arab villages. The important Roman city of Timgad has been abandoned since about 250 AD, while beside it is the clearly marked channel of a now vanished river. The Roman metropolis of Leptis Magna (near modern Tripoli, and now a ruins in the desert) was one of the primary commercial centers of the Roman Mediterranean, and the birthplace of emperor Septimus

196. Evenari, et al, 1982, ibid., p.353.

197. Pliny, as quoted in Daniels, 1969, ibid., pp.48-50.

198. Murray, 1949, ibid., pp.102-3.

199. Huzayyin, 1973, ibid., p.318.

200. Butzer, 1961, ibid., p.45.

Severus... Numerous Roman mosaics from these and other North African sites depict fauna now found only in tropical Africa. Ruins of great aqueducts and reservoirs dot the almost uninhabited plains of North Africa inland from the relatively better watered littoral. Roman remains in the oases of the Libyan and Egyptian deserts show that there was formerly a cultivated area many times larger than the present, together with large buildings and extensive irrigation works. Well-traveled roads, whose traces are now nearly obliterated by drifting sand, connected all of the Roman towns with each other and with the sea or the Nile. Roman Africa was a flourishing area which contributed an important part of the imperial capital's food supply, as well as supporting a sedentary population greatly in excess of the present." [201]

"Abandoned Roman wells are far more numerous in all of the oases than currently used wells. In some cases these older disused wells have become silted up or their sidewalls have collapsed, but in many case they have apparently been abandoned only because they have lost their artesian head and the water must be lifted. The present Arabs do not use water lifting devices, and they refuse to draw water from wells which no longer flow naturally." [202]

Huzayyin also raised a number of questions regarding relatively recent climate changes posed by archaeological evidence.

"In dealing with this still unsettled question of relatively recent climatic changes, archaeological evidence may be of particular value. Roman cisterns found on the Mediterranean coast of Egypt west of Alexandria and in several places on the edge of the Syrian Desert are not fully examined or even properly known. Their abundance and the large capacity of some of them may be taken as a sign of greater precipitation of rainfall at that time... In North Africa and the interior of the Sahara archaeological evidence is still lacking, but the abundance of remains in many of the desolate areas may be taken as an indication of the somewhat more favorable climate lasting until Roman times." [203]

As the moister conditions which persisted into Roman times gave way to aridity, North Africa came under the influences of the theology and laws of the Near East, which were increasingly of an antifemale and antichild character. Such influences at first came indirectly through Rome and Byzantium, which, as shall be demonstrated, were themselves under considerable influence from Semitic and Indo-Aryan concepts imported from Persia and the Near East.

For instance, Philo (c.30 CE), an Alexandrian Jew, taught that sex was justified only for procreation, even in marriage, and gave voice to a growing hostility toward both women and sexual pleasure. He advocated genital mutilations as a means for:

"...the excision of passions, which blind the mind. For since among all passions that of intercourse between man and woman is greatest, the lawgivers have commended that that instrument, which serves this intercourse, be mutilated, pointing out, that these powerful passions must be bridled, and thinking not only this, but all passions would be controlled through this one." [204]

Philo's teachings considerably influenced later Christian theolo-

201. Murphy, R.: "The Decline of North Africa Since the Roman Occupation: Climatic or Human?", *Annals, Assoc. Am. Geographers*, 41(2):120-2, 1951; cf. Knight, M.M.: "Water and the Course of Empire in North Africa", *Quat. J. Economics*, 43:44-93, 1928-29; Shaw, B.D.: "Climate, Environment, and History: the Case of Roman North Africa", in *Climate and History*, T.M.L. Wigley, et al, Eds., Cambridge U. Press, 1981.

202. Murphy, 1951, ibid., p.127.
203. Huzayyin, 1973, ibid., p.319.
204. deMause, 1974, ibid., p.24.

gians.[205] In later years Clement of Alexandria (c.200 CE), a Christian scholar steeped in the mystery religions of the East, preached against the remaining female influences in Egypt, proclaiming that "every woman should blush at the thought that she is a woman".[206] He interpreted Jesus' advise to "become as little children" as meaning "uncontaminated, without knowledge of sex".[207] Tertullian of Carthage (c.200 CE), another Christian theologian of the severe Montanism sect, opposed the use of judicial torture, but railed against women, marriage, and the chemical/ pessary birth control practices which the ancient Egyptians had developed.[208] In later years the Roman Emperor Diocletian attempted to crush the Christian movement, ordering the execution and burning of tens of thousands of Christians in Egypt.[209]

By contrast, it was during this early Christian period that many aboriginal North African groups, such as the Berber and Tuareg, were still clinging to fertility amulets and matristic cultural influences, some aspects of which persist into the modern era.[210] Indeed, the Alexandrian Greeks considered the early inhabitants of North Africa as "barbarians", from which the term *Berber* derives. The Berbers resisted the Romans as they had the Greeks and Phoenicians before them, and the Moslems and Europeans afterward, by raiding their settlements and fortifications from hidden mountain enclaves deep within the Sahara.

Christian influences spread across coastal North Africa and up the Nile to Ethiopia, whose leaders were then engaged in subduing the pastoralists of rapidly desiccating Kush.[211] Northwest Africa did not long remain in Christian hands, however, given the invasion of that region by the Vandals. The Vandals, whose name means "wanderer" but who have been more clearly associated with the term "vandalism", were a Germanic group whose roots were in Central Asia; they swept across Gaul, Spain, and entered the Maghreb (Morocco-N.Algeria-Tunisia) in the 400s CE. Sacking Rome in 455 CE, the Vandals threatened the western border of the Byzantine Christian empire in both Europe and Africa.

During this turbulent period, Christian theologians in North Africa pushed for clerical celibacy and ordered mobs to burn the "pagan" library at Alexandria.[212] Scholarship and learning in general ground to a halt, and hostility towards women increased. In c.430 CE the monks under St. Cyril, Bishop of Alexandria, murdered Alexandrian intellectuals including the beautiful Hypatia, a learned mathematician and astronomer and daughter of the scientist Theon; Cyril's monks dragged Hypatia into a church, stripped her naked and scraped her to death with oyster shells.[213]

The rights and status of females continued to erode under the Christian philosophers until the Islamic invasion of North Africa following c.641 CE. While the reforms of the Prophet Muhammed may have stabilized and possibly even raised the status of some women among the nomadic warrior caste Bedouin peoples of Arabia, his laws brought new repressions to women in Egypt and other parts of "pagan" North Africa, not the least of which was the military conquest and enslavement of North Africa itself. The great library at Alexandria was again burned, this time by Moslem armies.[214]

North Africa was invaded entirely between CE 640-700 by Moslem Arab armies who pushed through the Maghreb into Spain, which was dominated and provoked into a constant state of war for several centuries. By this time, most of the remaining forests had vanished from coastal Libya and Egypt, but enough wood and scrub remained in Northwest Africa, compared to Arabia, such that Moslem armies called it a "land of continuous shade".[215]

Other Islamic armies invaded North Africa between CE 1000-1100, followed by a great influx of Bedouin nomads. This influx occurred at a

The Murder of Hypatia by the monks of St. Cyril, Bishop of Alexandria (c.430 CE).

205. Bullough, 1976, ibid., p.169.

206. Lewinsohn, 1958, ibid., p.95.

207. deMause, 1974, ibid., p.47.

208. Himes, N.: *Medical History of Contraception*, Williams & Wilkins, Baltimore, 1936, pp.61-6; Bullough, 1976, ibid., pp.185-6; Scott, 1939, ibid., p.134.

209. Scott, 1939, ibid., p.143.

210. for instance, see Pond, A.: *The Desert World*, Greenwood Press, Westport, CT, 1975, p.205.

211. Davidson, 1966, ibid., pp.38-9,129.

212. Bullough, 1976, ibid., pp.168-72,183-7; Jeans, J.: *The Growth of Physical Science*, Premier, 1958, p.99.

213. Thompson, P.: *Secrets of the Great Pyramid*, Harper & Row, 1971, p.4; Jeans, 1948, ibid., pp.99-100.

214. Jeans, 1948, ibid., p.101.

215. Mikesell, M.W.: "The Deforestation of Northern Morocco", *Science*, 132:441-8, 1960.

time when the Near East was itself being invaded by Central Asian nomads. Drought also was occurring in the Nile Valley around CE 1069 when flood waters failed for several years. Widespread famine followed, which was probably also exacerbated by the invasions of nomads.[216]

After 1100 CE, the Fatimid rulers of Egypt were ousted by Saracen Muslims of a more puritanical and militant character, and female status declined even further. The Saracens, whose origins were in the deserts of Syria and Arabia, fought with Christian crusaders in the Near East, and also with the Nubians of the Upper Nile region.[217] During this same period, Norman armies were active in Europe and the Mediterranean, and raids were also made along coastal North Africa.[218]

While North Africa became exceptionally arid following the Roman period, Nicholson identified semihumid trends in the Southwestern Sahara between CE 800-1200, and 1500-1750, the latter of which roughly corresponds with the "Little Ice Age" period of Europe. These semihumid periods were not very widespread, nor as intense or humid as the period before 3000 BCE, however. A number of African states did develop in the Southwest Sahara during these moist interludes, partly in response to these moister conditions, but also partly due to the growth of new trading opportunities, and increased central state power following prior invasions and migrations into the region. Nicholson has summarized the paleoclimatic and historical evidences for humid phases during these later periods.[219]

The major African kingdoms which formed during the years CE were those of Ghana (CE 700-1200), Mali (1000-1350), and Songhai (1350-1600), which developed in the region between Senegambia and the Niger River; the Kanem-Bornu kingdom (CE 800-1800) of mounted Moorish knights also developed and persisted in the region around Lake Chad.

The African Kingdoms of the Sub-Saharan *Sahel* (Arabic for "edge" of the Sahara Desert) were partly stimulated by migrating tribes of Berber immigrants, forced south by invasions occurring across North Africa. Where peoples of the Sahel did adopt Islam, it was usually under the force of armed might, and then generally mixed with vestiges of their previous animistic religions. Islam was moderated and shaped according to African custom, the greater the distance from Mecca, the greater the moderation. Indeed, Ibn Battuta, a Berber scholar and Muslim theologian, was somewhat shocked by the peoples of the Islamic empire of Mali; he wrote:

> *"The women there have 'friends' and 'companions' amongst the men outside their own families, and the men in the same way have their 'companions' amongst the women of other families. One day at Walata I went into the judge's house, after asking his permission, and found with him a young woman of remarkable beauty. I was shocked and turned to go out, but...the judge said to me: 'Why are you going out? She is my companion'. I was amazed... for he was a theologian and a pilgrim to boot."* [220]

216. Davies, 1981, ibid., pp.163-4.
217. Davidson, 1966, ibid., p.40.
218. Grove, A.T.: *Africa*, Oxford U. Press, 1978, p.88.
219. Nicholson, 1975, ibid., pp.96-7.
220. Davidson, 1966, ibid., pp.82-3.
221. Davidson, 1966, ibid., pp.61,82-3.
222. Davidson, 1966, ibid., pp.17,24.
223. see appropriate sections in Chapter 5.

African Islamic Kings were purportedly "known" for their tolerance, justice, and lack of absolutism, at least towards other males, and their territories were touted as being safe for travelers.[221] It was during these times that Europeans first learned of the great universities, books, and scholars of Timbuktu.[222] Nevertheless, Islam for females in sub-Saharan Africa usually meant virtual slavery: forced marriages, polygamy, concubinage, loss of economic, property and inheritance rights, general subordination to males, plus painful and deforming genital mutilations, such as clitoridectomy and infibulation.[223] Arab culture in general had (and still has) a predatory relationship towards darker complexioned Africans of either sex, often enslaving them to the lighter-colored warrior caste. The Tuaregs and Berbers violently resisted the Arabs; they adopted Islam only nominally, if at all, with their women rarely adopting the veil.[224]

The Berbers were pushed south during the more vigorous Moslem invasions of the 1100s, possibly stimulating the subsequent formation of the Songhai empire in the Niger-Chad watershed. Moslem armies established strongholds in the Maghreb, and in Spain. Spain and Portugal later dislodged the Moors, and themselves took control of some coastal regions in Northwest Africa. By that time, however, Saharasian customs regarding females and children were deeply imbedded into the Spanish character, and championed as their own; in the next sections, the rooting of European Christian doctrines in similar Near Eastern customs will be discussed. A special aspect of the Spanish cultural transition is that, after the Moors were driven out of Spain, the horrors of the Inquisition and Code of Torquemada were instituted to deal with various non-Christian peoples. The New World was also discovered around the same time, giving the warrior caste of Spanish society a new land to colonize; as if to emphasize the psychological damage done to Spanish culture by hundreds of years of warfare against the Saharasian Moors, *Matamoro*, meaning "kill the Moors" (or alternatively to "kill the unbaptized" non-Christians), thereafter appeared as a common place name in Spanish settled America. In North Africa, the Portuguese were eventually dislodged by the Moors, but the Spanish remained until the coming of the Ottoman Turks some two hundred years later.[225]

The Sahel region was repeatedly in conflict with regions to the north, across the Sahara, a conditions which has lasted into modern times. In 1586, a Moorish army invaded south from Morocco to grab and control the lucrative trade routes and gold mines of the Songhai. The Moors routed the Songhai and established a center at Timbuktu. Their military action further stirred up trouble and turmoil among Berber, Tuareg, and other groups south of the Sahara. Timbuktu later broke away from the Shereef of Morocco, but declined in influence as desiccation progressively diminished pasture and water resources in the region.[226]

The Sahara rapidly depopulated with increased desiccation. Peoples either migrated to the moister regions of the south, or wherever secure waters remained available: along the Nile, Niger, Senegal, and Gambia rivers, around Lake Chad, on the coastal side of the Atlas mountains, and along the narrow coastal zone adjacent to the Mediterranean. The regions south of the Sahara were consequently subject to increasing influence from the north, by both land and sea, from both Arabs and Europeans, and also from peoples to the East.

For instance, the Zimbabwe stone ruins of Central Southern Africa were built sometime between 1100-1400 CE, probably by Bantu groups in response to trade between the interior and the coastal port city of Sofala, to the east. Chinese porcelain and Indian beads have been found there, along with other foreign goods suggesting widespread overland and linked oceanic trade.[227]

Other peoples, starting with the Greeks, Romans, and Arabs, sailed through the Red Sea to ports as far south as Mozambique, where trade with India is evident. Chinese porcelain dating from 1100 CE has also been found along the African coast. Cushitic settlers also constructed agricultural terraces along the coast between Ethiopia and Mozambique. Peoples from Southeast Asia, probably Indonesia, also settled in Madagascar and coastal Africa. The yams, bananas and rice which they brought eventually reached Central Africa.[228] As was the case with Rome and Byzantium, it will later be demonstrated that the cultural influences

224. Turnbull, 1976, ibid., p.170.
225. Grove, 1978, ibid., p.88; Scott, 1939, ibid., p.69; Davies, 1981, ibid., p.245.
226. Grove, 1978, ibid., pp.87-8.
227. Grove, 1978, ibid., pp.86-7.
228. Grove, 1978, ibid., pp.86-7.

transmitted from the Indian Ocean and Indonesia into Africa were themselves a product of antifemale and antichild influences, which originated in Saharasia.

The African kingly states of the sub-Saharan region formed partly in response to overland or sea trade with, or invasion from North Africa, the Mediterranean, or Arabia. Not only did dominant males come to power through the stimulus of invasions and warfare, but the philosophies of the foreign groups dictated, more or less, the subordinate nature of women and children. In more recent times, the Arab and European slave trade exacerbated these tendencies by stimulating raiding between bordering states or tribal/ethnic groups.

The Arabs exploited Africa as a source of slaves for the harems of the Near East, eventually establishing large slave trading centers along the Red Sea coast, and later on the island of Zanzibar. Sub-Saharan Africa, particularly the Sudan, became a prime capture zone for sex slaves, who would be sold as either concubines or eunuchs. The Arab, Turkish, and later European slave trade contributed to and exacerbated the military turmoil of Africa, placing various ethnic groups at each others throats as never before.

Regarding the Arab-Turk slave trade, the peoples of the Nubian regions of Southern Sudan to this day:

Arab slavers (above) forcing a group of black Africans to the coast for transport to Near Eastern destinations. Kidnapping for slavery began in Africa several thousand years ago, as part of the general epoch of warfare, conquest and empire-building across North Africa and the Near East. The capturing of Africans for ship-transported slavery to the New World (below) was equally brutal, but by contrast lasted only a few hundred years. In either case, the socially-disruptive effects of slavery were tremendous, lasting for generations.

"...[have] a picture of the Northern Sudanese as people who abducted their women and property, who regarded them as naturally inferior, and referred to them as slaves." [229]

This antipathy between dark-skinned animistic or Christian peoples of sub-Saharan Africa with light-skinned Moslem, Arabic peoples of North Africa underlies much of the social turmoil and warfare which currently exists across Nigeria, Cameroon, Chad, Sudan, Ethiopia and Somalia.

Regarding the later European slave trade in West Africa, Grove comments that:

"Nearly everywhere, it seems, social life was disrupted by the [slave] trade. Some coastal peoples, armed with European firearms, raided for slaves far into the interior. States like Dahomey and Ashanti grew strong and wealthy; but for most of the west-coast people, the years between 1550 and 1850 must have been years of turmoil, with armed bands and mercenary troops a constant threat to settled life." [230]

229. Godfrey Morrison, as quoted in Downing, D.: *An Atlas of Territorial and Border Disputes*, New English Library, London, 1980, p.66.

230. Grove, 1978, ibid., p.91.

As demonstrated in the maps in Chapter 5, male circumcision appears in the Sub-Saharan region, but with a declining incidence the farther south one moves from North Africa. This suggests a southward spread and dilution of the practices with pre-existing, non-mutilating cultures. Similarly, female genital mutilations center on the Red Sea coast, in Sudan, and have spread west across the subtropical bush and grasslands. Infant cranial deformation, which appeared in Egypt, also spread south up the Nile and across the subtropical bush and grasslands.

The metallurgical connections of the Bantu with regions to the north, as well as their cultural similarities regarding women and children, suggests that the Bantu played a key role in the spread of circumcision, cranial deformation, and other armored, patriarchal Saharasian institutions and beliefs into Sub-Saharan Africa. The Pygmy, Bushman, and Hottentot peoples were often subordinated by the Bantu, but themselves did not adopt such practices, or did so only nominally, as a compromise, survival measure. They occupied areas deemed less desirable by the larger Bantu peoples, such as the desert or rainforest, and themselves remained mostly peaceful, monogamous, and female oriented, with a high status for women and children. Their behaviors and beliefs today more closely reflect the presumed original cultural makeup of Africa, before events following c.3500 BCE which dried up the Sahara and stimulated invasions from the Near East.[231]

Later influences from colonial Europe were mixed in character, being exploitative and authoritarian on the one hand, but opposing tribalism and the worst excesses of a few bloody African tyrants on the other. While racial superiority was a fairly dominant theme among all the European colonial powers, some European nations in general attempted to educate and train native Africans in colonial administration and technical skills (waterworks, railroads, schools, clinics, telegraph-telephone, etc.) while others did nothing in this direction — these differences had profound influences which can be seen today.

For example, one may observe a sharp difference between former Portuguese colonies and former British colonies. Portugal did not attempt to train, educate or assist native Africans in any significant manner; they arrived as conquerors and demanded subservience. In the Portuguese colony of Angola, after centuries of exploitative colonialism, slaving and war with discontented Africans, and facing considerable civil discontent with the Angolan war at home, the Portuguese colonials left, taking with them everything which was not nailed down — telephones, generators, waterworks, etc. — leaving behind gutted cities and a totally chaotic situation into which various African warlords stepped in to fill a power vacuum. The chaos continues in Angola until today. The British, by contrast, ended slavery much earlier than other European nations, and did in general bring native Africans into their infrastructure, establishing clinics and schools; Kenya and Tanzania by comparison experienced a *relatively* smooth transition from British colonialism to native African rule, and social conditions there remained stable and thriving.

These few sentences only serve to briefly contrast the affects of two different European colonial powers, which constituted but another layer of culture-contact and influence on top of the larger historical-migratory considerations previously discussed. African patrism is much older than European colonialism, and persists on it own among the new nation-states of Africa, being generally more intensive in the north and north-east than in the west and south-west.

231. Hallet, J.P. & Relle, A.: *Pygmy Kitabu*, Random House, 1973, pp.5, 24-7; cf. Turnbull, C.: *The Forest People*, Touchstone, 1961.

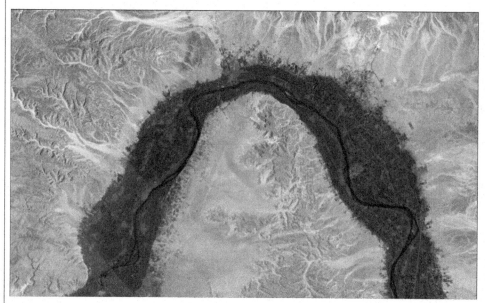

An aerial view of the Great Bend in the Nile River (Egypt) as seen today where it flows through the hyper-arid Sahara Desert. Where there is water, there is life. Without water, only barrenness.

Watercourses within Saharasia, such as the Nile River, increasingly became focal points for dominance and conflict, and a source of wealth and power for patristic divine-king Empires.

B. ARABIA

The Arabian Peninsula lies in the same climatic belt as the Sahara Desert in North Africa; it possesses a similar hyperarid climate. The Peninsula slopes from highlands along the Red Sea coast on the west, to lowlands along the Persian Gulf to the east. Only the southwestern highlands and the adjacent coastal strips obtain any significant moisture. However, these same highlands act as an additional barrier to moisture moving toward the interior.

At the south end of the Arabian Peninsula is the hyper-arid *Rub'al-Khali*, or "empty quarter", which is the hard core of the Arabian deserts. It is composed of over 100,000 mi^2 (259,000km^2) of mobile sand dunes with regions of deflated pebble, and until recently was one of the least explored regions on Earth.[1] Sand dunes stretch north from the Rub'al-Khali through the interior of Arabia, reaching to the Syrian Desert which dominates the north end of the Peninsula. The entire Arabian Peninsula is devoid of a single flowing river or significant body of water.

A variety of evidences point toward moister conditions in Arabia in Neolithic times. Fossil springs, streams, lakes, and wadis exist across the Peninsula dating from c.8000-4000 BCE, suggesting ecological changes which parallel those of North Africa.[2] Masry states that:

> "...the wadis...were once probably very active streams. In northeast Arabia, the al-Batin and Sahba systems dried up after their interior charges from basin catchments were affected by drought. Another geomorphic feature is the gravel plain, several of which are located in the area near wadi beds. These plains are presumed to be associated with early Quaternary water runoff from the former active streams. Sabkhas or salt-flats constitute yet another feature indicating climate changes...where water accumulated from abundant rain, forming lake-like basins which fed several streams. These basins were transformed into the modern sabkhas through evaporation caused by extreme aridity... The eolian sand deposits, most salient of all the topographic features, may be the product of very recent transformations, since they appear to overlie the water-transported gravel beds in many places. Finally, there are remains of many fresh water lakes along the periphery of the Empty Quarter. A carbon 14 date from one gave a determination of 4,700 BC."[3]

Even the Rub'al-Khali has revealed evidence of water and water fauna, to include gravel beds from flowing water and freshwater shells. Early European travelers into the region reported "immense deposits" of freshwater shells, flint implements, and abandoned wells of more recent times.[4] More recent explorations of these same ultra-dry regions:

> "...have also yielded (in profusion on their surface) some stone artifacts of a 'Neolithic' facies. The complete aridity of the area at the present (and probably also in historic times) is incompatible with the existence of a culture using these implements."[5]

By 7000 BCE, agriculture spread into Arabia from the north during a moist period when conditions were favorable for a broader subsistence pattern. This led to widespread Neolithic settlements and subsequent interaction with early farming communities to the north, in Mesopotamia. By 6000 BCE domestication of cattle, sheep, and goat was underway, practices which also spread into Arabia from the north. Spring formations

1. Wadia, D.N.: *Post Glacial Desiccation of Central Asia*, Nat. Inst. of Sciences of India, Delhi, 1960, p.5.

2. Masry, A.H.: *Prehistory in Northeastern Arabia: The Problem of Interregional Interaction*, Unpublished dissertation, University of Chicago, Anthropology, 1973, pp.72-4; Hawkes, J.: *Atlas of Ancient Archaeology*, McGraw Hill, 1974, p.167.

3. Masry, 1973, ibid., pp.73-4, cf. p.255.

4. Philby, H.J.: "Rub' Al Khali: An Account of Exploration..", *Geographical Journal*, 81(1):11,17 Jan. 1933.

5. Huzayyin, S.: *The Place of Egypt in Prehistory, A Correlated Study of Climates and Cultures in the Old World*, Memoires, L'Institute D'Egypte, Cairo, 1941, p.105.

related to gullys and wadis have also been found dating to this period of increased population.[6]

After 5000 BCE, the direction of interaction between Mesopotamia and Arabia was no longer primarily from north to south. Trade relations and migrations from Arabia to Southern Mesopotamia occurred. With passing time, peoples of interior Arabia developed migratory adaptations and an emphasis upon nomadic herding. However, Northeast Arabia from the Jabrin region to Qatar remained extensively settled.[7]

Hamblin reviewed the archaeological materials in Saudi Arabia, and made the following conclusions:

> "...the Arabia Deserta of legend is not, and never has been, either all desert or all deserted. Lost in the wastes are cities with defense walls and towers, elaborate palaces with murals and shrines to forgotten gods. There are still-sturdy dams across now-dry streams, and networks of stone-lined cisterns and canals which once served fields of grain, herds of sheep, and the needs of the slowly plodding caravans. There are literally tons of Stone Age tools, and thousands of incised or pecked-out pictures of hunters and their game on the sandstone scarps where once were Pleistocene lakes."[8]

Additional evidence for a moist period in Arabia comes from deep-sea cores made in both the Persian Gulf, and the Gulf of Aden, which reflects the character of the Red Sea region. These cores indicate a moist period centered on 4000 BCE, a much later date for the Arabian humid phase than that determined by others.[9]

Changes After 4000 BCE:

In Arabia, c.4000 BCE marks the start of a progressive desiccation, which intensified and accelerated under the pressures of human actions, such as overgrazing, wood grubbing, and burning.[10] The Rub'al-Khali may have dried out first, followed by surrounding regions.[11] The onset of this arid phase also led to a progressive abandonment of settlements, and a general dispersal of Semitic peoples out of Arabia, in all directions. The Arabian settlements of this period were:

> "...successively displaced from southern interior latitudes to northern coastal ones with a concomitant progressive reliance on marine resource adaptation vs. generalized herding and broad-spectrum hunting and gathering. This was probably related to shifting ecologic and climatic conditions."[12]

A migratory, nomadic adaptation was increasingly undertaken. Various Early Bronze Age settlements to the north of Arabia, primarily Mesopotamia, began to be harassed by newcomers from Arabia, and also by immigrants from the Zagros Mountain region to the north.[13] As previously discussed, this period also saw the arrival of a Semitic culture in the Upper Nile Valley.

By 3500 BCE, Semitic elements appeared in Sumerian cuneiform texts,[14] and the language of Egypt developed Semitic aspects.[15] Evidence of the behaviors and beliefs of Semitic peoples in Arabia for this period is generally lacking, however it is inferred that these people were carriers of the beliefs and social institutions which were adopted in Egypt following its invasion from the east during this same period. Changes in Mesopota-

6. Masry, 1973, ibid., pp.72-4, 103, 175, 223, 255.

7. Masry, 1973, ibid., pp.97-8, 219-20.

8. Hamblin, D.J.: "Treasures of the Sands", *Smithsonian*, Sept., 1983, pp.43-53; quote p.45; cf. Masry, 1973, ibid., for survey of region.

9. Butzer, K.W.: "Patterns of Environmental Change in the Near East During Late Pleistocene and Early Holocene Times", in *Problems in Prehistory: North Africa and the Levant*, F. Wendorf & A.E. Marks, eds., S. Methodist U. Press, Dallas, 1975, p.398.

10. McClure, H.: "Ar Rub'al-Khali", in *Quaternary Period in Saudi Arabia, Vol. 1*, S.Al-Sayari & J.Zotl, eds., Springer Verlag, NY, 1978, p.257; Hotzl, H. & Zotl, J.: "Climatic Changes During the Quaternary Period", in Al-Sayari & Zotl, 1978, ibid., pp.301-3; Anton, D.: "Aspects of Geomorphological Evolution; Paleosols and Dunes in Saudi Arabia", in *Quaternary Period in Saudi Arabia, Vol. 2*, A.Jado & J.Zotl, eds., Springer Verlag, NY, 1984, pp.288-96.

11. Hotzl & Zotl, 1978, ibid., p.303.

12. Masry, 1973, ibid., p.220.

13. Hawkes, 1974, ibid., pp.11,167; Masry, 1973, ibid., pp.74, 225-6.

14. Hawkes, 1974, ibid., p.168.

15. Hoffman, M.: *Egypt Before the Pharaohs*, Knopf, NY, 1979, pp.290-1.

Ruins of the Marib Dam (top) and nearby Sabaean "Pillars of the Moon" (bottom) of Southern Arabia. These both date to a moist period between 750 BCE to 500 CE, located in what is today an extremely dry, desolate area.

16. Masry, 1973, ibid., pp.97,172-4; Hawkes, 1974, ibid., p.187.

17. Grun, B.: *Timetables of History*, Touchstone, NY, 1982; .
Mikesell, M.: "The Deforestation of Mt. Lebanon", *Geographical Review*, 59(1):11, 1969; Hawkes, 1974, ibid., p.167.

18. Field, H.: *Contributions to the Anthropology of Saudi Arabia*, Field Research Projects, Coconut Grove, 1971, pp.30-3.

19. Anati, E.: *Rock Art in Central Arabia, Vol 2*, Institut Orientaliste, Louvain, 1968, p.47.

20. Anati, 1968, ibid., p.66.

21. Gupta, S.P.: *Archaeology of Soviet Central Asia and the Surrounding Indian Borderlands*, Vol. 2, B.R. Publishing, Delhi, 1979, pp.257-8; Masry, 1973, ibid., p.221.

22. Hawkes, 1974, ibid., pp.11,168; Grun, 1982, ibid.; Masry, 1973, ibid., p.222.

23. Evenari, M., et al: *The Negev*, Harvard U. Press, Cambridge, 1982, pp.350-1.

24. Butzer, K.W.: "Climatic Change in Arid Regions Since the Pliocene", in *A History of Land Use in Arid Regions,* L.D. Stamp, ed., UNESCO, 1961, p.45.

mia of a similar character will be discussed shortly. It appears that Arabia was a region from which successive displacement of peoples of a radically different cultural mode occurred: that of increasingly militarized nomadic herding, with a greatly subordinated status for females and children.

Prior to 3000 BCE, the Persian Gulf coast was a broadly settled region, whose peoples engaged in hunting, herding, marine resources, and trade.[16] Peoples displaced out of central Arabia were settling along the coasts, and migrating towards both the Levant and Syria, to form what would later be known as the Kingdom of Aram.[17] Egypt and Mesopotamia were also influenced by these migrations.

As was the case in North Africa, Arabia also contains rock art from its ancient peoples. These rock art have not been quantitatively dated, nor do they compare in a qualitative sense to the fabulous frescoes and carvings of North Africa. However, they do contain images of antelope, gazelle, oryx , ibex, ostrich, cattle, horse, and camel. The Arabian rock art appears to follow the same changes in motif found in Saharan art. The oldest art depicts animals, hunting scenes and people, while newer art includes depictions of mounted men on horse and camel.[18] Anati identified a style of rock art dating from c.3000-1000 BCE which was dominated by adult men carrying more than one weapon, and was distributed throughout Arabia, Egypt, Palestine and Jordan.[19] Only a small percentage of the pictures of this particular style included women and children. However, Anati identified other groupings of Arabian rock art which contained a large percentage of women.[20]

Increasing numbers of Semitic peoples, using domesticated camels and horses were migrating from Arabia to the north and west after c.2500 BCE. Around the same time Indo-Aryan peoples using the chariot and iron weapons were abandoning the Iranian Plateau and Central Asia, migrating to the south. Numerous settlements in East Arabia were abandoned at this time, which also saw a major shift in trade routes from the land to the sea, and a cessation in cultural interactions.[21]

Similar changes were occurring at the time across the entire region stretching from the Nile to the Indus Valley. The Semitic Akkadian Empire of Sargon was founded in Agade around 2400 BCE.[22] Egypt, as previously discussed, was invaded by Asiatic-Semitic peoples around 2300 BCE. The Egyptians initially retained their territory against the militant nomads who were plundering the Near East. However, they too were finally overwhelmed by the Asiatic/Semitic Hyksos after c.1800 BCE.

The southern region of the Arabian Peninsula is presently slightly moister than the interior. Here, in present day North Yemen, the Bronze Age Kingdoms of Qataban (c.2000 BCE) and Saba (1500 BCE to CE 600) developed, based upon control of trade routes and extensive runoff agriculture. Considered to be master builders and hydrologists, the Sabaeans constructed a large dam in 750 BCE at Marib, their capital city.

The Marib dam was part of a complex water diversion system, with a catchment area of 10,000 km^2. The 400 million m^3 capacity dam diverted water into giant sluice gates which irrigated 20,000 hectares; the system had the potential of supporting an estimated 300,000 people. The dam and waterworks lasted until c.575 CE when a large flood destroyed it, greatly diminishing the agricultural output of Sabaea. This, in turn, prompted another dispersal of Arabian peoples. About 1700 hectares are still irrigated at the same site today, and plans have been discussed to restore the dam.[23] Butzer identified declines in the South Arabian water table of 3 to 4 meters since Sabaean times, with declines in other Arabian wells of 1 meter or more.[24]

After c.1000 BCE, Bronze and Iron Age semi-nomadic peoples engaged in hunting and gathering were still widely scattered across Arabia.[25] Runoff agriculture was also practiced, with settlements located near the mouths of wadis. Conditions were definitely drier than in the past, but runoff agriculture technology allowed mixed agricultural/pastoral/hunting settlements here and there. Where rains were sufficient, settlements were generally widely dispersed and not solely at the mouths of Wadis. However, the great energy expended in such water catchment and control systems, is another indication of the gradual drying up of the region.

Architectural designs in Bronze Age Arabia give evidence of an historical decline in moisture. Protective walls and fortresses for securing vital but dwindling water supplies were increasingly erected, particularly along the lucrative trade routes. The fortifications at Taima Oasis, with walls several meters thick and many kilometers long, are an example here.[26] After 1000 BCE, Northern Arabia sequentially came under the control of various groups, to include the Egyptians, Hebrews, Babylonians, Assyrians, and Persians. Central and Southern Arabia remained autonomous, however, under the control of nomadic tribes or fortified city-states based upon runoff agriculture, pastoralism, hunting, and trade.

Some portions of Arabia which were sustained by trade and agriculture appear to have maintained a relatively high status for women, in spite of the cultural transitions occurring among the nomads. The Biblical Queen of Sheba ruled a southern Arab state with connections to Ethiopia and identified with ancient Sabaea. Trade with many nations occurred, including with Solomon's Israel. Sheba's "kingdom" allowed more freedom and a higher status for females than Solomon's once-nomadic Hebrew state — the Patriarchs and Rabbis of Jerusalem blamed the Queen of Sheba for Solomon's "wicked ways".[27] It should be noted that numerous Arabian cities along the Red Sea coast today sit upon ancient *tells* (settlement mounds) which have never been excavated. This is true for much of Arabia, whose prehistory remains largely unknown. Arabian archaeology is in such a young situation, with so many uncertainties remaining, that it was possible for the scholar Salibi, to recently propose — on the basis of ancient place names — that the cities along the Arabian Red Sea coast actually constituted the ancient Kingdom of Israel, to include ancient Jerusalem.[28] More will be said on the ancient Hebrews in the next section.

Regarding more recent climate change, Huzayyin felt that physiographic evidence indicated a wet phase in Southern Arabia which persisted into the 400s CE.[29] McClure also viewed the physiography as evidence of moisture lasting until early historic times.[30] This view is supported not only by the presence of ruins of Bronze Age cities in Arabia, but also by historical records of Arab scholars, such as Yakut, who wrote of a land of milk and honey in southern Arabia called Ubar:

> "...it was a land richly afforested, with copious water, with perennial streams, and thickly inhabited".[31]

Philby narrated the end of the city of Wabar (Yakut's Ubar). He found the site of Wabar to be devoid of structures, but contained numerous impact craters, fused slag heaps, iron-nickel meteorites, fulgurites (lightning tubes), and jet-black tektites:

> "the city of a wicked king [was] destroyed by fire from heaven and thenceforward inhabited only by semi-human monomembrous monsters...[and containing] the pearls of the numerous ladies who...had graced the court of King 'Ad and perished in the flames."[32]

25. Masry, 1973, ibid., pp.104-7.

26. Masry, A., ed.: *An Introduction to Saudi Arabian Antiquities*, Dept. Antiquities & Museums, Ministry of Education, Saudi Arabia, 1975; Hamblin, 1983, ibid., p.48.

27. Stone, M.: *When God Was A Woman*, Dial Press, NY, 1976, pp.55,173 .

28. Salibi, K.: "The Bible Came from Arabia", unpublished manuscript, 1984, Archaeology Dept., American University of Beruit; cf. "Jerusalem in Saudi Arabia", *Newsweek*, 10 Sept. 1984, p.74.

29. Huzayyin, 1941, ibid., pp.125-6.

30. McClure, H.A.: *The Arabian Peninsula and Prehistoric Populations*, Field Research Projects, Miami, 1971; cf Masry, 1973, ibid., p.250.

31. as paraphrased by Bertram Thomas in Philby, 1933, ibid., p.23 .

32. Philby, 1933, ibid., pp.12-3, 25.

A 4800-pound iron meteorite has been found in the Empty Quarter near Al Hadi-dah, and Wabar is now recognized as an authentic meteorite crater.[33] The myths suggest an impact during times when Southern Arabia was moist, following which desiccation occurred. However, it is not known what geology has to say about the age of the Wabar craters. One researcher hopes to use the Space Shuttle's imaging radar to look for Ubar beneath the sands of the Rub'al-Khali, at a location near other ruins and roads mentioned in the legends.[34]

The recent unearthing of Qaryat al-Fau, a pre-Islamic Bronze Age city and trading center on the northwest edge of the Rub'al-Khali, further supports the idea of moister conditions into historic times. It also suggests the presence of a culture at least somewhat different from the nomads who later developed the legends of "wicked" King 'Ad and his ladies. The finds at Quaryat al-Fau date from 200 BCE to 500 CE. As the capital of Kinda, a trading state, the city was invaded periodically from 100 CE until its final abandonment around 500 CE. The site is presently a grassless, rocky wasteland dotted here and there with small shrubs. However, date trees, vines, and grains were grown and a large number of cattle, goats, sheep, and camel were raised at the time of its occupation. Abundant remains of wildlife, to include gazelle, deer, and ibex, indicate that rainfall was much greater than at present. Even so, rainwaters must have been less than desired for this urban center as its inhabitants developed extensive subterranean canals to tap groundwater and surface canals to tap now dry streams from nearby mountains.[35] Also significant are the finds of statues of nude children, the Goddess Minerva, and unveiled women, indicating connections with Mediterranean cultures relatively unaffected by nomadic influences.[36] Nomadic tribes increasingly opposed these settled traders, however, and defensive fortifications at Quaryat al-Fau increased in size toward the end of its habitation.[37]

Huzayyin also discussed relatively recent changes at the extreme north and south ends of the Arabian Peninsula:

> "In the desert east of the Jordan there is some evidence that the underground water level, which is ultimately affected by the supply of rainfall, has fallen by 2 meters since Roman times. Other and perhaps more conclusive evidence comes from such remote parts of Arabia as the highlands of Yemen. There we know that the earlier historic civilizations were established on the lower step of the table land, that of the Yemen Jauf, lying at about 1000 meters elevation. Gradually, the ancient Mineans and Sabaeans shifted ther capitals to the higher slopes of Yemen until, finally, the capital was transferred to San'a at some 2,200 meters. Also, if we trace the ancient cisterns of the Sabaeans and Himyarites in Yemen, we find that invariably they were situated on the crests of hills rather than in collecting basins, where we find the present-day cisterns. At present, the ancient cisterns are never more than partly filled with water, while the low-lying new ones are usually filled to the rim. It is evident, however, that precipitation and runoff on the crests of hills must have at one time been enough to fill these Sabaean and Himyarite cisterns (c.1000 BCE-AD 500) with water." [38]

The various ruins in Arabia dating from c.1500 BCE into the Christian era thrived given a greater moisture supply than at present, but also obtained at least as much from their hydraulic technology as from the moister subhumid episode. However, these Bronze Age settlements were also eventually abandoned, due to increasing desertification. Between CE 100-600, settled populations increasingly turned to nomadic ways in a

33. Snead, R.: *Atlas of World Physical Features*, J.Wiley, 1972, p.142; Abercrombie, T.J.: "Saudi Arabia: Beyond the Sands of Mecca", *National Geographic*, January 1966, pp.33-5.

34. Eberhart, J.: "Radar From Space", *Science News*, 9/22/84, p.187.

35. al-Ansary, A.R.: *Qaryat al-Fau, A Portrait of Pre-Islamic Civilization in Saudi Arabia*, U. Riyadh, St. Martin's, 1982, pp.15-6,57.

36. al-Ansary, 1982, ibid., pp.24-5,95,100-1,113,121.

37. al-Ansary, 1982, ibid., p.16.

38. Huzayyin, S.: "Changes in Climate, Vegetation, and Human Adjustment in the Saharo-Arabian Belt", in *Man's Role in Changing the Face of the Earth*, Vol 1, W.L. Thomas, ed., U. Chicago Press, 1972, p.319.

process which has been called the 'Bedouinization' of Arabia.[39]

The pre-Islamic city of Mecca was established in c.400 CE, as a sacred enclave, or *haram*, to which pilgrimages would be made. Trade would occur at the haram only at certain times of the year, when a total suspension of inter-tribal raiding and bloodshed was mandated. Other harams were established at nodes in trade routes.[40]

The irruption of nomadic Arabs from Arabia after CE 630 closely followed the abandonment of Qaryat al-Fau (CE 500) and Sabaea (CE 500-600), which was related to both a general failure of the water supply systems and increasing pressure from nomadic groups.[41] Huntington also argued the abandonment of Arabian cities and out-migrations as being due to further desiccation between AD 600-700.[42]

It was upon these events that the prophet Muhammed gained his following, through conquest and massacre of cultures unwilling to adopt his new Islam, and via transfer of loyalties among militant nomads from a tribal focus toward a regional religious statehood. In short turn, Islamic Arab armies irrupted and eventually conquered all of North Africa, most of Spain, and much of the Near East and Central Asia (see Figure 20). These irruptions, dating from c.640-800 CE and c.1000-1100 CE, carried with them not only the Islamic faith, but also a family structure and set of beliefs about women and children which would reinforce, amplify, and codify into law the ongoing social transformations of the invaded territories.[43] Strong men of an Islamic Arabian background thereafter ruled from agricultural cities on secure water supplies (Nile, Euphrates) of conquered territories. Prostration, originally required only during prayer, was extended to Muslim secular affairs later on.[44] Pre-existing non-Muslim peoples of a lesser-patristic nature were either absorbed into Islam, reduced to slave status, or wiped out.

Muslim apologists often state that before Muhammed and Islam, women had virtually no legal status, being bought and sold, and possessed no property or inheritance rights.[45] However, there is no evidence to support this claim, and pre-Islamic archaeology suggestive of a higher status for women goes against it entirely. *Possibly*, this claim held some truth for women of the migratory warrior nomad cultures of Arabia, who were elevated a tiny bit by Muhammed's "reforms". Not so, however, for women of the independent agricultural communities or trading states conquered by the Arab Muslim armies. For those cultures where sexuality was more open and women held a higher status and more active role in society, the arrival of Islam and Quranic law was an unmitigated social disaster, including for those families forcibly converted to Islam.[46]

For instance, Muhammed's laws forbade female infanticide, but terminated worship of the Sun Goddess Allat; later, his law mandated the veiling and seclusion of women. Slaves were given specific legal status and rights, but slavery itself was not abolished and, indeed, was spread to many new territories by Muslim warriors. Women were guaranteed inheritance rights, but only as a fraction of what their male relatives would receive; they also were guaranteed property rights during divorce, but could not initiate divorce themselves except on grounds of the husband's erective impotence. And they lost the important rights to select their own husbands. Men could acquire multiple wives according to what they could afford economically, and divorce for any reason, by a simple public proclamation. Islamic heaven provided no place for women, except as *houris*, the voluptuous "sex angels" devoted to the pleasures of the male. Women were also forbidden to enter the Mosque or to pray. As the booty of war, captive women of conquered lands were absorbed as wedded slaves into the polygamous Arab harems and tents.

Muslim men could marry many times over to both Muslim and non-

Young woman being prepared for Islamic ritual murder by stoning, presumably for "family honor" and in accordance with the demands of Muslim *Sharia Law*. She is firstly tied up, then buried to the waist, after which her relatives and neighbors will gather in a cursing mob and throw fist-sized stones at her until she is dead. The offenses which demand this public sadism are typically sexual in nature, including the "crime" of being raped, or falling in love with an "unapproved" male. Male violators are also sometimes stoned, but far less frequently. Lesser punishments, such as cutting off the nose, throwing acid in the face, severe floggings or lengthy prison terms, are also frequently carried out in virtually every Islamic nation. *I speak here about current 21st Century conditions, and regrettably not of some ancient Dark Age!*

39. Eickelman, D.F.: *The Middle East: An Anthropological Approach*, Prentice Hall, 1981, p.254.

40. Moscati, S.: *The Semites in Ancient History*, U. Wales, Cardiff, 1959, p.34.

41. Heichelheim, F.M.: "Effects of Classical Antiquity on the Land", in Thomas, 1972, ibid., p.179.

42. Huntington, E.: *Palestine and Its Transformation*, Houghton Mifflin, Cambridge, 1911, p.327.

43. Beck, L. & Keddie, N., eds.: *Women in the Muslim World*, Harvard U. Press, Cambridge, 1978.

44. Wittfogel, K.: *Oriental Despotism*, 1957, p.154.

45. Coulson, N. & Hinchcliffe, D.: "Women and Law Reform in Contemporary Islam", in Beck & Keddie, 1978, ibid., p.37.

46. Beck & Keddie, 1978, ibid., p.26.

Islam means *Submission*, and Islamic *Sharia Law*, a central-core body of rules of behavior for obedient Muslims, regulates all aspects of life, from cradle to grave. Sharia Law demands multiple daily prayers, ritual absolutions against defecation, urination and other body functions, and a subordination of young people's lives, especially of women and sexual life, fully into the arranged polygamous rape-marriage system. It also demands sadistic punishments for offenses, such as public whippings, amputations, stonings and beheadings, as well as killing of apostates (those who leave Islam or convert to other religions) and subordination into dhimmitude slavery or killing of "infidels". Sharia's root meaning is *the way to the water*, suggestive of a set of obedience demands created by conquering warrior-cult nomads who control a given water supply in the middle of a harsh desert.[55]

Muslim women, but Muslim women could marry only one man, and could not marry outside the faith. Adulterous women, or sexually active unmarried girls could be killed on the spot by husband, father, or brother, without penalty to the murderer, and often with the full and open encouragements of the hapless girl's maternal kin as well. Indeed, preservation of "family honor" often demanded the death of such a female, at the hand of her father or brother.[47] A man who would *not* carry out such a murder of their daughter or sister or mother would be ridiculed by other men, and by older village women as well, as someone who could not control his own family. Younger men who shied away from their "obligations" to murder a "sinful" sister or mother would be unlikely to marry themselves, until their offending female relative was murdered. Through such mechanisms, of severe sexual segregation of the young, the seclusion of women, genital mutilations, marrying girls at a very young age, and murdering offending girls, a village would insure the absolute suppression of sexual feelings in the young. The adulterous man, or sexually active unmarried male was not punished in such a severe manner — but they surely would have a most difficult time finding a female partner willing to take such a risk.[48] Muslim law accorded the word of a man greater status than the word of a woman, or of several women.[49] Muhammed dictated that no man should take more than four wives, but himself married 14 women, leaving behind 9 widows. His second "wife", when Muhammed was over 50 years old, was a 9 year old girl, Ayesha, who moved into his tent with her toys.[50]

After Muhammed's death, armies of his followers swept out of Arabia, across North Africa, and into the Levant and Mesopotamia. The political center of Islam later shifted to Baghdad, on the secure waters of the Tigris-Euphrates. Arabia itself suffered from profound desiccation, and developed along distinctly puritan lines. For instance, Rhazes (c.900 CE), an Islamic scholar steeped in Persian philosophy, condemned passionate love and sexual activity as "foul".[51] Ghazaili's book, *Council for Kings* (c.1100 CE) listed the various pains and torments women were forced to suffer given Eve's transgression in the Garden of Eden:[52]

"1. Menstruation.
2. Childbirth.
3. Separation from mother and father, and marriage to a stranger.
4. Pregnancy.
5. Not having control over her own person.
6. A lesser share of inheritance.
7. Ability to be divorced and inability to divorce.
8. Lawfulness of man to have 4 wives but the wife only 1 husband.
9. Seclusion inside the house.
10. Head must be covered inside the house.
11. Testimony of two women to equal that of one man.
12. Must be accompanied by a near relative while outside the home.
13. Men pray on Friday and feast and pray at funerals while women don't.
14. Women are disqualified for rulership or judgeship.
15. Merit has 1000 components, only 1 of which is attributable to women, while 999 are attributable to men.
16. If women are profligate, they will get 1/2 as much torment on resurrection day.
17. Widows must wait several months to remarry, while widowers do not.
18. Female divorcees must wait several months to remarry, while male divorcees do not." [52]

47. "Honor killing" of sexually active girls by their male relatives, with full approval and encouragement or even *insistence* by female relatives, is still customary in many Near Eastern and Central Asian cultures; cf. Radbill, S.: "A History of Child Abuse and Infanticide", in *The Battered Child*, R.Helfer & H.Kempe, eds., U.Chicago Press, 1974, p.7; also see the account by Accad in Chapter 3 of this work.

48. As a consequence, homosexuality and pedophilia flourish within the Islamic world, at levels probably much higher than anything seen in the West, even in the face of severe taboos against homosexuality. See the section on homosexuality, p.154.

49. Lewinsohn, R.: *A History of Sexual Customs*, Harper Bros., 1958, pp.103-8; Lacey, R.: *The Kingdom*, Harcourt, Brace, & Jovanovich, NY, 1981, pp.116,220-59; Stone, 1976 ibid.; Beck & Keddie, 1978, ibid. .

50. Lewinsohn, 1958, ibid., p.161.

51. Bullough, V.L.: *Sexual Variance in Society and History*, J. Wiley, NY, 1976, p.237.

52. Tannahill, R.: *Sex in History*, Stein & Day, 1982, pp.33-4; These sentiments, and the historical conditions described in the paragraphs above still characterize the situation across most of present-day Islamic Saharasia, where veiling and extreme subordination of women, along with "honor" murders of women by their own husbands, brothers and sons is a continuing blight.

Islamic puritanism peaked in the mid-1700s CE and early 1800s under Wahhab, who led a movement to purge Arabia of its "pagan" remnants. In addition to warring against the Mameluke Egyptians and Turks, his followers destroyed temples and shrines, banned smoking, dancing and music, and reinstituted the stoning of females for adultery.[53] Wahhabist sentiments and creeds today are foundations for national laws within many Islamic nations of Saharasia, to include Saudi Arabia and Taliban Afghanistan, and are likewise foundational for virtually all of the Muslim Jihad terrorist groups around the world.

Throughout the centuries CE Arabia remained relatively isolated from foreign contact or conquest, and the Ottoman Turks captured most of the Arab lands in the Near East and North Africa, and parts of Northern Arabia. However, neither they, nor the Egyptians ever conquered the arid central or southern end of the Arabian peninsula. Northern Arabia was taken over by the Ottoman Turks, who allied themselves with the German Kaiser during the World War I, and consequently were driven out by British forces.

The World Legacy of Islam[54]

While much has been written to claim that Islam is a "religion of peace", authentic history shows it has spread almost uniformly by armed conquest, after which conquered inhabitants were either killed, enslaved as low-caste "dhimmis", or forcibly converted — the three choices given to non-believers as dictated by the Koran.[55] Whatever mildly democratic and egalitarian tendencies which existed in the conquered populations, especially the rights of women and more open attitudes about sexuality, were basically swept off the map as totalitarian Islamic states were established, generally led by debauched tyrants. Chronic jihad "holy war" thereafter ensued against the "inferior" non-Muslim peoples at the borders of the new Islamic states, an expression of the Koran's mandate to spread Islam to all corners of the world, and of the violence provoked within Muslim populations coming into contact with non-believers.[56] Slavery was particularly boosted and spread widely by Islam, with tens of millions of black slaves dragged northward, and millions of additional white slaves dragged southward by Muslim raiders into Slavic ("slave") and Western Europe, from as far away as Iceland.[57] While the giant territory dominated by the defeated Ottoman Caliphate was dismantled into smaller states after World War I, the family life and sexual-emotional configurations of Islamic society were scarcely altered. From a sex-economic perspective, this helps to understand why Islamic peoples and regions have remained in a near-perpetual state of warfare and conflict.

Islamic Saharasia — from Mauritania and Morocco eastward across North Africa, Arabia, the Middle East and across Central Asia as far as Western China — continues to be a violent region, suspicious of foreigners and almost wholly undemocratic, extreme patriarchal and totalitarian. Awash in trillions of petro-dollars, the anti-sex, anti-female ideology of Islamist supremacy[58] is today being spread world-wide though a system of Saudi/Wahhabi-funded mosques, creating social tensions and suicide-homicide bomber, mass-murder terrorism world-wide, directed against the hated non-believers, or Muslims of less-fanatical lifestyles. "Multicultural" openness in the West has allowed a massive immigration of Muslim peoples from Saharasian regions, but Islamic law and custom carried with them has proven nearly immune to significant change.[59] We may expect a continued and amplifying global clash of cultures as Islamic-Saharasian ideology and behavior slowly asserts itself through continued out-migration and demographic expansion.

53. Eickelman, 1981, ibid., pp.29-31; Lacey, ibid., pp.56-7.

54. For more details, see the author's forthcoming work, *Saharasia Since 1900*.

55. Ye'or, B., *The Dhimmi: Jews and Christians Under Islam*, Fairleigh Dickinson Univ. Press, 1985; Ye'or, B., *Islam and Dhimmitude: Where Civilizations Collide*, Fairleigh Dickinson Univ. Press, 2002; Spencer, R.: *The Politically Incorrect Guide to Islam (and the Crusades)*, Regnery Pub., Washington, DC, 2005.

56. North Africa and the Eastern Mediterranean were, before the arrival of Islamic armies, occupied by peoples who today constitute only remnant populations. Muslims of Arab, Turkic and Persian backgrounds today dominate the home regions of Berbers, Copts, Assyrians, Zoroastrians, Armenians, Greeks, Jews and hundreds of Byzantine Christian ethnicities, to name a few. All were crushed down and swept aside by Islam, wiped out in many cases. Today with the sole exception of Israel and Lebanon — which gained independence from prior slave-dhimmi status under Ottoman domination via international agreements (and to which Muslims reacted with open warfare), as respective Jewish and Christian homelands — they live as second-class citizens within their Islamic "host" nations.

57. Milton, G.: *White Gold: The Extraordinary Story of Thyomas Pellow and Islam's One Million White Slaves*, 2005.

58. The Islamic world never went through social transformations similar to the European Reformation, Enlightenment, Renaissance or the American or French Revolutions, whereby the power of priests and kings was checked. Consequently, outside of a few urban regions, there is no factual separation of Mosque and State. "Secularist" totalitarians such as Saddam Hussein murdered far more women and children than he ever "liberated" and never stepped outside the boundary of what the mullahs would tolerate. Even the half-way secularism of Turkey appears fragile, and is today being undone by a newly-elected Islamist regime — *people have voted for less freedom!* State-controlled newspapers and mosques across the Islamic world, including within the "Western allies" of Egypt and Jordan, pour forth a constant river of scapegoating hatred of the infidel, and of Jews and Americans especially. *Mein Kampf* is meanwhile a best-selling book, along with the *Protocols of Zion*, in multiple translations across the Islamic world (which allied itself with Hitler during World War II). Islam has thereby maintained both political-authoritarian and sexual-emotional ("spiritual") characteristics which are more correctly compared to mass-psychological totalitarian political movements than to secularized "religion" as seen in most other parts of the world. Nazism's racist nationalism merged with mystic occultism, or communism's "religion of atheism" which was equally violent towards "non-believers," are comparable.

59. Ye'or, B., *Eurabia: The Euro-Arab Axis*, Fairleigh Dickinson Univ. Press, 2005.

C. THE LEVANT TO MESOPOTAMIA

Physiography

The towering mountains of the Levant lie on a north-south axis at the eastern end of the Mediterranean Sea. These highlands squeeze moisture from winds moving inland off the Sea, and are relatively moist; their downslope and downwind portions are considerably dry, however. Presently, moist winds penetrate inland only to the north of these mountains where a gap exists between the Lebanese and Anatolian highlands. The downwind eastern side of the Lebanese Mountains is dominated by the arid Damascus Basin and Syrian Desert, whose dry conditions stretch east to the Persian Gulf. The region receives less than 250 mm (10") of rainfall per year. East from the Syrian Desert, additional moisture of significance is found only in the Mesopotamian rivers, the Tigris and Euphrates, which are fed primarily by runoff from the Anatolian-Iranian highlands.

> *"Dust storms occur in all parts of the year, but true sand storms are rare. Occasionally, great sandstorms do come off the Arabian Desert and cover Iraq for a day or several days, leaving behind a layer of sand that is sometimes red. The more usual 'sandstorm', however, is actually silt lifted from the desert. In winter the dust rises in front of an oncoming depression. In summer, the dust-storms are more severe, beginning at about ten o'clock and lasting until late afternoon. Five to eight days a month have duststorms, and few days between May 1 and September 30 are without haze."* [1]

In Israel (Palestine), at the southern end of the Levant, mountains also squeeze moisture from Mediterranean winds; these mountains and their adjacent valleys are relatively well watered. Mt. Carmel, a site of multiple occupations since Pleistocene times, is located here. The Jordan Valley lies on the downwind side of these mountains in a basin which captures significant stream flow. The Jordan River flows from the highlands near the Lebanon-Syrian border southward through the northern end of the Jordan Valley into the Sea of Galilee, which in turn overflows into the Dead Sea. Water does not flow out of the saline Dead Sea, as it lies at the lowest point of the Jordan Valley basin; the Valley itself does continue south from here, but only as a dry trench. The western side of the southern Jordan Valley is occupied by the Negev Desert, while its eastern side merges into the Syrian Desert.

In the Jordan Valley, ancient terraces and sediment accumulations evidence a former time much wetter than at present. Fossils of hippopotamus, crocodile, water tortoise, and rhinoceros are found, suggesting the past presence of ponds, lakes, and marshes. [2] The Jordan Valley presently contains the Sea of Galilee and the Dead Sea, which, given its prominent elevated beaches, was the first body of water to be described as "pluvial" in the geological literature. These two bodies of water were greatly enlarged and nearly joined as late as 16,000 BCE, forming Glacial Lake Lisan. The Dead Sea then stood about 200 meters above its present level of 393 meters below sea level. A pluvial precipitation 2.5 times that of today has been estimated for the region. [3] The basin region yields evidence of aridity after c.16,000 BCE, after which Lake Lisan shrank, and tectonic subsidence of the northern end of the basin is thought to

1. Gibson, M.: *The City and Area of Kish*, Field Research Projects, Coconut Grove, 1972, p.17.

2. Huzayyin, S.: *The Place of Egypt in Prehistory, A Correlated Study of Climates and Cultures in the Old World*, Memoires, L'Institute D'Egypte, Cairo, 1941, pp.95-103.

3. Farrand, W.R.: "Pluvial Climates and Frost Action During the Last Glacial Cycle in the Eastern Mediterranean — Evidence from Archaeological Sites", *Quaternary Paleoclimate*, W.C. Mahaney, ed., Geo Abstracts, Norwich, England, 1981, p.400.

have played a major role. Moister conditions occurred again, however, from around 3500-2500 BCE, after which aridity once more took hold.[4] Regardless of climatic pendulations, the unique physical characteristics of the Dead Sea basin, its great depression in elevation, made it generally drier than surrounding upland regions.

Butzer has argued for a moister Neolithic period from c.6000-2000 BCE for the region stretching from the Jordan Valley, across Mesopotamia (Tigris-Euphrates river valleys) and the Near East, and into Southwest Asia:

"...the evidence seems to indicate that there was a moist Neolithic period as well, although the area has not been subjected to intensive investigation. The Dead Sea, for example, has several postpluvial strandlines which are geomorphologically just as fresh and well-preserved as the Neolithic lake deposits of the Fayum. That the subpluvial was felt even as far as India can be seen from deposits in the Narbada Valley, east of Bombay..." [5]

A number of other smaller lakes and springs also existed in the Jordan and Hula Valleys, and the Jafr depression of Southeast Jordan contained a lake once covering 1800 km^2, with an "accumulation of over 25 m. of marls, silts, and limestones, with a rich freshwater snail fauna".[6] These lakes persisted into the late Pleistocene, being fed by precipitation and meltwaters greater than that of today.

"In the mountainous area most of the wadis flowed constantly throughout the whole year and the area was densely covered with a forest composed mainly of oaks... These environmental conditions were excellent for the settlement of hominids in the area of Israel; indeed the Wurmian sediments almost everywhere in Israel yield rich assemblages of artifacts, proving that the whole country was inhabited down at least to the central Negev." [7]

The eastern half of the Damascus Basin [Syria], now a salt swamp, was filled to its highest level during the Pleistocene, due to a "drastic lowering of temperature and obvious optimal precipitation".[8] Wright observed a number of great wadis carved by rains of the moister period which drained the Syrian Desert plateau toward the Euphrates. These wadis are now extinct with evidence indicating a cessation of water flow at some point in time before the Neolithic. A large expanse of wind-borne sand invaded and now covers one stretch of this wadi system.[9]

"The extinct wadi system of the Syrian Desert...is the most impressive record of Pleistocene pluvial climate that the writer has been able to discover in the desert regions of the East Mediterranean." [10]

Flora and Fauna

Fossils of hippopotamus, crocodile, water tortoise, and rhinoceros are found in the Jordan Valley. The mountains of the coastal Levant also yield evidence for vastly different conditions in ancient times. They once supported vast and "impenetrable" cedar forests; at present, however, they are as barren as the mountains of the Sahara, except for a small number of stands remaining in preserves and around holy shrines. These forests were heavily lumbered through the centuries by various human cultures for fuel, charcoal, temples, and ships. The Assyrians, Phoeni-

4. Hotzl, H., et al: "The Youngest Pleistocene", in *Quaternary Period in Saudi Arabia*, Vol. 2, A.Jado & J.Zotl, eds., Springer Verlag, NY, 1984, pp.316-7.

5. Butzer, K.W.: "Climatic Change in Arid Regions Since the Pliocene", in *A History of Land Use in Arid Regions*, UNESCO, 1961, p.41; cf Butzer, K.W.: "Patterns in Environmental Change in the Near East During Late Pleistocene and Early Holocene Times", in *Problems in Prehistory, North Africa and the Levant*, F. Wendorf & A.E. Marks, eds., S. Methodist U. Press, Dallas, 1975, pp.390-391.

6. Butzer, 1975, ibid., p.393; cf. Farrand, 1981, ibid., pp. 396-400.

7. Horowitz, A.: "The Pleistocene Paleoenvironments of Israel", in Wendorf & Marks, 1975, ibid., p.222.

8. Nutzel, W.: "The Climate Changes of Mesopotamia and Bordering Areas, 14,000 to 2,000 BCE", *Sumer*, 32, 1976.

9. Wright, H.E.: "An Extinct Wadi System in the Syrian Desert", *Bull. Res. Counc. of Israel*, 76:53-9, 1958.

10. Wright, 1958, ibid., p.56.

cians, Egyptians, Mesopotamians, Persians, Greeks, Romans, Arabs and Turks, from c.2600 BCE onward, exploited the forests relentlessly. In more recent times, Maronite and Druze exploited the remaining scrub in a similar fashion. Ecological changes here appear to be due to human actions at least as much as from climate change.[11]

A reconstruction of the original vegetation cover in Syria indicates the present plant cover (or absence of it) is greatly regressed from an original condition of open forest/herbaceous steppe. A similar deforestation has occurred across Iraq, on the Khuzistan lowland plains of Iran, and the adjacent Zagros region; the dense natural forest cover which once characterized the region has been almost completely destroyed up to about 1000 meters elevation. Aside from climate change factors which would inhibit regrowth of trees, natural processes which would lead to establishment of such a climax plant community are opposed by human activities such as tree-cutting, grazing, cultivation, wood-grubbing, and consequent erosion.[12]

In Mesopotamia, evidence of desiccation also abounds. Terraces exist on the banks of the Middle Euphrates at levels of 15, 30, 60, and 100 meters above the river. Torrential activity is further evidenced by gravel spreads found in its middle and upper parts.[13] Some of the depressions of now arid Western Iraq were lakes with human habitations during Paleolithic times, predating 6000 BCE. Flint artifacts have been found in such depressions, to include the Iraqi Western Desert.[14]

According to Wadia, the environmental history of Mesopotamia and the Levant parallels that of Central Asia:

"In Asia Minor and Syria, which is called the 'graveyard of a hundred cities', in Iraq which possessed not many centuries ago in the fertile plains, between the Euphrates and the Tigris, the granary of the ancient world, in parts of Iran, India, China, the same sequence of events has happened, increasing dryness, migration of the indigenous fauna and flora, erosion of the soil cover by wind and undisciplined rush of water across the fields during the few occasional rain storms and the loss of vegetation cover. These ravages of nature have been supplemented by the acts of man, his misuse and neglect of soil, overstocking and overgrazing of cattle, deforestation, political turmoil and the depredations of war." [15]

Butler, who observed Near Eastern deserts in the 1920's, gave reports of widespread desiccation and ruin which characterize the region:

"The results of the explorations of the last twenty years have been most astonishing... Practically all of the wide area lying between the coast range of the eastern Mediterranean and the Euphrates, appearing upon the maps as the Syrian Desert, an area embracing somewhat more than 20,000 square miles, was more thickly populated than any area of similar dimensions in England or the United States is today, if one excludes the immediate vicinity of large modern cities... An enormous desert tract lying to the east of Palestine, stretching eastward and southward into the country which we know as Arabia, was also a densely populated country. How far these settled regions extended in antiquity is still unknown, but the most distant explorations in these directions have failed to reach the end of ruins and other signs of former occupation... On ascending into the hills the traveler is astonished to find at every turn remnants of the work of men's hands, paved roads, walls which divided fields, terrace walls of massive structures.

11. Mikesell, M.W.: "The Deforestation of Mount Lebanon", *Geographical Review*, 59(1):1-28, 1969.

12. Pabot, H.: *Rapport au Gouvernement de Syrie sur l'ecologie vegetale et ses applications*, 1957 (FAO/ETAP report #663); cf. Whyte, R.O.: "Evolution of Land Use in Southwestern Asia", in *A History of Land Use in Arid Regions*, Arid Zone Research XVII, UNESCO, 1961, p.72.

13. Huzayyin, 1941, ibid., p.105.

14. Voute, C.C.: "A Prehistoric Site Near Razzaza", *Sumer*, 13(1-2):135-56, 1957; Whyte, 1961, ibid., p.105.

15. Wadia, D.N.: *The Post-Glacial Desiccation of Central Asia*, Nat. Inst. of Sciences of India, Delhi, 1960, p.1.

Presently he comes upon a small deserted and partly ruined town composed of buildings large and small constructed of beautifully wrought blocks of limestone, all rising out of the barren rocks which forms the ribs of the hills. If he mounts an eminence in the vicinity, he will be further astonished to behold similar ruins lying in all directions... Mile after mile of this barren gray country may be traversed without encountering a single human being. Day after day may be spent in traveling from one ruined town to another without seeing any green thing save a terebinth tree or two standing among the ruins, which have sent their roots down into the earth still preserved in the foundations of some ancient building... Passing eastward from this range of hills one descends into a gently rolling country that stretches miles away toward the Euphrates. At the eastern foot of the hills one finds oneself in a totally different country, at first quite fertile and dotted with frequent villages of flat-roofed houses. Here practically all the remains of ancient times have been destroyed through ages of building and rebuilding. Beyond this narrow fertile strip the soil grows drier and more barren, until presently another kind of desert is reached, an undulating waste of dead soil. Few walls or towers or arches rise to break the monotony of the unbroken landscape; but the careful explorer will find on closer examination that this region was more thickly populated in antiquity than even the hill country to the west. Every unevenness of the surface marks the site of a town, some of the cities of considerable extent... It is perfectly apparent that large parts of Syria once had soil and forests and springs and rivers, while it has none of these now..." [16]

Hughes also articulated a similar view for the region:

"The once-prosperous capital cities of Mesopotamia are now mounds of clay in the desert, where the courses of dry canals may be traced under dust and sand blown there by the winds. The famous 'fertile crescent' of early agriculture once arched from the borders of Egypt through Syria and Mesopotamia to the Persian Gulf, but pictures taken from space by the astronauts show that it is now a shrunken remnant." [17]

New Material 2005: See *Update on Saharasia* (Appendix B) for presentation of newer archaeological evidence for an early dry period in the Eastern Mediterranean, and related evidence for a temporary, early period of social violence. Also new material is presented on very early long-distance migration paths from this region across Southern and into Central Europe.

General Settlement Patterns and Culture Change, to Include Influences Transmitted to Mediterranean Europe:

Upper Paleolithic artifacts from cultures such as the Natufians, who founded Jericho, are found in the Jordan Valley, coastal plain, and Negev, indicating establishment of settlements between 10,000 to 8000 BCE. The region then supported large populations of a complex hunting and gathering society.[18] This early period in the Near East following the Pleistocene was one of rapid climatic and environmental changes during a time when the first attempts at settled agriculture were being undertaken in the Levant and Mesopotamia.[19] It appears that such climatic and environmental changes affected the Natufians of the region, and by 9500 BCE, drier conditions began to prevail.[20] These events also occurred around the same time that agricultural and animal herding peoples spread into the Levant and Mesopotamian region from the Zagros Mountains, continuing their migrations southward into the Negev and Northern Arabia.[21]

The site of Murybet in the river valley of Northern Mesopotamia still possessed, in 8500 BCE, a climate more favorable than that of today;

16. Butler, H.C.: "Desert Syria, the Land of a Lost Civilization". *Geographical Review*, Feb., 1920, pp.77-108.

17. Hughes, J.D.: *Ecology in Ancient Civilizations*, U. New Mexico Press, Albuquerque, 1975, p.2.

18. Ronen, A.: "The Paleolithic Archaeology and Chronology of Israel", in Wendorf & Marks, 1975, ibid., pp.243-244.

19. Henry, D.: "Adaptive Evolution within the Epipaleolithic of the Near East", in *Advances in World Archaeology*, Vol. 2, F.Wendorf & A.Close, eds., Academic Press, NY, 1983, p.99.

20. Mellaart, J.: *The Neolithic of the Near East*, Thames & Hudson, London, 1975, pp.32,36-8,44.

21. Masry, A.H.: *Prehistory in Northeastern Arabia: The Problem of Interregional Interaction*, diss., U. Chicago, Anthropology, 1973, p.103.

materials found in the settlement mound there include the bones of sheep, gazelle, cow, equids, birds, rodents, fish, freshwater muscles, plus grains of wheat, barley, and chickpea.[22] The Levantine mountains then possessed leopard, lion, wolf, hyena, gazelle, antelope, fallow deer, and bear, with hippos in the coastal plain; Syria possessed both onagers and elephant.[23]

Natufian Jericho of c.8350 BCE developed during a period of increasing post-Pleistocene aridity, and possessed a population of around 2000, probably due to its location on a secure water source at the north end of the Dead Sea basin. At this early date, however, it did not possess the defensive walls and towers which characterized it later on. Flint and obsidian were traded between Jericho and Anatolia during this period.[24]

By 8000 BCE, new peoples who established permanent agricultural settlements were spreading southward from the Zagros Mountains.[25] The Mediterranean forest which characterized much of the Levant began to vanish, retreating northward with changing climatic conditions. According to Mellaart, this period of peak aridity and ecological change saw the desertion of many Natufian sites, and "...may well have precipitated a crisis which affected Natufian man and his livelihood."[26]

After 8000 BCE, the city of Jericho expanded to about 10 acres, and fortification walls, towers, and tombs for dead kings were constructed. These deposits are the earliest of all evidences ever found anywhere for the presence of possible conflict and social stratification.[27] Pastoralism also increased across Jordan and Syria in the years following 8000 BCE, as did the use of elongated rock structures on the ground, called desert "kites", used to channel running game into traps.[28] Jericho was deserted by c.7500 BCE, and some *suggestions* of infant cranial deformation have been found. The desertion of the city left no traces of violence at the site, however, and appears to have been connected with the increasing aridity of the area.[29]

However, it appears that violence and social stratification were neither continuous nor widespread around Jericho or elsewhere in the Near East at this early date; the evidence at Jericho appears to reflect the unique geography of the city at a time when temporary local or regional desiccation was occurring, when an increasing number of people were moving south from the Anatolian-Iranian highlands into the Jordan/Dead Sea Valley.

> *"In a geographical sense, Jericho is one of the most remarkable of ancient sites. It lies in a hot and barren zone in the rain shadow of the Judean hills, 650 feet (approx. 200 m) below sea level, and relies for its existence on the water issuing from a powerful spring which maintains an oasis-like environment."* [30]

Only fleeting visions of military conflict, fortifications, social stratification, or cranial deformation occur in the Near East before c.5000 BCE, appearing here and there at isolated sites, and without any clear pattern or widespread distribution. And even after 5000 BCE, and for the next thousand years, evidence of social turmoil still seems to crop up only occasionally, perhaps incubating within some unknown culture or region before revealing itself in studied archaeological deposits. It is only after c.4000 BCE, when desiccation clearly became more widespread and intense, that these initial traces of disturbed human behavior begin to blossom in clear, unambiguous, and often organized, institutionalized forms.

To continue, Jericho was reinhabited after c.7350 BCE at the onset of wetter conditions, and numerous settlements and a circuit wall were

Ancient Jericho
Early photograph of the ruins,
one of the oldest city-states

22. Mellaart, 1975, ibid., p.55.

23. Mellaart, J.: *The Chalcolithic & Early Bronze Ages in the Near East and Anatolia*, Khayats, Lebanon, 1966, p.4-5.

24. Mellaart, 1966, ibid., p.9; Mellaart, 1975, ibid., p.39,48-51.

25. Hawkes, J.: *Atlas of Ancient Archaeology*, McGraw-Hill, 1974, p.168.

26. Mellaart, 1975, ibid., p.32,38.

27. Roper, M.K: "Evidence of Warfare in the Near East from 10,000-4300 BCE", in *War, Its Causes and Correlates*, M.A. Nettleship, et al, eds., Mouton Publishers, Paris, 1973, p.304; Jaynes, J.: *The Origins of Consciousness in the Breakdown of the Bicameral Mind*, Houghton Mifflin, Boston, 1976, pp.141-2; Hawkes, 1974, ibid., p.198.

28. Helms, S.: *Jawa: Lost City of the Black Desert*, Cornell U. Press, NY, 1981, pp.26-7,47.

29. Mellaart, 1975, ibid., pp.50-1.

30. Hawkes, 1974, ibid., p.198.

constructed, but no clear-cut fortifications were present. Trade and technology were significantly advanced over the earlier settlements, but neither shrines nor cult rooms were found. These new settlers were not Natufian, but related to cultures in Northern Syria; they engaged in mixed hunting, animal domestication and animal herding. Around the same time that Jericho was reinhabited, the city of Beidha, south of the Dead Sea, was established on top of a sand dune.[31] This new culture continued to spread across Syria, Palestine, and into the Negev and North Arabia.

Regarding the Negev desert, there are many sites in its relatively unexplored central region indicating Upper Paleolithic and early Neolithic societies dating from c.16,000 to 6,000 BCE.[32] The Negev is said to contain an "astonishing number of flint artifacts scattered widely" on its surface.[33] The peoples who colonized the Negev brought new methods of hunting, food collection, trade, and architecture.[34]

By 6000 BCE, agricultural settlements with cattle, sheep, and goat were established across Mesopotamia into the Levant and Arabia, also having spread into the region from the north. The cultural aspects of Jericho after 6000 BCE are found across the Near East into Iran and Central Asia.[35] Walled cattle impoundments or flood control barriers also appeared at Tell es-Sawwan on the Tigris River after this date.[36] Aside from the possible fortifications at Jericho and Tell es-Swwan, archaeological evidence for the broad region around the Near East indicates that this was a peaceful period.[37]

Mellaart's review of the evidence indicated that the period 6000-5500 BCE was a time of decreasing moisture across parts of the Near East, with declining tree pollens and general reduction in vegetation. During this drying period Jericho was again abandoned without signs of violence, along with other sites in Palestine, Syria, and the Zagros mountains. Migration characterized the region.[38]

"...a climatic oscillation affected the drier parts of the Levant, inland Syria, and Mesopotamia...destroying the previous pattern of agricultural settlements. The response of the population was two fold: emigration both northward and westward into the Mediterranean forest zone, which gradually led to the opening up of this previously shunned area. Others adjusted by turning to semi-nomadic pastoralism and stockbreeding, occasionally visiting the ancestral sites, and setting up their seasonal camps on the ruins of the villages of their fathers. For perhaps half a millennium or so these conditions prevailed until at last, enriched by new ideas and technological achievements...picked up in the north and east during their wanderings, they gradually filtered back into Palestine, conditions permitting, to found the various cultures of the 'pottery Neolithic' period with their strong semi-sedentary overtones." [39]

Evidence from the Zagros mountains for this period of aridity is contradictory, including both female figurines with bulging breasts and thighs (at Ganjdareh), which suggest a higher female status, and female cranial deformation (at Alikosh), which suggests the onset of nomadism and traumatic influences directed towards infants. Female figurines from Palestine of 5500 BCE also suggest the use of cranial deformation.[40] The unsettled period c.6000-5500 BCE also saw the settlement and later nonviolent desertion of the Neolithic Khirokitia site on Cyprus, where cranial deformation was practiced among a culture without female figurines.[41]

31. Mellaart, 1975, ibid., pp.51,57,59, 65.

32. Marks, A.: "An Outline of Prehistoric Occurrences and Chronology in the Central Negev, Israel", in Wendorf & Marks, 1975, ibid., p.361.

33. Evenari, M. et al: *The Negev*, Harvard U. Press, Cambridge, 1982, pp.11,343.

34. Yosef, O.B.: "The Epipaleolithic in Palestine and Sinai", in Wendorf & Marks, 1975, ibid., p.375.

35. Gupta, S.P.: *Archaeology of Soviet Central Asia and the Surrounding Indian Borderlands*, Vol 2, B.R. Publishing, Delhi, 1979, p.290.

36. Helms, 1981, ibid., pp.94-5.

37. Beaumont, P.: "The Middle East - Case Study of Desertification", in *The Threatened Drylands*, J.Mabbutt & S.Berkowicz, eds., 24th Intern. Geographical Congress, UNESCO/U. New South Wales, Australia, p.47; Roper, 1973, ibid., p.317.

38. Mellaart, 1975, ibid., pp.48,67-8,88, 90,135-6.

39. Mellaart, 1975, ibid., pp.68-9.

40. Mellaart, 1975, ibid., pp.82,88,239.

41. Mellaart, 1975, ibid., pp.98,130-2.

By 5500 BCE, as moist conditions began to return, new settlements appeared across Mesopotamia, into Anatolia and the Zagros region. These cultures possessed naturalistic female figurines.[42] The Halaf culture, noted for superior ceramics, thereafter established itself across North Syria and Upper Mesopotamia. After c.5000 BCE, moist conditions peaked and settlements also spread throughout the Negev and Sinai.[43] The number of artifacts dating between c.5000-4000 BCE indicate that the Negev, at least, was relatively well populated,[44] and this period also saw the development of the first large cities in Mesopotamia. Cultural contacts and trade were occurring from Egypt to the Iranian Plateau.[45]

The period following c.5000 BCE brought gradually increasing harassment, conflict, and destruction to Mesopotamian settlements, by hill and mountain peoples from the Zagros region to the northeast, and also from Semitic peoples to the south.[46] Fortifications appeared on the Anatolian Plateau, in Iraq near Baghdad, at Tell es Sawwan,[47] and at Ras Shamra in Syria shortly thereafter.[48] Other fortifications appeared after 5000 BCE in Syria and East Turkey, and numerous sites dated after 5000 BCE exhibit destruction layers.[49] The site of Arpachiyah in Mesopotamia, in which female figurines had been made, was destroyed and looted by an unknown enemy around 4700 BCE, followed shortly thereafter by the sudden end of the Halaf culture. A new culture, the Ubaid, expanded into the area from Arabia northward to the edge of the Anatolian-Iranian highlands. Unrest was apparent at sites in Anatolia, Cyprus, and Syria.[50] Of the disappearance of the Halafians, Mellaart said:

"One is tempted to blame...soil exhaustion, animal exhaustion, animal disease, drought, nomadic incursions and the like, factors that have often affected life in these areas in historical times, [given] the destruction and more often desertion of once prosperous settlements and the general impoverishment evidenced by the sites that survived." [51]

Changes After c.4000 BCE:

Mesopotamia appears to have maintained a relatively peaceful existence before 4000 BCE. A religion of the mother goddess prevailed, and women maintained a high status as independent stenographers, diviners, maids, cooks, hairdressers, weavers, and necromancers, while numerous "wise women" existed. Cuneiform writing and other technological developments were underway by this time, but the great temples, ziggurats, tombs and social institutions associated with a warrior and priestly caste, or divine kingship were absent.[52]

More so than either North Africa or Arabia, the Levant and Mesopotamia would later develop numerous central state empires in the centuries after 4000 BCE. Before this date, the region was diffused with various peoples who migrated in primarily from the north. As was the case with the Nile Valley, the coastal Levant and Mesopotamian river valleys retained their water supplies as desiccation set in, attracting other peoples who were abandoning surrounding arid regions.

The Semitic irruption from Arabia after 4000 BCE, the simultaneous period of desiccation in that region, and the arrival of Semitic/Asian peoples in North Africa and Arabia after c.3300 BCE have already been discussed. It is important to recognize, however, that the Asiatic peoples who so strongly affected Africa arrived in the Levant and Mesopotamia before their arrival in Northern Arabia or Africa. By 4000 BCE, these peoples were migrating south from the Anatolian/Iranian highlands, and

42. Adams, R. M.: *Heartland of Cities*, U. Chicago, 1981, p.94; Mellaart, 1975, ibid., pp.111,114-5,151,154-5.

43. Horowitz, ibid.; Evenari, 1982, ibid., pp.11,343.

44. Evenari, et al, 1982, ibid., pp.11,343.

45. Hawkes, 1974, ibid., p.168.

46. Hawkes, 1974, ibid., p.11.

47. Roper, 1973, ibid., pp.317-21.

48. Hawkes, 1974, ibid., p.200.

49. Roper, 1973, ibid., pp.318,324; Mellaart, 1975, ibid., 126; Mellaart, 1966, ibid., pp.99,108; Todd, 1976, ibid., p.137.

50. Mellaart, 1975, ibid., pp.166-7,236-7; Mellaart, 1966, ibid., pp.20,39.

51. Mellaart, 1975, ibid., p.236.

52. Stone, M.: *When God Was A Woman*, Dial Press, NY, 1976, pp.39-44; Tannahill, R.: *Sex in History*, Stein & Day, 1980, p.62; Roper, 1973, ibid., p.300.

similar types of pottery and skeleton are found from the Mediterranean into Central Asia dating from that time.[53] Indo-Aryan peoples from the north and Semitic peoples from the south would shape the cultural structure of the Levant and Mesopotamia henceforth:

> "The vicissitudes of the ancient Orient are dominated by the struggle for the possession of the 'fertile crescent', the Syro-Mesopotamian area which constitutes its point of greatest fruitfulness. In this struggle there take part, from opposite directions, the peoples of the mountains which bound the area to the northeast, and the peoples of the desert which forms its other boundary... The peoples of the mountains are made up of various stocks, but have in common and Indo-European element in their ruling classes; the desert peoples are homogeneous, being Semites. The history of the ancient Near East receives its fundamental impulse from the convergent pressure of these two forces. Success attends now one, now the other; and the fate of the two groups is different: in the area of the mountain peoples invasions constantly recur, bringing about a ceaseless modification of ethnic composition; in the desert there is no substantial alteration, and the fundamental Semitic unity is so preserved..." [54]

Mesopotamia developed an increasing number of city-states after 4000 BCE, to include Ur, Nippur, and Babylon. Irrigation agriculture rapidly developed, as did writing and metallurgy; some of these innovations were stimulated by the newcomers from the Asian steppe, but others were native to Mesopotamia and adopted by the northern invaders later on. The period 4000-3500 BCE brought widespread, severe flooding in the floodplain of the Tigris and Euphrates, probably related to the same climatic instability which was causing desiccation in Arabia at the time.[55] After c.3500 BCE, Mesopotamia also began to experience a general desiccation.[56] The period marks the lower boundary for many of the pastoralist and hunting groups which had inhabited Jordan and Syria more or less continuously since c.8000 BCE.[57]

By 3500 BCE, the influence of Arabian names and terms appeared in Sumerian cuneiform texts. Regarding these influences, which were brought by migrant nomadic populations abandoning Arabia, Hawkes states that "such infiltration has been a recurrent feature of Mesopotamian history since the earliest records..."[58]

Mesopotamia and the Levant were increasingly harassed by northern Indo-Aryan and southern Semitic nomads after 3500 BCE. And the various migrations of these peoples were increasingly characterized by militant conflict, involving highly organized armies of mounted nomads who challenged the various city-states for control of the secure water supplies upon which they were founded. Indo-Aryan and Semitic groups established various kingdoms across the Levant and Mesopotamia, the influences of which persist down to the present day. These invading groups include the Indo-Aryan Hurrians of Mesopotamia,[59] and the Semitic Phoenicians of the Levant.[60] Semitic groups also settled Assyria, on the plain of Akkad and Shinar.[61]

After c.3500 BCE, the end of the flood period, the Sumerian cities of Mesopotamia underwent a radical transformation. Massive city walls, temples, palaces, and tombs were erected and horse-drawn wheeled vehicles, such as the war cart, came into use. New products, such as linen, and ideas regarding the calendar, writing, agriculture and husbandry also developed after this date.[62] This was a period with a "widespread sharing of technologies, aesthetic and cultic knowledge and practices", from Egypt

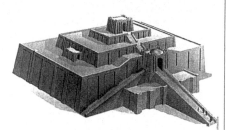

Model of the Ziggurat of Ur, near the Euphrates River.

The Zigurat of Ur, prior to its excavation and reconstruction.

Ruins of the Ziggurat of Eanna, one of many others marking ancient city-states of theTigris-Euphrates region. All of them are today found in desertified regions, swamped with desert sands and marking the presence of salinized soils.

53. Gupta, 1979 Vol 2, ibid., p.239; Mellaart, 1966, ibid., p.20.

54. Moscati, S.: *The Semites in Ancient History*, U. Wales, Cardiff, 1959, p.134.

55. Grun, B.: *Timetables of History*, Touchstone, NY, 1982; Hawkes, 1974, ibid., pp.167,173.

56. Nutzel, W.: "The Climate Changes of Mesopotamia and Bordering Areas, 14,000 to 2,000 BCE", *Sumer*, 32:23, 1976.

57. Helms, 1981, ibid., pp.26-7,47.

58. Hawkes, 1974, ibid., pp.167-8; Mellaart, 1966, ibid., p.54.

59. Hawkes, 1974, ibid., p.12.

60. Mikesell, 1969, ibid., p.11.

61. Grun, 1982, ibid.

62. Hawkes, 1974, ibid., p.175; Grun, 1982, ibid.

to Central Asia and the Indus Valley.[63]

Social institutions also underwent a radical transformation at this time, in the same general manner as that observed in Egypt after c.3300 BCE: the decentralized, female and child oriented agricultural and trading villages of Mesopotamia, and the hunters and pastoralists of outlying regions bearing a similar social makeup, were radically and violently transformed. Divine Kingship, the despotic central state, human sacrifice, slavery, polygamy, concubinage, female seclusion, infant cranial deformation, swaddling, and circumcision appeared in Mesopotamia in fully blown, institutionalized forms only after c.3500 BCE. As was the case in Egypt, these transformations occurred under the stimulus of foreign groups which entered the area from Arabia and Central Asia, at a time when the surrounding environment was undergoing desiccation. Palestine became a virtual crossroads, and "melting pot" of various cultures living in fortified cities, many of which display multiple destruction layers, and foreign elements from the north or south.[64]

The Mesopotamian world after 3500 BCE became dark with demons, the gods having abandoned the world, departing for "Heaven". Humans were divorced from the gods, and divination with a professional priesthood gained in importance and status. Natural phenomena, particularly childbirth, began to assume a mystical, demonic, and hostile character.[65]

The epics of Gilgamesh, similar in tone and plot to Homer's Odyssey, were placed during the turbulent times after 3500 BCE. Pottery lamented the death of Tammuz, the son of the mother goddess Innin, and festivals celebrated the victory of the god of spring over the goddess of chaos.[66] In Babylon, the god Marduk murdered his mother, the creator goddess Tiamat, to gain his favored position.[67] The goddesses' husband-consort Humban was derisively known as the "father of the weak" by peoples to the north.[68] The god Inshushinak, also known as the "father of the weak", became the "King of the gods", while the mother goddess was reduced to the "Great Wife".[69]

"She is dragged from the throne and threatened with death until she agrees to marry her assailant... who then kisses away her tears and becomes her husband and rules beside her." [70]

Images of a bull and lion, from the gates of Babylon. These virile and powerful animals were frequently depicted larger than life in Mesopotamian royal art, as desirable traits for emulation by their god-king dictators.

In Assyria, a region dominated by the northern warriors, a fear of castration by the female appeared in both myth and law. Queen Semiramis, the legendary founder of Babylon, was blamed for originating and introducing the castration of boys,[71] being called "the most beautiful, most cruel, most powerful, and most lustful of Oriental Queens" by Herodotus.[72] A woman who destroyed one of a man's testicles would be punished by having a finger amputated; if both testicles were damaged, she would have both breasts "torn off".[73] However, these fears did not stop the Assyrians from administering castration as a form of punishment.[74] Infant cranial deformation also appeared in Assyria, as it later did in Phoenicia, Sidon, and Tyre.[75]

Women were gradually reduced to being property, and female adultery was made a property crime. An adulterous woman could either be killed or have her nose cut off, while her male partner could either be killed or castrated; the wife of the offending man would be given to the father or husband of the offending woman, to do with as he wished.[76] A rapist could be punished by having his own wife raped by the original victim's father or husband. Abortion was punished by killing the offending woman.[77] Various female uncleanliness rituals connected with a vaginal blood taboo also began at this time.[78] Excavations of the period suggest the presence of the harem,[79] and later Assyrian codes give some of

63. Gupta, 1979 Vol 2, ibid., p.257; cf. pp.166-7,281.

64. Mellaart, 1966, ibid., pp.34-7,44-7,53-4,65-7.

65. Jaynes, 1976, ibid., pp.232,253-4.

66. Grun, 1982, ibid.

67. Stone, 1976, ibid., pp.41-3.

68. Jaynes, 1976, ibid., p.161.

69. Stone, 1976, ibid., pp.41-3.

70. Stone, 1976, ibid., p.39.

71. Brandt, P.: *Sexual Life in Ancient Greece*, AMS Press, 1974, p.510.

72. Tannahill, 1980, ibid., p.61.

73. Tannahill, 1980, ibid., p.69.

74. Wittfogel, K.A.: *Oriental Despotism*, Yale U. Press, New Haven, 1957, p.355; Brandt, 1974, ibid., p.994.

75. Dingwall, E.J.: *Artificial Cranial Deformation*, London, 1931, pp.83-5.

76. Bullough, V.: *Sexual Variance in Society and History*, J. Wiley, NY, 1976, p.53.

77. Stone, 1976, ibid., p.59.

78. Tannahill, 1980, ibid., p.66.

79. Lewinsohn, R.: *A History of Sexual Customs*, Harper Bros., 1958, p.115.

the earliest references to facial veiling.[80] Temple prostitution, under the supervision of male priests, was institutionalized.[81] Kings of the period introduced human sacrifice, butchering their favorite wives, concubines, and servants, who were taken to the grave with their master, sometimes being buried alive in obvious crouching positions.[82] Figurines appeared in Syria of circumcised males,[83] and Sumerian texts spoke of a "man in charge of the whip", assigned to beat schoolboys at the slightest hint of disobedience.[84]

These cultural changes gradually intensified throughout Mesopotamia and the Levant after 3500 BCE; however, some regions do indicate remnants of a lingering matrilineal kinship and inheritance, as late as 2000 BCE.[85] Some of the harsher practices like ritual sacrifice of widows also eventually diminished. Still, the general trend was toward an intensification and institutionalization of the various antifemale and antichild practices, many of which prevail in the region to this day.

Desiccation of the environment continued through this turbulent period. The first historically known dam was constructed in the volcanic rock desert of Jordan, across a wadi near the ancient city of Jawa in c.3250 BCE; runoff agriculture likewise developed in Jordan around 3000 BCE as a response to increasingly scanty and inadequate rainfalls. From Jordan, these new forms of technology gradually spread to other regions suffering the effects of aridity.[86]

In terms of depopulation and abandonment, the major impact of the period was upon settlements in the Syrian and Negev Deserts, which were considerably diminished. Mesopotamia and the coastal Levant retained their supply of moisture and remained populated, attracting peoples who were abandoning the desiccating regions. Neolithic and Early Bronze Age settlements vanished from the Negev after 3000-2500 BCE, with intensification of aridity.[87] Other settlements across Palestine and the Syrian Desert were also abandoned after 3500 BCE. In fact, the history of the Levant, Palestine, and Mesopotamia after 3500 BCE is an incredible mix of migration, invasion, war, rebellion, general social chaos and cultural change, all against the backdrop of a desiccating environment.[88]

Jawa, for instance, which lay on the migration route between Egypt and Mesopotamia, developed fortification architecture after c.3250 BCE, with thick layers of ash underlying its fortifications. Invaders from the surrounding desert destroyed the site in c.3200 BCE, a scant 50 years after its first major construction and occupation. Its subsequent occupants enlarged its water diversion and storage systems, and its fortifications; new hut circles then appeared outside the walls of the fort.[89] Cities across the region, including Mesopotamia, increased in size as aridity increased. Adams commented on the environmental and social changes of the period:

"...it can hardly be accidental that widespread tendencies to concentrate in urban centers of unprecedented size coincided with heightened competition over reduced volumes of irrigation water." [90]

Before 2600 BCE, Semitic Akkadian peoples had migrated from the desiccating northern part of Mesopotamia southeast to Sumeria,[91] forming the Akkadian Dynasty of Ur. Sixteen Royal graves of the period have been found, each containing evidence of prolonged ceremony and ritual, to include from 6 to 80 victims of ritual sacrifice, most of whom were women.[92] This period generally coincides with the onset of an "abrupt and widespread abandonment of sites" across the region from Mesopotamia to Central Asia, and the disruption of overland trade.[93]

The port city of Byblos saw tremendous growth between 2700-2300

Marduk, god of Babylon Murdered his mother, the creator-goddess Tiamat, to gain power. His myth reflected a reality which was to become increasingly widespread (post-3500 BCE).

80. Keddie, N. & Beck, L.: *Women in the Muslim World*, Harvard U. Press, Cambridge, 1978, pp.25,32.

81. Lewinsohn, 1958, ibid., p.29.

82. Jaynes, 1976, ibid., p.161.

83. Mellaart, 1966, ibid., pp.62-3.

84. Radbill, S.: "A History of Child Abuse and Infanticide", in *The Battered Child*, R.Helfer & H.Kempe, eds., U. Chicago Press, 1974, p.3.

85. Stone, 1976, ibid., p.41.

86. Evenari, et al, 1982, ibid., pp.350-4; Helms, 1981, ibid., pp.166-7.

87. Butzer, 1961, ibid., p.42.

88. Helms, 1981, ibid., p.57; Adams, 1981, ibid., pp.94,130,349.

89. Helms, 1981, ibid., pp.69,85,203-5.

90. Adams, 1981, ibid., p.94.

91. Butzer, 1961, ibid., p.42.

92. Davies, N.: *Human Sacrifice in History and Today*, Wm. Morrow, NY, 1981, p.289.

93. Gupta, 1979 Vol 2, ibid., pp.257-8; Mellaart, 1966, ibid., pp.67-8.

Phoenician warships of the Mediterranean. Battle shields are hung over the railing, in the manner of the later Vikings.

94. Mellaart, 1966, ibid., pp.68-71.

95. Mikesell, 1969, ibid., p.1.

96. Hawkes, 1974, ibid., p.196.

97. Mellaart, 1966, ibid., pp.92,168-9.

98. Payne, G.: *The Child in Human Progress*, Sears, NY, 1916,. pp.150-2; Mellaart, 1966, ibid., pp.71,91-5,174; Scott, 1939, ibid., p.153.

99. Whyte, 1961, ibid., p.106; McGhee, R.: "Archaeological Evidence for Climatic Change During the Last 5000 Years", in *Climate and History*, T.M.L. Wigley, et al, eds., Cambridge U. Press, 1981, p.168; Hawkes, 1974, ibid., pp.11,168; Mellaart, 1966, ibid., pp.71,95.

100. Jaynes, 1976, ibid., p.293.

101. Gumilev, L.N.: "Heterochronism in the Moisture Supply of Eurasia in Antiquity", *Soviet Geography*, 7(10):35, 1966.

102. Grun, 1982, ibid.

103. Huntington, E.: *Palestine and its Transformation*, Houghton- Mifflin, NY, 1911, p.383; Von Wissmann, H.: "On the Role of Nature and Man in Changing the Face of the Dry Belt of Asia", in *Man's Role in Changing the Face of the Earth*, Vol 1, W.L. Thomas, ed., U. Chicago, 1972, pp.296.

104. Grun, 1982, ibid.

BCE, as Mediterranean trade increased and building materials for monumental architecture and ships, namely large trees, came into great demand.[94] Mikesell identifies 2600 BCE as the onset of a period of intense lumbering and deforestation in Lebanon.[95] Shortly after this, between 2400-2300 BCE, the Levant saw a decline of urbanism in favor of a semi-nomadic subsistence, and similar changes occurred in Syria.[96]

The various city-states of Mesopotamia were conquered by Sargon of Akkad around 2400 BCE, and trade was established with the Indus Valley via Magan (Oman) and Dilmun (Bahrain). Seals with Indian humped cattle and water buffalo appeared in Mesopotamia, as trade in slaves and exotic animals proceeded. Meanwhile, a "heterogeneous nomadic and warlike people from the surrounding deserts", invaded Byblos, Jericho, and other towns in Palestine, which were burned, looted, and their inhabitants massacred or enslaved.[97] Naram-Sin, a Mesopotamian King, recorded the presence of "demoniac hordes", at the west end of the empire. These invader peoples, also of a Semitic background, settled on the coast where they were later called Canaanites, and in the hinterlands where they were called Amorites. They made pottery shapes with feet, constructed individual tombs, and possessed war armaments. Later, these same Semitic invaders dominated sea trade in the Mediterranean and were called the Phoenicians, founding settlements in Carthage, coastal Spain and France, and Sicily, where their antifemale and antichild cultural attitudes and social institutions were transplanted. Evidence of child sacrifice and female grave murder has been found in excavations of their period of habitation, the former practice being widespread among the Semitic strata; children were murdered and burned in large numbers, buried in special clay jars, while prisoners were flayed alive and crucified. Child sacrifice only declined later under Greek influence. For all this mayhem and slaughter, the Phoenicians became a much feared and hated people.[98]

Around the same time that these events were occurring in the Levant, the Akkadian empire of Mesopotamia was itself expiring, under the pressure of invading Asiatic/Indo-Aryan cultures from the Zagros region. Around 2300 BCE, wave after wave of these peoples swept out of the Anatolian/Iranian highlands and across the Mesopotamian region, destroying cities, enslaving peoples, establishing various new Kingdoms in their wake, and further transforming and entrenching their own antifemale, antichild social institutions.[99]

As discussed earlier, these Asiatic, Indo-Aryan and Semitic invaders often swept south and west into Egypt and North Africa (Asiatics, Hyksos, Persians, Arabs, Turks). It was the onset of a period of desiccation, migration, and destruction, whose aftermath can be seen in the present day social and political chaos of the region. The Fertile Crescent was flooded with what the Babylonians called *khabiru* (Hebrew) peoples, identified as "vagrants, refugees, and wanderers".[100] The Great Flood of the Babylonians and Hebrews was dated to this period, c.2300 BCE, which saw climatic instability plus widespread ruin and destruction.[101] Later epic poetry from Babylonia celebrated the recreation of the world.[102] Pastoralist nomads from Arabia and Asia continued to abandon their lands, and increasingly harassed settled peoples.[103]

It was against this chaotic environmental and social backdrop that the patriarch Abraham left the city of Ur for Canaan.[104] Stone has argued that Abraham, the first prophet of the male god Yahweh, came from an Indo-Aryan background, with similarities to the Brahmin castes of India. His own mother left to obscurity, he married Hittite women and was buried in a traditional Hittite grave. Indo-Aryan concepts lay at the foundation of Judaism, Christianity, and Islam, all of which claim Abraham as a common ancestor.[105] Still, there is evidence that the early Hebrews held women and sexual pleasure in relatively high esteem. The "flesh" was not seen as inherently evil, and matrilineal descent was observed.[106] No punishment was given for voluntary sexual relations between the unmarried,[107] and celibacy itself was considered a violation of the biblical injunction to "be fruitful".[108] However, a puritanical, antifemale, antisexual, and antichild Judaism would develop in later centuries.

By 2100 BCE, the forests of the Levant were relatively depleted.[109] The port city of Byblos was again troubled by invaders, and Ras Shamra was entirely destroyed, to be used only as a graveyard for several centuries afterward.[110] In Mesopotamia, invasions and fortifications of cities also increased.[111] The ruins of the Mesopotamian city of Eridu, populated since c.5000 BCE, yielded evidence of the desiccation which was progressing at the time:

> "...the ziggurat [of Ur-Nammu, c.2100 BCE] was never completed and there is little evidence of occupation after this date; Eridu, on the extreme south-western edge of the alluvial plain, was an inhospitable site and the excavations have shown that even during its occupation the encroachment of sand dunes was a recurrent menace. The mounds of the ancient city now extend over an area some 1300 feet by 1000 feet, and are dominated by the ruins of Ur-Nammu's ziggurat, but the excavated buildings have once more been covered by blown sand and nothing is visible of their plan."[112]

Semitic Amorites abandoned interior Canaan after c.2000 BCE, moving to Mesopotamia where they established a new dynasty in Babylon under Hammurabi.[113] The Amorites had been displaced from Canaan by a continued migration of other Semitic peoples from rapidly desiccating Arabia.[114] It is recalled that this was the same time that Canaanite (Phoenician) settlers also appeared in Tunisia.

The Babylonian Amorite Dynasty under Hammurabi followed the dictates of Enlil, the god of punishment and compulsion.[115] Hammurabi warred against other Mesopotamian cities, uniting them into a common empire. His famous "Laws" or "Code" contained 252 articles, 64 of which related to family law. In general these laws codified the compulsive antifemale and antichild views discussed above, abolishing any remnants of the instinctive nature of the family. Some new twists were also apparent, such as the ability of a husband to use his wife or daughter as security for a debt for a 3 year period, a law which later developed into a specialized branch of slave trade.[116] Children could be legally sold by parents, and ritual widow murder persisted; one tomb from the 3rd dynasty of Ur possessed 64 women with gold and silver hair ribbons, all of whom were butchered before being entombed.[117] Even so, legal wives under Hammurabi retained some property, inheritance, and divorce rights, and "unlawful" rapists were dealt with harshly.[118] Hammurabi's code also mandated selling into slavery anyone who neglected their irrigation system and thereby caused damage to fields.[119] In general, most of Hammurabi's codes reflected tribal values of the invading nomadic Semitic and Indo-Aryan cultures, and were present, more or less, throughout most of the Near East at the time.

Around the time the Amorites first arrived in Mesopotamia, the Euphrates shifted its course so as to favor Babylon.[120] Salinization first appeared in the Tigris-Euphrates floodplain at this time, along with evidences of desertification in other areas of Mesopotamia. Three major salinization phases have been identified, between 2400-1700 BCE, 1300-900 BCE, and around c.1200 CE. These phenomena have been variously

Excavated ruins of Tyre, on the Lebanese coast. Capitol of the violent Phoenician Empire c.1100 to 573 BCE, and from which brutal raids were launched all across the Mediterranean Sea. It was sequentially conquered and destroyed by armies of Assyria, Babylonia, Alexander the Great, Roman Empire, Muslim Empire, Crusader armies, and finally once more by Muslim armies, etc.

105. Stone, 1976, ibid., pp.xii,117-9.
106. Bullough, 1976, ibid., p.75.
107. Taylor, G.R.: *Sex in History*, Thames & Hudson, London, 1953, pp.163,241; May, G.: *Social Control of Sex Expression,* Geo Allen & Unwin, London, 1930, p.23.
108. Taylor, 1953, ibid., p.245.
109. Hawkes, 1974, ibid., p.196.
110. Hawkes, 1974, ibid., pp.199-200.
111. Hawkes, 1974, ibid., p.173.
112. Hawkes, 1974, ibid., p.171.
113. Hawkes, 1974, ibid., p.170; Whyte, 1961, ibid., p.106.
114. Butzer, 1961, ibid., p.42.
115. Wittfogel, 1957, ibid., p.138.
116. Lewinsohn, 1958, ibid., pp.22-6.
117. deMause, L.: "The Evolution of Childhood", in *The History of Childhood*, L.deMause, ed., Psychohistory Press, NY, 1974, p.32; Davies, 1981, ibid., pp.30-1.
118. Stone, 1976, ibid., pp.43,59.
119. Helms, 1981, ibid., p.170.
120. Grun, 1982, ibid.

Hammurabi's Code
Top of a stone pillar on which his laws were inscribed, for public viewing. Hammurabi is depicted receiving a "scepter of authority" from the sun-god Shamash.

attributed to excessive irrigation and poor drainage,[121] plus decreases in the fallow cycle.[122] However, such may simply have been complicating factors, brought about by social unrest and conflicts associated with the migration of peoples into the region from areas more completely desiccated. Jacobsen and Adams made the following observations:

"One indication of the ecological shift which took place in succeeding millennia is that permanent swamps today have virtually disappeared from the entire northern half of the [Tigris-Euphrates] alluvium... There is good historical evidence that devastating cycles of abandonment affected the whole alluvium. The wide and simultaneous onset of these cycles soon after relatively peaceful and prosperous times suggests that they proceeded from sociopolitical, rather than natural, causes... For example, the numerous Old Babylonian settlements...had been reduced in number by more than 80 percent within 500 years following, leaving only small outposts scattered at wide intervals along watercourses which previously had been thickly settled." [123]

The cycles of land abandonment apparently were not consistent in either time or space, however, as another moist phase occurred at least in the Negev region between 2100-1900 BCE. Peoples from other regions moved into the Negev and established widespread agricultural settlements of a Middle Bronze Age character.

"...traces are clearly visible to every traveler who follows the main road from Yeroukham to Makhtesh Ramon. If he pauses on his route to examine the outlines of hillsides and mountain ranges, he can discern large stone mounds silhouetted against the skyline, like huge warts. The traveler who exerts himself to climb to the hilltops will find that these mounds, called tumuli, are of sturdy construction... also found [are] grinding stones, quarns, flint blades, and olive pips... the remains of numerous villages and 'high places' dedicated to worship...which have been found in the Negev show that the Negev was populated relatively densely with sedentary, stable people around the year 2000 BCE. Very little detailed study has been made of this period." [124]

From the above, it would appear that the Negev region, far to the south of chaotic Mesopotamia and the Northern Levant, was a relatively safe and secure zone. However, this situation soon changed and another period of abandonment occurred in the Negev and Syrian Desert from 1900 to 1000 BCE, in part related to turbulent political events and warfare, but also due to reductions in the carrying capacity of the land.[125]

It was during the Negev moist phase of c.2100-1900 BCE that the Egyptian Pharaoh Kheti III (c.2050 BCE) finally drove the Asiatics out of the Nile Delta region. This event may be related to the destruction of Jericho around 2000 BCE by nomadic peoples. Jericho tombs after this date give evidence of mass murder, human sacrifice, ritual widow murder, caste, and social stratification like never before. It was also around c.1900 BCE that Jericho was again destroyed and abandoned.[126]

This latter destruction of Jericho also coincided with an earthquake disaster which destroyed several cities on the plain south of the Dead Sea.[127] Volcanism also continued around c.2000 BCE in the black volcanic rock desert of Jordan, where the city of Jawa persisted. Jawa was possibly the city identified in the Bible as Jebel Druze, a "place where oak trees grow and people in disgrace might find refuge in exile." Jawa of c.2000

121. Jacobsen, T. & Adams, R.M.: "Salt and Silt in Ancient Mesopotamian Agriculture", *Science*, 128:1251-8, 1958.
122. Gibson, M.: "Violation of Fallow and Engineered Disaster in Mesopotamian Civilization", in *Irrigation's Impact on Society*, T.E. Downing & M. Gibson, eds., Tucson, U. Arizona Press, 1974, pp.7-19.
123. Jacobsen & Adams, 1958, ibid., p.1254.
124. Evenari, et al, 1982, ibid., pp.11-3
125. Evenari, et al, 1982, ibid., pp.13-4.
126. Kenyon, K.: *Excavations at Jericho*, British School of Archaeology, Jerusalem, (Vol. I, 1960, pp.180-2,186-7,264-7, 301, 306-8, 454,488,502), (Vol. II, 1965, pp.171, 388); Mellaart, 1966, ibid., p.71.
127. Finegan, J.: *Light From the Ancient Past: The Archaeological Background of the Hebrew-Christian Religion*, Princeton U. Press, 1946, p.126; Harland, J.P.: "Sodom and Gomorrah", *The Biblical Archaeologist*, 6:3, 1943, and 5:2, 1942.

BCE sported a prison-like "citadel" at its summit, and other sites were spread across the basalt desert to the northeast. The city then obtained all of its water from the adjacent dam and wadi, but the site was abandoned soon after this time.[128]

This was also the time (c.1850 BCE) when Chedarlaomer and his Kings "smote" entire nations living in the Southern Negev.[129] Populations also sharply declined in Mesopotamia around 1750 BCE,[130] the time of "a vast ethnic movement which affected the whole of Western Asia and ultimately Egypt" in the form of the Hyksos invasion.[131]

Populations were able to persist in the Negev, however, but only by practicing extensive runoff agriculture. Evenari, et al, believe that all Negev settlements from 1900 BCE until the Moslem period were dependent upon extensive runoff agriculture, runoff-fed cisterns, deep wells, and/or control of lucrative trade passing through the region.[132]

After c.1700 BCE, Kassites from the Iranian Plateau/Central Asia region invaded and ruled Babylon and Southern Mesopotamia,[133] while Northern Mesopotamia was alternately invaded by the Indo-Aryan Hittites and Mittani, who struggled for control. The royal names of the Mittani, for instance, were Aryan, and Vedic names appeared in their documents.[134]

Other Indo-Aryan invaders, such as the Hurrians, divided up remaining parts of Syria and the Levant.[135] Each of these groups used the horse and chariot, emphasized military caste, and introduced a host of antifemale and antichild social changes in all of their conquered lands. The Hittites monopolized the use of iron weapons for several centuries, making general trouble for everybody. Meanwhile, the Egyptians finally overthrew the Hyksos after c.1570 BCE and in turn invaded Palestine and Syria up to the banks of the Euphrates.[136]

History records the period following c.1400 BCE as exceptionally chaotic. By 1400 BCE, desiccation was forcing Arameans and other peoples out of the Syrian Desert, in one wave after another toward Mesopotamia.[137]

The Hittites abandoned their homeland in the desiccating Anatolian plateau after c.1200 BCE, migrating through Palestine and Mesopotamia.[138] They again came into conflict with many groups, including the Egyptians.[139] The period saw Garamantians settling in Libya, Peoples of the Sea laying waste to the Eastern Mediterranean, and nomads of the interior assaulting various settled lands. Numerous sites reveal destruction layers at this time, the start of a general "Dark Age".[140] The Assyrian Empire, which married the iron weapons of the Hittites to merciless bloodlust tactics, emerged supreme over the Bronze Age Egyptians and Mesopotamians.[141] Their empire held sway over the Eastern Mediterranean after c.1200 BCE.[142]

The chaotic period around 1400 BCE also saw the Hebrew migration out of captivity from Egypt into Canaan under Moses. The Levite priests of the Hebrews derived their beliefs and legal codes from the Semitic and Indo-Aryan roots previously discussed, and these views appear to have been reinforced during their period of enslavement in Egypt. Hebraic writings of the period viewed the female as a lesser being, a dangerous seductress who brought about the fall of humanity from a state of grace. These beliefs, and their corresponding legal interpretations and social institutions, did not originate with the Hebrews, but with the Indo-Aryan and Semitic invaders who previously overran much of the Near East. However, the Hebrew and Assyrian codes of the Mosaic period were harsher and more compulsive than in previous times.

Subsequent Hebrew taboos regarding widows, divorced women, female virginity, and vaginal blood were similar to those of Indo-Aryan

128. Helms, 1981, ibid., pp.22,30-1,182.

129. Evenari, et al, 1982, ibid., p.13; Genesis 14:5-7.

130. Bowden, M.J., et al: "The Effect of Climate Fluctuations on Human Populations: Two Hypotheses", in Wigley, et al, 1981, ibid., p.488.

131. Edwards, I.E.S.: *The Pyramids of Egypt*, Viking Press, NY, 1972, p.181.

132. Evenari, et al, 1982, ibid., pp.11-13.

133. Phillips, E.D.: *The Royal Hordes*, McGraw Hill, London, 1965, p.46; Hawkes, 1974, ibid., pp.170,180; Stone, 1976, ibid., pp.42,58; Huntington, 1911, ibid., p.385.

134. Sastri, K.A.N.: *Cultural Contacts Between Aryans and Dravidians,* Bombay, 1967, pp.28-9; Jaynes, 1976, ibid., pp.211-2; Hawkes, 1974, ibid., pp.113,170; Phillips, 1965, ibid., pp.41-2; Gupta, 1979 Vol 2, ibid., p.312.

135. Phillips, 1965, ibid., p.46; Stone, 1976, ibid., pp.58-66.

136. Hawkes, 1974, ibid., p.11; Grun, 1982, ibid.

137. Butzer, 1961, ibid., p.42; Huntington, 1911, ibid., p.389.

138. Bryson & Murray, 1977, ibid., p.15.

139. Hawkes, 1974, ibid., p.159.

140. Jaynes, 1976, ibid., pp.213-4; Hawkes, 1974, ibid., pp.170,196; McGhee, 1981, ibid., p.168.

141. Jaynes, 1976, ibid., p.214; Hawkes, 1974, ibid., p.13.

142. Hawkes, 1974, ibid., p.170.

Assyria. A menstruous woman was unclean for 7 days, and all she would touch or look at was likewise unclean. Following childbirth, she was required to undergo three ritual cleansings, the same as for leprosy. Death by stoning or burning was proscribed for blasphemy or various sexual "crimes", such as a female's loss of virginity before marriage, intercourse with one's mother-in-law, intercourse with the daughter of a priest, or for sodomy, bestiality, or adultery. Mosaic law called for the death of a raped woman as well as her rapist, unless the woman agreed to marry the rapist.[143]

"Since the moral resentment of the woman may gradually turn into an instinctual consent, it is next to impossible for a woman to be raped." [144]

Widows were placed in the same category as harlots and profane women, similar to Indo-Aryan Brahman India.[145] Daughters and wives were essentially the property of their male relatives, being bought and sold. Unlike the Egyptian or Babylonian agriculturalists, Hebrew kinship was by this time traced entirely through the male line, and women possessed no inheritance, property, or divorce rights.[146] Mosaic law severely punished transgressors, stoning the offender for minor offenses, and sometimes stoning the wives and children of male offenders as well.[147] These early Hebrews, like other Semitic and Indo-Aryan groups of the Near East, also practiced child sacrifice and human sacrifice. Following the Exodus, the firstborn of all women and animals belonged to Yahweh as potential sacrifice victims.[148]

According to the Old Testament, following the death of Moses, circumcision, which was widespread in Egypt, was made compulsory for all Hebrew males by Joshua: "And the Lord said to Joshua, This day I have taken away the reproach of the Egyptians from you...", the reproach being that the Hebrews were previously uncircumcised, and the Egyptians had mocked the Hebrew men for their former natural condition.[149]

The Hebrews warred with numerous peoples in the Levant and Mesopotamia, particularly those who continued to worship the mother goddess.[150] Joshua's troops massacred the inhabitants of entire cities across Palestine,[151] during a generally chaotic "Dark Age" around c.1300 BCE which extended across the Mediterranean and Near East.[152] Joshua was equally harsh with Hebrew transgressors, as seen in the example of Achan, who was stoned to death along with his entire family, for keeping Babylonian war booty.[153]

Assyrians under Kings such as Tiglath Pileser I, the "butcher of innocents", also went on campaigns of bloodlust, mass executions, and horrible tortures. Assyrians also practiced self-castration during frenzied religious ceremonies.[154] These events signaled an anarchic departure from the previous periods of disciplined rule under Pharaonic Egypt and the Babylonia of Hammurabi.[155] The period saw the first amulets for protection against evil spirits, incantations to protect the individual from demons, and exorcism through medicine.[156]

After 1000 BCE, following such turbulent years, an expanding network of trade developed in the Old World, stretching from China and India, across Arabia and the Sinai, into Egypt and beyond. The Sinai and Negev were important crossroads for support and control of these trade routes, and the Kings of Judea constructed fortresses to control them. Highlands were gradually abandoned in favor of lowlands, where large settlements based upon extensive runoff agriculture developed. This period coincided with the transformation of the Hebrews from loose nomadic tribes into a centrally controlled state.[157]

Offering of human heads to an Assyrian general.

143. Stone, 1976, ibid., pp.55,59,117-8; Gage, M.J.: *Woman, Church, and State*, Persephone Press, Watertown, Mass., 1980, p.29; Bullough, 1976, ibid., pp.76-81; Scott, 1939, ibid., pp.157-8,82.

144. Bullough, 1976, ibid., p.81.

145. Gage, 1980, ibid., p.158.

146. Stone, 1976, ibid., p.55.

147. Pitt-Rivers, G.H.L.: *Clash of Cultures and the Contact of Races*, Routledge & Sons, London, 1927, p.215; Joshua 7:24-25, Numbers 15:31.

148. Davies, 1981, ibid., pp.60,61,65.

149. Lewinsohn, 1958, ibid., p.35; Joshua 5:3-9.

150. Stone, 1976, ibid., p.196; Joshua 10-13, I Kings 18:40, II Kings 10.

151. Stone, 1976, ibid., pp.169-72; Joshua 10-13.

152. McGhee, 1981, ibid., p.168.

153. Davies, 1981, ibid., pp.60-1.

154. Taylor, 1953, ibid., p.234.

155. Jaynes, 1976, ibid., p.214.

156. Jaynes, 1976, ibid., pp.232-3.

157. Evenari, et al, 1982, ibid., pp.14-7.

Under Saul, the first Israeli King, the Hebrews adopted some of the ways of the goddess worshiping city states they had conquered, no doubt in part due to influences from the large number of slave girls they brought back as war booty.[158] King Solomon, who ruled Israel after c.960 BCE, was called the "wisest man who ever lived". His harem included some 300 wives and 700 foreign princess concubines (gifts from allies and trading partners) and slave girls from conquered lands.[159] Influences of the mother goddess increased and, much to the dismay of the male priesthood, Solomon turned to worship Ashtoreth, Queen of Heaven.[160] These conditions did not persist, however, and other Israeli kings invoked the letter of Mosaic law. The Queen of Maacah was dethroned for worshiping Asherah instead of Yahweh.[161] Jezebel, wife of King Ahab, was thrown from a high window by several Hebrew eunuchs and trampled by horses for her refusal to worship Yahweh.[162] King Ahab also worshiped the mother goddess Asherah. After his death the lives of his 70 sons were demanded by the Yahweh-worshiping King Jehu, who used deceit to massacre others not committed to Yahweh.[163] Athaliah, the daughter of Jezebel was likewise deceived and murdered for continuing to worship the Queen of Heaven.[164] In spite of such attacks, the Queen of Heaven continued to be worshiped in Judea until the Arab invasions of c.630 CE.[165] The Hebrews themselves recited the profound changes in environmental and social conditions which occurred at the time that the Queen of Heaven was given up for Yahweh:

"...to burn incense unto the queen of heaven, and to pour out drink offerings unto her, as we have done, we, and our fathers, our kings, and our princes, in the cities of Judah, and in the streets of Jerusalem; <u>for then had we plenty of victuals, and were well, and saw no evil. But since we left off to burn incense to the Queen of Heaven, and to pour out drink offerings unto her, we have wanted all things, and have been consumed by the sword and by the famine.</u>" [166] [emphasis added]

Puritanical followers of Yahweh also sacrificed humans, a practice which peaked out in the area between 800-700 BCE. According to Davies, burnt offerings were made of infants in large pits, similar to the method of the Phoenicians. King Manasseh (c.723 BCE) made human sacrifices, as did the prophet Samuel, and Ahaz (c.730 BCE) offered up his own children. Others, such as Ezekiel (c.597 BCE), however, condemned child sacrifice.[167]

Assyrians also warred extensively in the area, ruling Syria, parts of Israel, and Egypt.[168] Jerusalem was spared in c.700 BCE when King Ezekiah bought off the Assyrian army with 800 talents of silver, 30 talents of gold, plus "his daughters, his harem, his male and female musicians".[169] However, the Assyrian Empire crumbled from within after c.600 BCE when their desiccating homeland in Syria finally completely converted into desert.[170] Jaynes reviewed various Assyrian letters and texts dating from this phase of severe desiccation, contrasting them to the previous era of Hammurabi:

"...the State letters of Assyria...are imbedded in a texture of deceit and divination, speaking of police investigations, complaints of lapsing ritual, paranoid fears, bribery, and pathetic appeals of imprisoned officers, all things unknown, unmentioned, and impossible in the world of Hammurabi." [171]

" The cruel Assyrian Kings, whose palaces are virile with muscular

Assyrian king Tiglath-pileser III assaults a city, his troops cutting throats and impaling prisoners.

158. Stone, 1976, ibid., p.172.

159. Stone, 1976, ibid., p.181; May, G.: *Social Control of Sex Expression*, Geo Allen, London, 1930, pp.22-3; Tannahill, 1980, ibid., p.63.

160. Stone, 1976, ibid., p.173.

161. Stone, 1976, ibid., p.173.

162. Stone, 1976, ibid., pp.57-8; II Kings 9:30-7.

163. II Kings 10:1-7 & 18-25.

164. Stone, 1976, ibid., p.58; II Kings 11:13-21.

165. Stone, 1976, ibid., p.174; cf. II Chronicles 34:3-7.

166. Jeremiah 44:17-8.

167. Davies, 1981, ibid., pp.61,64.

168. Hawkes, 1974, ibid., pp.11,204.

169. Tannahill, 1980, ibid., p.63.

170. Jaynes, 1976, ibid., p.248.

171. Jaynes, 1976, ibid., p.249.

depictions of lion hunts and grappling with clawing beasts, are in their letters indecisive frightened creatures appealing to their astrologers and diviners to contact the gods and tell them what to do and when to do it. These kings are told by their diviners that they are beggars or that their sins are making a god angry; they are told what to wear, or what to eat, or not to eat until further notice." [172]

Israel fell entirely in 641 BCE, when the Babylonian King Nebuchadnezzar attacked, massacring much of the population and carrying the remnants away into slavery. Under the influences of enslavement, Levite priests and Hebrew philosophers following c.500 BCE adopted an even harsher antifemale and antisexual tone:[173]

"And I find more bitter than death the woman, whose heart is snares and nets, and her hands as bands: whoso pleaseth God shall escape from her; but the sinner shall be taken by her." [174]

"Women are overcome by the spirit of fornication more than men and in their heat they plot against men." [175]

"Happy is the harem which is undefiled...and happy is the eunuch." [176]

Boys were forbidden from playing with girls, and it became a sin for a man to look at or speak with an unchaperoned woman. Masturbation was viewed as a crime warranting death, and touching one's own genitals, even for micturition, was discouraged. Mothers-in-law were forbidden from living with their daughters for fear they might seduce their daughter's husband. Women were expressly forbidden from attacking a man's genitals, and homosexual fears heightened. Fathers could no longer kiss their sons or appear naked in front of them, while mothers could no longer kiss their daughters.[177] Further, parents were encouraged to kill their own children should they revert to "pagan" ways.[178] These attitudes were not restricted to the Hebrews, however, but spread widely across the Near East. As will be discussed in the next section, these changes stretched from Egypt into Persia and from India into Greece and Rome.[179]

The Babylonian conquest resulted in a sharp decline in population of the region,[180] with abandonment of the Negev persisting until the Nabatean period of 400 BCE to CE 100. The Nabateans also built an empire based upon control of trade routes through Northern Arabia, the Sinai, Southern Palestine, and the Negev; their carved rock tombs are found scattered in Northern Arabia.

Before the Nabateans settled the region surrounding Petra, southeast of the Jordan Valley, habitation was possible only through careful allocation and storage of scanty rains, and through clever techniques of runoff irrigation. Hieronymus of Cardia, a contemporary of Alexander the Great,§ describes the early Nabateans of North Arabia (c.300 BCE), before they abandoned the nomadic lifestyle in favor of settlements based upon runoff agriculture:

"In the eastern part of the land situated between Syria and Egypt the Nabateans range over a country which is partly desert and partly waterless, though a small section of it is fruitful. And they lead a life of brigandage and, overrunning a large part of the neighboring territory, they pillage it, being difficult to overcome in war. For in the waterless region...they have dug wells...and have

King Assurbanipal of Assyria (c.650 BCE), typical of the many antifemale, anti-sexual, bloodthirsty thugs ruling the region. Shortly after his rule, an alliance of Babylonians and Scythians attacked and destroyed the drought-weakened Assyrian Empire, sacking the capitol city of Nineveh, and kidnapping away its entire surviving population as slaves. A similar fate befell the Hebrew population of ancient Israel, only a few decades earlier, at the hands of Babylonian King Nebuchadnezzar.

172. Hawkes, 1974, ibid., p.170.
173. For instance, see Jeremiah 5:7 & 3:6-10, Ezekiel 23.
174. Ecclesiastes 7:26.
175. Taylor, 1953, ibid., pp.244-5.
176. Taylor, 1953, ibid., pp.244-5.
177. Taylor, 1953, ibid., pp.244-6.
178. Zechariah 13:2-4.
179. Taylor, 1953, ibid., pp.247-51.
180. cf. population curve, Figure 21.2, in Bowden, et al, 1981, ibid., p.488.
§ Alexander's influences are also touched upon on pages 238-239 and 301.

kept the knowledge of them hidden from the peoples of all other nations... For since they themselves know about the places of hidden water and open them up, they have for their use drinking water in abundance". [181]

Diodorus called the Nabateans "nomads engaged in trade", and there is some indication that they came from South Arabia.[182] Several centuries later, however, the Nabateans settled in Southern Palestine, a now-arid region which Josephus (CE 75) described as being:

"...moist enough for agriculture, and very fruitful. They have abundance of trees, and are full of autumn fruits, both wild and cultivated. They are not naturally watered by many rivers, but derive their chief moisture from rain, of which they have no want. By reason of their excellent grass, their cattle yield more milk than do those in other places; and as the sign of excellence and abundance they are very full of people." [183]

The Nabateans established their magnificent capital city in Petra, southeast of the Dead Sea in the mountains of Edom. The water supply at Petra is scant at present, with no water at all during most summers; at its peak, however, at least 20,000 people lived in Petra. Ruins of hundreds of Nabatean villages are found spread across the region, out into what is now barren desert. Huntington described the region during a visit in 1900, a time when few settlements existed due to lack of water.[184]

"It is hard to realize how greatly Petra has changed. Today its ruins lie in a desolate valley... in spite of the rains of the last three days, water could be obtained only by going half a mile or more either above or below the ruins. Even the small village of Elchi, higher up the valley, was suffering for lack of water to irrigate part of the fields upon which the villagers depend for food. Yet in the past there was water enough not only for Elchi and its dry fields, and for other fields or orchards whose walls appear on every side of Petra, but for the city itself, which must have had at least 20,000 or 30,000 inhabitants, and possibly more... Its inhabitants were not poor...They were among the really opulent people of their day." [185]

When the Romans established alternative trade routes based upon shipping which detoured the Nabatean controlled regions (from India around the Arabian Peninsula to the Red Sea, and then overland to the Nile) the economic foundation of the Nabatean empire was shaken. Roman armies occupied the region without much resistance in CE 106.[186] Nabatea remained under Roman control until the early Christian era, after which it passed into the Byzantine Empire. Byzantium maintained the Nabatean agricultural works until losing the area to Moslem armies in the 600s, whereafter abandonment gradually occurred.

Recent experiments with rebuilt runoff agricultural systems in the Negev indicate that existing rainfalls are sufficient to grow a crop.[187] However, such experiments have not proven that such systems were or could have been the exclusive sustenance upon which the region's large population was based; such would have required a substantially greater quantity of readily available water. The fact that cleaned-out Roman wells in the area remained dry indicates at least a higher groundwater level during Roman times.[188] It is possible that Nabatea flourished at a time when groundwater levels were high, and declined as this resource was consumed. However, it is also possible that the higher groundwater of

Nabatean rock tombs, at Qasr al Bint (top) and Hreba (bottom), N. Arabia.

181. as quoted in Evenari, et al, 1982, ibid., p.345.

182. Evenari, et al, 1982, ibid., p.18.

183. Josephus, as quoted in Huntington, E.F.: *The Pulse of Asia*, Houghton-Mifflin, 1907, p.368.

184. Huntington, E.F. & Shaw, E.B.: *Principles of Human Geography*, John Wiley & Sons, NY, 1951, pp.619-23.

185. Huntington, 1911, ibid., pp.222-3.

186. Evenari, et al, 1982, ibid., pp.17-22.

187. Evenari, et al, 1982, ibid.

188. Butzer, 1961, ibid., p.45.

Nabatean temples (above) and amphitheater (below) at Petra, in the southern Jordan Valley.

189. Von Wissmann, 1972, ibid., p.297.
190. Davis, J.H.: "Influences of Man Upon Coast Lines", in Thomas, 1972, p.510; Evenari, et al, 1982, ibid., p.22.
191. Taylor, 1953, ibid., p.253.
192. Huntington, 1911, ibid., pp.39-40.
193. Huntington, 1911, ibid., p.134.
194. Huntington, 1911, ibid., p.135.

Nabatean times was not a fossil deposit, being derived from higher rainfalls of the day.

It must also be acknowledged that field experiments such as those of Evenari may eventually prove that present rains are sufficient to support a population as large as pre-Roman Nabatea. However, to my knowledge this has not yet been demonstrated. Even modern day Israel, with all its wells and hydrological systems, has not reached the same density of population in the arid regions of ancient Nabatea. Most of the Negev and the region east of the Jordan Valley and Dead Sea is still minimally populated or uninhabited, except where water is piped in from the Sea of Galilee, or where deep wells tap newly found aquifers.

The period 100-300 CE saw increasing Roman influence in the Levant and Palestine, and also increasing conflict between Rome and Persia, with a considerable spread of Bedouin nomadism into Syria and Mesopotamia.[189] Mesopotamian canal systems began to silt up and decline, while various outlying settlements were attacked as the trade-raid pattern of nomadic subsistence increased.[190] It was a period when numerous Persian (Indo-Aryan) and Semitic concepts were sweeping and altering the Roman world.[191]

Huntington raised a number of questions regarding desiccation in Palestine since Roman times. His observations, made in 1911, still retain their vitality:

"In all parts of Palestine unnumbered ruins show that once the population was more dense than now. In Judea, Samaria, and Galilee the terraces of vineyards, the cisterns of farmhouses, and the stones of villages strew scores of barren hillsides, or are scattered far beyond the limits of modern habitations. East of Jordan and in the low south-country of the Negev ruins are still more abundant than in the inhabited parts of the land. There, in places where not a vestige of settled life can now be seen, dozens of ancient ruins proclaim the former existence, not of petty villages, but of prosperous towns and rich cities so large that they were graced with splendid temples and theatres. The same is true around the Sea of Galilee, and for hundreds of miles to the north and northeast in Syria. In many places the sites of the old villages are waterless; elsewhere the limestone hills are so devoid of soil that a single farmer, and far less a whole village, could scarcely find land enough to raise crops. Something has clearly changed."[192]

"One wonders where the ancient people procured water. We found the Arabs drinking water which they had brought two or three miles from the lower part of the Wadi es Seba, where pools stand far into the summer. Wells might be dug, but no one feels like running the risk. As we rode along, our escort pointed out a place near the ruins of Khurbet Abu Khalyun, where an Arab chief is reported to have spent two hundred pounds in digging a well. When it was finished the water was too salty to drink."[193]

"Today the southernmost border of profitable agriculture runs from the Wadi Sheriyah eastward to Debir and the other villages on the southern border of the Judean Plateau. This, too, is the boundary between the domains of the nomadic Bedouin and the agricultural fellahin. In former days the boundary ran farther south, certainly fifty miles away and possibly more. The change in the boundary does not seem to be due to any human action unless man has somehow changed the amount of rainfall."[194] [emphasis added]

Similar examples come from the Syrian Desert. Huntington observed a small Arab village squatted in the ruins of ancient Bosra, which occupied nearly a square mile. The Roman era population of ancient Bosra was estimated at 35,000 to 55,000 inhabitants. Ancient Bosra sported a hippodrome, or sport arena, with seating for 25,000, and a theatre with seating for 9,000. However, the Arab village there in c.1900 supported a population of only 1500 inhabitants, and this at the height of the dry season when people crowded the village to use its spring waters. The village occupied a territory of 1200' by 750', less than 1/25 that of ancient Bosra. Huntington observed other large ruins with small villages crowded around shrunken springs and wells in the Syrian Desert region.[195]

Huzayyin also believed the evidence indicated a lowering of subterranean water levels in Syrian Desert wells since Roman times.[196] The Roman foggara (qanat) at Palmyra, in the Syrian Desert, no longer function via natural spring activity as they once did. Water must now be pumped up from the fallen aquifer through boreholes. Other renovated foggara, wells, and cisterns in Syria evidence a fallen water table since Roman times.[197] Huntington gave the following description of the ruins at Palmyra and the Syrian Desert:

> *"In the early centuries of the Christian Era Palmyra was a great city as large as modern Damascus, which has a population of 150,000 [c.1950 CE]. Ancient writers speak with enthusiasm of its sweet water and beautiful gardens. Its caravans traveled all over western Asia, and it grew so wealthy that its rich citizens took pride in adorning it with wonderful colonnades and temples. Today [c.1950] Palmyra is a vast desolate ruin in the midst of the desert, and harbors only a village of about 1,500 people. Its water is still derived from the old aqueducts, but instead of being sweet and abundant, it has a disagreeable odor of sulphur, and is so scanty that it will irrigate only a few hundred acres of palm trees and gardens."* [198]

The ruins of Palmyra,
Eastern Jordan Valley.
"...a vast desolate ruin in the midst of the desert..."

> *"Syria abounds in dry stream beds, large and small. The winter of 1904-05 was the wettest for decades; but, even then, scarcely a drop of water flowed in most of the channels in the interior of the country. Yet in the days of the Romans it was deemed necessary to build bridges to span these dry streams. For example, at Burak, forty or fifty miles south of Bosra, a sturdy bridge stretches its arches across a channel of gravel. Farther east in the desert the branches of the Wadi Butm and Wadi Rajil, lying south of Jebel Hauran at a distance of about twenty miles from Sulkhad, are always dry. Yet three Roman bridges were built there. In North Syria a bridge spans the dry bed of the Dahna, thirty miles northeast of Homs. Under present conditions so sane a people as the Romans would scarcely build bridges in such locations. More conclusive than the bridges over dry ground are the spring-houses where no springs exist."* [199]

> *"In Syria [Roman baths] are scattered from north to south, and from the desert to the coast. Many are now waterless. One of the most remote is the 'Hamam' or 'Bath' es Serakh in the desert... Today the place is absolutely dry. Even in the unusually moist season of 1904-1905 the great well, a hundred feet deep and eight feet in diameter, was merely a waterless hole yawning among the bones of dead beasts in a dreary expanse strewn with flints."* [200]

195. Huntington, 1911, ibid., pp.292-5.

196. Huzayyin, 1941, ibid., p.107.

197. Butzer, 1961, ibid., p.45; Huntington, 1911, ibid., pp.361-2.

198. Huntington & Shaw, 1951, ibid., p.619; cf Huntington, E.F. & Visher, S.S.: *Climatic Changes, Their Nature and Causes*, Yale U. Press, New Haven, 1922, p.66.

199. Huntington, 1911, ibid., pp.288-9.

200. Huntington, 1911, ibid., pp.289-90.

Aramc

Strong female images persisted in a few places in Southern Palestine up until the Christian and Islamic period, but no later. Noses (and breasts, genitals) of beautiful ancient statuary were routinely smashed by later Christian and Moslem fanatics, as a method to "destroy the evil spirits". At top is a Nabatean "Goddess of Destiny" in a zodiac wheel (c.200 BCE); at bottom is a statue from Palmyra (c.150 AD).

201. Simkhovitch, V.G.: *Toward the Understanding of Jesus and Other Historical Studies*, Macmillan, 1921, NY, p.8.

202. Simkhovitch, 1921, ibid., p.5.

203. Scott, 1939, ibid., p.144.

204. Bullough, 1976, ibid., p.176.

205. Bullough, 1976, ibid., pp.184-6; May, 1930, ibid., p.33; deMause, 1974, ibid., pp.11-3,28; Gage, 1980, ibid., pp.38,255.

206. Lewinsohn, 1958, ibid., p.96; Bullough, 1976, ibid., p.176; May, 1930, ibid., p.41.

The Romans were generally opposed to the austere antifemale and antisexual philosophies which prevailed in the Near East, and they viewed male circumcision as a ghastly, barbaric rite. The heavy-handed attempt of the Antiochus Epiphanes to prohibit Jewish religious practices and circumcision backfired, however, provoking the Maccabees rebellion of 167 BCE wherein Judea became independent from Rome. When Rome regained control of Judea in 63 BCE, their rule was characterized by the same brutal methods of previous Near Eastern kings; as will be discussed later, Rome itself was being transformed through its increasing contacts with Saharasia.

For instance, under Imperial Roman overlordship, Herod slaughtered the children of Bethlehem. Both Jews and early Christians alike were viewed with suspicion by the Romans, who annexed Judea to Roman Syria in 6 CE. The Romans ruled with an iron hand, and brutally put down the Jewish Zealot cult, which often practiced ritual suicide and family murder.[201] The Emperor Titus, representing an increasingly authoritarian Rome, sacked a rebellious Jerusalem in 70 CE, killing or selling into slavery over a million inhabitants by one estimate.[202] Other Roman governors of Palestine were equally brutal sadists: Urbanus had young men castrated on whim, and turned virgin girls over to brothels; Firmillianus mercilessly tortured suspected heretics and criminals.[203]

According to the New Testament, Jesus attempted to bring a teaching of love and compassion into the region, and to reform the Jewish priesthood. However, his teachings were largely rejected, and he was murdered. Still, like most philosophers of that era, he preached against sexual freedom and divorce, calling marriage to a widow the same as fornication or adultery, while praising the virtues of a celibate life.[204] His teachings attracted followers who in later years would amplify the preexisting antifemale, antisexual, and antichild Near Eastern philosophies.

For instance, Justin Martyr (c.153 CE) approved of Christian youth who would undergo castration in order to protect their purity, and approved of the renunciation of marriage. Martyr opposed the "exposing" of infants (infanticide), but only because the parent later ran the risk of violating the incest taboo when the child, saved but sold into bondage, might be met by the parent in a brothel. Cassianus, a Gnostic philosopher, taught that Jesus had been sent to save men from copulating, while other Gnostic writers cast Jesus as a severe hater of women and sex. Tatian, a pupil of Martyr, taught that sex was invented by the devil, and that marriage was corrupt, having been invented by Adam, not god; he counseled newlyweds against the "filth" of intercourse. His influence upon the Syrian Christian church, which later refused to baptize married people, was great. Similarly, Dion Chryostomus (c.390 CE) viewed sex a "cancer" upon the world, reflecting a widespread view that celibacy protected the self against "infestation with demons". His view of children was similarly negative, and he approved of telling them stories about horrid cannibalistic demons, to make them "less rash and ungovernable". Such terrifying stories, he felt, helped to deter children "...when they want food or play or anything else unreasonable."[205]

The Christian theologian Origen (c.250 CE), who viewed women as "the daughters of Satan", castrated himself to avoid sexual temptation, creating a self-castration cult known as the Valesians. Believing they were serving god, the Valesians would forcibly castrate or mutilate the sexual organs of anyone unfortunate enough to fall into their hands.[206]

The philosopher Mani (c.260 CE), founder of the Manichaean cult, taught that sexual intercourse chained the soul to the devil. His sect combined elements from Gnostic, Christian, Zoroastrian, and Greek ideas, and Manichaeanism spread across the Near East into Greece and

Rome.[207] These and similar Eastern concepts flooded both Greece and Rome during the early centuries of the Current Era (CE).

Rome itself gradually absorbed various Eastern philosophies, and its social institutions and laws changed accordingly. During the rule of Septimius Severus (c.200 CE), a Roman Emperor native to Syria, the State was made owner of all arable land. Severus instituted policies of confiscation, torture, and slaughter, demanding to be called dominus, or "master". He also pushed for strict enforcement of adultery laws in Rome, an effort he later gave up on. Of his rule, Wittfogel said "it was as if the spirit of ancient Assyria had taken possession of the palace".[208]

The "orientalization" of Rome was complete by 300 CE with the start of the Papacy.[209] The first Christian Emperor Constantine (c.320 CE) persecuted the followers of Ashtoreth in Canaan as "immoral".[210] In 392 CE the Emperor Theodosius mandated Christianity in the Negev, where the "ecclesiastical population must have been very large indeed".[211] The aesthetic monk Marun preached in Syria at the time, and his followers later founded the Maronite sect.[212]

Rome's territories in the Levant were eventually ceded to Byzantium, while Persia dominated Mesopotamia. Arab armies seized both after 640 CE. Following Arab conquests in the Negev, agricultural settlements declined or were abandoned, and remained abandoned until the present day.[213] The Moslem Caliphate established in Damascus and Baghdad adopted many Indo-Aryan customs previously not held by Semitic nomads, such as the veil, massive harems, and widespread use of eunuchs in state administration and the military.[214] These practices were in turn spread, more or less, to wherever the Moslem faith was instituted.

Egypt and Baghdad fought in the 900s, and another great irruption of Bedouin nomads occurred from Arabia after 1000 CE; these nomads swept across North Africa, Mesopotamia, and the Levant.[215] These were true nomadic herders with no use for settled agriculture; both towns and fields were decimated.[216] Mesopotamian cities thereafter experienced a very sharp population decline, from 1,500,000 down to 300,000.[217] In 1055, the Seljuk Turks swept in from beyond the Iranian Plateau and conquered Baghdad, establishing an even more militant and severe form of Islam which still characterizes the region.[218] The European Crusades also began shortly after these events, with seven major expeditions between 1100-1290. After the crusaders captured the Levant and Jerusalem, contact with the Arab world indirectly fostered a revival of learning and scholarship in Europe.

The centuries immediately following 1200 CE saw climate changes across the Northern Hemisphere toward cooler conditions in northerly latitudes, the start of what has been called the "Little Ice Age". From what has been given above, it would appear that Arabia and Central Asia were once again being abandoned by nomadic cultures, who irrupted into surrounding regions.

Mesopotamian canals were again subject to siltation and salinization around 1200 CE, coinciding with the Mongol invasions. Iraq suffered heavily under the Mongols; however, Jacobsen and Adams attributed much to climatic phenomena:

"By the middle of the 12th century [CE] most of the Nahrwan region already was abandoned. Only a trickle of water passed down the upper sections of the main canal to supply a few dying towns in the now hostile desert. Invading Mongol horsemen under Hulagu Khan, who first must have surveyed this devastated scene a century later, have been unjustly blamed for causing it ever since." [219]

Massive fortification walls of Jerusalem.

207. Bullough, 1976, ibid., pp.189-90.

208. Wittfogel, 1957, ibid., p.211f; Scott, 1939, ibid., p.144.

209. Wittfogel, 1957, ibid., p.211.

210. Stone, 1976, ibid., p.194.

211. Evenari, et al, 1982, ibid., p.22.

212. Mikesell, 1969, ibid., p.21.

213. Evenari, et al, 1982, ibid., p.27.

214. Lewinsohn, 1958, ibid., pp.110-3; Wittfogel, 1957, ibid., p.358.

215. Turnbull, C. *Man in Africa*, Doubleday, 1976, p.173; Wittfogel, 1957, ibid., p.358.

216. Turnbull, 1976, ibid., p.173.

217. Bowden, et al, 1981, ibid., p.488.

218. Bullough, 1976, ibid., p.237; Pitcher, D.E.: *An Historical Geography of the Ottoman Empire*, E.J. Brill, Leiden, 1972, p.23; Davidson, B.: *African Kingdoms*, Time-Life, NY, 1966, p.40.

219. Jacobsen & Adams, 1958, ibid., p.1257.

Jacobsen and Adams would appear correct regarding climate changes in the region, but they appear equally incorrect about the nature of the Mongol invasions, which were of "uncommon magnitude in the history of blood"; the Mongols were the "atom bomb" of their day, massacring millions of people, to include an estimated 5,000,000 souls in only four Near Eastern cities, one of which was Baghdad.[220] As happened so many times in the past, these invaders from the steppes of Central Asia were halted in their march toward Africa only by the Egyptians, this time under the Mameluke Caliphs. Iraq suffered 300 years of devastation under Mongol rule.[221] The Mamelukes controlled the Levant and Syria until the 1500s CE, when it was invaded by the Ottoman Turks. The harsh treatment of agricultural peoples by the Turks led to further land abandonment and degradation, a situation which lasted until the defeat of the Turks during the first World War.[222]

220. Wiencek, H., et al: *Storm Across Asia*, H.B.J. Press, NY, 1980, pp.77,88.
221. Wiencek, 1980, ibid., p.85; Whyte, 1961, ibid., p.108.
222. Whyte, 1961, ibid., p.100.

All photos, Aramco World

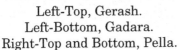

Left-Top, Gerash.
Left-Bottom, Gadara.
Right-Top and Bottom, Pella.

The Decapolis Region of the Eastern Mediterranean

Ten major city-states of antiquity are found scattered across modern-day Jordan and Syria. They are Philadelphia (modern Amman), Scythopolis (Beisan), Gadara (Umm Qais), Hippos (Qal'at al-Husn), Pella (Tabaqat Fahl), Gerash (Jerash), Canatha (Kanawat), Damascus, Dium and Raphana. Nearly all of these ancient cities thrived on more abundant rainfall and surface water runoff than is available in modern times; each city was in turn surrounded by a widespread agricultural hinterland composed of numerous smaller villages and estates with orchards, crops and pasturelands. The cities peaked during Roman times, with the building of a major road network, along with amphitheaters, aqueducts and baths, but the region was well-populated for many thousands of years before the Romans. Increasing aridity — along with complement warfare, deforestation and land-abandonment — eventually dried up their sustenance.

D. ANATOLIA TO THE INDUS VALLEY

The Anatolian plateau, which composes the modern nation of Turkey, with the Kurdistan and Zagros mountains, act as a barrier to moist air moving northeastward out of the Mediterranean. A tongue of moisture penetrates inland along the southern edge of these mountains bringing light winter rains and snow, and a number of streams and rivers flow from the highlands north and south into surrounding dry basin regions. Moist air also penetrates inland beyond the highlands to the southern parts of the Caspian Sea. The central portion of Anatolia is semiarid to arid, while most of the mountains of Iran receive seasonal rains above 250 mm (10"). Summers are short and hot, and winters are long and bitterly cold.

Anatolia itself possesses two east-west mountain chains, the Pontic in the north, and Taurus in the south, with an elevated plateau in between. The plateau descends to meet the Aegean in the west, but retains a high altitude toward the east, where it merges with Transcaucasia and Iranian Azerbaijan. The region possesses many recent lava flows, and a number of large and small lakes, most of which are salty. Lake Van is charged with washing soda, while Lake Urmia (Rezaiyeh) contains "evil smelling salt and black mud".[1]

To the east of the Anatolian-Iranian mountains lie the desert basins of central and eastern Iran, Afghanistan, and the Indus Valley. The Iranian Plateau is today largely salt desert punctuated here and there with oases, and is without a major river. The Iranian Deserts are given names which aptly describe their character: *Dasht-i-Kavir* (Desert of Salt), *Dasht-i-Naumed* (Desert of Despair), *Dasht-i-Margo* (Desert of Death).

At the far eastern end of Iran, on the border with Afghanistan and Pakistan, lie the treeless wastes and salt flats of the Seistan basin, of which Fairservis gives the following description:

"Seistan is perhaps the most extreme of the Iranian plateau... The milder winter period has the greatest rainfall [annual av. of 65 mm or 2.5"]. Snow is rare; temperatures frequently fall below freezing during the night. The smaller irrigation channels and rain pools freeze very frequently... Summer, from May to October, is extremely arid... Temperatures may rise above 115° [F, 46 °C] in the shade. The extreme heat bakes the soil to great hardness... In winter, windy and calm days alternate, but, beginning in May and ending in September, the so-called bad-i-sad-o-bist ruz, or 120-days wind, blows virtually without cessation. The wind is very violent [and]...has a marked effect upon the topography. Dunes, for example, build up with remarkable speed... There are advantages gained in that the wind also blows away the swarms of flies..."[2]

Vicious sandstorms accompany the 120-days wind which may blow up to 160 kmph (100 mph), in a belt stretching from Herat on the Iranian border, through Afghanistan to Pakistani Baluchistan.[3]

Afghanistan also occupies a zone which is relatively dry at present, lying to the northeast of the Iranian Desert, northwest of the Rajasthan Desert of the Indus Valley, and south of the Central Asian arid zone. Farming is largely dependent upon meltwater from the mountains of the Hindu Kush and Himalayas. Hot, waterless, barren deserts, some strewn with varnished pebbles and scattered lenses of volcanic lavas and ash,

1. Mellaart, J.: *The Neolithic of the Near East,* Thames & Hudson, London, 1975, pp.91,195.
2. Fairservis, W.A.: "Archaeological Studies in the Seistan Basin of Southwestern Afghanistan and Eastern Iran", *Anthro. Papers Am. Mus. Nat. Hist.,* 48(1):14-6, 1961.
3. Dupree, L: *Afghanistan,* Princeton U. Press, 1973, pp.4,31; Fairservis, 1961, ibid., p.16.

and some with shifting sands, are found in lower altitudes to the south, towards Iran and Pakistan. They are to this day relatively unexplored.[4] Summertime dust storms occur, sweeping south into Afghanistan from Central Asia:

"The loess often seems to hang in the air, and penetrates every-thing, skin and clothing, with an almost oily consistency, and sometimes blocks out the sun in the afternoon" [5]

At present, the entire region of Pakistan and Northwest India is an arid zone, called the Rajasthan Desert; the central part of the Rajasthan Desert, called the Thar, receives less than 125 mm (5") annual precipitation.[6] The Rajasthan is presently characterized by "extremes of temperature, severe drought, high velocity winds, and low relative humidity".[7] Bryson and Baerreis feel the aridity of the area may be sustained by the great quantity of atmospheric dusts borne into the air, which interfere with normal precipitation processes. As the dust appears to come from the devegetated desert surface itself, important questions are raised regarding the role of human activity (burning, wood cutting, herding, etc.) in affecting atmospheric processes. Plants which protect soils presently grow only at the fringe of the desert, though they used to grow in its central parts prior to being assaulted by humans and herd animals. The observed dust layer stretches across at least the central portions of Saharasia.

"Deep, dense dust over North Africa and Arabia in the spring appears to thin eastward along the southern coast of Iran and Baluchistan...but becomes very dense and deep over the Rajasthan Desert. From there it appears to diminish southward and east-ward..." [8]

Some 60 tons of dust is borne northeastward out of the Sahara and Arabian deserts across the Arabian Sea into Pakistan and India per year, a quantity which is equal to about 1/3 of the yearly sediment transport of the Nile.[9] The dust layer over the Rajasthan is exceptional:

"Over each square mile of Rajasthan sometimes hangs more than five tons of suspended dust. That is several times the average density over Chicago; Rajasthan air is turbid indeed. During pollution episodes the air in large cities does not become as turbid as the average air over the Indus area, but to lesser heights. The daytime Rajasthan sun is a hazy red or is completely hidden. At night dust veils the stars." [10]

Following the glacial epoch, a humid phase occurred in Anatolia, Iran, Afghanistan, and in the Indus Valley; such conditions extended into Central Asia as well.

"In the historical and archaeological records of several Asian countries — Asia Minor, the Aral-Caspian region, Iran Plateau, and Eastern Turkestan — there are well authenticated facts of wet periods marked by normal or abundant rainfall, punctuated by dry arid periods. But these cycles have eventually given place to more or less continuous progressive aridity during the last 2000 years." [11]

Not only did a different and more abundant flora and fauna exist, but more surface water ran off into streams and rivers, and numerous large and small lakes were formed.

4. Dupree, 1973, ibid., pp.4,31.

5. Dupree, 1973, ibid., pp.25-8.

6. Bryson, R.A. & Baerreis, D.A.: "Possibilities of Major Climatic Modification and their Implications: Northwest India, a Case for Study", *Bull. Am. Meteorol. Soc.*, 48(3):136, March 1967.

7. Singh, G.: "The Indus Valley Culture", *Archaeology & Physical Anthropology in Oceania*, 6:180, 1971.

8. Bryson, & Baerreis, 1967, ibid., pp.137,140.

9. Hare, F.K.: "Connections Between Climate and Desertification", *Environmental Conservation*, 4:96, 1977.

10. Bryson R. & Murray, T.: *Climates of Hunger*, U. Wisconsin Press, 1977, p.112; cf. Bryson, R.A.: "Is Man Changing the Climate of Earth?", *Saturday Review*, 1 April 1967, p.112.

11. Wadia, D.N.: *The Post-Glacial Desiccation of Central Asia*, Natl. Inst. of Sciences of India, Delhi, 1960, p.2.

"More or less similar vicissitudes of wet and dry cycles ending in aridity is the history of large numbers of...lake basins in Turkey, Khorasan [NE Iran], Sinkiang [Tarim Basin], and Mongolia."[12]

Moister conditions lasted into Neolithic times possibly down to the early historic era stretching uniformly throughout Southwest Asia, from Palestine to India, and from there into Central Asia, with a generally similar climatic chronology.[13]

Indeed, the entire region stretching from the Eastern Mediterranean to the Indus valley appears to have remained relatively moist until around 2000 BCE. The region then suffered from both climate change and widespread destruction and ruin. Cultural centers in Egypt, the Aegean, Anatolia, Mesopotamia, and the Indus Valley declined, related to the widespread desiccation, famine, migrations and invasions of the period.[14]

While climate change outside the Saharasian belt is beyond the scope of this work, some have attributed the decline of Greece[15] and Rome,[16] to climate changes of a similar chronology. These regions are in the same latitude zone as Anatolia and the Indus, and also are close to the northern edge of the Sahara.

Ancient Glaciers, Rivers, and Lakes:

The easterly sections of Saharasia lie at higher latitudes and hence possess a slightly different climatic history than the western sections, being generally frigid during winter months. During the Pleistocene, these subsections of Saharasia were closer to the great ice sheets, and were subjected to higher rain and snowfalls, lower temperatures and lower evaporation. Traces of former glaciations have been found as far south as the Balkans, Caucasus, and Iranian mountains, as well as in Anatolia, Kurdistan, Lebanon, and Ethiopia.[17]

During the Pleistocene, many of the lakes in and bordering Saharasia stood at levels much higher than today. The examples of lakes in Chad and Ethiopia have already been discussed, as has Glacial Lake Lisan, the remnant of which is the Dead Sea. A number of other shallow but extensive pluvial lakes are found in the Near East, notably in Turkey and Iran; however, the chronology of these lakes is not well known. The largest of these pluvial lakes, Lake Van in Turkey and Lake Rezaiyeh (Urmia) in Iran, have been studied somewhat indicating expansion of depths from 100 to 200 meters and volume increases of 3 to 8 times that of today. Lake Tuz, the "Great Salt Lake" of Turkey, presently stands at 2 meters depth; it once had a volume 56 times that of present.[18] During glacial times, the central desert of the Iranian plateau was an immense lake. The lake shrank during subsequent periods, leaving behind fertile savanna and pasture land.[19] Farrand has estimated the presence of pluvial precipitation 1.6 times that of today for Lake Rezaiyeh.[20]

More recently, Butzer has summarized lacrustine evidences from interior basins of Iran:

"Although there is generally no absolute dating, the sedimentary sequences of the intermontaine basin floors [in Iran]... indicate overall semiarid conditions that were periodically ameliorated by lower temperatures, reduced evaporation ratios, and increased spring runoff from expanded mountain snowpacks. Existing shallow lakes were deeper and greatly expanded; present salt flats harbored shallow sheets of water during winter and spring while higher water tables inhibited wind erosion... At Kerman...deposits

12. Wadia, 1960, ibid., p.13; cf. p.6.

13. Butzer, K.W.: "Climatic Change in Arid Regions Since the Pliocene", in *A History of Land Use in Arid Regions*, UNESCO, 1961, p.41.

14. McGhee, R.: "Archaeological Evidence for Climatic Change during the Last 5000 Years", in Wigley, T.M.L. et al, eds.: *Climate and History*, Cambridge U. Press, NY, 1981, pp.167-8.

15. For Greece, see: Huntington, E.: "The Burial of Olympia", *Geographical Journal.*, 36:657-86, 1910; Carpenter, R.: *Discontinuity in Greek Civilization*, W.W. Norton, NY, 1968; Bryson, R.A. et al: "Drought and the Decline of Mycenae", *Antiquity*, 48:46-50, 1974; Bryson & Murray, 1977, ibid.

16. For Rome, see: Huntington, E.: "Climatic Change and Agricultural Exhaustion as Elements in the Fall of Rome", *Quar. J. Economics*, 31:173-208, 1917; Baynes, N.H.: "The Decline of the Roman Empire in Western Europe. Some Modern Explanations", *J. Roman Studies*, 33:29-35, 1943; Kagan, D. ed.: *The Decline and Fall of the Roman Empire: Why did it Collapse?*, Heath, Boston, 1962.

17. Huzayyin, S.: *The Place of Egypt in Prehistory, A Correlated Study of Climates and Cultures in the Old World*, Memoires, L'Institute D'Egypte, Cairo, 1941, pp.58,95,105-6,119; Butzer, K.W.: "Patterns in Environmental Change in the Near East During Late Pleistocene and Early Holocene Times", in *Problems in Prehistory, North Africa and the Levant*, F. Wendorf & A.E. Marks, eds., S. Methodist U. Press, Dallas, 1975, pp.390-1; Farrand, W.R.: "Pluvial Climates and Frost Action During the Last Glacial Cycle in the Eastern Mediterranean - Evidence from Archaeological Sites", in *Quaternary Paleoclimate*, W.C. Mahaney, ed., Geo Abstracts, Norwich, England, 1981, p.394.

18. Farrand, 1981, ibid., p.396.

19. Whyte, R.O.: "Evolution of Land Use in Southwestern Asia", in *A History of Land Use in Arid Regions*, UNESCO, 1961, p.100.

20. Farrand, 1981, ibid., p.400.

include a rich mushroom flora, abundant ferns... [and] a host of mollusca requiring permanently moist conditions... Saline soils with a completely different, xerophile molluscan fauna are present today." [21]

The Seistan depression itself once contained a very large lake and abundant ground water which supported a large population. Elevated strands mark the old lake level where only a few small scattered marshes and lakes presently exist. Megalithic monuments, abandoned irrigation works, and dry wells of considerable depth may also be found.[22] Huntington discussed a shallow lake in Seistan with ruins on its elevated beaches, about which local inhabitants had preserved legends of its drying up.

"Many, perhaps most, of the facts can of course be explained individually upon other theories than that of climatic change. No other theory explains all the facts. The most significant feature of the history of Seistan is this: A comparison of physiographic, archaeological, historical, and legendary data shows that all these lines of evidence agree in proving that the water supply of Seistan has fluctuated during historic times. The fluctuations agree in time and character with the climatic pulsations of Chinese Turkestan." [23]

In the northwest portion of the Rajasthan are numerous desiccated river valleys which once reached the sea; these rivers now maintain flow only in their upper portions, and often loose themselves in the sand.[24] Allchin, et al's study summarized the large amount of evidence for wetter conditions in the area between c.8000-3000 BCE. Following this period, the region underwent oscillatory climate change toward desiccation.[25]

"One consequence of the formerly increased humidity in Rajasthan is that there are numerous old river courses now blocked by sand deposition, and identifiable on air photographs and by the presence of coarse fluvial gravels in the field. Associated with these old courses are lines and clusters of saline depressions... Such river courses...are often associated with Middle Stone Age artefacts.." [26]

Similarly, the plain stretching eastward from the Indus to the Ganges, on the southern edge of the Himalayas, were both wetter and cooler at the time that the Himalayas were covered with ice.[27] Today, much of India is either arid, or suffers from too frequent failures of Monsoon rains. Failure of rains and famine have been recorded in central India since Moghul times (c.1520 CE).[28]

Ancient Vegetation:

The vegetation history of Anatolia is not well known, but it does appear that a great deal of deforestation has occurred. An analysis by Birand indicated that 70 percent of the countryside was once covered by forests; today only 13 percent remains forested.[29] These forested regions of Anatolia appear to have stretched eastward into the Iranian mountains, according to the distributions of 5000 BCE determined by Bobek.[30] They were decimated by semi-nomadic peoples engaged in extensive grazing, dry farming, and charcoal burning.

Bobek's work in the barren Iranian Plateau indicated that a cover of open forest/herbaceous steppe vegetation existed around 5000 BCE, prior

21. Butzer, 1975, ibid., pp.391-2.

22. Wadia, 1960, ibid., pp.6,12.

23. Huntington, E.F.: *The Pulse of Asia*, Houghton Mifflin, NY, 1907, p.322.

24. Bryson & Baerreis, 1967, ibid., p.136.

25. Allchin, B., et al: *The Prehistory and Paleogeography of the Great Indian Desert*, Academic Press, NY, 1978.

26. Allchin, 1978, ibid., p.59.

27. Huzayyin, 1941, ibid., p.127.

28. Whyte, 1961, ibid., pp.112-3.

29. Birand, H.: "Vue d'ensemble sur la vegetation de la Turquie", *Vegetatie*, V-VI:41-4, 1954; Birand, H.: "La vegetation anatoliennen et la necessite de sa protection", *Report for Sixth Congr. Internat. Union Conserv. Nature*, Athens, 1958.

30. Bobek, H.: "Beitrage zur Klimaokologischen Gliederung Irans", *Erdkunde*, 6:65-84, 1952.

to its degradation into desert.[31] At the time of Bobek's 1952 study, only a fraction of Iran's previously forested areas remained, and probably far less than this exists today:

25% of the Caspian coast humid forest
50% of the Caspian mountain humid forest
    (with severely degraded remnants)
15% to 20% of Zagros oak and oak/juniper forest
5% of Elburz and Khorasan juniper

Afghanistan and Baluchistan were also once forested to a much greater extent than today.[32] The territory, as discussed below, was also home to some of the earliest agricultural and pastoral peoples who now thrive only in select places where seasonal precipitation or runoff from mountain snowpack is regular. Some forested areas persisted in Afghanistan until the Islamic Arab invasions, after which massive deforestation occurred.[33] Deforestation in India was underway by 4000 BCE.[34]

Analysis of pollen from cores out of Rajasthan salt lakes indicate scrub burning by humans with a transition to grassland around 9000 BCE, and a moist period clearly starting by 8000 BCE. The pollen record further indicates a sudden transition to even wetter conditions around 3000 BCE, with a trend towards intense dry conditions by 1800 BCE.[35] These dates generally coincide with the rise and decline of the Harappan civilization;[36] the invasion of the Indus region by Indo-Aryan peoples from the Northwest also began by 1800 BCE.[37]

Naturalistic Petroglyphs of Anatolia – "bison rock" c.8000 BCE

Ancient Artwork:

Ancient petroglyphs are found in the southeastern portion of Anatolia near the saline waters of Lake Van, in the almost inaccessible Tirsin Plateau and Gevaruk Valley. The most ancient rock carvings and paintings of Anatolia date to c.8000 BCE and consist of naturalistic figures of animals similar to the early phase of North Africa and Arabia. Only "much later" do the rock art degenerate into stick-man abstract figures, including evil spirits, demons, and witch doctors. Combat also appears in the later motifs. Uyanik, a Turkish expert on the rock art, feels that this relatively unexplored area contains the most extensive prehistoric petroglyphs in the world.[38]

Farther east, across Iran, Afghanistan, the Indus Valley, and Central Asia, various female figurines shaped out of clay are found, dating from almost the earliest cultural layers. As shown in Figure 59, early female figurines are of a gentle, sensuous quality, displaying prominent breasts and vulva. However, figurines of later periods changed from naturalistic realism to the abstract, in a manner very similar to that observed in the rock art. Figurines from the centuries after 3000-2500 BCE, shown in Figure 60, are more abstract and uniform, depicting females with severe features, small sexual organs, sharp angular bodies, and totally lacking of their previous sensual characteristics. Male figurines with exaggerated phalli also appear for the first time in these cultural layers. This early subtle change in the figurines suggests a relationship to similar but more profound changes in bona-fide goddesses of later centuries, where the "Queen of Heaven" is transmuted into the murderous "Destroyer of the Earth". Such later goddesses often possessed snarling, ferocious, or grotesque faces with exaggerated bosoms, as with the fierce Hindu Goddess Khali.[39]

Gupta's work indicates this stylistic change in the figurines was

31. Bobek, 1952, ibid.

32. Pabot, H.: *Rapport au Gouvernement d'Afghanistan sur l'amelioration des paturages naturels*, 1959 (FAO/ETAP Report #1093); cf. Whyte, 1961, ibid., p.73.

33. Dupree, 1973, ibid., p.19.

34. Hawkes, 1974, ibid., p.206.

35. Singh, 1971, ibid., pp.177-9; Singh, G.: "Stratigraphical and Palynological Evidence for Desertification in the Great Indian Desert", Special Number of the *Annals of the Arid Zone* on Desertification, Jodhpur, India.

36. Agrawal, D.P. et al: "Radiocarbon Dates of Archaeological Samples", *Current Science*, 33(9):266, 1964.

37. Wheeler, M.: *The Cambridge History of India: The Indus Civilization,* 3rd Ed., Cambridge U. Press, 1968, p.127; Allchin, B. & Allchin, R.: *The Birth of Indian Civilization*, Harmondsworth, Middlesex, Pelican Books, pp.144-56.

38. Uyanik, M.: *Petroglyphs of Southeastern Anatolia*, Akademische Druck-u.Verlagsanstalt, Graz, Austria, 1974, pp.12-3; cf. Akpinar, A. & Alok, E.: "The Petroglyphs of Anatolia", *Aramco World Magazine*, 35(2):2, March-April, 1984.

39. Gupta, S.P.: *Archaeology of Soviet Central Asia and the Surrounding Indian Borderlands*, Vol 2, B.R. Publishing, Delhi, 1979, pp.94-5, 125-30, 273-4, 281; Dupree, L.: *Afghanistan*, Princeton U. Press, NJ, 1973, pp.268-9.

reflected in Southern Central Asia, across the highlands of Anatolia, Iran, Afghanistan, and Pakistan, and into Mesopotamia as well. Regarding the style change itself, he felt that:

> *"When human and animal figures are also set in geometric patterns and perfect standards are evolved, the spontaneous creative impulse is lost."* [40]

40. Gupta, 1979 Vol 2, ibid., p.93.

41. Masson, V.M.: "Altyn-depe during the Aeneolithic Period", in *The Bronze Age Civilization of Central Asia: Recent Soviet Discoveries*, P.Kohl, ed., Sharpe, NY, 1981, p.90.

Others have also noted a relative abundance of female figurines at the lowest levels of sites in Iran, with a near absence of such figurines in later periods.[41] Females predominate during times of peace and plenty; males in times of warfare and chaotic displacement.

Figure 56: Chalcolithic Terracotta Figurines (Predominantly Female), Moist and Peaceful Period c.4000-2500 BCE. From sites in Iran, Turkestan, Afghanistan, and Pakistan. (Gupta, 1979, pp.70-129)

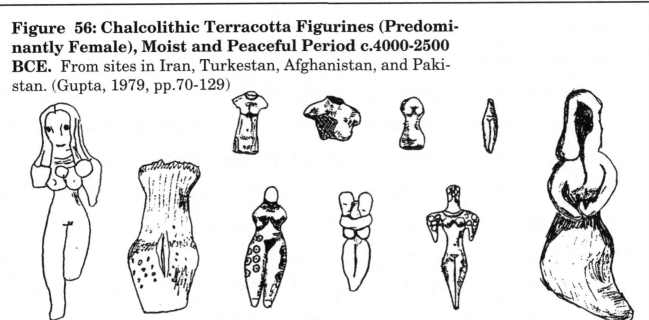

Figure 57: Bronze Age Terracotta Figurines (Female and Male), After the Start of Aridity and Conflict, c.2500 BCE. From sites in Iran, Turkestan, Afghanistan and Pakistan. (Gupta, 1979, pp.114-170)

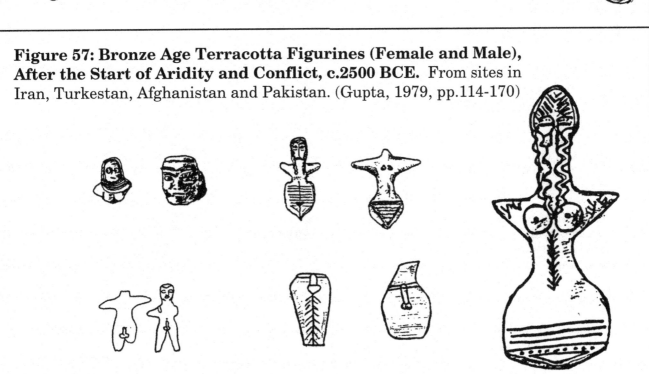

General Settlement Patterns and Culture Change, to Include Influences Transmitted into India and Mediterranean Europe:

The highland region stretching from the Eastern Mediterranean, through Anatolia, Northern Iraq, Iran, Afghanistan, and into Turkestan was moister with light, workable soils around 9000 BCE. Some of the first farming communities appeared there, based upon use of the digging stick and hoe. In the nearby Zagros region, communities based upon animal herding and domestication appeared around or after 9000 BCE. This same date saw the onset of shrub burning in the Indus region, presumably by grain agriculturalists.[42]

By 8000 BCE, the highlands saw the development of mixed pastoral and agricultural communities. One of the earliest Neolithic wheat/barley sheep/goat complexes developed in Afghanistan, later spreading westward across the highland plateaus of Iran, Anatolia, and into the Aegean.[43] Peoples from this broad highland region also began to spread south into Palestine, Mesopotamia, and Arabia, and trade relations developed.[44]

Cattle and the horse appeared in Northern Afghanistan, which then had close ties to Central Asia.[45] By c.7000 BCE, mature Neolithic agricultural communities existed from Anatolia into Iran, with domesticated animals, pottery, weaving, crafts, mudbrick dwellings and female figurines.[46] These highland cultures, particularly in Anatolia, developed and traded with peoples in the Levant, Mesopotamia, and Central Asia. In Anatolia, large communal apartment buildings appeared, similar in appearance to the American Pueblos.[47] These developments paralleled those in Mesopotamia, which have already been discussed.

After c.6500 BCE, the settlement at Catal Huyuk (Anatolia) was developed, based upon irrigation agriculture, cattle breeding, and wild game hunting. Evidence of olive, almond, pistachio, apple, juniper, hackberry, and oak have been found, along with depictions of bull, red deer, fallow deer, boar, wild ass, bear, dog, wolf, and lion. The residents at Catal Huyuk made pottery and wove fabric. They cold-hammered copper ornaments and tools, and cast others in lead. Wood was widely used in construction early in their history, which, when taken with other evidences, suggests a moister climate than today. Use of wood declined in later years. Social conditions in Catal Huyuk were peaceful, with a considerable female influence. Female figurines with pronounced breasts and genitalia abounded, and cranial deformation was not practiced. A retaining wall composing the outer wall of the apartment complex existed, possibly for stock enclosure or flood protection, but no clear evidence of fortifications existed. Great social organization clearly existed, but no clear evidence of either hierarchy or stratification was found. A room was found with mounted bull horns on the walls and floor, but no evidence of a male deity existed.[48] Stern detailed the social situation:

> *"The masculine symbolism of the bull was important in Catal Huyuk, but the long established mother goddess was still supreme, as the many cult images of her show. Some of them look like the fat ladies of Malta. James Mellaart, who was in charge of the University of London expedition that did the excavation, thinks that the local religion was dominated by women."* [49]

Some have argued that the "bull cult" room is actually a male "god's room",[50] but there is no independent or unambiguous evidence for this interpretation, particularly since other evidences point towards strong female influences. Neither the presence of bull horns in a room, nor the

42. Hawkes, J.: *Atlas of Ancient Archaeology*, McGraw Hill, 1974, p.10; Singh, 1971, ibid., pp.184-5; Mellaart, 1975, ibid., pp.70-1.

43. Dupree, ibid., pp.257,263.

44. Mellaart, 1966, ibid., p.9; Mellaart, 1975, ibid., p.39.

45. Gupta, 1979 Vol 2, ibid., p.266.

46. Stone, M.: *When God Was a Woman*, Dial, NY, 1976, p.44; Hawkes, 1974, ibid., pp.131,188; Gupta, 1979 Vol 2, pp.266-7.

47. Hawkes, 1974, ibid., p.139.

48. Todd, 1976, ibid., pp.25-7,30,42-5,74,105,116,131-3,222; Mellaart, 1975, ibid., pp.98-9,101,108.

49. Stern, 1976, ibid., p.223.

50. cf. Jaynes, J.: *The Origins of Consciousness in the Breakdown of the Bicameral Mind*, Houghton-Mifflin, Boston, 1976, p.152.

existence of clay female figurines, are alone sufficient to claim the presence of either a "god" or "goddess". There is no clear evidence these early cultures had a religion as we view it today, with temples and a caste of religious specialists devoted to worship of the deity and apart from the everyday work of the culture. Such evidence certainly appears in later centuries, however.

After 6000 BCE, sites in Anatolia, Syria, and Iraq yield evidence of drought, decline, and/or disturbed social conditions. Aridity appeared to increase from the Levant to Khuzistan. Jarmo, Buran, Sarab, Alikash, and other sites in the Zagros region were deserted as agriculture declined and pastoralism increased.[51] Hacilar in Anatolia yields evidence of the earliest known fortified or animal-impoundment farmstead, also dated after 6000 BCE, with the first destruction evidence, of conflagration, dating to c.5200 BCE. Bodies of victims, mostly children, along with a greasy black material, burnt brick and charcoal were found.[52] Female figurines of a naturalistic style, including animals and mother-child motifs were found in the ruins, and similar artifacts have been found elsewhere in the Anatolian-Iranian highlands.[53] Beycesultan, astride the best route from the Anatolian plateau to the Mediterranean sea, saw the development of either a defensive wall or a stock enclosure at this time.[54]

Other major destructions appeared between 5000-4300 BCE in other Anatolian and Syrian settlements along overland trade routes.[55] It was also around this time period that Mesopotamia was being harassed by Semites from Arabia and hill tribes from the Zagros. Cultural contacts between the highlands and the steppes of Central Asia were established by this time, which was a period of movement and unrest.[56] As discussed by Roper:

"It was not in Palestine, therefore, but notably in Turkey that subsequent developments in the history of warfare can be traced. Events in Syria and Iraq also seem to have played a part." [57]

Mellaart observed that "everywhere were dismal traces of destruction, violence, and semi-barbarism."[58] Catal Huyuk, Mersin, and Can Hasan were either abandoned or burned and destroyed after 4800 BCE, and figurines suggesting cranial deformation were found in Can Hasan and Arukhlo.[59] Mersin suffered from repeated destructions and rebuilding of fortifications, followed by settlement of peoples who later became known as the Hittites.[60] Sites in Mesopotamia and the Levant were also greatly disturbed, as discussed in the previous section, but none permanently so.

Changes After 4000 BCE:

Perhaps even more so than Mesopotamia and the Levant, the highland region between Anatolia and the Indus has been subject to repeated conflict, invasion, foreign domination, and all the associated depredations of war. Such events accelerated after 4000 BCE with a general change toward desiccation which has seriously affected the region.

Fortress architecture appeared more frequently in Anatolia after c.4000 BCE, and female figurines declined in abundance or vanished altogether. The first royal tombs appeared in Alaca Huyuk after this date, as Indo-Aryan peoples continued to arrive from regions to the north. Artifacts from Iran and the Indus region indicate the sharing of artistic and cultic elements with Central Asia.[61]

As determined through artifacts in the Rajasthan, the period 4000-1500 BCE saw a transition from hunting to more or less settled agricul-

New Material 2005: See *Update on Saharasia* (Appendix B) for presentation of newer archaeological evidence on this early dry period in the Eastern Mediterranean and Anatolia (Turkey), with parallel evidence for a temporary, early period of social violence at the same time. Also new material is presented on very early long-distance migration paths across this region into Southern and Central Europe.

51. Mellaart, 1975, ibid., pp.48,88,90, 135.

52. Roper, M.K.: "Evidence of Warfare in the Near East from 10,000-4300 BCE", in *War, Its Causes and Correlates*, M.A. Nettleship, et al, eds., Mouton Publishers, Paris, 1973, pp.322-3,329; Mellaart, 1966, ibid., p.124; Helms identified Hacilar's fortifications as dating after 5000 BCE: Helms, S.W.: *Jawa: Lost City of the Black Desert*, Cornell U., NY, 1981, p.95.

53. Mellaart, 1975, ibid., pp.111,114-5,151,154-5.

54. Mellaart, 1966, ibid., pp.110-11.

55. Roper, 1973, ibid., pp.318,324.

56. Phillips, E.D.: *The Royal Hordes*, McGraw Hill, London, 1965, p.17; Gupta, 1979 Vol 2, ibid., pp.272,302.

57. Roper, 1973, ibid., p.318.

58. Mellaart, 1966, ibid., p.99.

59. Todd, 1976, ibid., p.137; Mellaart, 1966, ibid., pp.102,108; Mellaart, 1975, ibid., pp.122,126,202-5.

60. Mellaart, 1966, ibid., pp.102,105; Mellaart, 1975, ibid., p.126.

61. Mellaart, 1966, ibid., pp.124,151-60.

ture, then to pastoralism and eventually to nomadic herding.[62] These human cultural changes parallel the ecological changes towards desertification in the region. Great interaction, trade, and competition ensued among peoples from the Balkans to the Hindu Kush, and from the Caspian to Egypt.[63]

Various Indo-Aryan peoples began pushing southward out of Southern Central Asia en masse after 3000 BCE, invading the Anatolian-Iranian highlands. A dramatic increase in migration and conflict occurred across the highlands and Central Asia. A "dramatic shift in trade routes and interaction spheres" occurred, along with a "total break in trade and significant cultural interactions", and "abrupt and widespread abandonment of sites".[64] In general, the period saw an abandonment of Near Eastern overland trade in favor of sea routes, which benefited the growth of trade dominance by Minoan Crete.[65] As mentioned previously, these movements of Indo-Aryan peoples powerfully affected Mesopotamia, the Levant, and Egypt. Fortified urban settlements spread across the highlands and connections with Central Asian cultural centers increased. While many cultural influences were traded across the highlands between peoples of the Eastern Mediterranean and Central Asia, the primary direction of migration was southward.[66]

The southward Indo-Aryan migration was widespread: The Mittani pushed south across the Caucasus; the Hittites moved into Anatolia; The Hurrians invaded Mesopotamia; Dorians moved into Greece; Italic speaking peoples arrived in Italy; other Aryan groups spread from Iran to India. Other groups driving horses and wielding battle axes invaded West and Central Europe, where the earliest fortified settlements date to after 3500 BCE.[67] According to Hawkes:

"What caused their migrations southwards, eastwards and to the west is uncertain, but the fact that...they caused upheavals across the Old World from Europe to India is well enough established." [68]

These militant peoples maintained the female in a position subordinate to the male, and they possessed characteristic social institutions. The Celtic, Germanic, Slavic, and Latin (French, Spanish) tongues were derived from these invaders, who arrived with new metallurgical knowledge, a military caste, strongman leaders, royal tombs, an emphasis upon the horse in both myth and sacrifice, and horse-drawn wagons and chariots.[69] The camel also occupied a central place in myths and cults in Afghanistan, Iran, and Turkmenia after c.3000 BCE, the Avesta of the later Persians attributing to the camel a lofty position of great strength and power.[70]

A cultural transition similar to that previously discussed in other regions of Saharasia occurred in the broad region across Anatolia, Iran, Afghanistan, and the Indus Valley. The decentralized, female and child oriented agricultural and trading villages, and the artistic, generally peaceful hunters and pastoralists of this broad, relatively moist highland region, were violently destroyed, or radically transformed toward the antifemale and antichild beliefs and practices of the invading cultures. Divine kingship, the despotic central state, slavery, polygamy, concubinage, female seclusion, the harem, infant cranial deformation, swaddling, circumcision, and other painful and traumatic treatments of infants and children clearly appeared only after c.4000 BCE.[71]

The foreign invaders who stimulated these changes in the Anatolia-Iran-Indus region came first from Central Asia, and later from Arabia; in both cases, the migrations and invasions appear to have been stimulated by the rapidly desiccating environment. In later centuries, the same

Realistic animal impression-seals of early Harappan culture, Indus Valley, found scattered by the hundreds in the ruins of abandoned cities which today are characterized by harsh desert.

62. Allchin & Allchin, 1978, ibid., pp.86,330.

63. Hoffman, M.: *Egypt Before the Pharaohs,* Alfred Knopf, NY, 1979, pp.201-5,293-4.

64. Gupta, 1979 Vol 2, ibid., pp.257-8.

65. Gupta, 1979 Vol 2, ibid., p.258.

66. Masson, "Altyn-depe during the Aeneolithic Period", 1981, ibid., pp.63-4,91-2; Kohl, 1981, ibid., p.xxxii; Gupta, 1979 Vol 2, ibid., p.118; Whyte, 1961, ibid., p.101; Hawkes, 1974, ibid., p.132.

67. Hawkes, 1974, ibid., pp.12,60; Mellaart, 1966, ibid., p.80.

68. Hawkes, 1974, ibid., p.12.

69. Hawkes, 1974, ibid., p.12; Gupta, 1979 Vol 2, ibid., p.311; Grun, 1982, ibid.

70. Sarianidi, V.I.: "Seal-Amulets of the Murghab Style", in Kohl, 1981, ibid., p.246.

71. Cranial deformation, fortifications and other evidences of conflict do appear earlier in Anatolia and Palestine, as previously discussed, but not persistently, from one archaeological period to the next; neither do all of such elements appear at once in these areas until after c.4000 BCE.

migrant cultures who stimulated change in the highlands would stimulate or directly force similar influences westward to Greece, Rome, and Europe, and east to India, Burma, Thailand, and ultimately to certain islands of the Southwestern Pacific.

For instance, the centuries surrounding c.3000 BCE saw the first fortified Neolithic settlements in Greece, a period when Central Asian groups were pushing south across the Caucasus. The most impressive of these fort cities was at Dimini, which yielded evidence of an aristocratic caste society with an immigrant colony. The town plan was significantly different from the open villages of the earlier Neolithic. Knossos and other settlements of Minoan Crete also yielded influences from the East around this time. The Megaron fortifications of Troy similarly appeared, to include stone towers and a substantial enclosure with a single gate.[72]

The Iranian Plateau of c.3000 BCE also yields evidences of great cultural diversity, and a continuing slow infiltration of Indo-Aryan nomadic peoples from Russian Turkestan; among the new groups were the Mittani, who would later harass Mesopotamian settlements. The relationship of the newcomers to urban centers was characterized by a trade-raid pattern, acted out against a rapidly desiccating environment.[73] For instance, one relatively unexplored urban settlement in Northeast Iran, dating to around 3000 BCE, suffered from decline and was covered by a crust of clay, sand, and salt at the time of its rediscovery in more recent times.[74]

In Afghanistan, the southern city of Mundigak also saw the arrival of newcomers; mutilated, decapitated bodies appeared in unkept, poorly constructed graves, revealing an uncharacteristic lack of care or concern for the dead.[75]

It was after 3000 BCE that the Harappan culture suddenly appeared in fully mature form in the Indus Valley. This period was exceptionally moist in the Indus region,[76] and Harappan cities possessed an elaborate drainage system to handle an increased rainfall. Bones of wetlands animal life such as elephant, tiger, rhinoceros, and water buffalo have been found along with pictures of these same animals on Harappa seals. Camels were absent. Marsh and forest are thought to have been extensive, supplying habitat for the animals and wood for the Harappan peoples.[77] Harappan peoples held the females in relatively high status,[78] and had cultural contacts with Mesopotamia,[79] as well as with Dravidian peoples, who had previously entered the Indian subcontinent from Central Asia.[80] Cultural contacts by sea occurred between the Indus Valley culture and coastal India, Bahrain, and Mesopotamia.[81] While the Harappan peoples appear to have had some contact with nomadic cultures to the north, they can not be traced to the northern Indo-Aryans who were at the time harassing peoples farther west.

Around 2600 BCE, Troy was destroyed by peoples from the Russian steppe. Its new inhabitants spread their culture across Western Anatolia and Macedonia.[82] Troy was thereafter re-occupied, enlarged, expanded and newly fortified.[83] Lerna, in Greece, and other sites were similarly "leveled off" and abandoned for a period.[84] From the Aegean across Anatolia, there occurred "...almost wholesale desertion of the coast from Troy to Izmir...a desertion shared by the Ciacus Valley and island of Lesbos opposite".[85]

Iran also witnessed an expansion of steppe peoples from the north, who continued to invade the plateau region, their arrival marked by a characteristic grey pottery.[86] Settlements in Baluchistan and parts of the Indus Valley also saw destruction by fire or flood, though urbanism increased elsewhere in the region.[87]

By c.2400 BCE, Indo-Aryan horse nomads were vigorously pushing

Bronze woman of Mohenjo-daro, Harappan culture, Indus Valley

72. Hawkes, 1974, ibid., pp.113-6,132.

73. Whyte, 1961, ibid., p.101; Hawkes, 1974, ibid., p.188.

74. Aykroyd, W.: *The Conquest of Famine*, Chatto & Windus, London, 1974, p.165.

75. Gupta, 1979 Vol 2, ibid., pp.112-5.

76. Singh, 1971, ibid., p.177.

77. Bharadwaj, O.P.: "The Arid Zone of India and Pakistan", in *A History of Land Use in Arid Regions*, L.D. Stamp, ed,, UNESCO, 1961, p.151.

78. Stone, 1976, ibid., p.72.

79. Hawkes, 1974, ibid., p.206.

80. Ammal, E.K.J.: "Introduction to the Subsistence Economy of India", in *Man's Role in Changing the Face of the Earth*, W.L. Thomas, ed., U. Chicago Press, 1972, p.326; Masson, V.M.: "Seals of a Proto- Indian Type from Altyn-depe", in Kohl, 1981, ibid., p.158.

81. Hawkes, 1974, ibid., pp.11-2; Masry, 1973, ibid., p.222.

82. Thomas, H.L.: "Archaeological Chronology of N. Europe", in *Chronologies of Old World Archaeology*, R.W. Ehrich, ed., U. Chicago Press, 1965, p.391.

83. Hawkes, 1974, ibid., p.134.

84. Hawkes, 1974, ibid., p.117.

85. Mellaart, 1966, ibid., p.140; cf. pp.67-8,148.

86. Thomas, 1965, ibid., p.391.

87. Gupta, 1979 Vol 2, ibid., p.275.

out of Central Asia, across the Anatolian-Iranian highlands, and into Syria and Mesopotamia. The Hurrians, a Caucasian people with a ruling caste of Indo-Aryans, pushed south out of Northern Iran from the mountains at the south end of the Caspian Sea, and established settlements in Assyria.[88] Large royal tombs appeared in Anatolia after this date, tombs which contained both Egyptian and Central Asian paraphernalia, as well as rulers with a cranial structure different from local peoples.[89]

Their influence was also felt in Mesopotamian cities, such as Babylon and Nippur, which at the time were being transformed via Semitic peoples pushing out of Arabia. The Semitic Akkadian dynasty of Sumer was harassed by the nomads from the north, but they were initially subdued. However, these same nomads later overwhelmed the Sumerians.[90]

Numerous sites across Greece and Anatolia reveal violent destruction and fire from the period between 2300-2000 BCE. Tarsus, Beycesultan, Troy, Polath, Lerna, and other sites in the highlands were annihilated. Destruction was widespread and complete, with very low rates of reoccupation. The cultural identity of these peoples is uncertain, as they left few traces behind. Other sites across Eastern Anatolia and Iran, however, revealed the intrusion of Central Asian peoples, and Asian characteristics on pottery. New waves of invaders also swept across Palestine and Mesopotamia at this time, a period when Jericho was destroyed, when the ancestors of the Semitic Phoenicians attacked in the Levant, and Asiatics attacked Egypt, as discussed in previous sections.[91]

The Harappan culture of the Indus Valley was subject to major hydrological changes in c.2200-2100 BCE as the Saraswati river shrank into an insignificant stream, and the Indus changed its course under the influences of flooding and tectonic activity.[92] By c.1750 BCE, the Harappa civilization decayed and vanished in a rapid fashion, leaving behind evidence of massacres and destruction in what once were grand unfortified cities, with irrigation works, fabrics, ceramics and numerous Bronze Age artifacts. While it is certain that both climate change and Indo-Aryan invasions occurred in the region, the relative roles of these forces in the decline of the Indus settlements continue to be widely debated.[93]

Undated settlement mounds from prehistoric times also dot the Baluchistan Hills, to the immediate west of the Indus Valley. The region now only supports a meager agriculture.[94] Contemporary rainfall in the Indus region is believed to be about 1/3 of that which occurred during the early Holocene.[95] Two major rivers mentioned in classical history have since disappeared, and the groundwater table has dropped 75-100 m. [250-350'] in some areas.[96] Taken together, these events speak for sweeping and powerful environmental change.

Wadia reviewed various evidences for desiccation in the region, pointing out the discovery of 39 deserted cities on the dry banks of the Ghaggar (the ancient Saraswati), which now loses its path in the sand dunes of the Rajasthan.[97]

"The same sequence of events appears to have happened here as in Central Asia — increasing dryness in an area of enclosed drainage, scanty precipitation and strong evaporation, leading to eolian action, sand drift, salinity and alkalinity of soil and decay of forests. The geomorphology of the Rajputana [Rajasthan] bears testimony to the prolonged action of these forces." [98]

The subsequent Indo-Aryan settlements of the Indus Valley were not only fewer in number, but they lay on river channels where remaining water supplies were concentrated.[99] Indeed, following 2000 BCE, desicca-

Sudden deaths at Mohenjo-daro, Indus Valley, following the arrival of violent invaders from the north. This well-developed and previously peaceful Harappan city was not significantly rebuilt afterward, given declining water resources and failure in agriculture, with a shift to a nomadic way of life.

88. Stone, 1976, ibid., p.75; Wiencek, et al, 1980, ibid., pp.41-2; Kohl, 1981, ibid., p.xxxii.

89. Mellaart, 1966, ibid., pp.151-60.

90. Hawkes, 1974, ibid., p.168; Whyte, 1961, ibid., p.106.

91. Weinberg, S.S.: "Relative Chronology of the Aegean in the Stone and Early Bronze Ages", in Ehrich, 1965, ibid., p.305; Gimbutas, M.: "Relative Chronology of Neolithic and Chalcolithic Cultures in E. Europe North of the Balkans and Black Sea", in Ehrich, 1965, ibid., p.485; Hawkes, 1974, ibid., pp.117,170; Mellaart, 1966, ibid., pp.83,88,91-5,122-3,174,176.

92. Walls, J.: *Man, Land, Sand,* Macmillan, NY, 1980, p.181; McGhee, 1981, ibid., p.168.

93. see Singh, 1971, ibid., pp.178-9 for summary.

94. Bharadwaj, 1961, ibid., p.151.

95. Singh, 1976, ibid.

96. Wadia, 1960, ibid., p.6.

97. Wadia, 1960, ibid., pp.4,6.

98. Wadia, 1960, ibid., p.6.

99. Bryson, & Baerreis, 1967, ibid., p.139.

Hindu gods, as well as the caste system and a fanatical antisexual, antifemale philosophy, were brought to India by conquering Indo-Aryan warriors from the north. Such gods were often characterized as gigantic fornicators, with many wives, concubines and handmaidens. In this respect, they were similar to the real-life kings and priests who created and worshiped them, kidnapping their endless supply of female slaves from conquered territories. The mythological Krishna, above, had 16,000 wives who bore him 180,000 sons. That is 11 sons per mother, with nothing recorded in myth about the fate of his female children.

100. Tannahill, R.: *Sex In History,* Stein & Day, 1980, p.200; Lewinsohn, R.: *A History of Sexual Customs,* Premier, 1964, pp.36-7; Hawkes, 1974, ibid., pp.12,206; Sastri, 1967, ibid., p.29; Phillips, 1965, ibid., p.42; Gupta, 1979 Vol.2, ibid., pp.119-20,175,230; Davies, 1981, ibid., p.103.

101. Sastri, K.A.N.: *Cultural Contacts between Aryans and Dravidians,* Bombay, 1967, p.39; Hurlimann, 1967, ibid., p.12.

102. Elwin, V.: *The Muria and Their Ghotul,* Bombay 1942; Lewinsohn, 1964, ibid., pp.36-7; Tannahill, 1980, ibid., p.200.

103. Aykroyd, 1974, ibid., p.48; Davies, 1981, ibid., pp.79-82; cf. pp.75,240.

104. Thompson, E.: *Suttee,* George Allen, London, 1928.

105. Boserup, E.: *Women's Role in Economic Development,* Geo. Allen & Unwin, London, 1970, p.49.

106. Thompson, 1928, ibid., pp.19,24, 74; cf. Dupree, 1973, ibid., pp.268-9.

107. Thompson, 1928, ibid., p.29.

108. Shutter, R. & Shutter, M.: *Oceanic Prehistory,* Cummings Pub. Co., Calif., 1975, p.79; Wada, S.: "The Philippine Islands as Known to the Chinese Before the Ming Dynasty"; *Memoirs, Tokyo Bunko, research dept.,* Tokyo, #4, 1929, p.138; Thompson, 1928, ibid., pp.25,127-8; Davies, 1981, ibid., pp.22,75,104,129,135,154,159-60,166-9,185-93.

tion accelerated all across the Anatolian-Iranian-Indus highlands. Invasions from the north continued, and peoples of the highlands continued to harass settled agriculturalists in Mesopotamia, their own region subject to desiccation. Peoples from Mesopotamia, ruled by either native or foreign kings, would periodically dominate portions of the highlands, but never for very long given the continual migratory pressure from Central Asia.

Sanskrit speaking Indo-Aryan migrants settled in Iran, becoming the dominant caste and ancestors of later Persian leaders. Persia experienced an urban decline and abandonment of settlements following 1800 BCE. Indo-Aryans pushed through the Indus Valley, dominating or pushing Dravidian and pre-Dravidian peoples south from Northern India. They brought with them the seeds of Hinduism, the caste system, infant cranial deformation, a general view of the woman as inferior, vaginal blood taboos, and the practice of ritual widow murder (*suttee* or *sati* which in Sanskrit means "chaste woman") or mother-murder.[100] The non-Aryan inhabitants of India became known as *dasyu,* a term which was probably derived from *dasi,* which means slave. The Hindu god-king Indra, according to legend, conquered various fortified cities and was given the name *Parumdara,* meaning "destroyer of cities".[101] Cultures which escaped Hindu-Aryan domination survived only by retreating far to the south, east, or into the rainforests, as was the case with the peaceful Muria.[102] Hindu scriptures reflect environmental conditions of the period, being filled with stories of drought, famine, and prayers for rain; some of the ritual human and horse sacrificial practices of India, also learned from the northern nomads, were connected to rainmaking.[103]

Ritual widow murder was most widely practiced in Northern India, which was firmly under the control of various Indo-Aryan groups, while Southern India, with a Dravidian and pre-Dravidian background, did not adopt it to any significant extent. Such antifemale blood rituals were not limited to merely burning a widow to death, which is horrid enough, but also contained powerful sadistic elements in which the unfortunate woman's own relatives (sons, brothers) were forced to directly participate. A widow might have her throat slit by her own son or brother at the dead husband's gravesite, or be buried alive with his corpse.[104] In later years, antifemale attitudes would spread to infant care practices; preferential female infanticide was greater in the north of India than the south, and northern folklore held that breast milk was good for boys but bad for girls.[105]

Hindu ritual widow murder was in some areas related to worship of the Terrible Goddess Khali which, as previously discussed, appeared as an Indo-Aryan degradation of the original gentle mother goddess.[106] Its roots to Brahmanism and the caste system were discussed by Thompson:

"The rite [widow-murder, mother-murder] was an aparage of rank, but was fostered and spread by priestly influence. It was introduced into Southern India with the Brahmanical civilization... In Malabar, the most primitive part of South India, the rite is forbidden. In Malabar, a matriarchal system prevailed, which may account for the absence of the rite." [107]

Over the centuries, taboos against vaginal blood, widows, and female sexuality in general spread east through India, across the fertile Ganges Valley and delta region, into Burma, Thailand, Malaysia, Cambodia, and later into Indonesia, Bali, the Philippines, and parts of Island Asia and Oceania, where elements of Hindu (and later Buddhist) ritual, to include widow murder and human sacrifice, persisted over the centuries.[108]

The history of the region across Greece, Anatolia, Iran, Afghanistan, and the Indus after 2000-1500 BCE is entirely dominated by mounted warrior nomadism, archers, and the chariot. A widespread decline in urbanism and abandonment of settlements occurred.[109] New empires were built by invading groups abandoning the Central Asian steppe, or by similar groups which previously established kingships on secure water supplies, such as in Mesopotamia. Indo-Aryan peoples firmly entrenched themselves across the highlands, either in colonies or as the rulers of native peoples. Migrations, conflicts, and destruction were widespread.

Ritual widow murder spread across the highlands with the mounted Indo-Aryan warriors, including to Thrace, in Southern Bulgaria, where the Thracians established a new kingdom around 1900 BCE; according to Herodotus:

"Each man among them has several wives; and no sooner does a man die than a sharp contest ensues among the wives upon the question which of them all the husband loved most tenderly. The friends of each eagerly plead on her behalf, and she to whom the honor is adjudged, after receiving the praises both of men and women, is <u>slain over the grave by the hand of her next of kin,</u> and then buried with her husband." [110] [emphasis added]

While mainland Greece was invaded by northern peoples, Minoan Crete retained its original trading culture which possessed a very high status for women. Minoan Crete focused upon the female, and women were not secluded, their dress often exposing the bare breast. These freer conditions were gradually eroded under pressures similar to those experienced upon the mainland.

"Cretan men and women were everywhere accustomed to seeing a splendid Goddess queening it over a small and suppliant male god, and this concept must surely have expressed some attitude present in the human society that accepted it...the fearless and natural emphasis on sexual life that ran through all religious expression...was made obvious in the provocative dress of both sexes and their easy mingling - a spirit best understood through its opposite: the total veiling and seclusion of Moslem women under a faith which even denied them a soul." [111]

Minoan Crete, which encompassed many of the islands to the north (such as modern-day Santorini, or Thera), by c.1800 BCE became a central trading power in the eastern Mediterranean, during the period of warfare and chaos on the mainland. Art and technology flourished in Minoan areas, as evidence by the growth of grand and unfortified Minoan cities. Minoan motifs and technology appeared subsequently, as transplants, across the eastern Mediterranean. The Egyptians viewed the Minoans with respect, benefited from their culture, and newer evidence suggests much of the architecture, ship-building, and other technological innovations of later Greek and Roman periods owes much to ancient Minoa. Not only did the Minoans have multi-storied apartments, but also running water and sewers inside their cities. Steam vents from the volcanic isle were channeled through pipes in the walls of their homes, to heat them, and primitive flush toilets inside the homes connected with an exterior covered sewage system. Primitive steam power, probably learned from the pressure inside volcanic vents, as well as primitive bi-metallic batteries have been suggested in archaeological findings, along with clay water pipes, elegant metal and obsidian cutting tools, woven fabrics,

Terrible Goddess Khali (Hindu) Grotesque female image from a Balinese bronze c.1700 CE. According to myth, Khali squats over her dead consort Shiva and devours his entrails, while her vagina devours his penis. A symbol of anxiety about female sexuality, and object of sacrifices at Hindu temples, this aspect of Khali "the destroyer" stands as a sharp contrast to the possibly older pre-patrist vision of Khali as "Treasure-House of Compassion, Giver of Life to the world, Mother of the Universe".

109. Phillips, 1965, ibid., pp.46-8; Gupta, 1979 Vol 2, ibid., pp.175,230; Dupree, 1973, ibid., p.269; Hawkes, 1974, ibid. p.188.

110. Herodotus, as quoted in Thompson, 1928, ibid., p.21; cf. Phillips, 1965, ibid., p.42.

111. Hawkes, as quoted in Stone, 1976, ibid., p.48.

movable type-faces for imprinting clay with their primitive symbolic alphabet, metallurgical innovations, and delicate ceramic plates, pottery and cups. Minoan ships were second to none, with square-rigged sails, multiple oars and steering tillers. Minoan cities contained massive storage jars, several meters high, containing farm produce from across the region. Socially, the art suggests cooperative and social work arrangements, without fortifications separating a ruling elite from ordinary folk.

The Minoan site of Acrotiri on Santorini was covered by a thick layer of ash, around 1630 BCE, preserving the city from total destruction by a cataclysmic explosion of the volcanic island shortly thereafter — only a crater remains of the island today, its core sunk hundreds of meters deep and filled with Mediterranean waters. Other Minoan cities on the north coast of Crete were destroyed by vast tidal waves produced by this same volcanic explosion. It has been suggested that, had not the Minoans been so destroyed, their rate of technological development was so rapid they might have developed rockets and landed on the moon by the time of Jesus.§ Certainly, the Minoan technology represented the high point of Mediterranean development for many thousands of years, and stands as a proof against the pseudo-scientific dictum that "technology and civilization are dependent upon sex-repression", or that a central-state culture dominated by warriors, kings and priests is somehow more technologically creative and ingenious than decentralized, democratic social groups.

§ This particular idea comes from Charles Pelegrino, whose marvelous book *Unearthing Atlantis: An Archaeological Odyssey* (Random House, NY 1991) gives an overview of recent archaeological findings of Minoan cities. In 1994, about 10 years after the major findings presented in this book were already made, I visited the major Minoan sites and museums on Santorini and Crete, confirming many of Pelegrino's observations, and gathering even more evidence of the profound cultural transformation stimulated in Minoan areas by patristic warriors migrating out of Saharasia.

Grand Unfortified Cities of Minoa

At top is an avenue and plaza from Acrotiri, best preserved of the Minoan sites, on the volcanic isle of modern-day Santorini. The island was called *Thera* – fear – by those who survived its ancient explosion, which was of an historically unprecedented magnitude. Acrotiri is slowly being excavated from under tens of meters of volcanic ash which fell when Thera exploded (c.1630 BCE). The city is so large, it will require an estimated 200 years to complete the painstaking excavations.

At bottom is Knossos, one of several grand unfortified cities on the larger island of Crete, just to the south of Santorini. Knossos, Phaistos, and other Minoan cities on Crete served as central administrative centers for large agricultural hinterlands and trading ports. They were all destroyed by earthquake and/or gargantuan tidal waves which accompanied the explosion of Thera. *All of these cities, their technology and artifacts significantly predate classical Greece and Rome, both of which drew heavily from Minoan influences.*

Figure 58: Artistic Images and Artifacts Of Ancient Minoa (c.2000 - 1700 BCE) Reflecting Early Periods of Relative Peace and Isolation from Mainland Chaos

Minoan frescoes from the early and middle periods of Thera and Crete reflected conditions generally before the arrival of mainland Mycenaean and Asian-Aryan influences, suggesting a society where females had leading social roles, equal to men, and where sexuality was openly expressed, neither cloaked in mystery nor displayed in exaggerated manners (as with phallic architecture and bas-reliefs in Egypt and later Rome). The wall paintings and pottery are expressive and joyful, filled with color and life. Realistic female statues were also found, but are often mystified by contemporary historians and archaeologists. A well-known statue of a bare-breasted woman holding a snake in each hand is often quickly dismissed as a "snake goddess" but the obvious relationship between breast-milk and snake-milk (venom) also suggests the statue represents a respected female healer, handling snakes for the medicinal value of their venom. Other artifacts such as apparent doll-houses (or architect's models) and dancers squashing grapes with their feet in a large tub (suggesting knowledge

of wine-making) are often mystified as "cult objects" needlessly portraying the Minoans as a spiritually-preoccupied "goddess" culture. Likewise, a Minoan bath-tub, decorated with fish, is mis-identified as a burial sarcophagus based upon the similarity in shape to a coffin, even though no bodies have ever been found in such a Minoan tub. In short, there is a subtle evasion of the obvious, a reluctance among contemporary scholars to consider that the peaceful matristic Minoans could have been technologically superior to the patristic, war-dominated societies of the region. Evidence suggests Minoan society lacked major patrist influences until very late in its history. There were no fortifications, war-weapons and other trappings of authoritarian rulers until a very late period, and then only at a few of their cities (i.e. Knossos). They were technologically sophisticated, in ship-building, metallurgy, architecture, sanitation, art, agriculture, medicine, materials-science, and social administration. Egyptian and later mainland Greek and Roman societies obtained much from the Minoans.

Photos of pottery courtesy of Angelos Rethemiotakis

Figure 59: Artistic Images and Artifacts Of Ancient Minoa, From Late Period (c.1700-1630 BCE) of Mainland Patrist-Warrior Influences.

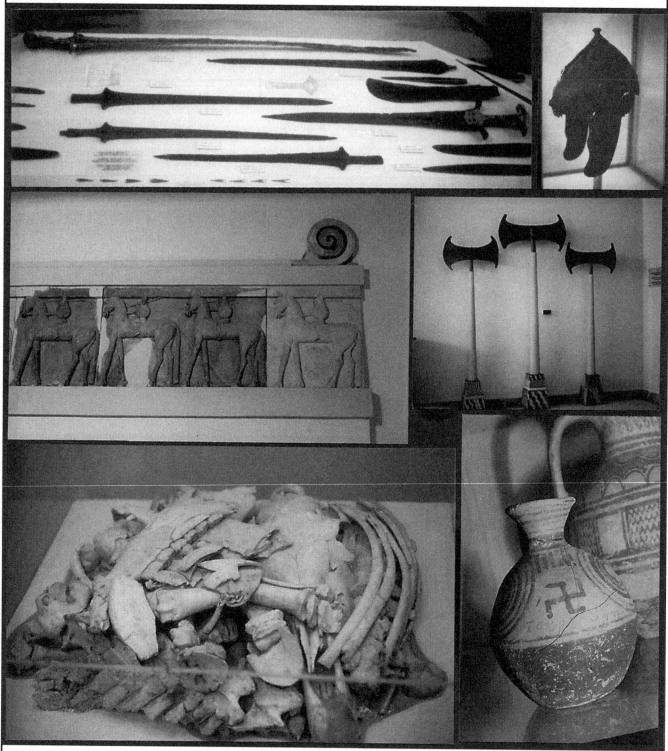

Late-period Minoan culture came to be dominated by distinctly patristic warrior influences from mainland Asia. Pottery quality and artwork degraded into colorless abstract symbols, abundant with the Indo-Aryan swastika. At Knossos and a few other Minoan sites, weapons of war make their first abundant appearance: swords, shields, battle-axes and helmets, etc. The battle-axe would similarly be used as a dominating symbol of central authority in those mainland-infected areas, along with images of giant bulls and horse-mounted warriors carrying lances and shields. Significantly, the remains of a Central Asian (Scythian) type ritual horse-

sacrifice was found on Crete at this time. Female images were degraded, into a stiff, emotion-less, deadly and de-sexualized "goddess", a fitting female companion to the authoritarian blood-drenched "gods" of the period. Gone are the bright, spiraling, sparkling images of life and nature, of the soft and natural sexuality of earlier periods. The volcanic eruption of Thera (c.1630 BCE) finally destroyed Minoan society, but preserved several major cities under ash layers before they were totally transformed by the mainland invaders. Minoa lay undisturbed for thousands of years, only now being uncovered and preserved by archaeologists.

Photos of pottery courtesy of Angelos Rethemiotakis

Minoan Greece eventually came under patriarchal authoritarian influences from the mainland, and central state kingships were established, with large and massive palaces and forts, boasting of virile bulls, horses and giant battle-axes. Gentle and sexual female images were then transformed into rigid, stiff-faced, deadly-appearing "goddesses" (see images and sidebar discussion, previous pages). Knossos was the focus of the Mycenaean Palace Age in Crete, and its occupants dominated the earthquake-prone island. Under such mainland influences, which came very late in Minoan history, both horse-sacrifices and human sacrifice occasionally surfaced.[112] By late Minoan III times, infant cranial deformation appeared on both Crete and Cyprus.[113] Minoan Crete was wiped out following the eruption of Thera in c.1630 BCE.[114] After this disaster, Mycenaean and other mainland Greeks, such as the Dorians, forcibly established themselves upon the devastated ruins, enslaving the survivors, establishing male gods and patrilineal succession, and otherwise lowering the status of women.[115]

The period saw destruction and desertion of settlements on the mainland, possibly in association with the economic collapse which certainly would have followed the tidal-wave destruction of the Minoan trading fleet, and Minoan trading centers. An increase in the building and strengthening of fortifications occurred afterward.[116] Signs of increased fortifications of settlements stretched across the Mediterranean at this time, to as far west as Spain, where burial practices were altered and "all traces of the mother goddess cult disappeared."[117]

112. cf. *National Geographic*, Vol. 159, #2, Feb 1981, p.205; Davies, 1981, ibid., p.55.

113. Dingwall, E.J.: *Artificial Cranial Deformation*, London, 1931, pp.81-2,226.

114. Hawkes, 1974, ibid., p.114.

115. Stone, 1976, ibid., pp.51-2; Jaynes, 1976, ibid., pp.214,255-6.

116. Hawkes, 1974, ibid., p.114.

117. Hawkes, 1974, ibid., p.89.

Spiral-Forms, Life-Energy and Matriarchy Vs. Abstract Forms, High Gods and Patriarchy

Less Armored Decorative Motifs   **More Armored Decorative Motifs**

The time-line of history, after the onset of aridity in Saharasia.

This *generalized* pattern in ancient pottery and artwork motifs was first observed and described by the Swiss scholar Hanspeter Seiler (1992), who argued that early matristic peaceful societies were more in touch with the forces of nature than later, armored patristic cultures; consequently they understood the fundamental relationships between spiral-forms and natural processes, and also could see with their eyes the visible spiral movements of orgone (life-energy) in the atmosphere; spiral forms therefore dominated the artwork of the earliest peoples, while abstractions were dominant among later more armored cultural groups. The spiral movements of life energy phenomena were firstly described scientifically by Wilhelm Reich (1951) who also argued that animistic peoples were generally more in touch with life energy processes in nature. Later armored cultures lost this visual capacity, lost organic sense contact with nature, and began to mystify natural forces into various metaphysical abstractions, high gods, etc. These same armored peoples, it is argued here, were very anti-child, anti-sexual, and violent, arising firstly within Saharasia. (H. Seiler: "Spiralform, Lebensenergie und Matriarchat", *Emotion* 10:137-167, 1992 (reprinted in *Nach Reich*, J. DeMeo & B. Senf, Eds., Zweitausendeins, Frankfurt, 1997); W. Reich, *Cosmic Superimposition*, Wilhelm Reich Foundation, Maine, 1951 (reprinted by Farrar, Straus & Giroux, NY, 1973.)

Indo-Aryan Hittite, Hurrian, and Kassite elements clearly appeared in Anatolia after 2000 BCE, where horse-dominated warfare and conflict again was a major feature. Around this same time, the Kassites, Hurrians and Mittani occupied Mesopotamia, while the Hittites expanded into Assyria. The Hittites were a military caste society which also practiced infant cranial deformation.[118] They were tall, used horse-drawn chariots, worshiped male gods, and rose to power based upon a monopoly of iron-smelting knowledge. They adopted the older Anatolian Sun Goddess of Arinna, but married her to a male storm god. Hittite kings possibly forcibly married the priestesses of the Sun Goddess.[119] After 1200 BCE, however, the Hittites declined rapidly under increasingly arid conditions in Anatolia. Where they once were powerful enough to challenge Egypt over control of the Near East, by 1200 BCE they had been reduced by famine to the point of requesting food aid from Egypt.[120]

Iran and Afghanistan also saw more invasions by nomadic warrior Indo-Aryan horsemen after 2000 BCE. Large numbers of settlements were abandoned or burned in a social decline which stretched into Central Asia as well. Use of the chariot and horse sharply increased as the nomadic lifestyle overtook settled agriculture.[121]

After 1200 BCE, Dorian invasions from Central Asia flooded Greece, and the male-dominated states of Corinth and Sparta were formed.[122] Phrygians displaced the weakened Hittites from Anatolia at this same time, dispersing them and liberating the knowledge of iron metallurgy upon which the Assyrian Empire would thereafter rise. Most of Iran was likewise conquered by new tribes from the north in 1200 BCE.[123] Other Central Asian tribes, such as the Scythians, Cimmerians, and Lydians, continued to pressure and invade the territory from Europe to India during a period of social turmoil and desiccation. A general "dark age" prevailed.[124]

This period centered on 1200 BCE also saw southerly invasions into Italy, Greece and Anatolia by groups from Central and Eastern Europe, who themselves were the ancestors of Central Asian "Kurgan" groups (discussed in the section on Soviet Central Asia). These groups assisted or figured prominently in the destruction of the Mycenaean Greeks, Hittites, and other peoples. Gimbutas believes these were the "Peoples of the Sea" who ravaged the Eastern Mediterranean at the time.[125]

Minoan society represented the high point of Mediterranean development for many thousands of years, and stands as a proof against the pseudo-scientific dictum that "technology and civilization are dependent upon sex-repression", or that a central-state culture dominated by warriors, kings and priests is somehow more technologically creative and ingenious than decentralized, democratic social groups.

"The catastrophic finale to the Mycenaean culture, the collapse of the Hittite empire, the destruction of Kode in Cilicia, Charchemish on the Euphrates, Arasa (Alasia) in Cyprus, Arvad on the Phoenician coast, and the raids on the Egyptian delta are documented not only by the archaeological evidence, but are mentioned by Egyptian sources of Merenptah and Rameses III, Assyrian inscriptions of Tiglat-Pileser I, and the Bible. The Egyptians remembered the fearful events with horror. Their inscriptions say that no lands in Anatolia and Syria could stop the aggressors." [126]

The years after 1000 BCE saw recurrent drought and famine in India, and also the start of cooperation between secular and religious authorities, with a consequent strengthening of caste structure.[127] This was the time of the early Indian law books, which gave fathers a king-like power over other members of his kin group, particularly over his wife and children.[128] This was also the era of *Avesta*, the sacred book of Iran, which codified the relations between master and slave.[129] Iran remained split up among warring groups of Urartians, Mannaeans, Medeans, Elamites, and Persians for many centuries following 1000 BCE.[130]

The years after 800 BCE saw the rise of the despotic central state

118. Dingwall, 1931, ibid., pp.82-3; Hawkes, 1974, ibid., p.132; Phillips, 1965, ibid., pp.7,39,41-2; Mellaart, 1966, ibid., pp.192-4.

119. Stone, 1976, ibid., pp.95,44.

120. Bryson & Murray, 1977, ibid., p.15.

121. Dupree, 1973, ibid., p.269; Stamp, 1961, ibid., p.101; Phillips, 1965, ibid., pp.42,47-8; Gupta, 1979 Vol.2, ibid., pp.119-20,312.

122. Grun, 1982, ibid.

123. Hawkes, 1974, ibid., pp.133,188.

124. Hawkes, 1974, ibid., p.170; Jaynes, 1976, ibid., p.213; Phillips, 1965, ibid., pp.50-2.

125. Gimbutas, 1965, ibid., pp.23,329-35,340-2.

126. Gimbutas, 1965, ibid., p.334.

127. Wittfogel, K.A.: *Oriental Despotism*, Yale U. Press, New Haven, 1957, p.97; Aykroyd, 1974, ibid., p.48.

128. Wittfogel, 1957, ibid., pp.117-9.

129. Gupta, 1979 Vol 2, ibid., p.230.

130. Hawkes, 1974, ibid., p.188.

power in Central Asia. The nomadic, warrior-caste Cimmerians and Scythians fought against each other, the former of which invaded Anatolia, and the latter, Iran and Iraq. Additional turmoil spread across the highland belt stretching from Greece to Iran, where local despotic central states rose to meet the new military challenges from Central Asia.[131]

By c.600 BCE, the Achaemenian Persians under Cyrus emerged supreme in Iran, forming an Empire which dominated the region from the Mediterranean to the Indus, and from Egypt into Central Asia nearly to China. The Achaemenians were a "genuflecting" society, with absolutism, slavery, and extreme subjugation of females and children through devices such as arranged marriages, polygamy, the harem, concubinage, and eunuchism. Eunuchism in particular flourished with the Achaemenid Persians. When Darius the Great conquered Babylon, he ordered a yearly tribute of 1000 talents of silver plus 500 castrated boys.[132] Other conquered nations gave tribute to Persia in material wealth, plus girls and castrated boys for use as sexual slaves.[133]

Persian concepts of the female and the sexual instinct, which were essentially Indo-Aryan in origin, flooded the Near East. The Persian Pahlavi texts of 600 BCE, which related a 7 day creation myth similar to Judeo-Christian Bible, accused the first woman of having intercourse with the devil; she was called "Queen of all whore demons", and the text stated that "since women are subservient to the devil, they are the cause of defilement in men."[134]

Fundamental changes also occurred in Greek institutions as military matters grew in importance, particularly as Persian power was projected across the Near East into the Mediterranean. Following the First Olympiad, the tribal gentile society was abolished and a central state aristocratic society was established. The Greek *Basileia*, a democratic institution based upon military, priestly, and judicial functions, was likewise replaced with the lifetime office of the *Arconship*, held by the upper classes of wealth and property.[135]

The freedoms of women were simultaneously restricted in many areas of Greece, particularly Athens, where married women were no longer allowed to watch the Olympic games.[136] Solon's Arconship was characterized by codification of class structure based upon wealth, and erosion of inheritance through matrilineal gens. He established and stocked the first brothels in Athens, and passed laws attempting to keep women at home.[137] At this time, the father possessed the power of life and death over members of his family, and honoring one's parents was the most important duty next to fearing the gods. Punishments for transgressions were harsh, as when Aesop was hurled from a high cliff for accused theft, and sacrifices of prisoners and condemned criminals to the gods occurred.[138]

A plethora of new world views and religions came into being after 800 BCE, most of which were founded directly upon the antifemale preachings of the Indo-Aryans. Some have argued these beliefs and behaviors evidence a major psychological shift in the human race, involving a universal fear of "pollution" from vaginal blood, or from genital intercourse with the female, with a complement craving for ritual purification or forgiveness, with homosexual overtones, from a male figure of fatherly authority.[139]

"...a remarkable psychological change crept over the Classical world. It was a change marked first by an increase in the amount of guilt felt, and second, with a sudden preoccupation with the after-life." [140]

Greek Acropolis
Fortified Imperial Athens

131. Phillips, 1965, ibid., pp.50-2,55.

132. Wittfogel, 1957, ibid., pp.154,357; Tannahill, 1980, ibid., p.247; Brandt, 1974, ibid., p.498; Gupta, 1979 Vol 2, ibid., p.xviii.

133. Brandt, 1974, ibid., p.496.

134. Stone, 1976, ibid., pp.110-1.

135. Morgan, L.H.: *Ancient Society*, Charles Ken, Chicago, 1877, pp.25-6,267-8.

136. Brandt, P.: *Sexual Life in Ancient Greece*, AMS Press, 1974, pp.93-4.

137. Morgan, 1877, ibid., pp.238,357-66; Tannahill, 1980, ibid., p.104; Brandt, 1974, ibid., p.30.

138. Dodds, E.R.: *The Greeks and the Irrational*, U. California Press, Berkeley, 1951, p.60; Scott, 1939, ibid., p.186; Davies, 1981, ibid., p.55.

139. cf. Reich, W.: *Function of the Orgasm*, Farrar, Straus & Giroux, NY, 1973; Taylor, G.R.: *Sex in History*, Thames & Hudson, London, 1953, p.245; Bettelheim, B.: *Symbolic Wounds*, Free Press, Glencoe, IL, 1954.

140. Taylor, 1953, ibid., pp.243-4.

The Persian Zoroaster (c.550 BCE) preached dualism, condemned nonprocreative sex, and considered the loss of male seed a great sin.[141] Jainist monks of India (c.550 BCE) were forbidden to even look at a woman: "They are to monks what a cat is to a chicken".[142] Widow murder was also prevalent in parts of India around 400 BCE, as were other Indo-Aryan practices.[143] Buddha opposed the rigidities of the caste system and human sacrifice, and viewed women as equals to men, but, as was the case with Jesus, his asceticism attracted followers who later translated his doctrine into one of female inferiority and pleasure-hatred.[144] Laws in India of the period c.400 BCE punished heterosexual transgressions more than homosexual ones.[145]

Far to the west in Rome, these influences were at first minimal. A military democracy developed under Etruscan influences, the upper classes of which came from the Caucasus and possessed a proto-Indo-Aryan background.[146] The early Roman King Severus Tellus (c.550 BCE) organized people into classes based upon wealth, following the example of Solon.[147] Unlike either Persia or Greece, however, Roman women maintained a relatively high status; love-match, common law marriages prevailed, divorce was easy, and there were no general taboos against young men and women sleeping together.

"The Romans regarded the sex-instinct as a natural force, not to be restricted even by the state except under extreme necessity. The enjoyment of it was mankind's natural right, woman's no less than man's. Rome was therefore not very punctilious about either virginity or marital fidelity. If man and wife proved ill-assorted, or could not satisfy one another, let them change partners. Even if complications occasionally arose, that was still better than letting sex starve." [148]

Regarding early Roman beliefs about the natural world:

"...ancient Roman religion knew no mythical histories of personal gods, no worship of legendary heroes, no cosmogony, no conceptions of life in the underworld." [149]

"So little of distinct personality had the spirits and powers of the fields and the farm-house, that the early Romans generally regarded them simply as forms or functional expressions of numen [spirit-energy, similar to Polynesian 'mana'] to which descriptive or personal names were to be assigned only to distinguish them from each other and little more. Consequently, they made no anthropomorphic images of them, had no pictures of them in their minds that they cared to draw on a wall or paint on a vase. It was only later that they learned from the Etruscans and the Greeks how to visualize and humanize their gods." [150]

The ancient Roman lawbook known as the *Twelve Tables* granted married women the right to keep and maintain their own property, and the right to leave an unhappy marriage; "marriage" here is defined by the fact that two people were living together, and the ancient Romans did not look to anyone to sanctify their relationships. The Twelve Tables codified even more ancient practices which prevailed across Europe. Efforts to prohibit the intermarriage between the Plebeian and Patrician classes were overwhelmingly opposed by the Romans, and the Roman King Decemviri (c.445 BCE) was overthrown by army revolt after passing such a law, which was repealed.[151] By law, the Roman father had the right to

A skeletal Gauptama Siddharta (Buddha, c.500 CE) fasting to extremes, a primary method for extinguishing the sexual urge. Buddha's asceticism attracted followers who translated his doctrine into one of female inferiority, pleasure-hatred, submission to priestly authority, and life-negation.

141. Bullough, V.L.: *Sexual Variance in Society and History*, J. Wiley, NY, 1976, pp.68-9.

142. Deo, S.B.: *History of Jaina Monachism*, 1956, p.208.

143. Thompson, 1928, ibid., pp.19-22.

144. Hurlimann, M.: *India*, Thames & Hudson, London, 1967, p.19; Davies, 1981, ibid., p.73.

145. Bullough, 1976, ibid., p.263.

146. Heichelheim, F.M.: "Effects of Classical Antiquity on the Land", in Thomas, 1972, ibid., p.166.

147. Morgan, 1877, ibid., pp.271-3.

148. Lewinsohn, 1958, ibid., p.63.

149. Clemen, C., as quoted in Noss, J.B.: *Man's Religions*, Macmillan, NY, 1965, p.95.

150. Noss, 1965, ibid., p.95.

151. Kiefer, O.: *Sexual Life in Ancient Rome*, Barnes & Nobel, NY, 1951, p.10; Gage, J.: *Woman, Church, & State*, Persephone Press, 1980, pp.129-30; Lewinsohn, 1958, ibid., p.64.

do with other family members as he wished, similar to the Greeks and Persians. However, during the early Roman Republic the rites of *patria postestas* were rarely invoked,[152] giving the appearance that such laws arrived from the Near East with the Etruscans, but went against the feelings of the average Roman citizen to such a degree that they were essentially ignored. In later years, however, as Near Eastern social institutions began to transform the Roman character structure, this would not be the case.

Relatively free conditions prevailed over most of Europe, far removed from events in Saharasia. Young people were allowed to sleep together without penalty, and to choose their own mates. Social institutions developed in which young people interacted, played, and loved. These social institutions were called the *kilbenen* in France, the *kirchgang* in Switzerland, and the *Dorfgehen* in Germany, and persisted until the 1500s CE. The Welsh custom of bundling is a remnant of these free attitudes toward adolescent sexuality.[153] As Rome came under increasing influence from the Near East, these ancient social institutions would come under attack, particularly from the Emperors of Imperial Rome, and later under the Christian Church.

Cultural conditions in Greece, however, were rapidly following the Near Eastern example, particularly as warfare with the Persian Empire intensified. The Pythagoreans (c.500 BCE), such as Empedocles, were celibate, holding chastity in great esteem; the repudiation of inner passion was their foremost goal.[154] Democritus (c.440 BCE), originator of the atomic theory of matter, disapproved of women and sexual activity, stating that the brave man was one who overcame "not only his enemies but his pleasures"; he felt the ultimate disgrace was for a man to be "slave to a woman".[155] The writer Euripides (c.430 BCE) taught that "to bury a woman is better than to marry her".[156] Epicuris (c.300 BCE), the disciple of Democritus, explicitly condemned "sexual intercourse with women", which "never benefited any man".[157] The Greek Sophists, from whence the word "sophisticated" comes, were allied with the Pythagoreans regarding their revulsed attitude toward women and heterosexuality.[158]

Plato (c.400 BCE), who's views would provide additional philosophical underpinnings for later Christian theology, thought sexual desire came from the "lowest" part of the psyche, and that copulation lowered man to the level of the animal.[159] Still, Plato argued against harsh treatment of women and children; he opposed the use of corporal punishment on children, and was for women's equality, even if only within the confines of the home and the compulsive marriage.[160] At the time, at least one Greek general sacrificed a young girl to the gods before a battle.[161]

The antifemale attitudes of the Greeks led to a complement increase in homosexuality, notably the use of young boys for sex by older men, a practice widespread in Persia and other places where taboos against vaginal intercourse with females prevailed. Growing up in Greece at this time often meant, for boys, being used sexually by older men; Athens actually possessed boy brothels and "rent-a-boy" services. Boys were often flogged on the altars of the goddess Diana, who had lost her original gentle character, and sexual attacks upon schoolchildren by teachers and pedagogues were relatively common.[162] At the Athens Museum, for example, there are life-sized head busts of Classical-period Greek teachers sitting on meter-high stone pedestals, carved by their students, with devotional sentences carved on the face of the pedestal — the pedestals also have *detailed carvings of the genitals of the teacher*, just below the devotional inscriptions. Indeed, the sexual confusions of the Greeks, which were strongly institutionalized, grew to such a point that, by around 400 BCE, statues of various goddesses began to display hermaph-

Greek Man-Boy Pedophilia Institutionalized and socially accepted, but nevertheless founded upon a pathological fear and hatred of female sexuality, and of the opposite sex. This same mechanism underlies pedophilia irrespective of gender, and no matter where it is found.

152. Radbill, S.: "A History of Child Abuse and Infanticide", in *The Battered Child*, R.Helfer & H.Kempe, eds., U. Chicago Press, 1974, p.6.

153. Kovalevsky, M.: *Modern Customs and Ancient Laws of Russia*, Burt Franklin, NY, 1891, p.13.

154. May, G.: *Social Control of Sex Expression*, Geo Allen & Unwin, Ltd., London, 1930, p.32; Bullough, 1976, ibid., pp.162-3.

155. Bullough, 1976, ibid., p.165.

156. Brandt, 1974, ibid., pp.72-3.

157. Bullough, 1976, ibid., p.166.

158. Taylor, 1953, ibid., p.135.

159. Bullough, 1976, ibid., p.165.

160. Brandt, 1974, ibid., pp.34-5; Lewinsohn, 1958, ibid., p.32; Radbill, 1974, ibid., p.4; Davies, 1981, ibid., p.56.

161. Radbill, 1974, ibid., p.3.

162. deMause, 1974, ibid., pp.43-4.

rodite characteristics, such as beard and male genitalia.[163] Meanwhile, the Romans were busy passing laws against celibacy, to oppose the growing influences of the Greek philosophers, such as the Pythagoreans who lived in colonies in Southern Italy.[164]

The hierarchal view of the male as superior to female developed parallel to the view of the human as superior to all of the natural world. Aristotle (c.350 BCE), a pupil of Plato and tutor of Alexander the Great, saw man as the highest of all creatures, superior to the female, with all the natural world having been created for him. Aristotle also saw the Earth as the immovable center of the universe. These themes were later adopted by Christian philosophers.[165] As a sign of the times, deMause has indicated that:

"Aristotle's main objection to Plato's idea that children should be held in common was that when men had sex with boys they wouldn't know if they were their own sons, which Aristotle says would be 'most unseemly'." [166]

Heracleides Ponticus, also a pupil of Plato, saw luxury in life and sensuality a right reserved only for the governing classes, with work and toil falling to slaves and the poor.[167] Zeno (c.300 BCE), founder of Greek Stoicism, which was similar to the earlier Cynic school of thought, viewed salvation as reserved to the sage alone; he felt men should be free from passion and calmly accept all occurrences as the unavoidable result of divine will.

Stoic philosophy encouraged indifference to either pleasure or pain.[168] Wherever the relationships between men and women became less spontaneous or instinctual in nature, being laden with fears, guilts, and angers, stoicism or similar philosophies and beliefs found fertile ground, and were both adopted and codified into law. Subsequently, the rights and privileges of women and children declined considerably.

Persia was conquered by Alexander the Great in c.330 BCE, and various Greek Satrapies were established. Indian, classical Mediterranean, and Central Asian ideologies and concepts were thereafter mixed and spread by the Greeks. A new Hellenistic dynasty, the Seleucids, was formed. After Alexander, remnant Greek democratic institutions were severely eroded, the state becoming absolute.[169] Prostration before Greek rulers became commonplace.[170] Democratic institutions only persisted afterward in isolated regions, as with the hill tribes in Afghanistan, where open councils similar to the early Greeks remained active.[171]

Alexander's campaign in India provoked several episodes of massive ritual murder/suicide among Hindus. At Agalassi, 20,000 survivors, mostly women and children, are claimed to have set the city on fire and thereafter jumped into the flames, rather than be captured. Other examples of massive murder/suicide were reported centuries later, when Moslem armies invaded India.[172]

After Alexander, and possibly before, preferential female infanticide permeated the Greek family. Milesia then possessed an excess male population, in spite of the warfare; Poseidippos claimed that "even a rich man always exposes a daughter." Polybius, the Greek historian, thought the problem of infanticide so great that he blamed the killing of legitimate children for the depopulation of Greece.[173]

It appears that Alexander's military incursions into Persia took place under moister climate conditions as compared to today. Wadia identified a subhumid phase in the Indus Valley from before the time of Alexander into the Christian era (c.400 BCE to CE 500), after which a general deterioration in climate set in.[174] Strabo, chronicling Alexander's cam-

Confusions of male and female. Above, an effeminate Greek Ganymedes with male genitals and female breasts being embraced from behind by Zeus, disguised in eagle feathers. Below, a female with penis dancing for Dionysus.

163. Brandt, 1974, ibid., p.126.
164. Tannahill, 1980, ibid., p.127.
165. Glacken, C.J.: "Changing Ideas of the Habitable World", in Thomas, 1972, ibid., p.72.
166. deMause, 1974, ibid., pp.43-4.
167. Brandt, 1974, ibid., p.9.
168. Morgan, L.H.: *League of the Iroquois*, Corinth Books, NY, 1922, pp.153-4; Bullough, 1976, ibid., p.166.
169. Wittfogel, 1957, ibid., p.208.
170. Wittfogel, 1957, ibid., pp.153-4,208.
171. Dupree, 1973, ibid., p.278.
172. Davies, 1981, ibid., p.102.
173. deMause, 1974, ibid., p.26.
174. Wadia, 1960, ibid., pp.4,6.

paigns in 330 BCE, described the area as a productive, fertile land with large trees and rivers.[175] Huntington has discussed the barrenness and ruin found in these same arid basin regions which indicates their fairly recent desiccation, a process which continued into historic times:

> "Ruins are incredibly numerous. Mighty cities of the dead crowd such places as Seyistan, the province of Kirman, the piedmont region of Afghanistan, and the northern border of the great desert of Dash-i-Lut. Those who have followed the track of Alexander declare that today it would be utterly impossible to travel as the conqueror did with a huge army in regions where now a small caravan of twenty or thirty camels can scarcely find water and forage." [176]

> "The 'desert stretch of more than 150 miles' along the north side of the Jaz Morian salt swamp, according to Syke's account, was once thickly populated, as is shown by numerous ruins, and by the remnants of kariz [qanat], to the reported number of 200, which are now dry. Many of the kariz have probably been abandoned because of wars, but that does not explain how Alexander procured water for an army where there are now merely salt pools: nor how he procured forage for all his baggage animals [including elephants! - J.D.] where today a few score can hardly subsist." [177]

These climatic and historical events took their toll upon the land as well as upon the fabric of human society. Rampant deforestation and soil erosion occurred across Greece, Cyprus, and elsewhere, as the demands for charcoal to make iron weapons increased.[178] Massive armies devastated the countryside in search of supplies, and in general promoted a disintegration of irrigation systems.[179]

The Hellenistic Seleucid dynasty remained established over Persia until c.250 BCE, when new groups of nomadic invaders arrived from Central Asia to found the Parthian dynasty. These new groups of invaders, who destroyed the Greek cities, were themselves pushed out of Central Asia by other militant groups in a chain reaction of events which started in China.[180]

Other Central Asian nomad armies pushed across parts of Afghanistan into the Indo-Gangetic plain of India, forming the Mauryan dynasty. Asoka "the Great", a Mauryan emperor, slaughtered hundreds of thousands during his conquests of Southern India. Only later was he touched with guilt over his deeds. Converting to Buddhism, Asoka "the hypocrite" ordered his kitchen to cease killing animals, and had "pillars of morality" erected across the land. While nominally Buddhist, his philosophy was firmly rooted in Indoaryan Brahmanism, with all the deadly trappings of the caste system and hatred of the female.[181] Buddhism mixed with the widow taboos and vaginal blood taboos of Brahmanism spread east from India into Burma, Thailand, and Cambodia, where central state rulership, plus shrines and temples were fostered.[182] Female status declined in these moist regions also, in proportion to the power exerted by strongman leaders and priests. Vaginal blood taboos were exacerbated such that only a Buddhist monk could safely deflower a virgin bride, in a rite similar to the later European *jus primae noctis*, where Kings, Lords, or monks had the right to sleep with new virgin brides on their wedding night:[183]

> "Blood flowing in the first intercourse of man and woman has in it a venomous poison, a root of all evils. Man can die on the spot." [184]

175. as quoted in Huntington, 1907, ibid., p.320.

176. Huntington, 1907, ibid., pp.315-6.

177. Huntington, 1907, ibid., p.317; cf. Huntington, E. & Visher, S.S.: *Climatic Changes: Their Nature and Causes,* Yale U. Press, New Haven, 1922, pp.88-90.

178. Glacken, ibid., 1972, pp.70-1.

179. Whyte, 1961, ibid., p.107.

180. Dupree, 1973, ibid., pp.297-9; Phillips, 1965, ibid., pp.110-2.

181. Hawkes, 1974, ibid., pp.206-8,212; Dupree, 1973, ibid., pp.286-8,296; Whyte, 1961, ibid., p.112.

182. Iwai, H.: "The Buddhist Priest and the Ceremony of Attaining Womanhood during the Yuan Dynasty", *Memoirs, Tokyo Bunko, Research Dept.,* #7, 1935, p.124; Hodgson, B. & Stansfield, J.: "Time and Again in Burma", *National Geographic,* 166(1):91, July 1984; Shutter & Shutter, 1975, ibid., p.79; Wada, 1929, ibid., p.138; Thompson, 1928, ibid., pp.127-8; Davies, 1981, ibid., pp.22,75,104,129,135,154,159-60,166-9,185-93.

183. Iwai, 1935, ibid., p.118; regarding practices in Europe, see Bullough, 1976, ibid., pp.424-5.

184. Iwai, 1935, ibid.

Buddhism mixed with Brahmanism diffused from India overland to Southeast Asia and by sea lanes previously developed by Hindu navigators. Buddhist images in association with ritual widow murder have been found as far east as the Philippines.[185]

Far to the west, stoic influences which opposed the free expression of emotion had completely permeated the Roman character structure by 350 BCE. After the lengthy Punic Wars against Phoenician Carthage (264-146 BCE), the Roman Republic was seriously eroded, a great military expansion began and the first bloody arena-game spectacles appeared. Rome conquered Spain, Gaul, and parts of Britain and the Alps, plus the entire North African coast. Rome displaced Greece in the Eastern Mediterranean after c.200 BCE, but Roman influence never penetrated beyond Eastern Anatolia given continued wars with an independent Persia. Besides acquiring much material wealth, Roman armies absorbed additional aspects of the intensely antifemale, antisexual Greek and Persian philosophy, which previously had been successfully resisted by the Romans.[186] Various mystic cults of the Orient were imported into Rome, such as those of *Cybele*, *Bacchus* (*Dionysus*), *Ma* of Cappadocia, *Adonis* of Syria, *Isis* and *Osiris* (*Serapis*) of Egypt, and *Mithras* of Persia. The relatively passive and inactive gods of old Rome were overthrown, and philosophies of disillusionment, self-denial, and emotionlessness set in.[187]

After the conquest of Syracuse, Sicily, the Romans adopted a legal system found there, which was based upon the absolutism of Egypt and Alexandrian Greece, and rampant slaving.[188] Mithraism, a Persian religion, was widely adopted by Roman soldiers,[189] and the Republic entered a period of strict authoritarianism, with harsher punishments, slavery laws, and ensuing civil wars. Myths of a "Golden Age" appeared in Rome, which was also undergoing intense deforestation and a decline in soil productivity.[190] Some efforts were aimed against the trend towards despotism, however. In 97 BCE, the Roman Senate enacted laws forbidding human sacrifice to the gods, and both Cicero (c.60 BCE) and Seneca (c.50 CE) spoke out against the use of judicial torture.[191] Unfortunately, such efforts at reform were not successful.

Aesthetic Persian, Greek, Hebrew, and Christian philosophies regarding women, children, and sexual conduct were increasingly adopted in Rome, or by Roman armies abroad. By 50 BCE, the Roman Republic collapsed entirely, being replaced by the Imperial Roman Empire of Caesar. Massive numbers of slaves poured in from conquered lands, and farmland increasingly fell into the hands of a privileged few.[192]

During the period, Pliny (c.50 CE) and others taught on the evil influences of menstruous women, who were blamed for all sorts of social ills and plagues, to include dead plants, soured fruits, blackened linen, tarnished mirrors, and miscarried livestock.[193] Epictetus (c.50 CE), a stoic philosopher, taught that the ideal teacher was unmarried.[194] His sick attitude towards children may be surmised from the following quote:

"What harm is there if you whisper to yourself at the very moment you are kissing your child, and say 'Tomorrow you will die'." [195]

The neo-Pythagorean Plutarch (c.100 CE) opposed the use of the whip on schoolchildren, but advocated homosexual relations between boys and older men; he also advocated abstaining from both wine and love in order "to honor God". Numenius, a scholar who traced Pythagorean and Platonic teachings to the Brahmins, Moses, the Egyptians, and the Magi, taught that individual salvation was achieved through the cessation of sexual activity. Chastity and abstinence were widely held to be prerequisites for philosophy, much as they were later for schoolteachers in Europe

Voluptuous female sculptures (above and left), of an erotic and/or exaggerated nature, are abundant within temples and palaces of patristic Hindu India, made to please dominant males (kings, priests, etc.) to whom women were subordinated or enslaved. Such statuary existed within a cultural background of intensive female sexual taboos, giant harems and ritual widow murder. The goddess *Devi* (above) was worshiped by the murderers of the Thuggee cult, who ritually strangled men, women and children in her name.

185. Wada, 1929, ibid., p.138.
186. Taylor, 1953, ibid., pp.252-3.
187. Noss, 1965, ibid., pp.104-5.
188. Wittfogel, 1957, ibid., p.208.
189. Taylor, 1953, ibid., p.252.
190. Glacken, 1972, ibid., pp.71-3; Heichelheim, 1972, ibid., pp.167-8,170; Kiefer, 1951, ibid., pp.77-8.
191. Davies, 1981, ibid., p.47; Scott, 1939, ibid., p.134.
192. Simkhovitch, V.G.: *Towards the Understanding of Jesus and Other Historical Studies*, Macmillan, 1921, ibid., p.85.
193. Brandt, 1974, ibid., p.367.
194. Discourses III, 22.
195. deMause, 1974, ibid., p.30.

and the United States, and medical scholars generally prescribed against too frequent sexual intercourse.[196]

Plotinus of Lycopolis (c.225 CE), a neo-Platonic scholar, felt sensuality was a sin which hindered spiritual knowledge. Plotinus rejected meat, and in general taught the shamefulness of the human body; he had considerable influence upon the Roman Emperor Gallienus.[197] Porphyry (c.280 CE), a pupil of Plotinus who would greatly influence St. Augustine (c.400 CE), wrote "On Abstinence", condemning all pleasures as sinful, to include horse-racing, theater-going, dancing, eating meat, and any form of sex.[198]

Children in Imperial Rome did not fare well either. Imperial Rome saw the use of castrated boys for sexual purposes, many of whom were castrated "in the cradle" for later use in brothels. Images of the god *Pirapus*, who held a sickle in one hand and an erect penis in the other, indicated a widespread and quite rational castration anxiety. And stalks of the giant fennel were used by Imperial Roman schoolmasters to whip schoolchildren into absolute obedience. As was the case in Greece under growing Persian and Near Eastern influences, sexual abuse of children by older men was increasingly common in Imperial Rome.[199]

These views were mirrored by the Christian philosophers of the day. St. Paul (c.50 CE), whose "spirit versus flesh" doctrine would provide the foundation upon which the Papacy was to rest, advocated celibacy over marriage, feeling the latter was a compromise with sin. Blaming Eve for the downfall of man, Paul preached the inferiority of women, forbade their singing in church, preached against the remaining goddess worshipers, and prescribed the strict subjugation of females:[200]

"It is good for a man not to touch a woman" [201]

"He that is married careth for his wife, but he that is unmarried careth for the Lord." [202]

By 250 CE, Christianity was replacing Mithraism in the Roman Empire. The doctrine of the mortification of the flesh brought about episodes of intense fasting, flagellation, and brutal masochistic actions among early Christian devotees: Ammonius tortured his own body with hot irons until covered with burns; Nacarius went naked in a mosquito-infested swamp until unrecognizable; Simeon ulcerated his flesh with an iron belt; Ponticus spent freezing nights under a water fountain; Ignacio, one of many Romans who engaged in a form of Christian "devotio", stated that he would provoke the lions in the arena such that they would be especially cruel; the Christian cult of Valesians initiated self-castration, and forcible castration and mutilation of kidnapped misfortunates.[203]

While Christianity was not adopted by the Roman Emperors until after Constantine, in c.310 CE, the Romans and their Emperors were greatly influence by the many of the same Semitic and Indo-Aryan beliefs which were later held by the Christian theologians. Divine kingship, ritual human sacrifice in the arena, and increasing restrictions on the rights and privileges of women characterized the Roman Empire, which was usually led by megalomaniac tyrants whose names are today synonymous for brutality, sadism, and extremes of sexual pathology.

For instance, Julius Caesar (c.45 BCE) was Rome's first Divine King, with a temple and priests dedicated to him. He ordered the construction of the first large amphitheater in Rome in which gladiator and sacrificial spectacles took place.[204]

Octavius Augustus (c.10 BCE) sacrificed over 300 men on the alter of Caesar, and enacted various laws to "protect the family". Women who

Adam and Eve, evicted from the Garden of Eden by an angel for sexual sins, according to popular interpretations. The myth appears rooted in the factual existence of a real garden-like environment in those parts of Saharasia which later dried up and became harsh deserts. Judaism, Christianity, Islam and Hinduism were all developed in the first instance by peoples who lived for generations under the harsh conditions of desert nomadism.

196. Bullough, 1976, ibid., pp.167,170-1; Radbill, 1974, ibid., p.4; deMause, 1974, ibid., p.45.

197. Kiefer, 1951, ibid., pp.141-3; Brandt, 1974, ibid., p.295; Bullough, 1976, ibid., p.171.

198. Bullough, 1976, ibid., p.172.

199. deMause, 1974, ibid., pp.43-4,46; Radbill, 1974, ibid., pp.3-4.

200. Taylor, 1953, ibid., pp.260-1; Bullough, 1976, ibid., pp.178-9; Stone, 1976, ibid., p.194; Gage, 1980, ibid., pp.53-4.

201. Stone, 1976, ibid., p.194.

202. Gage, 1980, ibid., p.80.

203. Taylor, 1953, ibid., p.257; Davies, 1981, ibid., p.130.

204. Bullough, 1976, ibid., p.135; Gage, 1980, ibid., pp.177-8.

were found guilty of adultery would loose their dowries, with 1/2 going to her husband, the other 1/2 going to the State. Divorce was also made more difficult. Laws were enacted forcing remarriage of widows within 2 months, and divorcees within 18 months; unmarried men would loose their inheritances, while sterile couples would inherit only 1/2 of that left to them. A moralizer against adultery and sex between the unmarried, Augustus lived a lifetime of adultery, requiring his third wife to recruit young virgin girls from the lower classes to help him overcome his impotency; at the same time, he banished and murdered his granddaughters for their sexual transgressions. His rule saw the introduction of formal marriage contracts, more frequent application of various punishments, including the death penalty, plus sadistic executions, using lions and bears in the arena. He banished the popular poet Ovid to his death on the Black Sea for his heterosexual love poems. The Roman nobleman Pacovius sacrificed himself for Augustus, an act of self-sacrifice and emperor-worship, called "devotio". Voltaire appropriately called Augustus "a debauched murderer".[205]

Tiberius (c.35 CE) forbade human sacrifice on the altar of Saturn, but ordered men and women to be severely tortured and butchered in grotesque ways; he also sexually abused temple priests, children, and infants. Caligula (c.40 CE) was a debauched sadist of the worst sort, who fed people to hungry lions and bears, and ordered children executed in front of their parents. He impregnated his sister, and then butchered her. He was finally killed by his own officers for his sickness. Nero (c.60 CE), a bisexual sadist who had young boys who pleased his eye castrated, would rape male and female prisoners tied to stakes. Vitellius (c.65 CE), a debauched pedophile, was killed by his own soldiers for his sickness, as was Heliogabalus (c.222 CE), a transvestite homosexual who filled the palace with eunuchs. Trajan (c.100 CE) mercilessly persecuted Christians and Jews. Commodius (c.190 CE) instituted human sacrifice as part of Mithraism, and studied the entrails of children for prophesy. Severus (c.200 CE), Decius (c.250 CE), Gallus (c.252 CE), and Valerian (c.255 CE) were all short-lived tyrants of a bloody and sadistic nature. Diocletian (c.290 CE) massacred thousands of Christians, boasting of its eradication, and finally institutionalized the use of eunuchs in the Roman administration; certain positions were reserved for eunuchs, who were given preferential promotion, prompting many ambitious parents to have their own sons castrated.[206]

While other Roman Emperors, such as Domitan (c.90 CE) and Hadrian (c.130 CE), attempted in some ways to curb these trends in the Empire, they generally failed in their efforts, enacting their own harsh brands of antifemale and antisexual legislation in the process. Domitan, for example, tortured animals as a child; later, as Emperor, he set dwarves and women to fighting each other in the arena. The first Christian Emperor Constantine (c.325 CE) legislated against the wanton killing or castration of slaves, but stimulated the castration of slaves abroad by allowing purchase of previously castrated slaves from "barbarian peoples".[207]

Constantine's Bishops also warred against marriage for priests, and urged abstinence for those already married.[208] After the capitol city was moved from Rome to Constantinople, prostration was introduced,[209] and the Christian Empire became a haven for eunuchs who occupied important offices and generalships.[210]

Constantine ordered rape to be punishable by being burnt alive at the stake, or by being torn to pieces by wild beasts in the arena. This law was extended to include the "seduction" of unmarried girls under 25. If the girl said she consented, she too was likewise murdered. Marriage would save

Octavius Augustus (c.10 BCE)
Emperor of Imperial Rome
"a debauched murderer".

205. Cory, D.: *Homosexuality, A Cross-Cultural Approach*, Julian Press, NY, 1956, pp.106,252; Lewinsohn, 1958, ibid., pp.75-7; Kiefer, 1951, ibid., pp.95-6,307-10; Simkhovitch, 1921, ibid., p.128; Bullough, 1976, ibid., p.138; Tannahill, 1980, ibid., p.134; Davies, 1981, ibid., pp.47,50.

206. Mantegazza, P.: *The Sexual Relations of Mankind*, Eugenics Pub., NY, 1935, p.104; Kiefer, 1951, ibid., p.151; Lewinsohn, 1958, ibid., p.77; Cory, 1956, ibid., pp.256-7; Bullough, 1976, ibid., pp.133,327; Wittfogel, 1957, ibid., p.357; Davies, 1981, ibid., pp.47,50-1; Scott, 1939, ibid., pp.141-4; de-Mause, 1974, ibid., p.45.

207. Kiefer, 1951, ibid., p.84; Bullough, 1976, ibid., pp.135,326; Scott, 1939, ibid., p.303; Davies, 1981, ibid., p.47.

208. Gage, 1980, ibid., p.251.

209. Wittfogel, 1957, ibid., p.154.

210. Wittfogel, 1957, ibid., pp.357-8.

the pair from death, but they were then exiled and their property confiscated. Slaves who helped arrange a liaison between "illegal" lovers were murdered by having molten lead poured down their throats, and male slaves who had intercourse with free women were likewise punished. Constantine's rule saw the application of numerous new sadistic methods of torture and execution, and the list of possible infractions grew longer.[211] Such was the legacy of puritanical Near Eastern philosophy in the dying Roman Republic; unfortunately, things would not end here.

During these early centuries CE, events closer to the heart of Saharasia, in Iran and India, were once again determined by Central Asian powers. The Persians under Parthian domination again enlarged their territory, and other Central Asian groups maintained a grip upon Afghanistan, the Indus Valley, and parts of India.[212] The Central Asian Huns invaded parts of Iran and India around 200-300 CE, establishing kingships, and the Parthians were overthrown in Iran in c.224 CE by local rebellions. The native Persian dynasty of the Sassanids was thereafter formed. The Sassanids were also a genuflecting society of Indo-Aryan moulding, and they extended their empire over much of the Kushan territories in Central Asia and India.[213] Other local rebellions occurred in India, where the native Gupta dynasty pushed the Saka and Kushan nomads back north toward their Central Asian homeland.[214] During this period of intense warring and shifting of empires, the great forests of Afghanistan were decimated; according to Dupree, "man remained, but the forests never returned."[215] A similar ecological disaster occurred in India, where "vast stretches of territory, notably in the Swat Valley, once prosperous, wore an appearance of desolation."[216]

The Hindu *Code of Manu* was written during this same period, codifying preexisting beliefs on the legal and illegal forms of marriage and intercourse. The Code required absolutions for intercourse with unmarried women, women of lower castes, or menstruating women,[217] and it commended the practice of ritual widow murder.[218] Widow celibacy and ritual widow murder increased in importance under the Gupta leaders, and further restrictions were placed upon women. Rulebooks were generated which governed the love life of the masses, the apparent aim of which was to reduce the spontaneous, instinctual nature of sexuality and marital relationships.[219] Following 400 CE, Hinduism was reestablished and Buddhism largely vanished from India.[220] Hunnish invaders ruled from Punjab, in Northwest India, where infant cranial deformation was introduced.[221] The murderous cult of Thuggee also traced its origins to the Hunnish period, perhaps originally as an underground resistance to the invaders.[222]

Events in Rome continued in a similar pattern. After Constantine, all Roman Emperors were of the Christian faith. The philosophies of "spirit versus flesh" dualism, and mortification of the flesh, of female inferiority, the inherent sinfulness of women and children, the inherent sinfulness of all sexual matters, including married heterosexual intercourse, and other such beliefs were increasingly spread into Europe. Wherever the Christian faith spread, usually by force of arms, so too were these beliefs codified into new laws by which the "converts" were governed. The system of Bishops developed and clerical rule over secular affairs increased, particularly over "family" matters: childbirth, love, sexual behavior, education of children, marriage, death, and inheritance.[223]

The rule of the Christian autocrats and Bishops over such matters of daily life was harsh and sadistic in the extreme. For instance, by force of law Christianity entirely replaced all other religions and ended the system of female oracles, or Sibyls, in the Eastern Mediterranean.[224] Castration of young boys grew fast, as a means of political advancement

211. Pitt-Rivers, G.H.L.: *Clash of Culture and Contact of Races*, Routledge & Sons, London, 1927, p.187; Scott, 1939, ibid., p.158; Kiefer, 1951, ibid., p.95.

212. Dupree, 1973, ibid., pp.291-300; Tannahill, 1980, ibid., p.212.

213. Wittfogel, 1957, ibid., pp.104,154; Phillips, 1965, ibid., pp.112,120.

214. Dupree, 1973, ibid., pp.110,302.

215. Dupree, 1973, ibid., p.19.

216. Whyte, 1961, ibid., p.112.

217. Bullough, 1976, ibid., pp.247-8.

218. Thompson, 1928, ibid., p.19.

219. Lewinsohn, 1958, ibid., pp.37-9; Tannahill, 1980, ibid., pp.202,228.

220. Hurliman, 1967, ibid., p.19.

221. Dingwall, 1931, ibid., pp.79-80,86.

222. Bruce, G.: *The Stranglers: The Cult of Thuggee*, Harcourt, Brace, & World, NY, 1968, p.11.

223. Bullough, 1976, ibid., p.385.

224. Jaynes, 1976, ibid., pp.331-2,347; Taylor, 1953, ibid., pp.253-4.

within the Empire, and also with a new twist: as a cure for various "diseases".[225] The tortures of Constantine were perfected, and the list of crimes grew longer. Flagellation with a lead-weighted whip was proscribed for the new crime of heresy. Christian castration and suicide cults also developed.[226]

Europe and the Christianized Roman Empire also saw a frenzy of ritual female murder between 300-1800 CE which rivaled events in India or elsewhere. The death penalty for ecclesiastical offenses was enacted and vigorously enforced after 385 CE.[227] Various "sexual sins" were also progressively punished by death, and since women were considered to be the harbingers of sin, the penalties of "slow death by torture" fell disproportionately upon them.[228]

Neither did infants escape the cultural turmoil — crying, fussy babies were considered to be evil "changelings", an imposter who was exchanged for the real baby "by a fairy or the devil". Killing or neglecting a baby until death was murder, but not so with a "changeling" baby. Even St. Augustine, who opposed the use of torture, agreed that some babies: "...suffer from a demon...they are under the power of the devil...some infants die from this vexation."[229]

A new "holiday" was developed for children called "Holy Innocents Day" (*Childermass*), when all children were whipped to make them remember the massacre of Herod. This mirrored similar practices in later centuries, where epileptics and "the insane" were whipped by clerics, in order to expel demons[230] (similar technologically-disguised sadistic traditions have been preserved into the modern era by compulsory psychiatric institutions).

Justinian "The Great", Pope of Eastern Rome in c.530 CE, ordered mutilating torture for anyone who insulted a priest or bishop in a church; St. Augustine persuaded the Council of Orleans to deny funeral rites to anyone who committed suicide while accused of a crime, in spite of the fact that severe tortures were in use at the time. Suicides were called the "martyrs of Satan".[231]

Central Asian nomads continued to determine events in the Anatolian-Iranian highlands, as well as in East Europe. The Huns pushed the Goths westward out of East Europe into eventual conflict with Rome; Goths sacked Rome in 410 CE. Huns ravaged Europe as far west as Gaul in the 400s, almost sacking Rome between 410-413, and later in 451. Under the weight of these invasions, the central power of Christianized Imperial Rome was broken up and a "dark age" set in between c.400-900 CE. Huns also pressed into Iran and India.[232]

By 500 CE, the region stretching from Anatolia to the Indus was divided among various empires with a generally similar view of the female, as a creature to be either avoided, subordinated, and/or dominated. Byzantine Kings ruled Anatolia, Sassanid Persians occupied Iran, and Huns ruled in the Indus Valley. Persians warred against kingdoms in the Levant during the early 600s, after which the Islamic Arab armies pushed northward.

According to Vredenburg, the Baluchistan hills west of the Indus maintained their agricultural productivity into the Moslem period, after which the region declined.

"In all the valleys around Zara there are to be seen hundreds of stone walls, which are called 'gor-band', or 'dams of the infidels'. Sometimes they stretch right across the flat pebbly floors of the great valleys, which, for want of a better name, are termed 'rivers'. They also occur across the entrance to most of the tributary ravines and at various heights above the main valley. The country is quite

225. deMause, 1974, ibid., p.47.

226. Davies, 1981, ibid., pp.130-1.

227. Taylor, 1953, ibid., p.265.

228. Bullough, 1976, ibid., p.347.

229. deMause, 1974, ibid., pp.10,47; Scott, 1939, ibid., p.134.

230. Radbill, 1974, ibid., p.3; Howard, D.: *1066: The Year of Conquest*, Dorset Press, 1977, p.28.

231. Davies, 1981, ibid., p.130; Scott, 1939, ibid., p.45.

232. Kiefer, 1951, ibid., p.357; Tannahill, 1980, ibid., pp.124,136; Dupree, 1973, ibid., p.303; Phillips, 1965, ibid., pp.120-4.

uninhabitable for want of water..." [233]

By 700-900 CE, remaining small settlements in the Indus Valley also were abandoned with a turn towards nomadic lifestyles.[234] Just prior to this period of abandonment, frequent dust storms were recorded, and a great spreading of the desert had occurred by 1000 CE.[235]

Moslem Arabs conquered the Sassanids after 650 CE, sweeping across the highlands into India and Central Asia, establishing the Abbasid Caliphate.[236] The Abbasid Caliphs traced their descent to Abba, uncle of Mohammed. The Persian system of female seclusion and harem were fully adopted and amplified by the Abbasids, who spread such concepts around the Muslim world. Of 38 Abbasid Caliphs between CE 750-1258, all but three were sons of foreign slave girls. The intrigues of the Harem weighed heavily upon the males of the ruling class also, many of whom were imprisoned from childhood until the day they would either rule, or be executed. Executions and torture were standard features of the Abbasid court.[237] Moslem influences spread south into the Indian Ocean at this time, along the east coast of Africa, and eastward into parts of the Indonesian, New Guinean, and Philippine Archipelago, where trade relations, Mosques and Caliphates were established (see Figure 20, page 102).

Constantly at war with Central Asian groups, the Arab Abbasids finally lost Persia to the Ghaznavid Turks of Asia around 900 CE. Under the leadership of Mahmud, newly converted to Islam, the Ghaznavids warred upon India. After Mahmud's death, the Ghaznavids were displaced from Persia by the Seljuk Turks, who also took Anatolia from the Byzantine Emperors and established an empire from Anatolia to China (c.1000-1200 CE). Relying heavily upon Turkish slaves and mercenaries, the Seljuks continued to push south, into Mesopotamia and the Levant. Under the influences of the newly converted Central Asian empires, Islam took on a more puritanical view of sexuality and women, a view which was subsequently transplanted into their conquered territories.[238]

Muslim armies persisted in warring against Hindu India, which retrenched and formed new, powerful alliances. Centuries of unrelenting, bloody warfare occurred between Moslems and Hindus. One of the Hindu kingdoms, the Vijayanagar (c.1350-1550 CE), held women in even lower esteem than the Moslems, who had earlier outlawed ritual widow murder. The Vijayanagar practiced ritual widow murder to an unprecedented degree. Wives and widows of the period were often but children or teenagers given the prevailing practice of child marriage; this usually meant an old man of wealth marrying teenage or preteen girls. Such young girls were not spared from the fate of older widows and were burned alive, often drugged and hog-tied, thrown into flames with the corpse of their dead "husband", a man whom they may have scarcely known given the prevailing system of bloated harems.

"The sacrifice was often, and especially in the case of princes, compulsory, so that scores or hundreds of women might be, and actually were, burnt at the funeral of a single Raja, with or without their consent." [239]

Thompson properly points out the compulsory nature of the female murder, and the mass hysteria which gripped a community about to burn its "widows". Hundreds or thousands of spectators would crowd about in a carnival atmosphere, becoming enraged if a girl or woman tried to flee. He cites many instances where crowds would kill a woman who tried to flee, sometimes repeatedly throwing the unfortunate woman back onto the fire, or incapacitating her with a knife before doing so.

Romanticized misportrayal of "Sati" or "Suttee" (meaning "virtuous woman"). The more precise terms are *Ritual Widow Murder,* or *Mother Murder* (when performed by the woman's son), a compulsory public act carried out by brute force in a circus-like atmosphere.

233. quoted in Huntington, 1907, ibid., p.320.

234. Bryson, & Baerreis, 1967, ibid., pp.139-140.

235. Bryson & Murray, 1977, ibid., p.109.

236. Dupree, 1973, ibid., p.312.

237. Dengler, I.C.: "Turkish Women in the Ottoman Empire: The Classical Age", in *Women in the Muslim World*, L. Beck & N. Keddie, eds., Harvard U. Press, Cambridge, 1978, pp.228-44; al Sayyid Marsot, A.L.: "The Revolutionary Gentlewomen in Egypt", in Beck & Keddie, 1978, ibid., p.263; Wittfogel, 1957, ibid., pp.139,144,344; Bullough, 1976, ibid., p.231.

238. Bullough, 1976, ibid., p.237; Wittfogel, 1957, ibid., p.360.

239. Thompson, 1928, ibid., p.28.

"Ritual" female murder in rural Hindu India of more recent times, appropriately sanctified by a holy man. No romantic disguise here. Every angry man in the village gets to stab his knife. (J. Campbell, *A Personal Narrative of 13 Years Service*, 1864)

It was the sexual experience of a widow, the fact that she no longer was a virgin, "in need of protection", and that she lacked a husband to "guide" her, which made the widow a dangerous threat to a social structure founded upon the absolute negation of female sexuality and instinctually directed sexual pleasure.

"...women were unsafe, and it was best to preserve their honor by burning them when their protectors died." [240]

It is interesting to note that these periods of massive widow murder coincided with the construction of the erotic temples at Khajuraho and Sna Lingam. But these temples were hardly evidence of a free sexuality, no matter how devoted they were to the fantasy-pleasures of the male. [241] In later years, the Temple of Kamakhya, "Goddess of Love", would practice human sacrifice. [242]

Other examples of massive female murder, or suttee, may be mentioned. When Muslim troops prepared to storm the city of Jaisalmer in CE 1285, the men of the city slit the throats of all their women folk to prevent them from falling into Muslim hands. A total of 24,000 women of all ages, the entire female population of the city, were ritually murdered by their own menfolk who then launched a suicidal counter-attack upon the larger Muslim force. Men of other Hindu cities, such as Chitor, similarly butchered their female kinfolk before being attacked by the Muslims, feeling the women would be better off dead than in the tents and beds of the hated enemy. [243] Even during times of "peace", it was not uncommon for 2000 to 3000 women, of all ages, to be burned alive in a single ghastly funeral "ceremony". The death of one Indian Raja was commemorated with the slaughter of 11,000 women. Archaeological evidence of the mass burnings has been found, in the form of layered slag and cinder heaps mixed with calcined bone. [244] Hindu scriptures of this period clearly encourage ritual suicide: "A devoted wife, who follows her husband in death, saves him from great sins."[245]

As much as the Muslims considered widow murder a barbaric custom, they were themselves no angles of mercy; under generals such as Bahman, and later Muhammed Shah, they slaughtered countless Hindus: men, women, and children. The Vijayanagar in turn would revenge themselves. At one point, an agreement was reached between Muhammad Shah and Bukka, his Hindu rival, to spare noncombatants. Unfortunately, the agreement did not last. [246] The contradictory Muslim view of women was best given by the Persian Iskander (1082 CE):

"If you have a daughter, entrust her to kindly nurses and give her a good nature. When she grows up, entrust her to a preceptor so that she shall learn the provisions of sacred law and the essential religious duties. But do not teach her to read and write; that is a great calamity. Once she is grown up, do your upmost to give her in marriage; it were best for a girl not to come into existence, but being born, she had better be married or buried..." [247]

Moving west from India, Fairservis observed the Seistan region, between Iran and Afghanistan, to be "literally covered with the ruined cities of the Medieval Islamic period (CE 1000-1500), as well as earlier sites," possibly indicating milder conditions and absence of the 120-days wind. [248] Other evidence indicates the role of warfare in the depopulation of Seistan. Originally, the Seistan Basin contained a peaceful farming community which supplied surplus food to surrounding powers. Various Iranian, Greek, Mauryan, and Arabic armies marched across it over the

240. Thompson, 1928, ibid., p.48.
241. Hurliman, 1967, ibid., p.133.
242. Bullough, 1976, ibid., p.259.
243. Thompson, 1928, ibid., p.37.
244. Thompson, 1928, ibid., p.35.
245. Davies, 1981, ibid., p.108.
246. Hurliman, 1967, ibid., pp.87,179.
247. Bullough, 1976, ibid., p.238.
248. Fairservis, W.A.: "Exploring the 'Desert of Death'", *Natural History*, June, 1950, p.252; cf Fairservis, 1961, ibid.

centuries, as a crossroads to other regions. Still, Seistan appeared to flourish, more or less, its agricultural potentials being appreciated by whomever captured the region. However, a series of Mongol invasions between CE 1297-1407 again brought slaughter, ruin, and great depopulation.[249]

Around this same time, returning to events in the West, we observe the rule of Charlemagne (c.650 CE), a polygamous Christian butcher of the Holy Roman Empire. Charlemagne brought the first systematic use of torture against accused criminals and "witches" in Europe. His reign reinforced the power of the priesthood over the laity, and an attempt was made to introduce prostration to clergy. Clergy were made exempt from secular law, and were granted the right to oversee, and approve or disapprove of all marriages. Charlemagne campaigned against "sexual crimes" at home, and warred against "pagans" in Northern Europe, bringing much bloodshed. Men or women who violated church rules could be dragged naked through the streets, subject to mob cruelty. The pillory, stocks, flogging, and whipping were prescribed for "immorality". Ritualized rape, or defloration of virgins on the wedding night by Kings, Lords, and monks (*jus primae noctis*), spread widely in Europe following Charlemagne's rule.[250]

The years after 900 CE in Europe saw a series of debauched Popes and Byzantine Emperors,[251] with ecclesiastical sexual scandals persisting over the years.[252] The Eastern and Western churches formally divided and excommunicated each other after 1000 CE over the specific issue of clerical celibacy, the Eastern clerics wanting to maintain open relations with women, the Western clerics not.[253] New laws were passed in the increasingly celibate Western Church by which wives of priests were automatically made into concubines, their children perpetual serfs of the Church. Mobs were incited to attack the homes of priests who refused to put away their wives. Priests were forbidden to say mass if they lived with any female, be she wife, sister, or mother.[254]

Preferential female infanticide was rampant in Europe in the 800s, with an estimated male/female ratio of 156 to 100. The female population was further reduced by burning of female "witches", as well as by punishing young girls for infanticide. Infanticide was added to the list of crimes for which burning at the stake was mandated, regardless of the circumstances of the pregnancy. The man who impregnated the girl invariably went free.[255]

The Hildebrand "reforms" of 1059 CE further reduced democracy in Europe, taking the election of the Pope out of secular hands, allowing the decision to rest with a College of Cardinals. By 1096, the first of four major Crusades was launched against Palestine to reclaim Jerusalem. The Crusades were not successful in achieving their stated goal for any significant length of time, and largely reflected the interests in warfare, booty, and military matters which preoccupied the Holy See. In 1204, the fourth Crusade was launched by Pope Innocent III of Rome, who also spent 30 years in a war of extermination against the heretical Albigenses. This Crusade was aimed against the Saraceans of Palestine, but, through convoluted events, detoured instead to loot Constantinople. A "Children's Crusade" was also launched, the result of which was several thousand European children being taken up as slaves into Egypt.[256]

The period following the Crusades saw importation of various new Near Eastern ideas, such as the chastity belt, which was used as an alternative to the harsher practice of infibulation.[257] It was an era of sexual hypocracy and cruelty, when clerics railed against various sexual heresies, but themselves lived out the greatest licentiousness.

After 1215 CE, the Roman Church got down to the business of rooting

Celibacy

The condition of being unmarried; without a husband or wife — *not to be confused with chastity or abstinence*. One can be celibate, as with an unmarried priest or monk, and still have many sexual partners. Celibacy (not abstinence) was firstly instituted by the Christian Church for power motivations, to prevent widows (legal wives of priests) and "legitimate" children from inheriting the land, church buildings and wealth accumulated by individual priests. With the enactment of priestly celibacy laws, all property went *to the Church institution* after a priest's death. Priests were still allowed, however, to sleep with women and to have concubines. The institution of priestly celibacy ushered in a more sadistic type of woman-hating individual, as most married men refused to disown their wives and children to become priests.

Chastity or Abstinence

The total avoidance of sexual intercourse; to abstain from sex and to be chaste or "pure". A much more difficult undertaking than celibacy.

249. Fairservis, 1961, ibid., pp.30-36.

250. Gage, 1980, ibid., pp.95,175; Wittfogel, 1957, ibid., p.154; Lewinsohn, 1958, ibid., p.134; Taylor, 1953, ibid., p.31; Bullough, 1976, ibid., p.353; Scott, 1939, ibid., p.134.

251. Taylor, 1953, ibid., pp.38,141; Gage, 1980, ibid., pp.54,255,258; Bullough, 1976, ibid., p.327.

252. Bullough, 1976, ibid., p.430; Taylor, 1953, ibid., p.141.

253. Bullough, 1976, ibid., p.317.

254. Lewinsohn, 1958, ibid., p.119; Gage, 1980, ibid., pp.33-4.

255. Piers, M.W.: *Infanticide*, W.W. Norton, 1978, pp.45-6; deMause, 1974, ibid., p.29.

256. Howard, 1977, ibid., p.101; Godfrey, J.: *1204: The Unholy Crusade*, Grove Press, c.1980.

257. Dingwall, E.J.: *The Girdle of Chastity*, Clarion, 1959, p.14.

out heretics, and called in the Dominican Inquisitors to take charge of the matter. The Dominicans could seize, imprison, and torture heretics, claim their possessions and wealth, and try and convict; secular authorities then carried out the executions. Under the Dominicans, clerical celibacy was finally enforced; Pope Innocent III officially ruled for the disinheritance of wives and children of priests, the denial of their burial rights, their eternal damnation, and their direct abduction and sale into slavery.[258] The underlying philosophy of these laws was, of course, a belief in the satanic and evil nature of women and children, and the inherent sinfulness of sexual pleasure.

Various Church penitentials were drawn up by the celibate priesthood, focusing upon sexual sins, listing various stages of sinfulness, each of which was progressively more sinful:

"1. foulish looks,
2. foul words,
3. foul touchings,
4. foul kissings,
5. cometh a man to do the deed". [259]

Medieval European Woodcut
"Devil Seducing a Witch"

The penitentials listed the many ways "the deed" could be done, some being more "foul" than others. The tone of the celibate priests is revealing:

"Into the filth of the flesh, into the manner of life of a beast, into thralldom of a man, and into the sorrows of the world... to cool thy lust with filth of thy body, to have delight of thy fleshy will from man's intercourse; before God, it is a nauseous thing to think thereon, and to speak thereon is yet more nauseous." [260]

Bullough properly pointed out the sexual preoccupations and psychopathology of the celibate priesthood, where everything from leprosy to witchcraft was attributed to sexual sin, and where fear from contamination from heretics and witches dominated sermons. Great hypocrisy also abounded. The Inquisition kept prison harems of young women, who were incarcerated purely because of their good looks; they were subject to repeated prison-rape by inquisitors and other strangers with connections to the Holy See, who threatened them with grotesque tortures if they failed to submit. Concubines were increasingly used by the Priesthood such that a compulsory concubine tax law was passed, a law which also disencumbered the Priest of any legal responsibility to the involved women or any children which would be born.[261] The following quotation is taken from a speech by Cardinal Hugo given during his visit to Lyons with Pope Innocent IV in 1250 CE:

"Witches Dancing with Demons"

"Since we came here we have effected great improvements. When we came, we found but three or four brothels. We leave behind but one. We must add, however, that it extends without interruption from the eastern to the western gate." [262]

258. Bullough, 1976, ibid., pp.320,393; Gage, 1980, ibid., p.37.

259. Bullough, 1976, ibid., p.387.

260. Bullough, 1976, ibid., pp.386-7.

261. Gage, 1980, ibid., pp.35-6,42; Bullough, 1976, ibid., pp.389,393.

262. quoted in Taylor, 1953, ibid., p.37.

263.264. Lewinsohn, 1958, ibid., p.135; Gage, 1980, ibid., pp.35,41; Tannahill, 1980, ibid., p.279.

264. May, 1930, ibid. pp.105-17; Taylor, 1953, ibid., p.70.

The Christian Church also engaged in temple prostitution, and kept brothels of young girls who would service only Christian men. The girls were required to say prayers, however.[263] Church coffers overflowed with monies from brothel and "sin rent" payments, which were allowed in lieu of more painful forms of penitence.[264] This was the Europe of Boccaccio's *Decameron*, and which Martin Luther's Reformation was aimed at. While 1215 CE saw the foundation of the Dominican Order in Mediterranean Europe, in distant England the year was marked by increasingly secular and democratically oriented developments, in particular, the forcing of King John's signature upon the Magna Carta.

The Dominicans did their work well, however, aided by the fact that testimony of women in courts was forbidden. New offenses such as copulation with the devil, incubi, succubi, or other evil spirits were officially recognized as major heresies punishable by death.[265] Ecclesiastical courts pondered the question of what the devil's penis must be like, if it was unusually hard, covered with scales, or if his semen was ice-cold, etc. Terrified women would be accused of having intercourse with the devil, and tortured to extract confessions. Afterward, they were burned at the stake. Even virgins were accused of this crime and, since torture was routinely applied, few failed to confess.[266] A "witch" hunting craze ensued all across Europe, and both Catholic and Protestant peasant populations of Europe were afflicted and decimated by this terror. One source cites 9 million "witches" burned between 1485-1784, but this is probably a low estimate.[267] The bulk of the victims, some 85%, were women, both young and old. Of one group of 291 condemned witches, only 23 were men, 11 of whom were guilty of crimes connected with women.[268] Clearly, the hatred of women and of sexual pleasure was the source of the widespread sadism directed towards both.

The women who were tortured and burned were accused of crimes ranging from being a midwife, possessing a cat, or having intercourse with the devil. Often, the smartest, most outspoken, most beautiful, educated, or wealthy of women in a community would be singled out for murder by the female-hating celibate clergy. Teenage girls merely accused of abortion or infanticide were tortured to gain confessions, and thereafter burned to death. At other times almost the entire female populations of a town would be immolated on a single day. In one German city, 900 women were burned in one day; in one French city, 400 were burned in a single day. The Saxon Jurist Carpzof boasted he had read the Bible 53 times and executed 20,000 women, mostly for witchcraft and infanticide. An Inquisitor of Como, Italy was quoted as having burned 1000 witches in a single year. The murders were not carried out by disorderly mobs, but were well-ordered, following a well-defined legal process involving sworn testimony from local citizens, gathering of evidence, extraction of confessions, and official ecclesiastical judgements.[269]

"A Bishop of Geneva is said to have burned 500 persons in 3 months. A Bishop of Bamberg 600, a Bishop of Wurzburg 900. 800 were condemned, apparently in one body, by the Senate of Savoy. Paramo boasts that in a century and a half from 1404, the Holy office had burned at least 30,000 witches... In Spain, Torquemada personally sent 10,220 persons to the stake and 97,371 to the Galleys. Counting those killed for other heresies, the persecutions were responsible for reducing the population from 20 million to 6 million in two hundred years..." [270]

The Black Death which struck Europe in 1348 also appears to have been instrumental in developing an even harsher, more suspicious view of women. With up to 3/4 of the population killed in many areas, sleeping with a woman was called a mortal sin.[271] The Bretheren of the Cross cult was organized, being devoted to mass flagellation, a phenomenon which first appeared earlier in Italy; like the Jainists of India, the Bretheren were forbidden to even speak with a woman.[272]

Petrarch (c.1370 CE), called "the first modern man", shortly thereafter would teach that women were devils:

Women being burned alive in Lausanne, Switzerland in 1573 CE. The executioner stands nearby with pitchfork, to push anyone escaping back into the fire — gallows and torture-wheel depicted in the background. Nearly every city and village in Europe had its own ghastly toll of brutally-murdered heretics, on demands of Christian authorities. Between 1581 and 1620, for example, some 970 persons were charged with "sorcery" in Lausanne alone — 90% of whom were burned.

265. Gage, 1980, ibid., p.256; Lewinsohn, 1958, ibid., pp.125-6; Bullough, 1976, ibid., pp.420-1.

266. Lewinsohn, 1958, ibid., pp.123-4.

267. Gage, 1980, ibid., p.107.

268. Ehrenreich, B. & English, D.: *Witches, Midwives, and Nurses: A History of Women Healers*, Feminist Press, 1973, pp.7-8.

269. Ehrenreich & English, 1973, ibid., pp.6-14; Gage, 1980, ibid., p.107; Lewinsohn, 1958, ibid., p.127; Bullough, 1976, ibid., p.422; Tannahill, 1980, ibid., pp.273-4.

The Catholic Church has, to this day, arrogantly denied any past wrong-doing for the horrific tortures and mass murders it instigated across Europe over the centuries. The Christian Church, to include both Catholic and Protestant sects, certainly murdered more people than did the Nazis, and possibly even more than Stalin or Mao Tse'Dung — and yet, no acknowledgment of guilt or apologies have ever been given, not even for a single death. Even the murders of prominent males, such as Giordano Bruno (who was tortured and burned) have never been acknowledged by the Vatican.

270. Taylor, 1953, ibid., pp.126-7.

271. May, 1930, ibid. p.136.

272. Gage, 1980, ibid., p.29; Scott, 1939, ibid., pp.256-7.

Hidden Holocaust of the
Christian Church.

A probable minimum of 9 million
persons, mostly women but men
and children as well, were tortured
and murdered between 1485-1784
CE. This total does not include a
probable equal number murdered
from the time of Constantine (the
first Christian Emperor of Rome,
c.300 CE) until the 1400s, nor another great number of natives murdered in the New World after the
arrival of Columbus in 1492.

273. Tannahill, 1980, ibid., p.288.
274. deMause, 1974, ibid., p.29.
275. May, 1930, ibid. p.111.
276. Scott, 1939, ibid., pp.63-69, 158,
217, 278; Davies, 1981, ibid., p.245; deMause,
1974, ibid., p.10.
277. Piers, 1978, ibid., pp.68-9;
Bullough, 1976, ibid., pp.436-42; Lewinsohn,
1958, ibid., pp.167-70.
278. Kovalevsky, 1891, ibid., p.13.
279. deMause, 1974, ibid., pp.11,229.

"...an enemy of peace, a source of provocation, a cause of disputes, from whom a man must hold himself apart if he wishes to taste tranquility." [273]

The period saw preferential female infanticide across Europe, with a male to female ratio of 172 to 100.[274] Another intensification of antifemale and antisexual laws occurred after the return of Columbus from the New World; his discoveries not only shook the foundations of many Church teachings, but a syphilis epidemic thereafter occurred which swept Europe. Roughly 1/3 of the population died by 1520.[275] The 1400s saw the start of the Inquisition in Spain, the Code of Torquemada, and the publication of the *Malleus Maleficarum*, a bible of witch hunting. Numerous people were burned at the stake, buried alive, and/or mercilessly tortured in ghoulish ways. Even barnyard animals were burned if they displayed unusual characteristics, or were even remotely involved in the death of a person.[276]

The Carolina, a Counter-Reformation era code of criminal law, was passed at this time by Charles V (c.1550 CE), King of Spain and Holy Roman Emperor. The Carolina proscribed death for "sinful" women, either by burning at the stake, being buried alive, or by impalement or sacking (drowning). If the latter method were used, the victim was to be torn with glowing thongs beforehand. These sadistic ritual murders of women took place in a public spectacle scarcely different from that of Hindu India, often with singing choir boys and pealing church bells at hand, designed to drown out the screams of the dying. Under Charles, marriage outside the church was ruled invalid, requiring the consent of the parents. Secular marriages or marriages without parental approval would result in disinheritance; if parents did not approve of a marriage they could bring the man up on charges of rape, which carried the death penalty. A list of banned books was drawn up, and nudity in art was prohibited. Michelangelo's Sistine Chapel painting, the Last Judgement, had fig-leaves, loin-cloths, and robes painted over the genitals of the figures.[277]

It was during this period also that the last vestiges of the young people's communal societies, and socially approved cohabitation before marriage were finally and thoroughly stamped out.[278] Infants were subject to rigid swaddling, and children harshly abused, according to a popular rhyme of the period:

"The child whom the father loves most dear, he does punish most tenderly in fear." [279]

Prior to the Reformation, baptism and exorcism had been performed simultaneously, given the view of the child's essentially evil nature. The combined rituals would leave children in pain and hysterically crying, but this was thought to be a sign of the devil being exorcised. After the Reformation some of these harsh child treatments were mitigated. However, Reformation clerics were only slightly more interested in the welfare of children or women than was the Holy See. Vicious stories continued to be told to children to frighten them into obedience, that God would hold bad children "to roast over the pit of hell, much as one holds a spider, or some other loathsome creature over the fire", or others, like "The little child is in this red hot oven. Hear how it screams to come out...It stamps its little feet on the floor..." Aside from swaddling, infants of the 1500s would also sometimes be branded on the neck with a hot iron to prevent "falling sickness", a form of exorcism by physicians which bears a resemblance to modern-day circumcision mutilations, also allegedly performed for "health" reasons. Calvin favored burning of witches, and considered children to be laden with sin; he called children the "imps of darkness". Luther's views on children and women were similar. He, too, advocated the burning of witches, and claimed he would rather have a dead son than a disobedient one.[280] He also observed that demons:

> *"...often take the children of women in childbed and lay themselves down in their place and are more obnoxious than ten children with their crapping, eating, and screaming."* [281]

Besides women, a wide variety of other people were subject to persecution: writers, poets, scientists, Jews and homosexuals. It was also heresy to deny the reality of witchcraft.[282] The 1600s saw a total decline of representative institutions in Europe.[283] Giordano Bruno and numerous other scientists were tortured and burned at the stake. Others, such as Galileo, were tortured and imprisoned. In Italy, hugging a married woman or widow could bring a heavy fine, which would be doubled if she was a virgin; in some cities death was proscribed for such a hug.[284] In England, the maypole was banished, flogging of children became a standard of school curriculum, and 200 separate moral crimes, including adultery and incest, could bring the death penalty; fornication led to a fine, imprisonment, public penance, and a ducking in "the foulest pool in the parish".[285] Midwives were driven underground under threat of death, being replaced by church-approved physicians or barber surgeons, whose fumblings killed numerous pregnant and birthing women. Pregnant women, being considered sick with sin, were ordered into church-run sick houses where, as Ignaz Semmelweis later discovered, contamination from the truly sick killed them in massive numbers.[286] And, of course, the *auto da fe* ("act of faith") continued from the 1500s to the 1700s; such ritual mass murder of heretics, performed on Sundays for the amusement of church-going Christians, continued at a weary pace.[287]

During the period of European mass insanity and butchery, rumors of a new "savior" from Central Asia circulated, a savior who lived among the peoples of the Central Asian Khanate, where Turko-Mongol peoples became increasingly dominant; his name was Genghis Khan, a man about whom more will be said in the next section. After 1200 CE, the Mongols conquered most of the highlands between Anatolia and India, but were unable to press into Europe.

In Afghanistan, the Mongols were successful in their invasions, and cities such as Bamian, Herat, and Khwarizm were completely leveled, with all citizens slaughtered in a bloody frenzy that lasted days. The Mongols also pressed into Mesopotamia, with great destruction and slaughter; indeed, wherever the Mongols rode, great destruction and massacres generally followed.[288]

The Ottoman Turks wrested control of the western highland region from the Mongols after c.1300 CE, eventually capturing much of Central Asia, the Near East, and North Africa. The Turks exacted harsh tribute, to include slave women, castrated boys, and from 1/2 to 2/3 of all crops. Peasants abandoned the land, fleeing to cities. Trees were no longer planted and terraces degraded. The Ottomans warred against Europe between 1354-1402, and then later in the 1500s and 1600s, always pressuring Eastern Europe. They captured Constantinople, Bulgaria, Thrace, Romania, Hungary, and part of Russia. Various European armies opposed the Turks, and prevented their intrusion beyond the Balkan Peninsula.[289]

The Grand Turks of Constantinople variously bloated their harems with from 300 to 1200 slave women from the markets of the Mediterra-

Tamerlane, claimed son of Genghis Khan and ruler of the Mogul Empire ("Mogul" is a Persian derivation of "Mongol") which eventually stretched from the Black Sea to India (c.1450 CE). Tamerlane is depicted here with a "holy aura", typical for such blood-drenched Central Asian "god-kings" who slaughtered millions and demanded absolute genuflecting obedience.

280. deMause, 1974, ibid., pp.10-12,30; Davies, 1981, ibid., pp.249-50; Radbill, 1974, ibid., p.4.

281. deMause, 1974, ibid., p.10.

282. Taylor, 1953, ibid., p.113.

283. Kovalevsky, 1891, ibid., p.206.

284. Dingwall, 1959, ibid., p.28.

285. Taylor, 1953, ibid., p.171; May, 1930, ibid. p.152; Bullough, 1976, ibid., pp.479-80.

286. May, 1930, ibid. p.108; Ehrenreich & English, 1973, ibid.

287. Scott, 1939, ibid., pp.70-5.

288. Wiencek, et al, *Storm Across Asia*, H.B.J. Press, NY, 1980, ibid., pp.32-3,37; Dupree, 1973, ibid., p.315.

289. Whyte, 1961, ibid., p.100; Wittfogel, 1957, ibid., pp.64f,361; Pitcher, 1972, ibid., p.21.

nean and Black Sea. When Suleiman "the Magnificent" wedded the slave girl Roxelana in c.1550, it was the first legal marriage of an Ottoman Caliph in over 100 years. The Ottoman harem system brought intrigues and cruelties to men as well, the Law of Fratricide demanding that all brothers of a new leader be killed. Achmet I abolished the Law of Fratricide in c.1610, replacing it with the "cage", a special prison where his brothers were kept until it was time for their succession to power. Men kept in the "cage" were given female sex-slave companions, any children of whom were drowned.[290] Sultan Ibrahim (c.1630) was kept prisoner in the "cage" from the age of 2 until he was 24. Devoted to orgiastic abuses, in a fit of anger he once tied his entire harem of 280 concubines into weighted sacks, and drown them in the Bosporus. Sultan Suleiman II and Osman III spent 39 and 50 years respectively in the "cage", only to become absolute rulers of the Empire upon emerging for the first time since infancy.[291]

Such cruelty and madness was the true character of the Near Eastern harem system. While acknowledging that some harems were likely more humane than others, classic books about the harem, such as *The Perfumed Garden* by Nefzaou, have done a great disservice to human emotion by making the harem appear as a place of romance, erotic passion, and enormous sexual gratification. In truth, the harem system bred murderous intrigues, sadistic cruelties, eunuchs, extreme sexual frustration, pathologies, disease, suicide and enormous misery.

The Turks constantly pressured Europe, invading Austria twice, and pillaging the local populations in a Genghis-Khan fashion. After 1683, the Turks were defeated and driven out of both Austria and Hungary; they also engaged in three wars with Russia between 1699 and 1812. The Turkish Empire began disintegrating in the 1800s, but still had enough energy to massacre millions of Armenians in the early 1900s.[292] Prostration before leaders and clergy persisted in the Ottoman Empire until the end of the Sultanate in 1919.[293]

To the east, in Iran, Afghanistan, and India, Tamerlane, who claimed descent from Genghis Khan, formed an alliance between various Turk and Mongol bands and developed an empire from the Black Sea to India (c.1450 CE). Tamerlane invaded India on the pretext that its Muslim rulers were not zealous enough; his campaigns mirrored those of his supposed ancestor, with immense slaughter, looting, and slaving. His empire disintegrated upon his death in 1405, but it shook the grip of the Golden Horde from the region. Mongols, called "Moguls" by the Indians, continued to rule over India, while Afghanistan was variously fragmented by hill tribes or ruled by Asian powers until later years, when it was pinned between the Russians and the British.[294]

In India of the 1500s, the king owned all the land, and in some places operated brothels to finance the military. Literature of the period strongly urged ritual widow murder, which was still widely practiced. When the King of Narsynga died in 1516, for example, 500 men and women accompanied him on the funeral pyre. Mass suicide during the continuing Moslem/Hindu wars also continued. When Hindu Chitore was attacked by Moslems in 1533, it is said that no less than 13,000 women died in a massive blast of gunpowder, the mother of the King having the honor of applying the torch.[295] The kingdom of Vijayanagar fell after massive famine brought about by Moslem scorched-earth tactics, and 2/3 of the population perished.[296]

In India, the Muslim Mogul rulers kept large harems of slave girls, but abhorred the Hindu practices of widow murder. Emperor Akbar (1556-1605), for instance, attempted to curb widow murder, prostitution, and child marriage, establishing a minimum marriage age of 16 for men

Tamerlane conquers Delhi in 1399 CE, massacring its inhabitants and building a tower from the severed heads. Massive gravesites mark the battlefields of Mongol and Mogul territory, characterized by skeletons with severed heads and limbs.

290. Tannahill, 1980, ibid., p.243.
291. Tannahill, 1980, ibid., pp.240-4.
292. Pitcher, 1972, ibid., p.21; Scott, 1939, ibid., pp.271-2.
293. Wittfogel, 1957, ibid., pp.154,361; Scott, 1939, ibid., p.209; Pitcher, 1972, ibid., p.21.
294. Dupree, 1973, ibid., p.317; Wiencek, et al, 1980, ibid., pp.45,86,94; Hurliman, 1967, ibid., p.179.
295. Davies, 1981, ibid., pp.101-2,108; Hurliman, 1967, ibid., p.88.
296. Aykroyd, 1974, ibid., p.50.

and 14 for women. He revoked a tax on non-Muslims and took a Hindu bride. However, his reforms were not successful.[297] Jahangir, son of Akbar, continued his father's practices of religious tolerance, and forbade the cutting off of the ears and nose as a punishment. He forbade sati upon pain of death, but succeeded in stamping the practice out only near Delhi. Still, one has to wonder how vigorously his policies were enforced, given that he kept busy a squad of 40 hangmen, and liked to watch criminals be trampled by elephants.[298] Sati actually increased in other parts of India during Jahangir's rule, and one Mathura ruler had 700 women burned alive at his funeral.[299] Ritual widow murder, other forms of human sacrifice, and murder/robber cults ceased in India only begrudgingly under the iron-handed policies of British colonialism.

In general, the belt of territory stretching from Europe into Anatolia, across Iran, Pakistan, and India has, in the past, possessed the highest levels of military and social violence, and the harshest cruelties to women and children of any region on Earth. The status and treatments of women and children in Europe only began to improve following the Italian Renaissance, Reformation, Enlightenment, and Revolutionary movements in America and France. Murderous and pathological conditions persisted in Europe until humanistic, rationalistic renaissance freethinkers, artists, scientists, feminists, and scholars, with a lot of help from Napoleon Bonaparte's army (which demolished intractable Church authority across Europe), returned family law into increasingly democratic, secular hands.[300] However, this European transition towards social reform occurred with many reactionary social upheavals, especially the rise of authoritarian fascist conditions during and between World Wars I and II.

This time period was similarly a historical turning point for Turkey, Iran, Afghanistan, Pakistan, and India, following the break-up of the same fascist central state empires, and a significant intrusion of Western secular influences. However, political absolutism and extreme sadistic butchery has persisted across the region. The British withdrawal from India was followed by widespread "civil" massacres between Muslims and Hindus, claiming the lives of millions of men, women and children in bloody village-against-village combat. Reactionary movements have thrived across the Anatolia-Indus region, developing from fertile anti-sexual, antifemale soil, overthrowing more reform-minded leaders and instituting new authoritarian regimes which persist to the present day.

Regardless of the current situation, the above pages document profound effects from the combined forces of desiccation and warfare. Huntington traveled through the Iranian Plateau regions in c.1900, and contrasted several areas containing old ruins, some of which were wet at the time of his visit, and some dry. He observed that the ruins in the presently wet regions had been re-populated, while ruins in the dry regions had never recovered from the depopulation of former times:

> "If war and misgovernment are the cause of the decay of Persia, it is remarkable that the two provinces which have suffered most from war and not less from misgovernment should now be the most prosperous and least depopulated; while the two which have suffered less from war and no more from misgovernment have been fearfully, and, it would seem, irreparably depopulated. It is also significant that the regions which have suffered the greatest ruin are those where water is least abundant, and where a decrease in the supply would be most quickly felt. War and misgovernment do not seem invariably to cause depopulation, nor has the process gone on most rapidly where war has been most prevalent."[301]

297. Thompson, 1928, ibid., p.57; Wiencek, 1980, ibid., pp.111-19; White, E.H.: "Legal Reform as an Indicator of Women's Status in Muslim Nations", in Beck & Keddie, 1978, ibid., p.54.

298. Thompson, 1928, ibid., p.57; Wiencek, et al, 1980, ibid., pp.128-9; Davies, 1981, ibid., p.92.

299. Thompson, 1928, ibid., p.36.

300. Lewinsohn, 1958, ibid., pp.139-40; Tannahill, 1980, ibid., p.326.

301. Huntington, 1907, ibid., p.326.

E. CENTRAL ASIA

General Discussion:

Central Asia lies at a much higher latitude than North Africa, being closer to the former ice sheets of the Pleistocene glaciation. During the glacial period temperatures were cooler, with a reduced evaporation. Meltwaters from the great ice sheet fed into the enclosed basins of Central Asia, supporting large freshwater lakes. Glaciers also were more active in the Hindu Kush, Himalayas, and Tibetan ranges (Astin Tagh, Karakoram), draining northward into these same basins. The Tien Shan, Altai, and other Mongolian mountains which cut north/south through the Central Asian steppelands were also glaciated, providing additional meltwaters which flowed into the enclosed basins. The forested region of Asia was shifted south during the glacial period, while snow lines and timber lines were also lower. These factors combined to create conditions of reduced temperatures and evaporative stress with greater runoff.[1]

After the last glacial period ended, a transition which started at c.10,000 BCE, temperatures and sea levels began to rise. Fluvial action of streams and rivers dependent upon snowpack and glacial meltwaters gradually decreased. Glaciers persisted longer in the high mountain regions on the southern fringe of the dry belt, however. Von Wissman identified the period of the Thermal Maximum, 5000-2500 BCE, as being more humid in Central Asia.[2] Others have contradicted the extension of humid conditions to as late as 2500 BCE, claiming that drier conditions prevailed across the steppelands of Central Asia after c.3500 BCE.[3]

The Soviet researcher Petrov believed that the drylands of Central Asia experienced oscillating climate change over the last several thousand years, with a general trend toward dry conditions stretching down into historic times. He summarized the conditions in Central Asia evidencing a progressive desiccation, in which human activity also played a role:

"1) A general reduction in glaciated areas,
2) A predominance of deflation and sheet wash among geomorphic processes,
3) Partial conversion of concentrated rill runoff to sheet flow,
4) A reduction of lake areas and progressive salinization,
5) A degradation of vegetation in level watersheds, down to complete disappearance,
6) A suppression of species forming processes in plant and animal life,
7) A progressive salinization of soils under level watershed conditions and the conservation of salt crusts." [4]

"The moving sands of Central Asia are also, in part, the outcome of the irrational behavior of man in the desert and semidesert regions. It is common knowledge that these deserts have been inhabited since time immemorial and that they have been used for the grazing of livestock for thousands of years. Equally as ancient is farming in the central regions of China and Kashgaria [Tarim Basin]. During this extensive period, overgrazing, excessive use of shrubs and herbaceous vegetation for fuel, and plowing of overgrown sands for sowing, unaccompanied by the necessary soil preservation measures, have resulted in the mass destruction of natural vegetation, the deflation of sands fixed by vegetation and their transformation into moving barchans. Nearly all the ancient

1. Von Wissman, H.: "On the Role of Nature and Man in Changing the Face of the Dry Belt of Asia", in Thomas, W.L., ed: *Man's Role in Changing the Face of the Earth*, Vol. 1, U. Chicago Press, 1972, pp.278-9.

2. Von Wissman, 1972, ibid., pp.281-2.

3. Thomas, H.L.: "Archaeological Chronology of Northern Europe", in *Chronologies in Old World Archaeology*, R.W. Ehrich, ed., U. Chicago Press, 1965, pp.378,391.

4. Petrov, M.P.: "Once Again About the Desiccation of Asia", *Soviet Geography*, 7(10):22, 1966.

agricultural regions bordering on the sand deserts and semideserts and the wells in the deserts are surrounded by a broad belt of moving barchan sands, which always lie at the edge of the fields or around inhabited spots, i.e., where man's influence on the sands has been extensive. This phenomena is widespread in Central Asia." [5]

Petrov pointed out the lack of direct evidence for recent climate changes, given the short time period since measurement-taking began. However, even some of those Soviet researchers opposed to historical climate change in Soviet Central Asia agree that desiccation has occurred in China, east of the Tien Shan. The deserts of China are much drier than those of Soviet Central Asia, with salinized soils and evaporation many times greater than precipitation. [6]

The conclusions made by Soviet researchers on landscape transformations in Central Asia in part mirror those of Huntington on climate change. Traveling through the region shortly after 1900, Huntington identified a series of enclosed basins and lakes in Central Asia which evidenced climatic oscillations of similar direction and chronology.

"Including Gyoljuk, our survey of western and central Asia has dealt with six distinct basins. On the west lies Gyoljuk in Turkey; then come the Caspian basin in Russia, and that of Seyistan to the south in Persia; while far to the east we have Lop and Turfan in the heart of Asia forming part of China, and Kashmir south of the Himalayas in India. ...the limits of our six basins lie over sixteen hundred miles apart from north to south, and over three thousand from east to west. All this vast area seems to have been subject to the same great waves of climatic change." [7]

Huntington also felt the Central Asian climate change chronology to be similar to that of Mesopotamia:

"...almost innumerable facts seem to indicate that two or three thousand years ago the climate was distinctly moister than at present. The evidence includes old lake strands, the traces of desiccated springs, roads in places now too dry for caravans, other roads which make detours around dry lake beds where no lakes now exist, and fragments of dead forests extending over hundreds of square miles where trees cannot now grow for lack of water. Still stronger evidence is furnished by ancient ruins, hundreds of which are located in places which are now so dry that only the merest fraction of the former inhabitants could find water." [8]

Chappell has summarized the various arguments for and against climate oscillations in Central Asia during more recent times, drawing upon the works of Huntington, Stein, various Soviet sources, and others who visited the region around the turn of the century. His review also indicates a thawing in attitude among some Soviet researchers regarding the idea of climate change and environmental factors. [9] That changes occurred in the landscapes of Central Asia is no longer seriously challenged; however, the mechanism of such changes and their absolute chronology are still less certain. While evidence of climate change is present, such changes are also strongly linked to human actions. Wadia also discussed the issue:

"Everywhere in these six million square miles extent of middle

5. Petrov, 1976, ibid., pp.23-4.

6. Petrov, 1966, ibid., pp.15-22.

7. Huntington, E.: *The Pulse of Asia*, Houghton Mifflin, NY, 1907, p.356.

8. Huntington, E.F. & Visher, S.S.: *Climatic Changes, Their Nature and Causes*, Yale U. Press, New Haven, 1922, p.66.

9. Chappell, J.E.: "Climatic Change Reconsidered: Another Look at 'The Pulse of Asia'", *Geographical Review*, 60(3):347-73, 1970.

Asia, the story is of once copious and flowing rivers now struggling against growing sand dunes; of once flourishing human settlements abandoned; forests withering and sand-bounded lakes migrating due to encroaching dunes, shrinking in volume and precipitating rock-salt and gypsum deposits in their deserted beds." [10]

"The hundreds of lakes of Inner Asia are shrinking and although accurate data and chronologies are lacking, we cannot consider any theory about aridity of Asia satisfactory which does not take into account the testimony of the drying lake basins dotting the desert panorama from Peking to the Caspian." [11]

Sinitsyn also authored a standard Soviet work on Central Asian physical geography, and, as quoted by Chappell, is a supporter of desiccating climate change, and the views of Huntington:

"Fresh traces of desiccation, relating not only to the recent geologic past but also to the historic period, are observed everywhere in Central Asia. This involves the cutting back of outer watercourses and the shortening of rivers that in the past extended far into the desert but now dry up in the foothills (Keriya, Niya, Endereh, and others); the decrease in the amount of glaciation in mountain regions and the rapid retreat of the glaciers; the narrowing and disintegration of the strip of piedmont oases into isolated plots thanks to the destruction of great masses of tugai forests and of the vegetation on the dune sands; the reduction of the area of human settlement from the third to the nineteenth centuries; the abandonment of a great number of ancient settlements, the ruins of which are deep within the sandy desert; the lowering of the level of groundwater on the plains and the destruction, in connection with this, of the vegetation of the hilly sands (tamarisk, dzhuzgun), which as a result are subjected to being blown about and built up into barchans; the advance of arid landforms on the adjacent territory and the erasure of former geobotanic and zoogeographic zones, of which only here and there isolated relict areas are preserved." [12]

Such pulsations in climate, hydrological adjustments, and transformations of the landscape appear well documented, as are the outbreaks of militant nomads themselves. Factors such as politics, population growth, and economics have also played a role in these events, but such factors are not static. Indeed, they tend not only to interact dynamically with each other, but are stimulated in the first instance by climate changes. Once triggered into movement, these processes overlapped in a common gestalt of events which shaped the history of the region. The important point for this study is that various significant events, social chaos, migrations and invasions, transpired nearly simultaneously to a major shift in climate and environment.

Huntington identified three separate periods of dryness between 2700-1000 BCE, with a steady decline in moisture between 1000 BCE to the start of the Christian era.[13] Soviet researchers have generally confirmed Huntington's views, identifying several Central Asian wet and dry phases in the pre-Christian era. Gumilev identified 3400-2100 BCE as having been arid, being preceded and followed by a moist phase; other dry phases were centered on 800 BCE and 250 BCE.[14] Shnitnikov identified 1900-400 BCE as a generally arid period.[15]

10. Wadia, 1960, ibid., p.7.

11. Wadia, 1960, ibid., p.12.

12. Sinitsyn, V.M., as quoted in Chappell, 1970, ibid., p.368.

13. Huntington, E.H.: *Palestine and Its Transformation*, Houghton- Mifflin, Cambridge, 1911, p.403.

14. Gumilev, 1966, ibid., pp.34-45.

15. Shnitnikov, 1957, discussed in Gumilev, 1966, ibid., p.39.

Huntington felt the alternating humid and dry periods he identified extended beyond just Central Asia, being present also in the Near East and Arabia, and responsible for the historic irruptions of Central Asian and Arabic nomads. These dry phases were CE 200-400 (Huns), CE 600-700 (Arabs), and CE 1150-1300 (Mongols).[16] Again, Huntington's chronology for Asian aridity is in reasonably good agreement with work done by others in more recent years. A few Soviet researchers have recently come to general agreement with his position on the 'Geographic Basis of History', as given in Huntington's *The Pulse of Asia*.[17] Phases of desiccation forced various cultural groups to abandon the steppes, to migrate in all directions toward moister regions. During the earliest of these phases, settled agricultural peoples with a common culture were infiltrated, challenged, and replaced by powerful nomadic groups of an Indo-Aryan background.[18]

All of the Central Asian dates given here must be viewed with caution. Gumilev properly gave a 100 year uncertainty for all his dates; in most cases physiographic evidence, historical data, and archaeological typology are used without corroborative sedimentological or palynological study. The problem of access to published studies, discussed previously, also fuels uncertainty here. Even so, the dates given above for phases of desiccation generally correspond to periods of outward expansion from Central Asia of various nomadic groups, as determined by historians, and as the preceding sections have demonstrated.

SOVIET / RUSSIAN CENTRAL ASIA

From the Caspian Sea eastward one finds a number of large sand deserts occupying basins within a generally arid zone. Southeast of the Caspian and Aral Seas are the Kara-Kum and Kyzyl-Kum sand regions, covered by both fixed and shifting sands. Smaller sand deserts lie to the north and northeast of the Caspian Sea, between the Aral Sea and Lake Balkhash, and to the east of Lake Balkhash. Southeast from Lake Balkhash are the Tien Shan mountains; on the other side of these mountains lie the arid regions of China, the Takla Makan and Gobi, which will be discussed shortly.

North of the Central Asian dry belt is a broad belt of steppeland which gradually merges into the forests of East Europe and North Asia. The steppelands bordering the arid zone are prone to a particularly harsh drought, called the *sukhovei*, which is characterized by hot, dry winds laden with a fine dust. Sukhovei winds lasting 20 to 30 days in a row may possess speeds of 18-30 kmph (10-20 mph) and relative humidities as low as 4-7%. Even crops grown in moist soil wilt under such conditions. The "rainy" season of the Aral-Caspian depression, from April to September, may experience 50 to 80 days of sukhovei. Further, the frequency of such episodes appears to be increasing since the 1800s, when meteorological records were first kept.[19] Soils in these arid steppe regions are typically highly mineralized, with sparse vegetation and very little organic material, less than 1%; abundant soda salts such as carbonates, sulfates, and sodium chloride are found either in the soils, or as surface deposits.[20]

Much of the Central Asian desert belt lies above 600 m (2000') elevation, being bordered and in some cases nearly enclosed by mountain ranges. The Soviet portion of the dry belt is bordered on the south by steep mountains; those on the border with Afghanistan reach to over 7000 m (23,000') while the Tien Shan regularly rise above 3000 m (9840').

16. Huntington, 1911, ibid., p.327.

17. Huntington, 1907, ibid.; cf. Chappell, 1970, ibid.

18. Gumilev, 1966, ibid., pp.34-45; Gimbutas, M.: "Relative Chronology of Neolithic and Chalcolithic Cultures in East Europe North of the Balkans and Black Sea", in *Chronologies in Old World Archaeology*, R.W. Ehrich, ed., U. Chicago Press, 1965, pp.484-6.

19. Kovda, V.A.: "Land Use Development in the Arid Regions of the Russian Plain, the Caucasus and Central Asia", in *A History of Land Use in Arid Regions*, L.D. Stamp, ed., UNESCO, 1961, p.184.

20. Petrov M.P.: *Deserts of the World*, J. Wiley & Sons, 1976, p.29.

These mountains are the highest in the USSR, and yield evidences of "extensive and severe" glaciation during the Pleistocene ice age.[21]

Climate Change and Hydrological Adjustments:

A variety of factors conspired to prevent the establishment of any long term settlements in the Soviet Central Asian arid zone, and to promote nomadic adjustments. Though fresh water was present in several massive lakes (Caspian, Balkhash) and large rivers (Uzboi, Syr Darya, Amu Darya, Chu), their dynamic nature wrought disaster after disaster upon peoples who established their communities along their banks. Subtle, and not- so-subtle climate shifts radically changed vegetation patterns and sediment characteristics of rivers which repeatedly shifted their courses in the loessic steppe, abandoning carefully built irrigation channels. The great lakes periodically shrank, exposing alluvium upon which agricultural empires would be built, only later to rise and inundate the same territory, forcing peoples in whole regions to abandon their settlements and migrate. Often, lake level fluctuations would occur at the same time surrounding steppelands were desiccating, stimulating famine, and driving large populations of both settled and nomadic peoples into other areas.[22]

1. The Caspian Sea

During the Pleistocene the Caspian Sea stood 75 m (250') higher than its present day level of 26 m (85') below sea level; it then covered a much larger area, connecting with the Aral and Black Seas.[23] The Caspian shrank somewhat after the Pleistocene, and even more so when a general desiccation prevailed over the Central Asian steppes after c.3000 BCE.[24] Regarding the region to the east of the Caspian:

> "...one thing is certain: the streams discharged enough water for longer periods of time in a year...than in subsequent periods, proved by the discovery of burnt pieces of trees like ash, poplar, and karagach in the excavations..." [25]

Forests north of the Black Sea were essentially gone by c.2300 BCE, possibly from the joint effects of deforestation and desiccation. This period of desiccation in Central Asia coincides with the various "deluge myth" periods of great flooding in regions at the periphery of Central Asia (Babylon 2379 BCE; Palestine 2355 BCE; China 2297 BCE).[26]

Huntington discussed historical, physiographic, and archaeological evidence for both raising and lowering of the water level in the Caspian and Aral Seas in historic times. He and Davis were able to observe several elevated lake strands with adjacent ruins, plus submerged fortifications and ruins during trips through the area in the early 1900s. Huntington felt the Caspian stood at a high level around 500 BCE, after which it steadily declined to a low around CE 400-650.[27] This date is in good agreement with Gumilev's determination of an CE 500 Caspian low.[28]

According to Huntington, moisture in the region had gradually decreased since c.2000 BCE. However, he felt that a particularly acute arid phase occurred between CE 550-700, stretching across the Near East and Central Asia, and also stimulating the Islamic irruption from Arabia. After a return of moisture to the region the Caspian level rose to a maxi-

21. Movius, H.L.: "Paleolithic and Mesolithic Sites in Soviet Central Asia", *Proceedings, Am. Phil. Soc.* 97(4):383, 1953.

22. Gupta, S.P.: *Archaeology of Soviet Central Asia, and the Indian Borderlands*, Vol. 1: B.R. Publishing, Delhi, 1979, pp.14-9.

23. Wadia, 1960, ibid., p.12.

24. Thomas, 1965, ibid., p.391; Gumilev, 1966, ibid., pp.36-7.

25. Gupta, 1979 Vol 2, ibid., p.160.

26. Gumilev, L.N.: "Heterochronism in the Moisture Supply of Eurasia in Antiquity", *Soviet Geography*, 7(10):35-7, 1966.

27. Huntington, 1907, ibid., pp.329-49.

28. Gumilev, L.N.: "Khazaria and the Caspian", *Soviet Geography*, 5(6):55, 1964.

mum, between CE 750-1000. Gumilev dates the rise in lake level to the 900s, which is in reasonably close agreement with Huntington. Unlike Huntington, Davis attributed the rise in the Caspian not to moist conditions in the arid zone, but due to a northward shift of storm tracks out of the region into the Volga watershed, the Asian steppes drying out as storms passed to the north. Huntington identified a new phase of aridity between CE 1150-1300 with an uncertain Caspian low between CE 1050-1250. He believed this arid phase stimulated the major irruption of Genghis Khan's armies out of Central Asia, after CE 1220. Gumilev identified a Central Asian arid phase in the 1200s, coupled with a major rise in the Caspian.[29]

2. The Aral Sea

East of the Caspian Sea lies the saline Aral Sea, occupying the center of a basin classified as true desert. Two major rivers presently flow into the Aral, the Syr Darya and Amu Darya (the ancient Oxus river). Over the centuries these rivers have shifted their courses in the surrounding steppe, much to the chagrin of settled peoples.

At one time, the Amu Darya flowed farther to the south than at present, feeding Lake Sarykamysh; from this lake, the Uzboi River flowed through the Kara Kum desert toward the Aral Sea. The quantity of water in both Lake Sarykamysh and the Uzboi were greatly reduced following desiccation in the region after 2000 BCE. The Uzboi eventually vanished entirely when the Amu Darya shifted its course around 1000 BCE. The dry bed and settlement ruins of the Uzboi are now covered by shifting desert sands.[30]

The Aral itself has changed in size with pulsating climate, becoming a shallow swamp around CE 350, during a period of particularly intense desiccation.[31]

3. Lake Balkhash

The region around Lake Balkhash possessed many lakes and rivers until at least 1000 BCE, or possibly later; they have long since vanished, however. The fresh-water character of Lake Balkhash led Berg to speculate that it had completely dried out around 300 CE, only subsequently refilling with new waters. Maps of the region dating from c.300 CE confirm a much smaller lake, with a greatly reduced volume corresponding to its deepest parts. By the 800s CE, however, Lake Balkhash had risen to a level higher than that observed today, and other new lakes appeared in the region.[32]

General Settlement Patterns and Culture Change, to Include Influences Transmitted into Russia and Europe:

Paleolithic hunter/gatherers spread across Russia and the forested regions of Asia during the late Pleistocene. Some of the first agriculture spread into Turkestan from the developing complex in Turkey and Iran, possibly as early as 9000 BCE,[33] but certainly by 7000 BCE.[34] By 8000 BCE, sites in Northern Afghanistan demonstrate a developing culture based upon cattle, goats, sheep, and horses.[35] Paleolithic similarities were also shared across the region from Northwest India into Turkestan, Siberia, Tibet and Mongolia.[36] Agricultural communities spread from

29. Gumilev, 1964, ibid., pp.54-68; Huntington, 1907, ibid., p.349; Huntington, 1911, ibid., pp.327,403.

30. Gupta, 1979 Vol 2, ibid., p.17.

31. Kovda, 1961, ibid., pp.199-201; Gumilev, 1968, ibid., p.29.

32. Gumilev, 1968, ibid., pp.25,28-31.

33. Hawkes, 1974, ibid., p.10.

34. Hawkes, 1974, ibid., p.131; Phillips, 1965, ibid., p.17.

35. Gupta, 1979 Vol 2, ibid., p.266.

36. Gupta, 1979 Vol 1, ibid., p.176.

north of the Black Sea westward into Europe and eastward across the steppes. These Neolithic agricultural communities left behind settlement mounds containing figurines of pregnant women, as well as naturalistic cave paintings of hunting scenes and wild animals. Some of these paintings are found high in the Pamirs and are indicative of warmer, wetter conditions.[37]

Neolithic occupations dating from c.6000-2000 BCE exist across the zone stretching from arid North Afghanistan, north and east into the arid steppes of Siberia and Kazakhstan. Cultural aspects from Central Asia were also reflected in Near Eastern sites during this time, as in Jericho and Jarmo. Neolithic communities were established on the Uzboi and at other sites in Southern Central Asia, where female figurines predominated.[38] While agriculture dominated in Southern Central Asia, by 6000 BCE pastoralism developed along the grassland fringe of the forested zone.[39] This was also a period when Neolithic cultures were emerging in East Europe, possessing features similar to the farming and pastoral cultures of Western Asia.[40] A naturalistic style of rock scribings appeared across Northern Scandinavia, Finland, Northern Soviet Union, and Siberia by 5500 BCE, rendered by peoples who were "masters of the creation of life and movement by the simplest of means".[41] According to Dolukhanov, who studied archaeological sites in Turkmenia:

> *"The majority of paleogeographers...share the view that the climate during the Middle Holocene...was moister than that of today. The...opinion is further substantiated by evidence recently obtained for the whole arid belt of the Old World."* [42]

Agricultural villages spread across the region of South Russia between the mountains and the Kara Kum desert, with irrigation agriculture along streams, between 5800-5300 BCE. The period after c.5000 BCE saw the start of a gradual, progressive desiccation which stretched down to c.2000 BCE, reflecting similar changes in the Near East.[43] In Turkmenia, this desiccation:

> *"...reduced the agricultural productivity of arable land, causing relative overpopulation. This, in its turn, caused a decrease in population density, a shift from the proto-urban to the village type of settlement, and a general exodus of the population to the east and west."* [44]

In other areas of southern Soviet Central Asia, however, where secure water supplies from mountain runoff existed, cities of a Bronze Age character continued to develop under the influence of irrigation.[45] The tugai biome, an open woods vegetation of poplars, maples, elms and ash, also persisted.[46] Between 5000-2000 BCE, the Kelteminar culture spread across the region which is now occupied by the Kara Kum and Kyzyl Kum deserts; ruins of numerous villages are scattered in these deserts, where 25% to 50% of all archaeological sites are covered by sand dunes. The Kelteminar culture engaged in hunting, gathering, and fishing, with stable, widespread settlements. At this same time, Turkmenian culture possessed naturalistic female figurines, and shared traits with the Indus Valley and West Asia.[47] Animal domestication spread widely. By 4900 BCE, or possibly afterward, Kurgan agricultural peoples settled north of and between the Black and Caspian Seas, on the newly exposed sediments of the shrunken lakes, particularly on the lower parts of the Dnieper and Volga Rivers.[48]

In Southern Turkmenia, a complex irrigation society developed on the

37. Gupta, 1979 Vol 1, ibid., pp.100-2; Gupta, 1979 Vol 2, ibid., pp.ix,266-7.

38. Gupta, 1979 Vol 2, ibid., pp.14-22,50-2,290.

39. Gupta, 1979 Vol 2, ibid., p.270.

40. Hawkes, 1974, ibid., pp.76-7.

41. Hawkes, 1974, ibid., pp.69-70.

42. Dolukhanov, P.: "The Ecological Prerequisites for Early Farming in Southern Turkmenia", in *The Bronze Age Civilization of Central Asia, Recent Soviet Discoveries*, P.Kohl, ed., M.E. Sharpe, NY, 1981, pp.359-60.

43. Mellaart, 1975, ibid., pp.209-12; Walls, 1980, ibid., p.30.

44. Dolukhanov, 1981, ibid., p.382.

45. Lisitsina, G.: "The History of Irrigation Agriculture in Southern Turkmenia", in Kohl, 1981, ibid., pp.350-3.

46. Masson, V. & Kiiatkina, T.: "Man at the Dawn of Civilization", in Kohl, 1981, ibid., p.108.

47. Gupta, 1979 Vol 2, ibid., pp.18-9,21,88; Sarianidi, V.: "Margiana in the Bronze Age", in Kohl, 1981, ibid., p.165.

48. Gimbutas, M.: *Bronze Age Cultures in Central and Eastern Europe*, Mouton, the Hague, 1965, p.23; Thomas, 1965, ibid., p.378.

Tedjen and Murghab rivers; the region, called a "little Central Asian Mesopotamia" is today characterized by aridity and a desiccated remnant plant cover called *takyr*.[49] As rainfalls became less dependable, these irrigation settlements suffered from both flood and drought.

> *"The instability of water supply had forced the aeneolithic farmers to shift their settlements to the full-water arteries in the middle part of the delta, eg, to the Geoksyur Oasis. The same considerations induced the Tedjun settlers to devise more complicated irrigational systems to regulate the discharge pattern and to supply their crops with water throughout the vegetational period."* [50]

At the Geoksyur Oasis, a complex network of irrigation canals has been uncovered, extending several kilometers in length with numerous branches.[51] However, even here the desiccating conditions had severe effects.

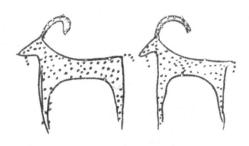

> *"By the end of the fourth millennium BCE, life in the Geoksyur Oasis had gradually faded away. One main factor was the instability of agriculture in the Tedjen delta. Several droughts in succession causing severe famines could have forced the entire population to abandon the area. Another reason is perhaps the result of primitive agriculture: an increase in the salinity of soils and destruction of irrigation devices. This last point might be at least partly connected with an increase in mud flows due to the deforestation of mountain slopes."* [52]

Around 4200 BCE, Central Asia reveals evidence of conflict and warfare. The Namazga II level at the Geoksyur Oasis site possessed fortifications and defensive walls; previous levels were not fortified. The artwork of the fortified levels was plaque-like and abstract, their figurines not possessing natural characteristics of other regions. Again, the pattern is of competition for access to water in a drying region, with fortification of a secure water site against attack by pastoralists and other groups.[53] Still, such characteristics were not yet widespread or common.

Changes After 4000 BCE:

The Amu Darya river, which flows into the southern Aral Sea, supported sedentary Neolithic fishing and hunting groups on its vast delta between 4000 to 2000 BCE.[54] Other Neolithic mixed farming and herding groups were established across Southern Russia, such as the Tripolye. Indeed, by 4000 BCE Neolithic similarities, expressed in tools, pottery and shared skeletal types, were spread across the Near East and Central Asia from the Mediterranean Sea, across the Iranian highlands to the Indus Valley, north across Central Asia into the Tarim Basin and North China.[55] Low stepped-pyramids and imprint seals were shared between Turkmenia and the Indus region.[56] These cultures had a strong feminine influence, with naturalistic female figurines; conflict and/or warfare was neither common nor serious.[57]

> *"...material equipment, as known from their graves, does not provide us with anything more than the rather humdrum products of peasant activity and skills."* [58]

49. Kohl, P.: "The Namazga Civilization: An Overview", in Kohl, 1981, ibid., pp.xi-xii.

50. Dolukhanov, 1981, ibid., p.377.

51. Dolukhanov, 1981, ibid., pp.379-80.

52. Dolukhanov, 1981, ibid., pp.379-80.

53. Mellaart, 1975, ibid., pp.208,219, 225.

54. Gupta, 1979 Vol. 2, ibid., p.16.

55. Phillips, 1965, ibid., pp.17-8; Gupta, 1979 Vol 2, ibid., pp.108-9,239,272,302.

56. Gupta, 1979 Vol 2, ibid., pp.273-4.

57. Phillips, 1965, ibid., pp.18-23; Gupta, 1979 Vol 2, ibid., pp.273-4.

58. Phillips, 1965, ibid., p.29.

Such generally peaceful conditions stretched across most of Central Asia into Europe as well, where:

"...the general absence of weapons of war among the grave furniture and burials provides...convincing proof of the absence of martial ideas in the hearts of the new peasantry." [59]

It is known that the settled peoples of Slavic and Russian territory were loosely governed by democratic tribal assemblies or *folkmotes*, similar in character to institutions elsewhere across ancient Europe. Slavic and Russian women of these early communities maintained considerable freedoms; marriage was based upon free choice, marked by a community festival, and divorce occurred at will. Kinship was matrilineal. Virginity had no importance, and their children were allowed to freely cohabit. Archaeology suggests that this same cultural character extended all across the Asian steppe prior to its desiccation. In later years, under pressures from invaders, male leadership was elected by the folkmote, which also retained the right to elect new leaders, as well as to make war or peace. [60]

By c.3500 BCE the desiccation identified in previous regional sections is more clearly observable in Central Asia. Displacement and migrations of peoples is apparent from archaeological sites, where new cultural materials were being laid down. Not only did new tools, irrigation methods, and houses suddenly appear, but art motifs and figurines underwent a significant transformation. In other areas the ornamental painting of pottery also ceased, a phenomena which appears related to the unpainted "grey ware" pottery of Indo-Aryan nomads of later periods. The first monumental architecture also appeared at this time. Oasis sites were abandoned and changes in river hydrology occurred, with a reduction in the number of smaller rivers. Migrations toward more stable water supplies occurred, with an increased emphasis upon pastoral nomadism. [61]

Eastern and Central Europe, after 3500 BCE, experienced demographic pressures from the shifting populations of Central Asia, and fortification of settlements occurred there as cultural variability increased. [62] New peoples carrying the battle-axe appeared in the archaeological record, intruding from South Russia and possibly beyond, replacing the more peaceful child- and female-oriented European peoples. [63]

"The influence of the Mother Goddess, who had been all powerful during the stone ages, now began to wane. Male deities, gods of war and conquest, were in the ascendant. Metal was the source of their might, and death-dealing armies equipped with Bronze, [and later] iron, and steel weapons were carrying their mandates on the tips of their swords." [64]

"...they were introducing violence to a part of the world that previously had been relatively peaceful. And along with ruthless invasions, undeclared warfare, and appropriation of women as their rightful spoils, they were developing a society in which masculinity was supreme. An insatiable desire for property and power, together with insensitivity to pain and suffering in themselves as well as in others, characterized everything they did. In Scandinavia, the medieval Vikings were to be their heirs, and even in more modern times, Germany was to adopt their belief that might makes right." [65]

Similar movements of peoples to the south also occurred, into the Near East and Mesopotamia, a feature of the subsequent history of those

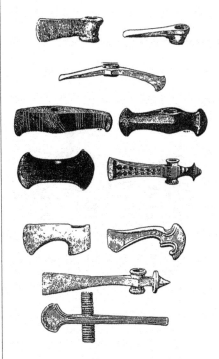

The Battle-Axe
Carried into a once-peaceful Europe by invading groups out of the Asian steppe.

59. Hawkes, J.: "Prehistoric Europe", in *History of Mankind*, Vol.I, UNESCO, London, 1963, p.265.

60. Kovalevsky, M.: *Modern Customs and Ancient Laws of Russia*, Burt Franklin, NY, 1891, pp.12-8,37,121,123,133.

61. Gupta, 1979 Vol 2, ibid., pp.90-103,139-40,152,274-5; Ehrich, 1965, ibid., p.391; Masson & Kiiatkina, 1981, ibid., p.109.

62. Hawkes, 1974, ibid., p.60.

63. Stern, P.: *Prehistoric Europe From Stone Age Man to the Early Greeks*, W.W. Norton, NY, 1969, pp.229-30.

64. Stern, 1969, ibid., p.302.

65. Stern, 1969, ibid., p.230.

areas which would repeatedly occur for thousands of years.[66]

Again, these events occurred against the background of an increasingly arid Central Asian environment, with oscillating lake and river levels, and the start of a general process of land abandonment. Waves of nomadic peoples would spill across Europe, the Near East, India, and China, introducing for the first time widespread violence, warfare, male dominance, castes, and kingships. They carried with them not only new technology and weapons, but also the seeds of a different cultural ethos which viewed the female in a highly subordinated position. Divine Kingship, male gods, the despotic central state, patrilineage, slavery, polygamy, concubinage, female seclusion, the harem and veil, ritual widow murder, infant cranial deformation and swaddling all appeared in Central Asia or in the territories conquered by Central Asian nomads after, and only after, this period.

Archaeological evidence from sites between the Black Sea and Caspian reveal evidence for this cultural transition in the form of female sacrificial victims in the graves of dead kings. This region saw the earliest royal tombs seen in Central Asia, dated to after c.3000 BCE, containing not only sacrificial victims but great tomb wealth.[67] In other places where sacrificial victims were not reported, female figurines have been found in graves.[68] Cemetery vaults of the period also revealed evidence of a growing patrilineage.[69] These were pastoral peoples, evidencing a highly stratified society which would eventually spread across the entire Central Asian steppe.

Central Asian Battle-Cart
Precursor to the Chariot

The Kurgan peoples between the Black and Caspian Seas, whose sites revealed a stable Neolithic agriculture since c.4900 BCE, were dispersing in all directions after 3000 BCE. Other cultures such as the Tripolye were in turn displaced. The relatively homogeneous Bronze Age cultures practicing agriculture with animal domestication, which once spread across the steppelands of Central Asia, were decimated.[70] Over the succeeding centuries, contacts and conflicts between settled agriculturalists and nomadic peoples increased, as did the number and size of fortifications built.[71] Of the Kurgan dispersal, Gimbutas has stated that:

"The Kurgan people came to the Black Sea area...when forested conditions over the present steppe area still prevailed. Perhaps the desiccation of the climate...in combination with their possession of horses, vehicles, knowledge of metallurgy, and social and economic structure as well, have to be reckoned with among the causes for their westward, northward, and southward expansion." [72]

Regarding Indo-Aryan peoples of the steppelands north of the Black Sea, between the Carpathians and the Caucasus, Hawkes has stated that:

"What caused their migrations southward, eastward, and to the west is uncertain, but the fact that...they caused upheavals across the Old World from Europe to India is well enough established." [73]

Desiccation of the Central Asian steppe intensified after 3000 BCE.[74] Cattle and pigs were replaced by goats and sheep. Also of importance was the appearance of the camel and cart in West Asia from the Gobi region far to the east.[75] The dispersals of Central Asian peoples and its associated cultural transition did not occur uniformly in either time or space, however. The Neolithic and Early Bronze Age hunting, farming, and herding tradition, with its natural view of the female persisted for several thousand years in the region between the Urals and the Black Sea, along the rivers to the south of the Aral Sea, and along the forest/steppe bound-

66. Masson, V.: "Altyn-depe during the Aeneolithic Period", in Kohl, 1981, ibid., pp.91-2; Kohl: "Overview", 1981, ibid., p.xxxii.

67. Phillips, 1965, ibid., pp.25-34.

68. Gupta, 1979 Vol 2, ibid., p.277.

69. Kircho, L.: "The Problem of the Origin of the Early Bronze Age Culture of Southern Turkmenia", in Kohl, 1981, ibid., p.29.

70. Gimbutas, 1965, ibid., pp.484-6; Gumilev, 1966, ibid., pp.35-7.

71. Gupta, 1979 Vol 2, p.16.

72. Gimbutas, 1965, ibid., p.32.

73. Hawkes, 1974, ibid., p.12.

74. Gumilev, 1964, ibid., pp.36-7; Gumilev, 1966, ibid., p.35.

75. Gupta, 1979 Vol 2, ibid., pp.147-9.

ary in Siberia.[76] Northward migrations of the dispersing peoples generally did not occur, perhaps because of the transport difficulties posed by difficult-to-penetrate marshlands and forests, and because of the less-desirable cold conditions and poor soils of the North.

The region east of the Caspian Sea also remained densely populated by Mesolithic and Neolithic peoples until after c.3000 BCE.[77] Bronze Age agricultural cultures were also established in the Amu Darya region after 3000 to 2000 BCE, and ultimately became the center of an irrigation farming civilization.[78] However, pastoral cultures characterized by the new, antifemale tradition soon displaced Neolithic settlements in Southern Central Asia. The Uzboi River experienced such a displacement after c.3000 BCE.[79] Sites in Afghanistan also reveal similar displacements and cultural transitions.[80]

The growth and decline of the Bronze Age civilization of Altyn-depe can be traced to this period as well. Altyn-depe is the oldest urban site in the Soviet Union, with a mound 22m (72') high, covering some 30 hectares (75 acres). Built on the intersection of alluvial fans formed by the Meana and Chaacka rivers, the site was harassed around 3000 BCE, as evident from the first construction of fortifications at that time. Altyn-depe contained irrigation works, bronze implements, ziggurat-style monumental architecture, and some class stratification and grave wealth, though violence was not a common feature of its peoples. The region continued to desiccate, however, and by c.1700 BCE, Altyn-depe and many surrounding Namazga sites were either abandoned or greatly restricted in size.[81]

Gumilev discussed these Central Asian settlement patterns:

> "The amount of food that nature provided for man was reduced, settlements decayed, and many were abandoned. The population either retreated into mountain valleys...or adopted a new mode of life suited to the new conditions"[82]

> "...the might and grandeur of the Eurasian nomads was preceded by the brilliance and charm of the settled peoples of the Bronze Age."[83]

The sequential diaspora of Central Asian steppe peoples has also been discussed by Gimbutas:

> "...a culture which, like a volcano, continued to erupt until its lava covered a great part of Europe and the Near East. The period ended with an increased differentiation of cultures north of the Black Sea and elsewhere in Europe..."[84]

It is recalled from previous sections that just before and after 3000 BCE, in both North Africa and the Near East, a period of social chaos occurred, with invasions by foreign cultures of an Indo-Aryan and Asiatic background. The specific cultural transition which occurred in those regions during the irruption of militant nomads from Central Asia can be seen happening in Central Asia itself around the same time period. Events in Central Asia, in turn, appear to have been triggered by changes in climate and hydrology of the Caspian Sea, as well as by arrival of camel and cart-using nomad groups from the Gobi region of China.

Groups continued to migrate into Soviet Central Asia from locations to the east, in China, displacing other groups. Europe, Greece, Anatolia, Iran, Afghanistan, and the Indus Valley were invaded by mounted nomads. Various central state kingships were established on secure water sources in Central Asia, and in conquered lands to the south and west.[85]

76. Phillips, 1965, ibid., pp.27-9; Gupta, 1979 Vol 2, ibid., pp.16-7,22; Kovda, 1961, ibid., pp.180,199.

77. Movius, H.L.: "Paleolithic and Mesolithic Sites in Soviet Central Asia", *Proceedings, Am. Phil. Soc.*, 97(4):383-421, 1953.

78. Gupta, 1979 Vol 2, p.16.

79. Gupta, 1979 Vol 2, ibid., p.17.

80. Dupree, 1973, ibid., pp.268-9.

81. Masson & Kiiatkina, 1981, ibid., pp.107-10,118,127-33; Masson, V.: "Urban Centers of Early Class Society" and "Seals of a Proto-Indian Type from Altyn-depe", in Kohl, 1981, ibid., pp.141,147,153; Dolukhanov, 1981, ibid., pp.380-2.

82. Gumilev, 1966, ibid., p.40.

83. Gumilev, 1966, ibid., p.44.

84. Gimbutas, 1965, ibid., p.488.

85. Gupta, 1979 Vol 2, ibid., pp.121,313.

"...the Kurgan pit-grave people from the Eurasiatic steppes, arrived no later than 2300-2200 BCE in the eastern Balkans, the Aegean Area, Western Anatolia, Central Europe, all of the Western and Eastern Baltic area, and Central Russia." [86]

Anatolia saw destruction of sites during the period of Asian migration, as previously discussed. Anatolian pottery, figurines, weapons, and religion reflected Central Asian motifs, and the first royal tombs appeared, also possessing Central Asian characteristics.[87] Central Russia and Eastern Europe also saw the arrival of Kurgan peoples, disturbing the previous cultures. Artifacts and ornaments of the Kurgan peoples are found stretching across both Europe and the Near East.[88] These lands were culturally transformed by the adopted social institutions of the Central Asian nomads, which were of an antifemale, antichild and antisexual nature.

For instance, ritual widow murder first appeared in Southern Germany and the Eastern Baltic around the time of these immigrants, followed in later years by cranial deformation of both male and female infants. These characteristics would increase and intensify in Europe during later years, as they did in the Near East. Central Europe, near Bohemia and the Carpathians, as well as the Eastern Baltic region, became centers for descendants of these Kurgan peoples in later years. The first clear evidences of warfare also appeared in Northern Europe after arrival of these Asian nomads, who worshiped the battle-axe, and built royal tombs in Central Europe.[89] Indeed, by c.2000 BCE the large South Russian settlements of the Tripolye culture had disappeared as the settled lifestyle was increasingly being abandoned.[90]

After c.2000 BCE, stone graves with grave wealth and sacrificed horses appeared around the Urals and Aral Sea region, and eastward across the steppes past Lake Balkhash and into the Altai Mountains.[91] This was an era of intense mounted warrior nomadism and chariot use. Increasingly, militant nomad peoples swept south out of Central Asia, across the Anatolian/Iranian/Afghan highlands, into Mesopotamia and Egypt, with artifacts being traded across Minoan territory, between Egypt and the Black Sea/Dnieper region. The previous urban tradition of Central Asia, Iran, and India suffered mightily under desiccation, hydrological changes, famine, migration, invasion, warfare, and social unrest. The settled, urban tradition declined and vanished in Central Asia, Iran, and India after 2000 BCE. Mounted nomadism and the significance of the horse increased.[92] The northern forested belt remained relatively immune from these events, however, and peaceful hunter-fisher people of Mongolian stock spread from North China to the Baltic. Similarly, the peaceful Andronov culture persisted in the region east of the Aral Sea.[93]

The ancestors of the warrior horse-using Scythian peoples were first observed on the Asian steppe around 2000 BCE, pushed on a migration towards the south and west by events in Chinese Central Asia. The Nomadic Cimmerians also appear in history around this same time.[94] The period saw invasions of horse and battle-axe peoples into Europe, but also Indo-Aryan nomad invasions south into Thrace, Anatolia, Iran, and the Indus Valley.[95]

In the Northern Carpathians of East Europe, the centuries between 1800-1250 BCE saw a widespread adoption of ritual widow murder in a high percentage of all graves. Horse sacrifices were also present.[96] The Central European Unetice culture, related to the previous Asian Kurgan culture, also developed during these same years; chiefly graves with rich grave-wealth appeared alongside poor commoners graves and mass burials. Skulls were often damaged, suggesting human sacrifice or execu-

New Material 2005: See *Update on Saharasia* (Appendix B) for presentation of newer archaeological evidence on influences transmitted into Europe from Central Asia, notably the discussion on the European *causewayed enclosures*, and on the early massacre sites at Ofnet, Talheim and Schletz.

86. Gimbutas, 1965, ibid., p.21.

87. Mellaart, 1966, ibid., pp.83,88,122-3,134,151-60.

88. Gimbutas, 1966, ibid., pp.23, 46, 340, 394, 585.

89. Grun, 1982, ibid.; Roper, 1973, ibid., p.300; Gimbutas, 1965, ibid., pp.23,394-5,402,423,455-6,551-2; Davies, 1981, ibid., p.33.

90. Phillips, 1965, ibid., p.24.

91. Phillips, 1965, ibid., pp.28-9,48; Gupta, 1979 Vol 2, ibid., p.312.

92. Gupta, 1979 Vol 2, ibid., pp.120, 175, 230; Phillips, 1965, ibid., pp.42-8; Murray, 1949, ibid., p.22.

93. Phillips, 1965, ibid., pp.14,28-9,48.

94. Gimbutas, 1965, ibid., pp.479,528, 533,577.

95. Whyte, R.: "Evolution of Land Use in S.W. Asia", in *A History of Land Use in Arid Regions*, L.Stamp, ed., UNESCO, 1961, p.101; Phillips, 1965, ibid., pp.7,39,41-2.

96. Gimbutas, 1965, ibid., pp.463-5.

tions.[97]

South Russia also revealed clear evidence of mounted horseback warrior nomads around 1500 BCE, a period when similar nomads were crossing the Caucasus into Iran and Afghanistan.[98] South Russia experienced an increase in fortifications and evidence of warfare. The horse became prominent in religion and in military cults. Human sacrifice and ritual widow murder became both widespread and "typical" of graves of the period. Dismemberment of sacrificial victims, as well as horrible tortures, were suggested by scattered human bones possessing knife-marks.[99] And far to the east, near Siberia, house plans began to suggest the presence of female seclusion.[100]

A network of trade routes spread throughout Europe, linking Central Asia with the British Isles and Sweden on the west, and with China on the east. The Hittite sky god Teshub was found from Lithuania to Anatolia; bronze spearhead similarities were found in sites in Britain, Poland, and China; amber was traded between Europe and Russia to the Urals and Siberia, as well as into the Near East through Greece or the Caucasus.[101]

With the spread of new technology and goods also went various Central Asian cultural practices. Graves in Central Russia and the Western Urals dating from c.1450 BCE yield dismembered, decapitated bodies during a period of invasion by the South Russian steppe people. Ritual widow sacrifice, and child sacrifice appear in the graves of Central Europe which saw the expansion of new cultures, the Tumulus and Urnfield, who were related to the widow-murdering Kurgan and Unetice cultures. These cultures spread across the upper Rhine, into Northern parts of Italy, the Adriatic, Yugoslavia, Greece, and Anatolia. Related peoples also expanded into the Baltic region, and proto-Celtic groups, also related to the Asian Kurgans, penetrated into France.[102]

The expansion of the Central European, old Kurgan peoples continued. Settlements to the south were destroyed, from Italy, across Greece and Anatolia, and spilling into the Eastern Mediterranean. Widespread and massive destruction and warfare followed in their wake.[103] Russian steppe peoples similarly continued migrating southward, into Iran, Afghanistan and India. And on the steppe itself, stockbreeding, male deities, horse sacrifice, and widow murder persisted and increased among groups such as the Turbino.[104]

The Uzboi River on the Aral Sea totally dried up after 1000 BCE, and its numerous settlements were abandoned when a shift in the course of the Amu Darya occurred.[105] Irrigation efforts increased thereafter along the deltas of the Aral Sea, which also was drying out.[106] Terracotta figurines of the period fell into "miserable" conditions.[107]

Still, there is some evidence of a moist interlude around 1000-900 BCE, when peoples began reoccupying the dry belt. At that time a new wave of Scythian migrants appeared, this time moving out of East Europe across the steppe. The light-skinned Scythian nomadic warriors traded with Rome and Greece, and brought a new phase of mixed agriculture and pastoralism to the northern steppe as far east as China.[108] Migrations of peoples back north from Iran occurred around the Aral Sea region, reoccupying previously abandoned sites. Several South Central Asian kingdoms grew to power at the time: Bactria, Margiana, Parthia, Sogdiana, the borders of which were constantly shifting. These kingdoms periodically coalesced from the various nomadic tribes of the steppe, who later were collectively called the Royal Horde. With shifting loyalties, these tribal groups variously formed alliances and constantly challenged each other for control of the Asian heartland.[109] The possible moist interlude of c.1000-900 BCE appears to have ended quickly, however, and

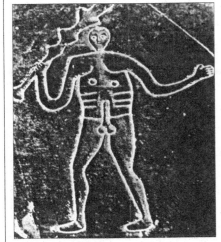

The Cerne Giant, Bronze-Age Celtic Britain (c.1000 BCE ?), carved into a chalk hillside, evidencing a territorial marker, or defensive reaction to an earlier arrival of warrior Saharasian immigrants — his club appears to be an inferior weapon to the Central Asian battle axe. The image stands 55m (180') tall, with an erect penis and war club. Often mis-identified as a "fertility image", the giant appears more clearly to issue a visual warning and threat to other cultures and strangers.

97. Gimbutas, 1965, ibid., pp.267,340.
98. Phillips, 1972, ibid., pp.42,46-8; Dupree,1973, ibid., p.269; Gupta, 1979 Vol.2, ibid., p.312.
99. P'iankova, L: "Bronze Age Settlements of Southern Tadjikistan", in Kohl, 1981, ibid., pp.291,303; Gimbutas, 1965, ibid., pp.539-46.
100. Gimbutas, 1965, ibid., pp.633,642.
101. Gimbutas, 1965, ibid., pp.47,112.
102. Gimbutas, 1965, ibid., pp.285,330,340,394,595-6,604-5,648.
103. see the previous section on "Anatolia to the Indus".
104. Gimbutas, 1965, ibid., pp.23,562,569,571-2,617,619,635-8; Hawkes, 1974, ibid., p.188.
105. Gupta, 1979 Vol 2, ibid., p.17.
106. Gupta, 1979 Vol 2, ibid., p.16.
107. Gupta, 1979 Vol 2, ibid., p.176.
108. Phillips, 1965, ibid., p.14; Chang, 1968, ibid., p.360; Kovda, 1961, ibid., p.181.
109. Phillips, 1965, ibid., pp.45-8; Gupta, 1979 Vol 2, ibid., p.x.

settlements remained only along secure water supplies. After 900 BCE, the Andronov and Karasuk cultures were pushed out of Kazakhstan by nomads, into fortified retreats at the forested fringe of the steppe. This date also saw militant, light-skinned nomads pushing east into the Tarim Basin of China. Other nomad groups spilled across the steppe after 800 BCE, when the arid zone demanded a true nomadic existence.[110] The entire steppe zone was then "unified into a pattern characterized by mounted warfare."[111]

The Scythians were one such mounted nomad culture who used Bronze and Iron weapons. A strongman military society, the Scythians murdered large numbers of women, slaves, and horses to put into the graves of their dead leaders, while the living smoked hashish and engaged in various self-mutilation rituals. They tortured and crucified enemies and criminals, took scalps, practiced concubinage, and worshiped male gods similar to the Dorian Greeks, including a war god later adopted by the Sarmatians and Huns. Griffins and clawing beasts were typical art motifs. The early Scythian period has been identified as the time when the matrilineal family system of Central Asia broke up, being replaced by patrilineage.[112] The Scythians and other nomadic steppe peoples practiced infant cranial deformation and prolonged swaddling, which were transmitted to many of the regions they conquered. Such practices had been present in the region even before the Scythians, however.[113]

Militant mounted nomads began displacing settled peoples of the eastern and western steppe. Settled peoples were variously conquered and enslaved. Often, they retreated into fortifications along the edge of the forested regions, or into mountain valleys where possible.[114] Equestrian equipment and artifacts uniformly appeared across Eastern and Central Europe, as well as other Eastern influences, originally dating from the Kurgans, and now taking on aspects of Central Asian Scythians, Cimmerians, and Sarmatians. The Danube region was "overrun by hordes of steppe horsemen", who continued to press into France, and Catalonia.[115]

After 700 BCE, Scythian and Sarmatian warriors ruled most of the Asian heartland from Hungary to Dzungaria, displacing other warrior nomad groups, such as the Cimmerians, southward into Thrace and Anatolia.[116] As mentioned previously, this was the time period when the Etruscans abandoned the region north of the Caucasus, migrating across Anatolia to eventually settle in Italy.[117]

The Khorezm State developed at this time, based upon irrigation agriculture and semi-nomadic stockbreeding in the lands south and east of the Aral Sea. Irrigation channels brought water to much of the land east of the Aral, linking the Amu Darya and Syr Darya deltas, which are now several hundred miles apart. At the time irrigation works were protected by fortresses and garrisons of soldiers, indicating conflicts with the nomadic groups.[118]

After 600 BCE, chariot warfare expanded across the steppe. Scythians again pressured Cimmerians, and unrest in Western Asia prompted growth of fortifications in Poland, Germany, and Switzerland. By 400 BCE, many of these same fortified European settlements were destroyed, either by flooding, or burning by Scythian and Celtic invaders. The Celts occupied territory in northeast Europe, and came into conflict with Scythian and Sarmatian nomads. They sacked Rome in 387 BCE.[119]

The Achaemenid Persians under Cyrus the Great also intruded into Central Asia after 600 BCE, displacing the Scythians who fled north to harass Eastern Europe.[120] The Persians conquered the nomads across Turkestan into Chinese territory. New nomadic groups appeared on the Chinese frontier moving westward, in a chain of displacement started by

Scythian burial mound, barrow or *Kurgan*, near Leninakan (formerly Alexandropol) Armenia, measuring around 90 meters (300') across and 18 meters (60') high. Such mounds contained heaps of treasure, and often murdered women, slaves and horses (c.800 BCE).

110. Phillips, 1965, ibid., pp.45-8; Gumilev, 1966, ibid., pp.40-1.

111. Chang, 1968, ibid., p.360.

112. Phillips, 1965, ibid., pp.50-2; Gumilev, 1966, ibid., pp.40-1; Scott, 1939, ibid., pp.153,211; Davies, 1981, ibid., p.53.

113. Phillips, 1965, ibid., pp.52-5,64-6,92-3; Chang, 1968, ibid., pp.50-2.

114. Phillips, 1965, ibid., pp.8,54-5,64-88; Chang, 1968, ibid., p.360; Kovda, 1961, ibid., pp.181-2.

115. Gimbutas, 1965, ibid., pp.159, 340 ,354-5.

116. Gimbutas, 1965, ibid., p.443; Phillips, 1965, ibid., p.52.

117. Heichelheim, 1972, ibid., p.166; Gimbutas, 1965, ibid., p.159.

118. Kovda, 1961, ibid., pp.199-201; Gumilev, 1968, ibid., p.29.

119. Phillips, 1965, ibid., pp.8,55.

120. Hawkes, 1974, ibid., pp.61,65,81; Heichelheim, 1972, ibid., p.167.

military action against nomads by the Chin Dynasty Chinese.[121] These groups were held in check by the Persians, who were later defeated by Alexander the Great. Alexander did not reclaim the Asian territory held by the Persians, being unable to overcome the ambush-encirclement tactics of the nomad calvary.[122] As the Greeks withdrew, mounted nomads again swept across Central Asia and southward into Iran, to form the Parthian Dynasty.

The warrior nomad Huns were active in displacing various groups within the Caspian region roughly between 200 BCE to 300 CE. They were then gradually moving west from the desiccating steppes of the Gobi and Ordos deserts of China.[123] The Chinese Yeuh Chi similarly pushed westward from China across the Aral Sea and Amu Darya region, destroying Greek cities in Bactria and founding the Kushan empire which encompassed the Caspian and Aral Seas, Tarim Basin, Indus Valley and Northern India.

Other groups fleeing from the Yeuh Chi, such as the Saka, fled across Iran, Afghanistan, and India to form new kingdoms of their own. In India, these peoples adopted Buddhism, and transmitted it across Central Asia, from where it spread to China, Korea, and ultimately Japan.[124]

The Sakas and Sarmatians were in turn displaced south and westward after 200 BCE, invading Afghanistan, Iran, and East Europe. In parts of Eastern and Central Europe, Germanic, Teutonic, and Gothic peoples with despotic, antifemale and antichild characteristics related to previous Battle-axe, Kurgan, Celtic, and Scythian groups ("Aryans") grew to dominance.[125]

Rome, also under control by peoples with a similar heritage, began its reign of conquest after 100 BCE, ultimately driving less powerful European groups, such as the Goths, eastward onto the steppe in the opposite direction. This eastward invasion of the western steppe would only be temporary, however.[126] When the Romans under Caesar entered Britain in 55 BCE, Celtic peoples and settlements along with megalithic monuments were found. Such megalithic monuments also existed across parts of France, Spain, Denmark, Southern Scandinavia, Malta, Sicily, and Ireland. The Romans made much out of the human sacrifice which existed there, but it is unclear if this was a custom introduced by the Romans themselves, as a by-product of the bloody warring which commenced after they arrived, or if such was introduced by the ancient Celts themselves as they migrated out of Eastern Europe with its Saharasian-influenced cultural circumstances. Nevertheless, the presence of brutality and sadism in Briton appears to have been a transplant from the mainland, related to the patristic Celts, Romans, or other immigrant groups who developed such traits by virtue of earlier Saharasian connections.[127]

During the period of increasing desiccation, sediments along the fringe of Lake Balkhash were exposed and provided pasture for nomadic groups. Chinese military accounts of 100 CE speak of Huns grazing herds numbering in the hundreds of thousands in the area. Other nomadic states lived in the area, grazing several hundred thousand horses, and presumably sheep and goats several times the number of horses. The region then possessed lakes and rivers, which are today vanished.[128]

Huns, Sarmatians, and other nomadic groups continued their westward movement, as well as the practices of ritual widow murder.[129] The Huns were particularly fierce, with skulls which appeared most frightening; male infants were also subject to scarification of the cheeks. With continued desiccation after CE 200, Huns defeated the Sarmatians and irrupted from Central Asia, disrupting overland trade routes, destroying cities, and enslaving peoples in the surrounding humid lands of East Europe and the Near East. In particular, infant cranial deformation and

Central Asian Warrior

121. Dupree, 1973, ibid., pp.297-8.

122. Dupree, 1973, ibid., pp.279-85; Heichelheim, 1972, ibid., p.167.

123. Gumilev, 1964, ibid., pp.54-65; Gumilev, 1966, ibid., p.43; Gumilev, 1968, ibid., p.25; Phillips, 1965, ibid., p.111.

124. Phillips, 1965, ibid., pp.98-9,110-2,126; Dupree, 1973, ibid., pp.xviii,297-9.

125. Gimbutas, 1965, ibid., p.355; Davies, 1981, ibid., pp.45-6.

126. Gumilev, 1968, ibid., pp.25,28.

127. Davies, 1981, ibid., pp.22,46-7; Scott, 1939, ibid., pp.86,90,93.

128. Phillips, 1965, ibid., p.112; Dupree, 1973, ibid., pp.xviii,299.

129. Phillips, 1965, ibid., pp.100-4,120.

swaddling were transmitted by the Huns to both Europe and India. Infant cranial deformation died out first, but swaddling persisted for centuries.[130] During this arid phase the steppe zones around the Aral and Balkhash declined in population. Military conflict and nomadic subsistence increased as various groups dispersed.[131]

During this period of aridity and unrest, semi-nomadic Indo-Aryan Khazar peoples migrated into the Caspian region. They began farming the sediments exposed by the shrinking lake, as had Kurgan peoples several thousand years earlier. Documents of the era testify to minimal rains but abundant rivers, fish, trees, wells, and a generally fertile and rich country. The Khazaria Khanate developed on the Volga delta during the Caspian low; it was a period of great wealth, with vineyards, orchards, fishing, farming, and livestock raising. Populations in the region were larger than that of the Cossacks in the 1800s. The arrival of the Khazars in the Caspian coincided with migrations of Slavs (which means *slave* in Greek and Latin) northward into territory previously occupied by Finno-Ugrian peoples.[132]

Meanwhile farther east, the Khorezm State on the Aral fell into a decline as desiccation turned the Aral into a shallow swamp. Persians, Huns, and Turks also fought over the territory after CE 350, initiating a series of blows from which the region never recovered.[133]

By 380 CE, Huns attacked and defeated the Goths, driving them westward into a temporary alliance with Rome. Commanding the conquered Sarmatians, the Huns pushed into East Europe and established headquarters in Hungary; they also fought continuous battles with the Sassanid Persians and Kushans of Iran. In 410 CE, the pressured Goths rebelled against Rome, sacking it. By 440, the Huns under Atilla, the "Scourge of God", ravaged Europe until their empire stretched from the Baltic to the Caspian, and eastward from there to the Tarim Basin. Atilla's death in 453 CE came as the Huns were initiating attacks against Gaul and Rome; disorder and rebellion thereafter fractured the Hunnish alliances, and their empire collapsed. Hunnish Kings continued to rule in parts of India while Visigoths and Ostrogoths, who were greatly devastated by the Huns, ruled in Gaul and Italy respectively. Hepthalite Huns also launched new attacks upon Iran.[134]

Central Asian Warriors

Pressures from the Huns and their allies, and from the growing Roman power, caused the Angles, Saxons, and Jutes to abandon North Europe and invade Britain; Frankish Law of the period codified for the first time the various forms of legal versus illegal torture. Legal application of torture was progressively expanded in Europe to include freemen as well as slaves.[135]

During the Hunnish period, Slavic and Russian peasants came into increasing conflicts with both Asian nomads and Christian Kings and philosophers, who greatly opposed the relatively free sexual life and high female status of these settled agricultural peoples. Christian clergy warred against adolescent cohabitation, and the secular marriage customs by which young people chose their own mates.[136]

Strongman Christian influences spread into Europe with the power of various Roman or Byzantine Emperors. Saint Patrick preached in Ireland against various remnant practices of the Celts, such as the (claimed) burning of firstborn children, but also introduced a host of other antifemale and antichild social institutions fast growing among Christians of Continental Europe. The myth about St. Patrick chasing all the snakes out of Ireland may be rooted in the "chasing out of sexuality" by Christian zealots. Germanic city-states grew to power during this period of urban decline, when the clergy were increasingly intruding into sexual and family matters. Ritual widow murder continued in East Europe, appear-

130. Dingwall, E.: *Artificial Cranial Deformation*, London, 1931, p.34; Phillips, 1965, ibid., p.126.

131. Kovda, 1961, ibid., pp.199-201; Gumilev, 1968, ibid., pp.29-30.

132. Gumilev, 1964, ibid., pp.54-65; Darby, H.C.: "The Clearing of the Woodland in Europe", in Thomas, 1972, ibid., p.206.

133. Kovda, 1961, ibid., pp.199-201; Gumilev, 1968, ibid., p.29.

134. Phillips, 1965, ibid., pp.120-2,124; Dupree, 1973, ibid., pp.302-3; Kiefer, 1951, ibid., p.357; Tannahill, 1980, ibid., p.136.

135. Wittfogel, 1957, ibid., p.146.

136. Kovalevsky, 1891, ibid., pp.7, 37,133; Bullough, 1976, ibid., pp.350-1.

ing among both Slavs and Serbs.[137]

The affect of the Huns upon Eastern and Central Europe might be further assessed from the Germanic epic of the *Niebelungenlied*, where victorious Germanic troops drank the blood of the fallen Hunnish foe. The Germanic legal code, of Visigoth inspiration, was codified from tradition and written on paper.[138]

This was a period when various new nomadic groups were settling in Eastern and Central Europe as a dominant ruling class, bringing new traditions which would rather easily mix with Christian antisexual philosophy (which also came from the same original source region, via the Near East, however). It was a period when the lower classes of Europe were being radically affected. With Church inspiration and approval, children were ritually beaten, women and "heretics" tortured and burned, and epidemics of self-flagellation appeared in the streets.[139]

In Southern Russia of 500 CE, the Khazars continued to flourish in the rich Caspian agricultural region, adopting a more settled lifestyle, and coming into conflict with other nomadic groups. Persians also instituted control over part of the Caspian region. The submerged Caspian ruins observed by Huntington and Davis were built during this period of Persian Sassanid influence (c.500 CE), when the Caspian Sea was lower than today. The Caspian fortifications included towers and walls 4m (13') thick and 18-20 m (60'-65') high.[140]

The first Turkish Empire appeared in history around 600 CE, occupying the old Hunnish territory; they originally hailed from the steppes of western Mongolia, as had the Huns. The Turks were forced to the northwest by the Arabs, who swept north across the Near East, invading Persia and Afghanistan after CE 625. The dispossessed Turks invaded Kievian Russia, a loose confederacy of Slavic principalities, and from there ruled an empire which stretched from the Caspian to Mongolia. During this process of invasion-displacement-invasion, settled agriculturalists suffered most.[141]

The 700s CE saw the development of the Viking cultures in the Baltic region. Descendants of the Battle-axe peoples, the Vikings sacrificed male and female adults and children to their gods, ritually murdered widows and female slaves at the gravesite, and displayed an incoherent form of "battle madness", during which they would go "berserk", losing all sense of fear or concern for self-preservation, making them an awesome enemy to deal with. To "go Viking" meant to go raiding, murdering and pirating. Between c.700-1050 CE, they would raid and invade Ireland, Scotland, England and Normandy from homelands in Norway and Denmark. Other Vikings pushed eastward into Russia after c.850, forming a ruling dynasty in Novgorod; they came to be known as the Varangian Rus, and dominated the region until the 16th century reign of Ivan the Terrible.[142]

The 700s CE also saw Arab armies sweeping eastward across the steppe of Southern Russia, into India and West China. They observed ritual widow murder on the Russian steppe, which had persisted from several thousand years earlier; a man's favorite wife would be strangled, and burned upon his funeral pyre.[143] Arab chroniclers also witnessed the ritual murder of a slave woman of a dead Varangian Rus chief; she was killed along with horses, cattle, and a dog, all of which were put on a boat with the corpse of the chief. The boat was set adrift, and burned.[144]

The southern part of Khazaria on the Caspian Sea, for instance, had been devastated by centuries of drought and invasions by nomadic warriors. Through the periods of chaos, the Khazars retained control over these previously Slavic regions. Unfortunately for them, a new threat occurred in the form of a rising Caspian Sea level. By CE 950, 2/3 of their lands were inundated by rising lake levels. In CE 965, the Khazars fought

137. Bullough, ibid., pp.347-8,352; Tannahill, 1980, ibid., p.216; Davies, 1981, ibid., p.46.

138. Bullough, 1976, ibid., p.349; Davies, 1981, ibid., p.45.

139. See the discussion in the previous section, "Anatolia to the Indus".

140. Gumilev, 1964, ibid., pp.54-68; Huntington, 1907, ibid., p.349; Huntington, 1911, ibid., pp.327,403.

141. Kovda, 1961, ibid., pp.182,199-201; Gumilev, 1968, ibid., pp.29-30; Dupree, 1973, ibid., p.312; Pitcher, 1972, ibid., p.23.

142. Jordan, R.: "When Vikings Sailed East", *National Geographic*, March, 1985, pp.282,294,303; Davies, 1981, ibid., pp.21,45,103; Howard, 1977, ibid., pp.61,110; Stern, 1969, ibid., p.230.

143. Pillsbury, B.: "Being Female in a Muslim Minority in China", in Beck & Keddie, 1978, ibid.; Davies, 1981, ibid., pp.103-4.

144. Jordan, 1985, ibid., pp.305-6.

with the Varangians of Northern Russia. The Varangians adopted the practices of female seclusion and polygamy from the Khazars, whom they defeated. Subsequently, the Khazars converted to Islam, as had the Turks before them, and joined the Royal Horde of the Great Khan.[145]

Kievian Russia repeatedly clashed with the Turks between 700-800 CE, after which Russian history recorded a shift to the north, to Novgorod, Vladimir, and Moscow. In spite of these invasions, however, Kievian Russia of c.800 CE did not possess a fully despotic character. Christianity and Islam had both been rejected by Kiev, given their austere views of festivity, celebration, and female sexuality. To the north, the rural peasantry and lower classes did not experience any major change in family structure or sexual life, and maintained a relatively high female status, at least until the time of conflict with the Varangians and the Khazars. Northern Russia retained many of its original, non-repressive social customs about females and children, and because of its relative isolation did not develop quite as harsh a view of women and children as the Central Asian nomads.[146]

During and after the 900s, additional changes occurred. Vladimir I (c.930 CE) forced Byzantine Christianity upon the Slavic peasants, bringing about the religious control of family matters, including the status of women and sexual behavior. While the Christian dictators put a stop to ritual widow murder, as they did in Eastern Europe, widows were still considered a taboo class. Byzantine Christians also persecuted and murdered "heretics", including many female "witches", though never to the same extent as the Roman Christian Church whose celibate and chaste priests always appeared more laden with sadism towards women and children. But Vladimir himself was no saint. He is said to have "shopped around" for a suitable religion, rejecting Islam because of its prohibition on alcohol; he worshiped idols and kept a personal harem of 800 concubines. Following forced Christianization of Kievian Russia, the Clerical and parental control of marriages resulted in spates of suicides among young girls condemned to unwanted marriages, and new laws were passed to penalize parents if their daughter killed herself following such a union. The peasantry gradually adopted severe Christian asceticism, and eventually believed that love-match marriages which went against the parental will would invoke the wrath of God and Heaven. Severe obedience training of children, female subservience, the sale of wives, and wife-beating progressively increased. Similarly, King Olaf (c.1030 CE), later to become "Saint" Olaf, began a campaign to "convert" the Norwegian peasants to Christianity — through armed invasion, massacre, enslavement, and confiscation of property.[147]

It was during this time period, of expanding Christian and Islamic influences, that the remaining democratic folkmotes disappeared from Southern Kievian Russia; they persisted, however, in Northern Russia near Novgorod.[148] Despotism in Southern Kievian Russia would increase.

> "Kievian society presents few features indicating a penchant for regimentation outstripping that of other medieval states. It is Tatar [Mongol] influence from 1240 onwards that seems to have given disciplinary trends their initial impetus, imparting despotic and traditionally 'oriental' features."[149]

As discussed in the previous section, much of Byzantine Christianity was founded upon "oriental" features adopted by Greece and Rome after 500 BCE; hence, the view of the Tatars as the sole source of Russian despotism must be tempered with the observation that influences of a similar character were forced upon Kievian society at earlier times by

145. Gumilev, 1964, ibid., pp.54-68; Huntington, 1907, ibid., p.349; Huntington, 1911, ibid., pp.327,403; Wittfogel, 1957, ibid., p.202; Gupta, 1979 Vol 2, ibid., p.xvii.

146. Hingley, R.: *The Russian Mind*, Bodleyhead, London, 1977, pp.27,46; Calverton, V.F.: *The Bankruptcy of Marriage*, Macaulay Co., NY, 1928, p.229; Kovalevsky, 1891, ibid., pp.132,142; Wittfogel, 1957, ibid., p.201; Kovda, 1961, ibid., p.182.

147. Kovalevsky, 1891, ibid., pp.36-7,142,149; Gage, 1980, ibid., p.382; Calverton, 1928, ibid., p.229; Jordan, 1985, ibid., pp.285,307; Davies, 1981, ibid., p.103.

148. Kovalevsky, 1891, ibid., pp.124, 127, 132, 141.

149. Hingley, 1977, ibid., p.162.

warlord Byzantine Christian emperors and other patristic Saharasian groups. On the Central Asian steppe, Moslem Turks captured Arab territories, eventually moving into Iran to form the Seljuk Dynasty of c.1000-1200 CE. The Seljuk Turks were dominant across Central Asia and the Near East, from Anatolia to India, until the Mongol period following c.1200 CE.[150]

Traces of the horse-riding warrior mentality, which originated in Asia, persisted in Europe in the form of knightly "chivalry". Chivalry was originally a quasi-religious cult of virginity-worship, horsemanship, and war, whose roots can be traced back to Germany. During its early centuries of existence, chivalry:

> "...was a social disaster. It produced a superfluity of conceited illiterate young men who had no ideals except to ride and hunt and fight, whose only interest in life was violence and the glory they saw in it. They were no good at anything else, and despised any peaceful occupation. In national wars they could be called on to fight by their feudal obligations, much like the Thanes in England, but just by existing, they created wars."[151]

Howard has contrasted the social conditions between Normandy and Briton just before the 1066 CE invasion of Briton by Norman despot, William the Conqueror. Normandy possessed characteristics transmitted to it by both the "chivalric" knights and other horse- and war-emphasizing peoples from Saharasia. The water barrier of the English Channel had allowed British cultural groups to survive in a greater degree of isolation from events on the mainland of Europe. The Celts and Romans had left a mark, to be sure, as had other groups, including the Vikings. But in the 1000s, British people lived under much freer conditions than existed on the mainland. Peoples in Normandy, by contrast, lived under absolutist kings with a strict hierarchy of power. Castles dominated a landscape prone to sporadic violence and warfare. The various knights always gave part of their booty to the Church, who forbade fighting on Lent, and from Wednesday evening to Monday morning. Selection of the Pope was no longer a secular affair after 1059, making the Holy See accountable only to its own Bishops. Church oversight of marriages and the sexual life of the average person was commonplace.[152]

The British, before the Norman invasion, lived in a class society with a king whose orders could be disobeyed if they violated unwritten laws regarding the rights of the peasant. Village meetings called the *moot* existed, similar to the previously-mentioned continental folkmote, with hierarchical levels of representation. While such moots were male dominated, they did allow the presentation of grievances and representation of all classes, and the embryonic stages of a parliament existed. The *witensmoot*, derived from the moots, advised the king, and selected his successor, as the folkmote once had the power to do on the continent. Nobody was above the law; Earls could be ejected by the peasants, and the witensmoot could dethrone a king. Britain also possessed secular marriages, which could be easily terminated. Premarital sex was common, and neither the clergy nor parents interfered much with the love life of young people. Women could own land, and bridegrooms commonly gave land to their wives as a wedding gift or dowry. Slavery was on the decline before the Norman invasion, given the absence of wars for the previous generation, and also due to church opposition to it.[153]

The Norman invasion cost Britain 20% of its total population. Norman troops, 2/3's of whom were mercenaries, looted, raped, murdered, and destroyed much of the countryside. William the Conqueror paid his

150. Wittfogel, 1957, ibid., p.360; Dupree, 1973, ibid., pp.313-5; Wiencek, et al, 1980, ibid., pp.16,91; Pitcher, 1972, ibid., p.26.

151. Howard, 1977, ibid., p.62.

152. Howard, 1977, ibid., pp.61-3,66,101.

153. Howard, 1977, ibid., pp.14,17,21-3,38-9,44-5,59,62.

troops partly with estates and land, and 200,000 of the French and Norman invaders settled on the English countryside as a dominant, ruling class. The Domesday Book was drafted and taxes were ruthlessly extracted, as fortified Norman castles appeared on the countryside.[154] The British monarchy in succeeding centuries became progressively more despotic, and worked to restrain the rights and freedoms of women and young people. Still, such traits never developed in Britain to the extent seen on the continent itself.

The forcing of King John's signature on the Magna Carta in 1215 CE evidenced the persistence of an independent spirit among the British peasantry. The middle 1200s on the continent, by contrast, saw the massive persecutions of the Holy Inquisition, in both Germany and France.

Britain was a major proponent in the Reformation, which opposed Papal tyranny, and also led the continent in later years with respect to legal reform over slavery, judicial torture, separation of church and state, limits to power of monarchs, development of democratic institutions, and greater rights and freedoms for women and children, to include divorce and marriage law reforms, and later legalization of contraceptives.

As a colonial power, Imperial Britain often as not transplanted vestiges of their own culture into regions that often were far more authoritarian and brutal in character than the British themselves, though this was certainly not always the case. And as reforms took place in Britain, such reform was often transmitted to its colonies, something not as clearly apparent in other European colonial powers. One major reason for this would appear to be the constant pressures from antifemale, antichild, warrior nomad groups which continued to press into Eastern Europe from Central Asia, and from which Britain was relatively immune, given its island character. Britain thereby managed to preserve some of its female and child orientation in the face of the Central Asian outbreak, something impossible in either Spain, Italy, Greece, Prussia, or Russia, all of which had been badly shocked by invasions from Saharasia. These latter areas had repelled the invaders, but later adopted many Saharasian treatments of females and children as their own. On this score, it is interesting to note the complaint of the German Queen Caroline, who said that the "English were not well bred enough because they were not whipped enough when they were young."[155]

In the 1200s, additional migrations toward the west from Central Asia affected Russia and Eastern Europe, this time under the leadership of the Mongols. Called "a confederation from one of the most forbidding places on Earth",[156] the Mongols (Tatars) under Genghis Khan established an empire from the Caspian to Pacific China and Korea. The Mongol nomad warriors had little but contempt for the hunters of the forest, or for urban dwellers; even their own game hunting was seen as preparation for killing men in war. They genuflected to the sun at dawn, and paid homage to the Great Khan as a God who was heaven-sent to unite them. Foreigners who refused to pay homage to his image were killed. As is well known, the Mongols perfected the meaning of barbarity, engaging in torture, murder, and bloodlust on such a scale that it appeared shocking even by Central Asian standards. Captured civilians and prisoners were driven in front of Mongol troops during battles, and forced to position siege equipment or fill trenches while under fire. Towns and villages were sacked, their inhabitants ruthlessly slaughtered in torture-murder frenzies which might last days or weeks:[157]

"Genghis Khan was the atom bomb of his day; and western Asia still bears the scars, still suffers from the economic impact... The

Mongol armies storm a town in Persia. Inhabitants of such conquered towns were generally slaughtered, or taken into slavery.

154. Howard, 1977, ibid., pp.103,146, 198-9.

155. Radbill, 1974, ibid., p.4; The official caning of English schoolchildren continues today, however.

156. Wiencek, H., et al, *Storm Across Asia*, HBJ Press, NY, 1980, p.12.

157. Wittfogel, 1957, ibid., p.219; Dupree, 1973, ibid., p.316; Wiencek, et al, 1980, ibid., pp.13-49.

silted canals and destroyed cities in western Asia sit as Genghis Khan's monuments." [158]

Mongol and Turk nomad armies established central states headquartered in humid regions surrounding the Central Asian arid zone, including in Kiev. The tenure of the Mongols in Kievian Russia and East Europe after 1240 was a severe blow which restructured beliefs and social institutions in a more powerful manner than the previous Byzantine influences. New Mongol words crept into the Russian language, such as *kabala* (bondage), *nagayka* (whip), and *kandaly* (fetters). The characteristic Asian nomad view of women, previously introduced by Scythians, Huns, and Byzantine Christians, was intensified.[159] The subservience and tribute which the Mongols demanded of the Russian Princes acted as a process of *unnatural selection*, by which:

> *"...only the most cunning, ambitious, and obsequious could survive... It was the princes of Moscow (who had to ingratiate themselves before the Great Khan in many ways), eventually promoted to Grand Princes, who revealed the greatest pertinaeity in ingratiating themselves with their masters and betraying their fellows."* [160]

Genghis Khan receives a supplicant at his court, allowing the man to actually touch his robe, which has an elongated sleeve, so the "god-king" does not have to touch ordinary mortals.

Puskin noted the ferocity of the Mongols, proclaiming them to have "...nothing in common with the Moors. When they conquered Russia, they gave her neither algebra nor Aristotle."[161] The Mongols pushed west in the 1200s, crushing Hungary and the German and Polish Knights; their invasion of Austria and the rest of Central and Western Europe was called off only when Ogadai Khan, their leader, died, forcing a competition for succession which immobilized the Golden Horde.[162] A by-product of this Mongol attack and retreat, as well as that of the Huns in previous years, was the retention of certain democratic, female and child oriented elements in Western and Central European culture which were mercilessly crushed in Eastern Europe and Russia.

Under Mongol rule, the Grand Duchy of Moscow developed based upon absolutism, prostration, State control of all land, and Mongol patterns of tax collection and military organization. All persons within the empire were considered the slaves of the Tsar, and their property his own. Flowing eastern robes were adopted, along with female seclusion. Restrictions upon foreign travel and foreign contacts were imposed in all areas except Novgorod, which though controlled by the Mongols was never entirely conquered by them. By the time the Grand Dukes of Muscovy gathered enough power to oust the Mongols, they had almost completely adopted the attitudes, behavior, and social institutions of their prior masters.[163]

> *"When the Golden Horde declined, and Moscow became strong enough to withhold tribute, no change was necessary: exactions previously passed on by Moscow to Saray now stayed in Moscow, while Muscovite Grand Princes continued to rule on principles absorbed from the Tatars: arbitrary despotic violence and indifference to the welfare of the subject."* [164]

> *"The bloody mire of Mongol slavery...forms the cradle of Muscovy, and modern Russia is but a metamorphosis of Muscovy."* [165]

Muscovy was assisted in its revolt against the Mongols after 1370, when Tamerlane's Turk-Mongol confederacy smashed the Golden Horde

158. Dupree, 1973, ibid., p.316.

159. Hingley, 1977, ibid., pp.26-8,134,148,162-3; Wiencek, et al, 1980, ibid., pp.86-8.

160. Hingley, 1977, ibid., p.162; cf. Kovalevsky, 1891, ibid., p.149.

161. as quoted in Wittfogel, 1957, ibid., p.219.

162. Wiencek, et al, 1980, ibid., pp.45,62,67.

163. Hingley, 1977, ibid., pp.9,114,163; Wittfogel, 1957, ibid., p.219.

164. Hingley, 1977, ibid., p.162.

165. Marx, K., as quoted in Hingley, 1977, ibid., pp.162-3.

in South Asia, conquering lands from the Black Sea to India.[166] Tamerlane's battles against the Golden Horde were no less brutal than the previous Mongol invasions, and evidence of massive slaughter has been found in excavations, where masses of skeletons indicate severed limbs and heads.[167] These events followed 3 successive Eurasian epidemics of the Black Plague, between 1346-1369, wherein 1/4 to 1/2 of the population of many areas died.[168]

After 1380, the Byzantine Emperor at Constantinople gave up the title "Tsar", and instead began to be called "Emperor of the Romans and Ruler of the Universe". The old title of "Tsar" was taken by the Duke of Muscovy.[169] Strongman, absolutist influences grew throughout the Russian lands bordering Saharasia during the Mongol and Tsarist periods, though some remote regions maintained vestiges of their original social makeup. The folkmote, for instance, persisted in the northern cities of Novgorod and Pscov, though it was wiped out in the south and west. Yearly festivals celebrating sexual freedom of youth were still carried out in the 1500s in Novgorod and Pscov, though the Christian clergy in those regions bitterly complained about the "corrupt young women and girls". The peasants resisted the priestly influences, often stoning and pelting the clergy when their ancient freedoms were threatened. Following old legal codes, women still chose their own husbands, divorced, and remarried.[170] Kovalevsky wrote:

> *"Local clergy were engaged in constant warfare with the shameful licentiousness which prevailed at the evening assemblies of the peasants, and more than once the clergy succeeded in inducing the authorities of the village to dissolve the assemblies by force."* [171]

Following the fall of Constantinople to the Turks in 1453 CE, Moscow took on the mantle of leadership for the Eastern Christian Church. The power of the Grand Duke increased, and the power of the independent Russian Knights, Dukes, and Princes declined. This was the period of Ivan the Third (c.1480), who defeated the Tatars, but turned about and ordered the abolition of the folkmote in Novgorod.[172]

> *"Successive rulers extended their power and territory in all directions. They also exerted irresistible pressure to erode the right of free movement whereby virtually all citizens, whether princes, boyars, or mere peasants, had been permitted to transfer allegiance from one master to another."* [173]

The Tsars progressively increased their absolutism, their empire spreading north, south, and east across the steppe. Notions of erotic or romantic love were crushed under the boot heels of the early Tsars, renown for their sadism.[174] Ivan IV, "the Terrible" (c.1570) was a murderous psychopath who should have been shut up in an asylum. His power exceeded any other monarch of the period, and it was sorely abused. He ordered the torture and murder of 20,000 persons in Novgorod, kept wild beasts for eating prisoners, and bashed out the brains of his own son with a club.[175] He wrote to Queen Elizabeth, calling her an "ordinary spinster" for yielding power to her "clodhopping tradesmen", and he called the Polish King Stephen Bathary an inferior for having a Parliament.[176] Ivan's rule:

> *"...rivals Stalin in the percentage of the then much smaller Russian population which he liquidated. Ivan even outdid the Georgian dictator in the element of apparent irrationality underlying the*

Ivan The Terrible
Early Tsar of Russia (c.1570) who initiated a reign of terror rivaling that of Russia's former Tatar (Mongol) masters. His sadistic absolutism generally characterized Tsarist Russia for hundreds of years. He declared all land, trade and industry to be his personal property, and exterminated a higher percentage of the Russian population than did Stalin in the 1900s.

166. Wittfogel, 1957, ibid., p.219; Dupree, 1973, ibid., p.317; Wiencek, et al, 1980, ibid., p.86.

167. Wiencek, et al, 1980, ibid., pp.76-77.

168. "Discussion: Subsistence Economies", in Thomas, 1972, ibid., p.419.

169. Kovalevsky, 1891, ibid., p.150.

170. Kovalevsky, 1891, ibid., pp.10-6,43,143-4.

171. Kovalevsky, 1891, ibid., p.12.

172. Wittfogel, 1957, ibid., pp.219-20; Hingley, 1977, ibid., pp.27- 8,107,114,163; Kovalevsky, 1891, ibid., pp.147-8,152,157.

173. Hingley, 1977, ibid., p.164.

174. Stern, M.: *Sex in the USSR*, Times Books, 1979, NY, p.14.

175. Scott, 1939, ibid., pp.146-7.

176. Hingley, 1977, ibid., pp.10,30,108, 132,164.

Serfdom in Russia
A life of poverty, beatings and crushing work, with all land and wealth owned by the Tsar and his underlings. Such conditions began to soften only with the first stirrings of social reform, in the 1700s.

177. Hingley, 1977, ibid., pp.28-9.

178. Kovalevsky, 1891, ibid., pp.37-8,153,210-17; Heichelheim, 1972, ibid., p.165; Gorer & Rickman, 1962, ibid., pp.101-2; Stern, 1979, ibid., pp.14-5.

179. Kovalevsky, 1891, ibid., p.11.

180. Wiencek, et al, 1980, ibid., p.21; Wittfogel, 1957, ibid., p.64; Scott, 1939, ibid., pp.63,69,158, 217, 245, 278; deMause, 1974, ibid., pp.10-2,29.

181. deMause, 1974, ibid., p.34.

182. Tannahill, 1980, ibid., p.326; Kovalevsky, 1891, ibid., pp.164,217-8; Hingley, 1977, ibid., pp.28,165.

183. Davies, 1981, ibid., pp.22,245; Scott, 1939, ibid., pp.70- 5,209,224,278; Radbill, 1974, ibid., p.9; Gage, 1980, ibid., p.27; Pitcher, 1972, ibid., p.21; Kovalevsky, 1981, ibid., p.164.

massacre of his subjects." [177]

Ivan was married 7 times, but rape was his preferred mode of sexual conduct. He would seize whatever woman grabbed his fancy, rape her, and then turn her over to his bodyguards who would do likewise. A bisexual, he also turned his brutal lust towards men; one, Basmanov, was his favorite, and enjoyed a long career in the Russian court. Ivan was also a moralist, who went unopposed by the Church. Indeed, his brand of absolutism was mirrored in Church Penitentials of the day, where the penitent kneeled in confession at the priest's feet, enduring cross examination and the imposition of severe penances. One Church official under Ivan's rule wrote a book counseling on proper use of the rod to keep one's wife, children, and servants in line. Strong measures were taken against young couples who married outside the Church, and freedom of migration was abolished.[178]

"[Ivan] took effectual measures for abolishing every vestige of paganism; amongst them, the yearly festivals held on Christmas Day, on the day of the Baptism of our Lord, and on St. John the Baptist, commonly called midsummer day. A general feature of these festivals, according to the code, was the prevalence of promiscuous intercourse between the sexes." [179]

In the late 1300s and 1400s, Europe saw increasing military conflicts with the Turks and increased sadism on the part of Christian autocrats, who furiously tortured and burned at the stake all who opposed their power and dogma.[180] The period saw widespread swaddling of infants, remnant traces of infant cranial deformation here and there, bizarre sexual taboos and grotesque fears of "pollution" by dangerous females, and the onset of taboos regarding breastfeeding of infants. Breastfeeding was considered a "swinish and filthy" habit; the rich hired out wetnurses, which usually meant death or malnutrition for the wetnurse's own child; infants of poor parents were generally malnourised.[181]

During the late 1500s and 1600s, Europe saw the first stirrings of freedom with the rise of humanistic philosophy, rationalism, and science; these would blossom more fully in the 1700s. Russia, however, remained in the grip of the Tsars during various periods of social chaos, famines, raids by nomads, and peasant rebellions, the latter of which were followed by government reprisals and increasingly restrictive laws. Border conflicts between Europeans and Russians occurred,[182] and Turks, having previously conquered Thrace, Bulgaria, and Macedonia, pressed into Austria and Hungary, bringing massacre, torture, rape-mutilations of women and children, and absolute havoc. The *auto da fe* ("act of faith") or ritual public mass murder of heretics by the Church on specified Sundays, increased in popularity, and children were sacrificed under the doorsteps of important buildings in Germany.[183] As in the past, events in Britain, while hardly democratic at the time, paled by contrast to events occurring on the continent or in Central Asia. Henry VIII (c.1540) established the Star Chamber, whose authority exceeded common law, but it was abolished within a century; likewise Queens Mary and Elizabeth tortured and burned at the stake a good number of heretics and opponents, but the total numbers of people affected and the time periods during which the persecutions were carried out were quite small as compared to events on the European continent, and did not make for total annihilation of the opposition. Ordinary people still possessed legal rights which even the Kings and Queens had to respect, due process of law could not be flagrantly violated for too long without eliciting a protest reaction from other factions competing for power, and this impeded the number of executions which could be summarily demanded by the various British autocrats. Scandinavia and France were characterized by similar trends away from absolutism, which would blossom to a fuller democratic social structure in later centuries; Russia, Austria, Greece, and Spain, lying geographically closer to Saharasian influences, would require additional centuries to experience such changes.[184]

Central state power and Russian territorial ambitions further increased under Peter the Great (1682-1725), who instituted a political police force and military conscription; a known wife-beater and adulterer, he maintained a harem of 400 concubines.[185] Peter the Great was not alone in the keeping of such bloated harems, but he certainly carried it to levels rarely seen in the West. Still, a pretty woman in Russia *or* Europe was at the mercy of any man of royal blood who happened to see her. The years between the 30-year's war and the French revolution (1648-1790) has been called the "Golden Age of Royal Mistresses", when the practice of keeping small or large harems became ubiquitous among European royalty. As harsh as these conditions appear from our present perspective, they marked a slow turning point from previous years when celibacy was glorified, and women ran the danger of being mutilated and burned alive by sex- and woman-hating Christian clergy simply because of their soft and sexual appearance, or for their intelligence or wealth. According to Lewinsohn:

> *"European courts were soon swarming Marquessas, Counts, Princes, and Dukes who owed their high-sounding names and their estates to the sexual needs of the Monarch."* [186]

Peter's rule also saw a slow turn towards the West on the part of the Russian Monarchs, given the developing role of European public health, science, and technology, but also due to the growing influences of French libertine and various new motifs in European arts, which were of great interest in the Russian courts.[187] Against the background of such conditions, official proclamations were made against alcohol consumption, and reports of widespread alcoholism among Russians regularly followed 1600, when travel in the countryside was slightly safer due to increased controls over nomadic groups, and increased central state power.[188]

Still, the Russian family maintained an orientation towards Central Asia, and increasingly reflected changes toward absolutism. Prolonged swaddling and other harsh infant and child treatments, the arranged marriage and bride price, and both wife-beating and child-abuse remained at epidemic levels.[189] Kostomarov has commented that "between parents and children reigned a spirit of slavery",[190] a spirit which also extended to relations between husband and wife. Domostroy wrote a widely-read manual of etiquette which counseled the Russian husband and father to "...teach his wife with love and sensible punishment", instructing him to beat his wife in a courteous and loving way so as to neither blind her or render her permanently deaf.[191] Regarding childrearing, the father was also advised:

> *"Punish your son in his early years and he will comfort you in your old age and be the ornament of your soul. Do not spare your child any beating, for the stick will not kill him, but will do him good; when you strike the body, you save the soul from death... Raise your child in fear and you will find peace and blessing in him."* [192]

An English visitor of the 1500s commented upon these practices:

Madame Lapuchin, a beauty of the Tsarist Court (c.1750), ordered publicly "knouted" by Elizabeth, Empress of Russia, for having a forbidden liaison with a foreigner. The whip-master flayed her entire back down to the bone and cut out her tongue, after which she was banished to Siberia. Such cruel and heartless punishments dominated the Tsarist period, as they did during the Tatar (Mongol) period before the Tsars. (W. Cooper, *History of the Rod*, London 1912)

184. Stern, 1979, ibid., p.15; deMause, 1974, ibid., pp.10,30; Scott, 1939, ibid., pp.56-8,62,81-2,87-9,135,147,200,224,278,280; Davies, 1981, ibid., pp.245-50; Radbill, 1974, ibid., pp.4-5; Gage, 1980, ibid., p.27.

185. Gage, 1980, ibid., pp.69,167; Hingley, 1977, ibid., p.165.

186. Lewinsohn, 1958, ibid., p.194.

187. Stern, 1979, ibid., pp.15-9; deMause, 1974, ibid., p.384.

188. Hingley, 1977, ibid., pp.46-8.

189. Gorer, G. & Rickman, J.: *The People of Great Russia*, W.W.Norton, NY, 1962; Kovalevsky, 1891, ibid., p.9.

190. deMause, 1974, ibid., p.390.

191. Hingley, 1977, ibid., p.154; Calverton, 1928, ibid., p.229.

192. deMause, 1974, ibid., p.393.

"If the [Russian] woman be not beaten with the whip once a week she will not be good...and the women say if their husbands did not beat them they would not love them." [193]

The 1700s saw massive social change in Europe. The arts, technology, and science blossomed, and in themselves challenged old beliefs about the nature of the universe, and the human relationship to it. A cry was heard for more humane treatment of wives and children, and the view of children as "evil imps" began to wane. However, masturbation phobias and toilet training pedagogy appeared, possibly given the fact that swaddling was increasingly being abandoned (again, starting first in Britain); the abandonment of swaddling allowed young children, for the first time, to explore their own bodies, and also to determine their own bowel functions more readily.[194] Corporal punishment in the schools and severe whipping of children was criticized in Britain, and some reforms were enacted. Foundling homes were set up for cast-off infants, and public hospitals established, their quality of care gradually improving.[195] Sex manuals, indicating a growing scientific understanding of procreation and the physiology of sex, were increasingly circulated, many discussing methods of workable contraception.[196] French women began to revolt against the stiff form of dress into which Spanish influences had cast them.[197] Such reform did not generally take place in Germany, Eastern Europe or Russia, where childhood remained a life-threatening ordeal up until the middle 1800s, and where severe corporal punishment and obedience training of infants, to include swaddling, persists more or less into contemporary times.[198]

The 1700s also saw the last trials in England and Scotland for witchcraft, as well as legal abolishment of burning at the stake, "pressing" to death, and judicial torture. These reforms were not completely instituted at once, but they eventually took hold.[199] France abolished the rack and flogging, as well as the torture of "boiling oil"; Italy likewise abolished torture. Frederick II, "The Great" ruled Prussia (1740-1792) in an era of "enlightened absolutism" which also saw the abolishment of torture, and Sweden entered the "Age of Freedom" (1720-1770). Slave revolts began to occur in European colonies abroad, where anti-colonial ideas were circulating as well.[200] Again, abuses continued for some time, with the legal reforms preceding actual cessation of the sadistic acts.

The 1700s also saw the challenging of Monarchy in America and France with revolutions. King George's iron-handed policies were successfully rebuked by the American colonists, which gave courage to the French reformers. Napoleon emerged out of the chaos of the French revolution, without much interest in reform himself. But he did smash the Holy Inquisition across Europe, destroying the Church's castles and torture-chambers, and he turned over vast holdings of Church land to the highest bidder. Clergy were again made accountable to secular courts, and ecclesiastical courts, which had previously dominated family law and offenses of the clergy, were abolished. Secular marriages again became widespread, and divorce was radically simplified.[201] Various guilds of craftsmen and traders also appeared, with their own councils, rules of order, and taxation methods, giving rise to the first forms of outside authority tolerated by either Monarchs or the Church since the early folkmotes had been smashed hundreds to thousands of years earlier.[202]

This was the era of Romanticism, the Enlightenment, a period of sweeping revolt against authority and tradition, with the first social, political, and sexual reforms. Such European changes appeared first in Britain and France, which were being greatly influenced by reforms and constitutional law in the United States. The names of the leading propo-

Prussian punishment for disobedience to authority, from a school book of the 1700s. Revolutionary social and legal reforms which occurred in Western Europe and America in the 1700s required several hundred years longer before they began to occur closer to Saharasia, in Eastern Europe and Russia.

193. Jenkins, A., as quoted in Hingley, 1977, ibid., p.154.

194. deMause, 1974, ibid., pp.38-40,48.

195. deMause, 1974, ibid., pp.5,29,42; Piers, M.: *Infanticide*, W.W.Norton, NY, 1978, p.66.

196. Bullough, 1976, ibid., p.472.

197. Lewinsohn, 1958, ibid., p.205.

198. deMause, 1974, ibid., pp.13,41,385-9; see appropriate sections in Chapter 5.

199. Scott, 1939, ibid., pp.91,100-1,136.

200. Midlarsky, M. & Thomas, S.: "Domestic Social Structure & International Warfare", in *War, Its Causes & Correlates*, M. Nettleship, et al, eds., Mouton, 1973, p.547; Pope, P.: "Danish Colonization in the West Indies", in Nettleship, et al, 1973, ibid., p.580; Scott, 1939, ibid., pp.135-6.

201. Lewinsohn, 1958, ibid., pp.221-3; Scott, 1939, ibid., pp.75-81.

202. Taylor, 1953, ibid., p.179.

nents of reform, or scientists whose work spurred reform, are a veritable "who's who" of founding figures for Western civilization, and the freedoms enjoyed by Westerners in the modern era: Tom Paine, Linnaeus, Jean Jacques Rousseau, Swedenborg, Wollaston, Adam Smith, Buffon, Benjamin Rush, Benjamin Franklin, Thomas Jefferson, Immanuel Kant, Beaumarchais, Mary Wollenstonecraft, Malthus, Beccaria, Voltaire, and so on.

Again, all of these reforms and changes were greatly delayed in Germany, Poland, and Russia, occurring first in Britain and France, and later in other European states. Reforms of various sorts continued through the 1800s and 1900s, granting greater and greater amounts of freedom to women and children, and legally exempting them from the sadistic punishments of either father or husband, who were increasingly subject to punishments themselves for inflicting physical harm upon their wives and children. These reforms, of course, continue to this very day.

Russia entered the renaissance only very late, and still lags behind Europe in essential reforms of family, sex, and child-care matters,[203] largely due to the lingering influences of the Central Asian Moslem warriors, who composed a large part of the Russian population. The Mongols and other Central Asian nomad groups were pushed back into their Saharasian enclaves only during the 1600s and 1700s, where they maintained a trade-raid relationship with surrounding lands. The Ottoman Turks filled the vacuum left by the disintegrating Mongol empire, engaging in many wars with the Russians and East Europeans, and persisting as a major power until c.1920. After that date, Central Asian power was dominated by the blood-drenched empires of Stalin and Mao Tse'Dung.

Present conditions in Central Asia continue to be quite arid, and contemporary news reports occasionally have noted massive sand storms and regional wild-fires in both former Soviet and Chinese areas. Russian scientists have discussed relatively recent environmental transformations which coincided with a change from settled agriculture to nomadic herding:

> *"Nowadays when one travels through the deserts, one encounters many old ruined fortresses, indisputable traces of what used to be irrigation systems, and some surviving artifacts. History gives witness to the fact that several hundred years ago [pre-Mongol?] these regions had flourishing towns and villages full of gardens and green fields, and that life there was quite active."* [204]

> *"According to the findings of geographers, soil scientists and archaeologists, this area [Caspian-Aral] was much better watered than others. We find here long strips of solonchak depressions which, according to popular legend, are sections of the ancient river of Sir Darya. Relics of the late Bronze Age indicate that this area was thickly populated in the past; and the name of these depressions (Kolgan-Sir: the channel of the Sir Darya) indicates...that the people still had memories of the time when this area was well watered... Lands formerly irrigated and fertile were transformed into deserts and became salinified. Parts of them were buried by shifting sands; extensive areas were transformed into takyry [clay hardpan]."* [205]

> *"Natural conditions in the black earth steppes and forest steppes deteriorated: the southern border of the forest land moved several hundred kilometers northwards as a result of tree felling; the forest*

Hercules prepares to kill Hippolyte, "Queen of the Amazons"

Patristic Saharasian Images from Ancient Greece infecting Europe, dated to the late 1800s, from a doorway into Old Vienna, Austria, symbolizing virtues the monarchy identified with (see opposite page).

203. "Child-care" centers violate their own definition when infants are routinely swaddled, ignored, or subject to severe obedience training. Similarly, "abortion rights" lose much meaning as a force for social or sexual liberation in the face of consistently unavailable or unreliable contraceptives, or where poor housing conditions with an utter lack of privacy prevail. Such is the case with much of former Soviet society, where women are free to work, but also must do all the housework and child-care chores, which their husbands will not generally help with. In these aspects, East Europe and the Soviet Union lag far behind Western Europe and the United States. The sale of children was outlawed in Russia only in the 1800s, and child-sacrifice under building foundations continued in Germany until 1843, practices which had been outlawed in Western Europe nearly a hundred years earlier. cf. deMause, 1974, ibid., pp.27,33.

204. Babayev, A.G.: "Principal Problems of Desert Land Reclamation in the USSR", in *Desertification: Environmental Degradation in and around Arid Lands*, M.H. Glantz, ed., Westview Press, Boulder, 1977, p.203.

205. Kovda, 1961, ibid., p.202.

steppes became completely de-forested, turning into anthropogenous steppe lands; the forest strips along the river terraces and on the sands were cut down; forest groves, and clumps of birches and blackthorns in the steppes disappeared altogether. The ploughed slopes of the high right banks of the Don, Dnieper, Donets and Volga suffered from water erosion: the plains began to be broken up by ravines; black dust storms began to occur regularly in the south and southeast of the country..." [206]

Hercules killing an unidentified male (above) and killing a serpent (below) — symbol of nature and the life-force.

Regarding the Soviet Union, established in the wake of the Russian Revolution, its early period of existence under Stalin's dictatorship was no less bloody than prior Central Asian states, largely because Soviet armoring and patrism was not touched or changed by the communist revolution. Stalin's death-toll was similar to the worst of the Tsars or Mongols, and even worse than that of his one-time allied friend Hitler, who rose to power on the heels of a failed attempt at democratic reform in Germany. Hitler and Stalin were allied against the more matristic democracies of Europe and America, and both had similar territorial and empire-building ambitions. Fascist Italy and Spain contributed to the Axis power, but those nations never established death-camps or gulags as seen in Nazi or communist-controlled lands. The open warfare between Germany and the Soviet Union which later occurred was not, at any basic level, a "struggle for freedom" by the Soviets, but rather a struggle between tyrants for land, resources and slaves. A contrast of the blood-soaked regimes of Hitler and Stalin shows the communists in the Soviet Union had a much higher death toll of murdering their own peoples, ethnic minorities and peoples in captive lands than did the Nazis, for all their atrocities (including the Holocaust). By contrast, for all their relatively small faults and occasional duplicity during this period of turmoil, the nations of Western Europe and America (Britain, France, Belgium, Scandinavia, Denmark, Netherlands, Switzerland, United States, Canada) — farther removed from Saharasia and possessing less-armored character structures — had nothing which could even remotely compare to the Soviet gulag or Nazi concentration camps.[207] This shows, once again, the considerable influence of the geographical and historical proximity to Saharasia in determining the behavior of people and political events, well into modern times.

Another example of this can be seen in the situation which developed following the break-up of the Soviet Union and Yugoslavia. With the ending of strong-handed central communist authority, those new nation-states which lay close to or within Saharasia have generally collapsed back into catastrophic turmoil, genocidal "civil" wars, or authoritarian dictatorships only slightly freer than the old communist system (ie., Bosnia, and the most southerly-Saharasian of the Commonwealth States: Turkmenistan, Uzbekistan, Tajikistan, Kyrgyzstan, Georgia, Armenia, Azerbaijan, Kazakhstan). By contrast, those parts of the former Soviet Union or its captive states farther away from Saharasia, have shown a more easy transition to democratic and freer conditions (ie., Lithuania, Latvia, Estonia, Belarus, Ukraine, Moldova, Poland, Hungary, Slovakia, Czech Republic). These factors are given additional documentation in Chapter 7, on "Expressions of Saharasia in Contemporary Demography".

206. Kovda, 1961, ibid., p.184.

207. Even in the cases of France in Algeria and Indochina, of other European powers in Africa or Asia, and of the United States in Vietnam, nothing remotely approaching the genocidal butchery of the Saharasian Islamists and communists, nor of the Nazis or Imperial Japanese (at the borderland of Saharasia) can be identified. While European colonialism had both enlightened and authoritarian tendencies, the Islamist and COMINTERN victories in Algeria and Indochina led to the deliberate genocidal massacres of civilians in hundreds of thousands to millions, dramatically greater numbers than who died from the French or American military, which by contrast mostly adhered to Western rules of warfare and deliberately avoided direct civilian casualties. When civilian protestors took to the streets in France and America, against their respective overseas wars, there were no mass-arrests with executions and mass-graves, nor mass-deportations to inland "Siberias" of the dissenting populations as was a typical response by modern Saharasian (and borderland) nations to similar opposition populations. The author's forthcoming work, *Saharasia Since 1900*, will deal with these issues in greater depth.

CHINESE CENTRAL ASIA

Like Soviet Central Asia, the Chinese portion of Saharasia is composed of a series of mountain rimmed basins and steppe zones. These include the Tarim basin, in which lies the Takla Makan, the Dzungaria basin, and the Gobi and Ordos basins.

The Tarim Basin contains one of the world's largest sand deserts, the Takla Makan, and is enclosed by mountains on all sides.[207] It is a nearly lifeless, barren wasteland, called the "dead heart of Asia", where:

"temperatures...go down to -30°F [-35°C] in winter and up to 125°F [52°C] in shade in summer. The diurnal range is equally high, more than 101°, and there is often [frost at] 50°. The relative humidity of air for long periods during the year remains at or below 5%." [208]

"...large masses of sand continually shift their position, driven by terrific winds and tornadoes. The sandstorms of the Tarim area are fierce storms raging for days, transporting enormous volumes of sand from one locality to another thus moulding and stamping new dune topography on the area." [209]

A yellow "dust-haze" is present over the Tarim and Lop Basins for more than half the year, probably with a rainfall-suppressing function similar to that observed and described by Bryson and Barreis over the Indus Valley.[210]

The lowest point of the Tarim Basin is found at its eastern end, where lies the Lop Nor salt lake and the Turfan Depression. Lop Nor is fed by the Tarim and Kuruk Darya rivers, which drain from the mountain ranges surrounding the basin, flowing from west to east at the borders of the Takla Makan. The various rivers which flow into the Tarim Basin are fed by glacial meltwater from the mountains of Tibet and to a lesser extent the Tien Shan; in the past, these glaciers must have been much larger, supporting a greater flow of water into the Tarim.[211] A plain of thick, rock-hard salt some 322 km (200 miles) wide surrounds the remnant lake at Lop Nor; ruins of villages on ancient shorelines presently lie among salt crusts and sand dunes of the Lop Nor environment, which is described as being without life and among the most barren of deserts anywhere in the world.[212]

North of the Takla Makan and the Turfan Depression is the elongated Dzungarian depression. The region is exceptionally arid today, being a sand and pebble desert basin which terminates in dry saline lakebeds. Southeast of the Takla Makan lies the highland Tsaidam Depression, a generally flat, pebble, sand and clay desert containing barren salt flats.[213] This salinized depression once contained the 31,000 km² (12,000 mi²) glacial Lake Tsaidam, which has greatly shrunk since the last glaciation. Tibet as a whole contains innumerable lakes possessing elevated beaches, which in some cases are reported to be 60-100 m. (200-330') higher than present lake levels. These lakes, as well as other landscapes in Tibet, are characterized by an increasing salinity; thick layers of wind-born sand are found mixed with ice at altitudes up to 4,575 m. (15,000'), the sand presumably transported from the Takla Makan and Gobi.[214]

East of the Takla Makan lies the vast Gobi Desert of North China, a broad, elongated basin between the highlands of Central China and Mongolia. Its level surfaces are generally covered with rock and pebble, alternating with hilly regions dissected by numerous dry stream beds.

207. Petrov, 1976, ibid., p.6.

208. Wadia, 1960, ibid., p.8.

209. Wadia, 1960, ibid., p.13.

210. Bryson, R. & D. Barreis: "Possibilities of Major Climatic Modification and their Implications: Northwest India, A Case for Study", *Bull. Am. Meteorol. Soc.*, 48(3):136-42, March 1967; Huntington, 1907, ibid., pp.92,134-5,147-8,157.

211. Wadia, 1960, ibid., p.9.

212. Huntington, E.: *Asia, A Geography Reader,* Rand McNally, Chicago, 1912, pp.169-70,239-61, map 388f.

213. Petrov, 1976, ibid., pp.11-6.

214. Wadia, 1960, ibid., pp.7, 10-1.

Ancient lake basins with strand lines and salt flats are also scattered about. Vegetation is "very sparse and almost useless for livestock" in many areas of the Gobi. Mineralized groundwaters feed small springs in a few locations, particularly as one moves eastward to regions of greater rainfall. Drainage is lacking throughout the entire Gobi Desert, and there is little or no surface water.[215]

> "The soils of the deserts of western China...are covered with a salt crust 30-80 cm [12-31"] thick, and sometimes 3-5 m [10-15'] thick. These salts are composed mainly of chlorides and sulfates and often of nitrates. Sometimes, dustlike forms of these salts create enormous dunes and barchans, tens of meters tall (pseudo-sand)."[216]

> "In the eastern Gobi, the southeast monsoons from the Pacific and the northwest winds from Siberia bring some amount of moisture to support thin scrubby vegetation for a few months each year, but in the west, the aridity progressively increases and human existence is precarious against the constantly drifting sand raised by violent tempests and storms, more dreaded than the snow-storms and blizzards of the Polar regions."[217]

South of the Gobi, nestled in the highlands which separate the arid steppe from the humid lowlands of China lies the Ordos Desert. The Ordos is presently covered with sand dunes and a stony plateau. The Yin Shan Desert lies to the west of the Ordos. Both regions were once quite green, though they are now devoid of grass cover. A similar transition occurred all across the steppes along the southern fringe of the Gobi.[218] Smil has summarized the more recent phases of desertification in China, as well as attempts to combat it.[219]

Chu felt that the climate of China and Chinese Central Asia had changed since c.4000 BCE, being warmer and wetter in the past. Near Hsian, sites from the Yangshao culture (c.3500 BCE) high up in the Yellow River Valley yield evidence of warmer and wetter conditions. Since that time, the northern limit of bamboo distribution has retreated southward by 3° of latitude, indicating the onset of cooler conditions.[220]

Chang summarized other evidence, from Taiwan, Beijing, and Manchuria, which date the postglacial "climatic optimum" in China between 6000-2000 BCE. Pollen analysis from near the Soviet border indicated a broadleaf forest maximally distributed between c.5700-500 BCE, following which the needleleaf Korean pine becomes dominant.[221]

> "...the beginning of the postglacial stage in China was marked by a gradual rise of the mean annual temperature, accompanied by an increasing vegetational cover and corresponding faunal and floral assemblages... The eastern low plains were wet and marshy...while the western high loesslands were clad with forests and dissected by watercourses."[222]

Gupta essentially agreed with Chang on warm, moist conditions in China, but moved the onset of such conditions back to 7000 BCE,[223] while Hawkes identified the "food producing Neolithic" at 6000-3000 BCE.[224] Chang also gave the following assessment of North China, to include the Gobi/Mongolian steppe:

> "These 'nors' [lakes] in Mongolia are now largely desiccated, but the wind blown cultural remains in this region 'occurred with such regularity in the various basins and hollows, large and small' as to

215. Petrov, 1976, ibid., pp.11-6.

216. Kovda, V.A.: *Land Aridization and Drought Control*, Westview Press, Boulder, 1980, p.79.

217. Wadia, 1960, ibid., p.10.

218. Gumilev, 1968, ibid., p.28.

219. Smil, V.: *The Bad Earth: Environmental Degradation in China*, M.E.Sharpe, NY, 1984.

220. Chu, K.: "A Preliminary Study on the Climatic Fluctuations During the Last 5000 Years in China", *Scientia Sinica*, 16(2):226-9, 1973.

221. Chang, K.: *The Archaeology of Ancient China*, Yale U. Press, New Haven, 1968, p.34.

222. Chang, 1968, ibid., p.35.

223. Gupta, 1979 Vol 1, ibid., pp.100-2.

224. Hawkes, 1974, ibid., p.215.

suggest that their formation took place under climatic conditions decidedly different from those of the present day and when these basins were filled with water. Similar climatic peaks during the early post-Pleistocene period are also indicated by peats at San-ho and Chi in Hopei in the plains area, and by the high water levels near prehistoric sites in Honan, as well as by the literary records of the existence of nors in Honan, Shansi, and Shensi in the western loesslands. Along with the moist climatic conditions in North China went a thick vegetational cover in areas that are now barren and semiarid, as indicated by a Black Earth horizon at some localities in the north... This black earth layer...probably represents an ancient forest cover. The existence of a thick forest in North China and on the Manchurian plains is further indicated by such cultural remains from prehistoric sites as the abundance of charcoal and woodworking implements (ax, adz, chisel, etc.) and by the frequency of bones of wild game. Some of these bones are definitely from forest-dwelling animals such as tigers and deer... The existence of rice, bamboo, and elephants is corroborated by the written records, while the presence of elephants and rhinoceri is further confirmed by sculptures and some zoomorphic bronzes from the archaeological site at Anyang." [225]

Chu estimated oscillations of temperature in China from various data, indicating warmer conditions from 3000-1000 BCE; fluctuations during the period were not ascertained. After this, a series of oscillations occurred, with colder phases at 100 BCE, and CE 400, 1200, and 1700. Chu attributed these cold temperature periods as being due to southeastward shifts in the Siberian High. [226] This being so, the highland steppes and interior basins of China would have been drier at these same times. These later dates for a cold dry phase on the Chinese steppes generally correlate to periods of nomadic outbreaks from the interior, namely the Hun and Turk-Mongol groups.

General Settlement Patterns and Culture Change, to Include Influences Transmitted to Coastal China and Japan:

From the earliest times, the western portions of Chinese Central Asia yielded Paleolithic similarities with cultures to the west, in Siberia and across the Pamirs in Northwest India. [227] The Gobi and Ordos Deserts have yielded microlithic stone tools and arrowheads as old as 20,000 BCE. These artifacts, which came from a prepottery culture that spread southwest across the steppes to as far as the Aral Sea and Indian Ocean, are indicative of moister conditions. [228] Late Paleolithic sites dating to after 6000 BCE stretch north from the Gobi to Lake Bikal and Manchuria indicating a hunting and fishing society. [229] The Ordos Desert in particular contains numerous Upper Paleolithic and Neolithic artifacts indicating a mixed hunting, fishing, and limited agricultural culture. [230]

By 4000 BCE, a highly developed Neolithic culture was spread across North China, and in the Yellow and Yangtze River valleys, with a population as large as any in the Near East. These peoples created fine painted bowls and vases, engaged in millet agriculture, and raised sheep, pigs, goats, and dogs. [231] Gupta has given a brief summary of the Neolithic materials in the Gobi region. While they have not been subjected to much systematic study, the region "...has yielded the richest collection of stone tools", to include scrapers, engravers, bifacial arrowheads, sickle-blades, axes and picks. [232] Similar Neolithic tools have been found in the Tarim

New Material 2005: See *Update on Saharasia* (Appendix B) for presentation of newer archaeological evidence on a c.5000 BCE Yangshao skeleton bearing the wound of a probable non-lethal hunting accident, but often misrepresented by advocates of early-violence theories as "evidence of war".

225. Chang, 1968, ibid., pp.32-3.
226. Chu, 1973, ibid., p.251.
227. Gupta, 1979 Vol 1, ibid., p.176.
228. Phillips, 1965, ibid., p.16.
229. Phillips, E.D.: *The Royal Hordes*, McGraw-Hill, London, 1965, p.16.
230. Chang, 1968, ibid., pp.162-3.
231. Hawkes, 1974, ibid., pp.10,215, 218.
232. Gupta, 1979 Vol 2, ibid., p.303.

Early Yang-shao (matristic)
ceramics and pottery designs

Basin, along with grey and black pottery.[233] These tools are suited for living in an environment much moister and richer in plant and animal life than present today in those regions.

Changes After 4000 BCE:

Trade and cultural interactions between China and regions to the west continued to develop over the centuries. By 4000-3500 BCE a common lithology and a multicolored ceramic pottery, originating in Russia, had spread across Afghanistan and the Tarim Basin.[234] Neolithic mixed farming cultures spread across the steppelands.[235] This was the time period of the early Neolithic "Sage Kings" of China, such as Shen Nung. Legends about these kings were written several thousand years later:[236]

"During the age of Shen Nung, people rested at ease and acted with vigor. They cared for their mothers, but not for their fathers. They lived among deer. They ate what they cultivated and wore what they wove. They did not think of harming one another." [237]

Archaeology has not directly confirmed the existence of Shen Nung, but has indicated an absence of militarism or significant social stratification among the earliest Chinese. The lack of significant caste and legends of high female status, plus textual prescriptions for abortion, suggests a fairly high female status for Neolithic China.[238] Chinese texts and legends have proven generally reliable for other ancient periods, such as the Shang period, confirmed in recent years by the discoveries at Anyang.

After c.3000 BCE, Chinese Central Asia experienced onset of drier conditions with a subsequent increase in the nomadic adaptation. Nomads progressively displaced the Neolithic peoples of the steppe, but not in the humid zones where Neolithic agriculture persisted into the historic period.[239] Domestication of the camel occurred by or before 3000 BCE, a time when peoples using both camels and carts left the Gobi region for destinations in West Asia.[240] These migrants came into contact with nomadic Indo-Aryan peoples of Soviet Central Asia, who were themselves migrating farther west and southwest. The primarily westward movement of the Central Asian migrations allowed the wetland coastal Chinese to preserve their original Neolithic traditions. Still, it was a period of some turbulence. Chinese legends captured part of the cultural transitions involved during the period of environmental transition in Central Asia.

"During the age of Shen Nung, men cultivated food and women wove clothing. People were governed without a criminal law and prestige was built without use of force. After Shen Nung, however, the strong began to rule over the weak and the many over the few. Therefore, Huang Ti administered internally with penalties and externally with armed forces." [241]

Huang Ti, known as the Yellow Emperor, appears as the first strongman ruler in China; he has been depicted as having had an artificially deformed crania, while the rebel soldiers he fought against were not so depicted. Infant cranial deformation, which was a characteristic of militant Indo-Aryan nomad invaders in other regions of the Near East, was reserved in China for similar strongman rulers. Actual artificially deformed crania have been found in Northwest China dating to c.2000 BCE.[242]

233. Gupta, 1979 Vol 2, ibid., pp.299-301.

234. Grun, 1982, ibid.; Gupta, 1979 Vol 2, ibid., pp.257,302.

235. Phillips, 1972, ibid., p.18; Gupta, 1979 Vol 2, ibid., pp.299-301.

236. Himes, N.: *The Medical History of Contraception*, Williams & Williams, Baltimore, 1936, p.109; Grun, 1982, ibid.

237. Chang, 1968, ibid., pp.78-9.

238. Himes, 1936, ibid., p.109.

239. Chang, 1968, ibid., pp.162-3.

240. Gupta, 1979 Vol 2, ibid., pp.147-9,238.

241. Chang, 1968, ibid., p.192.

242. Dingwall, 1931, ibid., pp.97-8.

Over subsequent centuries the desiccation of Central Asia proceeded. By c.2300 BCE, the "Great Flood" of Chinese legend occurred as the Yellow and Yangtze met, possibly related to vegetation changes and sediment adjustments in their drainage basins.[243] At this time, the Neolithic Yang Shao culture was established in much of Western China, including the Tarim Basin. These peoples viewed the female as being possessed of great knowledge and power, and superior in many ways to the male. Female shamanism predominated, as did female symbols on their pottery.[244] Still, they had much cultural contact with peoples in West Central Asia.[245] Also by this time a Neolithic hunter-fisher culture of Mongolian stock was spread across the northern forests of Asia, having migrated from North China to the Baltic.[246]

Northern China was the home of the Hsia dynasty which, according to legend, was founded after the Great Flood. Yu, a Chinese Noah, had saved mankind from the flood and founded the Hsia, ruling a Golden Age similar to that described for Shen Nung.[247]

However, migrations of militant nomad peoples from Central Asia into the humid portions of China did occur, increasingly after 2000 BCE. This date also marks the earliest period of Chinese literature, and the beginnings of an elaborate writing and counting system.[248] Gumilev pointed out various Indo-Aryan words, concepts, and practices which crept into Eastern China after c.2000 BCE:[249]

Indo-Aryan:	East China:
fee (fairy)	fei (emperor's concubine)
phoenix	feng-hua (legendary bird)
order (Latin=ordo)	orda (Turk-military camp)
baga (god)	bagadur (epic hero)
cohen (Semitic=high priest)	khan

It appears likely that these words were introduced by the militant authoritarian Shang culture, who made their capital at Anyang. The advanced Bronze Age Shang, like the militaristic groups who presided over Upper Egypt, Mesopotamia, and Central Asia, engaged in a plethora of authoritarian, antifemale, antichild practices: Divine Kingship, male gods, monumental tombs with massive grave wealth, human sacrifice, to include widow murder and interment of females and servants in the grave of the dead King, a military caste, slavery, concubinage, polygamy, infant cranial deformation and swaddling, and use of the horse and chariot. Sacrificial tombs often reveal evidence of great struggle on the part of victims, who were sometimes interred alive. Ritual widow murder predominated in Shang China, and children were butchered as foundation sacrifices as well, with hundreds of children being buried at the entrance to the Royal Palaces. These warlike peoples completely overran the generally peaceful, female and child oriented Neolithic communities of China. Farmers appear to have been enslaved, their skeletons revealing signs of malnutrition.[250]

The Shang arrived in Eastern China from the Indo-Aryan dominated steppe of Central Asia, which by 2000 BCE was rapidly desiccating. Aside from bringing similar cultural traditions, the Shang also brought chariots and bronze axes similar to those of West-Central Asia. Thousands of oracle bones have been found at Anyang, dated roughly to 1800 BCE, containing prayers for rain or snow and indicative of a dry phase.[251]

Events in Japan also appear related here. Following 2000 BCE, the Jomon culture first arrived in the south, and displaced northward the more peaceful Neolithic Ainu peoples of Japan.[252] Like the mainland Chinese Neolithic peoples, the Ainu possessed a culture firmly focused

Shang and Chou period artwork: (above and right). Note the similarity to Meso-American design.

243. Gumilev, 1966, ibid., p.35.

244. Chang, 1968, ibid., pp.103,121; Van Gulik, R.H.: *Sexual Life in Ancient China,* E.J. Brill, London, 1961, pp.8,13; Himes, 1936, ibid., p.109.

245. Chang, 1968, ibid., pp.168-9.

246. Phillips, 1972, ibid., p.14.

247. Van Gulik, 1961, ibid., p.10.

248. Grun, 1982, ibid.; Wittfogel, 1957, ibid., p.51.

249. Gumilev, 1966, ibid., pp.38-9.

250. Chang, 1968, ibid., pp.37,154,171-4,192-9,214-9,231-52; Roper, 1973, ibid., p.300; Hawkes, 1974, ibid., p.219; Jaynes, 1976, ibid., p.163; Phillips, 1965, ibid., p.43; Davies, 1981, ibid., p.38.

251. Hawkes, 1974, ibid., pp.216,219; Von Wissman, 1972, ibid., p.291; Chu, 1973, ibid., pp.226-9.

252. Grun, 1982, ibid.

Chou Step-Pyramid (above) Szechwan, China.

upon the female and child. Among the Ainu, sex was not viewed as shameful, young people cohabited, and females maintained a high status. Marriages were love-matches, decided by the young people themselves without parental interference. A matrilocal trial marriage known as the *yobai* existed, as did simple divorce. Only women were allowed to serve the gods and a woman's first menstruation was celebrated; there was no emphasis upon female virginity. It was an era of lovemaking, poems, songs, and dance, which lasted up until the predynastic era of c.400 CE. The Ainu, who were once widespread in Japan, are now found largely on the northern island of Hokkaido where some of these customs still persist.[253] The Jomon immigrants did not develop a strongman central state culture, however, and they adopted or possessed themselves many of the same peaceful social attitudes characteristic of the Ainu. Social conditions on Japan remained relatively uninfluenced by events on the mainland for many centuries to come.

After c.1500 BCE, a new migrant people even more warlike than the Shang appeared in West China, bringing with them new West Asian gods, weapons, artifacts, and ceremonial platforms similar to the step pyramid. These people founded the Chou Dynasty; they lived in forts among an enslaved peasant class who lived in huts. State ownership of all land, taxation, and corvee labor were introduced.[254]

"The Chou sovereigns behaved toward the territorial rulers not as the first among equals, but as supreme masters responsible only to heaven..." [255]

The Chou continued the Shang traditions of murdering a dead man's wife, concubines, slaves, and interring them in his grave. In the days of the early Western Chou, fully 5% of all graves contained human sacrifice victims, many of whom were buried alive. Later, the customs were modified such that all sacrificial victims were killed before being buried. Later still, people might be replaced with clay or burnt-paper effigies, though the practice of sacrificing the living did not fully die out. Foundation sacrifices of children were also continued by the Chou.[256]

Indeed, murder of widows persisted among the upper class Chinese for centuries, lasting even longer than in India. The practice of grave murder was so widespread in China that a special character was used to denote the burial of the living with the dead; the character "...is constantly to be found in ancient and even in more modern texts."[257] In one example, following the death of the favorite concubine of Mu-Yeng Hi, ruler of the state of Yen, Hi was so taken with grief that he ordered his sister-in-law into the grave with the dead concubine.[258]

The patriarchal caste system in China with its inferior status for females was amplified over the centuries. The following early Chou poem reflects the view of the female among the literate upper classes:

253. Levy, H.: *Sex, Love, and the Japanese*, Warm-Soft Village Press, Washington, D.C., 1971, pp.1-2,8,23-41,64-5; Hawkes, 1974, ibid., p.221.

254. Chang, 1968, ibid., pp.260-3,319-20,388-9,415-6; Von Wissman, 1972, ibid., p.291.

255. Wittfogel, 1957, ibid., p.33.

256. Chang, 1968, ibid., p.327; Van Gulik, 1961, ibid., p.14; Davies, 1981, ibid., p.38.

257. Davies, 1981, ibid., p.105.

258. Davies, 1981, ibid., p.39.

259. *Book of Odes*, #189, as quoted in Van Gulik, 1961, ibid., pp.15-6.

"When a son is born, he is cradled on the bed,
He is clothed in robes, Given a jade scepter as a toy.
His lusty cries portend his vigor
He shall wear bright, red knee-caps,
Shall be the lord of a hereditary house.
When a daughter is born
She is cradled on the floor,
She is clothed in swaddling bands,
Given a loom-worf as a toy.
She shall wear no badges of honor,
Shall only take care of food and drink,
And not cause trouble to her parents." [259]

Lower class Chinese did not immediately adhere to such views, and lower class women possessed much freedom, to cohabit with or marry whomever they pleased, or to break an engagement, or to divorce. They could dance and sing with men in public. However, upper class women of the ruling Chou caste were greatly restricted and slaves in every sense of the word.[260]

Over the centuries, the common folk entirely adopted the rules and customs originally practiced only by those of the ruling caste. Infant cranial deformation was spread to new regions, to include Korea, later becoming a mark by which settlers with nomadic, ruling caste ancestry could be identified, and in some cases killed by rebels. A father's control over his family, to include the marriage choices of his children was extended, acting to curtail love-match marriages. The harem system was adopted and widely practiced. Young boys were castrated en masse to provide eunuchs for the harems, while hundreds of females were enslaved to satisfy the lusts of Chou Kings and Princes.[261] Imperial harems would sometimes contain over 1000 women, and sexual attitudes were adopted with the goal to minimize sexual pleasure. *Cong fou*, the practice of extreme female passivity ("going limp") and contactlessness during sexual intercourse, was practiced by harem women, and considered highly desirable by the Chou Lords; this practice may have originated as a natural response of despondent female captives to being repeatedly raped by strange and ruthless men who might already have murdered their relatives, and who would likewise kill them for the slightest reason. Sexual prescriptions appeared in the ruling class literature advocating the avoidance of orgasm, plus multiple female partners, as a means of attaining great health benefits. Special "sexual accountants" appeared in harems to monitor who the Lord slept with on a given night.[262] Female desires were not considered in this period of sexual absolutism, which also saw an increase in general social stratification and introduction of prostration before rulers.[263]

Shang and Chou influences pushed southeast, leaving relatively unmolested the more peaceful Karasuk culture of northeast Asia; the Karasuk practiced mixed agriculture and herding of sheep and horses.[264] Fossil and archaeological evidence indicates warmer and wetter conditions in China after 1000 BCE. Chu estimated that during this wet phase the interior Yellow River Basin, which now receives 25-50 cm (10-20") of rain, received as much as the Yangtze, which gets 100-150 cm (40-60").[265] Still, this moist period does not appear to have been sufficient to cause a general settling down of various Central Asian nomad groups.

Nomadic groups such as the light-skinned Tocharians, known to the Chinese as the Yueh Chi, pushed across the Tarim, Mongolian steppe, and into northeast Asia where they displaced peaceful Neolithic Andronov and Karasuk communities. Eventually, the Chou Dynasty was also displaced eastward, and court records spoke of red hair and green eyes among the new western invaders The Yueh Chi were the first mounted warriors to reach the Gobi, and they had a connection of origins as far west as Persia.[266] This was a time when the Central Asian steppe was being generally overrun by mounted warrior nomads. Nomads were contained in Western China, however, by the slightly moister conditions which provided pasture, and also by the first of the Great Walls built by the Chou after c.700 BCE.[267] Indeed, Aykroyd assessed the situation in China:

"Between the major dynasties, six in number, each of which lasted from two to three hundred years, there were periods of disorder and bloodshed, with bandits ravishing the countryside, and a

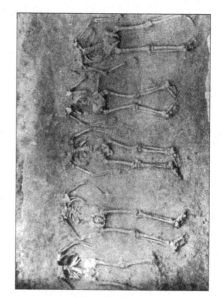

Human sacrifice (above) Shang period, An-yang; heads were buried separately. Male-female double burials (below) in most cases denotes ritual murder of widows.

260. Van Gulik, 1961, ibid., pp.14-5,21-25.

261. Wittfogel, 1957, ibid., pp.355-6; Van Gulik, 1961, ibid., pp.15- 25; Himes, 1936, ibid., p.112; Dingwall, 1931, ibid., pp.98-9.

262. Himes, 1936, ibid., p.161; Bullough, 1976, ibid., pp.290-9; Van Gulik, 1961, ibid., p.17.

263. Wittfogel, 1957, ibid., pp.33,153; Von Wissman, 1972, ibid., p.291.

264. Phillips, 1972, ibid., pp.43-4; Hawkes, 1974, ibid., p.216.

265. Chu, 1973, ibid., pp.226-9.

266. Phillips, 1972, ibid., pp.14,45-8,91,111.

267. Hawkes, 1974, ibid., p.220.

breakdown in civil administration. Famines were most frequent and severe during these interludes." [268]

The west end of the Tarim Basin was repeatedly influenced by the various Central Asian empires, though it never became attached to them; examples are the Sarmatian and Persian empires of c.800-500 BCE. The various antifemale and antichild concepts of the Central Asian nomads had been fine tuned by the Persians, and these influenced China also through the introduction of various new philosophies. [269]

The years after 800 BCE were known as the "period of the hundred philosophers", a condition which also seemed to prevail in the Mediterranean and Near East at the time. Early Taoist philosophy sought the worthy goal of uniting human behavior with Natural Law, with a return to the "Golden Age" of happiness and long life. But while Taoism venerated women in their procreative capacity, this was only in a strictly regulated manner, controlled by men. Taoist philosophers spoke little about love, and generally advocated the sexual enslavement of women, such that men could, through sex, obtain their female Yin essence. The result was a general reinforcing of sexual taboos and prescriptions, and an increase in the prevailing system of concubinage and sexual slavery. Sex handbooks were widely used by the literate classes, and these openly advocated different methods of orgasm avoidance (*coitus obstructus, coitus reservatus*) and use of multiple female sex slaves, as a means of accumulating great quantities of Yin; after a man had accumulated sufficient Yin, he was advised to ejaculate into his legal wife who was expected to then produce strong sons. The system demanded readily available female sex slaves, who were bought or otherwise taken from the lower classes, or from conquered lands. Such lower-class girls were freely sold, traded, or given away, and later became the basis of a system of state-run hereditary prostitution. [270] The Taoist sex handbooks, written by men, expressed a great anxiety, fear and hatred regarding women, who were called "the enemy":§

> *"The plain girl said: when engaging the enemy the man should consider her as worthless as a tile or a stone, and himself as precious as gold or jade... Exercising the coitus with a woman is like riding a running horse with a worn rein, or like tottering on the brink of a precipice bristling with bare blades and ready to engulf one."* [271]

The harem system, concubinage, and polygamy were widespread during Chou times, and the father's control over the love and marital affairs of his daughters was considered absolute. Once married, a woman fell under similar control by her husband; when widowed, she was controlled by her oldest son. [272] The bloated nature of the Imperial Chou harems led to numerous intrigues, and complaints were put to ink about the "meddling in state affairs" by women and eunuchs. [273] Sadistic abuses were also common, for example:

> *"The Prince of Wu orders his famous strategist Sun-tzu to demonstrate his strategic principles, using the 180 ladies of the Prince's harem. Sun-tzu divided them into two groups, each headed by a 'general', one of the favorite consorts of the Prince. When the ladies laughed and did not execute Sun-tzu's commands, he had the two 'generals' beheaded on the spot, disregarding the frantic protests of the Prince. Thus the Prince was made to understand the necessity of iron discipline in an army, and he gratefully appointed Sun-tzu as his generalissimo."* [274]

268. Aykroyd, 1974, ibid., p.86.

269. Van Gulik, 1961, ibid., pp.8, 13, 59, 157; Wittfogel, 1957, ibid., pp.139,151; Bullough, V.L.: *Sexual Variance in Society and History*, J. Wiley, NY, 1976, p.300; Denis, A.: Taboo, Putnam, NY, 1967, p.126.

270. Van Gulik, 1961, ibid., pp.28,35,45-9,65.

§ It is amazing to see these forms of Taoist sexual alchemy, which were born from ruthless warfare, ritual widow murder and sexual slavery, find fertile ground today among the mindless Western advocates of an orientalized "sexual freedom" or "New Age consciousness".

271. Van Gulik, 1961, ibid., p.157.

272. Van Gulik, 1961, ibid., pp.30-4.

273. Bullough, 1976, ibid., p.307.

274. Van Gulik, 1961, ibid., p.157.

When Prince Wu died around 677 BCE, a total of 66 people were killed to accompany him in his grave; when his nephew Muh died, 177 victims were butchered at the gravesite, prompting an unusual outburst of protest by the common people, from which the victims were selected.[275]

Chinese philosophers of the period in general stressed the need for segregation of young boys and girls, and the need for ceremonial behavior between the sexes; intimacy was discouraged, and family hierarchy with the eldest male on top was mandated. Lao Tse (c.550 BCE), a founder of Taoist thought, taught that youthful desire should be subordinated to age and the family, a view echoed by Confucius (c.500 BCE), who emphasized absolute obedience to one's teachers, loyalty to one's master, and the hierarchical family as the sound basis for a well-ordered state. Moti taught that the major reason for homes was to seclude women, and separate them from men. And Mencius (c.400 BCE) taught that the greatest of all sins was not to have a son.[276] These views later became the basis of an Oriental puritanism which spread across China, and into Japan. Not unexpectedly, ritual widow murder persisted throughout this period.[277]

Confucian puritanism developed slowly after 500 BCE, becoming ever more critical of women, who were viewed as sinister and dangerous, in great need of external controls and domination. Post-Confucian China became:

"A country shaped by old men, and they tried to drive love entirely from licit relations, even from marriage." [278]

Under the Chou Confucianists, "reform" came to mean the making of change via punishments, and the Ministry of Justice became known as the "Ministry of Punishments".[279] Stacey has discussed the effect of these practices upon young women, forced into unwanted marriages, and among whom high suicide rates prevailed:

"The young bride (usually between fifteen to seventeen years old) had to confront a sudden sexual adjustment for which she was typically unprepared, a household of strangers almost all of whom commanded her subservience, and inordinate pressure to contribute male progeny to her husband's lineage." [280]

The Chou pushed north and south during the time when nomadic pressures were occurring on the western frontier. Fragmentation of the empire eventually transpired, with rebellions against the central Kingship. China entered the Period of Warring States (470-220 BCE). Art motifs of the period included horse and combat scenes. Sarmatians were grazing herds in Dzungaria, Yueh Chi were in Mongolia and Transbaikalia, and the Pazyryk nomads, who smoked hashish and interred murdered widows and servants in their graves, were also on the Chinese steppe. The Chou themselves were still practicing ritual widow murder, but only in 0.5% of all graves, down from the earlier high of 5%.[281]

Between 400-100 BCE, the Ordos and Yin Shan region was described as covered with forests and grasses, abounding in wild game. Huns grazed their herds here during the period. Later, however, they and the Yueh Chi were forced westward out of the Ordos region through conflict with the growing power of the new Ch'in Dynasty. The Great Wall of China, built by the Ch'in, separates the Ordos from humid regions to the east. As the Huns and Yeuh Chi were pushed west, the Sarmatians appeared in South Russia on a westward migration. The Chinese scholar Mencius described

Taoist Sexual Alchemy emphasized a loveless sexuality and male orgasm-avoidance with multiple female partners, claiming health benefits for men — the needs of women were of no essential interest, and the entire system was dependent upon large harems and female sexual slavery for its continuance.

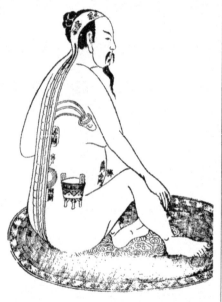

275. Davies, 1981, ibid., pp.38-9.
276. Bullough, 1976, ibid., p.301; Hingley, 1977, ibid., p.300; Himes, 1936, ibid., p.108; Wittfogel, 1957, ibid., p.151; Denis, 1967, ibid., p.132; Van Gulik, 1961, ibid., p.8.
277. Davies, 1981, ibid., pp.39,104.
278. Denis, 1967, ibid., p.126.
279. Wittfogel, 1957, ibid., p.139.
280. Stacey, 1983, ibid., p.43.
281. Glacken, 1972, ibid., p.70; Klimm, L.E.: "Man's Ports and Channels", in Thomas, 1972, ibid., p.531; Chang, 1968, ibid., pp.327,339,353,361; Phillips, 1965, ibid., pp.78-93.

the great deforestation of Eurasia in progress during this period.[282]

Ch'in leaders came from the drier western end of China, and brought superior iron weapons. They conquered all of China's various warring states to establish a single dictatorial Kingship. With respect to women and children, the Ch'in carried on most of the same traditions as the Shang and Chou. In addition to the Great Wall, the Ch'in initiated construction on the large road networks, irrigation works, and canal systems of eastern China.[283]

Nomads of the Gobi region subsequently formed a confederacy known as the Hsiung Nu to challenge the Ch'in. The Hsiung Nu were mounted warriors, expert in the use of the calvary and reflex bow; they expanded their empire from the Altai, across the Gobi, and into Korea. One leader of the Hsiung Nu, Shanyu, died and was buried with much grave wealth, to include his slaughtered wives, concubines, and ministers, "along with thousands of humbler followers".[284] This same period saw massive widow murder in the Wei state, along the Yellow River, where various Princes had their favorite females murdered. One grave in the region had 100 bodies in it, 99 of which were female.[285]

It was around 300 BCE that the Yayoi culture migrated from Korea into Japan, lording over or displacing the Jomon and Ainu. The Yayoi possessed a culture closer to that of the Central Asian Hsiung Nu. They constructed large tombs for their rulers, and their graves reveal more social stratification and strongman rule; clan power increased. Bronze weapons, iron metallurgy, and other mainland innovations also arrived in Japan with the Yayoi.[286]

Dzungaria was still used by the Huns for agriculture during c.200-100 BCE, but this was not to last much longer.[287] Chinese sources made increasing reference to droughts following 100 BCE, a time when the Huns and other groups were abandoning agriculture and the Gobi.[288] During this period of unrest, the Huns emerged victorious over other nomadic groups, whose loyalties constantly shifted as they moved about the steppe.[289] By 165 BCE, the peoples of north Asia rebelled against the Hsiung Nu, driving them west and southwest. Once this was done, human sacrifice ceased, and grave wealth diminished. The Hsiung Nu in turn displaced the Saka and Yeuh Chi westward, where they would harass Iran, Afghanistan, and India, forming new kingdoms. One of these new kingdoms, the Kushan (100 BCE-300 CE), spread control over much of Central Asia, the Indus, and Iran, transmitting Buddhism into China.[290]

Civil war erupted in China around 207 BCE and the Han overthrew the Ch'in. Confucian scholars were recruited for the new Han civil bureaucracy, and they generally reaffirmed female inferiority. New laws were instituted to segregate the sexes, forbidding public touching, even to hand items to each other.[291] The poet Szu-ma Hsiang-ju wrote of the Han Confucianists:

"...they refuse to attend a party when female entertainers are there, and they run away as soon as they hear the sound of songs and laughter..." [292]

These conditions led to the first significant increases in homosexuality in Chinese palace affairs, where eunuchs, boy lovers, and male transvestite companions abounded. Eunuchs in particular attained great positions of power. Many of the Han Princes and Kings were debauched sadists, who engaged in great abuses of male and female members of their courts. Palace and harem intrigues were incessant, with a large percent of the leadership being murdered, imprisoned, tortured, executed, or committing

Romanticized portrayal of a gentleman with his concubines. In reality, such women were obtained by kidnapping from conquered territories, or outright purchase from cultures considered "lower" than the ruling class. Disobedience or refusal on the part of a woman brought swift and often merciless punishment. Drawings and paintings such as these were made for the viewing pleasure of the upperclass.

282. Gumilev, 1968, ibid., p.28; Gumilev, 1966, ibid., p.42.

283. Van Gulik, 1961, ibid., p.55; Wittfogel, 1957, ibid., pp.32-8,66; Hawkes, 1974, ibid., pp.216,220; Kovda, 1961, ibid., p.207; Chang, 1968, ibid., p.37.

284. Phillips, 1972, ibid., p.112; cf. Dupree, 1973, ibid., p.298; Davies, 1981, ibid., p.38.

285. Davies, 1981, ibid., p.38.

286. Gregg, N.: "Hagi: Where Japan's Revolution Began" (Insert Map), *National Geographic*, 165:750-73, June 1984; Hawkes, 1974, ibid., p.216.

287. Gumilev, 1968, ibid., p.25.

288. Gumilev, 1968, ibid., pp.28-30.

289. Gumilev, 1966, ibid., p.43; Dupree, 1973, ibid., p.298; Phillips, 1965, ibid., pp.111-7.

290. Dupree, 1973, ibid., pp.xviii, 299.

291. Van Gulik, 1961, ibid., pp.57-9.

292. as quoted in Van Gulik, 1961, ibid., p.68.

suicide.[293] The Han dynasty also saw numerous cases where both wife and daughter would be murdered or commit suicide at the grave of their deceased father or husband.[294]

The Han period also saw spreading of desert conditions in Western China, and Han ruins can today be found deep within desert sand regions. Careful records of drought and famine were made at least from the Han period onward, being most frequent in the north and central provinces. Indeed, it has been estimated that from 108 BCE to 1911 CE, a total of 1828 famines occurred in China, almost one per year. One of the main jobs of the Emperor, it is recorded, was to pray for rain.[295]

It was during the Han period that Chinese navigators established contact with India, southern Asia and the Philippines,[296] and Japan came under increasing influence from the mainland.[297] Influences from mainland China, which were derived from Central Asia, soon spread to Japan, and greatly increased after 400 CE. The first Imperial clan and central state appeared shortly thereafter. Buddhism and Confucianism arrived in Japan from China and Korea, and land ownership fell into the hands of the State. Great burial mounds for strongman rulers were constructed, in which horse figurines were placed. The ancient Japanese love-match marriages declined, with marriages increasingly being decided upon by parents, particularly among the upper classes engaged in political maneuvering. Polygamy and the harem system appeared, with all its complement cruelties to women. One account from the period mentions a Japanese Emperor who prepared to decapitate a concubine for allowing a tree leaf to fall into his wine; she escaped this fate only by soothing him with a song in which she pleaded for her life.[298]

In another account from Japan from c.2 BCE, the Emperor ordered a halt to ritual widow murder following the death sacrifices of Prince Yamato-hiko. The Prince's entire retinue was buried alive in his tomb, but "...for several days they died not, but wept and wailed". Even the Emperor was moved by the cries of the dying women. While he did nothing to save them, he subsequently ordered that only clay images should be used thereafter.[299]

Late in the Chinese Han period, Wang Man usurped the throne and instituted reforms, abolishing male slavery and returning land to the peasants. He continued female slavery, however, surrounding himself with women taken from all over China. A brief turn from Confucian puritanism toward Taoism occurred, though the Confucian philosophers mightily objected:[300]

> *"If women are entrusted with tasks involving contact with the outside, they will cause disorder and confusion in the Empire, harm and shame on the Imperial Court... The Book of Documents cautions against the hen answering dawn instead of the cock..."* [301]

Confucian puritanism later emerged supreme over Taoism and Buddhism, one Emperor actually declaring a hierarchy of importance, with Confucianism first, Taoism second, and Buddhism third.[302] Confucianism and Taoism dominated the world views of the Chinese, particularly after the prolonged period of conflict with the Huns (150 CE). For many years the Huns tried but failed to penetrate the Great Wall or to dislodge the Han Chinese. The Huns were themselves driven westward by continued desiccation and conflict after 100 CE, the Hunnish empire spreading from Dzungaria and the Tarim all the way to Hungary, and south into Iran and India, as discussed in the preceding section. The combined effects of desiccation and conflict caused the northern steppes to be depopulated by 300 CE. Even after the Huns were defeated in the Gobi

293. Van Gulik, 1961, ibid., pp.28,48,61-3,65,167; Wittfogel, 1957, ibid., pp.338,356; Bullough, 1976, ibid., pp.303,309; Levy, 1971, ibid., p.11.

294. Davies, 1981, ibid., pp.105-6.

295. Songqiao, Z.: "Desertification and De-desertification in China", in *The Threatened Drylands*, J.Mabbutt & S.Berkowicz, eds., UNESCO/U. New So. Wales, 1980, p.84; Aykroyd, 1974, ibid., pp.82-3.

296. Wada, ibid., p.121.

297. Hawkes, 1974, ibid., p.216.

298. Hawkes, 1974, ibid., p.216; Levy, 1971, ibid., p.36; Cory, 1956, ibid., p.105; Gregg, 1984, ibid.

299. Davies, 1981, ibid., p.40.

300. Van Gulik, 1961, ibid., p.76-97; Bullough, 1976, ibid., p.291.

301. Van Gulik, 1961, ibid., p.87.

302. Van Gulik, 1961, ibid., pp.114-5.

by the Hsien-pi around this time, settlements were restricted only to the desert margins, the ecology of the region never recovering.[303]

The years after 300 CE saw increasing competition between Taoist and Confucian scholars for the ear of the Emperor, with Buddhism largely playing a minor role. The first Buddhist nunneries appeared in China only after 350 CE. Confucian, Taoist, and Buddhist philosophers all viewed women in a subordinated role, the laws of the day allowing a husband to execute his wife for moral crimes. The elder headwife in a polygamous household was placed in a similar position of authority over younger wives and concubines. Politically, this was an era of divided smaller states variously in alliances or at war with each other, with raids by nomads along the drier margins, plus various debauched and sadistic Lords, Princes, and Emperors.[304]

Using historical sources, Chu identified the period 300-700 CE as one of excessive dryness, followed by some moisture.[305] A period of land abandonment around CE 300 in the Tarim coincides with the phase of desiccation and conflict.[306] The Turks first appeared in history in the 600s on the slopes of the Altai Mountains in West Mongolia, being related to the earlier Hsiung Nu and later Mongols. The northern vegetated margin of the Gobi was much broader in the 6th and 7th centuries than today, and the Turkut Khanate of the steppe derived its power largely from local resources. Mongolia then still possessed forested areas; today it does not.[307]

The Turks in Mongolia were defeated by the T'ang Dynasty Chinese around 630 CE, though they maintained a vast empire over Soviet Central Asia. T'ang ruins have been found far out in the desert sands of present day China, but not as far out as those of the preceding Han period, indicating a progressive expansion of the desert. China saw a peaking out of central state power under the T'ang Dynasty, and all major official Chinese histories from this time on included special large sections on the military.[308] T'ang China was an exceedingly patriarchal authoritarian society where women were bought and sold like cattle into marriage, concubinage, or hereditary prostitution.[309]

"Palace agents used to scour the entire empire for beautiful and accomplished women, and apparently took them wherever they found them, not despising even commercial or government brothels. When a number of women had been so collected, the eunuchs and duennas sorted them out. The best were chosen for the Imperial Harem." [310]

T'ang society, which developed during a period of great Chinese militarism and central state power, organized a massive brothel system, in which women were kept as virtual slaves. Women were recruited from conquered territories, or from the lower classes. Female criminals could be sentenced to the brothels, as were the wives and children of males who were convicted of crimes. Pimps and bullies of various sorts kept the girls in line, and taxes were paid to the government; brothels were organized into classes, the lowest serving the military. During the period, marriages were entirely arranged by parents, the bride and groom often meeting each other just before the wedding. Only concubines, who could arrange their relations through negotiated contract, appear to have maintained a degree of freedom in their relationships.[311]

T'ang society appears to have institutionalized the previously random forms of debauchery and sadism of earlier years, behaviors which in large measure had been confined to the harems of the ruling classes.[312] The T'ang also continued use of the harem, where in the K'ai-yuan era every

303. Phillips, 1965, ibid., pp.113,120; Gumilev, L.N.: "Heterochronism in the Moisture Supply of Eurasia in the Middle Ages", *Soviet Geography*, Jan. 1968.

304. Van Gulik, 1961, ibid., pp.28,84-113; Bullough, 1976, ibid., pp.291-2.

305. Chu, C.: "Climatic Pulsations During Historic Time in China", *Geographical Review*, 16(2):274-83, 1926.

306. see Chappell, 1970, ibid., pp.361-4.

307. Gumilev, 1968, ibid., pp.28-30.

308. Wittfogel, 1957, ibid., p.63; Songqiao, 1980, ibid., p.84.

309. Levy, 1971, ibid., p.42.

310. Van Gulik, 1961, ibid., p.184.

311. Van Gulik, 1961, ibid., pp.170-83.

312. Van Gulik, 1961, ibid., pp.28,93-4,119-121; Wittfogel, 1957, ibid., p.40.

women who slept with the Emperor received a stamp on her arm for proof. Sadistic Imperial abuses also occurred, however, to include the severe mutilating torture of women.[313]

During this period of central state expansion and great female repression, various quasi-Buddhist and Confucian philosophies saturated China, adopting the general theme of female inferiority. The sex handbooks of the period actually described the physical characteristics, such as hair color, face shape, etc., of women considered most suited or unsuitable for coitus. Extensive new rules for lovemaking were formulated, to include an amplification of the above mentioned sexual prescriptions. These now included recommendations on the use of drugs and medicines during intercourse, and reflected interests in male immortality through multiple female partners. A central element of these prescriptions, however, was avoidance of the spontaneous, instinctual nature of sex, to include orgasm-avoidance (*coitus reservatus, coitus obstructus*).[314]

Buddhism arrived in Japan via Korea after 500 CE, around the same time the Yamato State began.[315] Family conditions and female status in Japan may be inferred from the Taika reforms of 645 CE, which exempted only pregnant women and new mothers from beatings, and attempted to limit the use of "court maidens" sent to the central palace to sway political decisions. Polygamy among the lower classes was later limited to three women, two wives and one concubine.[316] While some limited vestiges of female and sexual freedom continued to persist in Japan after 700 CE,[317] evidence of a cultural transition identical to that of China abounds after this date. Young girls and boys were procured to meet the sexual demands of the Buddhist clergy inside the temples, and children were eventually made to service the sexual appetites of the rulers and bureaucrats of the Imperial palace. Wives and daughters could be used as collateral for a loan, with free sexual access to them by the lender until the debt was paid. Women were virtually without legal rights or status of any kind, and lenders or tax collectors could drag them away from their families as slaves.[318] Hereditary prostitution developed along with an institutionalization of woman beating. Females were beaten in temples during "promiscuity confession rituals", where they confessed to sins. A special female "buttocks beating festival" evolved in Japan, persisting until the end of World War II.[319] Women also continued to be murdered as grave sacrifices at royal funerals.[320] Japanese literature of the period began to express a longing for the ancient days when even poor farm girls could pick their own husbands, and could even reject the amorous attentions of an Emperor.[321] Levy discussed the transition in Japanese morals:

"Eighth century records in prose and poetry...reveal that Japanese women of earliest recorded history must have enjoyed a far greater degree of sexual freedom than did Japanese females of later centuries, during which religious-philosophical dogmas prevailed. The punishment of adultery and illicit intercourse in Japan was due mainly to the foreign influences of Buddhism and Confucianism. With the diffusion of foreign concepts of Deity, the Japanese came to consider defilement of the gods a crime and sexual purity a matter of religious conviction. Women were even regarded as wives of the gods, and blood from defloration seemed holy. With these concepts in the background, there developed strict injunctions against a females losing her purity before the wedding encounter." [322]

West China saw increased conflict during the 700s, when Arab armies attempted to invade; a similar situation occurred in Mongolia as Turks again were active in the area. Both Mongolia and the Tarim Basin were

313. Van Gulik, 1961, ibid., pp.189-90.
314. Van Gulik, 1961, ibid., pp.149-50.
315. Hawkes, 1974, ibid., p.216; Gregg, 1984, ibid.
316. Levy, 1971, ibid., pp.35-6,71.
317. Levy, 1971, ibid., pp.22,26,33,67-8.
318. Levy, 1971, ibid., pp.31-6; Cory, 1956, ibid., p.105.
319. Levy, 1971, ibid., pp.39,41,67-8.
320. Davies, 1981, ibid., p.40.
321. Levy, 1971, ibid., p.22.
322. Levy, 1971, ibid., p.30.

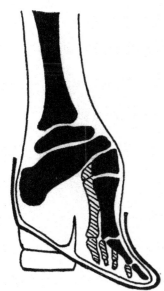

Upper-class Chinese woman, binding her feet. The procedure compressed the bones of the foot such that walking was greatly diminished, and running impossible.

323. Gumilev, 1968, ibid., p.30; Chappell, 1970, ibid., pp.361-4.

324. Huntington, 1907, ibid., pp.311-2.

325. Huntington, 1907, ibid., pp.239-61,280-94.

326. Huntington, 1907, ibid., pp.160-1.

327. Tannahill, R.: Sex in History, Stein & Day, 1980, p.195; Van Gulik, 1961, ibid., pp.219-24,253.

328. Stacey, J.: Patriarchy and Socialist Revolution in China, U. California Press, Berkeley, 1983, pp.40-1.

329. Van Gulik, 1961, ibid., pp.xii,108,219; Wittfogel, 1957, ibid., p.122; Davies, 1981, ibid., p.39.

330. Levy, 1971, ibid., pp.9-10, 28, 33, 38, 41, 70.

depopulated in the 800s.[323] However, the phase of land abandonment appears to have been short lived, the occupation of the largest ruin in the nearby Turfan Depression, Kara-Khoja, being established after 874 CE. Qanat systems were observed in the Turfan and also around Lop Nor in the Tarim. The region, which is now exceptionally arid, was then "renown for its library, its art, and its craft, as well as its might in war".[324]

Around 900 CE, the humid regions of China again came under attack from peoples abandoning the arid steppe. Northern China was invaded, and Sung Emperors retreated south. Turks increased their activity in the West. At the southerly edge of the Tarim Basin lie a number of ruins which were abandoned around 1000 CE, apparently due to water shortages. Lulan, one of the major cities on the lake, supported around 17,000 people, or about twenty times its later population of 1900 CE.[325] Choka, a city at the base of mountains forming the southerly rim of the Tarim Basin, supported 3000 - 5000 persons from a water source which in 1900 supported only 12 families.[326] Ruins of numerous other small villages and large towns lie on the southern edge of sand sea of the Takla Makan.

The period after 900 CE displayed growing tensions in desiccating northwestern China, with increased unrest among populations. Caste distinctions and puritanism increased in China afterwards, and female inferiority and strict separation of the sexes was emphasized even more than before. Female foot-binding was greatly emphasized, becoming mandatory for upper class women. Li Yu, the emperor who is said to have introduced this crippling deformity, was also known as a great Chinese "love poet". Yu saw the foot as most erotic, and considered touching a woman's foot as a most serious offense. Women's feet produced such sexual anxiety that they were rarely drawn or painted, even in erotic art of the period.[327]

However, it would be a mistake to view foot-binding as merely a class trait. Even farm girls had it done, by some estimates up to 14% of them, who would forever afterward work in the fields on their knees. The bandages would be started at age 6-7, and increased in tightness every few days until the desired effect was achieved (the ball of the big toe being crunched back to touch the heel). All women who underwent the mutilation became permanently disabled, and death from infection related to it may have been as high as 10%. The Confucians would say: "Why must the feet be bound? To prevent barbarous running around."[328]

Widow taboos continued through the period, with new wrinkles; a repudiated wife was considered forever disgraced, being unable to remarry; repudiated concubines were sent into forced prostitution. Confucian puritanism grew in China, with repressive controls enacted against erotic art and literature, and against Buddhist and Taoist institutions which continued to practice the various sexual prescriptions. The female body was considered indecent, and female dance and physical exercise was forbidden. Excavations of the period also indicate the persistence of ritual grave murder.[329]

Puritanism also spread into Japan during the 800s through the agency of Confucian and Buddhist scholars. They preached chastity for women, concubines for men, and placed taboos on remarriage of widows or female divorcees. Marriage formality increased, and marriage celebrations declined. Women were seen as the root of all evil, and vaginal blood taboos and cleaning rituals appeared. Vaginal blood was feared as a poison by the common man, and Buddhist priests were appointed to deflower virgin brides on their wedding night, demonstrating to any shy groom how the act was performed.[330] By the 1100s an increasing emphasis upon military force developed in Japan, possibly primarily as a reaction to the tide of events on the mainland.[331]

The outbreak of the Mongols from West China in the early 1200s has not been definitively linked with desiccation. However, Chu identified the period 1280-1367 CE in China as one of climatic turmoil, with excessive drought and flooding combined.[332] Initially thwarted by the Great Wall, the Mongols eventually broke into the humid zone, sacking Beijing and other Chinese cities to build an empire which stretched from the Caspian to the Pacific. Marco Polo, who visited China during the Mongol period and witnessed the opulence of the Chinese courts and palaces, discussed the widespread destruction of the countryside:

"The province named Thebeth [Tibet] was laid entirely to waste at the time that Mangu-Khan carried his arms into that country. To the distance of 20 days journey you see numberless towns and castles in a state of ruin." [333]

After this period of Mongol activity, the basins of West China were finally transformed into desert, if not purely from climate change, then certainly with the encouragement of human populations who, after the Mongols, increasingly took up the nomadic herding tradition.

Huntington concluded that depopulation of West China resulted from a progressive failure of the runoff and groundwater irrigation systems supporting their agriculture, rather than from human factors, such as war or decay of the irrigation works. He discussed several sites in the Tarim which had been abandoned only in 1834, when streams no longer discharged enough water to reach them.[334]

Huntington provided an interesting bit of evidence related to the Keriya River, a perennial stream flowing northward into the Takla Makan and terminating in the heart of the sand dunes. It flows toward the Tarim River, but never gets closer than 150 miles. Stein, a contemporary of Huntington, was told of a road along the Keriya by which one could cross the Takla Makan, to reach the city of Kucha on the Tarim River. This is the route which ancient Buddhists took to flee from Moslem invaders around 1000 CE. He says "a remark of Mirza Haidar, the Moghul leader and historian, makes it very probable that the Keriya River reached the Tarim as late as the sixteenth century."[335]

According to Huntington, other rivers along the southern rim of the Takla Makan give evidence of greater flow and human settlement in recent historic times, similar to that observed for the Keriya. Ruins of large settlements, such as the ancient Chinese kingdom of Tu-ho-lo (Tukhara), and fields of dead vegetation were found far out into the desert, often beyond the present day limits of river flow.[336]

One of these rivers, the Niya, was observed by Huntington to periodically vanish into the sand as it wound its way into the Takla Makan. Its final appearance at the surface was made some 50 miles into the desert, where a Moslem holy shrine was built. The photographs in Huntington's account show very large stands of oaks and poplars at the holy site, uncharacteristic for other parts of the region today, and indicative of a selective deforestation phenomena similar to that observed by Mikesell in Morocco and Lebanon. Other parts of the Takla Makan, contain large, old trees which have not been cut down, but which clearly suffer from the effects of a decreased water supply.[337] Various ruins along the elevated ancient shoreline surrounding modern-day Lop Nor indicate a greatly expanded size. The following legend on the origins of the Takla Makan would suggest that at one time, people remembered the formerly humid conditions and the social changes which accompanied the drying-up of the landscape:

331. Levy, 1971, ibid., p.42; Gregg, 1984, ibid.

332. Chu, 1926, ibid., pp.274-83.

333. Iwai, 1935, ibid., p.108.

334. Huntington, 1907, ibid., pp.169-73,187-9.

335. Huntington, 1907, ibid., pp.193-4.

336. Huntington, 1907, ibid., pp.199-314.

337. Huntington, 1907, ibid., pp.197-203,219-22.

"...a Mohammedan priest came to Kenan one day, long after the driving out of the former Buddhist inhabitants, and found no one at home. Men, women, and children had all gone out to work on the canals. The holy man was hungry and tired. Being accustomed to live off the fat of the land, he was irritated at finding the houses shut and empty. He offered a prayer, which can hardly be supposed to have been pious, and began to turn a hand-mill standing in a courtyard, whereupon sand rained down from heaven... From that time onward the water supply decreased..." [338]

According to the ancient Chinese, Lop Nor was a marshland, or "Lake of Reeds", while the Tarim Basin itself was in 1900 still called the Han Hai by the Chinese, meaning 'dried up sea'. [339] Huntington mentioned the ruins of ancient villages or remnants of ancient vegetation "far out in the midst of the sand" in the Takla Makan "where now there is no water". [340] Dating roughly from the Buddhist periods, many of these sand-buried towns lie on dried up watercourses. In some cases, ruins lay in abandoned oases, surrounded by fields of dead vegetation, such as reeds, poplars, and tamarisk. The above legend suggests that the final desiccation of the Takla Makan took place during Moslem times, after 1000 CE.

Huntington also discussed similar evidence for relatively recent climate change in the Turfan Depression, north of Lop Nor.

"Everywhere one encounters vast beds of dead reeds... Grum-Grshimailo, a Russian explorer, who visited Turfan in 1889, says that though now the playa of Bojanti is dry most of the year, Chinese records and an old song seem to indicate that formerly there was at least a large reed-swamp, if not a lake. From this and from the evidence of ruins he concludes that the water supply has greatly diminished. The plain is dotted with ruins not only in districts which are now inhabited, but in more remote regions, where no surface water is now available and the underground supply is saline." [341]

Regarding the arid zones of China, Huntington concluded that "The water supply throughout the whole region was formerly more abundant than now". [342] Stein also concluded that:

"...shrinkage of available water-supply has taken place in the Tarim Basin during historical times, and...it must be connected with a general desiccation period affecting the whole of Central Asia...". [343]

In addition to these dramatic environmental changes, Chinese history records equally dramatic changes in culture. Confucian puritanism in China received an additional boost under the Mongols, who established the ruling caste Yuan Dynasty, and stationed their troops across the land. Women were subjected to seclusion in the home and public veiling to shield them from the Mongols. A household system of demerits was established for sex-related sins, and foot-binding became more popular and nearly ubiquitous in the cities and among upper class women. [344]

Buddhist priests figured prominent in the Yuan court, and in the requisitioning of women for Kublai Khan; [345] according to Marco Polo, the Khan:

"...annually sent for 400-500 girls from a particular province, famous for its beauties, having them assessed according to a point

338. Huntington, 1907, ibid., p.178.
339. Wadia, 1960, ibid., p.7; Huntington, 1907, ibid., pp.239-61,280-94.
340. Huntington, 1912, ibid., p.167; Huntington, 1907, ibid., p.103.
341. Huntington, 1907, ibid., pp.309-11.
342. Huntington, 1907, ibid., pp.169-73,187-9.
343. Stein, M.A.: *The Ruins of Desert Cathay*, Vol. 1, London, 1912, p.257; cf. Stein, M.A.: *On Central Asian Tracks*, Pantheon, NY, 1964.
344. Bullough, 1976, ibid., pp.295,300; Tannahill, 1980, ibid., p.196.
345. Iwai, 1935, ibid., pp.155-6.

system. The top 30 or 40 shared his bed in rotation for a subsequent year, the others given to his noblemen." [346]

In Japan, which had been a cultural satellite of China, events followed a similar pattern. Mongols attacked Japan by sea in 1274 from their base in China, but were turned back at the beach when they ran out of arrows. The subsequent Mongol invasion of 1281 was also held to a beachhead and repelled. A third invasion force of 3500 Mongol ships was sunk by a storm, which the Japanese called *kamikaze*, or divine wind. [347]

These invasions spurred a growth of militarism in Japan during which love between men and women, and the status of women further eroded. Levy identified the period as characterized by a "tendency toward coolness and lack of sentiment, or an overt display of it...", where compassion and softness in the male was equated with weakness. [348] Anonymous child and teen betrothal were initiated, and a special class of taxed prostitutes and sex-slave girls was established for servicing the military. Japan of the 1300s saw a debauched military with massive harems. Peasant women were bought, sold, or given away as gifts; no woman was safe from kidnapping, rape, and enslavement. [349] Japanese Buddhism of the period invented female phantoms with great claws, and sequential hells for moral offenses. Hellfire awaited the adulterer after death, but only the female was punished for it while still alive. A husband had the right to decapitate his adulterous wife or concubine, or alternatively cut off her nose, fingers, or breasts. Sadistic crushing and tearing tortures were developed for females guilty of moral crimes, aimed at the sexual organs. Buddhist priests were forbidden from sexual indulgence, and could be killed if they went into a home where a woman lived alone; when found guilty, such priests were targeted for the severest of tortures, involving skin flaying and molten lead. [350] Under these antifemale and anti-heterosexual conditions, homosexuality flourished. [351]

The Yuan Chinese military also invaded Java and the Philippines, and armies marched across Tibet into Burma, Laos, and Cambodia, where they were turned back from the jungles by the Khmer peoples. Strongman militaristic influences, as well as Buddhism and Taoism, were left behind or reinforced in each of the areas invaded. [352]

The Mongols were expelled from China in 1370 by the Ming emperors, who proceeded to imitate their previous conquerors. The Ming maintained a lavish and cruel style, mixed with a resurgence of Confucian puritanism. Foreign dress, names, foods, and habits were banned, and the Great Wall was fortified with hardened bricks. Marriages were totally arranged in the upper classes, and only slightly less so in lower classes. A newly married couple would now see each other only at the threshold of the marriage bed. Sex handbooks were also censored and restricted. [353] One Ming notebook on "family instruction" gave the following advice:

> "1. *New concubines should be forced to watch coitus of the husband with the other wives.*
> 2. *If a woman makes a mistake she should be admonished; if she repeats the mistake, she should be caned 5 or 6 strokes on the bare buttocks.*
> 3. *Men should not tie their wives to a post and beat them until they are covered with blood.*" [354]

Men followed Confucian, Taoist, and Buddhist philosophy, but women were almost exclusively Buddhist, given its granting to women some small shelter and consideration from the larger anti-female culture. Women were less likely to be abused sexually by the generally homo-

346. as quoted in Tannahill, 1980, ibid., p.240.

347. Wiencek, et al, 1980, ibid., p.36.
348. Levy, 1971, ibid., p.42.
349. Levy, 1971, ibid., pp.42-3.
350. Levy, 1971, ibid., pp.71-3.
351. Levy, 1971, ibid., pp.10,33-4.
352. Van Gulik, 1961, ibid., p.258; Wiencek, et al, 1980, ibid., p.45; Iwai, 1935, ibid., p.116.
353. Van Gulik, 1961, ibid., pp.226,235,266-9; Tannahill, 1980, ibid., p.319; Hawkes, 1974, ibid., p.220; Pillsbury, 1978, ibid., p.653.
354. Van Gulik, 1961, ibid., pp.268-9.

sexual-oriented Buddhist male priesthood. Buddhist nuns, who came and went freely about the city and into women's quarters, were viewed with much suspicion by men, the nunneries often being a woman's only haven of escape from compulsory marriage to an unknown man, or from a cruel tyrant husband or mother in law.[355] Buddhism remained a minority influence, however, and ritual widow murder continued to persist.[356]

The desert continued to expand eastward during the Ming period, forcing widespread abandonment of lands.[357] As was the case with the Yuan, the Ming sent out naval expeditions into the Pacific, into the South Seas and to the coasts of Java and Ceylon.[358] Syphilis struck China for the first time around 1500, further intensifying puritan trends. By 1556, Portuguese missionaries observed a near absence of women in the streets of Canton, though prostitution based upon female slavery flourished in special sections of all major cities. Marriages were restricted within given social classes, though all men were free to buy or sell concubines as they wished. Puritanism in China continued through the 1600s into the 1900s, along with the complement subordinated status of women and harem system. Open public display of affection between males and females was taboo from the 1700s up to the present day.[359]

Regarding ritual widow murder in China, Davies made the following observations:

> *"The belief was firmly held that widows and even daughters were the property of the dead man, and logic demanded their sacrifice as an act of devotion to him and to the gods... The act was favored by public opinion and moralists were lavish in its praise... The methods of self sacrifice adopted by Chinese women were more varied than those found in India, where immolation by fire was the rule. The majority hanged themselves or cut their throats; but others took poison, or leaped into a chasm. Cases have been recorded of wives who threw themselves into burning buildings in which their husbands and parents were trapped in order to perish with them and of others who cast themselves into a fire that had been kindled to burn up the chattels of the dead man. Widows could also drown themselves."* [360]

Indeed, ritual widow murder/suicide by public hanging was popular in upper class China of the 1800s, being announced in advance via advertising, and watched by the public.[361] Again, other social conditions prevalent in China fit the overall pattern seen elsewhere in Saharasia.

By the 1900s, things had not changed much in China. In addition to ritual widow murder, a general pattern of sex-negation and female taboos persisted. In Mongolia of 1921, around 40% of the population of adult males were celibate Buddhist Lamas, who also owned most of the agricultural land which was farmed by serfs. The Lamas lorded over the population in a predatory manner, sharing power with a military caste. A similar situation of feudalistic repression existed in Buddhist-run Tibet, and other western parts of China before the revolution. The Lamas were dispossessed and often murdered during the communist insurgency after that date, and atrocities were widespread across China at the time. Secular military power developed, and wars occurred between the Japanese (who had invaded coastal mainland China) and the Chinese under the autocratic republican rule of Chaing Kai-shek, who previously had dispossessed the Ming rulers. The Japanese were eventually driven from China and Korea, following their defeat in World War II, while the communists, fighting from bases at the Saharasian fringe, in turn displaced the republican forces to Taiwan. The ascendancy of the commu-

Ming-era brothel
romanticized portrayal

355. Van Gulik, 1961, ibid., pp.267-8.
356. Davies, 1981, ibid., p.39.
357. Songqiao, 1980, ibid., p.84.
358. Van Gulik, 1961, ibid., pp.131,256,306.
359. Van Gulik, 1961, ibid., pp.xi,48-9,265,312,334-5; Tannahill, 1980, ibid., p.319.
360. Davies, 1981, ibid., pp.105-6.
361. Davies, 1981, ibid., pp.106-7.

nists to power in China was followed by a few genuine reforms, but in general the prior feudal masters were simply replaced by a new group of Communist Party task-masters, who began a policy of exterminating any individual or group which failed to fall in line with the new ideology. Reform of marriage law only occurred in China after 1950, when the "women's law" was passed to allow free and easy divorce. Over 800,000 divorces per year occurred thereafter, almost all of which were requested by women. Sadly, almost 10% of the women making requests for divorce were thereafter killed by their own families, which prompted a stiffening of divorce laws after 1953. This pattern, of firstly loosening and then stiffening social reform laws, was similar to that which occurred almost simultaneously in the Soviet Union.[362]

Similar situations prevailed in Japan throughout the Yedo era (1603-1867), which followed two centuries of war. Love came to be openly denigrated as it opposed the system of arranged marriages. Strict segregation of boys and girls over 6 was instituted, and child betrothal was practiced. Divorce could only be initiated by the male. Female status declined further as poor farmgirls could be sold into slavery or prostitution, to live a life of beatings, torture and abuse. Widows were pressured to take a vow of lifelong celibacy in deference to their deceased husbands, the word "widow" itself becoming a euphemism for "whore". Sexual contact with women over 30 was disadvised, and the suppression of orgasm was advocated as a key to longevity. Harems ballooned in size, and various drugs, rituals and rigid rules were developed to crush the emotions of the female. Prostitutes kept in cages were put on public display in special parts of the city. Newlywed girls would be deflowered by priests in the temples, or by high-level bureaucrats, vaginal blood being studiously avoided by the average man. Women were forbidden from acting in Kabuki plays, giving rise to a cultivated class of male transvestites; this later evolved into male prostitution which was later banned along with Kabuki plays as well. The Shogunate, meanwhile, developed lavish and infantile customs, such as specialized royal palace toilet attendants whose sole job was to "wipe the Majestic Buttocks" of the Emperor after defecation.[363]

Japanese courtesan, c.1840

The 1600s saw development of a Shinto cult in Japan, with mystic devotion to the Emperor, who was considered god's representative upon Earth. A caste of Samurai devotees was organized who pursued the martial arts, duty and devotion to Emperor and ancestors, "high" moral values, and readiness to commit *hara-kiri*, or ritual suicide. This ritual demanded that the devotee plunge a knife into his stomach, make a lateral slice, withdraw the knife, and then cut his own throat. The devotee's wife was allowed to forego the stomach cut, and merely cut her own throat. In one notable instance, the Shogun Nobunega, facing defeat at the hands of rebels, slit his wife's throat and then killed himself, after which 50 of his bodyguards did likewise.[364] Yedo Japan also saw campaigns to drive out foreigners, notably Christians, who were horribly tortured to death irrespective of age or sex.[365]

Japanese despotism, in both family and State affairs, increased after a period of industrialization and subsequent empire building, which began in the late 1800s. The branch of Shintoism known as *Bushido*, meaning the "way of the warrior", was elevated to a national religion by the god-Emperor, to whom total obedience was obligated. Imperial Japan invaded Korea, Formosa (Taiwan), Southeast Asia and China. Atrocities and torture abounded in the conquered regions, as in the 1937 rape of Nanking — a two-month orgy of slaughter which claimed the lives of over 300,000 Chinese civilians. Additional hundreds of thousands of captured prisoners of war and civilians from other nations, including over 150,000

362. Stacey, 1983, ibid., pp.3-4,178; Scott, 1939, ibid., pp.105,268-72; Allen, T. & Conger, D.: "Time and Again in Mongolia", *National Geographic,* Feb. 1985, p.253.

363. Levy, 1971, ibid., pp.29-53,73-5; Himes, 1936, ibid., p.127.

364. Davies, 1981, ibid., pp.126-9,285.

365. Scott, 1939, ibid., pp.107-10.

British and French civilians, were slaughtered by the Imperial Japanese army during the course of World War II,[366] while Japanese terroristic despotism and starvation in conquered territories indirectly led to additional millions of deaths. The United States came into direct conflict with the Japanese in later years when Japan began extending its empire into Southeast Asia, attacking American forces in the Philippines, Hawaii (Pearl Harbor) and elsewhere. After defeat of the Japanese by the Allies during World War II, the Japanese Empirate was dissolved, and various reforms were forced upon them by the Americans. Land was redistributed from feudal Lords to ordinary working people, hereditary prostitution was outlawed, parliamentary democracy was instituted, women were given the vote, and other changes appeared which had a significant influence upon the affairs of women and young people in Japan. Before the coming of the Americans, however, changes in Japanese society were clearly of the same character, direction and intensity as those which occurred in China, which were in turn derived from the nomad warrior peoples of Saharasia.

There is today a sharp contrast observable between the cultures of inland China and coastal China. The cities of Shanghai and Hong Kong have a considerably greater inclination towards freedom and self-regulation (democratic interests, a limited market-economy and freer flow of information) as compared to the interior, a contrast which is even more sharply reflected in the differences between mainland China and Taiwan, or between China and modern Japan. In all cases, the closer one gets to Saharasia, the more patristic and authoritarian is the culture and political climate. A similar difference is found between North Korea (closer to Saharasia) and South Korea (farther from Saharasia), though the influence of relatively free-wheeling American culture following World War II and the Korean War has been considerable. When the pro-democracy students erected a statue of the "Goddess of Democracy" in Tienamen Square (Beijing), in addition to requests for relaxation of many authoritarian rules, they asked for a few sex-economic reforms, related to marriage and contraceptives, to have friends of the opposite sex, and so forth. In response, the elderly communist leaders sent in troops from Central Asian divisions, well-known for their more ruthless capacities. The students, young males and females, were machine-gunned and crushed by tanks. New laws also appeared in China forbidding public hugging and kissing by young people, and also to curb "decadent Western influences" in the coastal cities, such as rock music and blue jeans. The war against public tenderness and love heats up — but the young Chinese are fighting back, even if only by holding hands in the park, and stealing a sweet kiss from under a portrait of the debauched mass-murderer, Chairman Mao. Love is the enemy of the authoritarian, world-wide, giving people strength and courage to stand up to the bully. Communist China hopefully will soon follow the path of Soviet communism: onto the trash-heap of history, with a breaking apart of the vast Chinese Empire into smaller autonomous regional states.

366. Scott, 1939, ibid., p.272; also see the new documentation gathered by Iris Chang, in *The Rape of Nanking: The Forgotten Holocaust of World War II*, Basic Books, NY 1997. Japan's direct responsibilities for, and awful conduct during World War II continues to be denied within much of Japan. Japanese school textbooks and history books have routinely censored critical information about Japan's unprovoked invasions of Manchuria, China, Korea, Formosa, the Philippines, New Guinea, Indonesia, Indochina, Burma, Pearl Harbor, etc., about the cruel manner in which its prison camps were run — with beatings, beheadings, gruesome medical experiments and high death rates — and other aspects of their deadly authoritarianism in conquered lands. These and other well-documented events of the period continue to be censored or uncritically misportrayed within Japan, with a general white-wash of massive war-time death tolls in Japanese-held territory and the merciless brutality of the Imperial Japanese army. Many former Japanese war criminals were never punished and allowed after the war to hold senior positions in Japanese industry and government, with no social stigma attached to their past behavior. The atomic-bombing of Hiroshima and Nagasaki (which together killed far fewer people than did the widely-denied rape of Nanking) by comparison are publicly discussed in graphic detail within Japan, with no mention of Japan's own massive a-bomb program (being developed with Nazi cooperation and by some accounts ready for mass production and use against the gathering American naval forces) often as a means to falsely portray Japan as a "victim" of World War II. Japan has to date paid nearly no wartime reparations to lands it devastated, even to the thousands of "comfort girls" kidnapped into sexual slavery for the Imperial Japanese army. This is a major contrast to modern-day Germany, which has paid large amounts of reparations to Jews and injured nations, which does not easily tolerate former Nazis in positions of power, and where every schoolchild is given a full education about Hitler and the Nazi crimes. Dissidents in Japan who have attempted to openly speak on these subjects have been subject to extreme social pressures, and may risk their lives for doing so — as was the case with the Mayor of Nagasaki, who as recently as 1990 was shot by a fanatic for daring to make even a mild criticism of Emperor Hirohito, that he bore partial responsibility for World War II. (Chang, 1997, p.12)

F. SUMMARY AND SYNTHESIS

From the materials presented in this Chapter, we may now make the following general summary and synthesis.

The archaeological record indicates the earliest settlements across all of Saharasia and its borderlands were characterized by relatively peaceful conditions, with minimal conflict before the onset of major desiccation and desert formation. The earliest people engaged in a mix of hunting, gathering, fishing, mixed agriculture, animal domestication and/or pastoralism. They were technologically sophisticated peoples, and possessed pottery, weaving, metallurgical and building construction knowledge, as well as skillful arts and various forms of technology needed to extract plentiful food from the environment. These peaceful cultural conditions prevailed at a time when Saharasia was a relatively wet and lush grassland biome. There is no evidence, in the form of destruction layers in settlements, weapons technology, or in graves, to suggest that peoples of these more fertile times went about warring upon each other, nor did they possess child-abusive practices which leave traces in bone structure, such as cranial deformation. The female principle was held in high status, or possibly revered, and in most cases female figurines of a naturalistic, sensual or erotic quality existed. Where rock art was the dominant form, the female is seen to have occupied an equal, or even central place in society, and was not viewed with anxiety by the artists. Royal tombs, excessive grave wealth, female grave murder, temple architecture, fortifications, weapons technology and a fully-blown nomadism were absent from these early cultures, as were male gods and phallic imagery.

While a few parts of Saharasia show signs of drying up as early as c.5000 BCE (see Table 9), these climatic turns towards dryness affected only relatively small geographical sub-regions, and did not appear to have lasting effects. A few traces of patrism which did appear coincidental to

Figure 60: Generalized Paths of Diffusion of Armored Human Culture (Patrism) in the Old World, after c.4000 BCE.[§]
1. Arabian Core Region   2. Central Asian Core Region

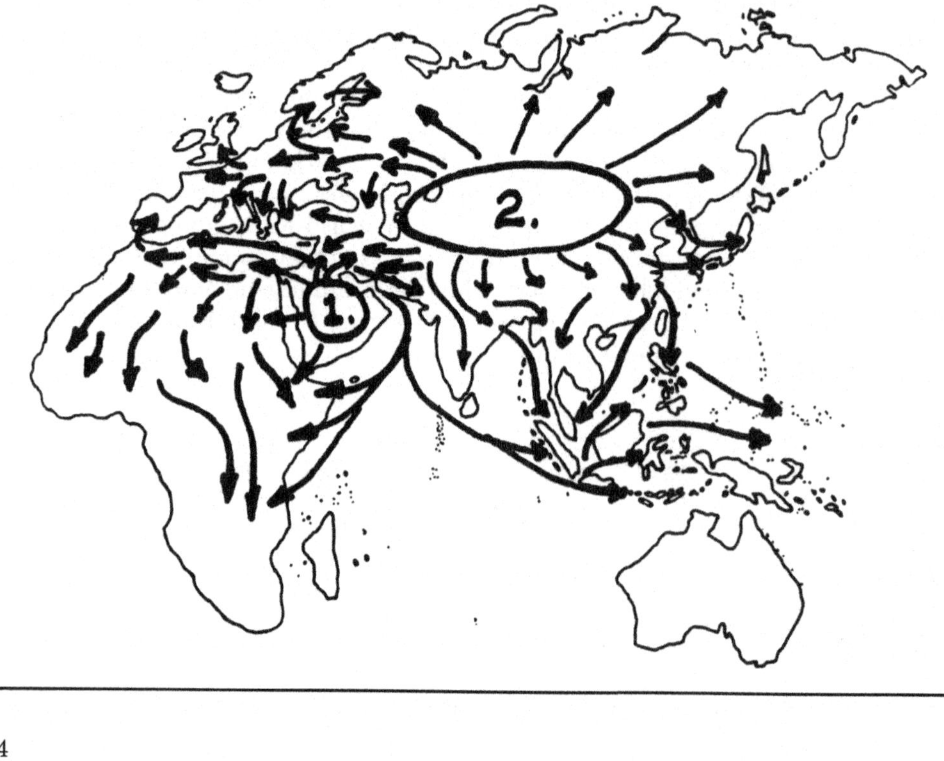

those climate changes (reported in the historical sections) were also regionally isolated, and not continuous from one archaeological layer to the next.[§]

Climate took its driest turn after c.4000 BCE, however, with widespread desiccation spreading across Arabia and Soviet Central Asia. The environmental change stimulated regional nomadic adjustments at the expense of fixed agriculture along water courses, which became increasingly undependable. Irrigation technology developed under such ecological change, allowing some peoples to maintain a settled agricultural lifestyle, which also persisted on exotic rivers, whose sources lay outside of the drying regions (ie, the Nile River, which is fed from moist highlands in Ethiopia). The onset of competitive, militant nomadism appears first in Arabia and Central Asia, generally simultaneous to the onset of the arid phase. In Central Asia, changes in the levees of the Caspian and Aral Seas and their tributaries worked to frustrate early agricultural efforts, and nomadism was further stimulated as population centers in those regions were abandoned, leading to major dispersals of previously settled peoples. In all likelihood, these environmental changes and mass migratory land-abandonments were shadowed by famine and even starvation conditions. In Arabia, Semitic peoples were increasingly abandoning their lands by 3500 BCE, moving north into the Near East and west into North Africa, with very militant and armored-patristic qualities in their social institutions and behavior. By 3000 BCE Semites became the dominant ruling caste in regions which possessed secure water supplies: the Nile Valley, Levantine Mountains, and Mesopotamia.

By 3000 BCE, nomadism also predominated across Central Asia, also with increasingly militant and armored-patristic qualities; Indo-Aryan nomads were pushing out of the Asian steppe into Europe, Anatolia, and Iran, and competing with Semitic warrior nomads for control of Mesopotamia. At this time desiccation also was spreading to North Africa, the Near East, and West China, increasing the area over which drought and famine would work their emotional-social damage, and the predominance of militant nomadism.

In Arabia and Southern Central Asia, desiccation occurred before the development of a patriarchal authoritarian, warlike nomadism, and was a central mechanism in stimulating such cultural changes in those areas. However, in much of North Africa, the Levant, Mesopotamia, and the moister fringes of Central Asia, these same warlike nomad groups generally invaded shortly before the onset of arid conditions. Once aridity took

[§] See the notation "New Material 2005" on page 368, which slightly modifies the dates presented here for "Onset of Major Patrist Phase" for the Middle East (Levant/Mesopotamia) and Anatolia.

Table 9: Summary of Dates for Ecological and Cultural Change[§]

Location	Onset of Desiccation	Onset of Major Patrist Phase
North Africa (West)	c.3000 BCE	c.2600-500 BCE
North Africa (East)	c.3000 BCE	c.3100 BCE
Arabia	c.4000 BCE	c.4000-3500 BCE
Levant/Mesopotamia	c.5000-2500 BCE*	c.3500-2500 BCE
Anatolia	c.5000-2000 BCE*	c.4000-3500 BCE
Iran	c.5000-2000 BCE*	c.4000-3000 BCE
Soviet Cent. Asia	c.5000-3500 BCE*	c.4000-3500 BCE
West China	c.3000-2000 BCE	c.2000 BCE
Indus Valley	c.2000 BCE	c.2000 BCE

* Oscillatory and regionally-isolated climate changes, with the most intense aridity occurring after c.4000 BCE

hold in these other areas, however, famine, starvation and mass-migrations also reinforced the trend towards armored and militant patriarchal authoritarian nomadism.

Mounted warrior nomadism, the war chariot, and a significantly more military oriented social structure thereafter developed across the open unforested lands of Saharasia. Remaining groups of once-peaceful hunter-gatherers and settled agriculturalists were either annihilated, enslaved, or pushed back to the forested fringes of the steppe, where cavalry and chariots were of limited use. Fortress architecture, to guard either the fringes of the moist regions or water supplies within Saharasia itself, was stimulated among settled peoples as a response to the cavalry tactics of the militant nomads. The methods of military organization and weaponry used by the nomads were eventually adopted by the settled peoples, either as a response to repeated assault, or the new weaponry predominated after they were annihilated and replaced by new inhabitants. More importantly for this study, the beliefs, behaviors, social structures and institutions of the patriarchal authoritarian nomads were transplanted into the newly conquered regions, irrespective of climate.

New practices and beliefs, such as divine kingship, the caste system, and male-dominated religions grew in the Saharasian regions conquered by the militant nomads. New kingdoms developed, the heartlands of which were generally located where water supplies were secure, or where larger populations with great agricultural potentials existed. The archaeological record, where such concerns have been addressed, documents the arrival of the conquering armies of the Indo-Aryan and Semitic groups with the sudden onset of female grave murder, along with royal tombs, massive grave wealth, and temple architecture. These new kingdoms include Egypt and Mesopotamia after c.3000 BCE, and Shang China and Brahmanic India in later centuries. During these later periods, infant cranial deformation, originally a mark of cradle-boards of nomadic peoples, also appeared as a high-caste distinction. Migrating or invading groups partly or entirely transplanted the Saharasian social fabric into Europe, Sub-Saharan Africa, India, China, Japan, and parts of Southeast Asia and Oceania.

These cultural changes are also clearly seen in the changing art styles and motifs of Saharasia, notably the rock frescoes of the Central Sahara and the female figurines of southern Central Asia. Females either vanished from such art motifs, or were changed in character towards less sensual, more abstract and non-erotic qualities.

Table 9 summarizes the dates for both climatic and cultural change across Saharasia, taken from the materials presented in this Chapter. Where the dates for onset of aridity and patrism coincide, drought and famine appear as the major triggers for patrism. Such a situation, it is argued, occurred in Arabia and Central Asia. However, where patrism appeared before the onset of major arid conditions, invasive migrations of armored, militant patrist peoples from Arabia and Central Asia appear as the first major mechanism for stimulating culture change towards patrism. In later years, increasing drought, famine, and conflict would have exacerbated and increased patrist tendencies. Saharasia's borderlands also underwent transformations in later centuries due to out-migrations or invasions of armored patrist peoples, though without the added disturbance of long-term drought and famine.

Figure 60 gives an exceedingly general graphic presentation of the migratory paths by which the patrist, Saharasian culture complex developed and spread through much of the Old World. Two spreading centers are identified, in Arabia and Central Asia. These were the first portions of Saharasia to dry up, leading to the first widespread historical epochs of

366

famine, social turmoil, land abandonment and outward migration. The arrows and pathways given are in general agreement with other studies on settlement patterns and migrations of cultures.[367] The map of migratory movement in Figure 60 also corresponds closely to the patterns of antifemale, antichild, and antisexual behavior derived from more recent anthropological data, on the World Behavior Map in Figures 1 and 24.

Discussion

A continuity clearly exists between the earliest changes found in the archaeological record with those of later historical periods. Early social institutions which nurtured children emotionally, promoting and protecting their self-regulatory natures, which venerated heterosexual love and allowed much freedom in sexual matters (for adolescents of both sexes, and the unmarried) once persisted across all of Europe, Asia, Africa, and India. In all cases, the arrival of militant armored nomad groups from Central Asia and Arabia initiated cultural transitions which destroyed the male-female and maternal-infant bond, and placed all family matters, to include choice of marriage partner, in the hands of dominant males. The early peaceful peoples were either exterminated and replaced by the armored newcomers, or they were enslaved, losing their own cultural identity and legal controls over their land, property and very lives.

The once self-regulated, free choice of sexual partner and mate selection was increasingly suppressed and subordinated to the demands of parents, priests and kings, and a newer cultural layer developed which denied sexual freedom, and nearly all other freedoms, especially to women and children. An emphasis upon military matters occurred with strongman leaders growing in power and veneration, to the point where the masses enthusiastically genuflected to divine kings. In all cases, change toward increasingly stratified political hierarchy and despotism was accompanied by institutionalized changes in family structure toward an increased male dominance over basic life decisions of females and children, and increased control of children by parents of both sexes.

Sadistic abuses within the family, in society at large, and in the military appear to have increased in direct proportion to these changes. Polygamy, the harem, veil, general seclusion of females, sexual slavery, male and female genital mutilations, and eunuchism all developed within the Saharasian zone as power of dominant males, divine kings, and the military caste grew. Their appearance in regions outside of Saharasia always came later, after repeated conflicts with Saharasian empires, or otherwise with adoption of Saharasian social institutions. These patterns were cross-cultural in character; spatially and temporally, such transitions occurred simultaneous to the spread of militant nomadism, first within Saharasia, and later outside of Saharasia.

As discussed in detail, a few Near Eastern sites do suggest, in the form of fleeting glimpses, the presence of grave wealth, the chiefly institution, possible fortifications, and other signs of disturbance centuries earlier than the date of general Saharasian desiccation at c.4000-3500 BCE. However, such evidence is not always clear, is not uniformly distributed or even regional in character, and does not appear in the archaeological record along with other aspects of fully-blown patrism. Neither is such evidence continuous, from one period to the next, even at a single archaeological site. Importantly, *evidence from these sites yielding "fleeting glimpses" of patrism prior to c.4000 BCE also indicates the presence of a simultaneous episodes of aridity, land abandonment, and nomadic pastoralism, making them exceptions which prove the rule of a desertifica-*

367. M. Edmonson, "Neolithic Diffusion Rates", *Current Anthropology*, 2(2):71-102, April 1961; M. Gimbutas, *Bronze Age Cultures in Central and Eastern Europe*, Mouton, The Hague, 1965, pp.19-22; C. McEvedy, *Atlas of African History*, NY, 1980, p.32; J. Spencer, *Oriental Asia, Themes Toward a Geography*, Prentice-Hall, NJ, 1973, pp.48-9; E. Isaac, *Geography of Domestication*, Prentice-Hall, NJ, 1970, pp.41,90-1; "The Appearance of World Religions", Map #11, *Historical Atlas of the World*, Rand McNally, NY, 1981, pp.28-9; "Migration and Race", Map, *The Atlas of Mankind*, Rand McNally, 1982, pp.14-5; Watkins, C.: "The Indo-European Origin of English", in *The American Heritage Dictionary of the English Language*, J. Trapper & G. Stuart, eds., Houghton-Mifflin, 1969, pp.xix-xx (cf. back cover diagram on "The Indo-European Family of Languages"); *People and Places of the Past: National Geographic Illustrated Cultural Atlas of the Ancient World*, Natl. Geog. Soc., 1983, pp.38, 211, 234, 279, 290, .305, 326.

tion-famine "trigger" for the genesis of armored patrism and the general *Saharasian theory.* Certainly, there does not exist any widespread, signifcant or persisting evidence of infant cranial deformation, female grave murder, divine kingship, temple architecture, militant nomadism, warfare, or other key aspects of the Saharasian cultural complex anywhere in the Old World prior to c.4000 BCE. After 4000 BCE, however, such practices and behaviors are increasingly evident in Saharasian archaeology and history, appearing first on the fringes of Arabia and Central Asia, and spreading outward from those regions with time. Also importantly, once armored patrism appeared in the ancient history of any given world region, with few exceptions it persisted in those same region over the centuries, right up into modern times.

The archaeological and historical survey given here — alone and by itself, and without reference to anthropological studies of the last 100 years discussed in prior chapters — confirms the past presence of a ubiquitously higher status for women, greater autonomy for children and adolescents, with a much more fluid and pleasure-oriented social fabric. Males had a solid role in the family as help-mates, lovers and partners, but did not dominate the basic life decisions of either the wife or children. Childhood and adolescent sexuality was not subordinate to dominating parental controls or priestly-shamanistic verbots, and was both instinctual and self-regulating in character. The survey also confirms that *destructive human aggression and sadism in its worst forms, to include despotism, warfare, ritual murder, and the brutal subjugation of females and children for sexual purposes, are a relatively recent development in human history, of less than 6000 years duration,* traceable back in time to Central Asia and Arabia through the temporal and spatial geographic patterns of the Saharasian cultural complex.

Further, and most importantly for this study, the archaeological and historical data independently support and confirm the spatial, geographic patterns in the anthropological data on the World Behavior Map, constructed from more recent data. By tracing the movement of peoples back in time across the surface of the map, and by focusing upon their social structure as it relates to children, sexuality and females, the times and locations of origin of such behavior, in the Old World at least, have been generally determined. Let's now turn our attention to the New World.

New Material 2005: The Appendix B article *Update on Saharasia* contains a review of more recent archaeological findings not available when my original research was undertaken in the 1980s and 1990s, and before the first edition of this book was published in 1998. These newer findings suggest the dates for the first appearance of unambiguous armoring, patrism and violence within the Middle East (Levant/ Mesopotamia) and Anatolia may be earlier than the indicated date of c.4000 BCE, and in fact may be closer to c.5000 BCE, or a bit earlier, timed closely to the previously identified sub-phase of aridity and desertification. A subsequent migration-diffusion of patristic violence into Europe also developed at this very early date, from the desiccating region of the Levant, as detailed in new maps of agricultural diffusion developed by European scholars and reproduced in the Appendix *Update*. These new findings suggest the spreading centers for patristic violence as identified here in Arabia and Central Asia (Figure 60) should be supplemented with a similar dry zone across Anatolia and the larger Middle East. This does not, however, undermine any of the basic conclusions presented in this work, and in fact further reinforces and supports them: *These new findings of pre-4000 BCE early social violence, clustered as they are in specific regions also suffering under severe desert-formation and famine-starvation conditions, or being the product of migration-diffusion of human social groups out of those same desertified regions, fully confirm the overall Saharasian discovery, of human armoring, patrism and social violence being the product of biological reactions to severe and prolonged drought-famine-starvation conditions, rather than something genetic and innate to the human species.* Additional new evidence is also presented in the *Update* for peaceful social conditions in ancient times across most other world regions.

9. Patrism in Oceania and the New World

Anasazi pottery design,
Southwestern USA,
peaceful conditions.

This study has, so far, not significantly addressed archaeological findings or historical events in the New World, or Oceania, to include Southeast Asia and Island Asia. However, the maps presented in prior chapters do show the presence of partial or significant patrist tendencies in some portions of Oceania, and in the Pacific Northwest of North America, in Mesoamerica, and in Peru. If the general theory on the origins of armored patrism in the process of desert formation is truly a global theory, these other regions of patrism will also have to be explained. Ideally, a systematic reconstruction of historical and prehistorical events, similar to that already completed for the Old World, would have to be done for these areas. This is a step which has been undertaken only partially. A number of archaeological and historical facts, as well as the geographic, spatial patterns on the maps, suggest processes at work in Oceania and the New World similar to but not as widespread as those seen at work in the Old World, namely desertification, vast migration patterns, the sudden appearances of warlike peoples, cultural changes toward increased infant and child trauma, disturbances in heterosexual love relations, decreasing status of women, and the diffusion of nomadic, warlike peoples among surrounding peaceful cultures, followed by the establishment of isolated despotic central states.

OCEANIA

If we assume that peoples on Earth were migrating about in a generally peaceful manner since the Pleistocene, exploring and shifting their locations, extending their domain from region to region, island to island, such peoples and migrations before c.4000 BCE, when Saharasia began to dry up, would have been of an almost purely matrist character. However, after Saharasia began to desiccate, initiating the massive cultural shift towards patrism in Central Asia and Arabia, migrations around the world would increasingly be characterized by patrism.

From these starting assumptions, it would follow that all patrist cultural characteristics in Oceania arrived there via transplantation from the Old World mainland after the onset of drier conditions. Also, whatever patrist tendencies were carried into Oceania would have diluted over time and distance, with increasing contact and intermarriage with preexisting matrist cultures. Such a dilution was observed with male genital mutilations, a distinctly patrist practice which was mapped on Figures 26-27, and the attribute "patrism" itself gradually diluted with passage across Oceania, as seen on the World Behavior Map. The male genital mutilations themselves stretch across a wide range of Oceania, through the eastern end of Melanesia and into the central Polynesian Islands to as far east as 100°W longitude, and eastward from there into Mesoamerica. Similar distributions exist for female genital mutilations.[1] Many of these same groups also practiced swaddling and infant cranial deformation.[2] None of the practices were uniformly distributed (for instance, they were generally common in Melanesia and Polynesia, but uncommon in Micronesia) and they are not the kind of thing one anticipates a culture would simply "invent" out of the thin air, unless they firstly developed similar characterological motivations, or were subject to clear-cut diffusion and

1. See Chapter 5, and A. Montagu, "Infibulation and Defibulation in the Old and New Worlds", *American Anthropologist*, 47:464-67, 1945; A. Montagu, "Ritual Mutilation Among Primitive Peoples", *Ciba Symposium*, October 1946, pp.421-36.

2. E. Dingwall, *Artificial Cranial Deformation*, London, 1931, pp.94-149, 238.

Figure 61: Diffusion Maps of G. Smith (1915-1933) and E. Loeb (1923)

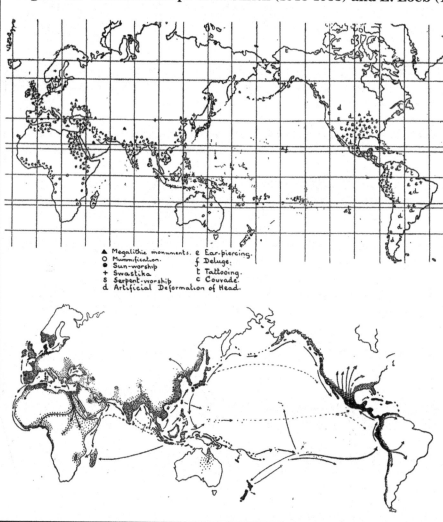

Megalithic monuments.   ℓ Ear-piercing.
O Mummification.   f Deluge.
● Sun-worship
+ Swastika   t Tattooing.
s Serpent-worship   c Couvade.
d Artificial Deformation of Head.

Insufficient data provided only limited support for Grafton Smith's[4] original bold hypotheses of a "heliolithic culture-complex", a problem similar to that encountered with E. Loeb[3] and his "blood sacrifice complex". The idea of global cultural diffusion was thereafter prematurely abandoned by most scholars. However, both Smith and Loeb appear to have touched upon parts of the "Saharasian cultural complex" (armored patristic culture) but neither addressed any significant number of variables on child-treatment, sexuality or the status of women. Only in the 1970s did it become possible to make more detailed and meaningful maps of cultural diffusion, with the creation of larger and more detailed anthropological data bases. As discussed in the Introduction, the present findings on Saharasia were the first to make use of these larger cross-cultural data bases for geographical analysis.

contact with a dominant group that forcibly introduced the practices.

For instance, in 1923 Loeb argued for a cross-ocean distribution of a "blood sacrifice complex", which included both circumcision, human sacrifice, and cannibalism. His study included world maps of their distribution, but the small number of cultures studied — 98 for cannibalism and human sacrifice, and 58 for circumcision — failed to reveal any telling pattern at all.[3] Still, the association between circumcision and despotic forms of central state and extreme hierarchy, cultural forms which lend themselves to human sacrifice, was found in both the historical reconstruction and cross-cultural analysis given in this work. Similarly, Smith argued for the diffusion of many cultural traits and artifacts into Oceania and the New World, primarily from Egypt via the Red Sea.[4] Some of his maps did reveal structured patterns, but none approached the level of detail on the maps given in Chapters 4 and 5 of this work, constructed from large, cross-cultural data bases. By the time these data bases came into use, the diffusionist theories had fallen into disfavor, and global maps were generally no longer made of trait or artifact distributions. Loeb, Smith and other diffusionists were clearly on the right track, but they lacked adequate data by which to test their theories.

It is known that Hindu, Islamic, Chinese, and Japanese sailors traveled around Southeast and Island Asia, into the Philippines and Indonesia, and they progressed eastward to unknown distances through

3. E. Loeb, "The Blood Sacrifice Complex", *Memoirs, American Anthropological Association*, 30:1-40, 1923.

4. G.E. Smith, *The Diffusion of Culture*, Kennikat Press, NY, 1971 (reprint of a 1933 work); G.E. Smith, *The Migrations of Early Culture*, Manchester U. Press, London, 1915; Also see W. Perry, *The Children of the Sun*, Scholarly Press, Michigan, 1968.

the various island chains of Oceania.[5] These groups all possessed strongly patrist attitudes at the time of their voyages, and would have acted as a socially disruptive force in the peaceful societies they visited. Patrism could thereby have been implanted among local peoples of Oceania, and subsequent trade interactions would have reinforced such tendencies. By imitation or force, the cultural institutions of the patrist navigators would have been adopted.

Specific cultural connections between Saharasia and Oceania can be determined or inferred from the presence of Buddhism, Hindu, and Islamic beliefs, rituals, and customs — and one map of Islamic/Moslem-influenced regions has already been presented, demonstrating the Moslem influence, at least, extended well into the island regions of Indonesia and Borneo. However, other maps suggest a significant carrying of patrism across much larger parts of Oceania. Along these lines, Elkin has argued for entry of circumcision into Australia from the Northwest coast, and similar practices appeared in the New World as discussed previously.[6] If such practices were brought to the New World across the Pacific from the Old, they would have left some traces across Oceania. However, neither Dingwall, Montagu, nor others studying such traits in detail produced maps of sufficient resolution to assist in such a line of inquiry. Only the poor maps of the early diffusionist ethnographers exist, which understandably have not been widely accepted. Even my own maps presented here, which are among the most detailed to be produced since the 1930s, can only be considered as a first approximation. In short, patrist tendencies could have come to Island Asia and Oceania either wholly or partly from the Old World mainland. But if only partly, then from where else?

One possible source would be the deserts of Australia, which may have acted as a locus for the independent genesis of patrist behaviors and attitudes, via mechanisms similar to those given for Saharasia. It is known, for instance, that some of the Australian Aborigine cultures possessed aspects of male domination, as well as severe genital mutilations and vaginal blood taboos. As pointed out earlier, Davidson's study of Australian genital mutilations[7] suggests that the harshest and most severe forms existed in the heart of the desert, being less severe towards the coast. His study further suggests that, like Saharasia, the severest form of mutilation, subincision, possessed a rather restricted distribution contained inside a wider region in which only circumcision was practiced. This certainly suggests a Saharasian-like mechanism for at least this one Australian trait. And Australia has, aside from Saharasia, one of the largest areas of harsh desert in the world.

Further, the Aboriginal population of Australia possessed a distribution which closely matched the Australian average annual rainfall, and access to water holes by the aborigines was considered a fighting matter.[8] Birdsell commented on this distribution:

"The high degree of correlation between rainfall and [population] density indicates that the Australian aborigines are subject to a rigorous environmental determinism..." [9]

Hence, it appears that the same principles at work in Saharasia could also have been at work in the Australian deserts, though probably to a lesser degree.

If the desert-famine hypothesis was at work significantly in Australia, then other parts of Oceania might have been influenced by diffusion of both Australian *and* Saharasian forms of patrism and associated institutions, to include later Hindu, Moslem, and Chinese voyagers. However,

5. For an interesting viewpoint on pathways across Oceania, see M. Levison, et al., *The Settlement of Polynesia, A Computer Simulation*, U. Minnesota Press, 1973; Also see citations in Chapter 8.

6. Discussed in A. Montagu, "The Origin of Subincision in Australia", *Oceania*, 8(2):202-3, 1937; cf. section on "Male Genital Mutilations" in Chapter 5.

7. D. Davidson, *The Chronological Aspects of Certain Australian Social Institutions, As Inferred from Geographical Distribution*, U. Pennsylvania, Philadelphia, 1928.

8. E. Graham, "Yuman Warfare: An Analysis of Ecological Factors from Ethnohistorical Sources", in *War, Its Causes and Correlates*, M.A. Nettleship, et al., eds., Mouton Pub., Paris, 1973, p.455.

9. J.B. Birdsell, *American Naturalist*, 39:171-207, 1953.

the idea that such practices were wholly transplanted into the Australian desert from Saharasia, along with other patrist attitudes and behaviors, cannot be dismissed *a-priori*. Armored patrism might have been carried into Australia by groups arriving after 3500 BCE, when Saharasia began to desiccate, or at any time thereafter. If this were the case, then patrist tendencies would probably have been introduced into both Australia and other parts of Oceania starting around the same time, intensifying thereafter.

Another interesting clue comes from Trobriand society, discussed previously, which was generally peaceful and matrist in character, but did possess a few patrist traits such as a polygamous headman of limited power. It is recalled that Reich felt such patrist traits had "invaded" their culture through mechanisms unclear and from unknown sources.[10] For instance, the headman's functions included those of war ritual. For the Trobriander, however, war was of a restrained character, designed mainly to humiliate, being full of various rules of etiquette: one could fight only at certain times of day, and the defeated opponent was left a path by which he could escape unharmed.[11]

Given a number of facts, I argue that the Trobrianders adopted such a headman and military apparatus purely for defensive purposes. Historically, the Trobriand Islands had been harassed by the intrusion of other, more militant peoples into Melanesia. Destructive antisocial aggression and sadistic impulses were not generated within Trobriand culture in significant enough quantities, such that an organized military structure would spontaneously develop. As Reich pointed out, the small amount of patrism which existed among the Trobrianders were "invasive" traits which were not deeply rooted in the character structure, much less in the instincts or "genes".

Warfare had been entirely given up by the Trobrianders by 1917 when Malinowski studied them.[12] According to Malinowski, warfare had never been a central aspect of their daily living or social structure, and the Trobriand headman was greatly limited in his authority. However, the early Trobrianders did have an occasional but very clear need for an organized defense against other warlike peoples in the region, primarily the warlike *Dobuans*. Trade routes once stretched southwest from the Trobriand Archipelago towards the homelands of the Dobuan and Amphlettan peoples, and from there to New Guinea and Australia.[13] The peoples of the south and west were more patrist and sex-repressive than the Trobrianders, and some groups in New Guinea were even more patrist still.[14] This being the case, the overseas trade routes would have been favored paths for any invasions, and suggest particular geographic directions (southwest, towards New Guinea and the Australian coast) from which patrist influences first arrived in the Trobriand Archipelago. I argue that it was the pressures of these more warlike peoples to the southwest which explains the presence of a limited-function headman and defensive military apparatus, and other "invasive" aspects of patrism among the Trobriand Islanders.

Malinowski made the following comparison between the Dobuans and the Trobrianders:

> *"The Trobrianders represent the enlightened, light-hearted, easygoing civilized tribes of North-West Melanesia. Not so their neighbors, the inhabitants of the fascinating yet gloomy, beautiful yet treacherous, 'mountain', the koya, as the Trobrianders call the southern district. The Dobuans as well as their landscape are an object of superstitious awe and attraction. The Koya, has always been and still is an El Dorado, a land of promise and hope, to*

10. See Chapter 3 for a discussion on this.

11. B. Malinowski, "War and Weapons Among the Natives of the Trobriand Islands", *Man*, January 1920, pp.10-12.

12. "Fighting and all that is connected with it is a thing of the past in the Trobriand Archipelago, as in practically all the districts of British New Guinea.": Malinowski, 1920, ibid., p.10.

13. R. Shutler & M. Shutler, *Oceanic Prehistory*, Cummings Pub., CA, 1975, pp.55-6.

14. Malinowski, 1932, ibid., pp.220, 234-5f,262; Malinowski later stated that the Dobuans were less sex-repressive than he first thought, based upon the study by R. Fortune (*Sorcerers of Dobu*, Routledge & Kegan Paul, London, 1963; see "Introduction" by Malinowski, p.xxiv). However, Fortune's own work states explicitly that the Dobuans displayed extreme anxiety and fear toward their environment and other people. They were prone to fits of murderous sexual jealousy, and engaged in the beating of wives and children in a manner not seen in the Trobriand Islands. Furthermore, they placed many barriers of a compulsive and fear-producing nature in the way of their young people's sexual relationships. The Dobuans very much fit the pattern of an authoritarian, antifemale, antichild, sex-repressive, and warlike culture, though not necessarily of the most extreme form. For instance, see pp.xxi,21-4,29-30,52,63-5,76-8,265 of Fortune, ibid.

generation after generation of sailors and adventurers from the Northern islands. For the two districts are united by an interesting intertribal trade, the kula. In the past as now the Trobrianders sailed year after year to the southern district on kula expeditions. They regarded the Dobuans as their envied superiors in some ways, as despised barbarians in others — the Dobuans who ate man and dog, but could produce more deadly witchcraft than anyone else; who were mean and jealous, but could fight and raid till they held in terror the whole koya. The Southerners were to the Trobrianders their partners and competitors, their foes and also their hosts — this latter in more than one sense, for at times a whole crew of Trobriand sailors were caught and eaten by their southern neighbors." [15]

The objections of Ahrens regarding cannibalism aside,[16] it would appear that the Trobrianders had to develop some form of defense against the more hostile Dobuans, who would periodically raid, steal women and property, and kill their more peaceful neighbors to the north. Perhaps for these reasons the Trobriand natives considered the southern, Dobuan-dominated part of the Archipelago with suspicion: the southern parts were said to be inhabited by malignant spirits, who caused epidemics, a belief not found in the north; Trobriand myth identified the spot where human immortality was lost as being on the lagoon shores of the southern part of the main island; also, the Trobriand after-death spirit world, or "heaven", was found in nearly the opposite direction,[17] to the north-west on Tuma Island where:

"...undisturbed by the troubles of the world, the spirits lead an existence very much like that of ordinary Trobriand life, only much more pleasant." [18]

These feelings of the Trobriand natives about compass directions toward which happiness or evil were to be found appear to be rooted in real historical events, namely the locations farthest removed from where enemy war canoes first appeared on the horizon, and from which the Trobriand shores were first ravaged by other peoples of a more brutal, patrist nature.

But if the Trobrianders acquired their patrist tendencies from the Dobuans, where did the Dobuans acquire their own patrist social structure? This could have come about from an invasion of their homeland in the distant past by other, even more warlike, patrist peoples, or the Dobuan culture could itself have been a late arrival into the region, carrying with it a social structure already patrist in character. Fortune states that the history of the Dobuans is not well known, but cites the work of MacDonald regarding a *proto-Semitic* linguistic connection between the Dobuans and Southwest Asia, near the Saharasian borderlands.[19] Indeed, MacDonald demonstrated a wide variety of *Arabian* words in the "Oceanic languages", including names for parts of a sailing craft, aspects of religious worship, slavery status, and other social characteristics which fit very well with the idea of patrist ocean voyagers.[20] His analysis of Oceanic alphabetic characters in the Malay Archipelago also indicates a strong *Phoenician* element, an aspect of the region which has also been discussed in more recent times by Fell, who argued that the Polynesian languages contained ancient Libyan, Anatolian, and Asian elements.[21]

MacDonald's map of the extent of Semitic linguistic elements in Oceania included almost all of the Indian Ocean, from Madagascar to

15. B. Malinowski, as given in the "Introduction" to Fortune, 1963, ibid., pp.xvii-xviii.

16. W. Arens, *The Man-Eating Myth: Anthropology and Anthropophagy*, Oxford U. Press, 1979.

17. Malinowski, 1932, ibid., pp.147f, 360-1.

18. Malinowski, 1932, ibid., p.361.

19. Fortune, 1963, ibid., p.xi.

20. D. MacDonald, *Oceanic Languages, Their Grammatical Structure, Vocabulary, and Origin*, Henry Frowde, London, 1907, pp.iii,ix-xi.

21. MacDonald, 1907, ibid., pp.4-6; Fell, B.: *America B.C.*, Demeter Press, NY, 1977, pp.180-7.

Malaysia, the Philippines, southern Taiwan, Indonesia, Borneo, the Celebes, New Guinea, New Zealand, and all of the Pacific Archipelagos eastward to Hawaii and Easter Island, with Australia excluded.[22] Murdock has discussed more recent studies which confirm a high degree of linguistic correlation, cultural interaction, and trade across this same broad region.[23]

Of additional interest on MacDonald's map is the *exclusion* of a small region lying at the eastern end of New Guinea from those regions classified as possessing Semitic influence; here, distinct non-Semitic linguistic elements persisted to an extent that he was compelled to point them out on his map.[24] This is precisely the area where the Trobriand Islands are located, islands whose peoples display behaviors of a most unarmored, matristic, *non-Saharasian* character. These observations are entirely in keeping with the theory of the origins of patrism in Oceania as diffusing *from Saharasia after c.3500 BCE*. My World Behavior Map shows most of the area surrounding New Guinea and adjacent islands to be of an "intermediate" matrist-patrist behavior, this is an averaged situation, resultant from the existence of many matristic and peaceful groups like the Trobrianders, and other more warlike and patristic groups, such as the Dobuans. An enlarged close-up of Oceania from the World Behavior Map is provided in Figure 62, showing these intermediate regions as encompassing and surrounding Indonesia, New Guinea, Borneo, the Southern Philippines, and a few other isolated island groups far to the northeast of Australia.

It is also known that a warlike, patrist culture engaging in human sacrifice was present in the southeastern end of Melanesia sometime before 1200 CE. At least one elaborate burial with human sacrifices has been found in the New Hebrides, dating from 1265 CE.[25] Other cultures in Island Asia and Oceania also possessed varying degrees of patrism, in a manner which on the surface appears to be distributed in a nonrandom manner.[26] However, these data have not yet been systematically correlated or geographically identified, as was done for the continental Old

22. MacDonald, 1907, ibid., see "Sketch Map of the Indian & Pacific Ocean", p.xvi.

23. G. Murdock, "Genetic Classification of the Austronesian Languages: A Key to Oceanic Culture History", *Ethnology*, III(4):124-6.

24. MacDonald, 1907, ibid., p.viii, map on xvi.

25. Shutler & Shutler, 1975, ibid., pp.67-8.

26. See various maps of individual cultural characteristics, particularly on infant cranial deformation and genital mutilations in Chapter 5; also see section on "Oceania" in Dingwall, 1931, ibid., and Davies, 1981, ibid.

Figure 62: Close-Up View of Oceania, World Behavior Map

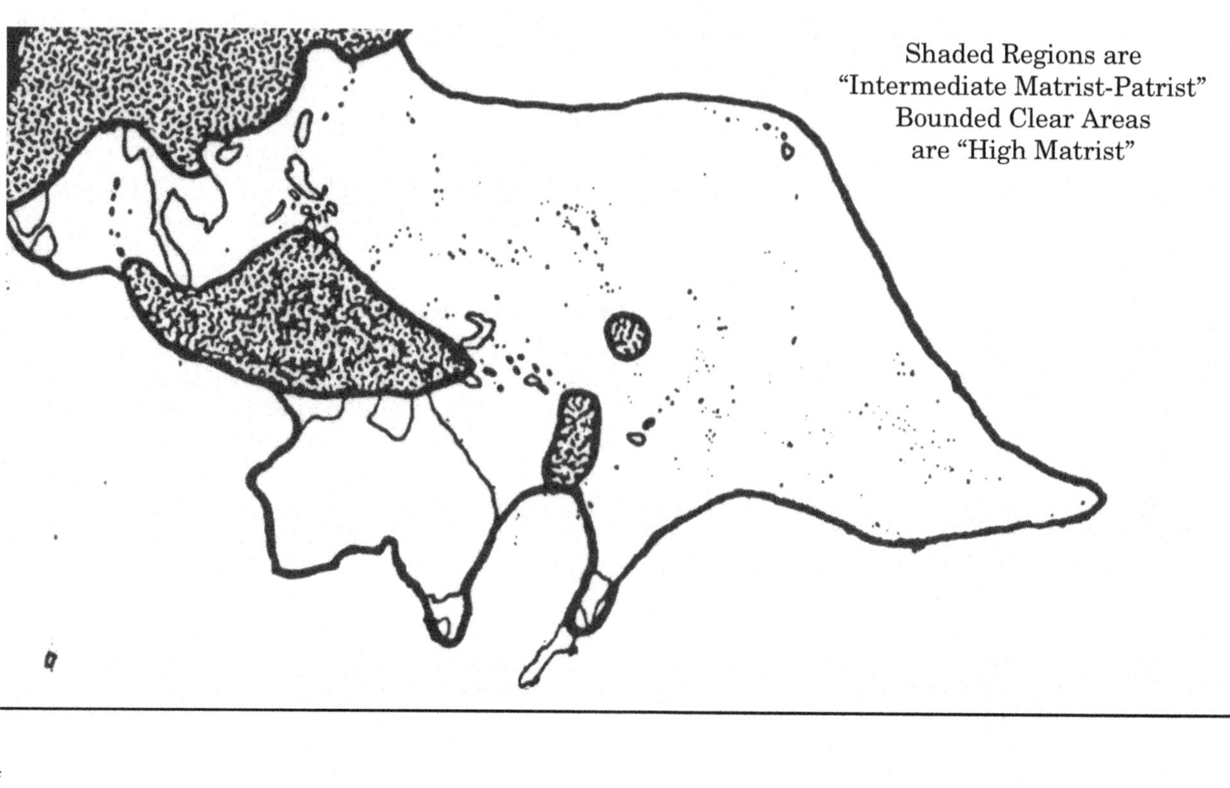

Shaded Regions are
"Intermediate Matrist-Patrist"
Bounded Clear Areas
are "High Matrist"

Peaceful North American natives living in large unfortified apartment complexes (above, Wupatki and Chaco Canyon) were forced to move into defendable cliff-dwellings (below, Montezuma's Castle and Mesa Verde), following the arrival of warlike, nomadic tribal groups in the Great Basin.

27. Discussed in Chapter 8.

28. Dingwall, 1931. ibid., pp.162-97.

29. H. Wormington, *Prehistoric Indians of the Southwest*, Denver Museum Nat. Hist., 1978, pp.59-60.

30. Graham, 1973, ibid., p.451; J.R. Ambler, "The Anasazi Abandonment of the San Juan Drainage and the Numic Expansion", *North Am. Archaeologist*, 10(1):39-53, 1989.

31. Wormington, 1978, ibid., p.40; F. Barnes & M. Pendleton, *Prehistoric Indians*, Wasatch Publishers, 1979, p.48.

World Saharasia. Additional research is needed to trace back in time and space the points of origin of each culture in Oceania to see how their various histories fit within the general Saharasian theory.

From these preliminary sketches and ideas, it can be seen how Oceania might have obtained its patrist cultures purely from diffusion alone. Or, if Oceania were invaded only at one end of a chain of closely spaced but separate islands, a cultural displacement and turbulent interaction could have occurred over great distances, much as was seen among nomadic cultures migrating and warring across the steppes of Central Asia.[27] An invading culture could also have gradually conquered the various island chains, setting up vassal states and alliances. The original patrist tendencies of the invaders would, however, have gradually been diluted through intermarriage with surrounding matrist cultures. Once so diluted, decentralization and breakup of the original forced alliances would have occurred, but lingering traces of patrism would have remained. Social institutions and behavior would thereby have been changed toward that of the invaders, namely patrism, but also would have been diluted with previous matrist culture; the greater the distance from the patrist homeland, the greater potential for such dilution.

These thoughts are offered as suggested points for more detailed research. It is predicted, however, that such research will demonstrate a much later arrival of patrism in Oceania than matrism, with a general absence of patrism before c.3500 BCE, the date of patrist origins in Saharasia; other connections should exist between patrism in Oceania and patrism in the deserts of Australia.

THE NEW WORLD

Saharasian theory suggests that cultures migrating to the New World across the Bering Strait (or by any other route) *before* c.3500 BCE would have been of an entirely peaceful, matrist character. And the behavior maps do indicate that matrism dominated the behaviors and social institutions of most native North and South Americans, with notable exceptions. Distinct patrist tendencies existed among some tribes of the Pacific Northwest, central Mississippi Valley, Mesoamerica, and Peru. Each of these regions acted as a locus of relative militant patrism amid a background of generally peaceful matristic culture.

For instance, patrist cultures from the Pacific Northwest spread infant traumatizing influences such as swaddling and cranial deformation southward into the Great Basin, as well as along the California coast,[28] long before the time of Columbus. Cranial deforming warrior groups moving south in the Great Basin also appear to have displaced the Anasazi or early Basketmaker culture of Arizona and New Mexico.[29] Groups such as the nomadic warrior Athapascan Apache and Navaho, and the warlike irrigator-farmer Mohave and River Yuma tribes, moved on a generally southerly course from uncertain northern regions, harassing, displacing and killing previous peaceful or less-warlike cultures. Similarly, patristic warrior groups such as the Numic speaking Paiute and Ute tribes also spread eastward from Southern California in Pre-Columbian times.[30]

The effects of such migrant patrist peoples on the previous nonviolent, matrist peoples was substantial; evidence of massacres have been found among some groups of Basketmaker peoples who themselves did not make or carry weapons.[31] The migration of patrist peoples into the Great Basin region occurred around the time that a well-developed farming culture crumbled to a lower state of social organization and technology;

the changes which occurred give the appearance of intruding armed, nomadic warrior cultures similar in character to those previously discussed for the Old World:

> *"This new cultural wave entering the southern part of canyon country [Arizona, New Mexico] was still essentially in the Archaic, or non-agricultural, phase. Even so, the tough tribesmen who moved into the territory vacated by the Anasazi were better armed than the peaceful, agrarian Anasazis. Further, the invaders were quite aggressive and warlike, with a curious cultural propensity for 'borrowing' from other tribes. Their borrowing included food, women and any cultural traits that seemed useful."* [32]

Also, at the far south end of the Great Basin, in Mexico, the nomad warrior Aztec became a dominant and bloody power. They possessed many patrist tendencies, as was also true of the Maya and their predecessors.[33] Cranial deformation and swaddling trauma were dominant in the upbringing of children in these Mesoamerican central-state empires, as were certain other forms of ritual mutilations, including incision of the genitals.[34] Many Mesoamerican groups, particularly the Aztecs, were originally migratory within a desert environment at least once in their history. Although I have not sufficiently researched the chronology, full cultural background, or migratory movements of these and other cultures in North America, common elements in directions of migration, subsistence patterns, environments, and behavior are suggested. Other pre-Aztec influences in Mesoamerica appear to have occurred from the Caribbean coast, and these will be discussed momentarily.

North American migrations in the Great Basin appear to have been in two main directions, firstly from north to south, as discussed above. However, Mesoamerican groups of a patrist character also sent influences back northward into the Arizona/New Mexico region, where ball courts and other evidences of contact exist. Mesoamerican influence also pressed northeast along the Gulf coast, up the Mississippi River into the Ohio Valley, where cranial-deforming, swaddling, patrist warrior communities with great earthen pyramids existed.[35] Such cultural influences also pressed into the Caribbean Islands and Orinoco region of South America, where shallow step-pyramids have also been found. These findings probably bore some dynamic relationship with Peru, which also possessed a number of extremely patrist societies, including or especially the Inca.[36] Inca Peru swaddled and deformed the crania of its high-caste infants, a trait which, as Dingwall has shown, was copied in South America, leaving remnant, though diluted practices throughout the Andes region and parts of the Amazon and Orinoco basins.[37]

In all the above cases, however, such patrist influences, which were spread through migration, trade, borrowing, or raiding, were diluted over both time and distance from their points of origin. The most severe forms of patrism in the New World were located in Mesoamerica, Peru, the Pacific Northwest, and Gulf Coast/Mississippi Valley.

The above points, derived from archaeology, history, and the behavior maps, indicate some nonrandom geographic pattern to patrism in the New World. However, such a pattern does not explain how these regions of relative patrism got that way in the first place. If the starting assumptions of this study are to retain meaning for Homo sapiens as a whole, namely that antichild, antifemale, and antisexual attitudes and behaviors cannot derive from biological instinct alone, but are rooted in social or environmental phenomenon, passed on from one generation to the next, then the source of patrism in the New World must be sought in

Mesoamerican sadism and bloodlust was as bad as anything which occurred in the Old World. Above, an Aztec priests cuts the hearts out of living captive prisoners, which are offered to the "god" Chac-mool. Below, a similar rite by a Mixtec priest. Warfare and a hunger for captive sacrifice victims were epidemic in Mesoamerica.

32. Barnes & Pendleton, 1979, ibid., p.20.

33. F. Anton, *Woman in Pre-Columbian America*, Abner Schram, NY, 1973; F. Guerra, *The Pre-Columbian Mind, A Study Into the Aberrant Nature of Sexual Drives, Drugs Affecting Behavior, and the Attitude Towards Life and Death, with a Survey of Psychotherapy, in pre-Columbian America*, Seminar Press, NY, 1971.

34. Dingwall, 1931, ibid., pp.152-9; Montagu, 1945, 1946, ibid.

35. Dingwall, 1931, ibid., pp.183-91.

36. Anton, 1973, ibid.; Guerra, 1971, ibid.; Dingwall, 1931, ibid., pp.152-9.

37. Dingwall, 1931, ibid., pp.201-25.

Above, Aztec form of self-abusive blood ritual, to appease the "god of the dead". Below, an Aztec statue of a man wearing the flayed skin of a sacrifice victim, in the manner of the Phoenicians. Wearing the skin, the torturer-murderer becomes the "god" Xipe Totec.

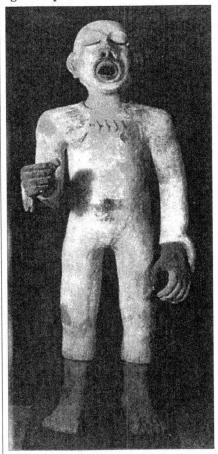

either processes similar to those of the Old World (desertification and famine) or in the physical diffusion of patrist peoples from the Old World to the New. Since patrist traits were found not to significantly exist in the Old World before the desiccation of Saharasia in c.3500 BCE, this implies that *diffusion of patrism to the New World must have occurred after that date, but before the voyages of Columbus*. This, unfortunately, projects the Saharasian thesis, controversial enough within itself, into the debate on pre-Columbian trans-oceanic contacts between the Old and New World. Indeed, the Saharasian maps presented in this work appear to oppose traditional ideas regarding the Bering Strait as the only route of crossing into the New World — the most severely patrist peoples in the New World are not to be found in the region of Beringia, but much farther south, in those very same regions for which various claims have been made by others regarding pre-Columbian contacts with the ocean-navigating Saharasian empires.

Before considering such ideas, however, the issue of desertification in the New World should first be addressed. Desertification as an explanation for patrism in the New World appears to merit study with respect to those groups within or along the edge of the Great Basin of North America, such as the Aztec, Apache, and other nomadic-warrior groups. Likewise, study of the roots of Inca society in the coastal Atacama desert also merits further exploration. And if such an approach would prove fruitful, then the patrism of the Mississippi, Amazon and Orinoco Basin cultures might be explained via a more localized diffusion. However, there are a number of problems which, at least at this preliminary stage of the research, oppose a direct desertification mechanism for patrism in the New World.

The deserts of the Great Basin, like those in Saharasia, appear to have been moister in the past. This might on the surface support the idea of a cultural effects from desertification. However, even today the Great Basin is absolutely wet and lush as compared to the deserts of Saharasia. In the 1980s, I traveled extensively through the United States portion of the arid Great Basin, and was amazed at the variety and quantity of plant and animal life there. In the past, plant cover was even greater in many areas, as was the case with wild game.

In 1980, I had been to the some of the driest parts of the Eastern Sahara, in Egypt, where rains were not even seasonal. A short distance away from the Nile River, vegetation was entirely and utterly absent. However, even in California's Death Valley, one of the hottest and harshest of the North American deserts, vegetation exists along upper canyon slopes and on alluvial fans. In short, the desert regions of North America bear no realistic comparison, in either intensity or size, to those of Saharasia. The analysis of dryness index and carrying capacity given in Chapter 4 confirms these observations, which appear partly applicable also for the coastal Atacama desert in South America. Regarding the Atacama, its coastal and geographically narrow characteristics would have allowed the exploitation of marine resources, or simple out-migration to new areas; famine would not have been a severe problem for peoples living in that area. Hence, a famine-related Saharasian-desertification mechanism for the origins of patrism in the New World cannot easily be invoked. Food is and was too abundant, even in "desert" regions of the Americas.

Even in most New World regions of minimal, scanty vegetation, much of the terrain is dissected by arroyos, ravines, and canyons. These downcut places provide much shade, cooler temperatures, and water storage, along with many forms of vegetation and game not present on the hotter upper plateaus. In fact, native North Americans built homes in these canyons, demonstrating their preference for the more pleasant and food-abundant conditions. Later, when patrist peoples were invading, the choice of canyon homes was favored even more so given their easy defense against attack. Indeed, my observations suggest that there were very few areas one could be in North American deserts where water and game was more than a day's journey away. Contrarily, in North Africa and Central Asia, there are many places where water and food would be more than a one-week journey away, by foot or by camel. This difference is most crucial.

Patrism in the Great Basin does not exist as a ubiquitous trait among all inhabitants of the region, as is the case for most of Saharasia. Only a few of the native North American cultures of the Great Basin were emphatically warlike; others were peaceful. The uneven distribution and wide cultural variance, which was also true in Oceania, would more clearly result from the invasion of warrior groups, rather than from any ubiquitous effect of instinct or climate. Also, the distributions in the Great Basin and Oceania would most clearly result from the invasion of patrist groups who would periodically attack *but not have sufficient numbers to completely overwhelm* the previously existing matrist cultures. It appears that the more peaceful, matrist groups retreated into canyons, or up on top of mesas and buttes. Fortifications were built, battles fought, and rivalries established, but everybody more or less kept their own cultural traditions intact. This is very different from Saharasia, where throngs of militant horse-mounted nomads overran entire regions in successive waves, butchering and enslaving all inhabitants, changing customs, laws, and religions, and establishing relatively permanent kingly states.

These points weaken the idea of desert-induced famine and starvation as the source of patrism in the regions surrounding the Great Basin. Instead, it appears that the nomadic warriors who invaded the Great Basin carried the first traces of North American patrism when they first arrived. Hence, the source of patrism would have to be traced back to the point of origins of those early immigrants, and this is a problem which remains to be tackled systematically for the New World. However, one major clue I have found is the *predominantly southerly migration of these cultures, starting in the Pacific Northwest region.* If a direct desert-culture effect is on shaky ground with respect to the Great Basin, however, it fails entirely as an explanation with regard to patrism in the Pacific Northwest, which is composed of very wet and lush, Pacific coast forests and meadows.

In the Pacific Northwest, infant cranial deformation and swaddling on cradle boards was practiced by groups such as the Kwakiutl, who appear *not* to have had a strong nomadic tradition. Warfare did occur, and slaves were taken; females were subordinated, though to a lesser extent than in the Old World. Cranial deformation was more prolonged and severe for females, however. A direct desertification impact cannot be invoked to explain their patrist tendencies. The environment was coastal forest, excessively wet, and not dry at all.

A similar difficulty exists with explaining the genesis of patrism in the narrow coastal Atacama desert. Like the Great Basin, this region also underwent a desiccation which stimulated movements of peoples. Coastal upland areas dried out, forcing people to concentrate their habitats in fog meadows, coastal river valleys, or in the highlands. Trading centers based upon irrigation and terrace agriculture formed, but were slowly decimated by increasing desiccation, and later by invasions from the highlands.[38] Patrism also appeared in the region, eventually reaching its peak intensity in the highlands with the Inca.[39] Still, it does not appear that patrism developed from the pressures of desertification. Indeed, patrist culture

Mesoamerican pyramids, such as those at Teotihuacan (c.100-600 CE, above) are reflected in the design of similar mound-structures in the Mississippi Valley, the largest of which is found at the Cahokia complex (c.900 CE) where the largest mound is 30m (100') high (below). At bottom is an ancient Mesoamerican ball court, in Arizona. Such structures mark a diffusion of Mesoamerican patristic influences, before the arrival of Europeans, but only within the last several thousand years.

38. E. Lanning, *Peru Before the Incas*, Prentice-Hall, NJ, 1967, pp.4-5,51,56.
39. Anton, 1973, ibid.; Guerra, 1971, ibid.

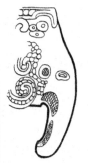

Stone carvings of Indian elephants in Central America (above). Memories from the Old World? In a similar manner, stone carvings exist in India of American maize.
(D. Mackenzie, *Myths of Pre-Columbian America*, Gresham, London 1932)

South America also has its pyramids, built by patristic cultures with possible origins in the Old World. Above, a pre-Inca Mochica pyramid, Chicama Valley, coastal Peru. Below, the colossal interlocking granite-block walls of Inca Cuzco.

40. Lanning, 1967, ibid., pp.57-9; cf. pp.65,77.

41. Lanning, 1967, ibid., p.102.

42. Anton, 1973, ibid.; Guerra, 1971, ibid.; Dingwall, 1931, ibid.; Davies, 1981, ibid.; N. Davies, "Human Sacrifice in the Old World and the New; Some Similarities and Differences", in *Ritual Human Sacrifice in Mesoamerica*, E.Benson & E.Boone, eds., Dunbarton Oaks Res. Lib., Washington, D.C., 1980, pp.211-26.

appears to have literally exploded on the scene in a fully formed manner, as if coming from the outside, as was the case in Dynastic Egypt and Shang China.

For instance, after 2500 BCE, agricultural communities along coastal Peru took a back seat to the development of an Ocean-oriented, caste-stratified, widow-murdering, irrigating, pyramid- and temple-building society; this new highly patristic culture spread out along the coast, and into the highlands. Lanning has described the events:

"The brief period from 2500 BCE to about 1800 BCE saw a remarkable development of many of the main features of ancient Peruvian culture. Most of the coastal valleys were cultivated. The total crop list expanded to include all of the major coastal cultigens except manioc and peanuts. Permanent settlements sprang up all along the coast, and some of them grew to considerable size. Public structures appeared in the form of temples, pyramids, and altars. Technology and art flourished in a number of media. The beginnings of formal art presaged the great styles of later periods. Burials were made in concentrated cemeteries, with the body wrapped in many layers of cloth and accompanied by elaborate grave goods. Skull deformations showed the application of aesthetic standards to the human body. Construction in stone, adobe, and packed clay was common. There are even some hints of the existence of stratified societies and of sociopolitical organizations which transcended the level of the village and perhaps even that of the single coastal valley.

Remarkably, these developments took place among people who were not primarily farmers, but shore-dwelling fishermen. To the best of my knowledge, this is the only case in which so many of the characteristics of civilization have been found without a basically agricultural economic foundation. Equally noteworthy are the speed with which these patterns spread along most of the Peruvian coast and the burgeoning of population that accompanied them. Some thirty Period VI villages are known in the area between Chicama and Nazca, and these undoubtedly represent only a small fraction of the villages that existed. None of the known sites can be dated earlier than 2500 BCE, yet all but one of them had been founded by 2000 BCE or shortly thereafter." [40]

Lanning left as an open question the ultimate resolution of the origins of these rapid cultural changes, and identified other similar ones which further intensified patrist tendencies in Peru, notably that of the Chavin cult around 850 BCE.[41] From our perspective here, however, such observations strongly suggest the arrival of ocean-navigating invaders from another extreme patrist society far away.

Another possible point of origins of patrism in the Americas is indicated in eastern coastal Mesoamerica, particularly in La Venta, home of the Olmecs (c.800 BCE). The Olmecs were the oldest of the Mesoamerican empires, and their cultural traditions can be found in later Toltec, Aztec, and Mayan culture. These groups constructed large pyramids and temples; female subordination, severe child obedience training, a priestly caste, military despotism, human sacrifices, flaying of prisoners, genital mutilations, cranial deformations, and other blood rituals formed important parts of their character and social structures.[42]

As was the case in Peru, this patrist cultural complex could not have arisen spontaneously, from instinct or genes. The late date of the Olmec center invites the speculation that it may have developed from influences

Figure 63: Close-up View of the Americas, World Behavior Map

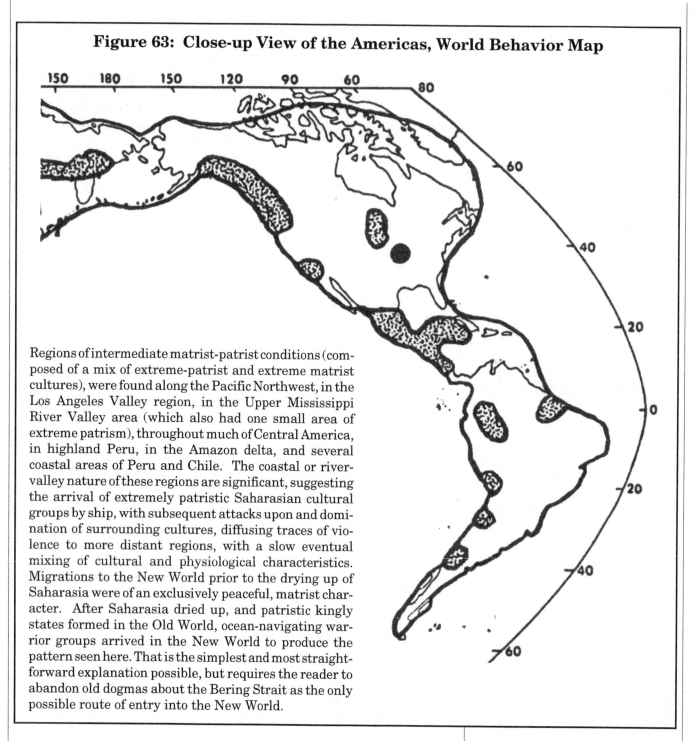

Regions of intermediate matrist-patrist conditions (composed of a mix of extreme-patrist and extreme matrist cultures), were found along the Pacific Northwest, in the Los Angeles Valley region, in the Upper Mississippi River Valley area (which also had one small area of extreme patrism), throughout much of Central America, in highland Peru, in the Amazon delta, and several coastal areas of Peru and Chile. The coastal or river-valley nature of these regions are significant, suggesting the arrival of extremely patristic Saharasian cultural groups by ship, with subsequent attacks upon and domination of surrounding cultures, diffusing traces of violence to more distant regions, with a slow eventual mixing of cultural and physiological characteristics. Migrations to the New World prior to the drying up of Saharasia were of an exclusively peaceful, matrist character. After Saharasia dried up, and patristic kingly states formed in the Old World, ocean-navigating warrior groups arrived in the New World to produce the pattern seen here. That is the simplest and most straightforward explanation possible, but requires the reader to abandon old dogmas about the Bering Strait as the only possible route of entry into the New World.

transmitted over great distances southward from patrist centers in the Pacific Northwest, or northward from Peru. However, it also appears to have developed relatively independently of other American regions.

The above discussions suggest that New World patrism can be traced back to at least three major points of origin, namely the Pacific Northwest, Pacific coastal Peru, and the Caribbean coast of Mesoamerica. In each of the above cases, the cultural transformations which occurred, from matrist to patrist, *must* have come from without and not from "spontaneous" internal cultural transformations alone. Given that desertification in the New World appears not to hold much hope for having stimulated the genesis of patrism, the question remains: Where was the ultimate origin of armored patrism in the New World?

New Material 2005: *Update on Saharasia* (Appendix B) presents Figure 63 with the addition of New World archaeological sites where solid evidence exists for violence and warfare, well before Columbus. There is a close geographical match between my identified regions of high patrism, and archaeological sites showing social violence, warfare and even mass-murder.

The cultural changes in the regions discussed above bear important similarities to matrist-patrist cultural changes which, under the influence of militant Saharasian nomads, occurred in the Old World. A detailed discussion has already been given for such changes in Mesopotamia and the Levant (post 3500 BCE), Egypt during the first dynasty (c.3100 BCE), and Shang China and Jomon Japan (c.2000 BCE).[43] If the basic argument of the origins of patrism in non-instinctual and non-genetic mechanisms is correct, then we must look elsewhere, outside of the Americas, for the ultimate origins of the peoples who first exhibited such patrist behaviors.

I have not researched the chronology of onset of patrism in the Pacific Northwest, but the direction of migration, to the south, suggests that its origins was quite separate from southerly regions. To my knowledge it has not been suggested that these groups came to the Pacific Northwest out of the Great Basin to the southeast. However, it has been suggested that they came from *Asia*, during the Shang or Chou period when patrism was clearly breaking out of the Asian arid zone into China. Both native Americans of the Pacific Northwest and dynastic Chinese of the period subordinated the female, repressed the child, and otherwise possessed a patrist character. They both used backpack cradleboards and swaddling to purposefully deform the crania of their infants. Similarities in linguistics, art design, clothing, drums, and diet between the two regions have also been noted.[44] In particular, Fell has come to similar conclusions, arguing that the Apache and Navaho were late arrivals from eastern Siberia, possessing recognizable Turkmenian roots in their language.[45]

Regarding patrism in Mesoamerica, a connection to Old World Mediterranean groups is possible. The connections between Phoenicia and the West African kingships — both of which were extremely patrist, engaging in human sacrifice and flaying alive of prisoners — have previously been discussed by Fox and others, who also pointed out the possible transoceanic connections.[46] Although these authors argued for a Mesoamerican/Mediterranean connection by virtue of artistic, linguistic, and artifactual evidence, my geographic analysis indicates that they also shared common patrist behaviors and social institutions.

With respect to coastal Peru, as previously discussed, the onset of patrism there appears to have arrived in fully-blown form and its obvious possible connections to Oceania have been argued by a number of authors.[47] Most interesting are the distributions of plant types, such as amaranth and sweet potato, used by Polynesian, South American, and Mesoamerican cultures alike.[48]

Aside from these similarities, and the geographical patterns demonstrated in my behavior maps, there is other evidence which further supports a connection between the patrist centers in the Pacific Northwest, Caribbean Mesoamerica, and Pacific coastal Peru.[49] These include similarities of pattern and design in artwork and pottery, similar myths and historical anecdotes, linguistic connections, and artifactual evidence. The list of such similarities in the diffusionist literature is quite long, and even a cursory review of it would be beyond the scope of this work.[50]

The great distance from the Old World to the New World has generally been seen as a major stumbling block to any diffusionist theory which would try to explain culture in the Americas. The most widely held viewpoint declares that Homo sapiens came into the Americas over the Bering Strait at the close of the Pleistocene, with independent cultural invention occurring ever since. However, it must be said that what once was merely a conservative viewpoint has now become a dogma. Beringia's monopoly on the question of New World origins is too often uncritically cited in textbooks, and young scholars are too often discouraged from seriously looking into the matter. The facts are, simply put, that the ships

43. See appropriate sections in Chapter 8.

44. J. Edwards & K. Johannessen, "Hypothesis of Taoist - New World Pre-Columbian Contacts", *AAG Program Abstracts*, San Antonio, 1982, p.120; O. Von Sadovszky, "The Discovery of California: Breaking the Silence of the Siberia to America Migrators", *The Californians*, Nov/Dec 1984, pp.9-20; S. Jett, "Pre-Columbian Transoceanic Contacts", in *Ancient North Americans*, J. Jennings, ed., W.H.Freeman, 1983; G. Ekholm, "The New Orientation Toward Problems of Asiatic-American Relationships", in *New Interpretations of Aboriginal American Culture History*, Cooper Square, 1972, pp.95-109.

45. B. Fell, *Bronze Age America*, Little, Brown & Co., NY, 1982, p.216; a similar argument has been developed by Von Sadovsky (1984, ibid.), who documented Hungarian linguistic traces among tribes of the North American West Coast.

46. H. Fox, *Gods of the Cataclysm*, Dorsett, NY, 1976; C. Irwin, *Fair Gods and Stone Faces*, St. Martins, NY, 1973; I. Van Sertina, *They Came Before Columbus*, Random House, NY, 1976; A. Von Wuthenau, *Unexpected Faces in Ancient America*, Crown, 1978; W. Smith, *Ancient Mysteries of the Mexican and Mayan Pyramids*, Kensington, NY, 1977.

47. See appropriate sections of C. Riley, et al., eds., *Man Across the Sea: Problems of Pre-Columbian Contacts*, U. Texas Press, 1971; Also Jett, 1983, ibid.

48. C. Sauer, "The Grain Amaranths: A Survey of Their History and Classification", *Annals, Missouri Bot. Garden*, XXXVII(4): 561-632, 1950; Shutler & Shutler, 1975, ibid., p.85; Also see other citations in ref. 50 below.

49. A fourth possible entry point is across the North Atlantic into Northeast Canada. However, it is argued that this route would have less likely been used due to unfavorable ocean currents and winds. Some influences could have been transmitted, however, and B. Fell has made such an argument for both Celtic and Norse influences via this route (B. Fell, *America B.C.: Ancient Settlers in the New World*, Demeter, 1977; B. Fell, *Saga America*, Times, NY, 1980).

50. See next page.

of the ocean-navigating Old World kingships were of sufficient size and seaworthiness to successfully make these long voyages.

The above ideas are offered only for suggestion, and I will resist presenting additional evidence for a diffusionist connection between the Old and New Worlds.[50] However, a map of suggested diffusionist patterns, a first-approximation derived from the behavior maps and archaeological/ historical review developed in this work, is given below in Figure 64.

The maps of behavior provide an unexpected independent verification of general Pre-Columbian diffusionist theories. Artifacts and linguistics have long been used to argue a Pre-Columbian connection between the Old and New Worlds. The fact that the behaviors and social institutions of the distant groups should also reveal common elements, with residues of similar cultural elements lying across the intervening geographical territory, is a most significant finding which cannot be underestimated.

As was the case with Island Asia and Oceania, the materials for the New World have not been subject to as complete and systematic a treatment as the materials for the continental Old World. Although this is a first-approximation of the problem of New World patrist origins, the geographical approach does support the general Saharasian thesis, and is pregnant with possibilities for bringing additional light on the subject.

50. For instance, each of the following works has its own extensive bibliography: Jett, 1983, ibid.; Fell, 1977, 1980, ibid.; Riley, et al., 1971, ibid.; Von Wuthenau, 1978, ibid.; Van Sertina, 1976, ibid.; Fox, 1981, ibid.; Irwin, 1963, ibid.; A. Mallery, *Lost America: The Story of Iron Age Civilization Prior to Columbus*, Overlook, 1951; J. Bailey, *The God-Kings & the Titans*, St. Martins, NY, 1973; C. Gordon, *Before Columbus*, Crown, NY, 1971; R. Jairazbhoy, *Ancient Egyptians and Chinese in America*, Rowman & Littlefield, NJ, 1974; R. Jairazbhoy, *Asians in Pre-Columbian Mexico*, London, 1976; C. Hapgood, *Maps of the Ancient Sea Kings*, E.P. Dutton, NY, 1979; N. Davies, *Voyagers to the New World: Fact or Fantasy?*, Macmillan, 1979; K.W. Butzer, *The Americas Before and After 1492: Current Geographical Research*, Special Issue of Annals of the Assoc. Am. Geographers, 82(3), 1992; J.L. Gardner, *Mysteries of the Ancient Americas, The New World Before Columbus*, Readers Digest, NY 1986.

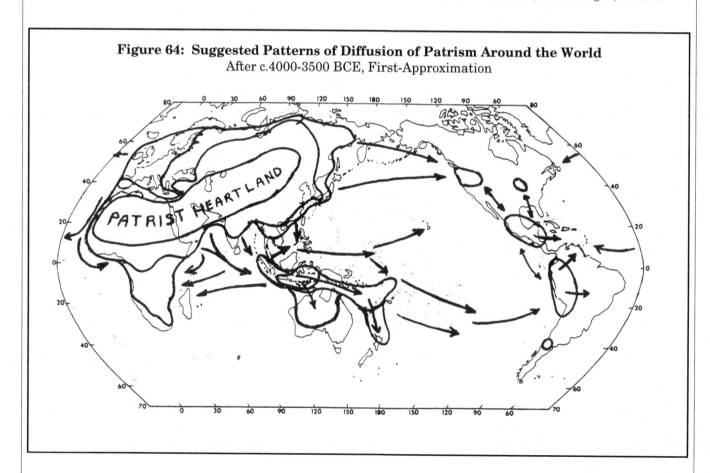

Figure 64: Suggested Patterns of Diffusion of Patrism Around the World
After c.4000-3500 BCE, First-Approximation

Patrist Influences After 1492 CE: The Arrival of Europeans

The affect of European culture upon native Americans, the so-called "Indians" as Columbus called them, was a disaster which lies beyond the scope of this work. However, for clarification it must again be stated explicitly that the maps provided here attempt to evaluate conditions in the Americas prior to the arrival of Columbus. One assumption applied to the anthropological data in the World Behavior Map, is that the scholars who studied the native Americans in the 1800s and 1900s did their very best to present the original, pre-European beliefs and social institutions of those cultures. The fact that so much of the Americas revealed a generally matristic cultural condition so different from the Europeans speaks to the validity of those cultural data — which has in any event already been subject to published critical review within the anthropological community. The fact that other regions of relatively high patrism in the New World correlate to areas with histories of violence and warfare (Central America and Peru), or with other mysteries of pre-Columbian contact theory, is itself enigmatic but also appears not to be accidental. These overlapping qualities further prove the validity of the discovery laid out in this work, of a major global geographical pattern imbedded within the data, revealed only when it is mapped.

Columbus is greeted by the Indians, with friendship
(Engraving from Bartolome de las Casas, *Spanish Cruelties*, 1609)

The enterprise of Columbus, if it occurred today, would without question be called genocide, with accurate comparisons to the atrocities of Hitler or any of the world's most cruel and thuggish brutes.[51] That he may have been a skilled navigator and sea captain will forever be overshadowed by the merciless enslavement, mutilation-torture and slaughter of just under a million Island Arawaks (Taino) at the hands of his followers, all within the few short years after his arrival at Hispanola (1492 to 1518). He approved of it all, and could have put a stop to it at any point, by invoking his authority as "Admiral of the Ocean Sea". He did not object, however, and generations of Europeans who settled the Americas followed similar patterns in their treatments of the native Americans, the most peaceful of which are today totally exterminated, partly from conflict with the European immigrants, and partly from epidemic European diseases to which the natives had no natural immunity. Only the more warlike and patristic native Americans, who fought back against new invaders, survived into the modern era: the Navaho, Apache, Comanche and others. Aside from the "Spanish Cruelties" recorded by las

Montezuma welcomes Cortez
(las Casas, 1609)

51. H. Konig, *Columbus: His Enterprise, Exploding the Myth*, Monthly Review Press, NY 1976.

Columbus' men chopping off the hands and gouging out the eyes of enslaved Taino Indians who failed to meet the quota of gold dust. (las Casas, 1609)

"They made bets as to who would slit a man in two, or cut off his head at one blow; or they opened up his bowels. They tore the babes from their mothers' breasts by the feet, and dashed their heads against the rocks... They made a gallows just high enough for the feet to nearly touch the ground, and by 13s, in honour and reverence of our Redeemer and the Twelve Apostles, they put wood underneath, and, with fire, they burned the Indians alive."

— *Bartholome de las Casas, 1609*

Columbus' soldiers burn Taino Indians alive in their homes, hanging others. (las Casas, 1609)

Casas several hundred years ago,[52] one can travel through large parts of both North and South America and visit places which are named after fully or relatively peaceful Indian tribes of whom only the smallest number of living descendants can be found, or which have altogether vanished from the face of the Earth.[53]

However big an error it has been for all Native Americans to be tarnished as "savages" it would be an equal error to characterize them as wholly peaceful. As identified previously, the dominant Native American groups of Mesoamerica were exceedingly bloody and cruel, predatory upon neighboring tribes. European immigrants often as not made alliances with subordinated Native American tribes, who sought their help with weapons and in battle against the more dominant ones.

The effects of the post-1492 clash of culture upon native North Americans raise other important questions, for surely in the ancient past, a similar culture-clash must have occurred. For example: What happened to local native peoples in these same areas of Central America, Peru, or the Mississippi Valley when the ocean-navigating warriors arrived from the absolutist divine-king states of Saharasia, from Shang-Chou China, or from the Persian Gulf or Mediterranean regions in the centuries between c.3500 BCE to 500 CE? It can hardly be imagined that those long-distance cultural contacts would have been any less violent than was the arrival of the Europeans. If any long-term or repeated contact occurred at the identified pre-Columbian landing-points along the coasts of the New World, then the result upon native peoples would have been similarly disastrous. The Europeans were not the first to bring armoring and patrism into the New World, and by the time the Europeans arrived, many extremely armored, patristic and warlike tribes already existed — large patristic empires preoccupied with war and human sacrifice, and led by blood-drenched god-kings similar in character to those of Old World Saharasia, from whence they appear to have originally derived.

For Oceania and the New World, the patterns on the World Behavior Map clearly appear to be the product of two different and separate sets of migratory influences:

Firstly, influences did come from over the Bering Strait during the years prior to c.8000 BCE, when the polar ice caps were larger and sea levels were lower. And this entry-route persisted along the Alaskan coast even after the sea levels rose at the end of the Ice Age. These earliest migrations, which covered most of Oceania and the New World, were of a peaceful and matristic social character. Patrism simply did not exist in any significant or enduring manner at those times — there is no evidence for it. These early peaceful people spread out and populated the islands of Oceania and the Americas as well.

Secondly, influences came from the Old World after c.4000 BCE, after Saharasia began to dry up, by way of ocean-navigating ships. In this latter case, the new immigrants reflected post-Saharasian conditions of armoredpatriarchal authoritarianism, with social institutions devoted to child-abuse, sex-repression, female subordination, warfare and social violence. Unlike so much of the Old World, however, the armored patristic newcomers to Oceania and the New World did not arrive in sufficient numbers to totally overwhelm their peaceful predecessors, except within smaller sub-regions — notably but not limited to the divine-king central states of Central America (Maya, Aztec, etc.) and South America (Inca). By contrast, the Europeans after 1492 carried terrible diseases for which they were immune, but to which Native Americans quickly succumbed.[53] And they came in sufficient numbers to overwhelm the remnant sometimes-peaceful, sometimes-warlike inhabitants of the Americas.

New Material 2005: See *Update on Saharasia* (Appendix B) for discussion of pre-Columbian archaeological evidence for warfare and social violence in the New World.

Navaho rock painting of horse-mounted Spanish soldiers carrying battle-lances.

52. Bartholome de las Casas, *Spanish Cruelties*, 1609.

53. One of the best summaries of this evidence, of vanished cultures in the New World after Columbus, is found in the work of Charles Mann *1491: New Revelations of the Americas Before Columbus* (Alfred Knopf, NY 2005). Mann argues, persuasively, that most of the deaths of indigenous Native Americans was due to inadvertent transmission of Old World diseases, against which the inhabitants of the New World had no natural immunities.

Part IV:

Review, Discussion and

Saharasia Today (c.1980)

10. Saharasia: Review and Discussion

This work has presented a radical thesis on the origins of child-abusive, sex-repressive and violent, armored human culture — patrism — in an area defined as *Saharasia*, with its subsequent diffusion to adjacent and distant regions. It has been argued that purposeful and socially-demanded infant trauma, child abuse, sex-repression, and female subordination, with their resultant antisocial violence and destructive aggression, developed and spread under a set of specific environmental and cultural processes which began in the Old World desert belt of North Africa, the Near East and Central Asia, starting around 4000 BCE. Some fleeting glimpses of patrism were identified in the centuries preceding 4000 BCE in regions of the Near East also suffering from temporary drought and attendant famine and migratory displacement[§] The year 4000 serves as a general marker for the time when aridity, famine and mass-migrations became relatively permanent fixtures on the geographical and human social landscape of Saharasia. This is factually established — the chapters presented in this work have provided solid and substantive new evidence: *The origins of patrism, defined and equated with patriarchal authoritarian or armored human culture, was originally rooted in the process of ancient desertification, specifically that which began in the Arabian and Central Asian portions of Saharasia. Famine, starvation and mass-migrations related to land-abandonment severely traumatized the originally peaceful and sex-positive inhabitants of those lands, inducing a distinct turning away from original matrism towards patristic forms of behavior. Once established within human cultures by new social institutions, armored patrism spread out beyond the borders of Saharasia by the new, trauma-inflicting social institutions, thereby recreating the older environmental traumas in each new generation.*

The above points are radical departures from conventional thinking, which itself is largely reflective of the general patristic orientation of the academic and behavior-science communities, and the larger armored patristic society in which they exist. A review of the evidence from which these conclusions are drawn is in order.

Review of Prior Chapters

Part I was a survey of the basic starting assumptions from which the overall findings developed. A detailed discussion was given of the roots of this study in the works of Wilhelm Reich, and the social/behavioral theory he developed known as *sex-economy*. The sex-economic viewpoint was described and independently supported from sources in the traditional behavioral science literature. Wilhelm Reich argued that antisocial violence and destructive aggression were rooted in specific, culturally determined infant and childhood traumas, adolescent sex repressions, and socioeconomic processes and institutions which worked to significantly subordinate children to adults, females to males, and ordinary people to class- or caste-structured hierarchies.

Specific *matrist* and *patrist* culture types were defined and identified, in accordance with Reich. A discussion was given of the logic behind Reich's speculations on the origins of armoring in desert regions, and my own research into the question — a link was anticipated between harsh

[§] **New Material 2005**: See *Update on Saharasia* (Appendix B) for further clarification.

389

desert climates and patrist culture. This discussion included a review of behavior in contemporary wet-region and desert-region cultures, and a detailed discussion of the human biological and social effects of long-term drought, famine and starvation. The effects of famine-related trauma upon infants and children, such as marasmus and kwashiorkor, were identified as major mechanisms by which severe drought and famine could substantively change adult behavior and character structure towards armoring and patrism. The influences of forced migrations, land-abandonment and nomadic adjustments, competition for scarce food and water, and of the harsh desert landscape and atmosphere itself were also identified. In short, a number of variables were identified by which the onset of widespread drought and aridity at c.4000 BCE would have destroyed previously peaceful and matristic human cultures, driving them towards a harsh and violent patrism.

Famine severely hurts and damages the child much in the same manner as somatosensory deprivation and disruptions of the maternal/infant bond. Such painfully traumatic deprivation *armors* humans against pleasurable sensation, depriving them of capacity for full contact with the environment or with other people, and furthermore implants a *pleasure-anxiety* which is experienced and expressed whenever potentially pleasurable situations confront such an individual. *Orgasm anxiety* blocks such an individual from full and deeper emotional contact with the opposite sex, preventing full discharge of bioenergy during the sexual embrace. From this comes the reservoir of undischarged internal sexual tension which feeds neurotic, psychotic, and sadistic-violent behavior, so often expressed in a sexual manner.

The biophysical armoring and all its social consequences are invariably transmitted from one generation to the next by new armored forms of behavior and social institutions, which are progressively turned away from pleasurable, biologically determined functions: childrearing becomes characterized by parental anxiety and contactlessness, pain may be willfully inflicted upon infants and children with irrational demands for obedience, the child's sexuality (which produces anxiety in the adult) is generally repressed, and the child's deeper emotional and physical needs are often ignored. *Sexual expressions of the older child and adolescent are routinely severely repressed, and made subject of the most central and repressive, pain-inflicting dogmas and pre-occupations of the sex-frustrated adult world.* This completes the self-reinforcing circle. Armoring is then securely implanted (no matter what the natural environment) and the bioenergy of youthful romance is impossibly crushed and channeled into socially-accepted institutions designed for repression-rationalization or sadistic discharge (fundamentalist religion, the military caste, political hysteria, arena games). This secures the continuance of yet another generation of "well-adjusted" and "socialized" armored Homo normalis.

From the above, the central question of ultimate origins — of how such self-defeating and dangerous modes of human conduct could have gotten started in the first instance — was addressed and hypothesized to have a connection with harsh desertification, famine and forced migrations. A systematic and quantitative testing of this hypothesis was then undertaken, independently, via several different methods.

A preliminary cross-cultural test was made, using 63 different variables taken from Textor's *Cross-Cultural Summary* [1] — a peer-reviewed collection of behavioral variables in a geographically-balanced global sample of 400 different human cultures. A larger *Correlation Table of Sex-Economic Factors* was also constructed using the Textor data (Appendix A). This explicit and systematic approach followed a methodology previously undertaken by Prescott, [2] and both Prescott's and my own cross-

1. R. Textor, *A Cross-Cultural Summary*, HRAF Press, New Haven, 1967.

2. J. Prescott, "Body Pleasure and the Origins of Violence", *Futurist*, April 1975, pp.64-74; also in *Bull. Atomic Scientists*, Nov. 1975, pp.10-20, and *Pulse of the Planet*, #3, 1991.

cultural analysis provided substantial supporting evidence for Reich's sex-economic views. The patristic cultural variables identified or anticipated from sex-economic theory showed from strong to moderately significant correlations with other patristic variables, 95% of which were positive in direction (that is, 95% of the 63 patristic variables showed a positive correlation with the other 63 patristic variables, demonstrating a firm and positive association between them). Only 5% of the selected variables showed negative correlations, and all of those could be explained as products of ambiguous codes within some of the data (as with the "segregation of boys" variable failing to address the "seclusion of girls" in Moslem regions, previously discussed on pages 135-37). In other words, Reich's sex-economic theory described human behavior in that 400-culture sample in a precise and robust manner.

An additional cross-cultural testing was then undertaken using a larger sample of 1170 different cultures, from Murdock's *Ethnographic Atlas* [3] data base. Both the Textor and Murdock data, when plotted on regional histograms, revealed a rough continental shift, from more patristic to more matristic conditions, as one moved outward from Saharasia towards Oceania and the New World.

A *World Behavior Map* was then prepared and presented from the Murdock data base, using 1170 different cultures. This map, composed of social/behavioral data from the more recently gathered anthropological reports, clarified the precise global distributions of patrism and matrism, within relatively recent times, with a minimizing of influences from European colonialism. *Saharasia* — North Africa, the Near (Middle) East, and Central Asia — *was the major locus of the most extreme patristic cultures, world-wide.* The farther one moved from Saharasia, the less patristic and more matristic social conditions became. The Saharasian borderlands showed an "intermediate" condition between extreme patrism and matrism, while much of Oceania and the New World showed up as matristic. The few exceptions to this general pattern appeared as the product of historically-documented migrations (such as patristic Moslem culture transplanted into Indonesia) or as historically-controversial migrations (such as Pre-Columbian contacts in the New World).

The following cannot be emphasized strongly enough: *No other theory on human behavior has such a powerful and systematic explanatory capacity for these same, widely used and peer-reviewed cross-cultural data; only Reich's sex-economic theory has exhibited such a powerful explanatory capacity.* The fact that the highly-structured geographical patterns appearing on the World Behavior Map were later produced from these same peer-reviewed data constitutes yet another strong proof of the overall theory and basic starting assumptions. One does not ever see such highly-structured patterns on maps and in nature, without a causal principle at work. The fact that the World Behavior Map is controversial, that it inadvertently forces into discussion other hotly-debated issues (such as adolescent sexuality, family issues, or pre-Columbian contacts) cannot be helped, and no genuine scholar will shrink from an open discussion of these incredible spatial correlations *which by themselves provide yet another form of independent supportive evidence.*

Part II addressed the more recent historical dimensions of armoring and patrism, with a series of environmental and cultural maps. *Chapter 4* presented a detailed analysis of environmental conditions within Saharasia, showing the pattern on the World Behavior Map was generally matched by the exceptionally harsh living conditions, and by specific vegetational and cultural characteristics of this same large territory. Specific world maps were presented for each of the following variables:

3. G. P. Murdock, *Ethnographic Atlas,* U. Pittsburgh Press, 1967.

Highest precipitation variability
Highest average temperatures
Highest peak temperatures
Highest dryness ratios
Lowest precipitation amounts
Largest contiguous expanse of barren land
Largest contiguous expanse of uninhabited land
Largest contiguous expanse of desert soils
Largest contiguous expanse of lowest primary productivity
Largest contiguous expanse of nomadic herding
Core region for Islamic empires
Core region for Turkish empires

No other place on Earth was found to possess the above characteristics in unison, and Saharasia appears on the maps as the central core of these same environmentally-extreme or historically related features. The spatial pattern on the World Behavior Map was thereby given additional support, within the context of the original theory and basic starting assumptions.

Chapter 5 provided a separate map and discussion of each of the patrist variables used to construct the World Behavior Map, demonstrating generally similar Saharasian distributions. Where possible, each variable was also discussed within the context of the *Correlation Table of Sex-Economic Factors*, rooting the maps into the real-world of observed human behavior. Additional behavioral variables not used in the World Behavior Map were also discussed and mapped, also in *Chapters 6 and 7*, and shown to possess similar Saharasian distributions. Many of the individual variables strongly hinted at specific pathways by which cultural diffusion might have occurred from Saharasia into island groups in the Pacific Ocean, and possibly even into the New World. A number of other cultural variables were discussed but not mapped; in such cases, with further geographical analysis, a Saharasian-related distribution was predicted. The variables mapped and discussed included:

Infant Cranial Deformation and Swaddling
Breastfeeding / Denial of the Breast
Male Genital Mutilations
Phallotomy and Eunuchism
Female Genital Mutilations
Unnecessary Hysterectomy and Mastectomy in Western Hospitals
Scarification of the Body
Female Premarital Sex Taboos
Segregation of Adolescent Boys
Incest and Incest Taboo
High Bride Price Marriage
Marital Residence
Polygamy
Contraception and Abortion
The Couvade, and Similar Practices
Post-Partum Sex Taboos
Homosexuality
Prostitution
Descent-Kinship
Cognitive Kin Groups
Inheritance Rules: Land and Movable Property
Ritual Widow Murder = *Mother Murder*
High God Religion and the *Mother Goddess*

Class Stratification
Caste Stratification
Slavery
Hydraulic Society and "Oriental Despotism"
Contraceptive Plants Used by Native Peoples
Status of Women Index
Contraceptive Use (contemporary nations)
Political-Social Freedoms
Press Freedoms

In each of the above cases, the geographical region of Saharasia with its more extreme patristic conditions could be identified on the mapped data. The behavior maps therefore confirmed and reinforced what was previously presented only in the Correlation Table and regional histograms, and on the World Behavior Map. Each of the individual patristic variables composing the World Behavior Map were shown to have their own individual Saharasian distributions. Additionally, some patristic variables which were not included in the World Behavior Map (such as swaddling and infant cranial deformation) were also shown to have a Saharasian distribution.

Part III addressed the question of ancient origins, reviewing published archaeological and historical data for Saharasia and its Old World borderlands, and for Oceania and the New World as well. The goal was to identify the dates when armored patrism was unambiguously observable in the archaeological record, to identify those world regions which showed the earliest conversion to patrism, and to trace the ancient origins of patrism into the modern era. Also of interest was the environmental conditions in different world regions at the times when armoring and patrism became firstly apparent in the archaeological record.

Chapter 8 addressed these fundamental questions for the Old World. In much of the prior research on the environmental questions, evidence was gathered suggesting the entire contemporary Saharasian desert belt was, prior to c.4000 BCE, a relatively wet and lush semi-forested grassland-savanna, completely different from the modern arid conditions. My research also suggested a similar time of origins for human armoring and patrism, c.4000 BCE, *before which only fleeting, temporary or ambiguous evidence could be identified in published archaeological reports, correlated with equally temporary sub-phases of aridity.*[§] The evidence, as reviewed and assembled from over 100 different scholarly sources, demonstrated a common time and regions of origins, of between c.4000 to 3500 BCE in Arabia and Central Asia, for the first-onset of *both* desert conditions *and persisting* armored patrism. The modern-day spatial pattern of Saharasia, as seen in both the harsh environment and armored social conditions, therefore, had *common regions and time-epochs of origins.* This review of archaeological and historical evidence required several years to complete, and was undertaken for each of the following subregions of Saharasia:

North Africa
Arabia
the Levant and Mesopotamia
Anatolia eastward to the Indus Valley
Soviet Central Asia
Chinese Central Asia

[§] **New Material 2005**: See *Update on Saharasia* (Appendix B) for further clarification.

The review included evidences from ancient glaciers, streams, rivers, lakes, and sands, fossil plants and animals, and ancient art work. General settlement patterns and cultural changes were reviewed for each subregion as determined from published archaeological and historical materials. The literature review confirmed what had earlier been only suggested: a radical environmental change in Saharasia, from a relatively wet semi-forested grassland-savanna to arid or hyperarid conditions at c.4000-3500 BCE. This change occurred across almost the whole Saharasian region, with its most intense and widespread phase starting in Arabia and Central Asia around c.4000 BCE, and proceeding in the rest of Saharasia after c.3500 BCE. The far western and eastern extremes of Saharasia, in West Africa and Chinese Central Asia appeared to dry out last. Minor climatic pulsations back toward wetness occurred here and there, but never with enough strength or duration to interrupt the general long-term trend towards hyperarid conditions.

A general change in subsistence lifestyle occurred in these same Saharasian regions, as desiccation took hold. Cultural groups dependent upon sedentary or semi-sedentary hunting, gathering, fishing, agriculture, and/or pastoralism, were decimated and otherwise forced towards nomadic pastoralism. Complement to these changes were increased competitive tendencies, to include developments in the technology of warfare, and construction of protective fortifications. A form of mounted, militant pastoral nomadism eventually developed across Saharasia, amplified by horse domestication. These events started first in Arabia and Central Asia, from which migrations and invasions spread across the desiccating grasslands to the rest of Saharasia; a pattern of land abandonment and mass-migrations followed the general pattern of desiccation.

The earliest sedentary peoples who lived in Saharasia during its formerly wet period were characterized by a general lack of weapons technology, as well as the absence of infant cranial deformation, ritual widow murder, significant grave wealth, royal tombs, large temples, or fortifications. As conditions became drier and more socially turbulent, these peoples changed in character towards more patristic conditions. Often, this change was accomplished by one culture invading and conquering another — but the militant invading cultures could, in each case, be traced back in time to the original inhabitants of Arabia and Central Asia, who were the first to suffer through dramatic widespread drought and famine conditions, which forced them from their original homelands under culturally traumatic conditions.

These migrating peoples, of a Semitic-Arabian and Indo-Aryan-Asian back ground, were characterized by militancy, strongman leaders, and anti-female and anti-child behaviors and social institutions. Infant cranial deformation, genital mutilations, ritual widow murder, significant grave wealth, royal tombs, large temples, slavery, the caste system, advanced military technology, and tendencies toward extreme violence and organized sadism in warfare were characteristically present and institutionalized in their cultures. Large empires were erected on conquered lands. Militant nomads invariably set themselves up as a ruling class in the conquered Saharasian regions with a stable water supply, notably along the Nile River, the Tigris and Euphrates, and in the moister regions of the Levantine and Anatolian-Iranian highlands. At later periods, similar cultural changes occurred along the Niger and Indus Rivers. Significant hydrological changes, soil salinization and continued desiccation thwarted large settlements in the Caspian and Aral regions, but the wetlands of both Russia and China were finally invaded and dominated by militant nomads.

The archaeology and history of Europe, Russia, sub-Saharan Africa, India, China and Japan were also reviewed with respect to influences they received from various peoples and armies who migrated or invaded out of Saharasia. A pattern of cultural change was observed in these borderland regions which possessed the following characteristics:

A: Before the onset of desiccation in Saharasia, the peoples in regions outside of Saharasia were similar to those inside Saharasia, namely *without* significant weapons, fortifications, royal tombs, cranial deformation, ritual widow murder, and other aspects of extreme or even mild patrism.

B: Following the onset of drier conditions in Saharasia, migrations occurred wherein cultures of an increasingly militant, nomadic character developed. Saharasia proper was incessantly overrun and criss-crossed by various nomad warrior groups who struggled with each other for dominance of the desiccating landscape. Migrations also occurred which were almost uniformly from the driest regions of Saharasia toward the wetter borderland regions, or toward regions with relatively secure water supplies. Peoples in the wetter regions were increasingly displaced, provoked to defend themselves, and/or conquered by the militant nomads. Hence, a military apparatus along with strongman leaders, weapons, and fortifications eventually appeared in the wetter regions. New empires were founded in the wetter regions by the invading peoples, who constituted a ruling class. Examples here are:

1) the Battle Axe, Kurgan, Scythian, Hunnish, and Turko-Mongol peoples who invaded and transformed Europe (Central Asian origins).

2) the Shang, Chou, and later Mongol peoples who invaded and transformed China (Central Asian origins).

3) the Indo-Aryan, Hunnish, and Mongol peoples who invaded and transformed India (Central Asian origins).

4) the Semitic, Arab, Indo-Aryan, Turkic, and Mongol peoples who invaded and transformed much of the Near East (Arabian and Central Asian origins).

5) the Semitic, Arab, Asiatic (Hyksos, etc.), and Turkic peoples who invaded North Africa and parts of sub-Saharan Africa (Arabian and Central Asian origins).

C: Over time, through intermarriage and mixing with original peoples of the wetter Saharasian borderlands, the migrant-invaders from Saharasia lost their emphasis upon nomadism, and tendencies toward militancy weakened in some cases. However, where new Saharasian armies continually pressured and invaded the borderlands, authoritarian central states continued to grow. The new kingly states of the Saharasian borderlands grew in size and power, and control was often extended over portions of the adjacent Saharasian desert, as well as over other wetter regions even farther away from Saharasia. In this manner, cultural influences which originated in Saharasia were transmitted to distant lands and peoples. The Imperial Greek and Roman states, Vikings, Mauryan India, Shang state of China, the Jomon and later empires of Japan, Bantu kingdoms of Africa, as well as the Germanic and Norman peoples of Europe all acquired their infant and child traumatizing, female suppressing, sex-negating, and violent armored patrism through a process characterized by these same long-distance invasive and migratory elements. And each, in turn, further spread their acquired armored social institutions to even more distant regions.

D: Lack of a written history, as well as an insufficient archaeology have inhibited addressing many events in sub-Saharan Africa. However, the diffusion of the Bantu iron culture, with its male-dominated family structure similar to that of Saharasia proper, appears also to have had a similar Saharasian source and influence.

E: Unarmored matrism persisted in a few major regions of the Saharasian borderlands, notably in Minoa (Crete, Santorini, until c.1700 BCE) and Harappan areas (Indus Valley, until c.1800 BCE). Both of these cultures had high levels of technology and social organization, but little direct evidence for armored patrism (weapons of war, fortifications, cranial deformation, royal tombs, grave murder, etc.) except in the last periods when they came into conflict with Indo-Aryan and/or Semitic invaders — which in any case marked the end of their civilizations. The very existence of these well-developed cultures, in and of themselves, refutes the worthless assertions that "technology and the central state are incompatible with a freer sexuality", or that "technology proceeds faster under the strong hand of military organization".

F: Unarmored matristic cultures lasted into modern times only within a few very isolated regions, away from migratory pathways and trade routes. Mostly they are documented in anthropological reports. Notable examples are the Muria of the Indian rainforest, the M'buti Pygmy of the African rainforest, and the Trobriand Islanders of Oceania. Other matristic groups exist, mostly inhabiting isolated Pacific Ocean islands, and among natives of the Americas. Geographical isolation from deadly patrism appears to be the major reason for their persistence into modern times.

G: The relatively recent Old World distributions of the Islamic and Turko-Mongol states appear to have been but the most recent phases of a migration-invasion process of great antiquity, rooted in Arabia and Central Asia. The same can be said of the large communist states of the Soviet Union and China, whose totalitarian aspects were not significantly different from what preceded them in history. The Soviet Union has broken down into smaller regional nations, but the areas closest to Saharasia remain the most patristic and politically authoritarian. China remains a totalitarian communist state at the time of this writing, though the areas farthest from Saharasia appear to have the greatest potential for democratic and freedom-oriented social reforms. Events in World War II, with the growth of fascist power in Germany and Japan (merging Saharasian brutality with the technological innovations and bureaucratic efficiency of the West), also appear as relatively recent consequences of this same Saharasian process.

Chapter 9 extended the above analysis, on the origins and diffusion of patrism, into Oceania and the Old World. The archaeological and historical review undertaken for these more distant regions was not as systematic or detailed as with the Old World, but a number of central considerations were raised as to how various patrist cultures and attributes arose in those regions very distant from Saharasia.

A final map of the global diffusion of patrism was prepared, as a first-approximation, developed from the World Behavior Map and other maps of individual cultural variables, and from a review of historical materials on ancient settlement patterns and migrations. Several major points of evidence supporting the patrist-diffusion hypothesis can be summarized:

A: The Saharasian cultural complex appeared in many areas of the New World in a sudden manner akin to that which occurred much earlier in dynastic Egypt, Shang China, and Kurgan Europe: a widow-murdering, temple-building, divine-king worshiping, male dominated, military caste society suddenly appeared in coastal Peru and Mesoamerica on top of cultural layers almost or completely *devoid of similar patrist characteristics.*

B: Some of these same patrist peoples emphasized shipbuilding and connections to the ocean, and possessed artifacts which cultural diffusionists have long argued came from across the ocean, from regions within Saharasia.

C: Linguistic links existed between Saharasia and Oceania containing a Semitic characteristic. Linguistic traces were also found between Central Asian groups and groups in the coastal Pacific Northwest of North America. It was argued that the patrist characteristics found in Oceania and the Pacific Northwest were transmitted by the same cultural groups who also left behind traces of their language.

D: Infant cranial deformation and swaddling were found among peoples of the Pacific Northwest and elsewhere in the New World, argued to have come from Saharasia where such practices were in more widespread use. Diluted forms of genital mutilations also occurred among isolated Central and South American groups possessed of a patrist, "Saharasian" cultural structure.

Discussion

Taken together, the prior Chapters demonstrate the earliest appearance of armored patrism within Saharasia, with a subsequent diffusion to distant lands. An unbroken chain of events has been demonstrated, from those earliest times of the first-appearance of patrism, up into the modern era. In this manner, the ancient archaeological and historical materials are connected, logically and spatially, with the more recent cross-cultural, anthropological data. Modern expressions of patrism are thereby shown to have ancient connections to a process which began c.4000 BCE, but no earlier.

Ancient Innate (Primary-Unarmored-Matrist) versus 6000-Year-Old Acquired (Secondary-Armored-Patrist) Behavior

From what has been given, it logically follows that the various armored and patristic cultural characteristics (originally identified in Part I) cannot have been the product of "spontaneous independent invention" or any immutable "genetic" form of inheritance. The only way such a mechanistic causal viewpoint can now be maintained is to ignore the cross-cultural and spatial components of human behavior and history, Saharasian in nature with strong elements of cultural diffusion, as presented in this work.

The deepest layer of archaeological materials, like the deepest layer of the human character structure, has been demonstrated to be of a nonviolent, egalitarian, child-affirmative, and sex-affirmative character. Such is what might be expected from self-regulated emotional patterns following the pleasure-principle, as first articulated by the young Freud, and later carried forward by his student, Reich. The primary core drives, as Reich described them, are self-regulated, pleasure-directed and matristic in nature, with distinct survival advantages; *they are innate and inherited*, reflecting a long-term evolutionary process. By contrast, chronic armoring, the secondary drives and social facade, as described by Reich, develop only as *later biological reactions* to prolonged trauma and repression.

A description of human sexual behavior among the earliest hominoids, as derived from physical anthropology and primatology, has been given by Fisher which also is in good general agreement with Reich's sex-economy and the Saharasian discoveries,[4] namely, *the structuring of family units based upon mutually pleasurable biological impulses, which are also egali-*

4. H. Fisher, *The Sex Contract: The Evolution of Human Behavior*, Wm. Morrow, NY, 1982

tarian, sex-positive, and child-positive in nature. Her descriptions of early family life are in agreement with the earliest of archaeological materials reviewed in this work, and therefore provide a bio-evolutionary basis for the natural origins and organization of matrist culture, worldwide.

A theory advocating biologically-inheritable components to behavior would therefore not be entirely at odds with the Saharasian discovery, *so long as the inheritable portions are limited to child-nurturing, sex-positive, pleasure-supporting, non-violent behaviors, of an exclusively unarmored matrist character.* It is absolutely clear, however, that genetic theories of inheritance can explain neither the genesis of the various facets of armored patrism, nor the demonstrated *Saharasian* historical-geographical patterns in human psychopathology given here. Armored patrism exists as a surface layer only 6000 years old, passed on through the generations via socially-approved trauma and repression, and imposed upon a much older and deeper foundation of unarmored matrism which was both worldwide and pan-cultural in ancient times. Matristic remnants of unarmored social conditions are clear and apparent on the various maps, particularly in those regions farthest removed from events in Saharasia.

Historically, unarmored matristic ways of raising children — natural birth, absence of genital mutilations, breastfeeding on demand, a positive view of childhood and adolescent sexuality, etc. — imparted distinct survival and health advantages to the evolving Homo sapiens. There was support and protection of the bonding functions between newborn babies and mothers, the developmental needs of children, and the bonds of romantic love and mutually-pleasurable sexual attraction between male and female which keep families together in a non-compulsive manner. From these pleasure-directed biological impulses developed other cooperative tendencies, as well as life-protecting and life-enhancing social structures. Technological innovations along peaceful lines also proceeded in those earliest of cultures, who were neither "primitive" nor "backwards" in any sense of the word. This has all been discussed and documented in the various chapters. The various pro-child, pro-female, and pleasure-oriented behaviors, have been demonstrated to occur in more recent times predominantly outside the bounds of Saharasia, though they once were present in stronger form both within and outside the geographic area of Saharasia, before the great drying-up occurred. The patrist behaviors, attitudes, and institutions which developed within Saharasia after it dried up are best explained as late biocultural adaptations and responses to the pressures and trauma of desertification and recurring famine.

These points have been demonstrated here in the cross-cultural analysis, in the spatial distributions plotted on maps, as well as in the archaeological and historical materials. The data used for the study were composed in tabular form by top-notch scholars, then published in mainstream journals and peer-reviewed for accuracy. They reflect cultural material contained in every major university library around the world. My role was to apply a set of powerful basic starting assumptions to the data, and to map them. The patterns on the map then emerged quite spontaneously. A subsequent review of archaeology and history later provided necessary confirmation.

Saharasia stands as an independent line of argument for the genesis of armored patrism in environmental, bio-cultural processes, originally stimulated by widespread climate transitions toward aridity and famine conditions in ancient times. The climate-change which created the Saharasian desert belt was *the greatest single global environmental change and catastrophe to occur since the end of the Pleistocene Ice Age,* and it had profound consequences for humans and their developing cultures.

A Few Predictions

Several predictions have also been made, based upon the overall Saharasian thesis. First, it is fully anticipated that further research on the question of patrist origins in Oceania and the New World will vindicate and extend most of the points given in Chapter 9. Patrism either developed wholly within Saharasia, diffusing to Oceania and the New World in later centuries, or possibly some aspects of patrism may have developed independently through *similar mechanisms* within the deserts of Australia, or possibly in the arid Atacama or Great Basin regions of the Americas.[§] In either case, the general Saharasian theory is confirmed. Second, a Saharasian-related distribution is expected to be found when global maps are constructed for the following variables: prostitution, homosexuality, infanticide, couvade, ritual widow murder, scarification of the body, and female figurines. A specific discussion of those variables, and the manner in which the data should be structured, is given in the respective sections of prior chapters. Third, suggestions have been given for restructuring of world data on demography, status-of-women, and other freedom-oriented factors, such that Saharasia would show up even more clearly on contemporary maps. With the break-up of the old Soviet empire, cultural data will be gathered separately on the diverse new independent nations along the southern part of Russia, and new cultural maps should become available in the next several years. A discussion of the anticipated distributions, and the manner in which data should be handled on each variable, has been given in the appropriate sections of Chapters 5 and 7.

[§] **New Material 2005**: See *Update on Saharasia* (Appendix B) for presentation of newer archaeological findings which indicate the temporary development of armored patrism and social violence in North Africa and Australia, during very early sub-phases of aridity.

Armored Patrism and Environmental Destruction

Another important question which this work has only touched upon is the influence of the human emotional desert of patrism upon living nature itself. This work has focused upon changes in human character and social structure as a response to the traumas of desert-spreading and related famine, starvation, mass-migrations and conflict. However, the other side of the antithesis requires future research. The desert landscape can also be created, maintained, and spread by human land-use practices which themselves are related to the patrist social structure. Just as patrism and armoring works to kill the biologically necessary spontaneous and self-regulatory elements of human social living, it also kills or assaults everything which is wild or uncontrolled in nature. Armored "civilizations" are, at this very moment, often destroying the beauty, bounty and grandeur of wild nature, either for the sake of "nation-building", or simply for short-term financial "profits". Overpopulation, itself the product of the patristic suppression of contraceptive knowledge, is a fundamental aspect of this problem, and ecosystems destruction is often at its worst in lesser-developed nations with massive populations. As a species, we are destroying the life-layer of the planetary biosphere — age-old forests vanish in favor of sparse grasslands, which then are attacked and converted into scrub and desert; toxic materials are spread into the air, water, and even on our foods — in so doing, we expand the size of already existing environmental deserts, and become more emotionally dry and deadly as a species.

It is clear that a given culture's view and treatment of the *natural environment* is rooted in the same mass characterological factors governing the way we treat our infants and children, the way we were treated ourselves in infancy. While an unarmored or less-armored individual may stand in the middle of a vast and ancient forest, or on a pristine shoreline or mountain-top and feel a powerful and expansive bioenergetic connection to nature, a feeling of being part of the larger web of life, another more armored individual feels only anxiety in such a situation, or they feel nothing at all, and may even start to calculate the amount of money which can be made from cutting down all the trees, or "developing" the last untouched spot on the coast or mountain range. These issues have been discussed more fully in the section on "High God Religion" — why one culture views nature with awe and respect, while another has only destructive poisonous contempt for the natural world. Hierarchical social structures established by extreme armored patristic groups tend to place male gods and men at the topmost positions of importance, while women, children, social groups lacking economic power, wild nature and non-human animal life are relegated to the bottom of the scheme of things. A negative perception of environment, and subsequent massive environmental destruction, are concrete aspects of patrism and armoring. They exist within a given social structure in direct proportion to the amount of trauma and repression experienced by infants, children and adolescents. Cultural groups (and individuals) which tolerate and even encourage free expression of pleasure-directed biological impulses, which celebrate and protect mother-love and sexual love, throughout history have always been the ones which viewed wild nature and nonhuman animal life with a deep respect and feeling of kinship. By contrast, those who have been most deeply infected with the *emotional desert of patrism* (or *emotional plague*, as Reich identified it in extreme cases) appear compelled to transform all of lush and wild nature into a *geographical desert* as well. This process has, in turn, reinforced the emotional desert of patrism, and a vicious cycle continues, century after century. Saharasia created a new kind of Homo sapiens, a dried-up life- and love-hating specimen who is emotionally better adjusted to existence within a toxic deforested wasteland than within the throbbing and pulsing web of life. We are caught in a suicidal symbiosis — an *emotional conspiracy* — with the desert itself, expanding and enlarging the deserts even into our own emotional and social structures. *Homo normalis,* the ape who thinks more than feels, is busy turning the only living blue planet in the solar system into toxic, lifeless sand and dust! There is far more than metaphor to these relationships.

The Problem is Both Men *and* Women

It has been said that the "action" in a given culture is not to be found in the big noise of politics or in events focused upon by popular media — it is found in the tents, huts, or houses where families reside. This study has peered into the huts, indeed, into the bedrooms of various cultures, finding that where armoring and patrism exist within character structure and family structure, they are mirrored in social institutions and state structure as well.

But the problem is surely more than just that of "dominating males" — it is both males *and* females who have absorbed into their guts the same antipleasure, emotional/cultural structure. Both fathers and mothers may become contactless parents, whose "care" for new life is defined less by the biological needs of children than by cultural traditions which demand the exacting from youth of a certain quantum of pain, and the compulsive stifling of spontaneity, enthusiasm, and passion. Cranial tourniquets, swaddling bands, genital mutilations, sex-segregation of children and adolescents, and strict methods of "obedience training" are heaped upon infants and children by both father and mother. However, the infant is more biologically bound to its mother during the first years, and painful punishments directed at it during those formative years more clearly and prejudicially shape its attitude towards the female.

It would be a mistake, therefore, to view this work as a blank-check of support for any social movement which views the female as a wholly innocent victim of male aggression, or which seeks to rectify the pattern of socially dangerous patrism through only "curbing male violence" or economic restructuring favoring females with children. Yes, for the most part, this must be done, and already has been done in many nations, but other key social factors demand addressing. In those regions where women are the most severely repressed, where the genitals of young girls are attacked with razors, and where rocks are thrown at unveiled women daring to go out in public, it is the women themselves who are most directly involved, as genital-mutilators or rock-throwers. It is also shocking to hear some feminists apologize for or even openly supporting male circumcision — this attitude itself reveals a certain castrated rage directed towards men, perhaps as vengeance for their own damaging experiences — it suggests how so many otherwise rational and decent women in our own armored culture are caught up unconsciously in the very activities which virtually insure the perpetuation of violent armored patrism into the next generation.

As wives, mothers and sisters, women play a vital role in supporting their men to undertake nationalistic wars of aggression against other countries; the softer male who rejects the macho, warmaking attitudes of his patristic culture will be despised and isolated by the majority of both men *and women* in his home culture. In a few of the more extreme patristic groups, females play a central role in demanding their men *kill other women*, that they undertake "honor" murders, to punish other women who shrug off the deadly grip of patristic village life. Both men and women in every patristic culture also have their respective "gossip systems", whereby nonconformists of either sex are ridiculed, shamed and isolated, or otherwise brought to obey the cultural norm, or finally brought to severe punishment. Clearly, however, it is females who suffer the most in all these contexts, even if they are active participants in the completely cruel and insane antisocial norm.

For a world-wide feminist movement that wishes to gain economic and social equality — or to secure unharassed reproductive rights, or to have the simple pleasure of walking down the street without a burqa or veil, without being harassed, raped, or butchered by some traumatized and sexually thwarted madman — an understanding of the sex-economic dynamic, and the role which mothers, fathers, hospitals, schools and religious institutions play in it, is absolutely essential. These concerns are essential also for a world which hopes to tame the forces of fascism (political left or right wing, or Islamo-fascist variety), which are but the institutionalized, organized expression of mass characterological pathos.[5] We must never lose sight of the fact that it is *primarily men* who carry out so many acts of physical violence — the beatings, raping, murdering, honor-killing, acting as policemen, military, etc., and as a first priority this must be condemned and stopped — but the larger problem is truly one of *both men and women*. Unconscious elements of fear and rage underlay the surface-facade attitudes of both men and women regarding sexuality and childrearing, though in this latter area women generally play a more central role.

5. W. Reich, "The Biological Miscalculation in the Struggle for Human Freedom", *Int. J. Sex-Economy & Orgone Research*, 2:97-121, 1943; reprinted in *J. Orgonomy*, 9(1):4-26 & 9(2):134-144, 1975.

A Way Out of the Trap[§]

Any concerned individual who asks "What Can Be Done" only has to review Table 1, given on page 5, and consider what they can personally do to move themselves, their family, and the local community farther into the "Unarmored Matrist" side of the equation. It is not an easy task, but one can start small, within one's own sphere of influence: personally, at home, and at work. Educating one's self, family and friends on matters of sex-economic importance is often a time-consuming and difficult task, but is unavoidable and centrally-important.

Protection of the newborn against mutilations and trauma, and of the rights of children and adolescents to a safe, self-regulating, sex-positive upbringing, are — in addition to securing the rights of adults to genuine self-determination — the most essential components for long-term social change; without them, all efforts to expand and safeguard human freedoms are doomed to fail. In the Western world, major reforms of schools and hospitals are necessary, in addition to general changes in the treatments of infants and children at home. Barbaric practices of all sorts, lacking in any significant scientific basis, take place every day in nearly every Western hospital, in spite of other helpful things which modern medicine can offer. Such practices in the West, generally undertaken by male doctors and aimed at destroying the sexuality of women and children, have already been described and criticized. In this context, the profession of empirical-midwifery has always acted as a remedy, and there is a profound need for gentle and matristically-oriented physicians to add their voices to the criticisms of modern hospital medicine.

Authoritarianism in the school system also must be restrained, and the psychiatric community held accountable for its irrational and scientifically unjustified drugging of school-children. AIDS hysteria is also the leading anti-sexual brain-wash of the century, and education of ordinary people and youth about the widespread "official lies" and unscientific nonsense about AIDS is of inestimable value in liberating human emotion and sexuality.[6] None of these large and tough problems will be easy to change, but change they must, if patrism is to be restrained and defeated.

We also must become more willing to confront hate-filled authoritarians in our home cultures, whether it is someone beating a child in public, a preacher-pedophile in the community, bearded mullahs demanding their "rights" to polygamy and the veil, or plaguey journalists and politicians working to amplify patristic tendencies through rumor-mongering and pestilent legislation. Active citizen's groups have been historically essential in restraining patrism and restoring matrism in the West, but there now is a pressing need to carry these reforms directly into the more highly armored and totalitarian cultures, within Islamic Saharasia.

The Saharasian discovery, which grew from the fertile ground of Wilhelm Reich's sex-economic discovery, is explicit about what specifically can and must be done to make the world a more rational and peaceful place. We now know from where and when the insane irrational anti-child, anti-sexual violence of Homo sapiens began, how it spread around the world, and its contemporary expressions. It all started in Saharasia, some 6000 years ago, as described here, in painful detail. Let us hope this new knowledge further illuminates "the exit out of the trap".

> *"Wherever we turn, we find man running around in circles as if trapped and searching for the exit in vain and in desperation... The trap is man's emotional structure, his character structure. There is little use in devising systems of thought about the nature of the trap if the only thing to do in order to get out of the trap is to know the trap and to find the exit... The exit is clearly visible to all... yet nobody seems to make a move towards the exit, or whoever points towards it is declared crazy or a criminal or a sinner to burn in hell... It is the basic evasion of the essential which is the problem of man. This evasion and evasiveness is a part of the deep structure of man. The running away from the exit out of the trap is the result of this structure of man. Man fears and hates the exit from the trap. He guards cruelly against any attempt at finding the exit. This is the great riddle.*
> — Wilhelm Reich, *The Murder of Christ*.

[§] **Note 2005:** These paragraphs were written primarily for the Western democracies and America specifically. However, they clearly do not provide any short-term solution to the exceedingly difficult problem of the Islamo-fascist nations, who are far more patristic and violent, and actively venting their collective cultural rage against the democratic West, and against Jewish, Christian, Hindu, Buddhist and animist populations all along the periphery of Saharasia. They certainly need sex-economic reforms even moreso than the West, but it remains an open question if sufficient numbers of their peoples can break free from the icy death-grip of Islamic ideology. More will be said on this issue in a forthcoming work *Saharasia Since 1900*.

6. See the citation list on page 15.

11. Saharasia Today (c.1980):
Sex-Economic Conditions in Egypt, Israel, and Other Parts of the Islamic World

In summer 1980 I traveled through Egypt and Israel at the hottest time of the year, observing the social relationships between men and women, parents and children, and the vast Sahara desert itself. The materials presented here are extracted almost directly from my field notes, with some post-facto analysis. The subjective element of daily emotional life was my primary concern. Just what did it mean, on a daily basis, to live in a region where freedoms for women and young people were largely nonexistent? What types of day-to-day obstacles did females and youth face in an extremely patrist culture? And how were conditions in Egypt different from those in Israel, a nation derived mostly from European traditions? How did these differences translate into patterns of daily living? Did such cultural differences bear a relationship to the political situation in the region? Some answers and telling observations were made, but the reader should understand the conditions described here were in a state of subtle change; some of what is written here may be less accurate as it was in 1980 — conditions could be worse in some regions, better in others. Most of this report, however, continues to be a generally valid description of social and sex-economic conditions.

Egypt, 1980

I arrived in Cairo on a jet from Paris on 28 July 1980, the significance being threefold. First, it was the hottest time of the year, when one might expect to observe the maximum influence of harsh desert conditions on human behavior. Second, it was the holy month of Ramadan, when all faithful Moslems abstain from eating or drinking from sunrise to sunset. Third, I arrived on the same "secret" plane that carried ex-President Nixon and the funeral entourage of the then-recently deceased Shah of Iran. The Shah, as is known, was overthrown by a popular revolution, and this will be touched upon later. While airborne, someone on the plane made a joke about the Ayatollah Khomeini (sworn enemy of the Shah) planting a bomb on the plane; this prompted absolutely no laughter at all, but did generate some chilly looks from a number of neat-as-a-pin, suited men wearing dark sunglasses, standing in the aisles. I had a fantasy, to walk up the aisle into the first-class section of the aircraft, and introduce myself to Nixon, to inform him of some ideas I had been researching on the subject of desert-greening,[§] and ask for an introduction to Anwar Sadat, the benevolent dictator of Egypt. But this fantasy was restrained by my unwillingness to explain myself to the cold-faced guys in the aisles.

Upon landing, Nixon and his entourage went out the front of the plane to a red carpet 21-gun salute, being welcomed by Sadat. The rest of us were herded out the back of the plane to a three hour wait at customs inspection. Security was especially tight given the antipathy the Shiite Moslems had towards the Shah and his family, as well as toward Sadat, who had supported the Shah and made peace with the hated Israel. All new arrivals, particularly those of foreign nationality, were considered to be potential terrorists by the customs officials — and that included me, a young, back-pack toting student from the USA.

[§] J. DeMeo, *Preliminary Analysis of Changes in Kansas Weather Coincidental to Experimental Operations with a Reich Cloudbuster*, U. Kansas, Thesis, 1979; J. DeMeo, "OROP Eritrea: A 5-Year Desert Greening Experiment in the East African Sahara-Sahel", *Pulse of the Planet* 5:183-211, 2002. J. DeMeo, "Cloudbusting: Growing Evidence for a New Method of Ending Drought and Greening Deserts", *AIBC Newsletter*, American Inst. of Biomedical Climatology, Sept. 1996, #20, p.1-4; J. DeMeo, "OROP Israel 1991-1992: A Cloudbusting Experiment to Restore Wintertime Rains to Israel and the Eastern Mediterranean During an Extended Period of Drought", *Special Report,* Orgone Biophysical Research Laboratory, 1992; J. DeMeo, "OROP Arizona 1989: A Cloud-busting Experiment to Bring Rains in the Desert Southwest", *Pulse of the Planet*, 3:82-92, 1991; also published as a *Special Report,* Orgone Biophysical Research Laboratory, 1991; J. DeMeo & R. Morris, "Preliminary Report on a Cloudbusting Experiment in the Southeastern Drought Zone, August 1986", *Southeastern Drought Symposium Proceedings,* March 4-5, 1987, South Carolina State Climatology Office Publication G-30, Columbia, SC, 1987.

In the airport waiting area were peoples from all over the Near East. Saudis with specially tailored robes and neatly trimmed beards, migrant laborer Yeminis, with grizzly beards and sun-baked skin, black Africans with colorful robes and fez, and a collection of American and European students and tourists. About 30 minutes into the wait, a large number of women, many in full-length robe-like veils which covered them from the top of their heads to their ankles, were noisily ushered through the customs portals without inspection while a man loudly announced in English, several times over: "In Islam, women are treated special!" I thought it might have been the harem of some Arab dignitary, but a few of the women were Westerners who had arrived on the plane. In fact, all the women in the customs waiting area were being allowed to leave without delay or inspection. Of course, who could object to such favored treatment for women? Later, other faces of Islam would show themselves. I wondered if one of the women was smuggling machine guns under her veil, as Algerian women did during the 1950s revolt against the French.

Cairo is a city of Mercedes-Benz limousines pushing past donkey carts with wooden wheels. It has one foot in the twentieth century, and the other in 500 BC. It is a city of ancient Saharasian traditions with a Western technological veneer, which in any case is none too deep. Once one leaves Cairo, with its Western-oriented urban elite, the *fellahin*, or peasant farmers of the Nile Valley predominate. Here, along the Nile, many live much as they did a thousand years ago, in mud and thatch huts, from hand to mouth, as farmers. The farmers and working class poor do not generally welcome the social aspects of Westernization, though they marvel at and desire our technology.

Egypt has no social welfare infrastructure to speak of, with only a rudimentary health care system hardly designed to service the private needs of women. In the old days, tribal tradition took care of dispossessed, widowed, or immoral women, by returning them to families, marrying them off to brothers, or killing them. Today, the crush of life in Cairo, as in any big city, breaks apart such family ties which would give women a new place when their old marriage ties disintegrate — but Cairo does give women a place to flee, from potentially deadly family situations. Westernization has also meant an increasingly acted-out sexual interest among Egyptian youth, though not anywhere near that of Western youth. Severe penalties for infraction of the virginity taboo, namely the "honor killing" of young women by their brothers or fathers, appear to continue unabated, at least in the rural areas, though this is only a subjective impression based upon conversations and newspaper reports. Today, if all else fails, young women can escape an unwanted arranged marriage, or escape "honor" vengeance by running away to Cairo. However, large cities are not kind to young runaway girls, no matter where in the world one goes, and in Cairo conditions are only slightly more free than along the banks of the Nile.

Once, I inadvertently walked near a female beggars region in Cairo, where toddlers grubbed in the sidewalk dirt for food, and young mothers begged until 2:00 AM amid the incessant noise of cruising-by taxis. Many young mothers sat cross-legged on the ground, covered with a veil, holding tiny, tightly-swaddled infants between their legs. Their open hand rested palm-up and immobile on one knee; in a long row, they waited for some small favor from passers-by. The tiny infants were swaddled too tightly, or were too exhausted, to cry, but they held a frozen, closed-eye grimace on their faces which was nearly identical from one child to the next. One young woman, not older than 14, walked — if that is the right word —incoherently with her infant, begging. Another gaunt and hungry woman, whose eyes were withdrawn and yellow, slowly shuffled towards me with a half outstretched hand, a tiny infant at her breast. I dug out a few coins from my pocket and pressed them into her hand, and walked away to cry in private. Coming as I had just done from the large meals and comfortable apartments of the USA, this sight was an incredible shock, which set me reeling. Yet to those living here, it was a commonplace sight, and went largely ignored.

It is difficult, nearly impossible, to get people to talk about matters such as the homeless, outcast and downtrodden, for in any culture they represent the living, and not-so-living, social debris or precipitate resultant from the clash between biological needs and culture or economy, what might be called the *dark side of civilization*. To address such problems means to acknowledge they exist, and to acknowledge they exist is to be forced to address their cause, which can in turn bring penetrating, critical light to bear upon emotional attitudes, family histories, world views, and social institutions which people may dearly cherish.

Most local Egyptians only wish you to see the pyramids, or other wonders of ancient times, and taxi drivers are often offended to the point of attempting to obstruct if you try to take pictures of such human tragedy. This impoverished region of Cairo — formerly a cemetery and, as I later learned, called *The City of the Dead* — had been taken over by dispossessed and thrown-out women, runaway girls and unmarried mothers who somehow avoided being killed, and divorced, widowed, and without-a-place-to-go women of the poorer classes. They survived, if one can call it that, by begging, or worse. From their ranks probably came the women who worked Cairo's seamy night clubs.[1] Several people I met considered them lucky, in that the more liberal interpretation of Islamic law in Egypt had at least allowed them to live. Certainly, the West also has its abused and outcast women who, with backs to the wall, do what they must to survive — but nothing like this.

Egypt is more liberal regarding sexual matters than other Near Eastern Moslem states. One aspect of this can be seen in the presence or absence of women in certain public places, or the manners in which the veil is worn, or not worn. For instance, working class women in Cairo may wear colorful veils which only cover the head and not the face. To the rural peasant farmer, of course, this is considered a disgrace. In the rural countryside, only small girls or dark-skinned Nubian women are allowed to go about unveiled or in colorful cloth. In rural public, grown women almost always wear a thick black veil all the time, regardless of air temperature or water availability. Such rural women may allow the veil to fall off the face while walking alone or only among other females, but not off the head entirely. When a stranger approaches, particularly a male foreigner, the veil is immediately pulled up across the face. More liberal rural women may allow half of the face to show, by holding one side of the veil in their teeth, combined with a turning away of the head to conceal the exposed side of the face.

Once, when sharing a large taxi with an Arab family for a nighttime ride across the Sinai desert, a humorous episode occurred. The elder grandmother, covered head to toe in a thick black, carpet-like veil, kept rolling the window down for the cool nighttime air of the desert. To the men in the back seat, including myself, this cold air was absolutely frigid. We wore only short-sleeve shirts and no head covering. The men would complain about the cold air, and she would roll the car window back up. Later, however, she would begin to sneak it back down, bit by bit, until someone noticed how cold it was again in the back seat, and another complaint would be voiced; she would then roll the window back up. This battle repeated itself for the entire three hour trip, but the obvious solution for her to remove the hot carpet-veil from her head was not given

1. For another street-view description of Cairo, see J. Raban, *Arabia: A Journey Through the Labyrinth*, Simon & Schuster, NY, 1979, pp.253-97.

consideration.

As one moves south out of Cairo, the number of fully veiled women increases to nearly 100%, and such conditions persist until reaching the area around Aswan, at the first cataract of the Nile. Here live Nubian peoples of darker complexion, with cultural roots in both Egypt and Sudan. The Nubian women wear only a head veil and do not cover their faces. Their dress is bright and colorful, apparently irrespective of age or marital status. However, it would be a mistake to consider them less constrained or "freer". In the Nubian regions, more severe forms of genital mutilations prevail, including infibulation, which closes over the vagina entirely. Traditional Arab purdah, or seclusion, is not practiced as much in these regions, as the women are needed for work in the fields and elsewhere; purdah is practiced primarily in the urban regions. This mandates that some other method for insuring virginity be used. Perhaps as a result of this, infibulation is widespread in Egyptian Nubia and even more so in Sudan. While Nubian and Sudanese women have more freedom to go out in public, the virginity of young girls is absolutely "protected", by this horrific sewing-up of the vagina.

In the big city of Cairo, conditions are not so severe. Women can work outside the home, and go unveiled if they are brave or wealthy enough, or have tolerant family support. Most of the working class young women abandon the veil while at work, or wear only a partial veil which completely exposes the face and neck. However, they completely veil themselves when they walk outside from place to place, being covered from head to toe, though usually with colorful cloth. Such young women who do hold jobs outside the home, generally secretarial or clerical positions, are often looked down upon by the average man in the street of Cairo.

Such women generally take taxis to work, but may walk or take a bus from time to time. When they walk, it is generally in the street, on the street side of cars parked on the curb, given the fact that the sidewalks are generally crowded with young men. The young men are not "street thugs" in the New York sense, but the sexual hunger of these young men is acute. They have a burning interest in the young girls who observably walk to and from work, but etiquette demands that contact between the sexes be minimized. So the girls walk in the middle of the street, carefully avoiding the young men. In any event, the number of women on the street at any given time, other than in the begging areas or during "rush hour", is very small. But the streets are filled with young men, at almost all hours of day or night. Women are also largely secluded inside the various apartments, which are their major refuge in the city.

The young men on the street are mostly poor and unemployed. They hustle foreign tourists and work for the wealthy Egyptian elite in many ways, however. One will get a taxi for you, another a hotel room, and another a tour to such and such a place, each time taking a gratuity from the tourist, and probably from the taxi, hotel, and tour agency as well. The young men all speak of getting married, of having a home, children, and possibly even a car, which are simple, honest goals. However, the crush of the growing population, and the poverty, dim the chances of the average youth considerably. Jobless and minimally educated for the most part, they are constantly on the sidewalks, where a considerable social and sexual tension permeates the atmosphere. One gets the sense that gasoline had been poured in the street, needing only a spark to set off a social conflagration. Although their interest in women is burning and deep, the inhibition of the men and their outward concern for proper female modesty — plus the constant watching eyes of everyone else on the streets — insures that unauthorized male-female contacts will not take place. The majority will eventually enter into arranged marriages with strangers.

Female modesty and the separation of the sexes are apparent in a number of other ways. As mentioned, women rarely ride the Cairo busses, for the following reasons. First, the busses are exceedingly dirty, dusty, hot, and noisy. If given a choice, the men would not ride them either. But then all of Cairo is rather hot and dusty. It never rains, so the dust and dirt just blows about here and there, coating everything, and never washing away. The dust of the desert covers everything, including skin and clothing, which hangs heavy from perspiration. The buses are additionally crowded like sardine cans most of the time, with flesh pressing flesh, packed three deep in the aisles. If you expect to get off at a certain stop, you must begin edging toward the door at least two stops in advance or it will be missed. A small boy climbs about on the pipe seat supports and handrails, gathering piasters. When the bus slows for the stop, men climb in and out of both doors and windows. I hesitate to say stop, as the bus will pause only for a second before rumbling off in a cloud of soot, with a crowd chasing behind it, trying to get on — or *in* — the same way others got out. Miraculously though, everyone gets a hand or foothold in the doorway, and the bus drives on with a gaggle of bodies, arms, and legs hanging or sticking out. The body-contact, shared-breath, and intimate-smell environment of Cairo's public bus system does not lend itself to participation by females in a culture where unacquainted men and women of the working classes do not openly speak to each other on the street, much less touch.

Still, an older matron or two will ride them during slack hours, and even do so unescorted. These older women, covered with full black veils, push through the crowd with a commanding authority, muttering in Arabic: "Get back! Get back!". They wave a bony finger or implement in front as they advance against the retreating crowd of men, acting like sheep backing down from a lion, and someone soon enough gives her their seat. Once I did see a very lovely young woman get on the bus, with half veil and fully exposed face; but she was escorted by a man who held her fast and tight at the upper arm. He glared angrily at the other men, as if to say "Back off! She's mine!", and was immediately given additional breathing room. The young girl fixed her gaze at the floor the entire time, and the man maintained his grip on her arm, and his angry glare, for the entire journey. This was also the only time I ever saw a young man and woman touching in public, if "touch" is the right word.

Touching in public is largely restricted to those of the same sex. Young men will often walk down the street touching, or go hand in arm or hand in hand. Similarly, women may share intimate body contact with other women, though women on the street are rare in any case. However, one will look in vain for an opposite-sex pair walking or talking as such, except for Western foreigners.[2] Indeed, tourists from all over Europe are in Cairo. They are easily identifiable not only by height and hair color, but also by their more colorful, and scantier clothing. Because of the differences in dress, there is often a turbulence in the working class areas of town when Western women walk through.

Western women not only will go about unveiled, but will wear clothing revealing their face, arms, neck, and legs. Such foreign women may provoke outrage in some areas of Cairo, and be openly spit upon or have rocks thrown at them. This happened to several European women I met during my stay in Cairo. In such cases, it is generally the older matrons, in their full length black veils, who make the loudest protests and start the stone throwing and spitting.

At Karnak and Luxor, sites frequented by busloads of European tourists, I saw a number of European men and women walking about in

2. Raban mentions hand-holding opposite sex couples in Cairo, but this is the only such account I have come across; still, my observations fully contradict his account. (Raban, 1979, ibid., pp.257,273).

clothing designed for the beach. They constantly turned the heads of the locals. Egyptian women in full length black veils would look back over their shoulders at Europeans dressed as such, doing double and triple takes, even though they must have been a common sight at this location. Insults were often shouted at the Europeans by the black-veiled matrons, who must have considered them absolutely naked. At El Arish, on the Sinai coast, the beach was populated by large numbers of Egyptian men in bathing suits. They swam and enjoyed the waters, but Egyptian women did not. They sat on the shore, with full veils, while their children frolicked in the water. The women occasionally waded knee-deep, never removing their full-length veils, but went no further. Only one English woman wore a bathing suit on the beautiful beaches of El Arish during my two days there, and she always attracted a group of friendly Egyptian men wherever she went.

Once I visited a large geographical library in Cairo, which was housed in an old Mosque converted to the purpose. Tradition held that an old Mosque could not be used for other purposes unless some small corner of it was preserved as a shrine, to provide a place of solitude, as well as a home for several greybeard holy men. I entered the building and asked several people about the collection, but encountered language difficulties, as I could not speak Arabic. Eventually, a young girl with a head veil and exposed face, about 16, approached me with several of the older men in tow; they were shouting rather angrily at her from behind. One of the greybeards, as if saying "I give up!", threw his hands into the air and walked off down a hall, resonantly calling out prayers to Allah and the name of the Prophet. The young girl introduced herself, saying that she was the only one in the building who spoke English, and said the men were upset that she, an unmarried girl, should speak in a foreign tongue (which they could not understand) to a non-Moslem stranger.

I thanked her for the courtesy of bending the social customs. We gathered some maps and manuscripts, and walked to a large table, being followed by several of the greybeards. They kept an eye on us for some time, occasionally reciting a verse from the Koran out loud, as if to exorcize the building from my sinful presence. However, they soon bored with conversation they could not understand, and eventually wandered off. The young girl relaxed quite a bit after they left, and conversation drifted away from maps to social questions, hers about America, mine about Egypt. She wanted to come to America to go to college, but doubted if she would ever do so. Her father was a general in the Egyptian army, and was rather traditional about her upbringing, she said, though he had allowed her to go to school. She said he "would kill" her if he knew she was speaking with me. I commended her bravery, and said I understood the situation, though I hoped she was using the phrase in the same sense that an American teenager would, and did not mean it literally.

This young woman was delightful, curious, and a storehouse of cultural information. She exploded many misconceptions I had about women raised under such conditions, though I did take her to be exceptional. She wanted to know all about "dating" in America, displaying much interest about freedom in matters of marriage and love. She had many good things to say about Jihan Sadat, Anwar Sadat's wife, who was working to pass laws favorable to women.[3] She was in many ways just like a young American girl with exceedingly strict parents, who sought to secretly bend the rules as much as possible in order to make life a bit more pleasant. However, in Egypt, the average parent was just as strict, or even more so, than her own.

I asked if she, with or without chaperone, would accompany me to the pyramids some time; this was absolutely out of the question, she said. I

3. G. Steinem, "Feminist Notes: Two Cheers for Egypt, Talks with Jihan Sadat and Other Daughters of the Nile", *Ms.*, June 1980, pp.72-4,96.

did not understand the situation in Egypt. She could not do such a thing, even with an Egyptian man known to her family. She was bethrothed, she said, to a man her father had selected for her, but whom she had never met herself. She appeared to lose her sparkley quality on discussing this, casting a blank stare at the table. But, she continued, the man was educated, and probably would not object to the fact that she had not had "the operation", which I took to mean clitoridectomy. I had not asked her about this, nor even hinted at it; her reference to it was entirely voluntary and spontaneous, as was her interest in the questions of love and sexual relations. I found that such topics burned in the minds of Egyptian youth, in proportion to the extremeness with which they were banished from everyday social life.

Another female informant I spoke with was an Egyptian airline stewardess, a member of the wealthy urban elite. This woman dressed entirely in the Western manner, with make-up, perfume, and clothing which exposed her arms, legs, and neck. She did not wear a veil at all, except, as she said, when the airline stopped in Saudi Arabia. There, she was not allowed to leave the plane, unless totally veiled and accompanied by a chaperone. However, the Saudi national airline would regularly employ Egyptian stewardesses, as, while Saudi men wanted all their own women to be veiled, and forbade them from gaining free access to international travel, they did enjoy the open friendliness of the unveiled Egyptian women.[4]

In Egypt, by contrast to Saudi Arabia, this woman could go unveiled, unchaperoned, obtain a job, fly abroad, and even drive her own car about town without the permission of any man. Ten years before, she said, much of this would have been impossible, and she gave much of the credit to the efforts of Sadat's wife, Jihan. I met her younger sister, also dressed in the Western style, as well as the sister's young fiance, who came over to visit her unannounced while I was there, as if he did so quite often, without prior checking, much as in the USA. They all wished to come to America, either as tourists or to study. The stewardess and I sat and talked at length, unchaperoned at her house alone. She later drove me to a bus stop in public, not wearing a veil while outside, though I did detect some nervousness on her part in doing so. This woman, however, was a rare sight, and I never saw another there who was either as wealthy, free, or brave. She was definitely a member of a tiny minority.

My discussions with men were more numerous, and generally revealed a different attitude regarding sexual relationships. One male informant, a doctor in Cairo, invited me to his house on several occasions, for tea and conversation. His apartment was not unlike many middle income apartments in the West, except for the style of furnishings, which in any event were both comfortable and functional. The doctor's sister served us tea and sweets, but all the women left the room after I arrived, and I never got to speak with them. He called Sadat a dictator, and said he hated all the 'isms', naming them: "socialism, capitalism, communism, fascism". But he loved and wanted "democracy" in Egypt. When asked about the female vote, he said this was "out of the question", that females didn't know enough about politics to know how to vote.

I asked him about circumcision, and he said this was a good thing. It made a man "cleaner", and "more potent", and "cooled a woman off". "Women", he said, "cannot control themselves like men can. You need to remove the clitoris to save the woman from the burning heat". Society would crumble into sexual anarchy if the female sex drive was allowed to fully blossom, I was told. I said women did pretty well what they wanted in America and we didn't have the anarchy he described, but he disagreed with me, citing things he had heard about America in the newspapers or

4. "...It's a sin to drive an automobile in Saudi Arabia, if you're a woman, states Sheik Abdel-Aziz bin Baz, a Moslem religious lawmaker. Women, he says, run the risk of 'falling into incalculable sins' when driving, including 'being bare faced and alone with strangers'. A driving ban is already effective in cities, and if Baz's pronouncement is enforced, it will be particularly difficult for rural women, since hired drivers are scarce and expensive." (*Ms.*, June 1982, p.22).

over TV. Rape, he said, "did not exist" in Islamic countries. I did not challenge him on this, but it was apparent that he considered neither an arranged and possibly forced marriage to a total stranger, nor the economically forced prostitution of homeless women in the slums of Cairo as the equivalent of rape.

An American woman had wanted to marry him, he said, but he refused her, even though it would have meant instant access to America, and American medical education, his lifelong dream. This, I took to be a fanciful boast. Western women, he explained, liked Islamic or Egyptian men better than their own European men as Egyptians were circumcised, which made them better lovers. "And besides, we know how to treat women over here", he continued. Later, under some questioning from me, he confessed that his medical education included neither the basics of female sexuality or contraception, nor anything which contradicted a conservative interpretation of the Koran. Days later, in a moment of honesty, he admitted to being a virgin, saying he was saving himself for marriage.

This doctor was a devout, conservative Moslem who obeyed the strictures regarding consumption of food or water during the daytime period of Ramadan, even risking heat stroke one day when we walked at length on the desert near the pyramids, refusing to drink water though nearly faint. His view of the female, and of sexuality, were among the most conservative of any I encountered in Egypt. Yet he was among the most highly educated of the working class people. His view of the "West" came entirely from movies, TV, radio and the newspapers, as well as through popular gossip stimulated by contact with tourists. His interest in females was great, but the barrier of inhibition was even greater.

On another occasion, I met with four young Arab men who were on their way to morning prayer at the Mosque. I was sitting on a bench overlooking the Nile, recovering from the last stages of the "Pharaoh's Revenge", having two days earlier consumed a bottle of soda pop contaminated with raw Nile water. At the time, I was not particularly interested in talking with anyone, except perhaps a salesman for diarrhea medication. However, they came up and sat with me, uninvited, and began asking questions about where I was from, what I was doing, and so on. I wanted to steer the conversation towards females, and simply asked if they were married. This triggered a barrage of questions and comments from them about girls, marriage, and sex. They asked if I was married. No, I said, and again asked if any of them were. No, they weren't. I asked if they had girlfriends. "No girlfriends allowed in Islam." Did I have a girlfriend? Yes, and I showed them a picture of my unveiled female friend. They talked among themselves, and one volunteered the statement that "Without the veil, everyone looks at a woman and wants to sleep with her". Another spontaneously said "If my sister ever slept with someone I would kill her immediately". These free associations speak for themselves, and suggest the reasons behind the requirement that a "sinful" girl's brother is the one required to kill her when the family honor code has been violated: As a child, the brother has the most intimate, emotional contact with a young female child, but suffers under a most strongly formulated incest taboo.

None of these young men had ever been outside Egypt, and I attempted to explain that in the West the lack of veils made for a much more pleasant situation all around, but this just gave them a great deal of anxiety, and changed the character of the discussion. Unveiled women, I was told, were immoral, and probably slept with many men. "Only prostitutes have knowledge of sex" several of them said, followed by "All American and European women are prostitutes", and so forth. I got angry and suddenly stood up to my full height, telling them they didn't have the slightest idea of what they were talking about, and that they ought to go see for themselves before judging! I stormed off, angrily.

This definition of a prostitute, "an unmarried woman who has knowledge of sex", was repeated by many of the men I talked with. Hence, I began to view the reports from Iran, where revolutionary guards were torturing and executing "prostitutes" in large numbers, in a different light. If this definition of prostitute was valid for other areas of the Near East, then it was probable that the Iranians were not executing merely the poor, lower class women who scraped along by selling their sexual favors, but any educated, unveiled woman of Western orientation whom the Mullahs or their young celibate male footsoldiers would claim had some "knowledge of sex". This might mean either book-learning, the absence of virginity in the unmarried, or the possession of a contraceptive.[5] The Iranian-Saharasian "prostitute", it would appear, is defined similar to the Medieval Christian "witch", and is treated similarly.

Television shows from the West are regularly translated and shown in Egypt, and one may see television sets going at all times of day and night, from even the smallest hovels and shanties in Cairo. It appears that much of the common person's view of Western morality comes over the translated television shows. Everyone knows the Egyptian newspapers are censored, and television is much more "believable" anyhow. Even a show like "Ozzie and Harriet" would be a culture shock, with unveiled women and teenage dating; I wondered if there was an Arabic version of "Dallas", and shuddered to think that they would get their version of American life from Hollywood movies and TV programs.

However, if the popular image of the American is distorted for Egyptians by television and tourists, so too is our image of Egypt by Arab stereotypes, and stories of oil wealth, harems, and pyramids. Sadat was hailed as a great peacemaker by the Western media, but the depth of popular hatred against him by the Egyptian man in the street was great. Soldiers armed with machine guns, fixed bayonets, and steel helmets were found at every important intersection, and surrounded government buildings and banks on a 24-hour basis. I once was nearly arrested for sitting on a brick wall by a fence to change the film in my camera — someone had thrown a bomb over the fence only several days before. I was saved from military arrest only by the unsolicited assistance of several Egyptian businessmen who were passing by.

Sadat was openly cursed by most Egyptian men I met on the street, educated or not, but never by the few women I met. He was widely viewed by men as a traitor for making peace with Israel, and his photograph was defaced in every public place it was hung. The depth of the anger was a surprise to me, but it appeared to reflect more than simple hatred of the man for making peace with the Hebrew enemy. Always attached to the anger were allusions to the sexual and family impacts of Westernization, to the behavior of females, and erosion of "tradition". Many of the Mullahs made a great deal over the fact that Sadat's wife was working so hard to educate Egyptian women, going about unveiled, and so forth. And Sadat had taken it upon himself to pass, solely by personal decree, the new "personal status" amendment which granted significant property and child custody rights to divorced women, including the right of a woman to a divorce should her husband take a second wife. This law was informally called "Jihan's Law" given the efforts Sadat's wife had undertaken in its behalf.[6] Sadat jailed many of the most vocal Mullahs, and pressed on with his program of reforms, as had the Shah of Iran. The Shah, like Sadat, also had a Western-oriented wife who worked to educate the rural poor woman; both of their wives also fought for contraceptive reform. They

5. Also see the report by American Feminists who went to Iran after the "revolution" (K. Millett, *Going to Iran*, Coward, McCann & Geoghegan, NY, 1982).

6. R. Morgan, "World: Good News From 18 Countries", *Ms.*, September 1985, p.14.

angered the Mullahs and other traditionalists. Both Sadat and the Shah had cabinet officers who appeared in public, posing for pictures, with their unveiled wives. And as history had it, both of these leaders were removed from their seats of power by religious zealots, backed to a greater or lesser degree by popular sentiment. Within a year after my visit to Cairo, Sadat was shot to death by fundamentalist officers in his army.

Based upon such observations, and regardless of other valid factors and forces at work, it would appear that the popular uprisings against both Sadat and the Shah were fed by significant elements of sexual anxiety among the masses, *a rebellion against sexual freedom*. The liberation of the female, with its sexual implications, was deeply frightening and angering the average male, who in any event possessed a distorted view of the female almost wholly fabricated by sex-hating Saharasian philosophers, as recorded in the Koran and other holy books. Where objections against "Westernization" are voiced, it is primarily the sexual values of the West and its more liberated view of the female which underlie the various objections.[7]

Israel 1980

As one leaves Egypt, heading eastward into Israel, one makes a journey into a region where conditions are not only moister, but where the prevailing cultural ethos is Western. The population is smaller, some 3.5 million compared to Egypt's 42 million, and population growth is also slower, 1% compared to 2.5%.[8] Israeli women are not only unveiled, they walk in public with full legs, arms, heads, and necks exposed. Their modern clothing may also reveal bellies, the shape of breasts, buttocks and thighs. Women on beaches wear skimpy bikinis, halter tops, or go braless with T-shirts — and at a few Israeli beaches (such as Eilat), totally nude mixed-sex sun-bathing takes place. Israeli women may walk about nearly naked by comparison to women in Egypt, or other parts of the Near East where veiling and purdah are dominant.

Still, Bedouin Arab or Palestinian women in Israel may wear full veils. Bedouin camps are found in Gaza and the West Bank, as well as on the fringes of the Negev. Here, many live as before the coming of the State of Israel, in tents with their camels, sheep, and goats, the women in full black veils, and the men with robes, headcloth and curved daggers in their belts. However, even among the Bedouin marriage custom and the status of women are tempered by Israeli law, which allows the female a considerably greater amount of freedom from the dictates of either her father, husband, or brother. The sharp differences regarding female status, and the entire approach to the matter of sexuality, leads to a persistent cultural friction.

While this friction occasionally takes a more violent form, it is generally expressed in a subtle fashion. For instance, one may often see both Arab and Israeli women on the bus system. The Israeli bus system is more efficient and cleaner than the Egyptian busses; even when crowded, there is breathing room, and one is not crammed against other people. Women can therefore ride them without necessarily being pushed or pressed against the bodies of strange men. Arab women in full veils often ride the bus unchaperoned. When an Arab woman gets on the bus with full veil, it is not uncommon for the Israeli women to gawk, point, look, and openly talk about the "veiled one".

Once, on an exceptionally hot day, an Israeli woman asked a veiled Arab woman sitting across from her "why don't you take that thing off, dear, and cool down... your men aren't here." The veiled woman just

7. The encroaching Western ethos of a relative sexual freedom also appeared central to other rebellions, as in Kenya, where Jomo Kenyata was unable to rally support against the British until they outlawed clitoridectomy. After that, he had all the support needed to gain power. (J. Kenyata, *Facing Mt. Kenya*, Vintage, NY, 1965, pp.125-48; J. Murray, *The Kikuyu Female Circumcision Controversy...*, Diss., UCLA, 1974) For a broader view of Iranian psychosexual factors, see the chapter on "Masculine History" in R. Baraheni, *The Crowned Cannibals, Writings on Repression in Iran*, Vintage, NY, pp.19-84; cf. C. Consalvi, "Some Cross- and Intracultural Comparisons of Expressed Values of Arab and American College Students", *J. Cross-Cultural Psychology*, 2(1):95- 107, 1971.

8. "World Data Sheet: 1983", Population Reference Bureau.

turned away to look out the window, and didn't say anything. Oppositely, I have seen veiled women purposefully sit on two seats at once in order to prevent an unveiled Israeli woman from sitting next to them, even though the additional seat might be the only vacant one on the bus. Veiled Arab women sit together, and unveiled Israeli women sit together, in a form of self-imposed apartheid-of-the-veil.

Still, young Arab girls in Israel often take great pains either to avoid or abandon the veil. Theirs is not an easy task, however. Israeli society is far freer than anything previously experienced by Arab youth. The traditions of Arab youth in Israel are similar to those of Egypt, but the closeness and accessibility of Western influences, plus the shelter of Israeli law regarding the female, are gradually reshaping their culture. It is an influence which is publicly loathed, but often taken advantage of privately, especially by young people. It is probable that Arab children are even more tightly controlled by their parents under such circumstances, but I could not fully determine this. It is known, however, that there are special secret organizations, and a complete "underground railroad" in Israel to assist young Arab women to escape "honor killing", smuggling them out of the country.[9] As in Egypt, such killings are rare in the cities, but more frequent in the villages and countryside where the virginity taboo is much stronger. Among sexually-educated secular Israeli women, who constitute the majority, the virginity taboo is relatively weak; Arab women of Gaza or the West bank, however, follow Egyptian or Jordanian traditions in that they do not get sex education in the schools.[10] These points would suggest both strong patrist traditions along with a simultaneous breaking down of traditions; it also suggests an unspoken point of friction between Arab and Israeli of a most intense form.

Once, while waiting at a bus stop in Beersheva, I observed the interaction between Arab and Israeli youth which, I believe, characterizes the cultural differences and changes which are occurring. As I sat waiting, three Arab women approached to also wait for the bus. There was an elder grandmother, a younger mother, and a teenage daughter. Each of the three women wore two separate veils, a light inner one of white cloth, and a heavy, carpet-like outer one of black fabric.

The older woman had both veils firmly and snugly wrapped around her head, exposing only her eyes, in spite of the over 100°F temperature, and the presence of shade at the stop. The younger mother had both veils over her head, but only the lighter inner veil pulled over her face, and that in a loose manner. The teenage daughter had the heavy carpet-veil down on her shoulders, with the inner white veil over her head, but with her face fully exposed, even though I, a non-Arab male, was sitting nearby. They sat quietly until the approach of a dozen Israeli adolescents, who had just gotten out of school.

The Israeli girls were wearing cut-off jeans and halter tops, without bras, and rubber thongs on their feet — a comfortable dress in the desert assuming one has a continuous supply of water. One of the Israeli girls, who was quite handsome and buxom, showed a considerable amount of breast above and at the side of the skimpy halter top, and all the girls had fully exposed arms and thighs. They walked with their boyfriends, some of them holding hands, others alone, but all cutting up with jokes and the like in a manner familiar in the West. One teenage couple held hands, pressed against each other and publicly kissed, and another fellow carried a large "ghetto-blaster" on his shoulder, out of which strains of Led Zeppelin rock-n-roll music came forth.

At the approach of the Israeli teenagers, the Arab women became very animated, and upset. The grandmother reached over to adjust the veil across her teenage granddaughter's face, which was exposed. The

9. D. Torgerson, "Arab Honor Killings Fought in Secret", *Miami Herald*, 4 January 1981.

10. Raymonda Tawil, as quoted by Torgerson, 1981, ibid.

young girl leaned away, shouldering her grandmother's arm away from her face, as if saying "Come on, ma! Leave me alone!". This went on for a minute or two until the grandmother gave up. The young Arab girl uneasily shifted her gaze back and forth between the desert and the Israeli teens. She became distressed and anxious, not only because her mother and grandmother were exerting pressure upon her to "cover up, and be decent", but also because of her own clear interest in the lifestyle of the Israeli teens. Her body language, and emotional reaction to the situation conveyed the impression that she, too, wished to be rid of the veil to wear blue jeans and a halter top, to have and kiss a boyfriend, and listen to rock-n-roll. However, she was trapped, and knew it. Meanwhile, the Israeli youth stayed to themselves at a distance, but gawked and eyeballed the veiled women in a cautious manner.

The differences in sexual behavior between the Westernized Israelis and the Bedouin Arabs is like night and day, yet they are neighbors. Most Arab youth suffer from enforced celibacy and strict separation of the sexes to a degree unknown among Israeli youth, with exception of a few extremist Jewish sects. Most of the people in Israel whom I met and spoke with do not even attend the Synagogue. Although the nation is technically a democratic theocracy and the religious parties command a greater power than their absolute numbers would normally allow, most Israelis are in fact secular in outlook.[11] They are a hearty bunch, their men and women very sturdy and handsome — "Hebrew Tarzans" is what Koestler called them in 1947,[12] and the description still applies. The secular Israelis are different from their American-Jewish counterparts, being more energetic, with the don't-waste-my-time straightforwardness of a New Yorker.

The Israelis are currently engaged in transforming the desert into productive land, land which was laid to ruin by Bedouin and Turkish armies for centuries. Terraces which lay dormant since the times of the Romans are being brought back into productive use. They are at work transforming the desert back into a garden, but they also threaten to transform the prevailing Saharasian cultural ethos of the region back into that which prevailed before the onset of desiccation 5000 years ago, before the invasions of Asian and Semitic migrants. Israel also is currently the only truly democratically oriented state existing within Saharasia proper.

Access to fertile land and water rights have traditionally been and still are fighting matters in this region. However, regardless of the amount of anger and hatred which is generated regarding the issue of productive land, there is clearly a significant quantum of fear, anger, and resentment directed towards the Israelis purely because of the different cultural ethos they hold regarding females and sexual behavior. To many of the Bedouin Arabs it is a secretly desired, but unattainable and hence a hated ethos, in the same manner that frustrated love turns hateful, or unattainable grapes are believed sour. Such emotional and subjective factors shape the political relationships between the various cultures in the region. Were it not for such differences in family and sexual matters, the difference between being an Arab Moslem or a Westernized Jew would be minimal.

11. The power of these sects has been vastly over-estimated, however, even for many years by myself. Their abusive violence mostly falls upon the backs of members of their own sects, being absurdly ridiculous and irritating to others, rather than deadly. While some are almost as patrist as neighboring Moslems, they are dramatically restrained in how far they can go by secular Israeli law. The larger percentage of Israeli Jews appear to fully disagree with their agendas, as determined by political voting patterns which have never brought them anything more than a few seats in the Knesset. Nevertheless, at their demands, mixed bathing and dancing have been banned at a few Israeli hotels catering to their members, non-Orthodox Jews have been forbidden burial in certain ultra-orthodox cemeteries to prevent them from being "defiled", and many Sabbath laws have been extended nationwide. Jewish fundamentalists have also tried to disrupt archaeological digs at Jericho, which have revealed both Arabic and female fertility-goddess influences during Hebrew periods when Yahweh was supposed to have been the primary god. While several outspoken Rabbis once openly preached uncompromising racist anti-Arab doctrines, others of such minority cults also preach that Israel should not even exist, because "the Messiah has not yet come". And new anti-encitement laws passed in Israel have basically made such firey anti-Arab preaching illegal, with violators drawing lengthy prison terms. By contrast, virulent Jew-hatred on par with Nazi Germany is spewed forth daily, without restraint or penalty, from nearby Palestinian mosques and mass-media, as well as from children's school-books. While there have been a few incidents of anti-Arab terrorism committed by Jews, these are indeed a very tiny number, to be counted on the fingers of one hand over many decades, as compared to the *thousands* of deadly attacks, and attempted attacks upon Jews by Arab Muslim fanatics, to include several major wars of aggression by collective Arab-Muslim armies, bent upon total Jewish extermination. In short, while Israel is very much like the rest of the West in that it carries expressions of patrism and armoring within its cultural makeup, there is no rational comparison possible between Israel and the larger Saharasian Arab-Muslim world, the latter of which continues to be dominated by extreme patrism, armoring and violence, expressed against both fellow Muslims and non-Muslims alike. (cf. N. Angier, "Unholy War at the City of David", *Discover*, January 1982, pp.30-6; J. Tamayo, "Confrontations Turn Ugly in Israel's Battle of Beliefs", *Miami Herald*, 12 June 1986, p.22A.)

12. A. Koestler, *Thieves In the Night: Chronicle of an Experiment*, Macmillan & Co., 1947.

There is a great deal more which could be said about contemporary conditions in Saharasia and its borderlands, about such events as: the break-down of the Berlin Wall, and later, of the Soviet Union itself; the brutal Iran-Iraq war; Saddam Hussein's death-camps, poison-gas attacks against his neighbors, and subsequent invasion of Kuwait and the Gulf War; the fanatic Taliban movement in Afghanistan; the Intifada and the difficult search for peace between the Israelis and Palestinians; the on-going slave trade in Moslem fundamentalist Sudan and other Sahelan nations; the massacre of students in China's Tienamen square, the massacre of Western tourists in Egypt, the even more terrible massacres in Algeria, and so on. All of these events have a clear sex-economic dynamic related to a 6000 year-old pattern of Saharasian history. The modern expressions of this pattern could be, and should be more fully discussed and described. Also needing clear and open discussion are the "Saharasian" behaviors adopted within the Western democracies, notably the USA, since the end of World War II: ie, male circumcision; AIDS hysteria; "abstinence education"; pathologizing of adolescent heterosexuality and "normalizing" pedophilia and child pornography; the "war" on selected illegal "street drugs" with antithetical pushing of toxic but legal "medical drugs" to schoolchildren; the dramatic growth of the prison population in the USA; the Western addiction to the new narcotic, petroleum, imported from Saharasia. This latter phenomenon currently has the Western democracies trapped in a deadly St. Vitus dance with different Saharasian extreme-patrist dictatorships, who control the spigot. And it must be point out how advocates of both liberal and conservative politic have preferentially supported many of these pathologies, even while rejecting others. But this discussion must wait for another time, to do the topic the justice it deserves.

I have often thought about that sad and conflicted teenage Bedouin girl at the bus stop in Israel, who gazed alternately at the desert and at the kissing Israeli teenagers. Did she ever make some daring, bold move to break free of her situation? Did she negotiate with her parents her need for an education, and gradually become a member of the urbanized Arab population who in at least a few ways are beginning to adopt the Western view of the female? Did she ever dare to meet with a boyfriend in secret, a move which might have brought the wrath of her parents and brothers down upon her head, possibly with fatal results? Did she run away to Tel Aviv or Jerusalem, or to Paris or Berlin, to reappear with a new name? Or did she simply age, resign to her fate, enter into an arranged marriage with an older man, have a dozen unwanted children, and begin to hate that which she longed for, but could not become, and could never have?

There are millions of such young people across Saharasia who are similarly being confronted, on a daily basis, with an older system of more biologically-oriented foreign values and ideas which are on the one hand very much attractive and desired, but on the other are wholly at odds with their prevailing social structure. It is probable that real or potential social changes — in the status of the female, and the rights of youth to sex-education and love-match relationships — will continue to stimulate major political change across the region, of both a reformist and reactionary character. The manner in which the older generations respond to these coming changes, and the courage of the young people in Saharasia in their pursuit of greater freedom, will be major determinants of political events within the region for the next several hundred years.

415

Appendices

CORRELATION TABLE of Sex-Economic Factors — 63 total   (Panel one of three)

PANEL ONE	FC#	_DESTRUCTIVE AGGRESSION & VIOLENCE_ 477	474	473	472	421	420	419	417	SOCIAL STRUCTURE 110	109	102	RELIGION 428	424	423	KINSHIP & INHERITANCE 200	198	188	186
INFANTS																			
Physical display of affection is low	317				70B								88B						
Infant indulgence is low	318		65C							64C		57C							
Protection from environment is low	319											63C				63D	56C		
Attention to needs is low	320											51C		75B					
Speed of attention to needs is low	321			64D	72C														
Consistency of attention to needs is low	322		82C						79D										
Constancy of nurture is low	323			82D	78C		80D	74C	79D										
Pain infliction is high	324									73C							62C		
CHILDREN:																			
Average satisfaction potential is low	302						52D												
Oral satisfaction potential is low	303			73B															
Anal satisfaction potential is low	304																		
Sexual satisfaction potential is low	305						56C	85D											
Average anxiety potential is high	308																		
Oral anxiety potential is high	309							79D											
Anal anxiety potential is high	310																		
Sexual anxiety potential is high	311																		
Early toilet training is present	329																		
Early weaning is present	330													83C					
Childhood indulgence is low	334					71B												62A	69C
ADOLESCENTS:																			
Adolescent sex dissociation is high	366																		
Segregation of adolescent boys is high	370						58C									-57A		52A	51B
Severe male initiations are present	376																		
Male genital mutilations are present	377			67D						62A	57A	68B	53A				64D	65A	57A
Painful female initiation rites present	383													-75C					
Sex expression restrictions are present	386				72D														
Female premarital sex taboo is high	390		71D			72D						68C	52C		55C				66D
MARRIAGE & FAMILY:																			
Marital residence near male kin	204									87C	96B	86C						94A	98A
Marital residence near male kin	209			96C						88B	98B	88A				95A	98A	86A	99A
Unrestricted Polygamy is present	242		94C	89C			90C	87B		88A				-70C				83C	89A
Father has family authority	254																		
Grandparents have authority over parents	255																		
Grandparents authority over grandchildren	256											60C							
Bride price is present	264																		
Exchange of female relatives is present	269I																		
FEMALES:																			
Female status inferior or subjected	277													64C		60D			
Female uncleanliness belief is high	395													82C					
Menstrual taboos are high	396													60D					
PREGNANCY:																			
Barrenness penalty is high	283																		
Isolation of pregnant women is high	290																		
Abortion penalty is severe	295					80C				86C									
Post-partum sex taboo 6 month to >2 yrs.	300		63D			61D			58C	57C									
ADULT SEXUALITY:																			
Extramarital coitus is punished	393					78B	76B	74B		77B									
Sex disabilities are present	397						64D												
Sex anxiety is high	398																		
Castration anxiety is high	399					71C	68C												
KINSHIP & INHERITANCE:																			
Exclusive patrilineal descent is present	186									51A	65A	56C				62A	63A		
Exclusive cognatic kin groups are absent	188									74A	86A					-80B			
Land inheritance favors male line	198									93C	97D	100C				99A		100A	
Movable property inheritance favors male line	200																		
RELIGION:																			
Organized priesthood is present	423																		
Full-time religious specialists present	424													53C					
High God present, active, sup. human morality	428										90C						73C		
SOCIAL STRUCTURE:																			
Class stratification is high	102			75C					62C	75A	78A			81C					62C
Castes are present	109			73B						80B		80A	63C			77D	74D	81A	75A
Slavery is present	110		56C	56D	53D			53D	59C		63B	61A					59C	50A	56A
DESTRUCTIVE AGGRESSION & VIOLENCE																			
Warfare prevalence is high	417		100B		100B	94D	100A	96A		95C		95C							
Military glory emphasis is high	419		85A	83C	88A	92A	100A		96A	76D									
Bellicosity is high	420		67A	65C	69A	78A		77A	81A										
Killing, torture, mutilate enemy emphasized	421		65A		58C		78D	67A	70D										
Narcissism index is high	472		83A	91A		70C	76A	69A	69B	64D									
Insult sensitivity is extreme	473		50C		62A		50C	56C		51D	89B	51C							
Boastfulness is extreme	474	75D		65C	74A	70A	67A	63A	69B	61C									
Alcoholic aggression is strong	477		67D																
	FC#	477	474	473	472	421	420	419	417	110	109	102	428	424	423	200	198	188	186

Explanation and Discussion Follows   (Panel two of three)

FC#	ADULT SEXUALITY				PREGNANCY				FEMALES			MARRIAGE & FAMILY								ADOLESCENTS							FC#
	399	398	397	393	300	293	290	283	396	395	277	269	264	256	255	254	242	209	204	390	386	383	377	376	370	366	
317								79C																			317
318														100C													318
319										60C			70C									100D					319
320			63C							50C																	320
321				79D																			64D				321
322																											322
323																											323
324											100C						56D	64C	64D								324
302																						70D					302
303																											303
304											-100D														83C		304
305		56D		65D	100B															100B							305
306	85C	86B		79C																100C							306
309		69D																					82D				309
310																				86D							310
311	80C	81A		65D					82C											100C	61D						311
329																											329
330																	-55D										330
334	71B	71D																									334
366			50D																								366
370										-90C			55A	86D			56B	53C	53B		86C		57B				370
376																											376
377								54D					56A			57D	71B	65A	65A			56B				65B	377
383																											383
386	73D	83D	100D	76C		78D		83B									-52D										386
390	65B		50C																								390
204													91A				85A			95C		95A			91B		204
209												100D	98A				83C				-63D	95A			89C		209
242		100C			100C			93D			100D	100D	90A			100C		86C	86A			93B			88B		242
254													100C				75C						100D				254
255														63B													255
256		-86D													71B										50D		256
264																											264
269																											269
277																	55D								-90C		277
395																											395
396	52C	53C	60C																								396
283																											283
290																											290
293				70C				80B									61C						64C				293
300																											300
393	87A	80C			88C	90C															61C						393
397	50D								55C											56C	72D					56D	397
398	79B			67C					89C							-67D	59C			100C	67D						398
399		79B	80D	87A					80C											100B	69D						399
186							92D						63A			58A	54A	54A		51D		56C	63A		52B		186
188												88D	86A			66C	76A	76A					88A		87A		188
198																	92A						96D				198
200																											200
423																											423
424																					-87C						424
428																	63C			89C			91A				428
102												-100A	68A	86C			60A		60C	76C			67B				102
109													76A				80B		80B				67A				109
110				53B	55C								62A				52A	51B	51C				62A				110
417													100D														417
419	77C			92B									77D				73B										419
420			80D	83B									70B				55C			75D						68C	420
421	60C			78B	73C																	54D	100D				421
472													66D				60C			77D			71D				472
473													53C				56C	58C									473
474																											474
477																											477
FC#	399	398	397	393	300	293	290	283	396	395	277	269	264	256	255	254	242	209	204	390	386	383	377	376	370	366	FC#

CORRELATION TABLE of Sex-Economic Factors — 63 total (Panel three of three)

		CHILDREN											INFANTS							
	FC#	334	330	329	311	310	309	306	305	304	303	302	324	323	322	321	320	319	318	317
INFANTS																				
Physical display of affection is low	317	74A	56D				68B						59C		56D	64C	76A		78A	
Infant indulgence is low	318	66A					65D						71A		69A	68B	83A	67B		75A
Protection from environment is low	319														60C	71B		67B		
Attention to needs is low	320	51C					55C								50B	56B	54B		63A	57A
Speed of attention to needs is low	321						65C				67C		55D	58B	66A		74B		65B	62C
Consistency of attention to needs is low	322			-56D	74C								72C	74A		84A	84B	78C	85A	69D
Constancy of nurture is low	323			-53D							85C	85D	81A		84B					
Pain infliction is high	324														66C	67D			83A	68C
CHILDREN:																				
Average satisfaction potential is low	302		53D	55D				59C			72B		50D							
Oral satisfaction potential is low	303		67C				62C					76B	63C			71C				
Anal satisfaction potential is low	304				77B															
Sexual satisfaction potential is low	305				74A	64D		77A												
Average anxiety potential is high	308				79A	82A	76C		85A			76C								
Oral anxiety potential is high	309	72C						73C			70C				70C	76C	73C		68D	81B
Anal anxiety potential is high	310							82A		77B										
Sexual anxiety potential is high	311							86A	87A					-62D	60D					
Early toilet training is present	329		78D									86D								
Early weaning is present	330	66C		70D							67C	73D								67D
Childhood indulgence is low	334		64C				65C										71C		74A	86A
ADOLESCENTS:																				
Adolescent sex dissociation is high	366																			
Segregation of adolescent boys is high	370								56C											
Severe male initiations are present	376																			
Male genital mutilations are present	377					64D					53D					61D				
Painful female initiation rites present	383																			
Sex expression restrictions are present	386				69D															
Female premarital sex taboo is high	390				65C				58A											
MARRIAGE & FAMILY:																				
Marital residence near male kin	204												93D							
Marital residence near male kin	209												88C							
Unrestricted Polygamy is present	242		-74D										94D							
Father has family authority	254																			
Grandparents have authority over parents	255																	60C		
Grandparents h/author over grandchildren	256																			
Bride price is present	264																			
Exchange of female relatives is present	269																			
FEMALES:																				
Female status inferior or subjected	277									-100D			63C				100C	100C		
Female uncleanliness belief is high	395																			
Menstrual taboos are high	396				53C															
PREGNANCY:																				
Barrenness penalty is high	283																			
Isolation of pregnant women is high	290																			
Abortion penalty is severe	295																			
Post-partum sex taboo 6 mo. to >2 yrs.	300																	-62D		
ADULT SEXUALITY:																				
Extramarital coitus is punished	393				76D			73C	76D							61D	71C			
Sex disabilities are present	397																			
Sex anxiety is high	398	67D			87A			69D	80B	75D										
Castration anxiety is high	399	71B			63C			61C												
KINSHIP & INHERITANCE:																				
Exclusive patrilineal descent is present	186	53C																		
Exclusive cognatic kin groups absent	188	89A																		
Land inheritance favors male line	198												100C							
Movable prop. inheritance favors male line	200																100C			
RELIGION:																				
Organized priesthood is present	423																			
Full-time religious specialists present	424		56C															67B		64B
High God present, active, sup. human mor.	428																			
SOCIAL STRUCTURE:																				
Class stratification is high	102																71C	67C	65C	
Castes are present	109																			
Slavery is present	110												53C						52C	
DESTRUCTIVE AGGRESSION & VIOLENCE																				
Warfare prevalence is high	417							92D	100D				92D							
Military glory emphasis is high	419								88C			85D	72C							
Bellicosity is high	420												50D	54D						
Kill., torture, mutilate enemy emphasized	421	65B														65C				67B
Narcissism index is high	472												62C	67D						
Insult sensitivity is extreme	473										58B		58D					50C		
Boastfulness is extreme	474													54C						
Alcoholic aggression is strong	477																			
	FC#	334	330	329	311	310	309	306	305	304	303	302	324	323	322	321	320	319	318	317

About the *Correlation Table*

This *Correlation Table of Sex-Economic Factors* was constructed from data contained in Robert Textor's *A Cross Cultural Summary* (HRAF Press, New Haven, 1967). Textor's large volume of peer-reviewed cross-cultural data for a regionally-balanced sample of 400 different cultures, was evaluated for the presence of variables similar to those presented in this work on Table 1 (page 5). Table 1 was constructed *a-priori* according to Wilhelm Reich's *Sex-Economic* theory of human behavior. A total of 63 different variables were so selected from Textor's *Summary*, and organized as given in the left-hand column of the *Correlation Table*.

The FC (Finished Characteristic) numbers correspond to those given in the Textor *Summary*. FC numbers across the top of the *Correlation Table* have the same attributes as the corresponding FC numbers at the far left of the *Table*. The numbers in the *Table* are the percent of the 400 cultures, for which data was available, where a statistically significant correlation existed between the two behavioral variables. A negative sign indicates a negative correlation; all other values are positive, indicating a positive correlation.

The given percent values were indexed from the *Summary* in the following manner. The primary attributes, given across the top of the *Table* in column order, were first accessed from the *Summary* by FC number. The FC numbers for the secondary attributes, given at the left in row order, were then searched through to see if a correlation existed. If so, the percent and probability values for that correlation, whether positive or negative in direction, were recorded on the *Correlation Table*.

Underlined entries indicate inferred positive correlations from double-negative correlations (as given in the wording of the variables in Textor's *Summary*), as with an attribute which "tends less to be absent". The letter immediately following the percent values indicates the probability of the given correlations, as determined through either the Chi-Square or Fisher Exact Test. Statistically insignificant correlations (p > .10) are not printed in Textor's *Summary*, nor included in the *Correlation Table*.

$$A = \quad\quad p < .001$$
$$B = .001 < p < .01$$
$$C = \ .01 < p < .05$$
$$D = \ .05 < p < .10$$

A total of 95% of all recorded correlations were positive in character, only 5% being negative, fully supporting the sex-economic theory. A 50%/50% split between positive and negative correlations, with a greatly reduced absolute number of correlations, was forecast by chance alone. 520 correlations were observed (500 positive and 20 negative) out of a total of 3906 possible correlations. Textor's *Summary* winnowed out a number of duplicate correlations, and also failed to run some FC numbers in both directions. Further, data was missing for a large number of the 400 cultures for some FC numbers. These factors resulted in a lower absolute number of correlations than would have otherwise been recorded. These facts make the observed overwhelmingly positive nature of the correlations all the more significant. Wilhelm Reich's theory is dramatically confirmed by the Textor data, demonstrating its global cross-cultural accuracy and predictive value. More details are given in Chapter 3, pages 68-70.

About the Ethnographic Data Used for Maps in This Study

In 1962, the *Ethnographic Atlas Project* was initiated by George P. Murdock of the University of Pittsburgh, as a means to comprehensively and systematically address questions of human culture. The original installments of Murdock's *Atlas* data, which was not a geographical atlas with maps, appeared in 22 issues of the journal *Ethnology* (see below), and included alphabetically coded data on more than 50 different cultural topics for 1170 different cultures, worldwide. A later hardcover book of these same data, the *Ethnographic Atlas*,[1] duplicated those data for an 862 culture sample of the original 1170 cultures. A numerically-coded IBM-card version of the Atlas data was also developed for the entire 1170 cultures.[2] It was this latter set of the *Atlas* data which was used by the author in the 1980s to construct the World Behavior Map, and other cultural maps.

The maps were developed by establishing the latitude and longitude for each of the 1170 cultures, generally as provided in the "Classification by Clusters" section of the *Ethnographic Atlas*. The data was then accessed through a computer program developed by the author which printed the various codings on individual base maps. Since the original publication of Murdock's data in *Ethnology*, a few of the codings have been modified — the maps presented here do not reflect those small number of modifications. Nor does this work include an additional roughly 100 cultures added to the Murdock data base beyond the original 1170 cultures.[3] Additional details on these data, and the use of them in the Saharasian research, can be found in the author's original University of Kansas dissertation,[4] which also contains a print-out of the computer programs used for coding and mapping the data.

Issues of *Ethnology* containing data on the full 1170 cultures.

Issue	Date	Pages	Code Number
1 #1	Jan. 1962	113-134	1-100
1 #2	April 1962	265-286	101-200
1 #3	July 1962	387-403	201-300
1 #4	Oct. 1962	533-545	301-350
2 #1	Jan. 1963	109-133	351-400
2 #2	April 1963	249-268	401-455
2 #3	July 1963	402-405	456-457
2 #4	Oct. 1963	541-548	458-483
3 #1	Jan. 1964	107-116	484-522
3 #2	April 1964	199-217	523-609
3 #3	July 1964	329-334	610-628
3 #4	Oct. 1964	420-423	629-634
4 #1	Jan. 1965	114-122	636-667
4 #2	April 1965	241-250	668-697
4 #3	July 1965	343-348	635 + 698-702
4 #4	Oct. 1965	448-455	703-725
5 #1	Jan. 1966	115-134	726-838
5 #2	April 1966	218-231	839-915
5 #3	July 1966	317-345	916-1087
5 #4	Oct. 1966	442-448	1088-1105
6 #1	Jan. 1967	103-107	1106-1132
6 #2	April 1967	109-236	1133-1170

1. Murdock, G.P., *Ethnographic Atlas*, U. Pittsburgh Press, 1967.

2. Available from the Human Relations Area Files, Department of Anthropology, Yale University, PO Box 2054, Yale Station, New Haven, CT 06520.

3. These additional data may be found in the following *Ethnology* issues: 6(4):481-487, 1967; 7(1):106-111, 1968; 7(2): 218-224, 1968; 7(3): 327-329, 1968; 8(1): 122-127, 1969; 8(2): 245-248, 1969; 8(3): 367-385, 1969.

4. DeMeo, J., *On the Origins and Diffusion of Patrism: The Saharasian Connection*, Geography Dept., Univ. of Kansas, 1986.

Update on *Saharasia*
New Findings Since the First Printing*
by James DeMeo, Ph.D.

Introduction and Background

In 1992, I was invited to Vienna, Austria, to give lectures on my research, and while there visited the *Natural History Museum*, which at the time had a large collection of East European artifacts organized chronologically. The display cabinets lined a pathway, which allowed one to see recovered artifacts and scenes reconstructing daily life, starting with the most ancient down to modern times. I made my way through the earliest collections of primitive stone tools, through Neanderthal times, and into the epoch of early *Homo sapiens*. Simple villages were shown in the reconstructed scenes, along with agriculture and animal domestication, some early types of pottery, fabrics and copper implements formed into decorative shapes. Settlements slowly grew in size, naturalistic artwork developed along with what I call "mother-dolls" (clay figures of women, what some have interpreted — wrongly I believe — as a "mother-goddess"). Artifacts of simple clay, stone, ceramic, copper, and even woven fabrics appeared, along with simple, yet elegant architecture, and the technology associated with agriculture, animal herding and hunting progressively improved in sophistication. All in all, it basically recorded an ordinary, though certainly vital and exciting existence of hunting, farming, dancing, and peaceful human relationships.

When the collection arrived at the middle of the fourth millennium BC (c.3500 BCE, or Before the Current Era) a broad white stripe, interrupting the path, had been painted on the walls and floor of the Museum gallery, bearing bold dark letters "CIVILIZATION BEGINS". Upon walking over that line, the display very dramatically included all kinds of war-weapons, battle-axes, shields and helmets. Artifacts related to horse-riding warriors appeared, as did crowns, coins and tombs for kings and other big-man leaders. Fortifications, palaces and temples then appeared, with all the evidence for war-making, despotic, and murderous *Homo normalis*, as discussed in Wilhelm Reich's monumental clinical discovery of human armoring,[1] the biophysical source of neurotic behavior and impulses towards sadism and brutality, and the wellspring for virtually every authoritarian social structure which exists, or

which has ever existed.

This example from the Museum depicts "civilization" in a manner quite unflattering as compared to the usual definitions, and implies that warfare and social violence is a relatively recent invention by our species, of only around 6000 years duration. It also implies that we have become so accustomed to warfare and violence as the "norm" that we have difficulty even conceptualizing there might be, or might have been in our most ancient past, another mode of social existence free of the horrors of warfare and all but the most uncommon examples of interpersonal violence. This point of view, however unrecognized or unpopular, has much evidence to support it.

In the decade before my visit to Vienna, from around 1980 through 1986, I undertook one of the most systematic global cross-cultural investigations on human behavior and the origins of violence that has ever been undertaken, as an effort to evaluate and test these ideas. My dissertation on the subject, presented to the Geography Department of the University of Kansas, created a controversy, but was accepted and eventually published as *Saharasia: The 4000 BCE Origins of Child Abuse, Sex-Repression, Warfare and Social Violence In the Deserts of the Old World*,[2] with various summary articles published in journals.[3]

This work demonstrated a previously-unknown global geographical pattern in the archaeological-historical literature, and in several large and widely-used anthropological data bases. The newly discovered geographical pattern demonstrated a strong spatial correlation between the world's most harsh *patriarchal-authoritarian* modes of social structure (synonymous with Wilhelm Reich's definition of *highly armored* character structures) to the most harsh global desert regions — in North Africa, the Middle East, and Central Asia — to which I gave the term *Saharasia*. Areas most distant from Saharasia, in Oceania and the New World, showed the softest and most fluid and flexible democratic and egalitarian social structures (synonymous with Reich's lightly armored, or unarmored character structures). Figures 1 and 2 (on page 9) reproduce my World Behavior Map, and the correlated Dryness Ratio Map identifying the world's harshest contemporary desert regions. Table 1 (on page 5) presents the dichotomous social-cultural factors which were mapped in the original study.[2,3]

---

* Revised and expanded from a prior article: J. DeMeo, "Update on *Saharasia*: Ambiguities and Uncertainties about *War Before Civilization*", in *Heretic's Notebook (Pulse of the Planet #5)*, 2002, p.15-44.

Additionally, I developed a new archaeological-historical data base, which when mapped showed a very strong correlation between the first drying-up of Saharasia around 4000-3500 BCE, to the general origins of human social violence — the earliest regions to dry up within Saharasia, notably in Arabia and Central Asia and their immediate peripheries, showed some of the earliest clear and unambiguous signs of social violence apparent in the archaeological record. Figures 3 and 4 (on page 10) present cultural diffusion maps as derived from the archaeological and historical materials.[2,3]

In *Saharasia*, I made the following argument: Human violence and warfare were the products of social institutions which inflicted great pain and trauma upon infants and children, as well as intense repressive sex-frustration within adults and adolescents, giving rise to sadistic impulses which were then channeled back into those same social institutions. Painful trauma and sex-repression experienced by children within such armored-patristic societies was adapted to and psychologically defended, and hence repetitively inflicted upon each new generation as "tradition" by the older generations. Drought and famine, extremely traumatic and deadly by themselves, were the triggers which drove previously peaceful unarmored-matristic human social groups towards increasingly disturbed and violent-sadistic behaviors, whereupon new social institutions appeared to guarantee their persistence, even under moist environmental conditions of food abundance.[1,2]

At the time when I undertook the basic research for *Saharasia*, a review of available archaeological materials demonstrated only a few regions in the Middle East, ranging from Anatolia into the Levant, and as far south as Jericho, possessed "fleeting glimpses" of violence prior to my marker date of c.4000 BCE. These unclear traces of violence appeared to begin around 5000 BCE, but were also timed to sub-phases of drought, aridity and land-abandonment, suggesting a similar drought-desert causation for the genesis of violence as was presented and argued for the post-4000 BCE event. The drying up of Saharasia after c.4000 BCE was, I argued, *the most significant climatological change which occurred on planet Earth following the end of the last Ice Age* (which ended around 10-8,000 BCE). In any case, at the time, according to the knowledge at hand, it appeared that neither drought nor violent episodes starting at the earlier date of c.5000 BCE were widespread, continuous, or persistent in the archaeological record. Only after c.4000-3500 BCE did drought and violence grip entire regions across the whole of Saharasia, a situation which I argued has lasted over 6,000 years, to be expressed in the more recent anthropological data as seen in the World Behavior Map. Archaeology, history and anthropology all presented mutually agreeable and reinforcing patterns on the world maps.

The exacting details of my Saharasia discovery with full citations has already been peer-reviewed and published.[2,3] Aside from these introductory notes, I shall assume the reader has a general familiarity with the earlier findings and underlying theory.

New Evidence For Ancient Violence

By 1999, I was alerted to new archaeological findings and books which *claimed* evidence for very ancient human violence, dating to well before c.4000 BCE. The book *War Before Civilization*[4] by Lawrence Keeley, is perhaps the most representative and widely-quoted example of this new genre of books, which basically argue for the innate, genetic or human evolutionary causation of war and violence, in opposition to the environmental-social-emotional causation argued in my *Saharasia*. Keeley's book laid down two basic arguments.

Argument One: *Intertribal warfare of an extreme and ruthless quality, as well as social-familial violence, existed among so-called "primitive" cultures of the New World long before the arrival of European colonials.* To this argument, I give a qualified agreement. In *Saharasia*, I cited some of the same evidence noted by Keeley, such as the butchery and despotism present among the Aztec, Inca, and Maya culture, long before the arrival of Columbus, Cortez or Pizarro. Likewise, the despotism and savagery of other "primitive" subsistence-level cultures in other world regions were detailed in *Saharasia*, well back into history and prior to any contacts with the sometimes equally despotic and savage Europeans. The findings on this point, in both my *Saharasia* and Keeley's *War Before Civilization* defeated many widespread myths about the supposed uniformly "peaceful" nature of "primitive man", "living in harmony with nature" — certainly, there are many well-documented cases of violence and organized warfare among isolated "primitive" tribal groups. *This was never in question.* However, unlike my *Saharasia,* these examples are too-often presented in such a manner as to mischaracterize *all* primitive cultures as carrying the seeds of violence. And so *I do object to making any kind of widespread and global extrapolation of these signs of violence among some aboriginal cultures as "proof" of an assumed but unproven ubiquitous violence among all cultures, in all regions, at all times.* Also, the authors pushing this line of argument almost always fail to take a genuine cross-cultural approach, and rarely openly address the various *peaceful aboriginal societies* as documented in various anthropological studies from the late 1800s and early 1900s, as detailed in my *Saharasia*. As a consequence, this first argument articulated by Keeley did not undermine or challenge my work in any manner. In fact, some of the archaeological evidence cited by Keeley and others for violence among ancient peoples of the New World — and which *I did not know about or cite in the first printing of Saharasia* — were located almost precisely in those regions where my World Behavior Map predicted such evidence might be found. More on this last point is given below. With confidence, I can therefore report, archaeological evidence on the question of "primitive violence" in more recent times, but prior to the epoch of European colonialism, provides excellent additional supporting evidence for my Saharasia discovery.

Argument Two: *Archaeological evidence for warfare and massacres exist in some very old archaeological sites, as early as 12,000 BCE, well before my c.4000 BCE marker date.* Keeley and other authors on the subject specifically mention ancient fortifications and graveyards filled with victims of violent deaths, well before c.4000 BCE. These archaeological reports superficially appear to provide a serious challenge to Saharasian theory, mainly because of the early dates. However, a close look at the original citations from the archaeologists who did the field work, and from those who are intimately familiar with the details, resolves the question in favor of the environmental-social-emotional causation implicit in Saharasian theory. In short, archaeological findings are often misquoted and misrepresented in more "popularized" accounts on "ancient violence".

To better understand the context and specific details of these newer archaeological findings for violence and warfare prior to c.4000 BCE, I shall explicitly address the major points of evidence.

Spanish Archers, rock art from Morella la Villa, Castellon, dated to "late Neolithic" (c.3000 BCE?)[5] Hunters in a ceremonial dance, or warriors in a battle? None of the figures appear injured or dead, and archaeology of the region does not support the idea of warfare at this early period. Similar rock art in Australia, claimed as evidence of violence, is even more abstract and ambiguous.[7]

Spanish and Australian Rock Art: Dancing or Fighting?

The article "The Beginnings of Warfare" by Trevor Watkins[5] is often cited to support the idea of a very ancient violent humanity. But Watkins does not provide such support. Watkins says: *"The origins of warfare are hidden in the mists of human prehistory, but by 1200 BC there was a long tradition of armies, campaigns, pitched battles and siege warfare"*.[5] It is quite a leap from "prehistory" to 1200 BC, and the latter date would surely be in good agreement with the chronology for first-origins of violence published in *Saharasia*. Watkins also stated, after a long discussion of human hunting skills and tools:

"The difficulty lies in recognizing whether a heavy arrowhead or a large spearhead, superbly and skillfully chipped from flint, was used for the hunt or as a weapon in fighting among humans. Only in one or two rare examples of later rock-art from south-east Spain are there pictorial references to the use of bows and arrows in conflicts between groups of people. Even then one is entitled to ask if what we are shown is a skirmish between rival bands or serious, organized warfare." [5]

I would amplify this qualification to seriously question if the rock art depicts a battle at all, as it can equally be interpreted as a scene of hunters engaged in a ceremonial dance of some sort, possibly in preparation for a hunt. Without some other evidence of violence in this same region, such as fortifications or skeletons with imbedded arrowheads, the Spanish rock art can only be viewed ambiguously.

Even so, if we give the benefit of the doubt to those who argue the Spanish rock art are battle scenes, it still would appear to be in agreement with the chronologies for first-origins of violence as given in *Saharasia*. There is only one undated rock-art reproduction in Watkins' article, from Morella la Villa, Castellon; a wider selection of similar rock art of the period is found in the work by Beltran, *Rock Art of the Spanish Levant*,[6] and it does contain a few scenes which are more supportive of the argument for group violence — as with the claimed "battle scene" at the Les Dogues site — but even here, the art may only record a village dance anticipating or celebrating a hunt. Whether violent, or not, it is reasonable to assume nearly all of this Spanish rock art is "late hunter-gatherer" period, approximating the "late Neolithic" identified in *Saharasia*, which would date the artwork no earlier than c.3000 BCE, well into the epoch of intense desertification which gripped North Africa.

Rock art depicting highly stylized and abstracted humans has been found in northern Australia,[7] dated as far back as c.8000 BCE, but the Australian images are even more ambiguous. Rock art which is so intensively stylized and abstract, such that even the simple form of a human being is difficult to make out from the drawings, where a specialist is required to point out what is a head or arm or torso, cannot be easily held up to conclude much of anything — especially when a simple line bisecting such a drawing is then interpreted as a "spear". To my knowledge the Australian scenes have not been matched to evidence of violence in skeletal remains in the region at those early dates. Below, some discussion will be given to the issue of confirmed interpersonal violence among ancient Australians in more southerly regions, also at very early dates — but significantly, only in relationship to a period of intense aridity and probable episodic famine.

Jericho, Catal Huyuk and Anatolia: Occasional and Discontinuous Violence in a Region of Early Drought and Desertification

Two of the earliest cities, Jericho and Catal Huyuk, are often misrepresented as having been subject to episodes of warfare during their earliest occupation layers, which have been dated to c.8350 BCE and c.6500 BCE respectively. Both had early enclosure walls which have sometimes been argued as evidence for fortifications — but without other evidence to support the existence of warfare, this interpretation is not warranted: the walls could just as easily have been for corralling and protecting domestic animals from roaming lions, hyenas or other large deadly or nuisance predators which are known to have inhabited those regions.

As discussed in *Saharasia*, the earliest evidence for social violence appears in Catal Huyuk and other Anatolian sites only temporarily, during a period of drought and attendant social decline, at around c.5200 BCE.[8] Drought and violence spread across Anatolia, Syria and the Levant *as a dominant and unrelenting social character only after c.5000-4300 BCE*, and Catal Huyuk was only finally destroyed after c.4800 BCE. This is close to the time of the world's first documented fortress, at Mersin, which was destroyed around 4300 BCE.[9]

The successive settlements and abandonments of Jericho were also timed to episodes of drought and land-abandonment across the wider territory, and there are walls and towers apparent at the site very early in its history. However, the earliest walls could have been for containing or protecting domesticated animals from predators, or possibly to protect against water and mud flows during heavy rains. The large circular tower of Jericho is one widely-noted bit of archaeology which is claimed to be "proof" of warfare, given its obvious similarities to towers found on genuine defensive fortifications elsewhere at later dates. However, a tower by itself does not warfare make. It could just as easily have been a lookout for predatory animals, or for long distance signaling.[10,11]

Jericho did eventually develop clearly defensive fortification walls, towers, and tombs for possible "kings", constituting some of the earliest evidence anywhere for possible conflict and social stratification.[11] However, like Catal Huyuk, Jericho's architecture does not prove itself to be the product of a social response to violent conditions, at least not until much later in the archaeological sequence, during periods of relatively harsh environmental conditions. Only then does the architecture take on a fortress-like quality, and unambiguously serve the purpose of protection against human attacks. Roper provided support for this viewpoint, stating that no signs of violence could be found at early Jericho, aside from the ambiguous walls and tower.[9]

Archaeologist Bar-Yosef made an extensive evaluation of early Jericho and came to basically the same conclusions,[12] additionally finding *an absence of evidence for warfare in the entire Near East region between 12,000 - 6,000 BCE*. While hunting technology was well developed, evidence for warfare could not be found.

Ancient ruin of Jericho (above top) and its Neolithic tower (bottom). Construction features such as towers and enclosure walls are not, by themself, evidence of warfare or social violence. Large walls can be impoundments for domesticated cattle, or protections against water and mud flows during rainy periods, while observation towers have many civil purposes. (from Kenyon[11])

In *Saharasia* I acknowledge the early evidence at Jericho and surrounding regions, stating:

"[Early] Jericho was deserted by c.7500 BCE... [leaving] no traces of violence at the site...[and this was] connected with the increasing aridity of the area. ...the evidence at Jericho appears to reflect the unique geography of the city at a time when temporary local or regional desiccation was occurring... Only fleeting visions of military conflict, fortifications, social stratification or cranial deformation occur in the Near East before c.5000 BCE, appearing here and there at isolated sites, and without any clear pattern or widespread distribution.... It is only after c.4000 BCE when desiccation became more widespread and intense that these initial traces of disturbed human behavior begin to blossom in clear, unambiguous and often organized institutional forms." [10]

Watkins also mentioned a clay sling-shot found at Catal Huyuk, and clustered buildings with rooftop entries and other factors which he interpreted as evidence for violence and warfare — but as discussed in *Saharasia*, James Mellaart, the man who excavated Catal Huyuk, viewed the same evidence firsthand and came to nearly opposite conclusions. [13]

It will be useful to review one of the original tables from *Saharasia* (Table 2, page 365), giving general dates for the onset of desert conditions, and the onset of first-evidence for patrism and violence. The Middle East, Anatolia, Iran, and Soviet Central Asia show their earliest signs of climatic degradation towards aridity at c.5000 BCE. Jericho was affected by these oscillatory environmental pressures much earlier, perhaps as a chronic feature of its unique geography, close to the Dead Sea and Jordan Valley. [10,11] The arguments presented in *Saharasia* therefore anticipate some discontinuous and episodic signs of social turmoil and conflict starting at *approximately* those same dates, but without persisting or widespread effects.

European Causewayed Encampments

There are now excavated a whole series of *causewayed encampments* which existed as central gathering places across Western and Central Europe. Regrettably, these are too often misrepresented in popular accounts as "fortifications", through the error of mixing the dates of *first habitation* with the dates for appearance of *first violence*, without careful reference. The error is, *extrapolating the violence backward in time*, without evidence for doing so. The field archaeologists who excavated these encampments were not convinced the earliest habitations had any clear warfare or defensive functions. The encampments were composed of concentric rings of shallow earth hills and trenches posing no significant obstacle to climb — with only about one meter distance between hilltops and trench-bottoms — which were also repeatedly broken with wide openings or "causeways" to facilitate the free passage of people in and out of those encampments, from the periphery all the way into the core. They appeared more in the manner of an unusual village architecture with mounds for privacy screens or trenches for animal corrals, allowing for separate family encampments. They appear to have served the functions of a central place for trading and seasonal gatherings, and in some cases as cemeteries. Later in the archaeological sequences, many of these encampments were raided by warriors using bows and arrows, and the battle-axe. Only then were the encampments transformed into closed defensive fortifications which rapidly were destroyed and/or abandoned.

Keeley gives dates of "5000 BC" or "4000 BC" for the appearance of violence at these sites. My own review of his cited references could not confirm such early dates. As best as I have been able to determine, from various reports published in many different languages, the earliest violence is documented at those encampments farthest to the east, as in Bavaria (c.3200 BC) followed by later conflicts in France and Denmark (c.2800 BC), followed lastly by conflicts in England (c.2600 BCE) — if true, this would be excellent *confirmation for Saharasia*, suggesting the arrival of violent invaders from Anatolia or Central Asia, moving on a Westerly migration route. Details on a few of these specific sites will follow.

The primary source for the "causewayed encampments" is: *Enclosures and Defences in the Neolithic of Western Europe,* edited by Colin Burgess, et al. [14] The various contributing authors, all of whom were field archaeologists who excavated these sites, pointed to the general dates given above for the first onset of violent conditions. Evans has given a general overview of the causewayed enclosures:

"Few monument forms have undergone such frequent radical re-assessment in their interpretation. Even now, after twenty-one examples have been excavated, they still stubbornly frustrate neat categorisation, and we are left with the impression of the blind men encountering the elephant... Unlike other major 'ritual' sites of the third and second millennia bc, the status of causewayed enclosures as 'monuments' has been somewhat ambiguous; their morphology would link them superficially with both henges and hillforts, yet their segmented ditches have led to doubts about their defensive capability..." [15]

In speaking about "The Neolithic Höhensiedlungen (high settlements) of Central Germany", Starling states:

"It is suggested that these sites were the communal foci of groups who used them for a variety of symbolic and practical activities, rather than centres of political and territorial control." [16]

<u>Hambledon Hill, Dorset England</u> is considered to be the oldest causewayed enclosure in England dating to around 3500 BCE. It was located close to a river system connecting to the sea, suggestive of an optimal location for trading of regional agricultural and other goods by boat. Archaeologist Mercer stated:

"About sixty [enclosures] are now known to exist within the southern half of England, and they range in size from about 1 - 60 hectares and in location from seasonally waterlogged valley bottom sites to sites set on hilltop and promontory positions... from single ditched enclosures to sites with up to five concentric rings of ditches with...a wide range of function. As a class of site, however, they are united by one idiosyncratic constructional feature - the ditches consistently appear to be 'causewayed' or interrupted at frequent and irregular intervals, in a manner that suggests that they were not conceived by their build-

ers as barriers in their own right but simply as a linear quarry for the construction of an internal bank or rampart." [17]

The Hambledon site was progressively transformed into a fortification, its ditches containing macabre evidence of corpse disposal, with human skulls set upright

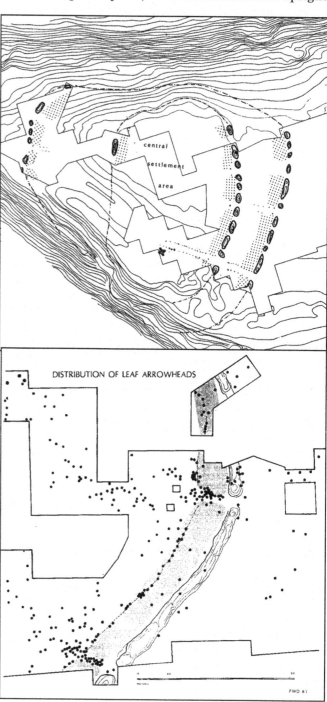

Hambledon Hill Stepleton Enclosure, Dorset, England (Top) the earliest such enclosure in the British isles, dating to c.3500 BCE. Bottom: Skeleton of a man, shot in the back with an arrow, with child, preserved under a collapsed and burned wall dating to c.2680 BCE, the earliest evidence for social violence in the UK. (from Mercer [17])

Crickley Hill, England (Top). Wide "causeways" allow free movement from the periphery to the core of the settlement. Constructed in the early Neolithic (c.3000 BCE?), it was attacked and eventually destroyed between c.2000-1000 BCE. Bottom: Distribution of arrowheads from an assault dated after c.2000 BCE. (from Andersen [18])

along periphery.[17] Also, many bones of young children were found in one section, giving the overall impression of slaughter and mayhem. The site was finally destroyed by fire during an attack by archers, with an arrowhead in one skeleton of a man carrying a child, found under a collapsed building wall. There is no question, this site showed violent events taking place. But at what dates? Carbon material found in trenches, which contained a vertical mixture of materials from different settlement periods, were dated by radiocarbon at 2610 BC, 2730 BC and 2890 BC, with errors of +/-150 years. Other radiocarbon dates were recorded at 2530 BC, 2650 BC and 2720 BC, with errors of +/- 130 years.[17] This is a rough average of 2686 BC, which suggests Hambledon Hill experienced perhaps *a thousands years of peaceful habitation before the unambiguous appearance of warfare.*

Crickley Hill, Glouchester England was first occupied in the "early Neolithic". The archaeologist Dixon reported the earliest occupation had a series of mounds organized into rings, with shallow trenches and causeways leading into the interior. No artifacts of any kind were found in those earliest phases. At the final enclosure (phase 1d) there were larger ditches with fenced roads leading into the interior. Dixon reports:

"The fate of this final enclosure was clearly shown by

the thick spread of flint arrowheads, over 400 of which choked the eastern entrance passageways and fanned out along the roadways into the interior. The enclosure had quite obvious been defended against archery attack..."[18]

After this attack, there was erected

"... a 70m long track leading up to a circular platform inside the enclosure, totally flat and clear of structures except at edges... we may consider it to have been the settlement's shrine" (p.84) and *"Like the Danish examples, the Crickley shrine was burnt down."*[18]

Clearly, this is evidence of warfare and violence — but at what date? Dixon says:

"The date of the end of the ritual phase 1e is still uncertain, though radiocarbon dates may eventually provide a guide. It occurred before the building of the hillfort, the latter perhaps early in the first millennium BC"[18]

This is the only mention of a date in the entire excavation report, but clearly demonstrates a very long period of peaceful conditions, from the "early Neolithic" (c.3000 BCE?) until the appearance of violence sometime after c.2000 BCE, with its final destruction after c.1000 BCE.[18]

A large number of similar causewayed enclosures

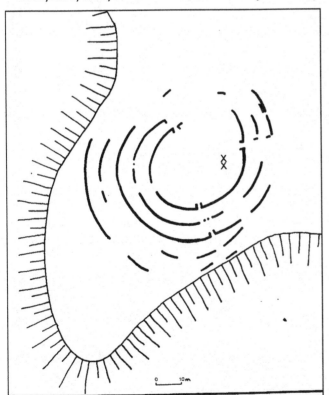

Causewayed Enclosure, Schalkenberg, Quenstedt, Germany, dated to c.3300 BCE. Sites in Bavaria have C14 dates of burning and abandonment ranging between c.3000 - 2000 BCE. (from Starling[16])

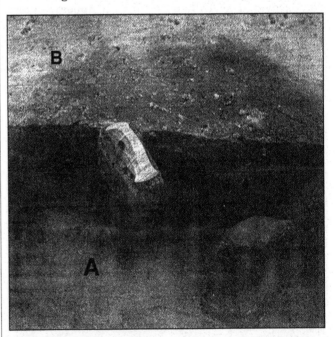

Battle-Axe at Causewayed Enclosure Camp in Sarup, Funen, Denmark. The lower part of the excavation (labeled A) is dated to c.3500 BCE, suggesting a more recent date for the upper strata (B) containing the battle-axe. (from Andersen[19])

are found scattered across continental Europe, as far east as Germany, all with generally similar architectures and probable social functions. As best as I can determine, all show peaceful conditions at their earliest dates of construction, with violence appearing only later on, in keeping with the generalized dates given in Saharasia for the onset of violence in Europe.

The biased "popularizers" of ancient human violence merely cite the approximate dates when these causewayed encampments were firstly constructed, and leave the reader with a clear impression those were the dates when violence first appeared. However, the dates for the onset of violence were in fact as I have given them above in these representative examples.

It should also be mentioned, that the Spanish rock-art depicting an apparent archery battle, mentioned above and identified with the "late Neolithic" or "late hunter-gatherer" period, may be chronologically connected to the same appearance of violence in the above causewayed encampments. Again, this is all support for the chronology and geography of violence as given in Saharasia.

Talheim, Schletz and Ofnet Cave, Germany: Massacres or Skull Burial Customs?

A collection of 50 human skeletons with evidence of trauma injuries was found at the Schletz site, dating to c.4000 BCE by radiocarbon determination. Another collection of 34 skeletons but with contradictory dates (of c.5500 to 4000 BCE depending upon dating method) was excavated at Talheim.[20,21,22] It appears likely, these people were in fact massacred, as determined from many trauma blows and the haphazard manner in

which the bodies were heaped into a shallow ditch. The dates suggest a connection to the social disruptions which led to the destruction of the causewayed encampments across Germany and the rest of Europe. The ambiguity in dating on these sites appears as a consequence of different dates being estimated by different researchers, and by different results being obtained from different radiocarbon laboratories. This problem also appears to affect the Ofnet site.

The Ofnet Cave in Bavaria is one of the most widely-cited "proofs" of early evidence for social violence. Claims have been made the site is proof of a single massacre, with the possible taking of heads as trophies of war.[4,21,23] From reading such accounts one would never know there were dissenting voices on the subject. The original excavation was undertaken in 1912 by Schmidt,[24] with all subsequent discussions on the finds focusing upon the remains of the skulls themselves. Grahame Clark described the site as follows:

"A more specialized form of collective burial is implied by nests of skulls found in caves and rock-

Ofnet Skull Nest, Germany, misrepresented as a group massacre, actually appears as a site for sequential skull burials which included grave offerings. (from Schmidt[24])

Massacre at Talheim, Germany, dated between c.5500-4000 BCE. (from Bahn[20])

Neolithic Skull Burials from Jericho, evidence of a burial custom, not massacres. (from Clarke[25])

shelters in south Germany, notably at Ofnet and Kaufertsberg near Nördlingen and at Hohlestein, Lonetal, near Ulm. Signs of cutting on the upper neck vertebrae suggest that the skulls had been detached from their trunks shortly after death. Their numbers, one nest at Ofnet comprising twenty-seven and another six skulls suggest that they relate to social groups comprising in all probability a number of hunting bands. Again their condition, those in the middle showing signs of having been pushed together and those on the periphery relatively intact and undisturbed, argues that, as in the later chamber tombs, they had been buried over a period of time." [25]

The Ofnet skulls — composed of four male, seven female, and 15 children — were coated with red ochre, and accompanied by personal ornaments and microliths. No mention of a violent massacre was made by Clark.[25] A more recent study of the skeletal materials by Jörg Orschiedt of the Archaeological Institute at the University of Hamburg confirms not only the *sequential* burial of the crania, but also refutes the theory of violence for all except a small sample of the skulls.

"A reexamination of the skulls from the Ofnet cave in southwest Germany showed that these and similar deposits should be understood as the expression of a special burial custom rather than head hunting practices from the late mesolithic. ... the reduction of group sizes in the late mesolithic as well as the demographic structure makes it unlikely that this deposit was a single event. The site was used several times as a burial place. As grave goods perforated canines of red deer and shells, probably necklaces, were placed on or around the heads. Red ochre was found around the heads and in the filling of the pits. The reexamination of the traumatic lesions on the Ofnet skulls showed that at least only six individuals had died from fatal blows. These heads were deposited on the northwestern rim of the larger skull pit and could possible represent a single event. The injuries were caused by a blunt, axe-shaped object. Most of the injuries are located in the occipital area. The only exception are two male individuals with several traumatic lesions which occur also on the parietal and frontal areas." [26,27]

These descriptions considerably tame down the descriptions of "massacre at Ofnet" from 32 individuals to a maximum of six.

The concept of skull-burial as a funeral custom, it must be noted, has a long history extending to sites beyond only Ofnet. Skull burials were found in Jericho, unrelated to any kind of violent death.[11,25] The city of Hallstatt, Austria, still has a display of hundreds of decorated skulls in a small church (now a museum) which the author visited most recently, evidencing a burial custom which ended only in the 1960s.

The dates for the Ofnet cave further confuse their relevance to the larger origins-of-violence question. Several radiocarbon dates of c.11,000 BCE were obtained in the 1980s, but these are today rejected in favor of newer dates of around 5500 BCE, obtained with newer methods said to be more accurate.[26,27] Assuming the six crania mentioned by Orschiedt were factually the consequence of deaths by violence, the date of c.5500 BCE would still place them too early to be explained by any invasion of warrior groups out of Central Asia at c.4000 BCE, when that region was being abandoned due to the pressures of desertification. However, we might postulate some kind of migratory invasion from Anatolia and the Levant, bringing social violence into Europe from that region at c.5500 BCE.

The geographical placement of the Ofnet site, in relative close proximity to the Talheim and Schletz sites, suggests a regional clustering of deadly events which, irrespective of chronology, are highly anomalistic and isolated in character, occurring as they do against a larger background of peaceful conditions across the wider geography of Europe for the greater part of prehistory. Only punctuated examples of violence seem to have occurred.

At the time of my writing and publishing of the Saharasia findings between 1986 and 1998, and even for the first "Update on Saharasia" article as published in Spring of 2002,[62] these findings at Talheim, Schletz and Ofnet, and a few other isolated examples across southern Europe dated from c.5500-4000 BCE remained a puzzlement. While it might have been possible to explain the violence seen at those sites as the consequence of the arrival of Central Asian, battle-axe and

Skulls on display in Hallstatt, Austria, in an old church.

Kurgan peoples, who entered Eastern Europe with much destruction after c.4000 BCE, violence dated to c.5500-4000 BCE could not be so readily explained. However, as mentioned in Saharasia (p.259), there always were *"fleeting glimpses"* of drought and aridity, coupled with military conflict, fortifications social stratification and cranial deformation in the Near East/Anatolia region *"before 5000 BCE, appearing here and there at isolated sites"*. This was the basis for my postulate, of some connection between Central European violence of c.5500 BCE, to the drought, land-abandoment and violence as seen in the Levant and Anatolia at around the same time.

During a trip to Germany in December of 2002, at a visit to an archaeological exhibition at the Martin-Gropius-Bau Museum in Berlin, I was stunned to see a large wall map on display which solved the mystery and supported my hypothesis. The map, first published by Andreas Zimmerman of the Universität zu Köln,[63] identified the diffusion pathways of agricultural technological development into Europe, starting from a point of origins in Anatolia and the Levant at around 9000 BCE, and from there over several millennia developing northwest into Europe via Mediterranean and Balkan pathways. By 6600-5500 BCE, this diffusion network had reached deep into Central Europe and Germany. Appendix Figure 1, shown here for the first time in my research, reproduces the essentials of Zimmerman's map. Viewing this map, one can easily imagine the development of desert-like conditions at c.5500 BCE (or earlier) across the Levant and Anatolia, being followed by dislocations of affected people northwest into Europe, following those same migratory pathways. In fact,

we might speculate that it was chronic drought and desertification which may have provided a major impetus for this particular migratory pathway, which late in the process came to be characterized by isolated bands of violent warrior-nomads. The consequence of their arrival, and clashes with local peoples of a more peaceful character, thereafter shows up in European archaeology, but only in a haphazard and isolated manner.

Ambiguous Evidence for Early Violence in China

Some skeletal remains found in China are also often cited as evidence for "very early violence", but again, the original archaeological report in question tends to undermine this interpretation, and confirm *Saharasia*. Underhill has written on the subject of warfare in Neolithic China,[28] and was cited by Keeley for his "early China violence" assertion. Underhill did discuss the finding of a skeleton of a man with an arrowhead in his thigh, dated to around c.5000 BCE, and found buried in a *Yangshao* archaeological strata, which is generally acknowledged to hold no clear or unambiguous evidence for warfare or violence. This single skeleton is the only recorded case of a Yangshao skeleton with an imbedded projectile point, to my knowledge, and the site where it was found holds *no other evidence for war or violence*. Taken together, the evidence suggests a hunting accident. This idea was also considered by the field archaeologists and written into their report, but rarely gets mentioned.

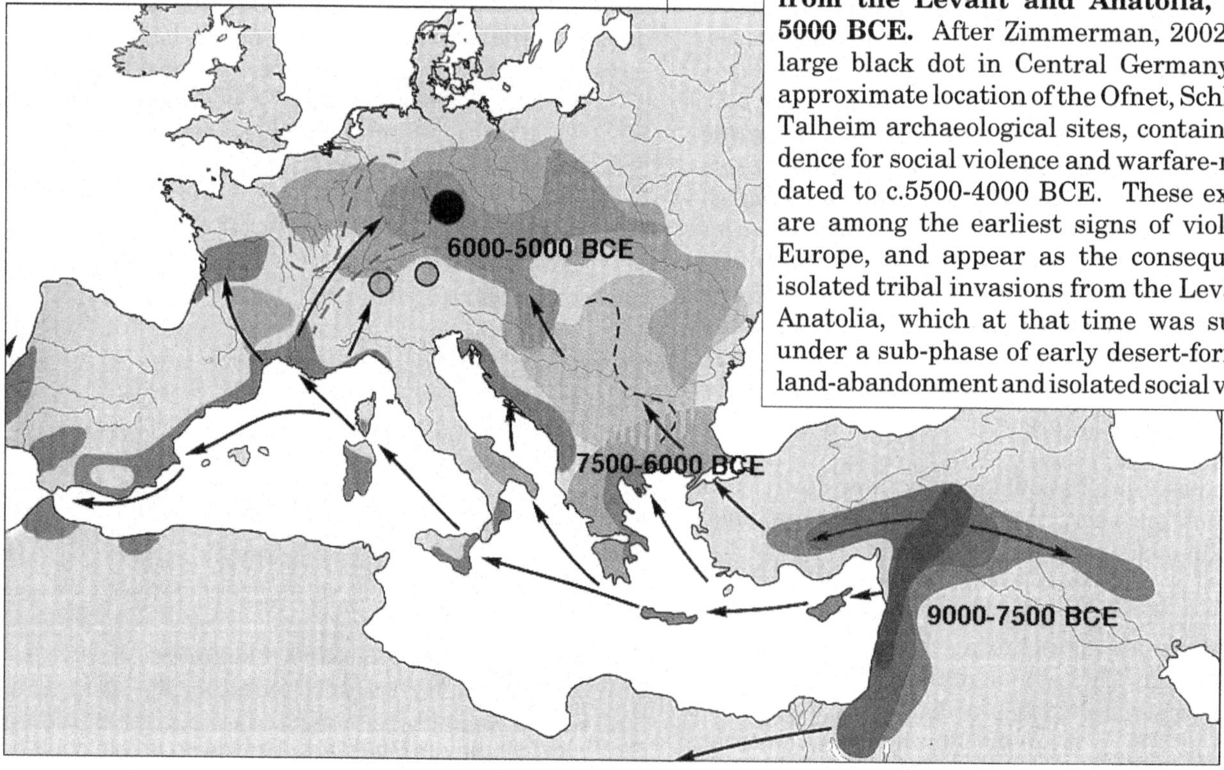

Figure 1: Pathways for Agricultural Diffusion (and Violence?) into Europe from the Levant and Anatolia, c.9000-5000 BCE. After Zimmerman, 2002.[63] The large black dot in Central Germany is the approximate location of the Ofnet, Schletz and Talheim archaeological sites, containing evidence for social violence and warfare-murder, dated to c.5500-4000 BCE. These examples are among the earliest signs of violence in Europe, and appear as the consequence of isolated tribal invasions from the Levant and Anatolia, which at that time was suffering under a sub-phase of early desert-formation, land-abandonment and isolated social violence.

6000-5000 BCE

7500-6000 BCE

9000-7500 BCE

Underhill also presented a chart for "defensive structures" in Neolithic China, and specifically identified two in the Yangshao period (before c.2500 BCE). However, both were marked as having a "debatable defensive function" — both were mere ditches surrounding habitations, or segments of ditches "possibly" joined by palisade-style fences. These are not conclusive by any means, and are at best ambiguous evidence for warfare and violence. One must ask, if these people had permanent settlements and domesticated animals, where did they keep them if not inside such an enclosed compound?

Later evidence for warfare in China is unambiguous. In discussing the subsequent *Longshan* culture, Underhill describes "...*evidence for a degree of violence not present during the pre-Longshan period.*"[28] These include grave evidence for mass executions, amputations, scalping, hacking of the limbs, and battle deaths, along with various weaponry (including jade battle-axes) not found in earlier times. Also present during the Longshan were child-sacrifice under or near foundations of buildings. Underhill also gives a chart identifying weapons found in various archaeological sites, such as axes, knives, spearheads, and arrowheads.[28] The earliest of this evidence is dated to c.2700-2100 BCE, and comes from Anyang, home of the earliest totalitarian Chinese society (Shang Dynasty), which was formed by invaders from the western, desertified regions of Central Asia.

All of these findings are in good agreement with what has already been written in *Saharasia*,[2] on p.345-348. The transition time from generally peaceful conditions to intensive warfare in China of c.2500 BCE given by Underhill, is in approximate agreement with my own figures for the first-time arrival of violence in Western China.

Jebel Sahaba, Egypt: Unambiguous Evidence for Social Violence and Warfare/Murder During an Early Period of Intensive Aridity

The ancient cemetery at Jebel Sahaba, on the desert highland plateau overlooking the Nile River Valley in Egypt, contains over 50 persons who were victims of a massacre, shot up with projectile points and showing other signs of violent death. The violence is unquestionable, and in Saharasia I had relied upon the chronological discussion by Michael Hoffman in his *Egypt Before the Pharaohs*,[29] which in keeping with other signs of violence in the region I had gathered, allowed placing Jebel Sahaba at c.4500 BCE.

Fred Wendorf's *Prehistory of Nubia*[30] presented the original field archaeological reports, which ambiguously placed the site between 12,000 BCE all the way down to 5000 or even 4500 BCE, based upon similarities between the flint projectile points imbedded in the skeletons to Qadan-era stone tools found at nearby

Jebel Sahaba Cemetery, unambiguous evidence of social violence and murder or warfare on the village level, during an epoch of intense aridity. (reproduced courtesy of Fred Wendorf, from the *Prehistory of Nubia*)[30]

sites. Wendorf originally openly expressed concerns about the ambiguous dates, but only in more recent years have radiocarbon evaluations been undertaken of the skeletons themselves. He discussed the newer findings, as follows:

"The Jebel Sahaba skeletons have only one post 1968 C14 date of 13,700 bp [11,700 BCE] on collagen from a human femur. It is discussed in the Conclusions to our book on Wadi Kubbaniya (1989) SMU Press. I wish we had more dates, but this agrees well with the Gadan artifacts imbedded in the skeletons. In 1968 the Gadan was not well dated, but subsequent work places that industry between 14,000 and 12,000 bp [12,000-10,000 BCE]. This was not the oldest evidence of violence in the Nile Valley. The Wadi Kubbaniya skeleton had a healed parry fracture, a partially healed wound with point imbedded in right humorous, and two points in the lower abdomen that killed him. This is dated by geology and the artifacts at greater than 20,000 bp [18,000 BCE]. There was some violence in the Nile Valley. Competition for limited resources?" [31]

This new information was somewhat eye-opening, as superficially it appeared to challenge the conclusions of *Saharasia* — in fact, upon deeper analysis, it provided a *confirmation* for the arguments given in *Saharasia*, for the environmental-social-emotional origins of violence. Jebel Sahaba could be dated to c.11,700 BCE, with yet other evidences for violence at c.18,000 BCE at nearby Wadi Kubbaniya. These dates, I noted, were certainly before my identified Saharasian transition dates of c.4000-3500 BCE — in fact, the dates were well before the Neolithic Wet Phase of North Africa, occurring at a time I had not even subjected to evaluation or review, given the widespread evidence for peaceful conditions during that Wet Phase.

Further investigation eventually resolved the question as follows: New research from the study of ancient climates is presented in Appendix Figure 2, revealing North Africa was extremely dry and arid during that early period of c.21,000-8000 BCE, similar to the modern condition of the Sahara Desert, but *well before the Neolithic Wet Phase.* The maps are from a larger global climate-mapping project directed by Jonathan Adams of Oak Ridge Laboratories, who prepared sequential maps of climate throughout the Quaternary, as based upon all available scientific evidence.[32] In fact, some of this evidence for a very early dry North Africa had been presented in *Saharasia*, though without discussion. Figures 50 and 51 in Chapter 8 of Saharasia,[2] p.221-222, depict this pre-8000 BCE dry period before the Neolithic Wet Phase.

These African maps of climate change show the transitions identified in graphs, but not discussed in my *Saharasia* research, regarding a very dry period in North Africa before c.8000 BCE, and prior to the wet and lush period which lasted from c.8000 BCE until at least c.4000 BCE. After c.4000-3500 BCE, dryness again gripped North Africa, and indeed all of Saharasia.

Taken together, these data demonstrate, *the vio-*

lence documented at Jebel Sahaba occurred during a very dry period in North African prehistory — a time of desert, low-vegetation and probable famine conditions — the violence did not occur during a time of plentiful food supplies. As such, the evidence from Jebel Sahaba and Wadi Kubbaniya supports the overall Saharasian discovery through validation of the environmental-social-emotional mechanism.

Another interesting factor which may be related here, is the existence of a few other skeletal remains suggestive of violent deaths in Sicily and southern Italy. Thorpe has reported:

"Two late Paleolithic bodies from about 11,000 BC have been found in Italy with flint points lodged in the bones. One from San Teodoro cave in Sicily, was a woman with a flint point in her pelvis. The other was a child with a flint point in its backbone, found in the Grotta del Fanciulli on the Italian mainland. Whether the points were spear-tips or arrowheads is unclear." [21]

While it is possible these were examples of hunting accidents, it is within the scope of the overall Saharasian theory that migrations from a more violence-prone desertified North Africa could have occurred, to bring social violence into the moister territory of Sicily and Italy — and perhaps even farther north into Europe — at that early period. However, if so, the patristic-violent influence must have withered away in both Europe and North Africa following the onset of the Neolithic Wet

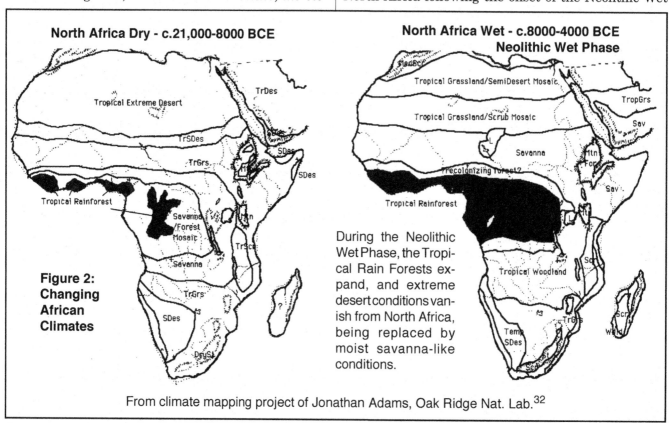

North Africa Dry - c.21,000-8000 BCE

North Africa Wet - c.8000-4000 BCE
Neolithic Wet Phase

Figure 2: Changing African Climates

During the Neolithic Wet Phase, the Tropical Rain Forests expand, and extreme desert conditions vanish from North Africa, being replaced by moist savanna-like conditions.

From climate mapping project of Jonathan Adams, Oak Ridge Nat. Lab.[32]

Phase in North Africa after c.8000 BCE. After that date, when moist conditions and food abundance returned to North Africa, there is little or no clearly identifiable social violence or warfare over the next 4000 years.

Ancient Artificial Cranial Deformation and a Cluster of Early Violence in S. Australia

Another challenge to my findings in *Saharasia* came in the claims for artificial infant cranial deformation among very early human cultures in Australia and elsewhere, shortly after the close of the last Ice Age, at c.10,000-8,000 BCE. Artificial cranial deformation was described in *Saharasia* as originating accidentally among nomadic peoples who used various kinds of infant head-bindings and cradle-boards, to secure the child in some kind of harness which was carried by adult caretakers on a long trek. Very harsh desert conditions were theorized to be underlying the infant cranial-deforming practices, especially where they appeared among a higher percentage of the population, eventually to become an admired group-identification feature. Deforming head-bindings were subsequently applied to infants as a "social custom", to continue the identifying marks even after the tribal group had settled down. Artificial cranial deformation therefore appeared as a trait which originally started by accident, but which spread with deliberation with the growth of nomadic lifestyles, and militant nomadism specifically. As discussed in a chapter in *Saharasia*, in Eastern Europe and Central Asia, at least, a deformed head often became a mark of the ruling class. The deformations were then undertaken more purposefully and with extreme measures indicative of a great deal of pain and agony for the infant. Mild forms of infant cranial deformation may therefore be associated with dry desert conditions and nomadic subsistence. More extreme forms as found in high-caste central-state societies surely were life-threatening ordeals for the infant, who also was swaddled tightly as an associated custom. Both of these practices, I argued in *Saharasia*, marked a *severe loss of emotional and nurturing contact between mothers and babies,* with generally low parenting skills combined with a buried anger towards the child (ie, a willingness to inflict painful trauma upon babies for the sake of "cultural tradition").

The existence of this painful practice at such very early archaeological periods superficially appeared to challenge the findings of Saharasia as the source of human armoring and child-abusive practices. However, a close examination revealed this was not the case.

Firstly, the most ancient examples for artificial cranial deformation appear to have little in common, in terms of the severity of the deformations, as compared to the more intensive and deadly practices of more recent historical periods. The examples of artificial cranial deformation given in *Saharasia* demonstrated adult skulls of frightening proportions, with foreheads towering upwards in a highly abnormal manner. The deformations were unmistakable, even to the non-specialist, based upon one's general observational knowledge of what the normal human crania looks like. By comparison, the late Pleistocene examples of artificial cranial deformations from Eurasia and Australia were quite minor, and even difficult to identify by the non-specialist.

The more severe deformations from more recent historical times surely produced a far more extreme infant trauma, with a more extreme disruption of the maternal-infant bond, and with more profound psychosomatic consequences as compared to the prehistoric examples. One can simply look at the skulls side-by-side, to get a sense of the greater amounts of pressure (using boards, tourniquets, metal bands, etc.) which must have been applied to the historical infant crania, and for longer time periods, to create their crania of more distorted and gigantic proportions. By comparison, the *prehistorical* infant cranial deformations could have been produced by simple cloth bands or flexible straps, for much shorter period. Some of the most ancient examples are today reclassified as "questionable", while others may be the consequence of adult activities, such as use of a forehead strap to carry heavy loads "Kikuyu style". Even so, some of the prehistorical deformations were of apparently sufficient severity as to correlate with episodes of social violence.

A very ancient Neanderthal crania (Shanidar 1, below) from c.53,000-42,000 BCE was once considered an example of artificial cranial deformation, but today this is considered highly questionable. Other skulls have been found in Jericho, Cyprus, Iraq, Lebanon and Syria, dating between 7-4,000 BCE.[31] As mentioned in *Saharasia* (and quoted above), these latter examples appear alongside correlated evidences of drought, land-abandonment and some isolated signs of social violence; a sub-phase of aridity existed, which spread across those same regions, strongly suggesting the genesis of this pain-inflicting ritual to the use of the nomadic backpack cradleboard.

A Chinese crania of early Homo sapiens (Shandingdong Upper Cave 102, below) was acknowledged as having suffered severe postmortem damage, but nevertheless is considered to represent an isolated early case of artificial deformation from adult use of the forehead strap. As such, it would represent a feature created not in infancy, but after the child was able to walk around and carry a heavy load. The location was near Beijing, dated somewhere between 30,000 to 8,000 BCE, a very uncertain time span.[33] In any case, this isolated example of *adolescent-adult* cranial deformation does not suggest infant trauma which might push an entire social group towards violent behavior. And as mentioned previously, no such violence has been found in the early Chinese archaeology.

As presented by archaeologist Peter Brown, the examples of cranial deformation from late-Pleistocene Australia appear better documented, with larger numbers of examples from sites such as Coobool Creek, Kow Swamp, Nacurrie and Cohuna. These sites are all found in SE Australia, dating from c.11,000-7000 BCE, where some additional evidence of tribal violence and conflict is also present:

"...well demarcated, single or multiple depressed fractures [exist] on the frontals or parietals of 59% of the females and 37% of the males. The majority of the fractures were located on the left side of the frontal and left parietal, which is consistent with a blow from a right-handed person, where the combatants are facing each other. In each instance there was bone regrowth associated with the fracture indicating that the people had survived what was often severe trauma." [34]

The above findings suggest a childrearing mode which tolerated a high degree of infant discomfort and trauma, in association with an adult culture infused with impulsive but generally non-lethal episodes of interpersonal violence. Given its non-lethal character, it is most probable that this violence was confined within existing social groups rather than indicating tribal warfare per se, though tribal conflict of a non-lethal nature cannot be ruled out. The fact that more female skulls showed depressed fractures than male skulls (59% versus 37%) demonstrates a significant social rage directed towards females,[34] who probably were the ones to whom the responsibility of culturally-demanded artificial cranial deformations was entrusted. If so, these Australian skulls may be the earliest evidence to exist showing the relationship between a harsh and pain-inflicting ritual directed at infants, which later produced a social violence directed more often than not towards the maternal figure.

What of the climatic conditions in SE Australia at this early time period, of c.11,000-7000 BCE? According to Adams' climatic reconstruction,[32] the period from c.16,000-10,000 BCE was extremely arid in most of SE and Central Australia, much drier than as seen in the modern times. After 10,000 BCE, conditions changed towards a slightly moister situation in those regions, more characteristic of the modern "outback" steppe or savanna-like climate, inland from the coastal zone. To quote from Brown again:

"Although there is an ethnographic account of cranial deformation from northern Victoria, there is no

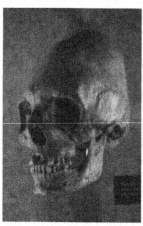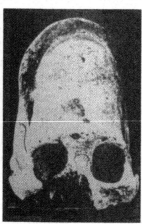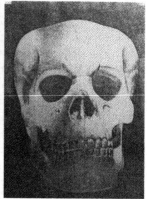

Figure 3: Comparison of Deliberate versus Accidental or Uncertain Artificial Cranial Deformation

Severe historical examples of Artificial Infant Cranial Deformation from the last several thousand years.
Left to Right: Russia, Peru, Mexico

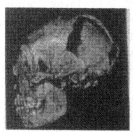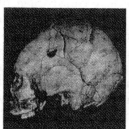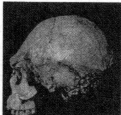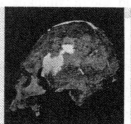

Ancient examples of mild or uncertain Artificial Infant Cranial Deformation,
Late Pleistocene/early Holocene periods.[33]
Left to Right: Shanidar 1 Neanderthal (Europe), Shandingdong Upper Cave 102 (China),
Coobool Creek (Australia), Kow Swamp 5 (Australia), and Cohuna (Australia).
(Reproduced courtesy of Peter Brown,[33,34] from http://www-personal.une.edu.au/~pbrown3/Deform.html)

evidence of the morphological pattern associated with deformation in the several thousand 'recent' crania from Victoria, South Australia and New South Wales in Australian museum collections. There is also no evidence of cranial deformation in the prehistoric samples from Roonka (7000 BP)... the mid-Holocene Barham series... or the Murray Valley group... and dated to 6000-750 years BP. In Australia, artificially deformed crania have only been recovered from Kow Swamp, Nacurrie and Coobool Creek. These sites are in close geographical proximity... The presence of artificially deformed crania in these three sites, and their absence from mid-Holocene and recent sites in the same area, suggests that they share a common cultural and chronological association."[34]

To summarize: The Australian sites mentioned here are located in the same general region. Extreme desert conditions existed across this region at the time when cranial deformations first appeared and were adopted; this suggests they developed from environmental pressures known to demand an intensive nomadism and use of back-pack cradles or similar apparatus for securely carrying babies around, which also deformed their crania. Social violence of a limited nature also developed around the same time, from the full complex of human responses to aridity and famine as noted in my Saharasia. Cranial deformations later became a social institution, and were purposefully recreated in later generations. Finally, both cranial deformations and social violence gradually disappear from the archaeological record following centuries of somewhat better environmental conditions and food supplies, disappearing entirely after c.7000 BCE.

Post-Saharasian Violence Among Pre-Columbian Tribal Cultures

One of the more controversial assertions made in *Saharasia*, was that the earliest migrants into the Americas were of a uniformly peaceful character, not prone to social violence because they held a more matristic and unarmored form of social organization. They attended to the needs of infants and children, and did not sex-repress their adolescents and adults. This argument was supported by the ethnographical evidence presented in the World Behavior Map, but also by the geographical locations of those cultures in the Americas which were of a more violent characteristic. Violence in the Americas, before Columbus, was found only in certain locations, and was not widely or randomly distributed on the map. The reader is referred back to my *Saharasia*[2,3] for full details on this question. Here, I am mainly interested in the following question: Do the locations of various archaeological sites recently dug up and showing clear evidence for violence among native

North American cultural groups, before the arrival of the Europeans, agree with the locations for violence in the Americas as determined by the World Behavior Map? Or not? This question can be directly answered by a locational comparison, as follows.

We can summarize some recent publications documenting either significant and ongoing interpersonal or intergroup social violence, as determined from skeletal remains, or even outright massacres suggestive of merciless and intensive tribal warfare, well before the arrival of Europeans into the Americas. The facts presented in the various papers are not in question. The point of interest for this paper, and for my Saharasian discovery, are the *locations* and *chronology* of the various archaeological sites, which are summarized in the following listing.

Major New World Sites of Violence and Warfare

1. SE Michigan, Riviere aux Vase, c.1000-1300 CE. Collection of several hundred skeletons showing signs of conflict and violence, predominantly against women.[35]

2. Illinois, Norris Farms, pre-Columbian. Substantial intergroup violence.[35]

3. South Dakota, Crow Creek, c.1300 CE. Site of a tribal massacre of around 500 individuals, men, women and children, but with a deficit of reproductive-age females.[20,35]

Crow Creek Massacre, South Dakota, a collection of 500 individuals killed in inter-tribal violence, c.1300 CE, in a region identified on the World Behavior Map as possessing isolated armored patrist groups within a background majority of unarmored peaceful matristic cultures.(from Bahn[20])

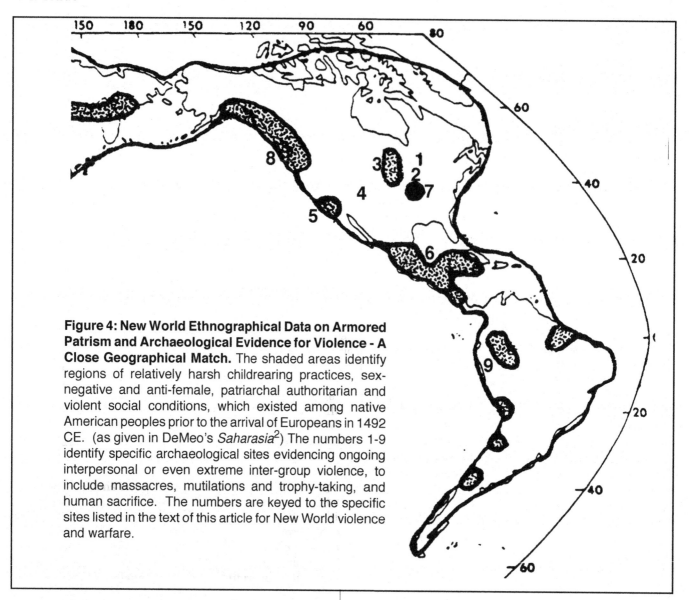

Figure 4: New World Ethnographical Data on Armored Patrism and Archaeological Evidence for Violence - A Close Geographical Match. The shaded areas identify regions of relatively harsh childrearing practices, sex-negative and anti-female, patriarchal authoritarian and violent social conditions, which existed among native American peoples prior to the arrival of Europeans in 1492 CE. (as given in DeMeo's *Saharasia*[2]) The numbers 1-9 identify specific archaeological sites evidencing ongoing interpersonal or even extreme inter-group violence, to include massacres, mutilations and trophy-taking, and human sacrifice. The numbers are keyed to the specific sites listed in the text of this article for New World violence and warfare.

4. La Plata River Valley, Four Corners, c.900-1300 CE. Substantial non-lethal interpersonal violence, especially against females.[36]

5. Santa Barbara Channel, S. California, c.1490 BCE or earlier to 1804 CE. Collection of 753 remains, demonstrating healed non-lethal cranial vault fractures in 128, or 17%,[37,38] with a similar high percentage of projectile point injuries and deaths. Males were more affected than females, children or the elderly, suggestive of combative roles.[38]

6. Central Southern Mexico, sites at Tetelpan, San Luis Potosi, and Mexico City, c.500 BCE - 1521 CE. Substantial interpersonal and intergroup violence with organized warfare, human sacrifice and possible cannibalism.[39]

7. W. Tennessee Valley, primarily late Archaic, c.2500-500 BCE, possibly earlier. Collection of several hundred skeletons showing signs of violent death and trophy-taking.[40]

8. SE Alaska, British Columbia, NW USA - Pacific NW Coast., c.3000 BCE - 900 AD. Substantial interpersonal violence with non-lethal skeletal injuries amplified eventually into organized warfare, defensive villages, especially after 1500 BCE.[41,42]

9. Peru, Coastal zone, Nasca and Ostra sites, c.3000 -1500 BCE. Ostra Site: Early (c.3000 BCE) ambiguous evidence of stone-weapons which might as easily have been used for other purposes.[43] Later unambiguous Nasca Artwork and mortuary evidence (c.1500 BCE) of warfare and headhunting, including mummified heads in the manner similar to later Jivaro and other head-hunting groups in adjacent regions.[44]

The above list of archaeological sites, when viewed geographically (Appendix Figure 4), shows a striking degree of correlation to those areas of the World Behavior Map identified from anthropological sources as containing high degrees of patriarchal authoritarian, violent culture. This suggests, *the social violence identified in those archaeological sites constitutes the historical underpinnings of the later social violence and patrism recorded in the ethnographical data.* Likewise, there is *a general absence of identified archaeological evidence for violence in most other regions of the Americas, with a similar absence of armored patrism in the ethnographical data for the unshaded parts of the map.* While this could simply reflect a lack of sufficient archaeological data for other parts of the New World, as discussed below there is much evidence in the archaeological record for peaceful conditions in the unmarked areas of the maps.

The above points are additionally in agreement with the pre-Columbian contact theory advocated in *Saharasia,* specifically regarding coastal arrival points of relatively violent invaders from the Old World, some of whom came from pyramid-building regions, and reproduced the same at their new homes in the New World. All are dated to time periods *well after* the 4000-3500 BCE origins of violence in the Old World, and well into the period of massive shipbuilding among the Old World kingly empires, who very likely transmitted violence into the New World according to the patterns given on the World Behavior Map.

Having said the above, I feel it important to also remind the reader, that *on average* the New World, Pre-Columbian cultures were nevertheless far more peaceful and genuinely social than were Old World cultures of the same time period. This is proven from my original cross-cultural evaluations, from *Saharasia* (p.73). Using those data, Appendix Table 1 (below) gives the average *percent patrist* values for the different regions indicated in my original Murdock *Histograms of Regional Behaviors,* as well as the number of cultures which appear in the upper and lower third of the percent-patrist categories, respectively. These data indicate, the Old World regions of Africa, Circum-Mediterranean and Eurasia contain *around 95% of all the world's extreme patrist cultures* (354 out of 368). By contrast, Oceania and the Americas held around 88% of all the world's *extreme matrist* cultures (259 out of 293). Native Americans, *as a generality,* were more peaceful and socially cooperative, unarmored and matristic, in spite of these terrible examples I have given above.[64]

Conclusions

The information contained in the above sections can be organized both temporally and geographically, into four major regional categories of prehistorical violence:

1. As discussed in *Saharasia,* and revisited in this article, there are a scattering of sites across Anatolia and the Middle East which showed "fleeting glimpses" of social violence as early as c.5000 BCE, and possibly even earlier. These are timed with a temporary episode of drought and aridity coincidental to the abandonment of many villages and sites across the region. This early evidence for land-abandonment and probable mass-migrations, with possible social violence appearing here and there, along with a few cases of infant cranial deformation, did not become epidemic, widespread or persistent in character. Drought appeared, followed by scattered and isolated signs of social disturbance. When wetter conditions reappeared in the region, settlements thrived once again under peaceful conditions.

2. A cluster of sites in southern Germany document violent conditions at several sites between c.5500-4000 BCE. These massacre sites, at Talheim, Schletz and Ofnet, may factually fall into the younger end of this range of dates, which would place them well within the time-line of events described in *Saharasia,* when Europe was transformed by invasions from Central Asia. If the older dates eventually prove to be correct, then they would appear to be somewhat anomalistic within the framework of Saharasian theory, but nevertheless also appear to have some relationship to the isolated, scattered and non-persisting signs of violence which spread across Anatolia and the Middle East — coincidental to a documented sub-phase of aridity and land-abandonment, as described in point #1, above. Whatever their dates, these massacre sites are not located in a formerly dryland region, and no obvious mechanism related to environmental pressures such as famine and starvation can be invoked to explain their "spontaneous" genesis of isolated violence. It appears certain, these sites are the consequence of cultural diffusion of warlike groups out of the neighboring drylands, either from Central Asia at c.4000 BCE, or more likely from Anato-

---

TABLE 1. Oceania/New World Cultures Were Less Violent-Patrist Than Old World Cultures (Murdock data, see p.73 of *Saharasia*)

Region	Average % Patrist Values	Number of Cultures Falling Within:	
		Upper Third Extreme Patrist	Lower Third Extreme Matrist
Africa:	65%	219	5
Circum-Mediterranean	67%	109	12
East Eurasia	55%	26	17
Insular Pacific (Oceania)	41%	11	31
North America	29%	1	166
South America	30%	2	62

lia sometime before or around c.5500 BCE, following the migratory pathways for agricultural diffusion previously identified in Appendix Figure 1.[63] The geographical clustering of the German sites does not support the assertion of any widespread or ubiquitous violence, but rather, the opposite, of isolated violence within a larger ocean of peaceful social conditions.

3. The violence in the Nile Valley at Jebel Sahaba, Wadi Kubbiyana and a few other sites at c.12,000 BCE does not fit within the original Saharasian chronology of drought and famine starting at c.4000 or even 5000 BCE, but nevertheless does occur during an earlier period of intense aridity, prior to the Neolithic Wet Phase of North Africa. As such, this very early violence in North Africa confirms the basic drought-famine mechanism for the genesis of violence as given in *Saharasia*. Whatever violence did exist at this very early time, however, was so scattered and isolated in its distribution, that it died out once the Neolithic Wet Phase developed. Once North Africa became wet and lush, supporting grasslands and trees with large herbivores, and numerous large rivers and lakes, evidence for human violence vanishes, only to reappear after c. 3500 BCE, when North Africa dries out again. In this latter case, the violent conditions persist, along with the harsh arid conditions, from c.3500 BCE all the way down into the modern era as a global phenomenon, to be recorded by ethnographers and anthropologists, and documented in *Saharasia* on the World Behavior Map.

4. In SE Australia, we have what appears to be an episode of "Saharasian"-type genesis of small-scale inter-group social violence — to include artificial infant cranial deformation, and generally non-lethal familial and tribal fights directed mostly at women — during an episode of unusually dry and possibly episodic famine conditions. The violence appeared during hyper-arid conditions starting at c.11,000 BCE, but died out and vanished by c.7000 BCE, after wetter conditions returned. This suggests the strong influence of desertification and aridity on social conditions, as detailed in *Saharasia*.

Appendix Figure 5 identifies these four locations of confirmed archaeological evidence for anomalous violence in the pre-Saharasian period, before c.4000 BCE.

After c.4000-3500 BCE, when all of Saharasia began declining into an intense and widespread aridity, the process of drought, famine, starvation and land-abandonment intensified, forcing the mass migratory events described in *Saharasia*. Violence then irrupted again, this time as a response to a more widespread and persisting drought-famine situation which forced the abandonment of entire regions. We have detailed here, the arrival of the new famine-affected and violent Central Asian migrants across the region of the European causewayed enclosures. They wreaked havoc among peaceful villages and trading centers, and ushered in the epoch of the battle-axe, Kurgan warrior nomads,

fortifications and warrior-kings, and were followed by subsequent waves of new immigrants who carried the seeds of violence in their desert-borne and desert-bred social institutions.

As argued in *Saharasia*, violence became anchored into human character structure, by virtue of the development of new social institutions for justifying and glorifying sadism and butchery, even when directed towards infants and children, and towards the opposite sex. The key for transmission of early famine-related violence outside of the dry regions is found in the development of new social institutions which re-create the violence generation after generation, irrespective of climate. The earliest episodes of human violence, specifically identified in the above four points, did not persist in such a manner, and this may be due to the fact that human social groups at these earlier dates had not yet developed either the size or the organizational complexity by which new social institutions could be readily preserved over the long term. One hypothesis which might explain the findings is, the conditions in Anatolia and the Middle-East generated some elements of social disturbance and violence within a small percentage of cultures, who then migrated into Southern Germany and committed massacres. A similar thing could have occurred in the region of the Nile, leading to the anomalous episode at Jebel Sahaba and Wadi Kubbiyana. At some point, these hypothesized violent cultural groups died off, or were assimilated into other peaceful cultures, or otherwise vanished. Peaceful social conditions then continued once rains and food supplies became abundant once again.

Much of the claims for violence in the archaeological record, described as "prehistoric" in the most general terms, really demands to be more critically reviewed and precisely reported in terms of both dates and locations. Human bones with cut-marks do not automatically constitute "evidence for cannibalism", given the existence of funeral rituals where the bones of the dead are cleaned of their flesh. Hunting accidents — where an occasional projectile point is found in an isolated human skeleton — cannot, by themselves, stand as evidence for widespread social violence and warfare, especially where the injured individual shows signs of bone-healing and sympathetic burial. Abstracted rock-art which claims to depict a person killed with numerous spears, but which requires a specialist to make the interpretation and to point out the details, falls down into the realm of ambiguous speculation at best. If the eye of an ordinary person cannot detect violence in the rock art scenes, it is likely that the violence existed only within the specialist's imagination. And in some cases, it surely is possible that later generations of violent people might have drawn spears on top of older rock-art of human subjects, just as people today add graffiti to "dress up" existing pictures of people — where archaeological digs fail to show violence in skeletons and struc-

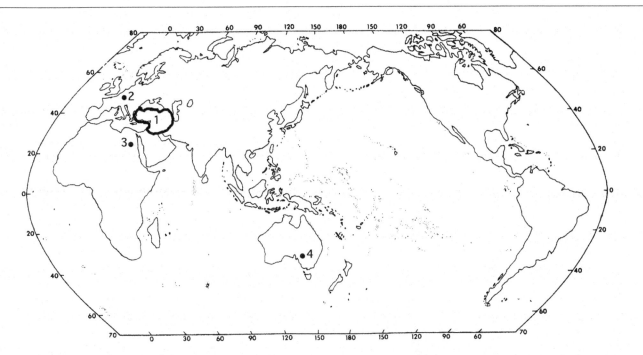

Figure 5: Confirmed Sites of Anomalous Violence in the Pre-Saharasian Period (Before c.4000 BCE)
1. Generalized region of isolated examples of violence and infant cranial deformation, Anatolia and Middle East, c.5000 BCE or possibly a bit earlier. **2.** Massacre sites at Ofnet, Talheim, Schletz, c.5500-4000 BCE. **3.** Massacre site at Jebel Sahaba, c.10,000 BCE. **4.** Region of early Australian infant cranial deformation and familial/intra-Group social violence, c.11,000 - 7,000 BCE. All of these regions, except for item #2 above, were characterized by harsh desert conditions at the time when social violence and warfare appeared. Widespread, massive and persisting social violence and warfare did not appear, however, until after c.4000-3500 BCE, when the larger part of Saharasia began to dry up, as discussed in the book *Saharasia*.[2]

tures, evidence from rock art can only be suggestive, at best. And, the date for the first-settlement of a location should not be confused with the date for the first clear and unambiguous evidence for violence. A site can be occupied for hundreds or thousands of years before the first clear signs of violence appear.

I have shown here, the violence in early China, in the causewayed enclosures of Europe, in Neolithic Spanish rock art, and in massacres of New World cultures before Columbus, all fit well within the parameters given in *Saharasia*, and these examples provide additional compelling support for the overall Saharasian theory. This is especially so for the Americas, where *most* of the evidence for village-scale massacres fits within those regions identified on the World Behavior Map as clusters of armored patrism. The close geographical associations are, in fact, striking.

What is at issue is: how, where, and under what conditions does human social violence and warfare develop. Is it something that can occur anywhere, under any conditions, something which lurks below the surface of the human character just waiting to spring forth to wreak social havoc? *Or does human violence conflict with and go against our basic biology, requiring only the most severe trauma to bring it forth; either trauma in the womb, in the crib, in the home and family, or the larger*

trauma of severe drought, land degradation, the disruption of food and water supplies, and the attendant famine and starvation conditions which follow?

All of these considerations were given focused discussion in *Saharasia*, and so will not be repeated here — but the issue is, to what extent has Saharasia's ancient historical components been challenged by these newer archaeological findings? From the discussion in this paper, I have shown that the larger Saharasian discovery and theory are not so easily challenged, due to the specificity of its construct — since the early violence identified at the c.4000-3500 BCE marker date is connected to the existence of severe drought, desertification, social displacement and famine within human populations, one can expect to find *similar social responses under similar environmental conditions, even if those conditions occur earlier than c.4000 BCE.* But more to the point, archaeology simply does not support the fantasy that ancient humans were just as warlike and bloody as either the historical or contemporary "civilizations". On the contrary, *the farther back one goes in time, before the c.4000-3500 BCE marker date, the more difficult it is to find clear and unambiguous evidence for human violence, and what does exist is observed to be regionally isolated and anomalous.*

Brian Ferguson, who has extensively reviewed the

archaeological record for evidence of human violence, has the following to say on the question, written as a conclusion for the book *Troubled Times: Violence and Warfare in the Past:*[45]

What does this evidence tell us? Paradoxically, by documenting violence and warfare and showing variations over space and time, these chapters highlight their absence in much of human prehistory. And this research is gathered together specifically to demonstrate the existence of violence. Another wideranging collection on "paleopathology at the dawn of agriculture' (Cohen & Armelagos, 198)[46] is striking for the relative absence of the sort of evidence presented here. Partly that may be neglect. But where trauma is specifically discussed, in many cases there is little or nothing to suggest any social pattern of violence. (Curiously, much of the evidence of trauma in Cohen and Armelagos comes from sites within the Mississippi drainage...

Other works similarly indicate a late emergence of violence and war. A survey of south Asian sites (Kennedy 1984: 178, 183)[47] finds limited skeletal evidence of trauma. Most of that appears in Harappan contexts, and even there earlier reports of massacres have been seriously questioned. In the Levant from the late Paleolithic well into the Neolithic, indications of violence and war are conspicuously absent from the abundant skeletal and settlement remains (Rathburn 1984; Roper 1974; Smith, Bar Yosef and Sillen 1984).[9,48,49]

A dedicated search for archaeological signs of war in South America (Redmond 1994)[50] produces little that is convincing and early. On the pre-ceramic Peruvian coast, any indication of violent conflict is late and limited to a few locations (Quilter 1989:65, 78, 85),[51] except for the highly problematic findings at Ostra (Topic 1989).[43] On the plains of western Venezuela, evidence of war only appears along with agricultural intensification and the rise of chiefdoms, post 500 AD (Spencer and Redmond 1992: 153).[52]

Europe and the Mesolithic and early Neolithic does produce some indications of personal violence (Meiklejohn et al 1984; Whittle 1985)[53,54] as discussed previously, but these are exceptional. The situation in China is similar: a very few signs of interpersonal violence (two skeletons with imbedded points) gives way to widespread evidence of war — fortifications, specialized weapons and multiple osteological signs — only in the final Neolithic, along with the development of economic inequality, not long before the rise of states (Underhill 1989).[28] A similar change occurred in prehistoric Japan, where evidence of violent death goes from about .002% of approximately 5000 skeletons from pre-agricultural Jomon times, to over 10% of all deaths in the subse-

quent, agricultural Yayoi epoch (Farris, n.d.).[55] In all these areas, war ultimately becomes entrenched and widespread, leaving unmistakable indicators. Again, it is difficult to understand how war could have been common earlier in each area and remain so invisible. ...

Roper (1969: 448)[56] calls into question some alleged instances of killing in the Paleolithic, but others remain convincing. The Australian rock art noted earlier (Tacon and Chippendale 1994)[7] indicates an early pattern of lethal violence, individual and then collective, but it stands as an exception that highlights the rule: individual killings seem rare and organized killing nearly absent throughout most of our collective past.

...if our ancestors were killing each other..dying after being stabbed, clubbed, or shot, we would see it in their remains. ...

The evident absence of warfare during most of our evolutionary past sinks a boat load of theories."[45]

Fuergeson is not alone in such an assessment. Consider the words also of Richard Gabriel, from *The Culture of War.*[57] Gabriel's statements are all the more illuminating given his basic belief in the roots of violence in our genetic-mental make-up:

"Using the Stone Age cultures of Homo sapiens and Neanderthal as a starting point, we find some remarkable data about the development of war. Man required thirty thousand years to learn how to use fire and another twenty thousand years to invent the fire-hardened, wooden-tipped spear; spear points would come much later. Sixty thousand years later, man invented the bow and arrow with transverse stone points. Ninety thousand years after the beginning of the Homo sapiens Stone Age, man learned to herd wild animals, and four thousand years later he learned to domesticate goats, sheep, cattle, and the dog. At about the same time there is evidence for the beginnings of systematic harvesting of wild grains, but it would take yet another two thousand years for man to learn how to transplant these wild grains to fixed campsites and another two thousand years to learn how to plant domesticated strains of cereal grain. It is only after this development, around 4000 BC, that warfare makes its appearance as a major human social institution. In sum, man has known war for only about 6 percent of the time since the Homo sapiens Stone Age began.

Once warfare had become established, it is difficult to find any other social institution that developed as quickly. In less than a thousand years, man brought forth the sword, sling, dagger, mace, bronze weapons, and large-scale fortifications. The next thousand years saw the emergence of iron weapons, the chariot, large standing professional armies, military acad-

emies, general staff structures, military training regimens, the first permanent arms industry, written texts on tactics, military procurement, logistics systems, conscription, and military pay. By 2000 BC, war had become the dominant social institution in almost all major cultures of the Middle East. ...

For the first ninety-five thousand years after the Homo sapiens Stone Age began, there is no evidence at all that man engaged in war on any level, let alone on a level requiring organized group violence. There is little evidence of any killing at all." [57]

These statements, from scholars intimately familiar with archaeological evidence, suggest a strong confirmation for the basic ideas presented in my *Saharasia*. This being the case, what are we to make of the various books and articles which continue to claim — without solid evidence — a violent and blood-drenched ancient history for our species? There are many books on violence in prehistorical periods which take great care in presenting archaeological evidence,[23,58,59,60,61] but none reviewed by this author was as bold in its unsupported claims and assertions of early violence as was Keeley's, which unfortunately tended to bias everything towards his own basic assumptions of the inevitability of war — that's not uncommon in today's world where "genetic determinism" dominates the sciences, and where the daily newspapers yield up plenty of evidence for the violent interpretation. Keeley and supporters are totally correct about violence among some "primitives" and their citations on warfare among Native American cultures has proven a treasure of additional evidence to support my Saharasian maps for the New World — and for much of the period of written human history, advocates for a deep rootedness of violence in the human species can draw from a wealth of evidence to support their viewpoints. However, this evidence becomes increasingly scarce the farther back into pre-history one digs, and it nearly vanishes entirely prior to c.4000 BCE. At a more basic level, in the assumption of the *innate* nature of violence, its "inevitability" and "genetic evolutionary roots" in our most ancient past, *the evidence simply does not support such a conclusion.*

The original conclusions given in my *Saharasia*,[2,3] first presented and published in the 1980s, are almost totally supported by the more recent archaeological evidence, even as articulated by the most staunch supporters of early-violence theory: *Generally peaceful social conditions existed worldwide, prior to the drying up of Saharasia after c.4000-3500 BCE. During the Saharasian wet period of c.8000-4000 BCE, peaceful social conditions prevailed as a world-wide phenomenon, with only the most isolated and even questionable of exceptions.* Where social violence did occur prior to 4000 BCE, it was in almost every case in association with the episodic appearance of harsh drought and

famine conditions — *only after such conditions became widespread and persistent does human social violence become a sustained and ongoing characteristic of the human animal.* Only *after* enduring the horrific and ongoing trauma consequent to massive drought and famine conditions, do the original peaceful and social human societies succumb and fall to the glory of violent warrior kings and patriarchal blood-lusting gods. Without the desert, without Saharasia, both history and humanity would today be entirely different.

Appendix References

1. Wilhelm Reich, *Character Analysis, Mass Psychology of Fascism, The Sexual Revolution, People in Trouble,* and *The Function of the Orgasm.*
2. James DeMeo, *Saharasia: The 4000 BCE Origins of Child-Abuse, Sex-Repression, Warfare and Social Violence, In the Deserts of the Old World,* Natural Energy Works, Ashland, 1998. (Revised from the dissertation: *On the Origins and Diffusion of Patrism: The Saharasian Connection,* Univ. Kansas, Geography Dept., Lawrence, Kansas, 1986.)
3. James DeMeo, "The Origins and Diffusion of Patrism in Saharasia, c.4000 BCE: Evidence for a Worldwide, Climate-Linked Geographical Pattern in Human Behavior", published in: *Kyoto Review* 23: 19-38, Spring 1990 (Japan) ; *Emotion* 10, 1991 (Germany); *World Futures: The Journal of General Evolution,* 30: 247-271, 1991; and *Pulse of the Planet* 3:3-16, 1991.
4. Lawrence Keeley, *War Before Civilization,* Oxford Univ. Press, NY, 1996.
5. Trevor Watkins, "The Beginnings of Warfare" in *Warfare in the Ancient World,* J. Hackett, Ed., Facts on File, NY 1989, p.15-16.
6. A. Beltran, *Rock Art of the Spanish Levant,* Cambridge Univ. Press, Cambridge, 1980.
7. P. Tacon & C. Chippindale, "Australia's Ancient Warriors: Changing Depictions of Fighting in the Rock Art of Arnhem Land, N.T.", *Cambridge Archaeological J.* 4:211-248, 1994. Also see: Lewis, D.: *Rock Paintings of Arnhem Land, Australia,* BAR Int. Series 415, 1988; Tacon, P.: "The Power of Stone: Symbolic Aspects of Stone and Tool Development in Western Arnhem Land, Australia", *Antiquity* 65:192-207, 1991.
8. DeMeo, Saharasia, 1998, ibid, p.284-295.
9. M.K. Roper, "Evidence of Warfare in the Near East from 10000-4300 BC" in *War: Its Causes and Correlates,* M. Nettleship, et al, Eds., Mouton, The Hague, pp.299-340, 1974.
10. DeMeo, ibid., p.259.
11. K. Kenyon *Digging Up Jericho,* Praeger, NY, 1958, p.127, 134.
12. O. Bar-Yosef, "The Walls of Jericho: An Alternative Interpretation" *Current Anthropology,* 27:157-162, 1986.
13. J. Mellaart, *The Chalcolithic & Early Bronze Ages in the Near East & Anatolia,* Khayats, Lebanon, 1966; J. Mellaart, *The Neolithic of the Near East,* Thames & Hudson, London, 1975.
14. Colin Burgess, et al, Eds., *Enclosures and Defences in the Neolithic of Western Europe, Vol.1 & 2,* BAR International series 403(i) and (ii), 1988.
15. Christopher Evans, "Monuments and Analogy: The Interpretation of Causewayed Enclosures" in Burgess, et al, ibid., p.47.
16. N. J. Starling, "The Neolithic Höhensiedlungen (high settlements) of Central Germany", in Burgess, et al, ibid., p.419, 427.
17. R. J. Mercer "Hambledon Hill, Dorset, England", in Burgess, et al, ibid., p.89, 95-96.
18. P. Dixon, "The Neolithic Settlements on Crickley Hill", in Burgess, et al, ibid., p.75-87.
19. Neils Andersen "The Neolithic Causewayed Enclosures at Sarup, on South-West Funen, Denmark", in Burgess, et al, ibid., p.354.
20. Paul Bahn, *Tombs, Graves and Mummies,* Barnes & Nobel, NY 1996, p.48, 50.

21. Nick Thorpe, "Origins of War: Mesolithic Conflict in Europe", *Archaeology* #52, April 2000.
http://www.britarch.ac.uk/ba/ba52/ba52feat.html
and "Origins of Violence - Mesolithic conflict in Europe"
http://www.hum.au.dk/fark/warfare/thorpe_paper_1.htm

22. M. Stuiver, A. Long & R.S. Kra, eds. "Die Datierung des Massakers von Schletz", *Radiocarbon* 35(1), 1993.
http://www.nhm-wien.ac.at/NHM/Prehist/Stadler/LVAS/QAM/14C/Schletz.html

23. David Frayer "Ofnet: Evidence for a Mesolithic Massacre", in Debra Martin & David Frayer, Eds., *Troubled Times: Violence and Warfare in the Past*, Gordon and Breach, 1997, p.181-216.

24. R.R. Schmidt *Die Altsteinzeitlichen Schädelgräber aus der Grossen Ofnet-Höhle und von Kaufertsberg*, J.F. Lehmanns, München, 1913.

25. Grahame Clarke, *Mesolithic Prelude*, University Press Edinburgh, 1980, p.62, 93.

26. Jörg Orschiedt, "Ergebnisse einer neuen Untersuchung der spätmesolithischen Kopfbestattungen aus Süddeutschland", *Urgeschichtliche Materialhefte* 12, 1998, p.147-160. Presented to the Paleoanthropology Society, 2000 Conference, abstract posted to: http://www.paleoanthro.org/abst2000.htm#orschiedt

27. Jörg Orschiedt, "Manipulationen an menschlichen Skelettresten. Taphonomische Prozesse, Sekundär-bestattungen oder Kannibalismus?", *Urgeschichtliche Materialhefte* 13, 1999.

28. Anne Underhill, "Warfare During the Chinese Neolithic Period: A Review of the Evidence", in Diana Tkaczuk and Brian Vivian, Eds. *Cultures in Conflict: Current Archaeological Perspectives*, Univ. of Calgary, Canada, 1989, p.221-223, 230-234.

29. Michael Hoffman, *Egypt Before the Pharaohs*, Barnes & Nobel, NY, 1979, p.98.

30. Fred Wendorf, *Prehistory of Nubia, Vol.2*, Southern Methodist Univ. Press, 1968.

31. Fred Wendorf, personal communication, Feb. 2001.

32. Jonathan Adams, *Sudden Climatic Transitions During the Quaternary*, with sequential climate maps
http://www.esd.ornl.gov/projects/qen/nerc.html

33. Patricia Lindsell, *Artificial Cranial Deformation* (from a dissertation in progress) 2001. More information is also available from Peter Brown and Patricia Lindsell at:
http://www-personal.une.edu.au/~pbrown3/Deform.html Also see: Don Brothwell, "Possible Evidence of a Cultural Practice Affecting Head Growth", *J. Archaeological Sci.* 2:75-77, 1975; K.A.R. Kennedy, "Growth, Nutrition and Pathology in Changing Demographic Settings in South Asia", in Cohen & Armelagos, ibid., pp.169-192, 1984.

34. Peter Brown, *Coobool Creek. A morphological and metrical analysis of the crania, mandibles and dentitions of a prehistoric Australian human population. Terra Australis*, 13. Department of Prehistory, Australian National University, Canberra 1989, p.71, 170. Also see: Brown, P.: "Artificial Cranial Deformation: A Component in the Variation in Pleistocene Australian Aboriginal Crania", *Archaeol Oceania* 16:156-167, 1981.

35. Richard G. Wilkinson, "Violence Against Women: Raiding and Abduction in Prehistoric Michigan", in Debra Martin & David Frayer, Eds., *Troubled Times: Violence and Warfare in the Past*, Gordon and Breach, 1997, p.21-43.

36. Debra Martin, "Violence Against Women in the La Plata River Valley, A.D. 1000-1300, in Martin & Frayer, 1997, ibid., p.45-75.

37. Patricia Lambert, "Patterns of Violence in Prehistoric Hunter-gatherer Societies of Coastal Southern California", in Martin & Frayer, 1997, ibid., p.84, 93-98.

38. Philip Walker, "Wife Beating, Boxing and Broken Noses: Skeletal Evidence for the Cultural Patterning of Violence", in Martin & Frayer, 1997, ibid., p.164

39. Carmen Ma, et al, "Evidence for Human Sacrifice, Bone Modification and Cannibalism in Ancient Mexico", in Martin & Frayer, ibid., p.217-239.

40. Maria Ostendorf-Smith, "Osteological Indications of Warfare in the Archaic period of the Western Tennessee Valley", in Martin & Frayer, 1997, ibid., p.241-265

41. Herbert Maschner, "The Evolution of Northwest Coast Warfare", in Martin & Frayer, 1997, ibid., p.267.

42. Gary Coupland, "Warfare and Social Complexity on the Northwest Coast", in Tkaczuk & Vivian, 1989, pp.205-214.

43. J. Topic, "The Ostra Site: The Earliest Fortified Site in the New World", in Tkaczuk & Vivian, 1989, p.215-228.

44. Donald Proulx, "Nasca Trophy Heads: Victims of Warfare or Ritual Sacrifice", in Tkaczuk & Vivian, pp.73-85, 1989.

45. Brian Ferguson, "Violence and War in Prehistory", in Martin & Frayer, 1997, p.332-334.

46. N. Cohen & G. Armelagos, Eds., *Paleopathology at the Origins of Agriculture*, Academic Press, Orlando, 1984.

47. K.A.R. Kennedy, "Growth, Nutrition and Pathology in Changing Demographic Settings in South Asia", in Cohen & Armelagos, ibid., pp.169-192, 1984.

48. T. Rathbun, "Skeletal Pathology from the Paleolithic through the Metal Ages in Iran and Iraq", in Cohen & Armelagos, ibid., pp.137-167, 1984.

49. P. Smith, et al., "Archaeological and Skeletal Evidence for Dietary Change during the Late Pleistocene/Early Holocene in the Levant", in Cohen & Armelagos, ibid, pp.101-136, 1984.

50. E. Redmond, *Tribal and Chiefly Warfare in South America*, Memoirs of the Museum of Anthropology, Univ. of Michigan, Ann Arbor, 1994.

51. J. Quilter, *Life and Death at Paloma: Society and Mortuary Practices in a Preceramic Peruvian Village*, Univ. Iowa Press, Iowa City, 1989.

52. C. Spencer & E. Redmond, "Prehispanic Chiefdoms of the Western Venezuelan Llanos", *World Archaeology*, 24:134-157, 1992.

53. C. Meikeljohn, et al, "Socioeconomic Change and Patterns of Pathology and Variation in the Mesolithic and Neolithic of Western Europe", in Cohen & Armelagos, pp.75-100, 1984.

54. A. Whittle, N*eolithic Europe: A Survey*, Cambridge Univ. Press, Cambridge, 1985.

55. W. Farris, *Sacred Texts and Buried Treasures: Essays in the Historical Archaeology of Japan*, undated manuscript, cited in Ferguson, op cit., p.348.

56. M.K. Roper, "A Survey of the Evidence for Intrahuman Killing in the Pleistocene", *Current Anthropology*, 10: 427-459, 1969.

57. Richard Gabriel, *The Culture of War: Invention and Early Development*, Greenwood Press, NY, 1990, p.20-21.

58. Jonathan Haas, *The Anthropology of War*, Cambridge U. Press, NY 1990.

59. Debra Martin & David Frayer, Eds., *Troubled Times: Violence and Warfare in the Past*, Gordon and Breach, 1997.

60. John Hackett, Ed., *Warfare in the Ancient World*, Facts on File, NY 1989.

61. Diana Tkaczuk and Brian Vivian, Eds. *Cultures in Conflict: Current Archaeological Perspectives*, Univ. of Calgary, Canada, 1989.

62. James DeMeo, "Update on *Saharasia*: Ambiguities and Uncertainties in *War Before Civilization*", *Heretic's Notebook (Pulse of the Planet #5)*, 2002, p.15-44.

63. Andreas Zimmermann, "Der Beginn der Landwirtschaft in Mitteleuropa", in *Menschen, Zeiten, Raum: Archäologie in Deutschland*, Wilfried Menghin, Editor, Konrad Theiss Verlag, Stuttgart, 2002, p. 133-134. (Map also reproduced on p.112 of: *Spuren der Jahrtausende: Archäologie und Geschichte in Deutschland*, Konrad Theiss Verlag, Stuttgart, 2002.)

64. James DeMeo, "Peaceful Versus Warlike Societies in Pre-Columbian America: What Does Archaeology and Anthropology Tell Us?" in *Unlearning the Language of Conquest: Scholars Expose Anti-Indianism in America*, Four Arrows (Don Jacobs) Editor, University of Texas Press, 2006 - in press.

BIBLIOGRAPHY for All Chapters

(Additional citations are found in the chapter on contraceptive plant materials on pages 195-196, and after the "Update on Saharasia" Appendix article, on pages 443-444.)

Anonymous citations are listed alphabetically by title.

Abercrombie, T.J., "Saudi Arabia: Beyond the Sands of Mecca", *National Geographic*, January 1966.

Accad, E., *Veil of Shame*, Editions Naaman, Sherbrooke, Quebec, 1978.

"Action Alert: Amnesty Protest Week", *Ms.*, December 1984, p.24.

Adams, J., *AIDS: The HIV Myth*, St. Martin's, NY, 1989.

Adams, R. McC., *Heartland of Cities*, U. Chicago, 1981.

Adamson, G. & Gare, D., "Home or Hospital Births?", *J. Am. Medical Assoc.*, 243(17):1732-6, 1980.

Agrawal, D.P. et al., "Radiocarbon Dates of Archaeological Samples", *Current Science*, 33(9):266, 1964.

Akpinar, A. & Alok, E., "The Petroglyphs of Anatolia", *Aramco World Magazine*, 35(2):2, March-April, 1984.

Al-Ansary, A.R., *Qaryat al-Fau: A Portrait of Pre-Islamic Civilization in Saudi Arabia*, U. Riyadh, St. Martin's, NY, 1982.

Allchin, B. & Allchin, R., *The Birth of Indian Civilization*, Pelican Books, Harmondsworth, Middlesex, 1968.

Allchin, B., et al., *The Prehistory and Paleogeography of the Great Indian Desert*, Academic Press, NY, 1978.

Allen, T. & Conger, D., "Time Catches Up With Mongolia", *National Geographic*, February 1985.

Al-Sayari, S. & Zotl, J., eds., *Quaternary Period in Saudi Arabia*, Vol. 1, Springer Verlag, NY, 1978.

Al Sayyid Marsot, A.L., "The Revolutionary Gentlewomen in Egypt", in Beck & Keddie, 1978, below.

Altman, I. & Wohlwill, J., eds., *Children and the Environment, Vol. 3 of Human Behavior and Environment*, Plenum Press, NY, 1978.

Altman, I., et al., eds., *Environment and Culture, Vol.4 of Human Behavior and Environment*, Plenum Press, NY, 1980.

Alvarez, L., et al., *Science*, 208:1095-108, 1980.

Ambler, J.R., "The Anasazi Abandonment of the San Juan Drainage and the Numic Expansion", *North Am. Archaeologist*, 10(1):39-53, 1989.

Ammal, E.K.J., "Introduction to the Subsistence Economy of India", in Thomas, 1973, below.

Anati, E., *Rock Art in Central Arabia*, Vol 2, Institut Orientaliste, Louvain, 1968.

Anders, T., "The Effects of Circumcision on Sleep-Wake States in Human Neonates", *Psychosomatic Medicine*, 36(2):174-9, 1974.

Andrews, A., "Toward a Status of Women Index", *Professional Geographer*, 34(1):24-31, 1982.

Angier, N., "Unholy War at the City of David", *Discover*, January 1982.

Anton, D., "Aspects of Geomorphological Evolution; Paleosols and Dunes in Saudi Arabia", in Jado & Zotl, 1984, below.

Anton, F., *Woman in Precolumbian America*, Abner Schram, NY, 1973.

Ardrey, R., *The Territorial Imperative*, Antheneum, NY, 1966.

Arens, W., *The Man-Eating Myth: Anthropology and Anthropophagy*, Oxford U. Press, NY, 1979.

Arms, S., *Immaculate Deception*, Houghton-Mifflin, Boston, 1975.

The Atlas of Mankind, Rand McNally, NY, 1982.

Axelrod, D.I., *Quaternary Extinctions of Large Mammals*, U. California Publications in Geological Sciences, Berkeley, 1967.

Aykroyd, W., *The Conquest of Famine*, Chatto & Windus, London, 1974.

Babayev, A.G., "Principal Problems of Desert Land Reclamation in the USSR", in Glantz, 1977, below.

Bailey, J., *The God-Kings & the Titans*, St. Martins, NY, 1973.

Ball, J., "Problems of the Libyan Desert", *Geographical J.*, 1927.

Baraheni, R., *The Crowned Cannibals, Writings on Repression in Iran*, Vintage, NY, 1977.

Barker, G., "Early Agriculture and Economic Change in North Africa", in *The Sahara: Ecological Change and Early Economic History*, J.A. Allan, ed., Middle East & North African Studies Press, England, 1981, pp.131-45.

Barnes, F. & Pendleton, M., *Prehistoric Indians*, Wasatch Publishers, Salt Lake City, 1979.

Barry, K., *Female Sexual Slavery*, Avon, NY, 1979.

Bayer, R., *Homosexuality and American Psychiatry: the Politics of Diagnosis*, Basic, NY, 1981.

Baynes, N.H., "The Decline of the Roman Empire in Western Europe. Some Modern Explanations", *J. Roman Studies*, 33:29-35, 1943.

Beaumont, P., "The Middle East - Case Study of Desertification", in Mabbut & Berkowicz, 1980, below.

Beck, L. & Keddie, N., eds., *Women in the Muslim World*, Harvard U. Press, Cambridge, 1978.

Bell, B., "The Dark Ages in Ancient History, 1: the First Dark Age in Egypt", *American J. Archaeology*, 75:1-26, 1971.

Berggren, W. & Van Couvering, J. eds., *Catastrophes and Earth History: The New Uniformitarianism*, Princeton U. Press, 1984.

Bergler, E., *Homosexuality, Disease or Way of Life?*, Collier, NY, 1956.

Berry, B., "Approaches to Regional Analysis: A Synthesis", *Annals, Assoc. Amer. Geographers*, 54:2-11, 1964.

Bess J.W., et al: "Microvesicles are a source of contaminating cellular proteins found in purified HIV-1 preparations". *Virol.* 230:134-144, 1997.

Best, A. & deBlij, H., *African Survey*, J. Wiley, NY, 1977.

Bethel, Tom: "The Cure that Failed: Did the AIDS lobby know what it was doing when it pressed the government to approve AZT?" *National Review*, 10 May 1993, pp.33-35.

Bettelheim, B., *Symbolic Wounds*, Free Press, Glencoe, IL, 1954; Collier Books, NY, 1962.

Bharadwaj, O.P., "The Arid Zone of India and Pakistan", in Stamp, 1961, below.

Birand, H., "Vue d'ensemble sur la vegetation de la Turquie", *Vegetatie*, V-VI:41-4, 1954.

Birand, H., "La vegetation anatoliennen et la necessite de sa protection", *Report for Sixth Congr. Internat. Union Conserv. Nature*, Athens, 1958.

Birdsell, J.B., *American Naturalist*, 39:171-207, 1953.

Biswas, M.R. & Biswas, A.K., eds., *Desertification*, Volume 12, United Nations Conference on Desertification, Pergamon Press, 1977, 1980.

Bobek, H., "Beitrage zur Klima-okologischen Gliederung Irans", *Erdkunde*, 6:65-84, 1952.

Bonaparte, M., "Uber die Symbolik der Kopftrophaen", *Imago*, 14:100-40, 1928.

Bongaarts, J. & Cain, M., "Demographic Responses to Famine", in Cahill, 1982, below.

Boserup, E., *Women's Role in Economic Development*, Geo. Allen & Unwin, London, 1970, p.49.

Bowden, M.J., et al., "The Effect of Climate Fluctuations on Human Populations: Two Hypotheses", in Wigley, et al., 1981, below.

Brackbill, Y., "Continuous Stimulation and Arousal Level in Infancy,: Effects of Stimulus Intensity and Stress", *Child Development*, 46:364-9, 1975.

Brackbill, Y. et al., *Birth Trap: The Legal Low-Down on High Tech Obstetrics*, C.V. Mosby, St. Louis, 1984.

Brandt, P., *Sexual Life in Ancient Greece*, AMS Press, NY, 1974.

Breggin, P., *Talking Back to Ritalin*, Common Courage, ME 1997.

Breggin, P., *Toxic Psychiatry: Why Therapy, Empathy and Love Must Replace the Drugs, Electroshock and Biochemical Theories of the "New Psychiatry"*, St. Martin's Press, NY 1991.

Broek, J., "National Character in the Perspective of Cultural Geography", *Annals, Am. Acad. Polit. Soc. Science*, 370:8-15, 1967.

Bruce, G., *The Stranglers: The Cult of Thuggee*, Harcourt, Brace, & World, NY, 1968.

Bryson, R.A., "Is Man Changing the Climate of Earth?", *Saturday Review*, 1 April 1967.

Bryson, R.A., et al., "Drought and the Decline of Mycenae", *Antiquity*, 48:46-50, 1974.

Bryson, R.A. & Murray, T.J., *Climates of Hunger*, U. Wisconsin Press, Madison, 1977.

Bryson, R.A. & Baerreis, D.A., "Possibilities of Major Climatic Modification and their Implications: Northwest India, A Case for Study", *Bull. Am. Meteorol. Soc.*, 48(3):136-42, March 1967.

Budyko, M.I., *The Heat Balance of the Earth's Surface*, N.A. Stepanova, trs., US Dept. of Commerce, Washington, D.C., 1958.

Bullough, V.L., *Sexual Variance in Society and History*, John Wiley, NY, 1976.

Burke, C., *Aggression in Man*, Lyle Stuart, NJ, 1975.

Burley, N. & Symanski, R., "Women Without: An Evolutionary and Cross- Cultural Perspective on Prostitution", in Symanski, 1981, below.

Burnett, C., et al., "Home Delivery and Neonatal Mortality in North Carolina", *J. Am. Medical Assoc.*, 244(24):2741-5, 1980.

Burton, R., *Personal Pilgrimage to Al-Medinah and Mecca*, Vol. 2, Dover, 1962.

Burton, R., *The Sotadic Zone*, Panurge Press, NY, c.1930.

Butler, H.C., "Desert Syria, the Land of a Lost Civilization". *Geographical Review,* Feb., 1920.

Butler, S., *Conspiracy of Silence: The Trauma of Incest*, New Glide Press, San Francisco, 1978.

Butzer, K.W., "Studien zum vor- und fruhgeschichtlichen Landschaftswandels der Sahara und Levante seit dem klassischen Altertum. II. Das okologische Problem der Neolitischen Felsbilder der ostlichen Sahara",*Abhandl. Akad. Wiss. u. Liter.* Mainz, Math.-naturwiss. Kl., 1:115-6, 1958.

Butzer, K.W., "Die Naturlandschaft Agyptens wahrend der Vorgeschichte und der Dynastischen Zeit", idem. III, *Abhandl. Akad. Wiss. u. Liter.* Mainz, Math.-Naturwiss. Kl. 2:96-109, 1959.

Butzer, K.W., "Climatic Change in Arid Regions Since the Pliocene", in Stamp, 1961, below.

Butzer, K.W., *Desert and River in Nubia*, U. Wisconsin Press, 1968.

Butzer, K.W., "Patterns of Environmental Change in the Near East During Late Pleistocene and Early Holocene Times", in Wendorf & Marks, 1975, below.

Butzer, K.W., *Early Hydraulic Civilization in Egypt*, U. Chicago, 1976.

Butzer, K.W. et al., "Radiocarbon Dating of East African Lake Levels", *Science*, 175:1074, 1972.

Byrk, F., *Circumcision in Man and Woman*, American Ethnological Press, NY, 1934.

Cahill, K., *Famine*, Orbis Books, Maryknoll, NY, 1982.

Cahill, K., "The Clinical Face of Famine in Somalia", in Cahill, 1982, below.

Calverton, V.F., *The Bankruptcy of Marriage*, Macaulay Co., NY, 1928.

Campbell, J., *A Personal Narrative of 13 Years Service*, London, 1864.

Campbell, R., "Personality as an Element of Regional Geography", *Annals, Assoc. Amer. Geographers*, December 1968, pp.748-59.

Camps, G., "The Prehistoric Cultures of North Africa: Radiocarbon Chronology", in Wendorf & Marks, 1975, below.

Cansever, G., "Psychological Effects of Circumcision", *British Journal of Medical Psychology*, 38:328-9, 1965.

Carlson, D., "Famine in History: With a Comparison of Two Modern Ethiopian Disasters", in Cahill, 1982, above.

Carpenter, R., *Discontinuity in Greek Civilization*, Cambridge U. Press, 1966; W.W. Norton, NY, 1968.

Carter, J., *Racketeering in Medicine: The Suppression of Alternatives*, Hampton Roads, 1992.

Carter, N., *Routine Circumcision: the Tragic Myth*, Londinium Press, London, 1979.

Catton, W., "Social and Behavioral Aspects of the Carrying Capacity of Natural Environments", *Human Behavior and Environment*, 6:269-306, 1983.

Chang, I., *The Rape of Nanking: The Forgotten Holocaust of World War II*, Basic Books, NY 1997.

Chang, K., *The Archaeology of Ancient China*, Yale U. Press, New Haven, 1968.

Chappell, J.E., "Climatic Change Reconsidered: Another Look at 'The Pulse of Asia'", *Geographical Review*, 60(3):347-73, 1970.

Chase, N.F. "Corporal Punishment in the Schools", *Wall Street Journal*, 11 March 1975.

Childe, G., *New Light on the Most Ancient Near East*, Praeger, NY, 1952.

Chu, C., "Climatic Pulsations During Historic Times in China", *Geographical Review*, 16(2):274-83, 1926.

Chu, K., "A Preliminary Study on the Climatic Fluctuations During the Last 5000 Years in China", *Scientia Sinica*, 16(2):226-9, 1973.

Clark, J.D., "Prehistoric Populations and Pressures Favoring Plant Domestication in Africa", in *Origins of African Plant Domestication*, J.R. Harlan, et al., eds., Mouton, the Hague, 1976.

Cocks, G., *Psychotherapy in the Third Reich*, Oxford U. Press, NY, 1985.

Cohen, N.W. & Estner, L.J., *Silent Knife: Caesarean Prevention and Vaginal Birth After Caesarean*, J.F.Bergin, South Hadley, Mass, 1983.

Colborn, T., et al., *Our Stolen Future: Are We Threatening Our Fertility, Intelligence and Survival?* Penguin, 1996.

Comas, J. & Marquer, P., *Craneos Deformados de la Isla de Sacrificios*, Veracruz, Mexico, Instituto de Investigaciones Historicas, Mexico, 1969.

Consalvi, C., "Some Cross- and Intracultural Comparisons of Expressed Values of Arab and American College Students", *J. Cross-Cultural Psychology*, 2(1):95-107, 1971.

Cooper, W., *History of the Rod*, London 1912.

Corea, G., *The Hidden Malpractice: How American Medicine Mistreats Women*, Wm. Morrow, NY, 1977.

Corea, G., "The Caesarean Epidemic — Who's Having This Baby Anyway — You or the Doctor?", *Mother Jones*, July 1980.

Corey, G., *Theory and Practice of Counseling and Psychotherapy*, Brooks/Cole, Monterey, CA, 1982.

"Corporal Punishment in the Schools", *Inequality in Education*, #23, Center for Law and Education, Gutman Library, Cambridge, MA, September 1978.

Cory, D., *Homosexuality: A Cross-Cultural Approach*, Julian Press, NY, 1956.

Coser, L., "The Political Functions of Eunuchism", *Am. Sociological Rev.*, 29:880-85, 1964.

Coulson, N. & Hinchcliffe, D., "Women and Law Reform in Contemporary Islam", in Beck & Keddie, 1978, above.

Cox, K. & Golledge, R., eds., *Behavioral Problems in Geography Revisited*, Methuen, NY, 1981.

Coyne, M., "Iran Under the Ayatollah", *National Geographic*, July 1985, pp.108-35.

Cravioto, J. & DeLicardie, E., "Neurointegrative Development and Intelligence in Children Rehabilitated from Severe Malnutrition", in Prescott, et al., 1975, above.

Croall, J., *Neill of Summerhill, the Permanent Rebel*, Pantheon, NY, 1983.

Daiches, S., *People in Distress: A Geographical Perspective on Psychological Well-Being,* U. Chicago Dept. Geography Research Paper #197, 1981.

Dalby, D. & Harrison-Church, R., eds., *Drought In Africa, Report of the 1973 Symposium*, Center for African Studies, U. of London, 1973.

Dando, W., *The Geography of Famine*, J. Wiley, NY, 1980.

Daniels, C.M., "The Garamantes", in Kanes, 1969, below.

Darby, H.C., "The Clearing of the Woodland in Europe", in Thomas, 1973, below.

Davidson, B., *African Kingdoms*, Time-Life, NY, 1966.

Davidson, D., *The Chronological Aspects of Certain Australian Social Institutions, As Inferred from Geographical Distribution*, Thesis, U. Pennsylvania, Philadelphia, 1928.

Davies, N., *Human Sacrifice in History and Today*, Wm. Morrow, NY, 1981.

Davies, N., *Voyagers to the New World: Fact or Fantasy?*, Macmillan, NY, 1979.

Davies, N., "Human Sacrifice in the Old World and the New; Some Similarities and Differences", in *Ritual Human Sacrifice in Mesoamerica*, E. Benson & E. Boone, eds., Dunbarton Oaks Res. Lib., Washington, D.C., 1980.

Davis, D., *Slavery and Human Progress,* Oxford U. Press, NY, 1984.

Davis, J.H., "Influences of Man Upon Coast Lines", in Thomas, 1973, below.

Dawson, W., *The Custom of Couvade*, Manchester U. Press, London, 1929.

deBeauvoir, S., *The Second Sex*, Vintage, NY, 1952.

deGrazia, A., *The Velikovsky Affair*, University Books, New Hyde Park, NY, 1966.

deMause, L., "The Evolution of Childhood", in deMause, ed., 1974, below.

deMause, L., ed., *The History of Childhood*, Psychohistory Press, NY, 1974.

deMause, L., "The Universality of Incest", *J. of Psychohistory*, 19(2):123-164, Fall 1991.

DeMeo, J., "Anti-Constitutional Activities and Abuse of Police Power by the US Food and Drug Administration and Other Federal Agencies, *Pulse of the Planet*, 4:106-113, 1993.

DeMeo, J., "Archaeological/Historical Reconstruction of Late Quaternary Environmental and Cultural Changes in Saharasia", unpublished monograph, 1985.

DeMeo, J., *Bibliography on Orgone Biophysics, 1934-1986*, Natural Energy Works, Ashland, 1986.

DeMeo, J., "Cloudbusting: Growing Evidence for a New Method of Ending Drought and Greening Deserts", *AIBC Newsletter*, American Inst. of Biomedical Climatology, Sept. 1996, #20, p.1-4.

DeMeo, J., "Cross-Cultoral Studies as a Tool in Geographical Research", Assoc. American Geographers, Annual Meeting, Louisville 1980, *AAG Program Abstracts*, p.167.

DeMeo, J., "Geographical Atlas of Ethnographic Data, Volume I: Human Behavior", unpublished monograph, 1985.

DeMeo, J., "Green Sea Eritrea: A 5-Year Desert Greening CORE Project in the East African Sahel-Sahara", in *Heretic's Notebook: Emotions, Protocells, Ether-Drift and Cosmic Life Energy, Pulse of the Planet #5*, J. DeMeo, Editor, p.183-211, 2002.

DeMeo, J., "HIV is Not the Cause of AIDS: A Summary of Current Research Findings", *Pulse of the Planet #4*, p.99-105, 1993.

DeMeo, J., *Preliminary Analysis of Changes in Kansas Weather Coincidental to Experimental Operations with a Reich Cloudbuster*, University of Kansas, Geography-Meteorology Dept., Thesis, 1979.

DeMeo, J., " 'Preventative' Mastectomy or Official Medical Quackery?" *Pulse of the Planet* 4:163-164, 1993; cf. "Genetics Does Not Equal Heredity", "Modern Horrific Medicine", "Infant Medical Experiments" & "The Mammogram", *Pulse of the Planet* 4:161-64.

DeMeo, J., "The Use of Herbs for Contraception by Primitive People", *AAG Program Abstracts*, Los Angeles 1981, Annual Meeting, Association of American Geographers, p.10; also see (by J. DeMeo): "Herbal Oral Contraceptives: Their Use by Primitive People", *Mothering* 5:24-28, 1977; "The Use of Contraceptive Plant materials by Native Peoples", *J. Orgonomy* 26(1):152-176, 1992; "Empfängnisverhütungsmittel bei Naturvölkern", *Emotion* 11:6-29, 1994.

DeMeo, J., *On the Origins and Diffusion of Patrism: The Saharasian Connection*, Univ. of Kansas, Geography Dept., dissertation, 1986.

DeMeo, J., *The Orgone Accumulator Handbook: Construction Plans, Experimental Use, and Protection Against Toxic Energy*, Natural Energy Works, Ashland, Oregon 1989.

DeMeo, J., "The Origins and Diffusion of Patrism in Saharasia, c.4000 BCE: Evidence for a Worldwide, Climate-Linked Geographical Pattern in Human Behavior", *Kyoto Review* [Japan] 23:19-38, Spring 1990; reprinted in *Emotion* [Germany] 10, 1991; *World Futures*, 30:247-271, 1991; and *Pulse of the Planet*, 3:3-16, 1991.

DeMeo, J., "OROP Arizona 1989: A Cloudbusting Experiment to Bring Rains in the Desert Southwest", *Pulse of the Planet #3*, J.DeMeo, Editor, p.82-92, 1991.

DeMeo, J., "OROP Israel 1991-1992: A Cloudbusting Experiment to Restore Wintertime Rains to Israel and the Eastern Mediterranean During an Extended Period of Drought", in *On Wilhelm Reich and Orgonomy, Pulse of the Planet #4*, J. DeMeo, Editor, p.92-98, 1993.

DeMeo, J., "Response to Martin Gardner's Attack on Reich and Orgone Research in the *Skeptical Inquirer*", *Pulse of the Planet* 1:11-17, 1989.

DeMeo, J. & R. Morris, "Preliminary Report on a Cloudbusting Experiment in the Southeastern Drought Zone, August 1986", *Southeast-*

ern Drought Symposium Proceedings, March 4-5, 1987, South Carolina State Climatology Office Pub. G-30, Columbia, SC, 1987.

DeMeo, J. & B. Senf, *Eds., Nach Reich, Neue Forschungen zur Orgonomie, Sexualökonomie, Die Entdeckung der Orgonenergie*, Zweitausendeins, Frankfurt, 1997.

Dengler, I.C., "Turkish Women in the Ottoman Empire: The Classical Age", in Beck & Keddie, 1978, above.

Denis, A., *Taboo*, Putnam, NY, 1967.

Dennis, W., "Infant Reactions to Restraint: an Evaluation of Watson's Theory", *Trans. NY Acad. Sci.*, Ser. 2, Vol. 2, 1940.

Deo, S.B., *History of Jaina Monachism*, Deccan College Post-Grad. & Res. Inst., India, 1956.

Devereux, G., *A Study of Abortion in Primitive Societies*, International Universities Press, NY, 1976.

Devitt, N., "The Transition from Home to Hospital Birth, USA", *Birth & Family J.*, 4(1):47-58, 1977.

Dingwall, E.J., *Artificial Cranial Deformation*, J. Bale, Sons, & Danielson, Ltd., London, 1931.

Dingwall, E.J., *The Girdle of Chastity*, Clarion, London, 1959.

Dodds, E.R., *The Greeks and the Irrational*, U. California Press, Berkeley, 1951.

Dolukhanov, P., "The Ecological Prerequisites for Early Farming in Southern Turkmenia", in Kohl, 1981, below.

Donovan, D., *Dublin Medical Press*, 19:67, 1848.

Doris, M., "Update: International Women's Decade: Ten Years After", *Ms.*, July 1985, pp.23-4.

Doughty, C., *Travels in Arabia Deserta*, Random House, NY, 1920.

Downing, D., *An Atlas of Territorial and Border Disputes*, New English Library, London, 1980.

Duesberg, Peter: "AIDS Acquired by Drug Consumption and Other Non-Contagious Risk Factors", *Pharmac. Ther.* 55:201-277, 1992.

Duesberg, P., *AIDS: Virus- or Drug-Induced? Contemporary Issues in Genetics and Evolution*, Vol. 5, Edited by Peter H. Duesberg (Reprinted from *Genetica* Vol. 95, No. 1-3, 1995), Kluwer Academic, NY 1996.

Duesberg, P., *Infectious AIDS: Have We Been Misled?*, North Atlantic Books, Berkeley, 1996.

Duesberg, P., *Inventing the AIDS Virus*, Regenery, NY 1996.

Dumont, H., "Relict Distribution Patterns of Aquatic Animals: Another Tool in Evaluating Late Pleistocene Climate Changes in the Sahara and Sahel", *Paleoecology of Africa and Surrounding Islands*, 14:1-24, 1982.

Dunn, P., "The Enemy is the Baby: Childhood in Imperial Russia", in deMause, 1974, above.

Dupree, L., *Afghanistan*, Princeton U. Press, NJ, 1973.

Eberhart, J., "Radar From Space", *Science News*, 9/22/84, p.187.

Eckholm, E.E. & Brown, L., *Spreading Deserts, the Hand of Man*, Worldwatch Paper #13, 1977.

Edmonson, M., "Neolithic Diffusion Rates", *Current Anthropology*, 2(2):71-102, April 1961.

Edwards, J. & Johannessen, K., "Hypothesis of Taoist - New World Pre- Columbian Contacts", *AAG Program Abstracts*, San Antonio, 1982.

Edwards, I.E.S., *The Pyramids of Egypt*, Viking Press, NY, 1972.

Edwards, P., "Wilhelm Reich", *Encyclopedia of Philosophy*, Macmillan/Free Press, NY, 7:104-15, 1967.

Ehrenreich, B. & English, D., *Witches, Midwives, and Nurses: A History of Women Healers*, Feminist Press, NY, 1973.

Ehrich, R.W., ed., *Chronologies of Old World Archaeology*, U. Chicago Press, 1965.

Eickelman, D.F., *The Middle East: An Anthropological Approach*, Prentice Hall, NJ, 1981.

Eisler, R., *The Chalice and the Blade*, Harper & Row, NY, 1987.

Eisler, R., *Sacred Pleasure*, Harper Collins, San Francisco, 1996.

Ekholm, G., "The New Orientation Toward Problems of Asiatic-American Relationships", in *New Interpretations of Aboriginal American Culture History*, Cooper Square, Towata, NJ, 1972.

Elwin, V., *The Kingdom of the Young*, Oxford U. Press, Bombay, 1968.

Elwin, V., *Maria Murder and Suicide*, Oxford U. Press, Bombay, 1942.

Elwin, V., *The Muria and Their Ghotul*, Oxford U. Press, Bombay, 1942.

Emde, R., et al., "Stress and Neonatal Sleep", *Psychosomatic Medicine*, 33(6):491-7, 1971.

Emery, W.B., *Archaic Egypt*, Penguin, London, 1972.

Espenshade, E.B. & Morrison, J.L., eds., *Goode's World Atlas*, 15th Edition, Rand McNally & Co., Chicago, 1978.

Esperandieu, G., "Domestication et elevange dans le Nord de L'Afrique au Neolithique ns la protohistoire d'apres les fiions repestres", *Actes 2e Cong. Pan-Afr. Prehist.*, Algiers, 1952.

European Workshop on the Impact of Endocrine Disrupters on Human Health and Wildlife, 2-4 Dec. 1996, Weybridge UK, Report of Proceedings [Report EUR 17549] Copenhagen, Denmark. European Commission DG XII, April 16, 1997.

Evans, H., *Harlots, Whores and Hookers*, Picture Library, NY, 1983.

Evenari, M., et al., *The Negev*, Harvard U. Press, Cambridge, 1982.

Fairbridge, R.W., "African Ice-Age Aridity", in *Problems in Paleoclimatology*, A.E.M. Nairn, ed., Interscience Pub., 1964, pp.356-63.

Fairbridge, R.W., *Quaternary Research*, 6:529, 1976.

Fairservis, W.A., "Archaeological Studies in the Seistan Basin of Southwestern Afghanistan and Eastern Iran", *Anthro. Papers Am. Mus. Nat. Hist.*, 48(1):14-6, 1961.

Fairservis, W.A., "Exploring the 'Desert of Death'", *Natural History*, June, 1950.

Faramarzi, S., "Honor Killings Persist Among Druze", Assoc. Press article, 20 Dec. 1995, *Medford Mail Tribune*, p.8-A.

Farrand, W.R., "Pluvial Climates and Frost Action During the Last Glacial Cycle in the Eastern Mediterranean - Evidence from Archaeological Sites", in *Quaternary Paleoclimate*, W.C. Mahaney, ed., Geo Abstracts, Norwich, England, 1981.

Fell, B., *America B.C.: Ancient Settlers in the New World*, Demeter, NY, 1977.

Fell, B., *Saga America*, Times, NY, 1980.

Fell, B., *Bronze Age America*, Little, Brown & Co., NY, 1982.

Field, H., *Contributions to the Anthropology of Saudi Arabia*, Field Research Projects, Coconut Grove, 1971.

Finegan, J., *Light From the Ancient Past: The Archaeological Background of the Hebrew-Christian Religion*, Princeton U. Press, NJ, 1946.

Fisher, H., *The Sex Contract: The Evolution of Human Behavior*, Wm. Morrow, NY, 1982.

Flint, R.F., *Glacial and Quaternary Geology*, Wiley, NY, 1971.

Flohn, H. & Nicholson, S., "Climatic Fluctuations in the Arid Belt of the 'Old World' Since the Last Glacial Maximum: Possible Causes and Future Implications", *Paleoecology of Africa and the Surrounding Islands*, 12:6-7, Rotterdam, 1980.

Foley, J., "The Unkindest Cut of All", *Fact*, July-August, 1966

Fortune, R., *Sorcerers of Dobu*, Routledge & Kegan Paul, London, 1963.

Fox, H., *Gods of the Cataclysm*, Dorsett, NY, 1976.

Frankl, G., *The Failure of the Sexual Revolution*, Khan & Averill, London, 1974.

Freud, S., "Sexuality in the Etiology of Neuroses" (1898); "Three Essays on the Theory of Sexuality" (1905); "'Civilized' Sexual Morality and Modern Nervous Illness" (1908), *The Standard Edition of the Complete Psychological Works*, Vol. III, Early Psychoanalytic Publications, London, 1961.

Freud, S., *Civilization and its Discontents*, W.W. Norton, NY, 1961.

Freud, S., *Beyond the Pleasure Principle*, Standard Edition, London, 1961.

Friedman, M., "A Morally Bankrupt Drug War", *Int. Herald Tribune*, 13 Jan. 1998.

Fromm, E., *Escape from Freedom*, Holt, Rinehart & Winston, NY, 1963.

Fromm, E., *The Sane Society*, Holt, Rinehart & Winston, NY, 1963.

Fumento, M., *The Myth of Heterosexual AIDS: How a Tragedy has been Distorted by Media and Partisan Politics*, Basic Books, NY 1990.

Fumento, M., "Teenaids, the Latest HIV Fib", *New Republic*, 10 August 1992, pp.17-19.

Gage, M.J., *Woman, Church, and State*, Persephone Press, Watertown, Mass., 1980.

Garcia, R. & Escudero, J., *The Constant Catastrophe: Malnutrition, Famines and Drought*, Vol. 2 of the Drought and Man series, IFIAS Project, Pergamon Press, NY, 1982.

Garcia, R., *Nature Pleads Not Guilty*, Vol. 1 of the Drought and Man series, IFIAS Project, Pergamon Press, NY, 1981.

Gardner, J.L., *Mysteries of the Ancient Americas: The New World Before Columbus,* Readers Digest, NY 1996.

Gibson, M., "Violation of Fallow and Engineered Disaster in Mesopotamian Civilization", in *Irrigation's Impact on Society*, T.E. Downing & M. Gibson, eds., Tuscon, U. Arizona Press, 1974.

Gibson, M., *The City and Area of Kish*, Field Research Projects, Coconut Grove, 1972.

Gimbutas, M., *Bronze Age Cultures in Central and Eastern Europe*, Mouton, The Hague, 1965.

Gimbutas, M., "Relative Chronology of Neolithic and Chalcolithic Cultures in East Europe North of the Balkans and Black Sea", in Ehrich, 1965, above.

Glacken, C.J., "Changing Ideas of the Habitable World", in Thomas, 1973, below.

Glantz, M.H., *Desertification, Environmental Degradation In and Around Arid Lands*, Westview Press, Boulder, 1977.

Gluschankof P, et al., Cell membrane vesicles are a major contaminant of gradient-enriched human immunodeficiency virus type-1 preparations. Virol. 230:125-133, 1997.

Godfrey, J., *1204: The Unholy Crusade*, Oxford U. Press, NY, 1980.

Gofman, J., *Preventing Breast Cancer: The Story of a Major, Proven, Preventable Cause of This Disease*, Committee for Nuclear Responsibility, San Francisco, 1996.

Goldberg, S., *The Inevitability of Patriarchy: Why the Biological Difference Between Men and Women Always Produces Male Domination*, Wm. Morrow, NY, 1973.

Goldenberg, N., *Changing of the Gods: Feminism and the End of Traditional Religions*, Beacon Press, Boston, 1979.

Goldsmith, D., *Nemesis: The Death-Star and Other Theories of Mass Extinction*, Walker, NY, 1985.

Golley, F.B., "Productivity and Mineral Cycling in Tropical Forests", in *Productivity of World Ecosystems, Proceedings of a Symposium: August-September, 1972,* National Research Council/National Academy of Sciences, Washington, D.C., 1977.

Gordon, C., *Before Columbus*, Crown, NY, 1971.

Gorer, G., "Some Aspects of the Psychology of the People of Great Russia", *The American Slavic and East European Review*, VIII(3):155- 67, October 1949.

Gorer, G. & Rickman, J., *The People of Great Russia; A Psychological Study*, W.W. Norton, NY, 1962.

Graham, E., "Yuman Warfare: An Analysis of Ecological Factors from Ethnohistorical Sources", in Nettleship, et al., 1973, below.

Graziosi, P., "Prehistory of Southwestern Libya", in Kanes, 1969, below.

Greenberg, M., et al., "First Mothers' Rooming-In with Their Newborns: It's Impact upon the Mother", *Am. J. Orthopsychiatry*, 43(5):783- 788, 1973.

Greenfield, J., *Wilhelm Reich Vs. the USA*, W.W. Norton, NY, 1974.

Gregg, N., "Hagi: Where Japan's Revolution Began" (Insert Map), *National Geographic*, 165:750-73, June 1984.

Griffiths, J.F. & Driscoll, D.M., *Survey of Climatology*, Charles Merrill Pub. Co., Columbus, Ohio, 1982.

Grossman, R., Ed., *The God that Failed*, Harper & Bros., NY 1949.

Grove, A.T., *Africa*, Oxford U. Press, NY, 1978.

Grove, A.T., "Desertification in the African Environment", in Dalby & Harrison-Church, 1973, above.

Grove, A.T. & Pullan, R.A., "Some Aspects of the Pleistocene Paleogeography of the Chad Basin", in Howell & Bourliere, 1963, below.

Grun, B., *The Timetables of History*, Touchstone, NY, 1982.

Gumilev, L.N., "Khazaria and the Caspian", *Soviet Geography*, 5(6):55, 1964.

Gumilev, L.N., "Heterochronism in the Moisture Supply of Eurasia in Antiquity", *Soviet Geography*, 7(10):35-7, 1966.

Gumilev, L.N., "Heterochronism in the Moisture Supply of Eurasia in the Middle Ages", *Soviet Geography*, 9(1):28-31, 1968.

Guerra, F., *The Pre-Columbian Mind, A Study Into the Aberrant Nature of Sexual Drives, Drugs Affecting Behavior, and the Attitude Towards Life and Death, with a Survey of Psychotherapy, in pre-Columbian America*, Seminar Press, NY, 1971.

Gupta, S.P., *Archaeology of Soviet Central Asia and the Surrounding Indian Borderlands*, Vol. 1 & 2, B.R. Publishing, Delhi, 1979.

Haire, D., *Childbirth in the Netherlands: A Contrast in Care*, monograph, Intern'l Childbirth Education Assoc., 1973.

Hallet, J.P. & Relle, A., *Pygmy Kitabu*, Random House, NY, 1973.

Hamblin, D.J., "Treasures of the Sands", *Smithsonian*, Sept., 1983, pp.43-53.

Hansson, C. & Liden, K., *Moscow Women: Thirteen Interviews*, (G. Bothmer et al., translators), Pantheon, NY, 1983.

Hapgood, C., *Maps of the Ancient Sea Kings*, E.P.Dutton, NY, 1979.

Hare, F.K., "Climate and Desertification", in *Desertification: Its Causes and Consequences*, United Nations, Pergamon Press, NY, 1977.

Hare, F.K., "Connections Between Climate & Desertification", *Environmental Conservation*, 4(2):81-90, 1977.

Harland, J.P., "Sodom and Gomorrah", *Biblical Archaeologist*, 6:3, 1943, and 5:2, 1942.

Harlow, H., *Love in Infant Monkeys*, San Francisco, 1959; reprinted in *Scientific American*, 200(6):68-, June 1959.

Harlow, H., *The Human Model: Primate Perspectives*, V.H. Winston, Wash., D.C., 1979.

Harman, R., "The Emotional Plague and the AIDS Hysteria," *Journal of Orgonomy*, 22(2):173-195, Nov. 1988.

Hawkes, J., ed., *Atlas of Ancient Archaeology*, McGraw Hill, NY, 1974.

Hawkes, J., "Prehistoric Europe", in *History of Mankind*, Vol.I, UNESCO, London, 1963.

Haynes, C.V., "Great Sand Sea and Selima Sand Sheet, Eastern Sahara: Geochronology of Desertification", *Science*, 217:629-33, 1982.

Hays, T.R., "Neolithic Settlement of the Sahara as it Relates to the Nile Valley", in Wendorf & Marks, 1975, below.

Heath, R., "Maternal-social Deprivation and Abnormal Brain Development; Disorders of Emotional and Social Behavior", in Prescott, et al., 1975, below.

Heichelheim, F.M., "Effects of Classical Antiquity on the Land", in Thomas, 1973, below.

Helfer, R.E. & Kempe, C.H., eds., *The Battered Child*, U. Chicago Press, 1974.

Hemmings, R., *Children's Freedom: A.S. Neill and the Evolution of the Summerhill Idea*, Schocken Books, NY, 1973.

Henry, D., "Adaptive Evolution within the Epipaleolithic of the Near East", in *Advances in World Archaeology*, Vol. 2, F.Wendorf & A.Close, eds., Academic Press, NY, 1983.

Helfer, R. & Kempe, H., eds., *The Battered Child*, U. Chicago Press, 1974.

Helms, S.W., *Jawa: Lost City of the Black Desert*, Cornell U. Press, Ithaca, NY, 1981.

Herbruck, C.C., *Breaking the Cycle of Child Abuse*, Winston Press, Minneapolis, MN, 1979.

Herman, J.L., *Father-Daughter Incest*, Harvard U. Press, Cambridge, 1982.

Himes, N., *Medical History of Contraception*, Williams & Wilkins, Baltimore, 1936.

Hinds, M., et al., "Neonatal Outcome in Planned vs. Unplanned Out-of- Hospital Births in Kentucky", *J. Am. Medical Assoc.*, 253(11):1578-82, 1985.

Hingley, R., *The Russian Mind*, Bodleyhead, London, 1977.

Historical Atlas of the World, Rand McNally, NY, 1981.

Hobler, P.M. and Hester, J.J., "Prehistory and Environment in the Libyan Desert", *S. Afr. Archaeol. Bull.*, 33:127-9, 1969.

Hochschild, A., "Is the Left Sick of Feminism?", *Mother Jones*, June 1983, pp.56-8.

Hodan, M., *The History of Modern Morals*, Wm. Heinemann, London, 1937.

Hodgkinson, N., *AIDS: The Failure of Contemporary Science, How a Virus that Never Was Deceived the World*, Fourth Estate, London, 1996.

Hodgson, B. & Stansfield, J., "Time and Again in Burma", *National Geographic*, 166(1):91, July 1984.

Hoffman, M., *Egypt Before the Pharaohs*, Alfred Knopf, NY, 1979.

Hooykaas, R., *Catastrophism in Geology, its Scientific Character in Relation to Actualism and Uniformitarianism*, North-Holland, NY, 1970.

Hopkins, K., "Eunuchs in Politics in the Later Roman Empire", *Proceedings, Cambridge Philological Soc.*, 189, 1963.

Horney, K., *New Ways in Psychoanalysis*, Norton, NY, 1966.

Horowitz, A., "The Pleistocene Paleoenvironments of Israel", in Wendorf & Marks, 1975, below.

Hosken, F., *The Hosken Report on Genital and Sexual Mutilation of Females*, 2nd Edition, Women's International Network News, Lexington, Mass., 1979.

Hosken, F., "Genital Mutilation of Women in Africa", *Munger Africana Library Notes*, 36, October 1976.

Hotzl, H. & Zotl, J., "Climatic Changes During the Quaternary Period", in Al-Sayari & Zotl, 1978, above.

Hotzl, H., et al., "The Youngest Pleistocene", in Jado & Zotl, 1984, below.

Howard, D., *1066: The Year of Conquest*, Dorset Press, NY, 1977.

Howell, F.C. & Bourliere, F., eds., *African Ecology & Evolution*, Viking Fund Publication in Anthro., #36, 1963.

Hufnagel, V., *No More Hysterectomies*, Inst. Reproductive Physiology, San Diego, 1990.

Hughes, J.D., *Ecology in Ancient Civilization*, U. New Mexico Press, Albuquerque, 1975.

Huntington, E.F., *Asia: A Geography Reader*, Rand McNally, Chicago, 1912.

Huntington, E.F., "The Burial of Olympia", *Geographical Journal*, 36:657-86, 1910.

Huntington, E.F., "Climatic Change and Agricultural Exhaustion as Elements in the Fall of Rome", *Quart. J. Economics*, 31:173-208, 1917.

Huntington, E.F., *Mainsprings of Civilization*, J. Wiley, NY, 1945.

Huntington, E.F., *Palestine and its Transformation*, Houghton-Mifflin, NY, 1911.

Huntington, E.F., *The Pulse of Asia*, Houghton-Mifflin, NY, 1907.

Huntington, E.F. & Shaw, E.B., *Principles of Human Geography*, John Wiley & Sons, NY, 1951.

Huntington, E.F. & Visher, S.S., *Climatic Changes, Their Nature and Causes*, Yale U. Press, New Haven, 1922.

Hurlimann, M., *India*, Thames & Hudson, London, 1967.

Huston, P., *Message from the Village*, Epoch B Foundation, NY, 1978.

Huzayyin, S., "Changes in Climate, Vegetation, and Human Adjustment in the Saharo-Arabian Belt, with Special Reference to Africa", in W. Thomas, 1973, below.

Huzayyin, S., *The Place of Egypt in Prehistory, A Correlated Study of Climates and Cultures in the Old World*, Memoires, L'Institute D'Egypte, Cairo, 1941.

Inhelder, B., "Early Cognitive Development and Malnutrition", in Garcia & Escudero, 1982, above.

Irwin, C., *Fair Gods and Stone Faces*, St. Martins, NY, 1973.

Isaac, E., *Geography of Domestication*, Prentice-Hall, NJ, 1970.

Iwai, H., "The Buddhist Priest and the Ceremony of Attaining Womanhood during the Yuan Dynasty", *Memoirs, Tokyo Bunko, Research Dept.*, #7, 1935.

Jacobsen, T. & Adams, R.M., "Salt and Silt in Ancient Mesopotamian Agriculture", *Science*, 128:1251-8, 1958.

Jacoby, R., *The Repression of Psychoanalysis*, Basic Books, NY, 1983.

Jado, A. & Zotl, J., eds., *Quaternary Period in Saudi Arabia*, Vol. 2, , Springer Verlag, NY, 1984.

Jairazbhoy, R., *Ancient Egyptians and Chinese in America*, Rowman & Littlefield, NJ, 1974.

Jairazbhoy, R., *Asians in Pre-Columbian Mexico*, Northwood, London, 1976.

Jakel, D., "Eine Klimakurve fur die Zentralsahara", in Sheel, 1980, below.

Jaynes, J., *The Origins of Consciousness in the Breakdown of the Bicameral Mind*, Houghton Mifflin, Boston, 1976.

Jeans, J., *The Growth of Physical Science*, Premier, NY, 1958.

Jett, S., "Precolumbian Transoceanic Contacts", in *Ancient North Americans*, J. Jennings, ed., W.H.Freeman, NY, 1983.

Jordan, B., *Birth in Four Cultures*, Eden Press Women's Publications, Montreal, 1980.

Jordan, R., "When Vikings Sailed East", *National Geographic*, March, 1985.

Jordan, T. & Rowntree, L., *The Human Mosaic*, Harper & Row, NY, 1979.

Kagan, D., ed., *The Decline and Fall of the Roman Empire: Why did it Collapse?*, Heath, Boston, 1962.

Kanes, W.H., ed., *Geology, Archaeology & Prehistory of the SW Fezzan, Libya*, Petroleum Exploration Society of Libya, 10th Annual Field Conference, 1969.

Kashala, O., et al., "Infection with Human Immunodeficiency Virus Type 1 (HIV-1) and Human T Cell Lymphotropic Viruses among Leprosy Patients and Contacts: Correlation Between HIV-1 Cross-Reactivity and Antibodies to Lipoarabinomannan", *J. Infectious Diseases*, 1994:169:296-304.

Kastner, et al., *Science*, 226:137-43, 1984.

Kay, M., ed., *Anthropology of Human Birth*, F.A.Davis, Philadelphia, 1982.

Keddie, N. & Beck, L., *Women in the Muslim World*, Harvard U. Press, Cambridge, 1978.

Kellogg, W.W., "Global Influences of Mankind on the Climate", in *Climatic Change*, Gribbin, J., ed., Cambridge U. Press, NY, 1978.

Kelman, H., *International Behavior*, Holt, Rinehart & Winston, NY, 1965.

Kenyata, J., *Facing Mt. Kenya*, Vintage, NY, 1965.

Kenyon, K., *Excavations at Jericho*, British School of Archaeology, Jerusalem, (Vol. I, 1960; Vol. II, 1965).

Kiefer, O., *Sexual Life in Ancient Rome*, Barnes & Nobel, NY, 1951.

Kircho, L., "The Problem of the Origin of the Early Bronze Age Culture of Southern Turkmenia", in Kohl, 1981.

Kitcher, P., *Vaulting Ambition: Sociobiology and the Quest for Human Nature*, MIT Press, Cambridge, 1985.

Klauber, G., "Circumcision and Phallic Fallacies, or The Case Against Routine Circumcision", *Connecticut Medicine*, 37(9):445-8, 1973.

Klaus, M.H., et al., "Maternal Attachment: Importance of the First Postpartum Days", *New England J. of Medicine*, 286:460-3, 1972.

Klaus, M.H., et al., "Human Maternal Behavior at the First Contact with Her Young", *Pediatrics*, 46(2):187-92, 1970.

Klaus, M.H. & Kennell, J.H., *Maternal-Infant Bonding: The Impact of Early Separation or Loss on Family Development*, C.V.Mosby, St. Louis, 1976.

Klimm, L.E., "Man's Ports and Channels", in Thomas, 1973, below.

Knight, M.M., "Water and the Course of Empire in North Africa", *Quat. J. Economics*, 43:44-93, 1928-29.

Koestler, A., *Thieves In the Night: Chronicle of an Experiment*, Macmillan & Co., London, 1947.

Kohl, P., ed., *The Bronze Age Civilization of Central Asia: Recent Soviet Discoveries*, M.E. Sharpe, NY, 1981.

Kohl, P., "The Namazga Civilization: An Overview", in Kohl, ed., 1981, above.

Konig, H., *Columbus: His Enterprise, Exploding the Myth*, Monthly Review Press, NY 1976.

Konia, C., "Orgone Therapy: A Case Presentation", *Psychotherapy Theory, Research, & Practice*, 12(2):192-7, 1975.

Koppen, W., "Da Geographische System der Klimate", Vol. 3 of W. Koppen & R. Geiger, *Handbuch der Klimatology*, Gebruder Borntrager, Berlin, 1936.

Korner, A., "Visual Alertness in Neonates as Evoked by Maternal Care", *J. Experimental Child Psychiatry*, 10:67-78, 1978.

Kovalevsky, M., *Modern Customs and Ancient Laws of Russia*, Burt Franklin, NY, 1891.

Kovda, V.A., "Land Use Development in the Arid Regions of the Russian Plain, the Caucasus and Central Asia", in Stamp, 1961, below.

Kovda, V.A., *Land Aridization and Drought Control*, Westview Press, Boulder, 1980.

Kron, R., et al., "Newborn Sucking Behavior Affected by Obstetrical Sedation", *Pediatrics*, 37:1012-6, 1966.

Lacey, R., *The Kingdom*, Harcourt, Brace & Jovanovich, NY, 1981.

Laing, R.D., "Liberation by Orgasm", *New Society*, 28 March 1968, p.464; also in *Pulse of the Planet*, 4:76-77, 1993.

Lagace, R., "Probability Sample Files", *Behavior Science Research*, 14:211-29, 1979.

Lamb, H.H., "Reconstruction of the Course of Postglacial Climate Over the World", in *Climatic Change in Later Prehistory*, A.F.Harding, ed., Edinburgh U. Press, 1982.

Lamberg-Karlovsky, C., "Afterword", in Kohl, 1981, above.

Lanning, E., *Peru Before the Incas*, Prentice-Hall, NJ, 1967.

Lane, H., *Talks to Parents and Teachers*, Allen & Unwin, London, 1928.

Lantier, J., *La Cite Magigue et Magie en Afrique Noire*, Librarie Fayard, Paris, 1972.

Lapon, L., *Mass Murderers in White Coats: Psychiatric Genocide in Nazi Germany and the United States*, Psychiatric Genocide Research Inst., Springfield, MA, 1986.

Lasar, T., "The Origins of the Emotional Disaster", *Orgonomic Functionalism*, VII(1):37-62, Jan. 1961; "...Part 2", VII(2):116-154, March 1961; "...Part 3", VII(5):283-329, Sept. 1961.

las Casas, B., *Spanish Cruelties*, 1609.

Lauritsen, J., *The AIDS War: Propaganda, Profiteering and Genocide from the Medical-Industrial Complex*, Asklepios, NY 1993.

Leavitt, J.E., ed., *The Battered Child: Selected Readings*, General Learning, Morristown, NJ, 1974.

Leboyer, F., *Birth Without Violence*, Alfred Knopf, NY, 1975.

Levison, M., et al., *The Settlement of Polynesia, A Computer Simulation*, U. Minnesota Press, Minneapolis, 1973.

Levy, B., et al., "Reducing Neonatal Mortality Rates with Nurse-Midwives", *Am. J. Obstet. Gynec.*, 109(1):50-58, 1971.

Levy, H.S., *Sex, Love, and the Japanese*, Warm-Soft Village Press, Washington, D.C., 1971.

Levy JA., "Infection by human immunodeficiency virus-CD4 is not enough", *NEJM* 335:1528-1530, 1996.

Lewinsohn, R., *A History of Sexual Customs*, Premier, Greenwich, CT, 1964.

Lewis, J., *In the Name of Humanity*, Freethought Press, NY, 1967.

Lewontin, et al., *Not In Our Genes: Biology, Ideology, and Human Nature*, Random House, NY, 1985.

Lipton, E., et al., "Swaddling, A Child Care Practice: Historical, Cultural, and Experimental Observations", *Pediatrics*, Supplement, 35, part 2, March 1965, pp.521-67.

Lisitsina, G., "The History of Irrigation Agriculture in Southern Turkmenia", in Kohl, 1981, above.

Loeb, E., "The Blood Sacrifice Complex", *Memoirs, American Anthropological Association*, 30:1-40, 1923.

Loewenberg, P., *Decoding the Past: the Psychohistorical Approach*, Knopf, NY, 1983.

Lowenthal, D., ed., *Environmental Perception and Behavior*, U. Chicago Dept. Geography Research Paper #109, 1967.

Mabbutt, J. & Berkowicz, S., eds., *The Threatened Drylands*, 24th Intern. Geographical Congress, UNESCO/U. New South Wales, Australia, 1980.

MacDonald, D., *Oceanic Languages, Their Grammatical Structure, Vocabulary, and Origin*, Henry Frowde, London, 1907.

Mackenzie, D.E., *Myths of Pre-Columbian America*, Gresham, London 1923.

Malcolm, J., *In the Freud Archives*, Knopf, NY, 1984.

Malinowski, B., "War and Weapons Among the Natives of the Trobriand Islands", *Man*, January 1920.

Malinowski, B., *Sex and Repression in Savage Society*, Routledge & Kegan Paul, London, 1927.

Malinowski, B., *The Sexual Life of Savages*, Routledge & Keegan Paul, London, 1932.

Mallery, A., *Lost America: The Story of Iron Age Civilization Prior to Columbus*, Overlook, NY, 1951.

Mantegazza, P., *The Sexual Relations of Mankind*, Eugenics Pub., NY, 1935.

Marks, A., "An Outline of Prehistoric Occurrences and Chronology in the Central Negev, Israel", in Wendorf & Marks, 1975, below.

Marsh, G., *Man and Nature*, Scribners, London, 1864.

Marshall, R., et al., "Circumcision I: Effects Upon Newborn Behavior", *Infant Behavior and Development*, 3:1-14, 1980.

Marsot, A.L., "The Revolutionary Gentlewomen in Egypt", in Beck & Keddie, 1978, above.

Masry, A.H., ed., *An Introduction to Saudi Arabian Antiquities*, Department of Antiquities & Museums, Ministry of Education, Kingdom of Saudi Arabia, 1975.

Masry, A.H., *Prehistory in Northeastern Arabia: The Problem of Interregional Interaction*, Unpublished dissertation, University of Chicago, Anthropology, 1973.

Masson, J., *The Assault on Truth: Freud's Suppression of the Seduction Theory*, Farrar, Straus & Giroux, NY, 1984.

Masson, J., "Jung Among the Nazis", in *Against Therapy*, Atheneum, 1988; cf. "Afterword to the Second Edition", edition by Common Courage Press, Monroe, ME 1994.

Masson, J., "The Persecution and Expulsion of Jeffrey Masson, as Performed by Members of the Freudian Establishment & Reported by Janet Malcolm", *Mother Jones*, December 1984.

Masson, V.M., "Altyn-depe during the Aeneolithic Period", "Seals of a Proto-Indian Type from Altyn-depe", "Urban Centers of Early Class Society" and "Seals of a Proto-Indian Type from Altyn-depe", in Kohl, 1981, above.

Masson, V.M. & Kiiatkina, T., "Man at the Dawn of Civilization", in Kohl, 1981, above.

May, G., *Social Control of Sex Expression*, Geo Allen & Unwin, London, 1930.

Mayer, L., *Views of Egypt from the Original Drawings in the Possession of Sir Robert Ainsley During His Embassy in Constantinople*, P. Bowyer, London 1804.

McBurney, C.B.M., *The Stone Age of Northern Africa*, Penguin Books, NJ, 1960.

McClure, H.A., "Ar Rub'al-Khali", in Al-Sayari & Zotl, 1978, above.

McClure, H.A., *The Arabian Peninsula and Prehistoric Populations*, Field Research Projects, Miami, 1971.

McEvedy, C., *Atlas of African History*, Facts on File, NY, 1980.

McGhee, R., "Archaeological Evidence for Climatic Change During the Last 5000 Years", in Wigley, et al., 1981, below.

McGinnies, W.G., et al., eds., *Deserts of the World*, U. Arizona Press, Tuscon, AZ, 1968.

McGregor, K., *Evaluation of Huntington's Ozone Hypothesis as a Basis for His Cyclonic Man Theory*, unpublished thesis, U. of Kansas, Geography Dept. 1976.

McHugh, W.P., *Late Prehistoric Cultural Adaptation in the Southeastern Libyan Desert*, Diss. Anthropology, U. Wisconsin, 1971.

McHugh, W.P., "Late Prehistoric Cultural Adaptation in Southwest Egypt and the Problem of the Nilotic Origins of Saharan Cattle Pastoralism", *J. Am. Res. Center in Egypt*, XI:12, 1974.

McQuarrie, H., "Home Delivery Controversy", *J. Am. Medical Assoc.*, 243(17):1747, 1980.

Mead, M., "The Swaddling Hypothesis, Its Reception", *Am. Anthropologist*, 56, 1954.

Mehl, L., *Scientific Research on Childbirth Alternatives & What it Tells Us About Hospital Practice*, NAPSAC Internat'l Publications, 1978.

Mehl, L., et al., "Outcomes of Elective Home Births: A Series of 1146 Cases", *J. Reproductive Medicine,* 19:281-99, 1977.

Meigs, P., "World Distribution of Arid and Semi-Arid Homoclimates", in *Reviews of Research on Arid Zone Hydrology*, UNESCO, Paris, Arid Zone Programme, 1:203-9, 1953.

Mellaart, J., *The Chalcolithic & Early Bronze Ages in the Near East & Anatolia*, Khayats, Lebanon, 1966.

Mellaart, J., *The Neolithic of the Near East*, Thames & Hudson, London, 1975.

Midlarsky, M. & Thomas, S., "Domestic Social Structure & International Warfare", in Nettleship, et al., 1973, below.

"Mid Revolutionary Mores", *Ms,* May 1983, p.28.

Mikesell, M.W., "The Deforestation of Northern Morocco", *Science*, 132:441-8, 1960.

Mikesell, M.W., "Review Article: Geographic Perspectives in Anthropology", *Annals, Assoc. Amer. Geographers*, September 1967, pp.617-34.

Mikesell, M.W., "The Deforestation of Mt. Lebanon", *Geographical Review*, 59(1):1-28, 1969.

Miller, A., *Thou Shalt Not Be Aware, Society's Betrayal of the Child*, Farrar, Straus & Giroux, NY, 1984.

Miller, A., *For Your Own Good: Hidden Cruelty in Child-Rearing and The Roots of Violence*, Farrar, Straus & Giroux, NY, 1983.

Millett, K., *Going to Iran*, Coward, McCann & Geoghegan, NY, 1982.

Mitchell, G., "What Monkeys Can Tell Us About Human Violence", *The Futurist*, April 1975, pp.75-80.

Monckeberg, F., "The Effect of Malnutrition on Physical Growth and Brain Development", in Prescott, et al., 1975, below.

Monod, T.:"The Late Tertiary and Pleistocene in the Sahara", in Howell & Bourliere, 1963, above.

Montagu, A., *Adolescent Sterility: A Study in the Comparative Physiology of the Infecundity of the Adolescent Organism in Mammals and Man*, Charles Thomas, Springfield, 1946.

Montagu, A., *Coming into Being Among the Australian Aborigines*, Routledge & Kegan Paul, London, 1974.

Montagu, A., "Infibulation and Defibulation in the Old and New Worlds", *American Anthropologist*, 47:464-67, 1945.

Montagu, A., ed., *Man and Aggression*, Oxford U. Press, NY, 1968.

Montagu, A., *The Natural Superiority of Women,* Macmillan, NY, 1953.

Montagu, A., "The Origins of Subincision in Australia", *Oceania*, 8(2):193-207, 1937.

Montagu, A., *Prenatal Influences*, Charles Thomas, Springfield, IL, 1962.

Montagu, A., "Ritual Mutilation Among Primitive Peoples", *Ciba Symposium*, pp.421-36, October 1946.

Montagu, A., "Social Impacts of Unnecessary Intervention and Unnatural Surroundings in Childbirth", in Stewart & Stewart, 1978, Vol. 2:589-610, below.

Montagu, A., ed., *Sociobiology Examined*, Oxford U. Press, NY, 1980.

Montagu, A., *Touching: The Human Significance of the Skin*, Columbia U. Press, NY, 1971.

Montgomery, T., "A Case for Nurse-Midwives", *Am. J. Obstet. Gynec.*, 105(1):3, 1969.

Morgan, L.H., *Ancient Society*, Charles Ken, Chicago, 1877.

Morgan, L.H., *League of the Iroquois*, Corinth Books, NY, 1922.

Morgan, R., "World: The Gandhi Assassination", *Ms.*, January 1985, p.93.

Morgan, R., "World: Good News From 18 Countries", *Ms*, September 1985.

Morgan, R. & Steinem, G., "The International Crime of Genital Mutilation", *Ms*, March 1980.

Morris, D., *The Naked Ape*, McGraw-Hill, NY, 1967.

Mori, F., "Prehistoric Cultures in Tadrart Acacus, Libyan Sahara", in Kanes, 1969, above.

Moscati, S., *The Semites in Ancient History*, U. Wales, Cardiff, 1959.

Moscati, S., *The World of the Phoenicians*, Praeger, NY, 1968.

Movius, H.L., "Paleolithic and Mesolithic Sites in Soviet Central Asia", *Proceedings, Am. Phil. Soc.*, 97(4):383-421, 1953.

Munson, P., *The Tichitt Tradition: A Late Prehistoric Occupation of the Southwestern Sahara*, Diss., Anthropology, U. Illinois, Urbana-Champaign, 1971.

Munson, P., "A Late Holocene (c.4500-2300 BP) Climatic Chronology for the Southwestern Sahara", *Paleoecology of Africa & Surrounding Islands*, 13:53-60, 1981.

Murdock, G.P., "Genetic Classification of the Austronesian Languages: A Key to Oceanic Culture History", *Ethnology*, III(4):124-6, 1964.

Murdock, G.P., *Ethnographic Atlas*, U. Pittsburgh Press, PA, 1967.

Murdock, G.P., *Atlas of World Cultures*, U. Pittsburgh Press, PA, 1981.

Murdock, G.P., and D.R. White: "Standard Cross-Cultural Sample", *Ethnology*, 8:329-369, 1969.

Murphy, R., "The Decline of North Africa Since the Roman Occupation: Climatic or Human?", *Annals, Assoc. Am. Geographers*, 41(2):120-2, 1951.

Murray, G.W., "Water from the Desert: Some Ancient Egyptian Achievements", *Geographical Journal*, London, 121:171-81, 1955.

451

Murray, J., *The Kikuyu Female Circumcision Controversy*, Diss., UCLA, 1974.

Murray, M.A., *The Splendor that was Egypt*, Philosophical Library, NY, 1949.

Murray, M.A., *The Genesis of Religion*, Routledge & Kegan Paul, London, 1963.

Mustafa, A., *J. Obstet. Gynaec. Brit. Cwlth.*, 73:302-6, 1966.

Neill, A.S., *Freedom, Not License!*, Hart, NY, 1966.

Neill, A.S., *The Problem Family*, Herbert Jenkins, London 1948.

Neill, A.S., *The Problem Parent*, Herbert Jenkins, London 1932.

Neill, A.S., *Summerhill: A Radical Approach to Child Rearing*, Hart, NY, 1960.

Neill, A.S., *That Dreadful School*, Herbert Jenkins, London 1937.

Nelson, A., "Orgone (Reichian) Therapy in Tension Headache", *Am. J. Psychotherapy*, 30(1):103-11, 1976.

Nettleship, M.A., et al., eds., *War, Its Causes and Correlates*, Mouton Publishers, Paris, 1973.

Nicholson, S.E., *A Climatic Chronology for Africa: Synthesis of Geological, Historical and Meteorological Data*, Diss. U. Wisconsin, 1975.

Nicholson, S.E., "The Climatology of Sub-Saharan Africa", in *Environmental Change in the West African Sahel*, National Academy Press, Washington, D.C., 1983, pp.71-92.

Nicholson, S.E. & Flohn, H., "African Environmental and Climatic Changes and the General Atmospheric Circulation in Late Pleistocene and Holocene", *Climatic Change*, 2:313-348, 1980.

Nitzschke, B., "Wilhelm Reich, Psychoanalyse und Nationalsozialismus", *Emotion*, 10:183-190, 1992.

Norman, C., "Environmental Destruction Hurts India's Development", *Science*, 218:1291, 1982.

Noss, J.B., *Man's Religions*, Macmillan, NY, 1965, p.95.

Nunberg, H., *Problems of Bisexuality as Reflected in Circumcision*, Imago, London, 1949.

Nutzel, W., "The Climate Changes of Mesopotamia and Bordering Areas, 14,000 to 2,000 BC", *Sumer*, 32, 1976.

O'Connor, D., "A Regional Population in Egypt to c.600 BC", in Spooner, 1972, below.

O'Leary, T.J., "Concordance of the Ethnographic Atlas with the Outline of World Cultures", *Behavioral Science Notes*, 4:165-207, 1969.

Oliver, J. & Lee, P., "Comparative Aspects of the Behavior of Juveniles in Two Species of Baboon in Tanzania", in *Recent Advances in Primatology, Vol. 1: Behavior*, D. Chivers & J. Herbert, Eds., Academic Press, 1978.

Ozturk, O., "Ritual Circumcision and Castration Anxiety", *Psychiatry*, 36:49-60, 1973.

Pabot, H., *Rapport au Gouvernement de Syrie sur l'ecologie vegetale et ses applications*, 1957 (FAO/ETAP report #663).

Pabot, H., *Rapport au Gouvernement d'Afghanistan sur l'amelioration des paturages naturels*, 1959 (FAO/ETAP Report #1093).

Pachur, H.J. & Braun, G., "The Paleoclimate of the Central Sahara, Libya and the Libyan Desert", *Paleoecology of Africa & Surrounding Islands*, 12:351, 1980.

Paige, K., "The Ritual of Circumcision", *Human Nature*, May 1978.

Paige, K. & Paige, J., *The Politics of Reproductive Ritual*, U. California Press, Los Angeles, 1981.

Papadopulos-Eleopulos, E., et al., "Is a Postive Western Blot Proof of HIV Infection?", *Biotechnology*, Vol.11, June 1993, p.696-707.

Papadopulos-Eleopulos, E., et al., "Has Gallo proven the role of HIV in AIDS?" *Emerg. Med.* [Australia] 5:113-123, 1993.

Papadopulos-Eleopulos, E., et al., "Is a positive Western blot proof of HIV infection?" *Bio/Technology* 11:696-707, 1993.

Papadopulos-Eleopulos, E., et al., "Reappraisal of AIDS: Is the oxidation caused by the risk factors the primary cause?" *Med. Hypotheses* 25:151-162, 1993.

Papadopulos-Eleopulos, E., et al., "Factor VIII, HIV and AIDS in haemophiliacs: an analysis of their relationship." *Genetica* 95:25-50, 1995.

Papadopulos-Eleopulos, E., et al., "A critical analysis of the HIV-T4-cell-AIDS hypothesis". *Genetica* 95:5-24, 1995.

Papadopulos-Eleopulos, E., et al., "The Isolation of HIV: Has it really been achieved?" *Continuum* 4: (No. 3). 1s-24s, 1996.

Parker, S., et al., "Father Absence and Cross-Sex Identity: The Puberty Rites Controversy Revisited", *Am. Ethnologist*, 1974, pp.687-706.

"Passage Rites", *Encyclopedia Britanica*, Macropaedia, Vol 13, 1982.

Payne, G., *The Child in Human Progress*, Sears, NY, 1916.

Peate, I., ed., *Studies in Regional Consciousness and Environment*, Books for Libraries Press, NY, 1968.

Pelegrino, C., *Unearthing Atlantis: An Archaeological Odyssey*, Random House, NY 1991.

Pensee Editors: *Velikovsky Reconsidered*, Doubleday, NY, 1976.

Perls, F., *In and Out the Garbage Pail*, Bantam, NY, 1969.

Perry, W., *The Children of the Sun*, Scholarly Press, Michigan, 1968.

Persinger, M., *The Weather Matrix and Human Behavior*, Praeger, NY, 1980.

Pesce, E., "Exploration of the Fezzan", in Kanes, 1969, above.

Peterson, G.H. & Mehl, L.E., "Comparative Studies of Psychological Outcome of Various Childbirth Alternatives", in Stewart & Stewart, 1978, below.

Petit-Marie, N. et al., "Pleistocene Lakes in the Shati Area, Fezzan", *Paleoecology of Africa & Surrounding Islands*, 12:293, 1980.

Petrov, M.P., *Deserts of the World*, J. Wiley & Sons, NY, 1976.

Petrov, M.P., "Once Again About the Desiccation of Asia", *Soviet Geography*, 7(10):22, 1966.

Pflanze, O., "Toward a Psychoanalytic Interpretation of Bismarck", *Am. Historical Rev.*, 76:419-44, 1972.

Philby, H.J., "Rub' Al Khali: An Account of Exploration..", *Geographical Journal*, 81(1):11,17 Jan. 1933.

Phillips, E.D., *The Royal Hordes*, McGraw Hill, London, 1965.

Piankova, L: "Bronze Age Settlements of Southern Tadjikistan", in Kohl, 1981, above.

Piers, M.W., *Infanticide*, W.W.Norton, NY, 1978.

Pillsbury, B., "Being Female in a Muslim Minority in China", in Beck & Keddie, 1978, above.

Pitcher, D.E., *An Historical Geography of the Ottoman Empire*, E.J. Brill, Leiden, 1972.

Pitt-Rivers, G.H.L., *Clash of Cultures and the Contact of Races*, Routledge & Sons, London, 1927.

Pollit, K., "Muscovites And Owenites", *Mother Jones*, August 1983, pp.55-6.

Pond, A., *The Desert World*, Greenwood Press, Westport, CT, 1975.

Pope, P., "Danish Colonization in the West Indies", in Nettleship, et al, 1973, above.

Population Dynamics of the World, Population Reference Bureau, Washington, D.C., 1981.

Prescott, J., "Body Pleasure and the Origins of Violence", *Futurist*, April 1975, pp.64-74; also in *Bull. Atomic Scientists*, Nov. 1975, pp.10-20, and *Pulse of the Planet*, #3:17-25, 1991.

Prescott, J., et al., eds., *Brain Function and Malnutrition*, Wiley & Sons, NY, 1975.

"Psychoanalysis and the Nazis: A Collaboration?", in *Tempo* section of *Chicago Tribune*, 29 August 1984.

Raban, J., *Arabia: A Journey Through the Labyrinth*, Simon & Schuster, NY, 1979.

Rackelmann, M., "Was War die Sexpol? Wilhelm Reich und der Einheitsverband für proletarische Sexualreform und Mutterschutz", *Emotion* 11:56-93, Berlin 1994.

Radbill, S., "A History of Child Abuse and Infanticide", in Helfer & Kempe, 1974, above.

Raknes, O., *Wilhelm Reich and Orgonomy*, St. Martin's Press, NY, 1970.

Ransom, C., *The Age of Velikovsky*, Kronos Press, Glassboro, NJ, 1976.

Rappoport J.,*AIDS Inc.: Scandal of the Century*, Human Energy Press, San Francisco, 1988.

"Reaping the Whirlwind: Vandalism and Corporal Punishment", unpublished monograph, EVAN-G, 977 Keeler Ave., Berkeley, CA 94708.

Reed, E., *Sexism and Science*, Pathfinder, NY, 1978.

Reich, W., *Selected Writings*, Farrar, Straus & Giroux, NY, 1973.

Reich, W., "Der Urgegensatz des vegetative Lebens (The Basic Antithesis of Vegetative Life Functions)", *Zeitschrift fur Politische Psychologie und Sexualokonomie*, I, 1934; English trans. in *Pulse of the Planet* 4, 1993, Reich, *The Impulsive Character*, 1974, below, and Reich, *Bioelectrical Investigation*, 1982, below.

Reich, W., *The Bioelectrical Investigation of Sexuality and Anxiety*, Farrar, Straus & Giroux, NY, 1982.

Reich, W., "The Biological Miscalculation in the Struggle for Human Freedom", *Int. J. Sex-Economy & Orgone Research*, 2:97-121, 1943; reprinted in *J. Orgonomy*, 9(1):4-26 & 9(2):134-144.

Reich, W., *Character Analysis*, Orgone Institute Press, NY, 1949; Farrar, Straus & Giroux, NY, 1971.

Reich, W., *Children of the Future, On the Prevention of Sexual Pathology*, Farrar, Straus, & Giroux, NY, 1983.

Reich, W., *Cosmic Superimposition*, Farrar, Straus & Giroux, NY, 1973.

Reich, W., "The Emotional Desert", in *Selected Writings*, Farrar, Strauss & Giroux, NY, 1973, pp.461-3.

Reich, W., *Ether, God and Devil*, Farrar, Straus & Giroux, NY, 1973.

Reich, W., *Experimentelle Ergebnisse über die Elektrische Funktion von Sexualitat und Angst (Experimental Investigation of the Electrical Function of Sexuality and Anxiety)*, Sexpol Press, Copenhagen 1937; English trans. in *Pulse of the Planet* 4, 1993, Reich, *The Impulsive Character*, 1974, below, and Reich, *Bioelectrical Investigation*, 1982, above.

Reich, W., *Function of the Orgasm*, Noonday, NY, 1961; Farrar, Strauss & Giroux, NY, 1973.

Reich, W., *Genitality in the Theory and Therapy of Neurosis,* Farrar, Straus & Giroux, NY, 1980.

Reich, W., *The Impulsive Character and Other Writings*, New American Library, NY, 1974.

Reich, W., *The Invasion of Compulsory Sexual Morality*, Farrar, Straus & Giroux, NY, 1971.

Reich, W., *The Mass Psychology of Fascism*, Orgone Institute Press, NY, 1946; Farrar, Straus & Giroux, NY, 1970.

Reich, W., *The Murder of Christ*, Orgone Institute Press, Rangeley, 1953; reprinted Farrar, Straus & Giroux, NY 1971.

Reich, W., "Der Orgasmus als elektrophysiologische Entladung (The Orgasm as an Electrophysiological Discharge)", *Zeitschrift fur Politische Psychologie und Sexualokonomie*, I, 1934; English trans. in *Pulse of the Planet* 4, 1993, Reich, *The Impulsive Character*, 1974, below, and Reich, *Bioelectrical Investigation*, 1982, above.

Reich, W., "Orgonomic Functionalism", *Orgone Energy Bulletin*, 2:1-15, 49-62, 99-123, 1950; and 4:1-12, 186-196, 1952.

Reich, W., *People in Trouble*, Farrar, Straus & Giroux, NY, 1976.

Reich, W., "The Problem of Homosexuality", in *The Sexual Struggle of Youth*, 1932, below.

Reich, W., *Reich Speaks of Freud*, Farrar, Straus & Giroux, NY, 1967.

Reich, W., *The Sexual Revolution*, Octagon Books, NY, 1971.

Reich, W., *The Sexual Struggle of Youth*, Sex-Pol Verlag, Berlin 1932; in Reich, *Children of the Future*, 1983, above.

Reich, W. & Neill, A.S., *Record of a Friendship: The Correspondence of Wilhelm Reich and A.S. Neill*, B. Placzek, ed., Farrar, Straus & Giroux, NY, 1981.

Remondino, P., *History of Circumcision*, F.A. Davis, Philadelphia, 1891.

Rhotert, H., *Libysche Felsbilder*, L.C. Wittich, Darmstadt, 1952.

Rice, R., "Maternal-Infant Bonding: The Profound Long-Term Benefits of Immediate, Continuous Skin & Eye Contact at Birth", in Stewart & Stewart, 1978, Vol. 2:373-86, below.

Richards, M., et al., "Early Behavioral Differences: Gender or Circumcision?", *Developmental Psychobiology*, 9(1):89-95, 1976.

Riley, C., et al., eds., *Man Across the Sea: Problems of Pre-Columbian Contacts*, U. Texas Press, Austin, 1971.

Robbins, M. et al., "Climate and Behavior: A Biocultural Study", *J. Cross-Cultural Psychology*, 3(4):331-44, 1972.

Ronen, A., "The Paleolithic Archaeology and Chronology of Israel", in Wendorf & Marks, 1975, below.

Root-Bernstein, R., *Rethinking AIDS: The Tragic Cost of Premature Consensus*, Free Press, NY 1993.

Roper, M.K., "Evidence of Warfare in the Near East from 10,000-4300 BC", in Nettleship, et al., 1973, above.

Roy, S.C., "Is the Incidence of Unusually dusty Weather Over Delhi... an Indication that the Rajasthan Desert is Advancing towards Delhi?", *Indian J. Met. & Geophy.*, 5(1):1-15, 1954.

Rush, F., *The Best Kept Secret*, Prentice Hall, NJ, 1981.

Saarinen, T., *Perception of Environment*, Assoc. Amer. Geographers Resource Paper #5, Wash., D.C., 1969.

Salibi, K., "The Bible Came from Arabia", unpublished manuscript, 1984, Archaeology Dept., American University of Beruit.

Saran, A.B., *Murder and Suicide Among the Munda and the Oraon*, National Pub. House, Delhi, 1974.

Sarianidi, V.I., "Seal-Amulets of the Murghab Style", in Kohl, 1981, above.

Sarianidi, V.I., "Margiana in the Bronze Age", in Kohl, 1981, above.

Sarnthein, M., "Sand Deserts During Glacial Maximum and Climatic Optimum", *Nature*, 272:43-46, 2 March 1978.

Sastri, K.A.N., *Cultural Contacts Between Aryans and Dravidians*, Bombay, 1967.

Sauer, C., *Land and Life*, U. Cal. Press, Berkeley, 1963.

Sauer, C., "The Grain Amaranths: A Survey of Their History and Classification", *Annals, Missouri Bot. Garden*, XXXVII(4):561-632, 1950.

Schatzman, M., *Soul Murder: Persecution in the Family*, Random House, NY, 1973.

Schlossman, H., "Circumcision as Defense: Study in Psychoanalysis and Religion", *Psychoanalytic Quarterly*, 35:340-56, 1966.

Schove, D.J., "African Droughts and Weather History", in Dalby & Harrison-Church, 1973, above.

Schulz, E., "Trends of Pleistocene and Holocene Research on the Sahara", *Paleoecology of Africa & Surrounding Islands*, 16:193, 1984.

Scott, G., *A History of Prostitution from Antiquity to the Present Day*, Medical Press, NY, 1954.

Scott, G.R., *A History of Torture Throughout the Ages*, Luxor Press, London, 1939; Sphere Books, London, 1971.

Sears, P.B., "Climate and Civilization", in *Climate Change: Evidence, Causes and Effects,* H. Shapley, ed., Harvard U. Press, Cambridge, 1953.

Seiler, H., "Spiralform, Lebensenergie und Matriarchat", *Emotion* 10:137-167, 1992; reprinted in DeMeo, J. & B. Senf, *Nach Reich*, Zweitausendeins, Frankfurt, 1997.

Semple, E., *Influences of Geographical Environment*, Holt, NY, 1911.

Shaw, B.D., "Climate, Environment and Prehistory in the Sahara", *World Archaeology*, 8(2):133-49, 1976-77.

Shaw, B.D., "Climate, Environment, and History: the Case of Roman North Africa", in Wigley, et al., 1981, below.

Sharaf, M., *Fury on Earth*, St. Martin's/Marek, NY, 1983.

Sheel, W., ed., *Sahara: 10,000 Jahre Zwischen Weide und Wuste*, Museen der Stadt Koln, c.1980.

Sheets, P., ed., *Archaeology and Volcanism in Central America,* U. Texas Press, Austin, 1983.

Sheets, P. & Grayson, D., eds., *Volcanic Activity and Human Ecology*, Academic Press, NY, 1979.

Shutter, R. & Shutter, M., *Oceanic Prehistory*, Cummings Pub. Co., Calif., 1975.

Siersted, E., "Wilhelm Reich in Denmark", *Pulse of the Planet*, 4:44-69, 1993.

Simkin, P., *Directory of Alternative Birth Services and Consumer Guide*, NAPSAC, 1978.

Simkhovitch, V.G., *Towards the Understanding of Jesus and Other Historical Studies*, Macmillan, NY, 1921.

Singh, G., "The Indus Valley Culture", *Archaeology & Physical Anthropology in Oceania*, 6:180, 1971.

Singh, G., "Stratigraphical and Palynological Evidence for Desertification in the Great Indian Desert", *Annals of the Arid Zone,* Special Number on Desertification, Jodhpur, India.

Smil, V., *The Bad Earth: Environmental Degradation in China*, M.E.Sharpe, NY, 1984.

Smith, G., *The Diffusion of Culture*, Kennikat Press, NY, 1971 (reprint of a 1933 work).

Smith, G., *The Migrations of Early Culture*, Manchester U. Press, London, 1915.

Smith, W., *Ancient Mysteries of the Mexican and Mayan Pyramids*, Kensington, NY, 1977.

Snead, R., *Atlas of World Physical Features*, J. Wiley, NY, 1972.

Sondhi, S., "Dowry Deaths in India", *Ms.*, January 1983, p.22.

Songqiao, Z., "Desertification and De-desertification in China", in Mabbutt & Berkowicz, 1980, above.

Sorokin, P., *Hunger As A Factor In Human Affairs*, U. Florida, Gainesville, 1975.

Spencer, J., *Oriental Asia, Themes Toward a Geography*, Prentice-Hall, NJ, 1973.

Spencer, R., "Cultural Aspects of Eunuchism", *Ciba Symposium 8*, 1946.

Spooner, B., *Population Growth, Anthropological Implications*, MIT Press, Cambridge, 1972.

Stacey, J., *Patriarchy and Socialist Revolution in China*, U. California Press, Berkeley, 1983.

Stafford, P., *Sexual Behavior in the Communist World*, Julian Press, NY, 1967.

Stallibrass, A., *The Self-Respecting Child*, Penguin, NJ, 1977.

Stamp, L.D., ed., *A History of Land Use in Arid Regions*, UNESCO, NY, 1961.

Stangl, J., "India: A Widow's Devastating Choice", *Ms.*, September 1984.

Stechler, G., "Newborn Attention as Affected by Medication During Labor", *Science*, 144:315-7, 1964.

Stein, M.A., *Ruins of Desert Cathay*, B.Blom, NY, 1968.

Stein, M.A., *On Central Asian Tracks*, Pantheon, NY, 1964.

Steinem, G., "Feminist Notes: Two Cheers for Egypt, Talks with Jihan Sadat and Other Daughters of the Nile", *Ms.*, June 1980.

Stephens, W., *The Family in Cross-Cultural Perspective*, Holt, Rinehart & Winston, NY, 1963.

Stern, M., *Sex in the USSR*, Times Books, NY, 1979.

Stern, P., *Prehistoric Europe From Stone Age Man to the Early Greeks*, W.W.Norton, NY, 1969.

Stewart, A. & Winter, D., "The Nature and Causes of Female Suppression", *Signs: Journal of Women in Culture and Society*, 2(3):531-53, 1977.

Stewart, D.: "The Conspiracy of Doctors Against Doctors", *NAPSAC News*, 6(1), Spring, 1981; "The Conspiracy of Doctors Against Midwives", *NAPSAC News*, 6(3), fall, 1981.

Stewart, D., ed., *The Five Standards of Safe Childbearing*, NAPSAC, Marble Hill, MO, 1981.

Stewart, L. & Stewart, D., eds., *21st Century Obstetrics Now!*, Vol. 1 and 2, NAPSAC, Marble Hill, MO, 1978.

Stewart, D. & Stewart, L., *Safe Alternatives in Childbirth*, NAPSAC, Chapel Hill, NC, 1978.

Stokes, N.M., *The Castrated Woman: What Your Doctor Won't Tell You About Hysterectomy*, Franklin Watts, NY 1986.

Stone, M., *When God Was A Woman*, Harcourt, Brace, Jovanovich, NY, 1976.

Street, F.A. & Grove, A.T., "Environmental and Climatic Implications of Late Quaternary Lake Level Fluctuations in Africa", *Nature*, 261:385-390, 3 June 1976.

Sullivan, D. & Beeman, R., "Four Years Experience with Home Birth by Licensed Midwives in Arizona", *J. Am. Public Health Assoc.*, 73(6):641-5, 1983.

Sulloway, F., *Freud: Biologist of the Mind - Beyond the Psychoanalytic Legend*, Basic, NY, 1979.

Sulman, F., *The Effect of Air Ionization, Electric Fields, Atmospherics and Other Electric Phenomena on Man and Animal*, Charles Thomas, Springfield, IL, 1980.

Symanski, R., *The Immoral Landscape*, Butterworths, Toronto, 1981.

Tamayo, J., "Confrontations Turn Ugly in Israel's Battle of Beliefs", *Miami Herald*, 12 June 1986, p.22A.

Tannahill, R., *Sex in History*, Stein & Day, NY, 1980.

Taylor, G.R., *Sex in History*, Thames & Hudson, London, 1953.

Textor, R., *A Cross-Cultural Summary*, HRAF Press, New Haven, 1967.

Theisenger, W., *The Marsh Arabs*, E.P.Dutton, NY, 1964.

Thomas, H.L., "Archaeological Chronology of Northern Europe", in Ehrich, 1965.

Thomas, W., ed., *Man's Role in Changing the Face of the Earth*, U. Chicago Press, 1973.

Thompson, E., *Suttee*, George Allen, London, 1928.

Thompson, H., et al., "Report on the Ad Hoc Task Force on Circumcision", *Pediatrics*, 56(4):610-11, 1975.

Thompson, P., *Secrets of the Great Pyramid*, Harper & Row, NY, 1971.

Thornthwaite, C.W., "An Approach Toward a Rational Classification of Climate", *Geographical Review*, 38(1):55-94, 1948.

Tickell C., *Climatic Change and World Affairs*, Harvard Studies in International Affairs, #37, 1977.

Todd, I., *Catal Huyuk in Perspective*, Cummings, Menlo Park, CA, 1976.

Torgerson, D., "Arab Honor Killings Fought in Secret", *Miami Herald*, 4 January 1981.

Trapper, J. & Stuart G., eds., *People and Places of the Past: National Geographic Illustrated Cultural Atlas of the Ancient World*, Natl. Geog. Soc., Washington, D.C., 1983.

Travis, C., "How Freud Betrayed Women", *Ms.*, March 1984.

Trewartha, G.T., *Introduction to Climate*, 4th Ed., McGraw Hill, NY, 1978.

Tromp, S., *Biometeorology: The Impact of the Weather and Climate on Humans and Their Environment*, Heyden, Philadelphia, 1980.

Turnbull, C., *Man in Africa*, Doubleday, NY, 1976.

Turnbull, C., *The Forest People*, Simon & Schuster, NY, 1961.

Turnbull, C., *The Mountain People*, Simon & Schuster, NY, 1972.

Ulrich, R., "Aesthetic and Affective Responses to Natural Environment", *Behavior and Natural Environment*, 6:107-20, 1983.

Uyanik, M., *Petroglyphs of Southeastern Anatolia*, Akademische Druck-u.Verlagsanstalt, Graz, Austria, 1974.

Van Buren, A., "Jerusalem, Reflections in an Empty Glass", *New Age Journal*, December 1983.

Van Gulik, R.H., *Sexual Life in Ancient China*, E.J.Brill, Leiden, 1961.

Van Sertina, I: *They Came Before Columbus*, Random House, NY, 1976.

Velikovsky, I., *Worlds in Collision*, Macmillan, NY, 1950.

Velikovsky, I., *Earth in Upheaval*, Doubleday, NY, 1955.

Velikovsky, I., *Ages in Chaos,* Doubleday, NY, 1952.

Velikovsky, I., *Oedipus and Akhnaten*, Doubleday, NY, 1960.

Velikovsky, I., *Mankind in Amnesia*, Doubleday, NY, 1984.

Verny, T., *The Secret Life of the Unborn Child*, Summit Books, NY, 1981.

Vita-Finzi, C., *The Mediterranean Valleys, Geological Changes in Historical Times*, Cambridge U. Press, 1969.

Von Cles-Reden, S., *The Realm of the Great Goddess*, Prentice-Hall, NJ, 1962.

Von Sadovszky, O: "The Discovery of California: Breaking the Silence of the Siberia to America Migrators", *The Californians*, Nov/Dec 1984.

Von Wissmann, H., "On the Role of Nature and Man in Changing the Face of the Dry Belt of Asia", in Thomas, 1973, above.

Von Wuthenau, A., *Unexpected Faces in Ancient America*, Crown, NY, 1978.

Voute, C.C., "A Prehistoric Site Near Razzaza", *Sumer*, 13(1-2):135-56, 1957.

Wada, S., "The Philippine Islands as Known to the Chinese Before the Ming Dynasty", *Memoirs, Tokyo Bunko, Research Dept.*, Tokyo, #4, 1929.

Wadia, D.N., *The Post Glacial Desiccation of Central Asia: Evolution of the Arid Zone of Asia*, National Institute of Sciences of India, Delhi, 1960.

Wallerstein, E., *Circumcision, an American Health Fallacy*, Springer, 1983.

Walls, J., *Man, Land, Sand*, Macmillan, NY, 1980.

Walton, K., *The Arid Zones*, Aldine Publishing Co., Chicago, 1969.

Watkins, C., "The Indo-European Origin of English", in *The American Heritage Dictionary of the English Language*, Houghton-Mifflin, Boston, 1969, pp.xix-xx.

Webster, H., *Primitive Secret Societies*, Octagon, NY, 1968.

Weinberg, S.S., "Relative Chronology of the Aegean in the Stone and Early Bronze Ages", in Ehrich, 1965, above.

Weitz, R. & Sullivan, D., "Licensed Lay Midwives in Arizona", *J. Nurse- Midwifery*, 29(1):21-8, 1984.

Welsh, R., "Severe Parental Punishment and Delinquency: A Developmental Theory", in *Psychology and the Problems of Today*, M. Wertheimer and L. Rappoport, eds., Scott, Foresman & Co., Glenview, IL, 1978; cf. earlier version of same title in *J. Clinical Child Psychology*, 5(1):17-21, 1976.

Wendorf, F., "The Paleolithic of the Lower Nile Valley", in Wendorf & Marks, 1975, below.

Wendorf, F. et al., "Late Pleistocene and Recent Climatic Changes in the Egyptian Sahara", *Geographical J.*, Roy. Geog. Soc., London, 143:211-34, 1977.

Wendorf, F., et al., "The Prehistory of the Egyptian Sahara", *Science*, 193:103, 1976.

Wendorf, F., et al., *Science* 205:1341, 1979.

Wendorf, F., et al., *Loaves and Fishes: The Prehistory of Wadi Kubbaniya*, Dept of Anthropology, Southern Methodist U., Dallas, 1980.

Wendorf, F., et al., *Science* 82, 3:68, Nov. 1982.

Wendorf, F. & Marks, A.E., eds., *Problems in Prehistory, North Africa and the Levant*, S. Methodist U. Press, Dallas, 1975.

Wendorf, F. & Schild, R., *Prehistory of the Eastern Sahara*, Academic Press, NY, 1980.

Wescott, R.W., "Review of: The Origins of Consciousness in the Breakdown of the Bicameral Mind", *Kronos*, III(4):78-85, 1978.

West, S., *The Hysterectomy Hoax: Why 90% of all Hysterectomies are Unnecessary*, Doubleday, NY 1994.

Wheeler, M., *The Cambridge History of India: The Indus Civilization*, 3rd ed., Cambridge U. Press, 1968.

Whitaker, J.I., *Motya: A Phoenician Colony in Sicily*, Bell & Sons, London, 1921.

White, E.H., "Legal Reform as an Indicator of Women's Status in Muslim Nations", in Beck & Keddie, 1978, above.

Whiting, J., "Effects of Climate on Certain Cultural Practices", in *Explorations in Cultural Anthropology*, W. Goodenough, ed., McGraw- Hill, NY, 1964, pp.511-44.

Whiting, J. & Child, I., *Child Training and Personality*, Yale U. Press, New Haven, 1953.

Whiting, J., et al., "The Function of Male Initiation Ceremonies at Puberty", in *Readings in Social Psychology*, E. Maccoby, et al., eds., Holt, Rinehart & Winston, NY, 1974.

Whyte, M., *The Status of Women in Preindustrial Societies*, Princeton U. Press, NJ, 1978.

Whyte, R.O., "Evolution of Land Use in Southwestern Asia", in Stamp, 1961, above.

Widstrand, C., "Female Infibulation", *Studia Ethnographica Upsaliensa*, 20:95-122, 1965.

Wiencek, H., et al., *Storm Across Asia*, H.B.J. Press, NY, 1980.

Wigley, T.M.L., et al., eds., *Climate and History*, Cambridge U. Press, NY, 1981.

Williams, W., "Dowry Murders in India", *Ms.*, April 1981.

Wilson, E., *Sociobiology: The New Synthesis*, Harvard U. Press, Cambridge, 1975.

Winick, M. & Rosso, P., "Malnutrition and Central Nervous System Development", in Prescott, et al., 1975, above.

Winkler, H.A., *Rock-Drawings of Southern Upper Egypt* II, Sir Robert Mond Desert Expedition, Oxford U. Press, London, 1939.

Winstanley, D., "Desertification: A Climatological Perspective", in *Origin and Evolution of Deserts*, S.G. Wells & D.R. Haragan, eds, U. New Mexico Press, Albuquerque, 1983.

Wittfogel, K.A., *Oriental Despotism*, Yale U. Press, New Haven, 1957.

Wolfe, T., *The Emotional Plague Versus Orgone Biophysics*, Orgone Institute Press, NY, 1948.

World Data Sheet: 1983, Population Reference Bureau, Washington, D.C.

Wormington, H., *Prehistoric Indians of the Southwest*, Denver Museum Nat. Hist., 1978.

Wright, J., *Human Nature in Geography*, Harvard U. Press, Cambridge, 1966.

Wright, H., "The Environmental Setting for Plant Domestication in the Near East", *Science*, 196:385-9, 1976.

Wright, H.E., "An Extinct Wadi System in the Syrian Desert", *Bull. Res. Counc. of Israel*, 76:53-9, 1958.

Yosef, O.B., "The Epipaleolithic in Palestine and Sinai", in Wendorf & Marks, 1975, above.

Young, F., "The Function of Male Initiation Ceremonies: A Cross-Cultural Test of an Alternative Hypothesis", *Am. J. Sociology*, 67:379-91, 1962.

Zeitlin, M., et al., *Nutrition and Population Growth: the Delicate Balance*, Oelgeschlager, Gunn & Hain, Cambridge, Mass., 1982.

Zimmer, P., "Modern Ritualistic Surgery", *Clinical Pediatrics*, 16(6):503-6, June 1977.

Zimmerman, F., "Origin and Significance of the Jewish Rite of Circumcision", *Psychoanalytic Review*, XXXVIII(2):103-12, 1951.

Zuckerman, M., "Physiological Measures of Sexual Arousal in the Human", *Psychological Bull.*, 75:297-329, 1971.

New Citations added into the Second Edition (in addition to the end-notes in Appendix B)

Bostom, A., *The Legacy of Jihad: Islamic Holy War and the Fate of Non-Muslims*, Prometheus Books, NY 2005.

DeMeo, J., "Update on *Saharasia*: Ambiguities and Uncertainties about *War Before Civilization*", in *Heretic's Notebook (Pulse of the Planet #5)*, 2002, p.15-44.

DeMeo, J.: "Editor's Postscript", *Heretic's Notebook (Pulse of the Planet #5)*, p.65, 2002.

DeMeo, J., *Preliminary Analysis of Changes in Kansas Weather Coincidental to Experimental Operations with a Reich Cloudbuster*, University of Kansas, Geography-Meteorology Dept., Thesis, 1979.

DeMeo, J. & R. Morris, "Preliminary Report on a Cloudbusting Experiment in the Southeastern Drought Zone, August 1986", *Southeastern Drought Symposium Proceedings*, March 4-5, 1987, South Carolina State Climatology Office Pub. G-30, Columbia, SC, 1987.

DeMeo, J., "OROP Arizona 1989: A Cloudbusting Experiment to Bring Rains in the Desert Southwest", *Pulse of the Planet #3*, J. DeMeo, Editor, p.82-92, 1991.

DeMeo, J., "OROP Israel 1991-1992: A Cloudbusting Experiment to Restore Wintertime Rains to Israel and the Eastern Mediterranean During an Extended Period of Drought", in *On Wilhelm Reich and Orgonomy, Pulse of the Planet #4*, J. DeMeo, Editor, p.92-98, 1993.

DeMeo, J., "Green Sea Eritrea: A 5-Year Desert Greening CORE Project in the East African Sahel-Sahara", in *Heretic's Notebook: Emotions, Protocells, Ether-Drift and Cosmic Life Energy, Pulse of the Planet #5*, J. DeMeo, Editor, p.183-211, 2002.

Lapkin, S., "Palestinian 'Honor'," *Frontpage Magazine*, 19 Jan. 2006.

Lewis, B.: "Europa wird am Ende des Jahrhunderts islamisch sein," *Die Welt*, July 28, 2004.

Milton, G.: *White Gold: The Extraordinary Story of Thyomas Pellow and Islam's One Million White Slaves*, Farrar, Straus & Giroux, NY 2004.

Spencer, R.: *The Politically Incorrect Guide to Islam (and the Crusades)*, Regnery Pub., Washington, DC, 2005.

Ye'or, B., *The Dhimmi: Jews and Christians Under Islam*, Fairleigh Dickinson Univ. Press, 1985.

Ye'or, B., *Islam and Dhimmitude: Where Civilizations Collide*, Fairleigh Dickinson Univ. Press, 2002.

Ye'or, B., *Eurabia: The Euro-Arab Axis*, Fairleigh Dickinson Univ. Press, 2005.

Index

CPSIA information can be obtained
at www.ICGtesting.com
Printed in the USA
BVOW04s2147170817
492244BV00004B/41/P